SUPERCONDUCTIVITY RES

SUPERCONDUCTIVITY

THEORY, MATERIALS AND APPLICATIONS

SUPERCONDUCTIVITY RESEARCH AND APPLICATIONS

Additional books in this series can be found on Nova's website under the Series tab.

Additional E-books in this series can be found on Nova's website under the E-books tab.

MATERIALS SCIENCE AND TECHNOLOGIES

Additional books in this series can be found on Nova's website under the Series tab.

Additional E-books in this series can be found on Nova's website under the E-books tab.

SUPERCONDUCTIVITY RESEARCH AND APPLICATIONS

SUPERCONDUCTIVITY

THEORY, MATERIALS AND APPLICATIONS

VLADIMIR REM ROMANOVSKII
EDITOR

Nova Science Publishers, Inc.
New York

Additional color graphics may be available in the e-book version of this book.

Library of Congress Cataloging-in-Publication Data

Superconductivity : theory, materials, and applications / editor, Vladimir Rem Romanovskii.
p. cm.
Includes index.
ISBN 978-1-61324-843-0 (hardcover)
1. Superconductivity. I. Romanovskii, Vladimir Rem.
QC610.92.S87 2011
621.3'5--dc23

2011017621

Published by Nova Science Publishers, Inc. † New York

CONTENTS

PREFACE

This book presents current research from across the globe in the study of superconductivity theory, materials and applications. Topics discussed include tunneling spectroscopy of novel layered superconductors; stability conditions of high-Tc superconductors; a study of the superconducting phase in metallic superconductors; numerical calculation of trapped magnetic field for bulk superconductors; ion modified high-Tc Josephson junctions and SQUIDS; and vortices in high temperature superconductors.

Chapter 1 - The discovery of novel high-T_c superconductivity in MgB_2 (T_c = 39.5 K) and $Li_{0.48}(THF)_yHfNCl$ (T_c = 25.5 K) initiated substantial progress in the field of superconductivity physics and its applications despite the fact that competing high-T_c cuprates remain the world leaders in almost all practically important superconducting parameters. This article describes electron tunneling and point-contact experimental studies of the indicated two materials and related substances, being crucial to elucidate the character of the quasiparticle energy spectrum both in superconducting and normal state. The account is based mostly on their own experiments, although works carried out in other laboratories are taken into account as well.

The authors studied superconducting gap structures by means of break-junction tunneling spectroscopy (BJTS), scanning tunneling spectroscopy (STS) and point-contact spectroscopy (PCS).

In the case of MgB_2, tunnel conductance $G(V) = dI/dV(V)$ reveals multiple-gap features. Here I is the quasiparticle tunnel current and V is the bias voltage. The two-gap model including the proximity effect (the correlated two-gap model) was used to describe the observed multiple-gap features in BJTS and STS. Three specific gap values can be identified as follows: $\Delta_S = 2 - 2.5$, $\Delta_M = 4.5 - 7.5$ and $\Delta_L = 10 - 12$ meV. The observed values $2\Delta_L(4.2\ K)/k_BT_c > 5 - 6$ constitute the largest known values of the superconducting gap to T_c ratio except for those appropriate to copper oxides and certain organic materials. Here k_B is the Boltzmann constant. On the other hand, ratios of the small gaps Δ_S to corresponding T_c's, $2\Delta_S(4.2\ K)/k_BT_c$, fall into the range 1.2 – 1.5, which is considerably below the Bardeen-Cooper-Schrieffer (BCS) value $2\Delta(4.2\ K)/k_BT_c \approx 3.5$ inherent to *s*-wave superconductors. The extrapolated highest gap-closing field B_c for the largest gap agrees with the upper-critical field, thereby indicating that this gap is predominant in MgB_2. Point-contact conductance $G(V)$ also demonstrates multiple-gap structures. Peak positions in the second derivative conductance of PCS are in a reasonable agreement with the phonon spectrum frequencies revealed by the inelastic neutron scattering measurements. The high-energy boron vibration modes (~ 75 meV) are suggested to play the important role in the overall electron-phonon interaction in MgB_2. The BJTS data of other AlB_2 type superconductors show the strong-coupling-size values of the ratio $2\Delta(0)/k_BT_c$, specifically, 4.2 – 4.5 for NbB_2 and 4.2 – 4.6 for CaAlSi.

Tunneling measurements have been carried out on layered nitride superconductors of the β(SmSI)-type $Li_{0.48}(THF)_xHfNCl$ (THF;C_4H_8O) ($T_c \approx 25.5$ K), $HfNCl_{0.7}$ ($T_c \approx$ 23-24 K) and $ZrNCl_{0.7}$ ($T_c \approx 14$ K). BJTS reveals Bardeen – Cooper - Schrieffer (BCS) - like gap structures with typical gap values of $2\Delta(4.2\ K)$ = 11-12 meV for $Li_{0.48}(THF)_xHfNCl$ with the highest $T_c \approx 25.5$ K. Their measurements revealed multiple gaps and dip-hump structures, the largest gap 2Δ (4.2 K) = 17 - 20 meV closing at T_c. It comes about that the highest obtained gap ratio $2\Delta/k_BT_c \sim 8$ substantially exceeds the BCS weak-coupling limiting values: ≈ 3.5 and ≈ 4.3 for *s*-wave and *d*-wave order parameter symmetry, respectively. For $ZrNCl_{0.7}$, two gap ratios $2\Delta/k_BT_c = 6 – 8$ and $2\Delta/k_BT_c = 3 – 4$ follow from the BJTS data. Such huge values of $2\Delta/k_BT_c$ are rather unusual for conventional superconductors

Chapter 2 - Macroscopic peculiarities of applied current penetration are theoretically investigated to understand the basic physical trends, which are characteristic for the stable formation of the thermo-electrodynamics states of high-T_c superconductors without stabilizing matrix placed in DC external magnetic field. The performed analysis shows that the definition of the current stability conditions of high-T_c superconductors must take into consideration the development of the interconnected thermal and electrical states when the temperature of superconductor may stably rise before instability from the temperature of coolant to its critical temperature. As a result, the thermal degradation effect of the current-carrying capacity of superconductor exists. The mechanisms of this effect are discussed. The boundary values of the electric field and the current above which the charged current is unstable are defined taking into account the size effect, cooling condition peculiarities, non-linear temperature dependences of the critical current density of superconductor and *n*-value of its *E-J* relation. It is proved that the allowable stable values of electric field and current can be both below and above those determined by a priori chosen critical current and electric field of the superconductor. The violation features of the steady current distribution in high-T_c superconductors cooled by liquid refrigerant (helium, hydrogen and nitrogen) are studied. The necessary criteria allowing one to determine the influence of the properties of superconductor and coolant on the current instability mechanism are written. The peculiarity formation of possible resistive states of high-T_c superconductor are investigated in the static approximation considering the temperature-decreasing dependence of the power exponent (*n*-value) of its voltage-current characteristics.

Chapter 3 - In this chapter, the theoretical investigation of five superconducting state parameters (SSP) viz; electron-phonon coupling strength λ, Coulomb pseudopotential μ^*, transition temperature T_C, isotope effect exponent α and effective interaction strength N_OV of metallic complexes using model potential formalism with the pseudo-alloy-atom (PAA) model is discussed. Various local field correction functions are used for the first time in the present investigation to study the screening influence on the aforesaid properties. It is observed that the electron-phonon coupling strength λ and the transition temperature T_C are quite sensitive to the selection of the local field correction functions, whereas the Coulomb pseudopotential μ^*, isotope effect exponent α and effective interaction strength $N_O V$ show weak dependences on the local field correction functions. The present results of the SSPs are found in qualitative agreement with the available experimental or theoretical data wherever exist. The T_C equation has been proposed, which provide successfully the T_C values of metallic complexes under consideration. Also, the present study confirms thc superconducting phase in metallic superconductors. A strong dependency of the SSPs on the valence 'Z' is found.

Chapter 4 – The authors present the results of the systematic comparative study of the thermopower, S, normal-state band spectrum, and superconducting properties for chain-free HTSC's of Bi-, Tl-, and Hg-systems. They have analyzed a large set of the experimental results on the $S(T)$ dependences in samples of all systems in case of increasing both number of copper-oxygen layers, n, and level of non-isovalent doping for different lattice positions. Based on systematization of all these data, they have revealed the common tendencies in $S(T)$ modification under doping both in the underdoped and overdoped regimes.

All the experimental data on thermopower were quantitatively analyzed within a narrow-band model allowing us to determine the main parameters of the band spectrum and charge-carrier system (the effective bandwidth, W_D, degree of the band filling with electrons, degree of state localization, and band asymmetry degree). They have identified the above parameters for all the studied samples and then analyzed the change in their values under variation of sample composition within different HTSC-families.

The authors have observed that for the samples of near-optimally doped compositions the values of the band parameters for all the systems are close to each other, that points to the fact that the band spectrum structures in Bi-, Tl-, and Hg-based HTSC's are rather alike in general. Besides, a correlation between the variations of the bandwidth and the T_c value with n was observed for all the studied systems.

They have also scaled the parameters of the band spectrum and charge-carrier system, as well as the T_c behavior under doping which level was estimated for samples with different types of impurities based on the thermopower value at T=300K. For all the three systems these values change in an analogous way, but differently in the underdoped and overdoped regimes indicating various mechanisms of the band spectrum modification in these two doping regimes. A qualitative distinction in influence of different impurities is discussed and explained with respect to their type and positions occupied in a lattice.

Analysis of the results has shown that the T_c value in all cases is directly related to the value of the density-of-states at the Fermi level, $D(E_F)$. They have discussed two different mechanisms of the band spectrum modification affecting the critical temperature in an opposite way. Fist, the disordering in the lattice induced by a rise of the defect number leads to a band broadening according to the Anderson localization mechanism that results in decreasing $D(E_F)$. Second, increasing number of copper-oxygen layers forming the band leads to a general rise in the density-of-states and, as a result, $D(E_F)$ increases. The relative contribution of these two mechanisms in band spectrum modification and T_c variation under increasing n and/or doping level is discussed based on all the results obtained.

Finally, the authors have observed the scaling behavior of T_c in the underdoped regime for each of studied HTSC-families. The variation of the correlation curve $T_c(W_D)$ with increasing number of CuO_2 layers, as well as with a rise of the maximal T_c in the system is briefly discussed.

Chapter 5 - Chemical substitution is a simple technological decision for the generation of a large number of nano-sized defects. Replacement of some atoms in the structure with other ones of different valence, ion radius and magnetic moment can result in deformation of lattice, charge distribution, appearance of oxygen and other atoms vacancies or disorder. Partial replacement of rare earth ion in $RBa_2Cu_3O_{7-\delta}$ by Ca^{2+} ion with similar ionic radius but lower valence value, provides additional holes and makes the overdoped region of the T(p) phase diagram accessible for study. Investigation of intra- and inter-granular effects in overdoped polycrystalline $R_{1-x}Ca_xBa_2Cu_3O_{7-\delta}$ (R=Y, Eu, Gd, Er and x=0, 0.025, 0.05, 0.10, 0.20, 0.30) samples was carried out by using different experimental technique. X-ray powder diffraction analysis and SEM were used for the examination of phase formation and microstructure. AC magnetic susceptibility measurements (of fundamental and third harmonics) as a function of temperature, DC magnetic field, frequency and AC magnetic field amplitude were exploited for the investigation of a large number of properties: differentiation between inter- and intra- granular effects, estimation of intergranular J_{cinter}, activation energy for TAFF, irreversibility line or non-linear dissipation processes. DC magnetization measurements were performed at fixed temperatures as a function of magnetic field and intragranular J_{cintra} was obtained. Transport measurements (resistivity vs. temperature and I-V characteristics) at different magnetic fields were used for the establishment of vortex-glass-vortex-liquid phase transition and scaling parameters.

It was established that low level overdoping leads to the improvement of intragranular critical current, flux pinning and irreversibility field at 77 K making it higher than in non substituted, fully oxygenated YBCO samples. Temperature dependence of intergranular critical current showed that it is governed by the S-I-S type joints between the grains. For highly overdoped samples the suppression of intragranular critical current and flux pinning has been observed. The intergranular critical current is characterized by S-N-S type. Indirect evidence suggests that this is a result of carriers' phase separation supporting the idea that the quality of superconducting condensate is strongly influenced by overdoping. The field dependence of activation energy for TAFF shows that 2D pancake vortices are characteristic of underdoped samples, while 3D vortex system exists in overdoped ones. Hole concentration is an essential parameter that controls many properties in HTS. They also investigated how the vortex dynamics was influenced by the doping effect. By using the third harmonics signal of AC magnetic susceptibility the irreversibility line was determined. The existence of vortex glass-vortex liquid phase transition was confirmed also by transport measurements. The scaling behavior of E-J data in Ca substituted samples is similar to the other polycrystalline YBCO samples. Previously established morphology dependence of dynamic exponent (**z**) was confirmed. However, **z** values are smaller than the usually reported for non-substituted YBCO. Static exponent (ν) shows a tendency for field dependence. These observations have been explained with the peculiarities of Ca substituted samples.

Chapter 6 - Bulk high-temperature superconductors have significant potential for industrial applications, such as magnetic bearings, flywheel energy storage systems, non-contact transport devices, levitation and trapped-field magnets. The term 'trapped-field magnet' describes a bulk superconductor supporting a trapped magnetic field after a magnetization process. The trapped magnetic field of the superconducting bulk magnet has been reported to be superior to that of a conventional permanent magnet.

For further enhancements of the trapped magnetic field of the bulk superconductors, it is important to increase the critical current density (J_c) and the size of the single domain. On the other hand, the trapped magnetic field depends also on several other parameters, such as the shape and the aspect ratio of the sample and the distance between the sample surface and the observation point, which is also important for the future design of electrical machines that utilize high trapped fields, like motors and generators.

For this reason this chapter summarizes the numerical calculation of trapped magnetic field. It is known that the calculation method using the sand-pile model and Biot–Savart law is useful for the determination of the magnetic characteristics of the sample. At the first stage, it was examined and extended earlier calculations of field distribution within the bulk superconductor for constant critical current density. At the second stage it was outlined calculation of field distribution within the bulk superconductor for field dependent critical current density

Chapter 7 - After the discovery of high-T_c superconductivity there has been considerable effort in developing Josephson junctions and SQUIDs from these materials. During this time only very limited numbers of fabrication techniques were developed, because of the difficult material properties of high-T_c superconductors. Therefore the most successful type junctions remain grain boundary junctions. Twenty years ago I proposed and successfully fabricated ion modified high-T_c Josephson junctions and SQUIDs. In this review all published results are summarized from their first demonstration in 1990 until the recent successfully fabrication in Berkeley of more then 15000 junctions with a critical current spread of only about 16 %.

Chapter 8 - Study on the vortex dynamics of type-II superconductors provides the information about the pinning ability and current-carrying ability of the superconductors. However, the vortex dynamics are currently described with multiple models. In this chapter, the authors presented a short review on the recently developed general mathematical models of flux relaxation and vortex penetration phenomena in the type-II superconductors. It is shown that the activation energy of vortices in flux relaxation and vortex penetration process can be expanded as series of current density (or internal field). The corresponding time evolution equations of current density (or internal field) can be obtained. These general expressions can be applied to arbitrary vortex systems. It is also shown that, by introducing a time parameter, one can convert a flux relaxation process starting with a current density below the critical value into a process starting with the critical current density. Similarly, by introducing some time parameters, one can convert a vortex penetration process with a non-zero initial internal field into a process with a zero initial internal field. Therefore, one can use a single formula to describe the flux relaxation (or vortex penetration) phenomenon with an arbitrary initial condition.

Chapter 9 - Y-based Superconductors(Y-Ba-Cu-O) is considered to be one of the promising superconducting materials owing to its good applications in many fields, and these applications need Y-Ba-Cu-O bulks with large dimension and complex shape. However, at present, the mean current density of large bulk decreases significantly compared with that of the small one and the shape of the bulk is very simple, usually in form of cylinder or hexagonal. In order to meet the need of the practical using of Y-Ba-Cu-O, joining technique has been used. Up to now, a lot of researchers had done a great deal of investigation on soldering, however, almost all of them needed a relatively long time to realize superconductive joining. In this paper, fast superconductive joining technology of Y-Ba-Cu-O was investigated.

Three kinds of solders, including sintered Yb-Ba-Cu-O, sintered Y-Ba-Cu-O/Ag and melt Y-Ba-Cu-O/Ag solder were synthesized and different soldering thermal cycles were designed respectively. The results show that the soldered bulk using sintered Yb-Ba-Cu-O can partially recover the superconductivity but accumulation of Yb_2BaCuO_5 (Yb211), which results in decrease in superconductivity, was observed in the bonding zone. In comparison, soldering with Y-Ba-Cu-O/Ag solders realized much better superconductive bonding owing to its thinner solder thickness and slower cooling rates during soldering. Moreover, soldering with the melt Y-Ba-Cu-O/Ag solder obtained better performance than soldering with the sintered solder. One plausible reason for the better performance of melt solder is that the dense structure owing to its pretreatment at higher temperature. The microstructure of the bonding zones was investigated by scan electron microscope and the images indicate that neither Ag particles nor Y_2BaCuO_5 (Y211) accumulated during the soldering process. The trapped field image also reveals that the supercurrent is almost not blocked by the bonding zone.

Chapter 10 - It is well known that magnetic flux can penetrate a type-II superconductor in the form of Abrikosov vortices. Many properties of the vortices are well described by phenomenological Ginzburg–Landau theory. High temperature superconductor also belongs to the class of type-II superconductors. These vortices (also known as flux-line lattice) form a triangular lattice in the low-temperature type-II superconductors. But experiments have shown that in the high-temperature superconductors, these vortices form an oblique lattice. This is attributed to various characteristics of high temperature superconductors, namely, pairing state symmetries, anisotropy, planar nature of the superconducting plane, symmetry of the unit cell modulated by the presence of CuO- chain, thermal fluctuations etc. The experimentally observed softening and subsequent melting of the vortex lattice is attributed to the small shear modulus of the vortex lattice. In this review the authors address these problems associated with the vortex lattice of high temperature superconductors. One of the central issues associated with the high-Tc superconductors is the pairing state symmetry of the order parameter components. Based on the experimental evidences a consensus could now be reached that these materials possesses mixed symmetry state scenario where the dominant order parameter is of d-wave symmetry along with which there is an admixture of a small s-wave order parameter component. The authors have studied the properties of the high-Tc superconductors involving mixed symmetry state of the order parameters in the framework of a two-order parameter Ginzburg–Landau (GL) model, over the entire range of applied magnetic field (Hc1 < H < Hc2) and wide range of temperature and for arbitrary GL parameter • and vortex lattice symmetry. Using the present model the limitations of the earlier theoretical works involving the GL theory could be overcome, wherein the studies were restricted to the upper (Hc2) and lower (Hc1) critical magnetic field regions and the problem was reduced to an effective single order parameter model by using an ansatz. The present theoretical model has been further generalized to take into account the effect of in-plane anisotropy. The authors calculate various properties of the high temperature superconductors including vortex core radius, penetration depth, vortex lattice symmetry, upper critical magnetic field, shear modulus of the vortex lattice etc. Furthermore, the effect of the admixture of the sub-dominant s-wave order parameter on the various properties of the high-Tc cuprates have been explored. The variations of these properties with applied magnetic field and temperature are also calculated and shown to be in very good agreement with experiments on high-Tc cuprate Y Ba2Cu3O7−± superconductors [1]- [5].

They have also used the same two-order parameter Ginzburg–Landau model appropriately modified to study the properties of the two-band inter-metallic high temperature superconductorMgB2.

The results of the analytical calculations are shown to agree very well with the experimental results of MgB2 [6].

Chapter 11 - The electronic band structure, charge distribution and hyperfine interactions, namely, electric field gradients (EFG) and contact terms of hyperfine fields (HFF) in stoichiometric GdBa2Cu3O7 (Gd123) high-temperature superconductor have been calculated on the first-principles basis using the full-potential linearized augmented plane wave (FP-LAPW) method. The generalized gradient approximation (GGA) was employed to treat the exchange and correlation effects. The Hubbard correction U was applied for 4f electrons to account for the strong on-site Coulomb repulsion. The calculated electronic structure and charge distribution is typical for (rare ¡ earth)123 class of compounds. It was shown that the evaluated strongly localized Gd magnetic moment, which is comparable with the one obtained from the neutron diffraction data, could not effectively influence the neighbors resulting in small itinerant moments in the CuO2 planes. The calculated EFG and HFF parameters are consistent with the M¨ossbauer measurements. Their results indicate that in the highly correlated Gd123 system the applied computational method yields reliable charge distribution to which the EFG and HFF are very sensitive.

Chapter 12 - High temperature superconducting BSCCO/Ag tapes have been fabricated commercially by PIT methods and many applications of the tapes need to join the tapes. There are several joining methods developed to implement the aim. Soldering technology is conventional joining methods and the joint owns larger resistance. Some superconducting joining methods also have been put out. First of all, the cold-press and post heat treatment technique according to the fabrication process of the tape is used to made superconducting joint with hundreds of hours and the joint is of lower superconducting property. Second, Diffusion bonding without an interlayer (directly) or with superconducting powders methods are developed to join the BSCCO/Ag tapes in shorter time, and these joints are of excellent superconducting properties comparing with the cold-press and post heat treatment method.

Chapter 13 - Recent progress in technology of melt-textured $REBa_2Cu_3O_y$ "RE-123" high-T_c superconductors has brought applications of superconducting permanent magnets in variety of industrial, medical, public, and research applications. When the superconducting pellet is magnetized to a high magnetic field, part of this field is trapped in the pellet and the authors get a superconducting permanent magnet or, shortly, super-magnet [1-13]. Such a name is fully justified as high-T_c superconductors can trap magnetic field by order of magnitude higher than the best hard ferromagnets nowadays known [14]. The trapped field depends on the critical current density, J_c of the melt textured material and on the size of the single grain. Therefore, to achieve a large B_T, one needs to enhance both J_c and the bulk material size. A further improvement of the critical current density and fabrication of large homogeneous good-quality single-grain superconducting disks capable of trapping high magnetic fields are fundamental issues for many industrial applications. In recent years, the major emphasis has been devoted to fabricating single-grain, high-performance LRE-$Ba_2Cu_3O_y$ "LRE-123" (LRE: Nd, Sm, Eu, Gd) pellets by means of the oxygen-controlled melt-growth process [15-40]. Melt-processed bulk disks as large as 140 mm in diameter were produced [41]. As a result of a continuous improvement in the pinning defect efficiency, this material showed very high critical current density even at 90.2 K, the boiling temperature of liquid oxygen [42-60]. Nowadays the performance reached the level necessary for industrial applications and they believe that bulk superconducting magnets will in a close future enter the market as basic parts of various industrial applications [61,62]. In all these cases a high number of pieces with equally high quality are required. Batch processing of LRE-123 materials with uniform properties is the necessary step in this process.

Batch processing is a principal step towards economical production of these materials. For a successful batch processing the following requirements are necessary: (i) the ability to process a large number of grains with the lowest possible loss fraction, (ii) For all processed blocks the final product quality must be satisfactory within each batch process. The seed crystal represents a key issue in this direction. In the batch melt growth a cold seeding process is used. The seed crystal is placed on the precursor pellet at room temperature. Normally, for the batch-processed melt-textured Y-123 pellets Nd-123 or Sm-123 crystals have been used as seeds [63-65]. However, in the case of LRE-123 the peritectic temperature is higher and these kinds of seeds get partially melted, degrading thus superconducting performance of the samples. In order to fabricate LRE-123 bulks using cold seeding method, the Cambridge group led by D. Cardwell introduced the Mg-doped LRE-123 crystals as a seed material [66]. These crystals possess higher melting temperature than pure LRE-123 ones. With generic Mg-doped Nd-123 seeds they succeeded in growing single-grain samples of 20 mm in diameter in air [67]. For elimination of a multiple nucleation in the case of LRE-123 bulk growth, one has to increase the maximum temperature that the seed can withstand to about 80 ^{o}C above the peritectic temperature of the particular LRE-123 [68]. For this, the recently discovered superheating effect of LRE-123 thin films has offered a genuine solution. Oda et al. [69] reported use of Sm-123 thin film seed grown on MgO crystal for growing Sm-123 block in the cold seeding process with a reduced oxygen atmosphere, taking advantage of the superheating phenomenon [70]. With this effect, the YBCO/MgO thin film grown by liquid phase epitaxy could be used as seeds for melt processing of high-temperature-melting NdBCO, [70] (in the bulk form YBCO has much lower peritectic temperature than Nd-123). Similarly, Sm-123/MgO thin films were shown to exhibit superheating effect, being stable up to 1100 ^{o}C, so that they can be used as seeds for growing any LRE-123 bulk [71]. Recently, they succeeded in batch processing of Gd-123 pellets 24 mm in size, 12 samples in one batch, using the Nd-123 thin film seeds [72-73].

In this contribution, the authors report on successful growth of several batches of LRE-123 pellets in air and partial oxygen pressure using the new class of Nd-123/MgO seeds. The emphasis was focused on producing high-performance large size single grains at a reduced price. The superconducting performance, microstructure, as well as the trapped magnetic field measurements at liquid nitrogen temperature and liquid argon temperature are reported. Finally, taking advantage of the batch processed material, the authors constructed home made child levitation disk, which is capable of levitating a mass greater than 35 kg. Their experimental results clearly indicate that using the new class of seeds, LRE-123 can scale up from laboratory to industrial production.

Chapter 14 - From accurate measurements of ac susceptibility, _, of a square YBa2Cu3O7□_ superconducting film of sides 2a = 4 mm and thickness t = 0:25 _m as a function of temperature T at different values of ac field amplitude Hm and frequency f, important superconducting properties are extracted and analyzed. The London penetration depth _(T) is determined from low-Hm _(T) measurements after the Meissner susceptibility of the square film is calculated numerically by minimizing relevant Gibbs potential.

The critical-current density Jc(T) is extracted from the measured _(T;Hm; f) based on the critical-state and flux-creep models. Having _(T) and Jc(T) determined, properly normalized _ vs Hm curves are further converted from the measured _(T;Hm; f) to be compared directly with the calculated critical-state curve. It is concluded that vortex dynamics is dominated by collective creep, Jc is practically independent of local flux density B, and the contributions from edge barriers to _(T;Hm; f) are important especially when T approaches Tc. For a more quantitative explanation of the observed _(T;Hm; f), type-II superconductivity theory for thin films in a perpendicular magnetic field is developed. Important anomalous phenomena are found and remain open for further study.

In: Superconductivity
Editor: Vladimir Rem Romanovskií

ISBN 978-1-61324-843-0

Chapter 1

TUNNELING SPECTROSCOPY OF NOVEL LAYERED SUPERCONDUCTORS: MgB_2, $Li_{0.48}(THF)_xHfNCl$ AND RELATED SUBSTANCES

***Tomoaki Takasaki*[1], *Toshikazu Ekino*[2*], *Alexander M. Gabovich*[3], *Akira Sugimoto*[2], *Shoji Yamanaka*[4], *Jun Akimitsu*[5]**

[1]Graduate School of Advanced Sciences of Matter, Hiroshima University
Higashi-Hiroshima, Japan
[2]Graduate School of Integrated Arts and Sciences, Hiroshima University
Higashi-Hiroshima, Japan
[3]Institute of Physics of the National Academy of Sciences
Nauka Avenue 46, 03680 Kiev, Ukraine
[4]Graduate School of Engineering, Hiroshima University
Higashi-Hiroshima, Japan
[5]Department of Physics, Aoyama-Gakuin University
Sagamihara, Kanagawa, Japan

ABSTRACT

The discovery of novel high-T_c superconductivity in MgB_2 (T_c = 39.5 K) and Li0.48(THF)yHfNCl (Tc = 25.5 K) initiated substantial progress in the field of superconductivity physics and its applications despite the fact that competing high-Tc cuprates remain the world leaders in almost all practically important superconducting parameters. This article describes electron tunneling and point-contact experimental studies of the indicated two materials and related substances, being crucial to elucidate the character of the quasiparticle energy spectrum both in superconducting and normal state. The account is based mostly on our own experiments, although works carried out in other laboratories are taken into account as well.

---

* E-mail address: ekino@hiroshima-u.ac.jp

We studied superconducting gap structures by means of break-junction tunneling spectroscopy (BJTS), scanning tunneling spectroscopy (STS) and point-contact spectroscopy (PCS).

In the case of MgB_2, tunnel conductance $G(V) = dI/dV(V)$ reveals multiple-gap features. Here I is the quasiparticle tunnel current and V is the bias voltage. The two-gap model including the proximity effect (the correlated two-gap model) was used to describe the observed multiple-gap features in BJTS and STS. Three specific gap values can be identified as follows: $\Delta_S = 2 - 2.5$, $\Delta_M = 4.5 - 7.5$ and $\Delta_L = 10 - 12$ meV. The observed values $2\Delta_L(4.2\text{ K})/k_BT_c > 5 - 6$ constitute the largest known values of the superconducting gap to T_c ratio except for those appropriate to copper oxides and certain organic materials. Here k_B is the Boltzmann constant. On the other hand, ratios of the small gaps Δ_S to corresponding T_c's, $2\Delta_S(4.2\text{ K})/k_BT_c$, fall into the range 1.2 – 1.5, which is considerably below the Bardeen-Cooper-Schrieffer (BCS) value $2\Delta(4.2\text{ K})/k_BT_c \approx 3.5$ inherent to s-wave superconductors. The extrapolated highest gap-closing field B_c for the largest gap agrees with the upper-critical field, thereby indicating that this gap is predominant in MgB_2. Point-contact conductance $G(V)$ also demonstrates multiple-gap structures. Peak positions in the second derivative conductance of PCS are in a reasonable agreement with the phonon spectrum frequencies revealed by the inelastic neutron scattering measurements. The high-energy boron vibration modes (~ 75 meV) are suggested to play the important role in the overall electron-phonon interaction in MgB_2. The BJTS data of other AlB_2 type superconductors show the strong-coupling-size values of the ratio $2\Delta(0)/k_BT_c$, specifically, 4.2 – 4.5 for NbB_2 and 4.2 – 4.6 for CaAlSi.

Tunneling measurements have been carried out on layered nitride superconductors of the β(SmSI)-type $Li_{0.48}(THF)_xHfNCl$ (THF;C_4H_8O) ($T_c \approx 25.5$ K), $HfNCl_{0.7}$ ($T_c \approx 23$-24 K) and $ZrNCl_{0.7}$ ($T_c \approx 14$ K). BJTS reveals Bardeen – Cooper - Schrieffer (BCS) - like gap structures with typical gap values of $2\Delta(4.2\text{ K}) = 11$-12 meV for $Li_{0.48}(THF)_xHfNCl$ with the highest $T_c \approx 25.5$ K. Our measurements revealed multiple gaps and dip-hump structures, the largest gap 2Δ (4.2 K) = 17 - 20 meV closing at T_c. It comes about that the highest obtained gap ratio $2\Delta/k_BT_c \sim 8$ substantially exceeds the BCS weak-coupling limiting values: ≈ 3.5 and ≈ 4.3 for s-wave and d-wave order parameter symmetry, respectively. For $ZrNCl_{0.7}$, two gap ratios $2\Delta/k_BT_c = 6 - 8$ and $2\Delta/k_BT_c = 3 - 4$ follow from the BJTS data. Such huge values of $2\Delta/k_BT_c$ are rather unusual for conventional superconductors.

1. Introduction

High-T_c superconductivity discovered in 1986 [1] was later found to occur in many copper oxides reaching its highest superconducting transition temperature of about 160 K [2]. It turned out, however, that cuprates are not unique as regards high T_c or other superconducting parameters. For instance, superconductivity with relatively high T_c was found in MgB_2 [3], [4], $Li_{0.48}(THF)_{0.3}HfNCl$ (THF; C_4H_8O) [5]-[7], sesquicarbide Y_2C_3 [8], and iron-based oxypnictide or pnictide compounds [9]-[12]. These discoveries were later reasonably recognized as "the next breakthrough in phonon-mediated superconductivity" [11].

In particular, discovery of superconductivity in previously overlooked binary MgB_2 became a remarkable step in the perpetual hunt for new classes of the high-T_c superconductors [13]. The observed $T_c \approx 39$ K is higher than that of the one of the first copper-oxide superconductors $La_{1.85}Sr_{0.15}CuO_4$ ($T_c \approx 35$ K) involving strongly correlated electrons [1]. The crystal structure is an hcp type (P6/mmm), consisting of alternating layers of Mg atoms and B honeycomb layers The band calculation predicts that the Fermi surfaces of this compound consist of two-dimensional cylindrical sheets arising from B p_{xy} orbital, and three-dimensional tubular networks arising from B p_z orbital [14]. Holes govern conductivity term determined by two-dimensional sheets, while both holes and electrons play an essential role in conductivity originating from three-dimensional tubules. Such an energy band structure near the Fermi level was inferred from angle-resolved photoemission spectroscopy (ARPES) measurements [15], [16]. Inelastic neutron scattering showed that high-energy phonon (E_{2g} mode) plays a crucial role in superconductivity [17]. Thus, Cooper pairing of the compound concerned is most probably induced by phonon exchange [18], [19], although Coulomb (plasmon) contribution might exist as well [20].

Another interesting high-T_c superconductor $Li_{0.48}(THF)_{0.3}HfNCl$ (THF; C_4H_8O) [5], [6] exhibits relatively high $T_c \approx 25.5$ K, which exceeds $T_c \approx 23$ K of the intermetallic compound Nb_3Ge, maximal in the pre-high-T_c epoch [21], [22]. Superconductivity in the compound $Li_{0.48}(THF)_{0.3}HfNCl$ (THF; C_4H_8O) is induced by electron doping to HfN double honeycomb layers through the intercalation of Li ions and organic molecules between Cl layers. Band calculations showed that strongly hybridized Hf $5d$ and N $2p$ orbitals form the conduction band [23]-[25]. Experimental studies of nuclear magnetic resonance (NMR) [26], muon-spin rotation (μSR) [27] and X-ray absorption [28] are consistent with those predictions. On the other hand, there is another way of electron doping of HfN layers, namely, de-intercalating (removing) Cl atoms from β-HfNCl [29]. The resultant materials $HfNCl_{1-x}$ and $ZrNCl_{1-x}$ exhibit superconductivity with $T_c \approx 23.5$ K and 13 K, respectively. Magnetic measurements in the related layered superconductor $ZrNCl_{0.7}$ revealed large upper critical field $H_{c2} \approx 27$ T [30].

We believe that it is crucially important to investigate the pairing mechanism (mechanisms) in the indicated above new layered compounds to find out another possible path to high-T_c superconductivity, different from that, which led to the discovery of varying superconducting ceramic oxide families [31], [32]. It is remarkable that contrary to all efforts all over the World and despite published and unpublished claims of final success, microscopic mechanisms of Cooper instability in high-T_c oxides are not known, which can be readily seen if one compares different trustworthy sources [33], [34].

Superconductivity is a complex cooperative phenomenon, being *non-universal* in the sense that various footprints of the underlying pairing interactions are revealed in such quantities as the ratio of the Cooper pairing-induced energy gap at zero temperature $T = 0$, $2\Delta(T = 0)$, to k_BT_c; the ratio of the electron specific heat jump ΔC at T_c to the electron normal specific heat C_n above T_c; or a tunnel electron density of states (DOS) [35]. Here k_B is the Boltzmann constant. Nevertheless, in the extremely successful weak-coupling Bardeen-Cooper-Schrieffer (BCS) scheme [36] these quantities become *universal*, since BCS theory is in essence the theory of corresponding states as in the famous van der Waals theory, the first of this kind [37].

In the framework of the BCS theory and in any of its strong-coupling [35] or more sophisticated [38] modifications the energy gap $2\Delta(T)$ serves as an inherent characteristic energy scale. Electron tunneling is especially suitable as the most direct probe to measure the energy gap 2Δ because the tunnel conductance $G(V)$ (V is bias voltage) is proportional to the quasiparticle DOS [39]. Thus, tunnel measurements can experimentally determine, in particular, the indicated above ratio $2\Delta(T=0)/k_BT_c$, which is $2\pi/\gamma \approx 3.52$ for weak-coupling isotropic s-wave superconductors and, say, $4\pi/\gamma\sqrt{e} \approx 4.28$ for d-wave superconductors [40]. Here $\gamma = 1.781...$ is the Euler constant, whereas $e = 2.718...$ is the base of natural logarithms (Napier number). In practice $2\Delta(T=0)/k_BT_c$ might be much larger than presented values, reflecting strong-coupling effects [35], an unconventional order parameter symmetry [33], [40], [41] or the influence of charge-density waves (CDWs) [42].

The superconducting gapped spectrum was probed in MgB_2 using various methods [15], [43]. Multiple-gap structures were found both in polycrystalline and single-crystal samples. Multiple (two, in most cases) gaps were attributed [19], [44]-[46] to the intrinsic two-band superconductivity [47], [48]. If the latter concept is valid, the inter-band scattering of quasiparticles by impurities should secure a merging of two gaps into a single one [49] (a direct consequence of the famous Anderson theorem [50]). However, in dirty polycrystalline or in impurity-substituted samples, distinct multiple-gap structures survive [43], contrary to expectations originating from the two-band model. It is possible to propose another totally different explanation of such multiple-gap structures on the basis of the spatial phase separation or intrinsic MgB_2 inhomogeneity [51]. Bearing in mind various feasible interpretations, we shall review the current situation in the field, as well as our own studies of the multiple-gap features in superconducting MgB_2.

Not so many experiments have been performed dealing with superconducting properties of another family of multiple-gap layered nitride superconductors $Li_{0.48}(THF)_xHfNCl$, $HfNCl_{0.7}$ and $ZrNCl_{0.7}$. The reason is the instability of these samples in the open air making the experiments very difficult. It is especially true as concerns tunneling data on superconducting gaps. On the other hand, less direct specific-heat measurements of the superconducting sample $Li_{0.12}ZrNCl$ with $T_c \approx 12.7$ K showed that the ratio $2\Delta(0)/k_BT_c \approx 5$ conspicuously exceeds the weak-coupling BCS value [52].

It is remarkable that recent experimental studies of $Li_{0.48}(THF)_xHfNCl$ and $ZrNCl_{0.7}$ revealed a negligibly small isotope effect $T_c \sim M^{0.07}$ (M here is the nitrogen atomic mass) [53], [54]. This result might be caused, e.g., by (i) the Born-Mayer repulsion of ion cores [55], crucially important for materials with noble metal atoms [56], [57], (ii) a purely electronic genesis of Cooper pairing [58]-[61], (iii) polaronic effects [62], (iv) anharmonicity of phonons [63]-[65], (v) van Hove singularity of the electron DOS because of the crystal lattice reduced dimensionality [66], [67], (vi) d-wave symmetry of the superconducting order parameter [68], (vii) or simply a non-equal contribution of different constituents into lattice vibration modes and Coulomb pseudopotential μ^* [68]-[72]. We emphasize that a small or inverse isotope effect is not a smoking gun of superconductivity being induced by a non-phonon bosonic glue, contrary to the statements concerning many materials, e. g., iron-based ones [73]. On the other hand, a discovered existence of the isotope effect in superconductivity cannot be considered as a crucial experiment proving the phonon background of superconductivity, since phonons (thermal) may participate in pair-breaking effects only (a well-known phenomenon *per se* [74], [75]), making no contribution to Cooper pairing [76].

At the same time, if one analyses high-pressure X-ray and Raman scattering results for $Li_{0.48}(THF)_xHfNCl$ and $ZrNCl_{0.7}$ crystals on the basis of the approximate strong-coupling McMillan formula taking into account only phonon-exchange interaction [77], a large electron-phonon coupling constant $\lambda > 3$ appears [78]. Thus, to reconcile experimental data of references [53], [54] and [78], it seems plausible (although not obligatory) to invoke a non-phonon superconductivity mechanism as a possible contributor to the actual Cooper pairing. Hence, one of the aims of this article is to analyze superconducting gaps of those layered nitride compounds, which would be helpful to uncover the relevant high-T_c superconducting mechanism (mechanisms).

The outline of this review is as follows. Section 2 introduces and describes the sample preparations, tunneling techniques and the way of data handling employed in the investigation of the physical properties of the compounds concerned. In Sections 3 to 5, we present the measured tunnel superconducting energy gap structures of MgB_2, other AlB_2 type superconductors (NbB_2 and CaAlSi), as well as layered nitride superconductors $Li_{0.48}(C_4H_8O)_yHfNCl$, $HfNCl_{0.7}$ and $ZrNCl_{0.7}$. In Section 3, data on multiple-gap structures of MgB_2 are considered for different kinds of samples and spectroscopies, namely break junction tunneling spectroscopy (BJTS), scanning tunneling spectroscopy (STS), point-contact spectroscopy (PCS). Peculiarities of the electron-phonon interaction in MgB_2 are discussed on the basis of the experimental results. In Section 4, the gap structures of NbB_2 and CaAlSi are presented and compared to those of MgB_2. Section 5 is devoted to the analysis of the superconducting energy gap structures for layered nitride compounds, obtained mainly by BJTS.

2. Experimental

2.1. Sample preparations

We used two kinds of MgB_2 polycrystalline samples, pellets and wires for BJTS, STS or PCS measurements. Synthesis was carried out in the following manner. MgB_2 pellets were fabricated from a mixture of $3N$ Mg powder and $2N$ amorphous B at 973 K in 200 atm Ar gas for 10 hours, accompanied by annealing at 1623 K under 5.5 GPa [3]. Wire segments of high purity MgB_2 were made by sealing boron filaments purchased from Textron (~ 100 μmϕ; $5N$) into a Mo tube with excess Mg ($3N$), using a ratio of Mg/B = 3. Reactions were carried out during 4 hours at 1223 K [79].

NbB_2 and CaAlSi to be used in tunneling measurements were produced along the following lines. Polycrystalline samples with the nominal composition $NbB_{2.1}$ were synthesized between 1873 and 2073 K under the pressure of 5.5 GPa in Ar atmosphere [80]. CaAlSi single crystal samples were prepared by RF heating at 1473 K in a BN crucible under Ar atmosphere [81]. The nominal composition of each reagent was Ca: Al: Si = 1: 1: 1, and the Ca metal was purified by a vacuum-distillation using a Ta crucible prior to be used.

Samples of $Li_{0.48}(THF)_xHfNCl$, $HfNCl_{0.7}$ and $ZrNCl_{0.7}$ for tunneling measurements were synthesized in the following fashion [5], [6]. Polycrystalline samples of alkali metal intercalated superconductor $Li_{0.48}(THF)_xHfNCl$ were prepared by reacting the β-HfNCl with Li naphthalene solution in tetrahydorofuran (THF) during 24 h. A molar ratio of β-HfNCl/Li corresponded to 1.0/0.48. The powder sample was compressed at a pressure of 1.8 GPa. All steps were made in an Ar filled glove box. Polycrystalline samples of Cl deintercalated superconductors $HfNCl_{0.7}$ and $ZrNCl_{0.7}$ were prepared by reacting β-HfNCl and β-ZrNCl with metal azides KN_3 and NaN_3 in a molar ratio of 1.0/0.3, respectively [29]. Each reaction process was carried out by annealing at 573 K and 593 K in vacuum during 24 h. The powder sample was compressed at a pressure of 1.8 GPa in an Ar filled glove box.

2.2. Experimental techniques

We have carried out the measurements by the four-probe *in situ* break-junction technique [82]. In this method, the thin platelet sample on a substrate is carefully cracked at 4.2 K to form a superconductor – insulator – superconductor (SIS) junction with a fresh and unaffected interface, which is obtained by applying the bending force parallel to the thickness direction. We believe it is the most effective way to fabricate tunnel junctions of this kind from a chemically reactive compound [82]-[85]. For an SIS junction, the characteristic peak-to-peak bias separation $V_{p\text{-}p}$ in the tunnel conductance, $G(V) \equiv dI/dV$, corresponds to $4\Delta/e$, where 2Δ stands for the superconducting energy gap between electron-like and hole-like quasiparticle branches, I is the quasiparticle current, V is voltage, and e is the (positive) elementary charge. Such an SIS junction is so sensitive that it can probe gap edge structures even at high T close to T_c because of the amplifying convolution in the integrand of the formula for $G(V)$ between electron gapped DOSes $N_{1,2}(E)$ originating from both electrodes [21], [86].

We also employed scanning tunneling microscopy/spectroscopy (STM/STS) measurements with PtIr tip in the ultra high vacuum ($< 10^{-8}$ Pa). The apparatus used in the present measurements was an OMICRON LT-UHV-STM system equipped with a low-T preparation chamber and other additional functions. The STM chamber temperature is kept at 5 K during the measurements. For the STM studies, the applied bias voltage $V = 15$ mV refers to the sample, and the tunneling current range is $I = 0.05$ - 0.3 nA. The typical scanning range is 200 nm × 200 nm. For the STS measurements, in which the superconductor - insulator (vacuum) - normal metal (SIN) junction is formed, the bias range is extended up to $V = \pm 30$ mV. The tunnel conductance dI/dV is calculated numerically from the I-V characteristics with the resolution better than 0.5 mV. In particular, a combination of break-junction and STS experiments allows one to unequivocally distinguish between realizations of SIS or SIN configurations.

We also performed PCS measurements by pressing a normal metal (Cu) wire on the cracked surface of MgB_2. The $dI/dV(V)$ characteristics of a barrier-less, metallic point contact (PC) formed between a normal metal and a superconductor can detect the superconducting gap feature, which stems from the Andreev reflection [87]. Therefore, this method can also be used as a complementary approach in addition to BJTS and STS measurements. The $d^2I/dV^2(V)$ characteristics of PC can reveal detailed spectral structures of the electron-phonon coupling ($\alpha^2F(\omega)$) at low T [88]-[90].

2.3. The way of data handling

To carry out a quantitative analysis, we compare experimental data with the well-known expression for the tunnel conductance across an SIS junction [21]

$$G(V) \equiv \frac{dI_{SIS}}{dV} = C\int_{-\infty}^{+\infty} N_1(E)\frac{dN_2(E+eV)}{dV}[f(E) - f(E+eV)]$$
$$+ N_1(E)N_2(E+eV)\left[-\frac{df(E+eV)}{dV}\right]dE \qquad (1)$$

Here C and $f(E)$ are the scaling parameter and the Fermi distribution function, respectively. Expression (1) is based on the concept of isotropic tunneling [21], [91]. If there is certain directionality, being important in the case of a significant electron spectrum anisotropy or an existence of several Fermi surface sheets, Eq. (1) should be substantially modified [91], [92], [129]. Unfortunately, in this situation the generality of the calculation is lost and one should make further specific *ad hoc* assumptions. Hereafter, to fit the results we use the complex density of states function $N(E, \varepsilon)$, which contains a phenomenological broadening parameter ε that has been originally introduced by Dynes *et al* [93], [94] as the lifetime quasiparticle broadening.

$$N(E,\varepsilon) = \left|\mathrm{Re}\frac{E - \mathrm{i}\varepsilon}{\sqrt{(E-\mathrm{i}\varepsilon)^2 - \Delta^2}}\right| \qquad (2)$$

The broadening most probably originates from the inelastic electron-electron scattering or the inverse proximity effect from the adjacent normal region [95], although other reasons are also possible [96] (see below). Hereafter Δ, ε, and C are considered as fitting parameters.

Taking C in Eq. (1) as a constant (normal state conductance) is an approximation valid for calculations of currents at small bias voltages, for which Δ is the relevant energy scale. This approximation corresponds to energy-independent tunnel matrix elements T_{12}. Harrison showed [97] that in the quasiclassical scheme T_{12} is a product of the reciprocals to the normal-metal quasiparticle densities of states on the left and on the right of the barrier times the tunnel integral. Therefore, those densities of states cancel out from the net expressions for the tunneling rate (see also Refs [21], [98]-[100]). On the contrary, the gap-induced square-root factors in the superconducting quasiparticle densities of states appear explicitly [101] in expressions for tunnel currents and conductivities in the same way as they enter Eq. (2). The same kind of approximation is used for calculation of Josephson currents between conventional superconductors [102], [103].

Eq. (2) describes the lifetime broadening, which always takes place even for superconductors with a spatially uniform order parameter. The existence of the parameter ε makes singularities smoother in the gapped superconducting electron DOSes. At the same time, the energy gap is filled with growing ε, which implies the existence of a leakage current.

The phenomenological Eq. (2) has been recently justified by a calculation of th e environment- (photon-) assisted high-T tunneling, which involves the inverse quasiparticle lifetime $e^2 k_B T_{env} R_{env}/\hbar^2$ [96]. Here T_{env} and R_{env} are the temperature and effective resistance of the dissipative environment. As a consequence, the probability density $P(E)$ for an electron to emit a photon of energy E in the limits of small R and E takes the form [96].

$$P(E) = \frac{1}{\pi\Delta}\frac{\varepsilon}{\varepsilon^2 + (E/\Delta)^2}. \tag{2a}$$

The effective DOS calculated on the basis of Eq. (2a) results exactly in the Dynes Eq. (2).

By no means should one forget about the existence of another significant smearing phenomenon: an inevitable gap amplitude distribution affecting the quasiparticle current, the scatter being either larger or smaller than the instrumental resolution. Lifetime broadening and the gap spread are complementary factors, and it is not a priori known, which of them is more important for a certain tunnel junction.

Multiple gap features were actually observed long ago even for low-T_c superconductors (see, e. g, [104]-[107]). Such observations became common for high-T_c oxides with short coherence lengths ξ [47], [108], [109]. The phenomena might be due to at least several reasons: (i) a true multiple-gap (e.g., two-gap) structure of the quasiparticle electron spectrum in the superconducting state [19], [48], [50], [110], [111] if the Anderson impurity mixing of different electron wave functions [112] is irrelevant; (ii) a more general kind of an energy gap anisotropy in the momentum (**k**) space [113]-[116] robust under the impurity influence; (iii) a spatial (**r** space) inhomogeneity of normal and superconducting electronic properties in the area, which determines the tunnel current of a specific junction [117]-[122]. Recently the two-gap model has been applied to pnictides, where the distinction between coherence factors was predicted for alternative phase differences between superconducting gaps on hole and electron Fermi surfaces [123].

Whatever this microscopic background, the simplest possible function $N(E, \varepsilon)$, averaged over a gap distribution, can be written down as

$$N(E,\varepsilon) = \int_{-\infty}^{+\infty} \left| \mathrm{Re}\left\{ \frac{E - i\varepsilon}{\sqrt{(E - i\varepsilon)^2 - \Delta^2}} \right\} \right| P(\Delta) d\Delta \tag{3}$$

where $P(\Delta)$ is the distribution function of the energy gap values. Since the existence of non-zero ε and the occurrence of the gap scatter $P(\Delta)$ are mutually independent, they are usually taken into account simultaneously [124]. To be specific, we adopt a random Gaussian distribution function, i.e. $P(\Delta)$ reads

$$P(\varDelta) = \frac{1}{\sqrt{2\pi}\delta_0} \exp\left(- \frac{(\varDelta - \varDelta_0)^2}{2\delta_0^2} \right) \tag{4}$$

Here δ_0 and $\varDelta_0$ are a half-width and a mean value of the energy gap distribution, respectively.

There are two types of superconducting states: intrinsic and induced. For the former case, the material becomes superconducting when the temperature drops below T_c. For the latter case, the material becomes superconducting (even if it is intrinsically a normal conductor) when it is in contact with another superconductor. There are two major ways of such a coupling between electron subsystems. One involves their spatial separation, which means the so-called proximity effect [125], [126]. The other situation is a separation in the momentum space corresponding to two-band (in a more general case, multiple-band) superconductivity [125], [127]. The mathematical expressions for energy gaps in the framework of the two-band model are formally the same as those describing the proximity effect [126]-[129].

$$\Delta_1(E) = \left\{ \Delta_1^{ph} + \Gamma_1\Delta_2(E) \Big/ \sqrt{\Delta_2^2(E) - (E - i\varepsilon_2)^2} \right\} \Big/ \left\{ 1 + \Gamma_1 \Big/ \sqrt{\Delta_2^2(E) - (E - i\varepsilon_2)^2} \right\} \quad (5)$$

$$\Delta_2(E) = \left\{ \Delta_2^{ph} + \Gamma_2\Delta_1(E) \Big/ \sqrt{\Delta_1^2(E) - (E - i\varepsilon_1)^2} \right\} \Big/ \left\{ 1 + \Gamma_2 \Big/ \sqrt{\Delta_1^2(E) - (E - i\varepsilon_1)^2} \right\} \quad (6)$$

where Δ_1^{ph} and Δ_2^{ph} ($\Delta_1^{ph} < \Delta_2^{ph}$) are the parent gap parameters, Γ_1 and Γ_2 are parameters describing the strength of the coupling between two superconducting subsystems, ε_1 and ε_2 are the phenomenological broadening parameters. The gaps Δ_1^{ph} and Δ_2^{ph} correspond to the bare order parameters in the relevant bands of the actual two-band superconductor or to the characteristics of a proximity sandwich. The sense of phenomenological parameters Γ_i might be twofold. They mean transition rates between bands in the case of a two-band superconductor. For the proximity sandwich they should be considered as intensities of the electron transfer between a normal state surface layer and a bulk superconducting phase.

In order to solve the equation system for $\Delta_1(E)$ and $\Delta_2(E)$ numerically, the iteration method was employed [126], in which the following convergence criterion was assumed to fulfill

$$\Delta_i^{(n+1)}(E) - \Delta_i^{(n)}(E) < 10^{-12} \quad (i = 1, 2), \quad (7)$$

where n denotes an iteration number. DOSes $N_1(E)$ and $N_2(E)$ for each band are as follows:

$$N_1(E) = \left| \mathrm{Re} \frac{E - i\varepsilon_1}{\sqrt{(E - i\varepsilon_1)^2 - \Delta_1^2(E)}} \right| \quad (8)$$

$$N_2(E) = \left| \mathrm{Re} \frac{E - i\varepsilon_2}{\sqrt{(E - i\varepsilon_2)^2 - \Delta_2^2(E)}} \right| \quad (9)$$

Figure 1 shows examples of calculated correlated $N_1(E)$ and $N_2(E)$, respectively. Note that the gap-peak positions and peak structures are sensitive to the inter-band coupling strengths Γ_1 and Γ_2. For $N_1(E)$ in Figure 1(a), the first-peak position corresponding to the effective gap value becomes larger than the bare quantity Δ_1^{ph}.

In the case of $\Delta_1^{ph} >> \Gamma_1$, the gap parameter can be estimated to be $\Delta_1 = \Delta_1^{ph} + \Gamma_1$. For small Γ_1, $N_1(E)$ is close to the separate BCS gap density of states.

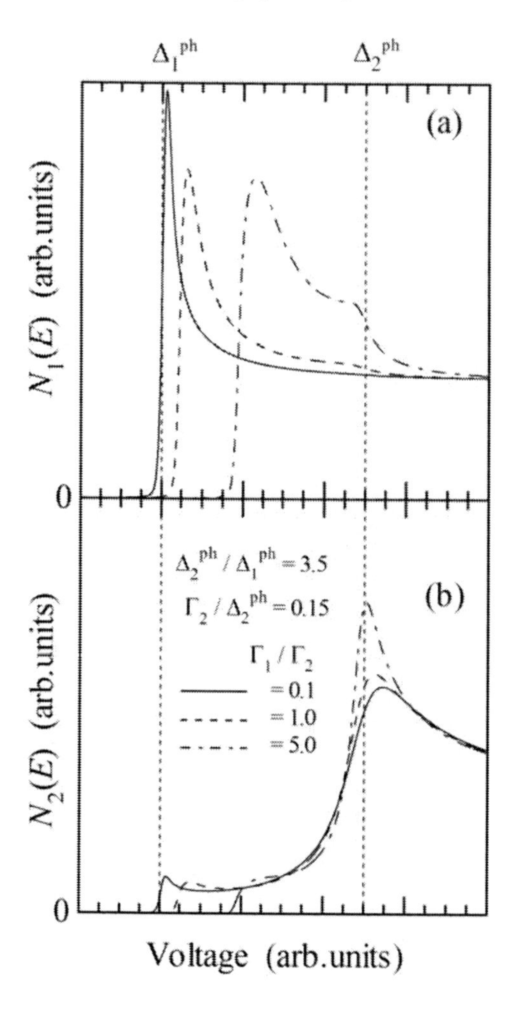

Figure 1. Partial band density of states (DOSes) $N_1(E)$ and $N_2(E)$ for the correlated two-gap model. Γ_1 and Γ_2 are parameters describing the strength of the coupling between two superconducting subsystems. The gaps Δ_1^{ph} and Δ_2^{ph} correspond to the bare order parameters in the relevant bands. Solid, dashed and dot-dashed curves correspond to $\Gamma_1/\Gamma_2 = 0.1$, 1.0 and 5.0, respectively. $\Gamma_2 = 0.15\Delta_2^{ph}$ and $\Delta_2^{ph} = 3.5\Delta_1^{ph}$.

The condition $\Gamma_1/\Gamma_2 >> 1$ means that the correlation influence of $N_2(E)$ on $N_1(E)$ is much stronger than that of $N_1(E)$ on $N_2(E)$. In $N_2(E)$ of Figure 1(b), the second-peak position corresponds approximately to Δ_2^{ph}. Therefore, the actual gap energy is always $\Delta_2 \approx \Delta_2^{ph}$ despite large variations of Γ_1. In both $N_1(E)$ and $N_2(E)$, the first peak position depends on the ratio Γ_1/Γ_2, and the larger is the Γ_1 value the broader becomes the coherent peak.

Depending on the relative orientation of the crystallographic axis to the current direction or the nature of the active surface layer, both $N_1(E)$ and $N_2(E)$ can contribute to the tunnel conductance. Therefore, as an estimate we use a linear combination of the constituent DOSes $N_1(E)$ and $N_2(E)$ to simulate the effective total DOS ($N_{total}(E)$).

$$N_{total}(E) = (1-x)N_1(E) + xN_2(E) \qquad (0 < x < 1) \tag{10}$$

We call this interpolation scheme the "correlated two-gap model". Figure 2 represents the calculated SIS conductance (formula (1)) for the limiting cases of separately acting superconducting components with DOSes $N_1(E)$ ($x = 0$) or $N_2(E)$ ($x = 1$) using the parameters of Figure 1. In Figure 2(a), the effect of the second gap on $N_1(E)$ is not noticeable for small Γ_1, while large Γ_1 produce clear dip-hump structures. On the other hand, in Figure 2(b), the large value of Γ_1 conspicuously enhances the primordial gap-edge peak in $N_2(E)$.

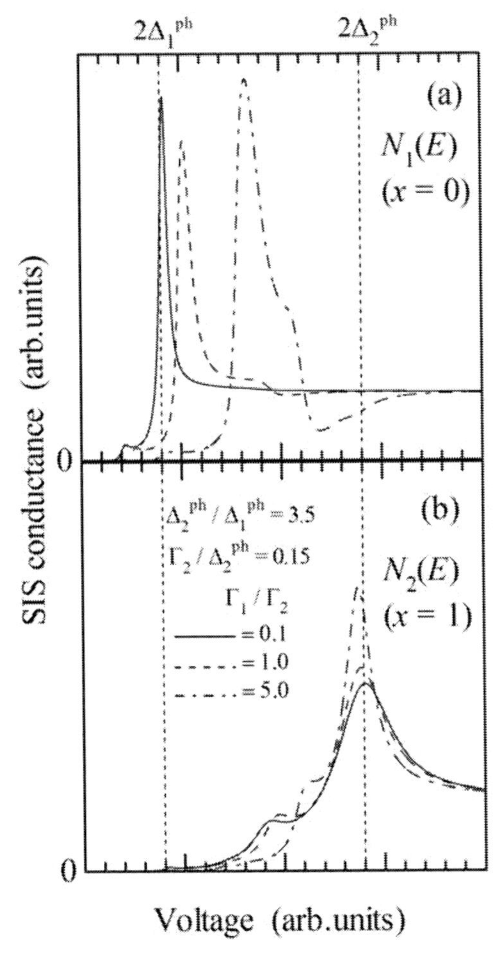

Figure 2. Calculated superconductor-insulator-superconductor (SIS) conductance for the limiting cases of the correlated two-gap model, namely, for constituent superconducting subsystems making separate contributions to the quasiparticle tunnel current. The total DOS for (a) and (b) corresponds to $x = 0$ and 1, respectively (see Eq. (10)). Parameters are the same as those of Figure 1. Solid, dashed and dot-dashed curves correspond to $\Gamma_1/\Gamma_2 = 0.1$, 1.0 and 5.0, respectively. $\Gamma_2 = 0.15\Delta_2^{ph}$ and $\Delta_2^{ph} = 3.5\Delta_1^{ph}$.

Figure 3 shows the overall SIS conductance taking into account partial contributions of superconducting subsystems into the overall quasiparticle current for x = 0.1, 0.5 and 0.9. Three pairs of peak-to-peak separations V_S, V_M and V_L are recognized. The correlation among the three-peak structures is determined by the relationship $V_M = (V_S+V_L)/2$. The peaks are located at voltages $V_S = 4\Delta_1/e$, $V_M = 2(\Delta_1 + \Delta_2)/e$, $V_L = 4\Delta_2/e$ using Δ_1 and Δ_2.

Quasiparticle scattering processes occur at the SN interface. If an electron with a momentum **k** in the N region with an energy $E < \Delta$ propagates in the direction of the interface SN it can not cross it because its energy is in the pairing-induced gap of the adjacent superconductor. Therefore, further free propagation of such an electron-like quasiparticle is impossible. Instead it becomes a component of the Cooper pair on the S side of the interface. The other component of the same pair should be an electron-like quasiparticle with the opposite momentum **-k** [36]. Hence, this electron is incorporated into the pair concerned and a *hole-like excitation* emerges in the superconductor with **-k**, i.e. moving into the depth of the normal metal in the direction opposite to the direction of the incident electron. Thus, the charge, the momentum and the energy are conserved in this process known as Andreev reflection [87].

A simple model of tunneling across an SN junction was developed by Blonder, Tinkham and Klapwijk (BTK model) [19], [130], according to which the infinitely thin energy barrier is located at the interface and the conductance takes the form

$$G = \left(\frac{dI}{dV}\right)_{SN} = C\int_{-\infty}^{+\infty}[1 + A(E) - B(E)]\left[-\frac{df(E-eV)}{dV}\right]dE \qquad (11)$$

Here $A(E)$ and $B(E)$ are the coefficients giving probabilities of Andreev and conventional electron reflections, respectively. For $E < \Delta$, $A(E) = \Delta^2/[E^2+(\Delta^2-E^2)(1+2Z^2)^2]$, and $B(E) = 1-A(E)$. For $E > \Delta$, $A(E) = u_0^2 v_0^2/\gamma_{BTK}^2$, $B(E) = \left(u_0^2 - v_0^2\right)^2 Z^2\left(1+Z^2\right)/\gamma_{BTK}^2$ and $\gamma_{BTK}^2 = \left[u_0^2 + Z^2\left(u_0^2 - v_0^2\right)\right]^2$. The quantity Z is a dimensionless barrier strength parameter and $Z = 0$ means no barrier at the SN interface. The parameters u_0 and v_0 are the Bogoliubov factors given by $u_0^2 = 1 - v_0^2 = \{1 + [(E^2 - \Delta^2)/E^2]^{1/2}\}/2$. Figure 4 is the calculated examples of BTK conductance for $Z = 0$, 0.5 and 5. The value $Z = 5$ corresponds to such a high barrier at the SN interface that one arrives at a SIN tunneling junction rather than a proximity sandwich.

It should be emphasized that the one-parameter BTK model is, in principle, a crude approximation to real junctions. In particular, it is necessary to take into account a possible directionality of superconducting Josephson and quasiparticle tunneling [131]. Moreover, the underlying physical picture behind the BTK simplified considerations is very involved especially for non-conventional Cooper pairing [132]-[134].

The validity of the adopted concept of the interface proximity-induced superconductivity was proved in the extreme case of a heterostructure involving a normal metal solid solution $La_{1.56}Sr_{0.44}CuO_4$ and an insulating oxide La_2CuO_4 [138]. Although neither of two layers were superconductors *per se*, the bilayer turned out to be a superconductor with $T_c \approx 30 \div 36$ K due to the hole accumulation on the insulating side of the sandwich.

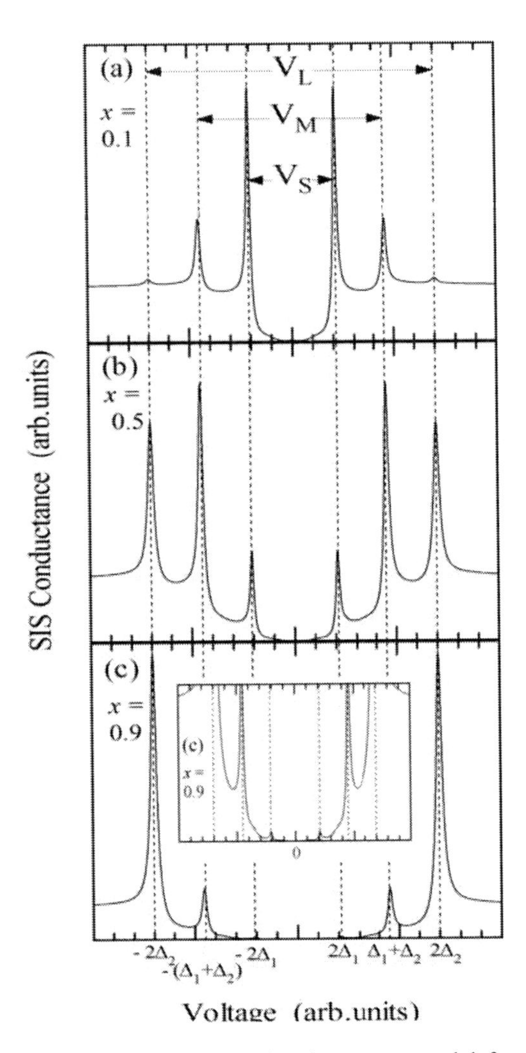

Figure 3. Calculated SIS conductance in the correlated two-gap model for x = 0.1, 0.5 and 0.9. The curves (a), (b) and (c) correspond to x = 0.1, 0.5 and 0.9, respectively. The peak positions V_S, V_M and V_L are estimated to be $4\Delta_1/e$, $2(\Delta_1+\Delta_2)/e$ and $4\Delta_2/e$, respectively, assuming $\Delta_1 < \Delta_2$. Inset in (c) is the escalerd central part of the figure.

One should bear in mind that, if a vibration mode with energy $\hbar\omega_0$ (= eV) is excited by the tunnel current I via the electron-phonon interaction, the current magnitude is reduced. Therefore, the point-contact resistance $R_{PC}(eV)$ is changed depending on the relevant relaxation time $\tau(eV)$ involving single phonon collision or emission processes. If the effect is relatively weak, the next relation is valid [90].

$$-\frac{d^2I}{dV^2}(V) \propto \frac{d}{dV}\left(\frac{1}{\tau(V)}\right) = \alpha^2(V)F(V) \tag{12}$$

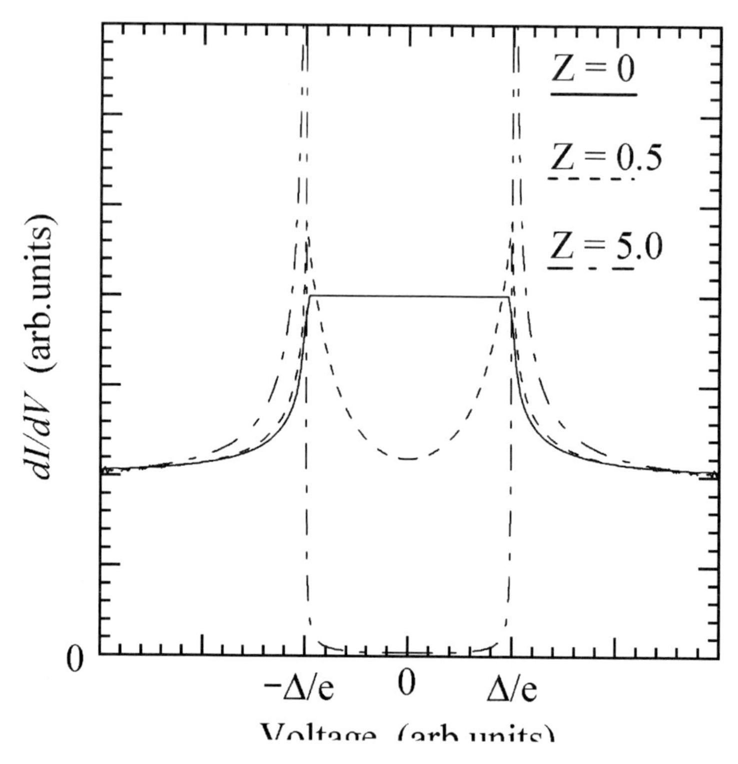

Figure 4. Calculated BTK conductance (see Eq. (11)). Solid, dashed and dot-dashed curves correspond to the dimensionless barrier strength $Z = 0$, 0.5 and 5.0.

3. SUPERCONDUCTING GAPS IN MgB_2

3.1. BJTS of sintered polycrystalline MgB_2

Polycrystalline samples of superconducting MgB_2 with the critical temperatures $T_c = 39$ K were studied by break-junction tunneling technique. The bias voltage-dependent $dI/dV(V)$ demonstrates varying multiple-gap features. All experimental curves exhibit underlying singularity patterns, which can be interpreted as a manifestation of two- (in the majority of instances) or three-peak structures. However, gap locations and the exact form of the coherent gap-edge peaks vary from curve to curve. Three typical gap intervals can be inferred from the data: $\Delta = 2 - 2.5$, $6 - 7.5$ and $10 - 12$ meV. In addition, minor sub-gap and outer-gap structures are sometimes observed. The correlated two-gap model can formally describe the observed multiple-gap spectra. All these gaps disappear above the same T_c (≈ 39 K). We have also observed the conductance spectra showing features inherent to normal-metal and/or low-T_c phases. The magnetic fields, which suppress superconducting gaps, depend on the zero-field energy gap value taken at 4.2 K. The extrapolated highest gap-closing field for the largest gap feature correlates with the independently measured upper critical magnetic field, thereby indicating that this gap is the main one, describing superconductivity in MgB_2.

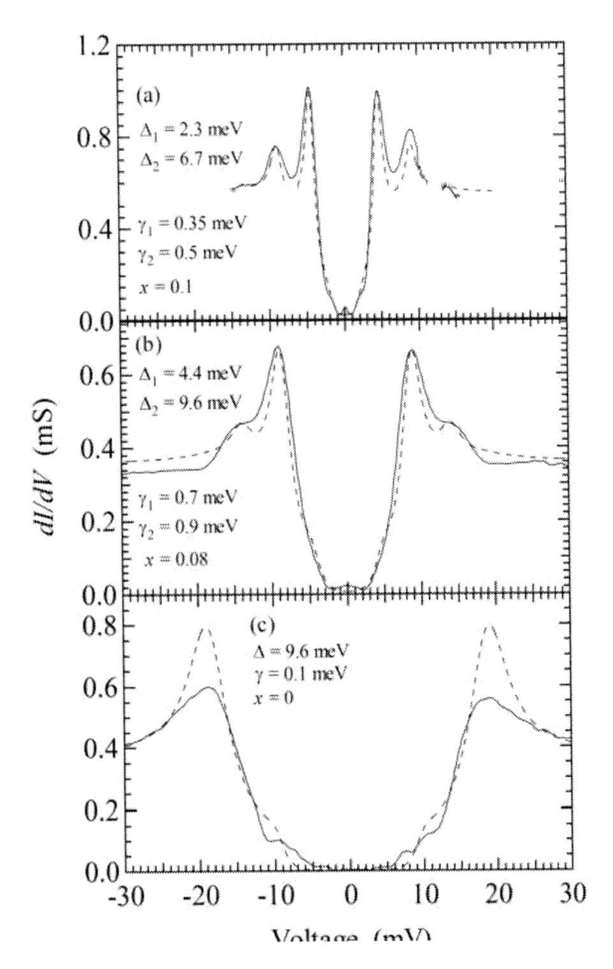

Figure 5. Break-junction tunneling spectra (BJTS) at 4.2 K are shown for three different representative measured data sets. The broken curves are calculated SIS conductances with the weighted sum of Bardeen-Cooper-Schrieffer-like (BCS-like) DOSes.

3.1.1. Energy gap structures at 4.2 K

Tunnel conductances *dI/dV* obtained for different break-junctions at 4.2 K can be classified as three typical ones presented in Figure 5. All of them are reproducible either among different samples or different junctions appropriate to the same sample. Additional fine structures around zero-bias suggest *n*-particle tunneling [21]. The BJTS data often accompanies the hump structures outside of the main peaks as shown in Figure 5(a). The intensities of hump structures vary among the spectra, while their peak-to-peak bias separations are almost the same.

Typical three-peak structures may be characterized by the relationship $V_M = (V_S + V_L)/2$. From this relationship, it comes about that two energy gaps for superconductors on each side of the SIS junction should exist. In order to comprehend the relation between the multiple gap-like features in SIS junctions, we fit the experimental spectra using a discussed above simple interpolation model [$N(\Delta_1, \Delta_2) = (1\text{-}x)N_1(\Delta_1) + xN_2(\Delta_2)$ (Eq. 3.15)], where the DOSes

$N_j(\Delta_j)$ (j = 1,2) are given by Eq. (2) [93] (see Section 2). The dashed curves in Figure 5 show the results of calculations. To reconcile the experimental conductance curves with calculated ones, a following identification should be used $V_S = 4\Delta_1/e$, $V_M = 2(\Delta_1 + \Delta_2)/e$ and $V_L = 4\Delta_2/e$. Note that the coherent peaks at the larger gap positions (V_L) can be easily reduced by a slight increase of ε_j. The calculated value of the smaller energy gap Δ_1 is about 2 meV, while the larger one is $\Delta_2 \approx 6.5$ meV.

The solid curve in Figure 5(b) shows a representative measured spectrum with the gap size twice as large as V_S of Figure 5(a). This kind of spectrum also involves the outer hump structures (shown by arrows), but their strength is weaker than that of their counterparts from Figure 5(a). One sees that the calculated SIS conductance can reproduce, in principle, the spectral features with the best fitting energy gap values Δ_1 = 4.4 meV and Δ_2 = 9.6 meV. It should be noted that the spectra with a single gap of $\Delta \approx 9.6$ meV are also observed as shown in Figure 5(c). This spectrum is roughly fitted by the SIS conductance with a single-gap BCS-like DOS, although the observed coherent peaks are smeared as compared with theoretical results.

3.1.2. Induced superconducting gaps in BJTS data

We indicated above that main multiple gap structures in dI/dV appropriate to MgB_2 can be imitated by the introduction of an SIS conductance with the weighted sum of the BCS-like DOSes. However the overall picture is much more involved. Namely, other gap-related features appear in many experimental BJTS $G(V)$ curves having varying shapes and locations. In particular, conspicuous dips or shoulders appear near V_M (and V_L), which are presented below. They can be hardly explained by the simple sum of the DOSes only. It is necessary to invoke also the concept of the induced superconductivity described, e. g., by the correlated two-gap model already presented above (see Eqs. (5) and (6)).

Typical tunnel conductances of BJTS made of MgB_2 are shown in Figure 6 to possess clear-cut multiple-gap structures. The peak locations V_S and V_M (or V_L) are almost the same as in the spectrum of Figure 5(a). Three pairs of peaks are indicated by arrows in Figures 5(a) and 5(b) ($V_S \approx 9$ mV, $V_M \approx 17$ mV, $V_L \approx 25$ mV). In Figures 5(b) and 5(d) well-developed structures are observed whereas the spectrum shown in Figure 5(c) contains only moderate shoulders in addition to the more or less conventional superconducting coherent peaks. The dashed lines correspond to the fitting curves of the correlated two-gap model discussed above. Fitting is good enough over the entire bias voltage range up to $\pm$20 mV.

Since the required fitting values x are of the order 0.1, the spectra in Figure 6 are dominated by the DOS $N_1(E)$ characterized by the small gap Δ_1. The calculated spectrum of Figure 6(a) were obtained using the intrinsic gap parameter Δ_1^{ph} = 2.2 meV and the ratio $\Gamma_1/\Gamma_2 = 0.4$ (see Sections 2 and 3), whereas the experimental spectrum depicted in Figure 6(b) is fitted well by taking $\Delta_1^{ph} = 0$ and relatively large ratio $\Gamma_1/\Gamma_2 = 10$. Both a relatively small x and large Γ_1 in Figure 6(b) result in the sharper gap edge than that appropriate to Figure 6(a) in spite of the negligible intrinsic pairing value Δ_1^{ph} for Figure 6(b).

The observed spectrum of Figure 6(c) is reproduced with $\Delta_1^{ph} \approx 1.5$ meV and the ratio Γ_1/Γ_2 = 0.7. In Figure 6(d), the features corresponding to V_M and V_S overlap. The fitting values Δ_1^{ph} = 0 and Γ_1/Γ_2 = 6.8 satisfactorily reproduce the experimental peak-dip structures similarly to the BJTS shown in Figure 6(b). The shapes of the fitting hump-dip-peak structures strongly depend on the value of the ratio Γ_1/Γ_2. Therefore, dip-hump structures should originate from the induced superconductivity. The resulting smaller and larger gap values are estimated to be $\Delta_1 = 2.2 \pm 0.3$ meV and $\Delta_2 = 6.0 \pm 1.5$ meV, respectively. The value of Δ_1 is kept almost constant among different junctions in spite of the varying proximity effect (induced superconductivity). On the other hand, the values of Δ_2 differ substantially for various junctions,

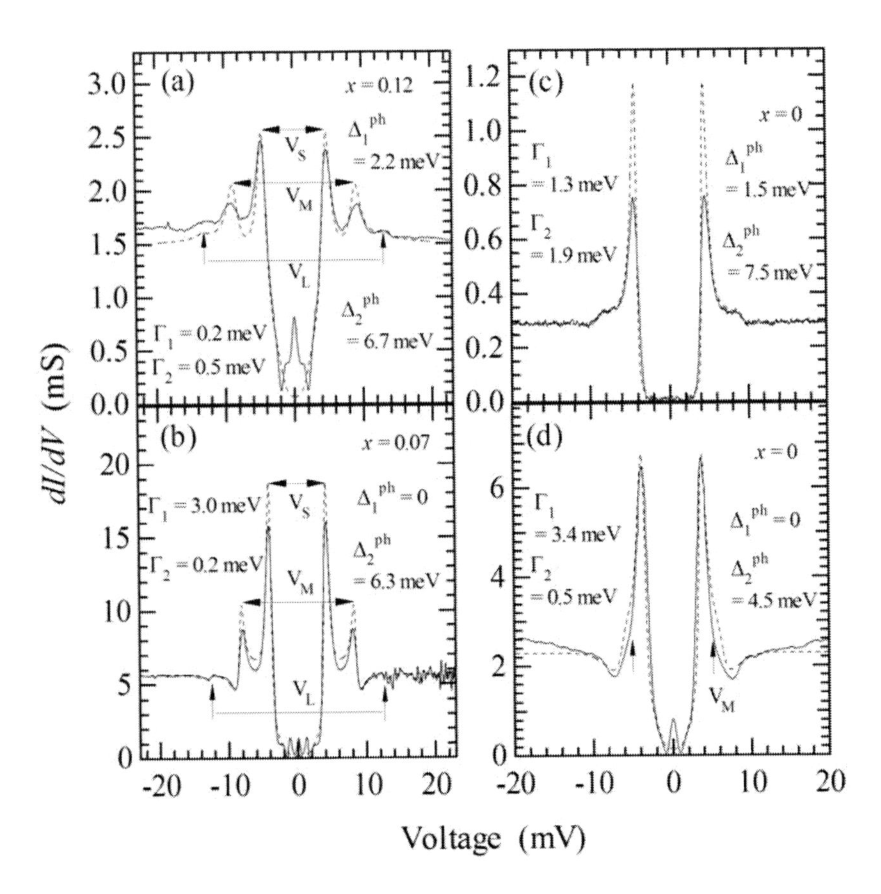

Figure 6. Multiple-gap tunnel conductance spectra for different MgB_2 break junctions measured at T = 4.2 K. The dashed curves are the results of fitting on the basis of the correlated two-gap model (see Eq. (10)).

Other types of tunnel spectra with sharp gap-edge peaks at smaller voltages, smeared features at larger gap locations $V_S \sim 20$ mV and very low-leakage conductance at zero bias are also often observed for MgB_2. Such spectra are shown in Figure 7. They are well reproduced by the correlated two-gap model similarly to the spectra depicted in Figure 6.

Since the values of model parameters in Figure 7 differ considerably from those appropriate to Figure 6, the DOSes are denoted as $N_3(E)$ and $N_4(E)$ rather than $N_1(E)$ and $N_2(E)$. The corresponding total DOS is now $N(E) = (1\text{-}x)N_3(E) + xN_4(E)$. The former DOS involves the smaller gap Δ_3 ($\approx \Delta_3^{ph} + \Gamma_3$), while is characterized by the larger one Δ_4 ($\approx \Delta_4^{ph}$, $\Delta_3 < \Delta_4$). Since the actual fitting value $x < 0.2$ in Figure 7, the spectra reveal mainly the small-gap component $N_3(E)$. The fit parameters are $\Delta_3^{ph} = 2.0 \pm 1.0$ meV, $\Delta_4^{ph} = 11.5 \pm 2.5$ meV, and $\Gamma_3/\Gamma_4 = 1.2\pm0.3$. Hence, $\Delta_3 = 4.7\pm0.7$ meV and $\Delta_4 = 11.5\pm2.5$ meV. The dispersion of Γ_3/Γ_4 is narrower than that of Γ_1/Γ_2. The fitting parameter Δ_3 is of the same order as Δ_2 from Figure 7, while Δ_4 is larger than any gap value in Figure 6. This indicates the existence of another kind of a gap Δ_4 in MgB_2 (larger gap) different from Δ_1, Δ_2 ($\approx \Delta_3$).

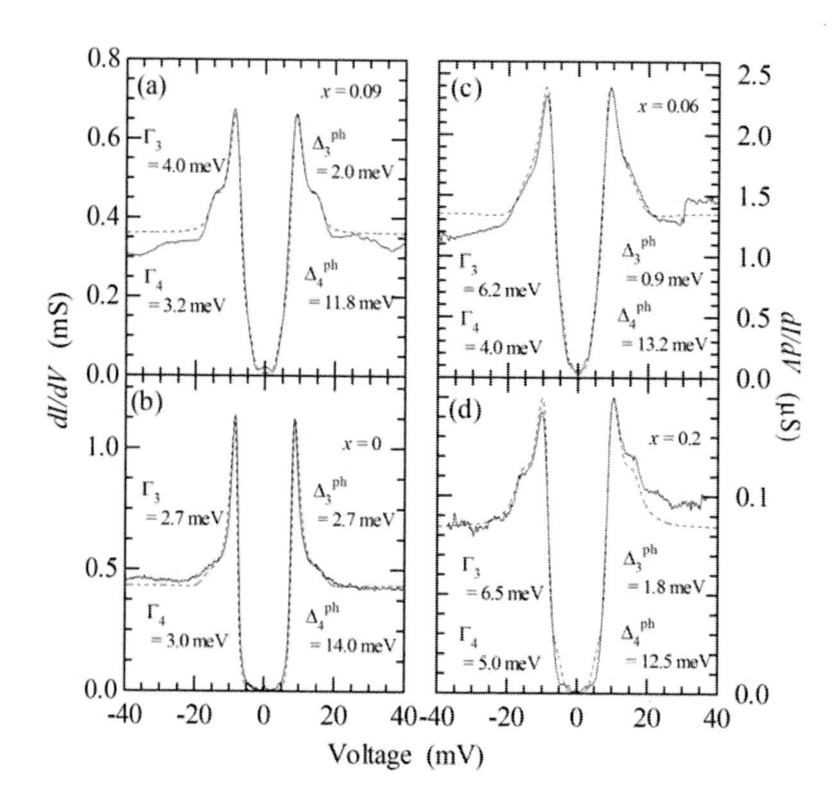

Figure 7. Multiple-gap structures at $T = 4.2$ K for different MgB_2 break junctions with the predominant peak-to-peak distance twice as large as those shown in Figure 6. The dashed curves are the calculation results in the framework of the correlated two-gap model.

Spectra with even larger apparent peak-to-peak separation of about 40 mV can be observed, their representatives shown in Figure 8. Gap-related features in Figure 8 are substantially broadened as compared to those from Figures 6 and 7. Zero-bias leakage is very low in Figures 8(a) and 8(b), while a narrow zero-bias dip is observed in spectra displayed in Figures 8(c) and 8(d). Low-bias fine structures in Figure 8(b) or zero-bias dips in Figures 8(c) and 8(d) are probably related to the smaller gaps discussed above.

The correlated two-gap model was used to fit the large-gap structure as shown by dashed curves. For spectra shown in Figures 8(a), 8(b) and 8(d) the values of Δ_4 are similar to those shown in Figure 7, while Δ_4 ($\approx \Delta_4^{ph}$) = 16 meV for Figure 8(c) is larger than fitting values for other spectra. For Figures 8(a), 8(b), 8(c), the ratio $\Gamma_3/\Gamma_4 \sim 0.8$ is consistent with that for Figure 7. The fitting parameter $x \approx 1$ in Figures 8(a), 8(b), 8(c) is in contrast to small $x < 0.2$ in Figure 7. It means that the tunneling spectra in Figure 7 are governed by the partial DOS $N_3(E)$, while $N_4(E)$ dominates the spectra of Figure 8.

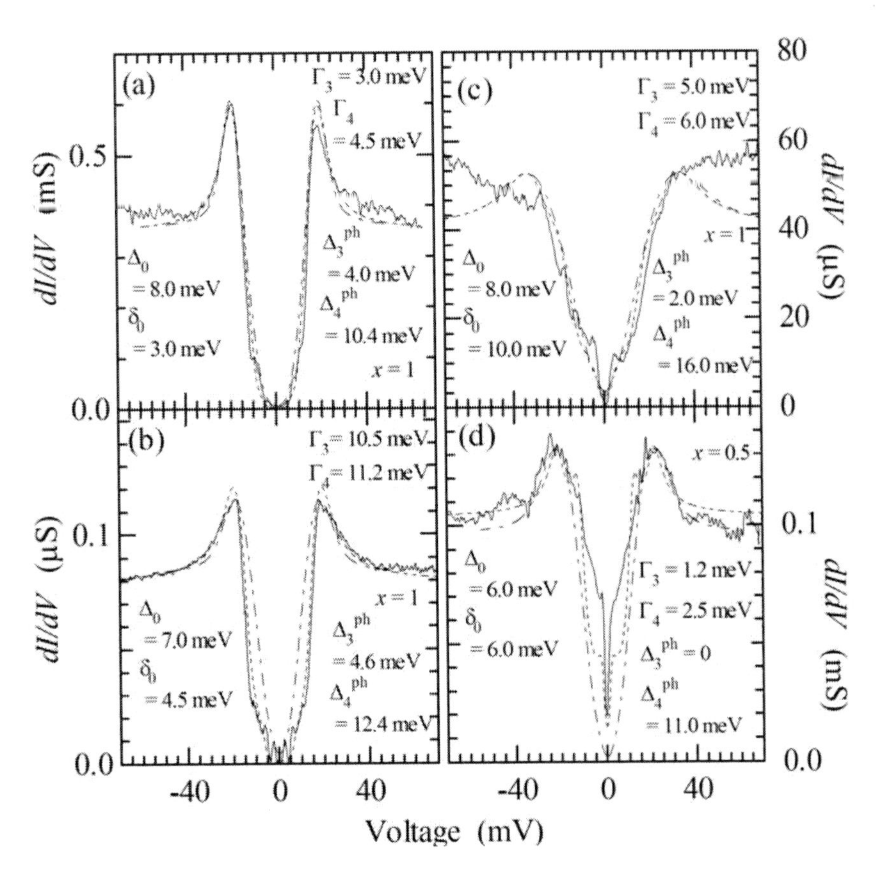

Figure 8. Multiple-gap structures at $T = 4.2$ K for different MgB_2 break junctions involving large gaps. Dashed curves are the fitting results using the correlated two-gap model, while the dot-dashed curves are the fitting results using the Gaussian gap distribution approach.

The broadened appearance of the gap structure in Figure 8 may be explained by an anisotropic nature of the superconducting gap [116] due to the multi-sheet anisotropic Fermi surface crossing p_σ bands [14], or by the energy gap spatial distributions connected to mesoscopic charge carrier density inhomogeneity [117], [118]. If one adopts the second origin of broadening, it is reasonable to use the simple Gaussian gap distribution in the approach given by Eq. (3) and Eq. (4).

Correspondent fittings of data presented in Figure 8 was carried out and the best fitting values are $\Delta_0 = 7.0 \pm 1.0$ meV and $\delta_0 = 4.5 \pm 1.5$ meV. For Figure 8(a), this model can describe the whole data set quite well, whereas for Figures 8(b) and 8(d) only the neighborhood of the gap-edge peaks are reproduced. For Figure 8(c), the sub-gap structure can be also fitted in this approach.

It turned out that some available low-T tunnel data for MgB_2 can be equally well treated by both the correlated two-gap model and the gap distribution approach. Therefore, it is difficult to make a reasonable choice between the models on the basis of those results only.

3.1.3. Temperature Dependence of the Tunnel Spectra

Thermal smearing inevitably wipes out all DOS-induced peculiarities of $G(V)$ provided T becomes high enough. Nevertheless, due to the convolution of two electron DOSes in SIS junctions (including BJ) the detrimental influence of temperature is not so significant [129], [278]. The gap features are especially distinct in the true tunneling regime without shunting when proximity effects *between different electrodes* are excluded.

Temperature variations of multiple-gap structures revealed by representative break junctions involving MgB_2 electrodes are shown in Figure 9. For conductances displayed in Figure 9(a) and 9(b), shoulder-like structures are visible for different T as is shown by thick arrows. These tunnel spectra can be well reproduced by the SIS fitting using the correlated two-gap model, and the gaps $\Delta_1(T)$ and $\Delta_2(T)$ can be followed up to the T_c neighborhood. Nevertheless, some discrepancies between the theory and experiment exist at low biases.

The fitting parameters at 4.2 K for Figure 9(a) and 9(b) are estimated to be Δ_1^{ph} = 1.6 (0.5) meV, Γ_1 = 0.9 (3.2) meV, Δ_2^{ph} = 6.4 (7.5) meV and Γ_2 = 1.2 (3.0) meV. These values are consistent with those of Figure 6. On the other hand, for Figure 9(c) distinct double-gap structures are visible at low T. The corresponding V_S = 5 mV at 4.2 K is half that of Figures 9(a) and 9(b). No fine structure is seen around zero bias in Figure 9(c) contrary to other junctions. This suggests formation of SIN contacts in the break junction [83].

In order to confirm whether a SIN appears, we compare calculated temperature variations of the suggested SIN conductance with those of the SIS. Figures 10(a) and 10(b) show the results of such calculations for different T assuming a textbook single-gap structure. For the SIN junction as shown in Figure 10(a) gap edges broaden with T, whereas the peak locations almost do not shift. Of course, there is no sub-gap structure near zero-bias. On the other hand, SIS junction as shown in Figure 10(b) exhibits the distinct gap-peak shift and zero-bias structure. The latter comprises a logarithmic peculiarity growing with increasing T according to the theory [119].

$$G(V) \sim \log\left(\frac{\min\left(k_B T, \Delta\right)}{eV}\right) \tag{13}$$

Matching of left and right gap widths for V → 0 is the origin of this feature. It is essential that singularity (13) appears only for genuinely symmetrical junctions. Therefore, it may vanish in break junctions where a spatial distribution of gaps always exists.

It is clearly seen that the spectrum of Figure 9(c) resembles that of an SIN junction rather than an SIS one. Indeed, as is shown by dashed curves in Figure 9(c) the correlated two-gap model for the SIN junction fits the spectrum well over the entire bias voltage range. The fitting parameters determined at T = 4.2 K are Δ_1^{ph} = 1.9 meV, Γ_1 = 0.6 meV, Δ_2^{ph} = 6.5 meV and Γ_2 = 0.1 meV. These two gaps disappear near T_c. These values are similar to $\Delta_1(T)$ and $\Delta_2(T)$ for the SIS junction found for Figure 9(a).

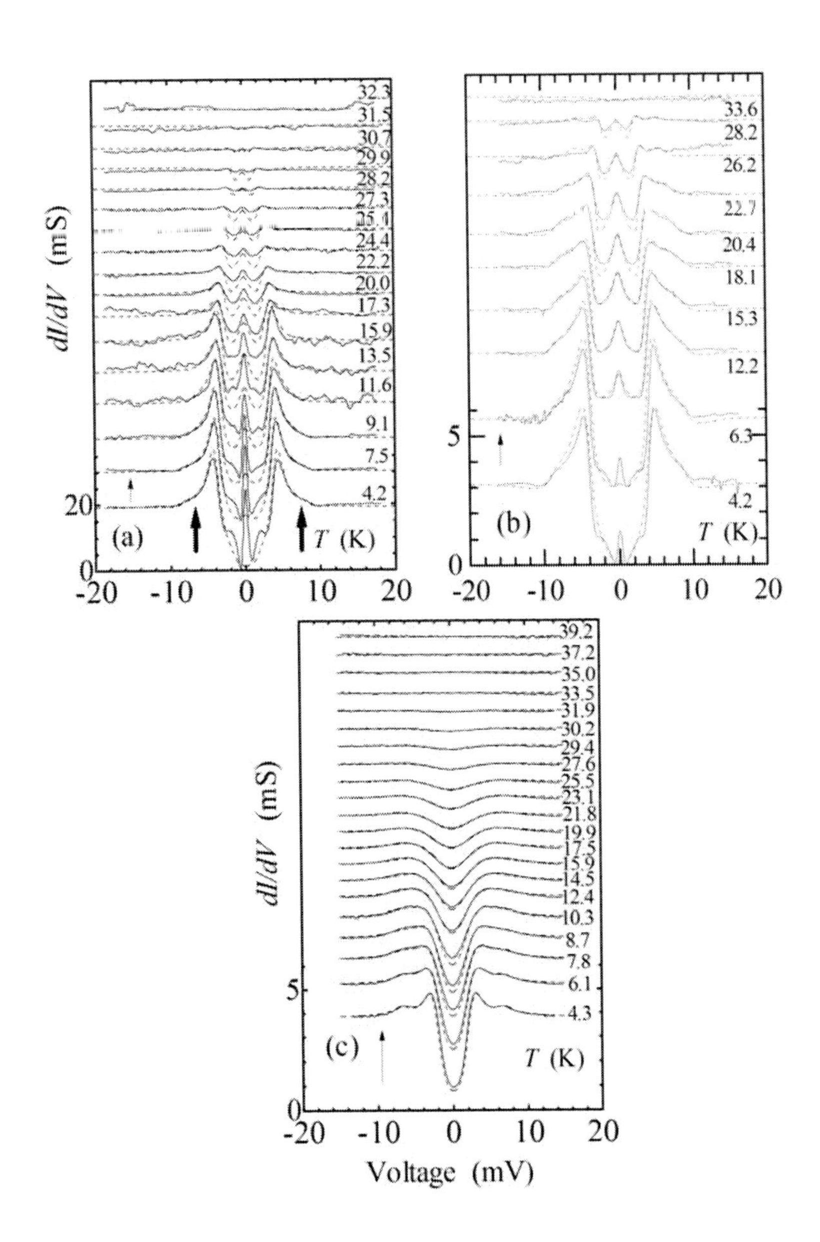

Figure 9. Temperature variations of $G(V)$ for different MgB_2 break-junctions. Fitting by the correlated two-gap model is shown by dashed curves. Figures (a) and (b) correspond to the SIS junction, whereas Figure (c) describes the SIN conductance formed while preparing the break junction. Dependences $G(V)$ for different T are shifted up for clarity, without any offsets between the experimental and calculated curves.

Figure 10(c) shows a SIS conductance calculated with the parameters characterizing the SIN conductance of Figure 9(c). The overall behavior of $G(V)$ for Figure 10(c) agrees with the SIS-like spectra of Figure 9(a) and Figure 6(a) and demonstrates similarity with theoretical guidelines presented in Figure 10(b). Different values of Δ_1^{ph} and Γ_1 appropriate to Figures 9(a) and 9(b) and the observation of SIN junction in Figure 9(c) suggest that some kinds of inhomogeneous electronic structure, e. g., a grain boundary region [139], [140], emerge near the cracked interface of the break junction. It is remarkable that in spite of the different values of fitting parameters and junction types in Figures 9(a) – 9(c), the small gap Δ_1 remains almost the same and is not closed up to T_c.

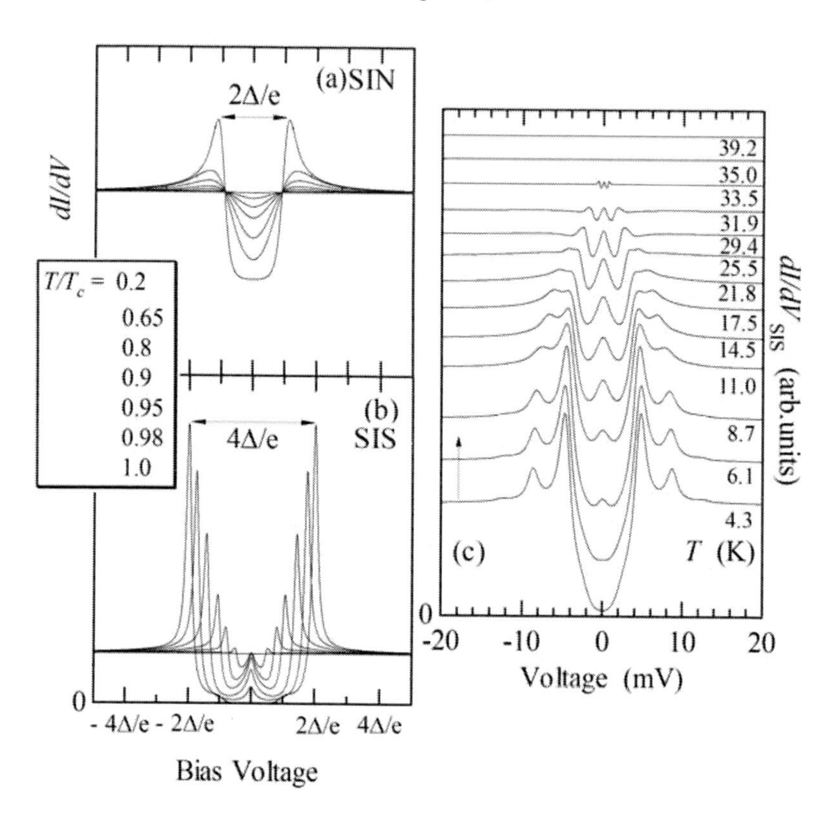

Figure 10. T dependence of $G(V)$ calculated for a conventional one-gap BCS superconductor: (a) for an SIN junction; (b) for an SIS junction; (c) the calculated SIS conductance with parameters taken from fitting curves of Figure 9 (c).

Some spectra can be regarded as the evidence for the normal conducting phase as is seen in Figure 11. Such $G(V)$ are often observed in the low-resistance ($R_J \sim 1\ \Omega$) break junctions. The plateau around zero bias voltage is characteristic of the Andreev reflection in the barrier-less SN junction (the localized barrier strength $Z = 0$ in original notations of Ref. [130]). In Figure 11 the plateau extends over the region $|V| \leq 10$ mV. The conductance has also minor peaks at $V \approx \pm 2.5$ mV. At higher voltages a number of stable features are seen. Such a structure can be attributed to the quasiparticle bound states in an SN junction [87], [134], 141] (or an SNS junction [133]-[135]). In the inset of Figure 11, a similar $G(V)$ feature is shown for a junction with $R_J \approx 0.8\ \Omega$.

In order to analyze the observed multiple peak structure, we tried to fit $G(V)$ by the weighted sum of two BTK conductances (see Eq. (11)) $(1\text{-}x)\sigma_1 + x\sigma_2$ for the SN junction [130]. The dashed curve in the main frame is the result of fitting. The enhanced conductance region is satisfactorily reproduced by BTK model. The fitting parameters for σ_1 and σ_2 are $\Delta_1 = 2.5$ meV (with $Z_1 = 0.2$) and $\Delta_2 = 8.3$ meV (with $Z_2 = 0.3$) with $x = 0.16$. The size of Δ_1 is similar to that of Figure 6, while Δ_2 is slightly smaller than Δ_4 in Figures 7 and 8. This result suggests that two peaks are related to the two energy gaps. The latter should be attributed to the correlated two-gap structures in Figure 6. We emphasize that the observation of SIN-like and SN-like junction spectra in certain MgB_2 break junctions confirms the idea that the normal conducting phase can appear near the cracked interface of a break junction.

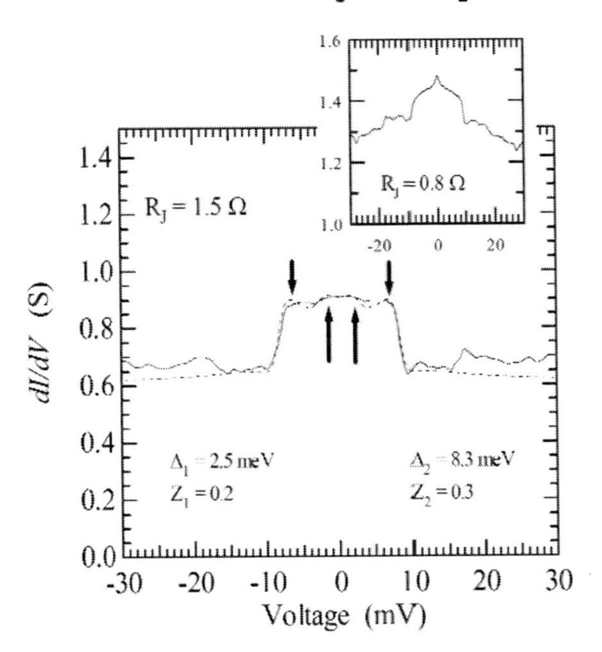

Figure 11. Low-resistant MgB_2 break junction tunnel spectra for T = 4.2 K (the main panel and the inset). The dashed curve represents the Blonder-Thinkham-Klapwijk (BTK) fitting [130].

It should be noted that the analysis of the MgB_2 differential tunnel conductance dI/dV carried both above and below is based on the assumption that magnesium diboride superconductivity is *isotropic*, which is consistent with the totality of various experimental data [4], [184]. For anisotropic, e. g., d-wave superconducting order parameter symmetry clear-cut zero-bias anomalies of $G(V)$ should exist [134], [138]. Indeed, such features (sometimes looking similarly [131] to those presented in our Figure 11) were found in tunnel and point-contact studies of high-T_c oxides [131], [134], [136] [138].

Figure 12(a) shows the T-dependence of the normalized zero-bias conductance (*NZBC*) for the junction described in Figure 9(c). As is well known, this quantity is proportional to the thermally smeared DOS at the Fermi level [21]. The *NZBC* value increases with T and flattens near T_c. Figure 12(b) shows the temperature dependence of the input normalized DOSes $N_1(0)$ and $N_2(0)$, which are determined by the fitting parameters appropriate to spectra displayed in Figures 9(a) and 9(c). The value $N_1(0)$ found for Figure 9(a) is very similar to that for Figure 9(c) at any T. Note that $N_1(0)$ almost coincides with that of BCS dependence (dashed curve) with $\varepsilon/\Delta_1(T) = 0.19$, while $N_2(0)$ increases rapidly above $T \approx 10$ K as compared to the BCS one [dashed curve, $\varepsilon/\Delta_2(T) = 0.07$].

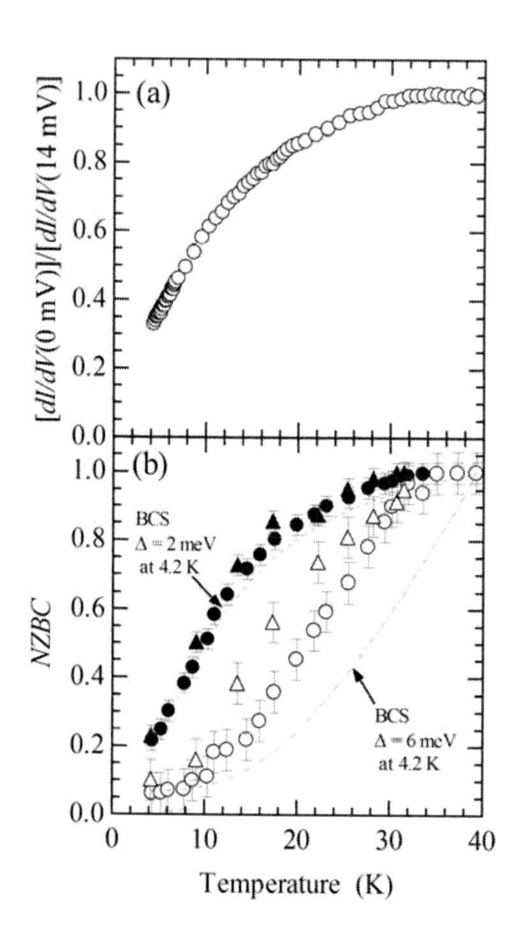

Figure 12. (a) Normalized zero-bias conductance (*NZBC*) corresponding to the data of Figure 9(c). (b) Calculated partial *NZBC* corresponding to $N_1(E)$ and $N_2(E)$ appropriate to Figure 9. Symbols (•) and (○) correspond to the partial *NZBC* proportional to $N_1(E)$ and $N_2(E)$ from Figure 9(a), while (▲) and (Δ) correspond to those appropriate to Figure 9(c). The dashed and dot-dashed curves describe the BCS dependences with $\Delta = 2$ meV and $\Delta = 6$ meV at 4.2 K, respectively.

Dependence of $G(V)$ on T for a junction with dominating largest gap (according to the adopted classification and similarly to the junction presented by Figure 8) is demonstrated in Figure 13. The critical temperature $T_c \sim 39$ K can be inferred from the T-dependence of the zero-bias conductance (*ZBC*) (if one ignores the *ZB* peak itself), which is shown in the right inset. In the left inset a similar dependence is displayed for another junction with the largest gap but gap features surviving at higher temperatures, which confirms truly bulk nature of this gap.

In Figure 14 T-dependences of $\Delta_1(T)$ and $\Delta_2(T)$ obtained by two-gap fitting of experimental spectra for three kinds of junctions from Figure 9 as well as from Figure 13. Both gaps decrease more rapidly than the conventional one-gap BCS theory predicts for the *normalized curve* $\Delta(T/T_c)/\Delta(0)$. It is especially conspicuous for $\Delta_2(T)$. The ratios

$\Delta_1(T/T_c)/\Delta_1(0)$ start to deviate from the BCS behavior above 20 K. The gap values $\Delta_1(0)$ are about 2.2 ± 0.3 meV and $T_c \approx 39$ K leading to $2\Delta_1(0)/k_BT_c \approx 1.3\pm0.2$, which is much smaller than the weak-coupling BCS value 3.53 [155]. Therefore, although $\Delta_1(T/T_c)/\Delta_1(0)$ is BCS-like in reduced variables this dependence should be regarded as a representative of the so-called α-class models with an adjustable phenomenological parameter $\alpha \equiv \Delta(0)/k_BT_c$ [142]. The deviation from the BCS theory in our case is caused mostly by the coupling between two gaps involved (see Eqs. (5), (6)).

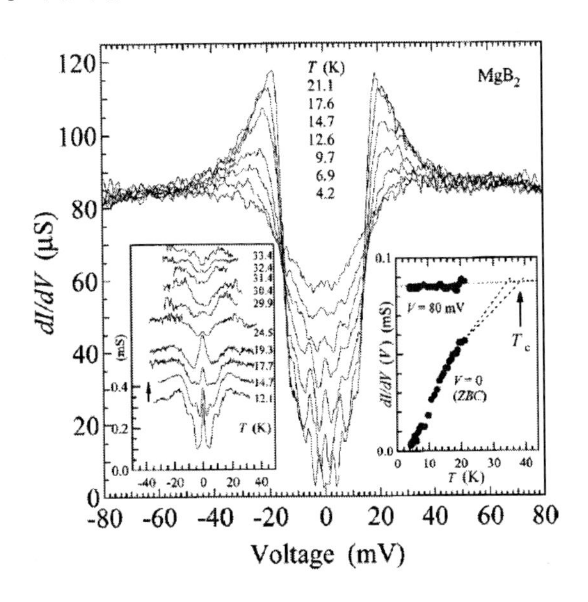

Figure 13. T-dependence of $G(V)$ measured for SIS junctions with the largest gap $\Delta = 9-10$ meV (main panel and left inset). Right inset: T dependence of $G(0)$ (ZBC) and $G(V = 80$ mV).

A large scatter in the values $2\Delta(0)/k_BT_c$ in one compound or even among members of the same class of materials, as in the case of MgB_2, is by no means trivial. For instance, in 122-type iron pnictides the ratio $2\Delta(0)/k_BT_c$ is quite robust and equals to ≈ 3.1 [156].

The function $\Delta_2(T)$ ($\Delta_2(T = 4.2$ K$) = 6.5$ meV) for junctions described in Figures 9(a) and 9(c) is linear over the entire T-range up to nearly T_c (~ 35 K). The same dependence for the junction corresponding to Figure 9(b) (($\Delta_2(T = 4.2$ K$) = 7.5$ meV) starts to deviate from the BCS behavior at about 20 K. This is probably due to the induced superconductivity. Since the T-dependences of tunneling DOSes and gaps are very similar for the genuine two-band and the proximity models (see above), it is difficult to distinguish between the phenomena from the spectrum fittings. An observation of more than two gaps with single T_c might be a required smoking gun.

This is exactly what is inherent to the totality of MgB_2 samples. The apparent set of observed gap values at 4.2 K includes three specific groups, namely Δ_S (= Δ_1) = 2 – 2.5 meV, Δ_M (= $\Delta_2 \approx \Delta_3$) = 4.5 – 7.5 meV and Δ_L = 10-12 meV. This is incompatible with the simple isotropic two-band superconductivity [44]. On the contrary, in the proximity scenario the function $\Delta(T)$ would undergo a step-like deviation a certain $T < T_c$, or continue to persist up to bulk T_c depending on the coupling between surface and bulk electron states [126], [143]. Note that the induced character of a gap can be much stronger for Δ_S (= Δ_1) than that for Δ_M (= Δ_2

$\approx \Delta_3$), although $\Delta_S < \Delta_M$. Various kinds of $\Delta_M(T)$ deviation from the BCS dependence shown in Figure 14 resemble the proximity-effect gap behavior in conventional superconductors [126].

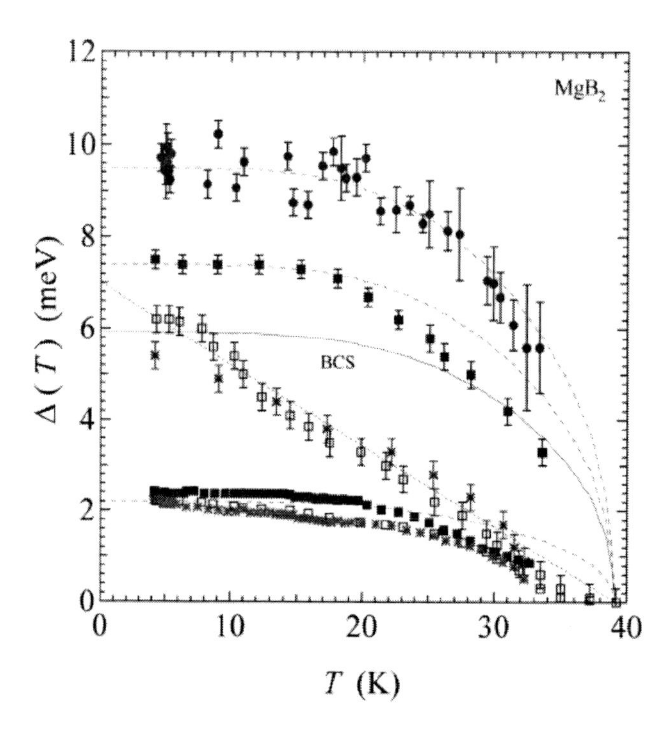

Figure 14. T-dependence of the energy gaps. Data with Δ_1(4.2 K) = 2.2 ± 0.3 meV, Δ_2(4.2 K) = 6.0 ± 1.5 meV denoted with open and closed squares correspond to the junctions described in Figures 9(b) and 9(c), respectively. The data marked by asterisks are taken from Figure 9(a). Closed circles correspond to the large gaps from Figure 13. Thin solid curve represents a conventional BCS dependence with T_c = 39 K, while the dashed curves are scaled to fit the experimental data. The dashed straight line represents a simple linear interpolation.

Large gap functions $\Delta(T/T_c)/\Delta(0)$ ($\Delta(0)$ = 9 – 10 meV) in Figure 13 almost follow the BCS T-dependence. The large gap to T_c ratio is in the range $2\Delta_L(0)/k_BT_c$ = 5 – 6. This is the largest ratio for superconductors except for copper oxides with $2\Delta(0)/k_BT_c > 8$ [144], [145]. The latter values are probably due to the interplay between superconductivity and CDWs [42].

In the multiple- (in particular, two-) gap approach [44] the compound MgB_2 with its multi-sheet Fermi surface consisting of boron p_σ and p_π bands might possess coexisting gaps in the clean limit including the large gap observed in our experiments. For conventional one-gap strong-coupling superconductors the relationship between the gap ratio $2\Delta(0)/k_BT_c$ and the logarithmic average phonon energy $\omega_{\ln}$ [146] is approximately given as follows (see also Refs. [298]-[300]).

$$\frac{2\Delta(0)}{k_BT_c} = 3.53\left[1 + 12.5\left(\frac{k_BT_c}{\omega_{\ln}}\right)^2 \ln\left(\frac{\omega_{\ln}}{2k_BT_c}\right)\right] \tag{14}$$

Hence, from the value $2\Delta(0)/k_BT_c \approx 5$ it comes about that $\omega_{ln} \approx 15$ meV. On the other hand, according to the neutron inelastic scattering measurements the averaged phonon frequency $\omega_{ln} = 57.9$ meV [17], which is much larger value. Moreover, the first principle calculation predicts that the strong electron-phonon interaction arises from the in-plane anharmonic stretching mode E_{2g} with $\omega_{ln} \approx 70$-75 meV [147]. These values correlate with the observed broad Raman-active E_{2g} peaks [148] [151]. One sees that the high value $2\Delta_L(0)/k_BT_c \approx 5$ can not be explained by strong-coupling gap effects.

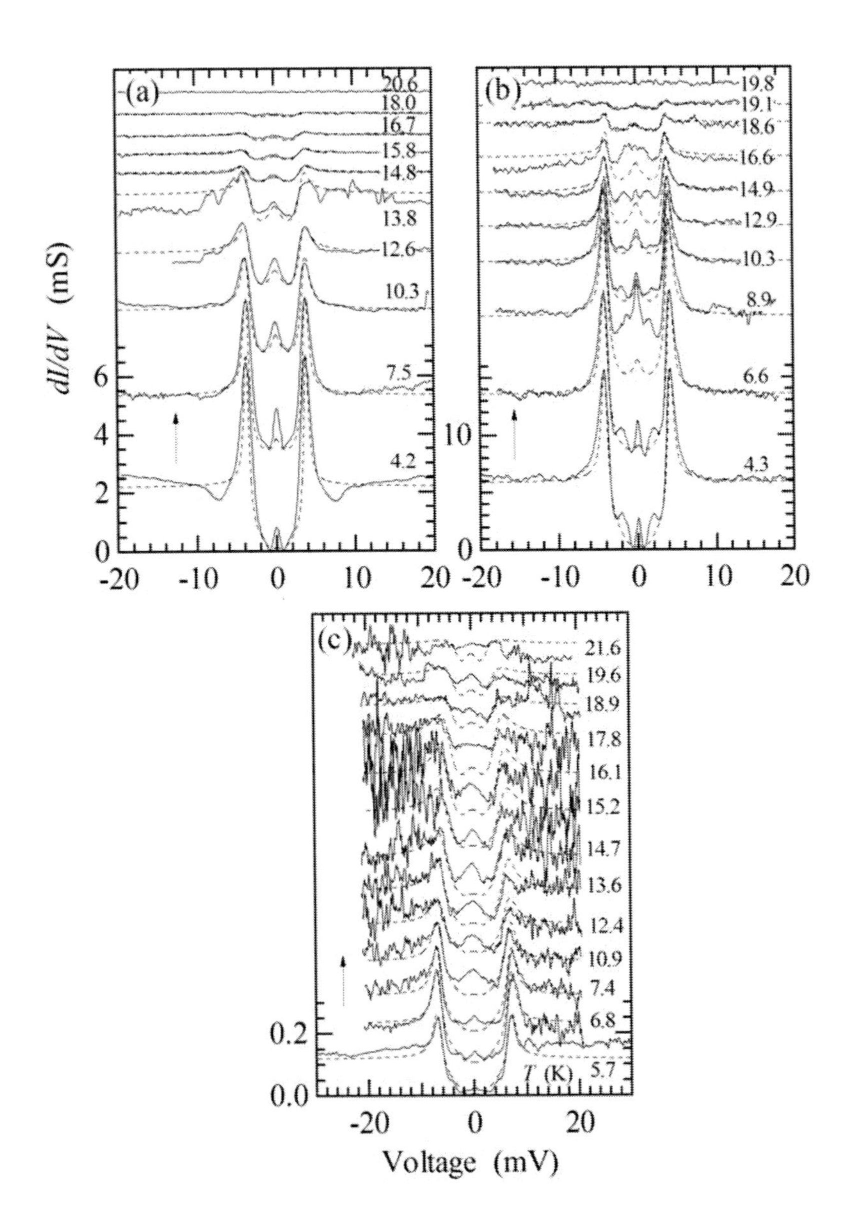

Figure 15. T variations of the tunnel $G(V)$. The data of (a) – (c) are inferred from different junctions. The dashed curves represent the fitting results using the BCS density of states for SIS junctions.

Our tunneling measurements have also revealed the existence of local low-T_c phase even for the samples with the bulk T_c = 39 K. Figure 15 shows corresponding gap features. For Figure 15(a) [this gap at 4.2 K is presented in Figure 6(d)] and Figure 15(b), the values $V_{p\text{-}p}$(4.2 K) are 7.4 mV and 8.4 mV, respectively. These values are consistent with those of Figures 6 and 9(a). For Figure 9(a), the well-developed gap structure is visible even at high temperature (~ 30 K). On the other hand, the gap features in Figures 15(a) and 15(b) vanish at much lower $T \approx$ 20 K. For Figure 15(c), the peak-to-peak distance $V_{p\text{-}p}$(5.7 K) = 14 mV is between those shown in Figures 6 and 7. In order to interpret the overall T variations of the gap structures, we fitted the data of Figures 15(a) – 15(c) using the BCS-like SIS conductance. The dashed curves represent the best-fit results. These fittings reproduce the spectral shape including the peak height. Substantial deviations around the dip structures in Figure 15(a) at low T can be reproduced by the correlated two-gap model [Figure 6(d)].

Temperature dependences of certain parameters inferred from Figure 15 are shown in Figure 16. In particular, the normalized conductance bottom is demonstrated in Figure 16(a) to reach unity already at $T \leq$ 20 K regardless of the fitting parameters. Moreover, the peak-to-peak separation in Figures 15(a) and 15(b) decreases slightly with T up to about 20 K. This is consistent with the behavior of $\Delta_1(T)$ below 20 K (see Figure 14).

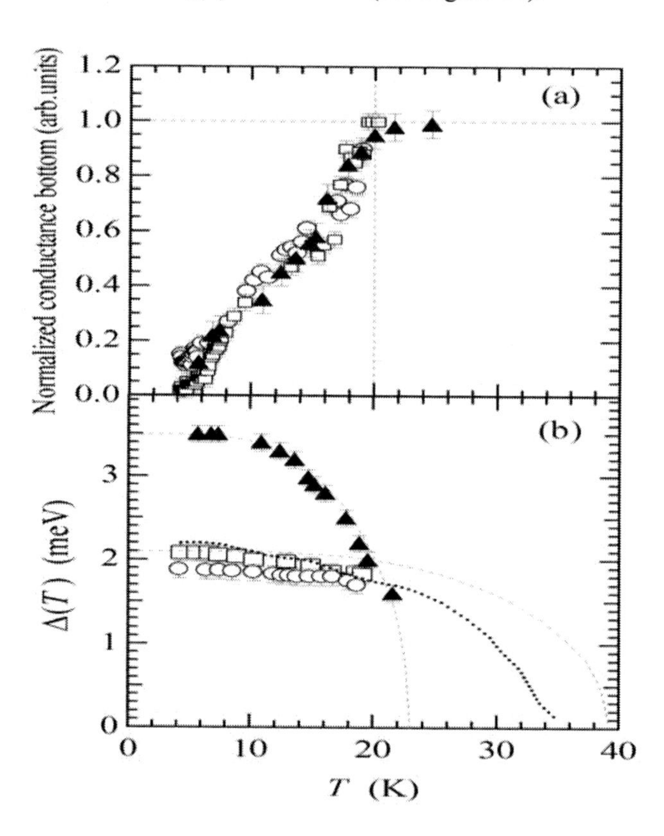

Figure 16. T-dependences of the normalized conductance bottom and the energy gap $\Delta(T)$ taken from Figure 15. Open squares, open circles and closed triangles correspond to the data of Figure 15(a) – 15(c), respectively. Dashed curves in (b) represent the scaled BCS dependences. The dotted curve in (b) is the representative $\Delta_1(T)$ taken from Figure 14.

The fact that $G(V)$ of Figures 15(a) and 15(b) become completely flat at 20 K is in contrast, e. g., to Figure 9. The flattening does not seem to have an extrinsic origin (such as junction instability) because the T-dependence of the conductance bottom is rather smooth as stems from Figure 16(a). Therefore, this behavior may be connected either to the proximity effect or the appearance of the local lower-T_c phase. In fact, the $G(V)$ pattern of Figure 15(a) at low T can be reproduced by the correlated two-gap (proximity effect) model as is shown in Figure 6(d). Since the intrinsic gap parameter Δ_1^{ph} is zero for Figure 6(d), the tunneling process should be dominated by the normal phase. Thus, vanishing of the apparent gap above 20 K may be due to the thermally induced reduction of the penetrating lengths in the proximity-induced junctions with small values of Δ_2^{ph}. In contrast to Figures 15(a) and 15(b), the gap value at 20 K for the junction corresponding to Figure 15(c) constitutes 60 % of $\Delta(T = 5.7\text{ K})$. This relationship agrees with the BCS theory for $T_c = 23$ K. For Figure 15(c) the ratio $2\Delta(T)/k_BT_c$ is about 3.5, which is just the BCS weak-coupling limit. Since in addition $\Delta(T)$ exhibits the conventional BCS behavior, the observed gap is most probably related to the local phase with lower T_c.

3.1.4. Tunnel Gap Spectra in the Magnetic Field

Magnetic field B is another powerful factor suppressing superconductivity by diamagnetic (Meissner currents) [155] and paramagnetic effects [157]. Figures 17 – 19 represent dI/dV at various fields directed perpendicular to the tunneling currents. Figures 17(a) and 17(b) describe two-gap structures including the small gap ($\Delta_S \sim 2$ meV).

Figure 18 involves a single middle gap pattern, while Figure 19 includes the single largest gap feature. Unfortunately, distinct gap features similar to those from Figures 6 – 8 are not stable for $B \neq 0$. Therefore, the B-dependences were studied for $G(V)$ with broader gaps, which are usually more robust against the applied magnetic field. Furthermore, we selected data exhibiting no zero-bias peak [except for the inset of Figure 17(a)], because such a leakage might cause an excessive local pair breaking summoned by the current-induced field in the junction.

The gap values in Figures 17(a) and 17(b) were obtained by the SIS-based fitting procedure using the weighted sum of the broadened BCS densities of states $N(\Delta_1, \Delta_2)$. The effectively zero-field $B = 0.1$ T for Figure 17(a) and zero-field for Figure 17(b) were carried out for fitting using $N(\Delta_1, \Delta_2)$. It happened to be satisfactory as is shown by the bottom curves, which correspond to the parameters $\Delta_1 = 2 - 2.5$ meV and $\Delta_2 = 6$ meV with $x < 0.2$ for both cases. These values are similar to those of Figure 6. The apparent shift of the multiple gaps in the field is demonstrated in the inset of Figure 17(a). These apparent coherent peak positions almost coincide with gap energies determined by the fitting.

Since MgB_2 is a type-II superconductor [155], [157] with a single Ginzburg-Landau parameter $\kappa \sim 5$ [4] or a system of two parameters ~ 10 in the two-band scheme [158], a gap for $B \neq 0$ probed by the break-junction tunneling is a spatially averaged order parameter in the vortex states [152]-[155]. With increasing B, the gap-edge peaks shift to lower biases and broaden. As the paramagnetic Zeeman energy (≈ 60 μeV/T) is much smaller than the gap, the observed gap closing should be attributed to the diamagnetic breaking of Cooper pairs in the condensate.

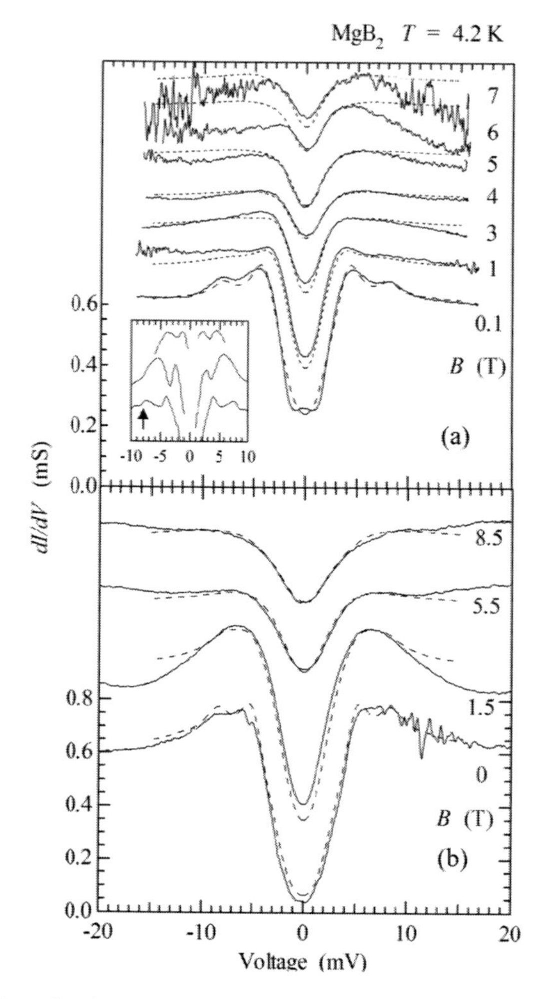

Figure 17. $G(V)$ for BJTS made of MgB_2 in the magnetic field B. Figures 17(a) and 17(b) show two-gap structures. Dashed curves describe fitting results. In the inset of Figure 17(a) multiple-gap spectra in the magnetic field B are shown for another break junction.

In Figure 20, theoretical dependences $\Delta(B) = \Delta(B = 0)[1 - B/B_c]^{1/2}$ for type-II BCS superconductors are presented (dashed curves) [154], [155]. B_c is the gap-closing upper critical field. The extrapolation of this expression to the larger gap have been made on the basis of the experimental data fitting by the calculated $\Delta(B)$ for the smaller gap. Therefore, an uncertainty remains about proper B_c value for the larger gap. Nevertheless, we adopt extrapolated values here because they decrease gradually as is shown in Figure 19, so that corresponding values of B_c can be inferred. Parameter B_c for Δ_M is determined from data of Figure 18. In Figure 20(a) Δ(B) curves are displayed on the basis of fitting using $N(\Delta_1, \Delta_2)$ and peak-to-peak separations $V_S = 8 - 12$ mV ($\Delta = 2 - 3$ meV) for certain junctions, whereas Figure 20(b) corresponds to peak-to-peak separation taken from Figures 18 and 19. For $\Delta = 9$ meV, ≈ 6 meV, ≈ 5 meV and 2 meV critical fields were determined as 18 – 20 T, 13 – 15 T, 5 – 9 T and 8 – 12 T, respectively.

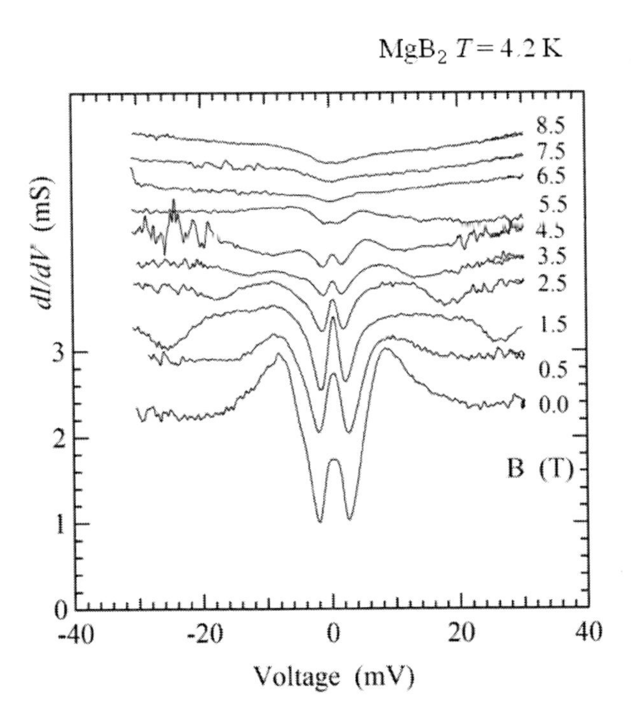

Figure 18. The same as in Figure 17 for a junction exhibiting a middle-gap structure.

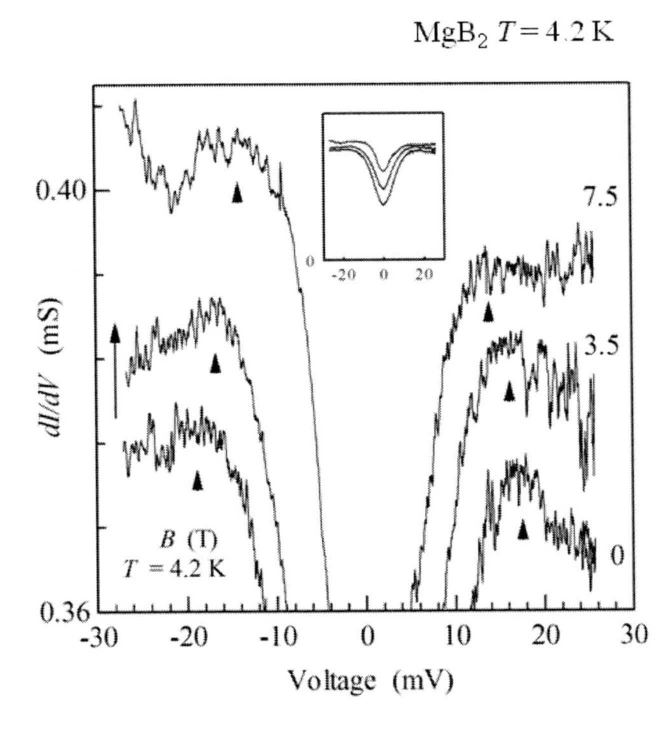

Figure 19. same as in Figure 17 for a junction exhibiting a large-gap structure. Overall conductance features are shown in the inset.

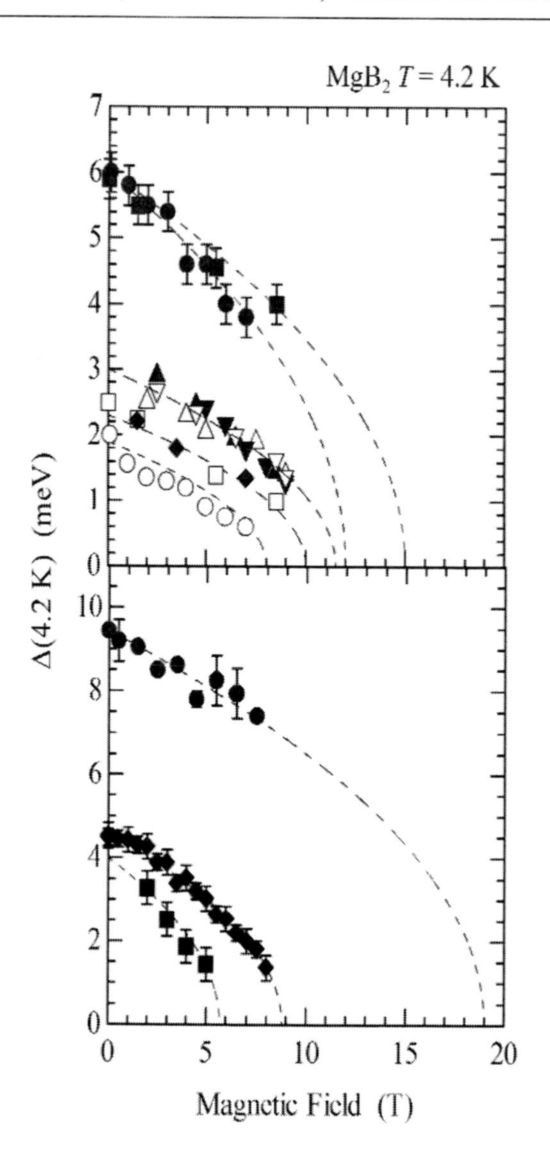

Figure 20. B- dependences of gap values. Dashed curves represent Ginzburg-Landau dependences $\Delta(B) = \Delta(B = 0)[1 - B/B_c]^{1/2}$, where B_c is the upper critical field. (a) double-gap values estimated from $N(\Delta_1,\Delta_2)$ (circles and squares) and peak-to-peak separations for several junctions (triangles), (b) peak-to-peak separations for middle- and large-gap structures corresponding to data from Figures 18 and 19, respectively. Closed squares are middle gaps from another junction.

One should bear in mind that MgB_2 and its relatives are highly anisotropic with the *T-dependent* upper critical magnetic field, so that a simple anisotropic generalization of the conventional Ginzburg-Landau theory is not applicable [159].

In Figure 21 the correlation between B_c and $\Delta(B = 0,\ T = 4.2\text{ K})$ is plotted using the data from Figure 20. Error bars describe differences between properties of break junctions with similar gap values. Variations are probably due to different flux penetration for varying geometries or manifestations of the gap anisotropy in magnetic measurements [158], [160], [163]. The extrapolated value $B_c \approx 18 - 20$ T for $\Delta \approx 9$ meV agrees with the upper-critical field 20 – 23 T in the *ab* plane (parallel to Mg-B sheets) found in magnetic measurements [161]-[163], thereby indicating that this gap dominating feature of MgB_2. The quantity B_c steeply increases with $\Delta(B=0)$ for $\Delta(B=0) > 5-6$ meV. We tried to fit for $\Delta > 5-6$ meV the observed dependence in Figure 20 by the following relationship given by Ginzburg-Landau-Abrikosov-Gor'kov theory [154], [155] for the second critical magnetic field: $B_c = a\Delta(B{=}0)^2$ [$a \approx 0.21$- 0.28 T/(meV)^2]. Let us assume the BCS value of the coefficient $a = c\pi^3/eh v_F^2$, where c is the light velocity, h is the Planck's constant (see above), and v_F is the Fermi velocity. Then from our data the latter is estimated to be $\approx 2.4-2.8\times10^5$ m/s, which is somewhat smaller than that given by band structure calculation [14].

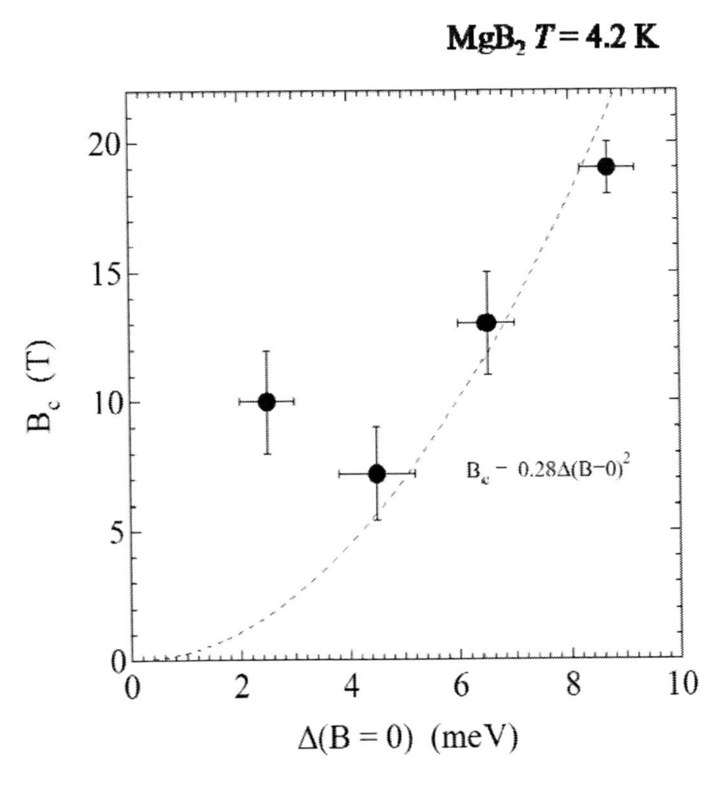

Figure 21. Dependence of the gap closing field B_c on the zero-field gap value $\Delta(B = 0)$. The fitting Ginzburg-Landau-BCS dependence $B_c \sim$ const. Δ^2 is depicted by a dashed curve.

3.2. BJTS data for high-purity MgB_2 wires

Polycrystalline pellets of MgB_2 have $T_c \approx 39$ K, residual resistivity $\rho_0 \approx 24$ μΩcm and residual resistivity ratio (RRR = $\rho(300K)/\rho$ (4 K)) ~ 4. Meanwhile, high purity MgB_2 wire segments show $T_c = 39.5$ K, residual resistivity $\rho_0 \approx 0.4$ μΩcm and RRR ≈ 25 [79]. The difference in ρ_0 and RRR values between two kinds of samples might be associated with the purity of boron. We have carried out break-junction tunneling measurements for the high-purity MgB_2 wire segments as well.

3.2.1. SIS conductance at 4.2 K

Typical spectra obtained for different MgB_2 wire break-junctions at $T = 4.2$ K are shown in Figure 22. They are reproducible and similar patterns can be observed either for different samples or different junctions corresponding to the same sample. Zero-bias peaks indicate Josephson effect and/or tunneling of thermally excited quasiparticles. Extra fine structures around zero-bias voltage are probably due to n-particle tunneling [21], [164]. Finally, conspicuous dip-hump structures are often observed outside the main superconducting coherent peaks.

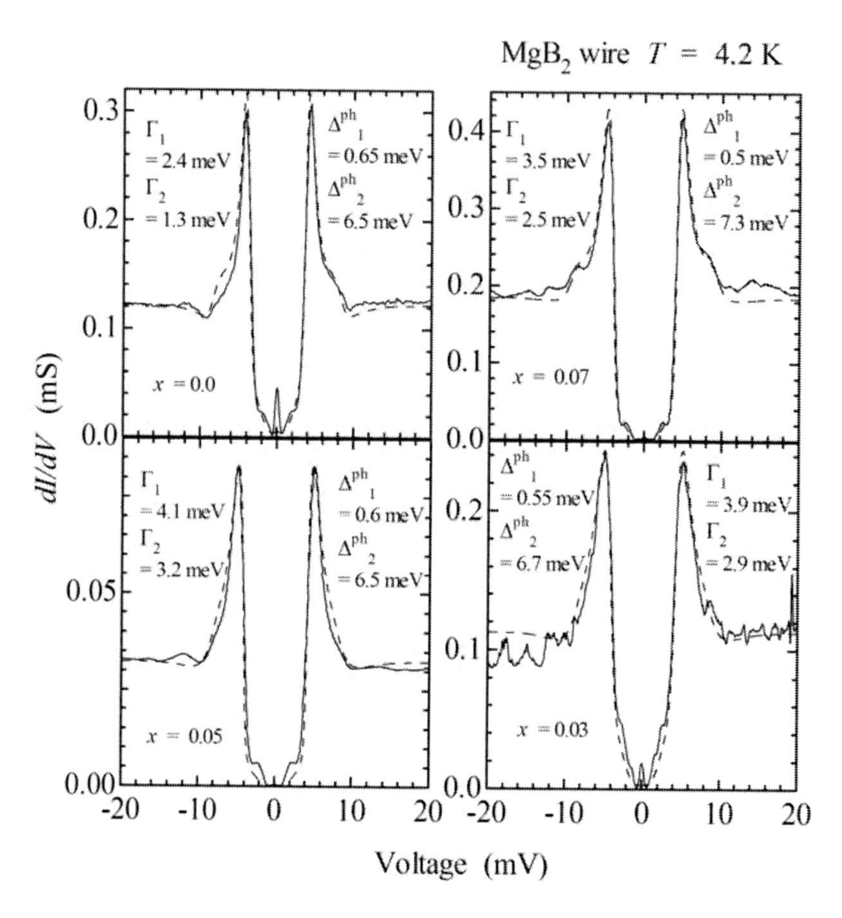

Figure 22. Representative tunnel $G(V)$ at 4.2 K for different break junctions made of MgB_2 wires. The dashed curves are fitting results using the correlated two-gap model.

Coherent peak structures have similar shapes and intensities for various spectra, their peak-to-peak separations being almost unchanged. The latter lie in the range $V_S = 8 - 10$ mV. Those values are close to characteristics of small gap spectra for pellets demonstrated in Figure 6. The dashed curves in Figure 22 are calculated results for SIS conductance in the framework of the correlated two-gap model. One sees that this model represents well main features of the experimental $G(V)$. Since $x < 0.1$ the spectra in Figure 22 are dominated by the contribution $N_1(E)$ involving the small gap Δ_1. Corresponding input parameters are $\Delta_1^{ph} = 0.5 \pm 0.1$ meV and the ratio $\Gamma_1/\Gamma_2 = 1.3 \pm 0.2$. The larger gap is estimated to be $\Delta_2 = 6.5 - 7.5$

meV. These values are kept almost constant among various junctions. On the contrary, spectra for pellets differed considerably from each other. The resulting intrinsic gap $\Delta_1^{ph} \approx 0.5$ meV is smaller than the correlation parameter value $\Gamma_1 = 3 - 4$ meV. It means that small gaps with $V_S = 8 - 10$ mV are caused mainly by the induced superconductivity rather than by the intrinsic one. The values of the larger gap parameter Δ_2 ($\approx \Delta_2^{ph}$) in different wire junctions do not change substantially, contrary to what is observed for pellets.

In Figure 23 other inherent conductances are shown for MgB_2 wire break junctions (solid curves). The spectrum of Figure 23(a) reveals outer hump structures. $G(V)$ is asymmetrical with respect to the bias sign, the distinct peak seen near – 13 mV as is indicated by an arrow. SIS conductance calculated on the basis of the correlated two-gap model satisfactorily reproduces the spectrum (dashed curve). The fitting small gap value and the correlation ratio are equal to $\Delta_1^{ph} = 0.4$ meV and $\Gamma_1/\Gamma_2 = 1.0$, respectively. The larger gap is Δ_2 ($\approx \Delta_2^{ph}$) = 12.5 meV. This value is equal to Δ_L in pellets. The spectrum in Figure 23(b) shows smeared peaks near ± 20 mV. A corresponding gap parameter is estimated to be $\Delta_L \sim 12$ meV on a basis of the observed peak-to-peak separation and assuming that SIS geometry is realized in this wire break junction. The two-gap model parameters $\Delta_1 = 2$ meV and $\Delta_2 = 12$ meV in Figure 23(a) are similar to those of Figure 8(d).

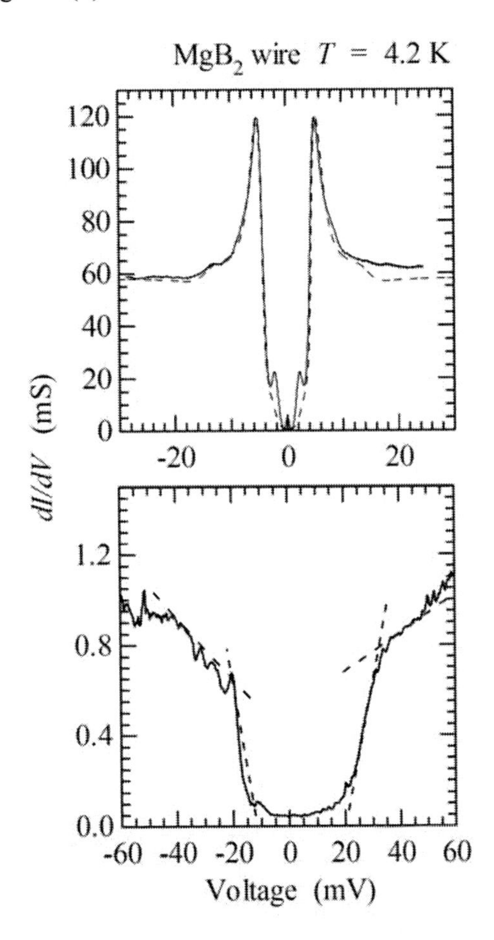

Figure 23. The same as in Figure 22 for other junctions.

3.3. Point-contact spectra of MgB_2

For comparison with tunnel data we studied MgB_2 samples by PCS [19, 88 - 90] as well. The conductance $dI/dV(V)$ demonstrates Andreev reflection features [87] at the superconductor-normal metal interface. The largest gap values turned out to be $\Delta(4.2\ K) = 9 - 12$ meV. We also observed clear evidence for the multiple-gap form of $G(V)$ with the smaller gap $\Delta(4.2\ K) = 1 - 3$ meV. Those gap values agree well with the previous break-junction data. The observed peak positions in the second derivative of the current ($d^2I/dV^2(V)$) are compatible with the phonon DOS revealed by inelastic neutron scattering measurements.

3.3.1. The differential conductance at 4.2 K

Figure 24 shows typical PC conductances $dI/dV(V)$ at 4.2 K. The spectra involve hump structures outside the inner coherent peaks. These hump structures vary among different contacts. Nevertheless, PCS measurements seem to confirm the validity of the two-gap picture for the superconducting MgB_2. Most probably, as has been indicated above, multiple-gap features of the PC spectra are due to the induced superconductivity. In order to properly analyze the situation we use (as before) the weighted sum for the overall conductance: $\sigma(\Delta_1, \Delta_2) = (1-x)\sigma_1(\Delta_1) + x\sigma_2(\Delta_2)$ ($\Delta_1 < \Delta_2$), where partial $\sigma_j(\Delta_j)$ ($j = 1,2$) are described by the Blonder-Tinkham-Klapwijk (BTK) conductance [19], [130], [134], and x is the weighting coefficient. Dashed curves in Figure 24 describe the results found in this way. It comes about from these fittings that the experimental conductance in Figures 24(a), 24(b) and 24(d) can be crudely expressed by $\sigma(\Delta_1, \Delta_2)$. Although fitting of the observed $G(V)$ in Figure 24(c) shows inner peak heights much larger than the experimental ones, the hump positions and zero-bias conductance are satisfactorily reproduced. The calculated value of the smaller energy gap is $\Delta_1 = 1 - 3$ meV, while the larger one is estimated to be $\Delta_2 = 3 - 12$ meV for different PC spectra.

The fitting parameters $\Delta_1 = 1 - 3$ meV and $\Delta_2 = 3 - 7$ meV in Figures 24(a) – 24(c) are consistent with those characterizing break junctions spectra of Figure 6. On the other hand, the larger $\Delta_2 \equiv \Delta_L$ (= 9 ~ 12 meV) in Figure 24(d) agrees with the BJTS data of Figure 8. These correlations indicate that both tunnel and PCS techniques probe the same electronic properties. In the BJTS case the T- dependence of the smaller gap Δ_2 (= 4.5 – 7.5 meV) resembles the proximity induced gap features found in more conventional Pb/PbO/Sn/Pb junctions [143]. Therefore the specified gap value could originate from the intrinsic superconducting gap reduced in the surface regions by the proximity effect [126]. It is quite reasonable that wide distributions of energy gaps Δ_2 in both PCS and BJTS correspond to the electronic phase separation or the proximity effect near grain boundaries [118]. In fact, electron energy loss spectroscopy (EELS) measurements revealed the existence of the oxide phase near the boundary region [139].

Δ_1 distribution turned out to be narrower than that for Δ_2. For the 2 meV gap the gap-closing field B_c is between 7 – 13 T as stems from BJTS results presented in Figures 20 and 21. Those values are similar to the upper critical field along the c axis determined by magnetic measurements [161]-[163]. Hence, a gap anisotropy [46], [116], [148], might also mimic the existence of the distinct smaller gap Δ_1.

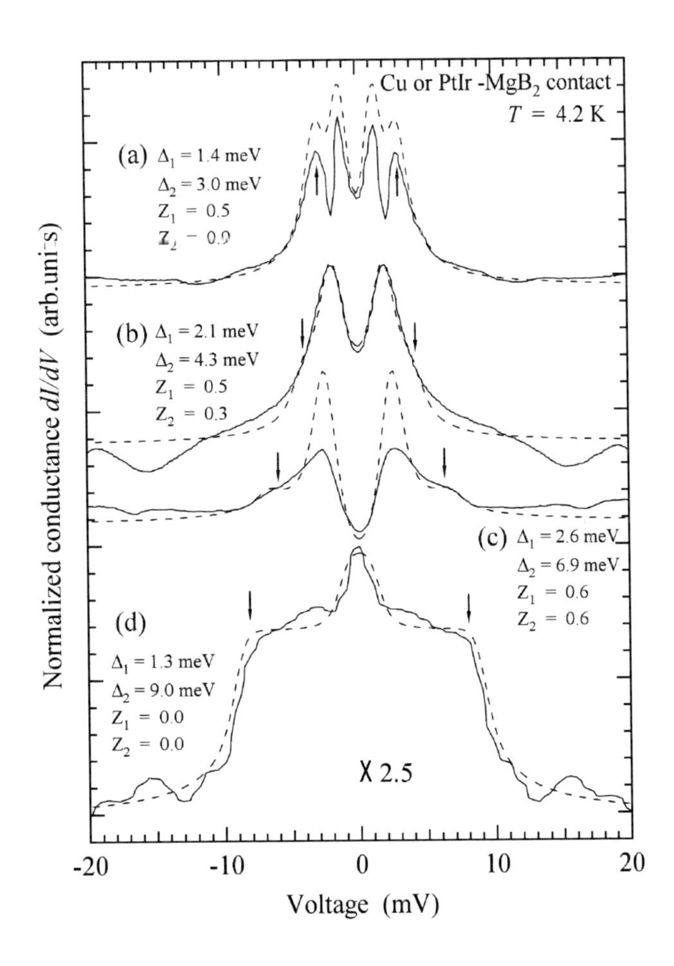

Figure 24. The point-contact (PC) conductance spectra $dI/dV(V)$ for Cu or PtIr wire-MgB_2 at 4.2 K. Dashed curves are the fittings results using the BTK model [130], [134]. The gap values and the barrier parameters are denoted by Δ_1, Δ_2 and Z_1, Z_2, respectively.

3.3.2. Second derivative $d^2I/dV^2(V)$ of the current at 4.2 K

Second derivatives $d^2I/dV^2(V)$, measured in two different Cu-MgB_2 point contacts are displayed in Figure 25 (solid curves). In Figure 25(a), we can see three pronounced peak structures: (i) in the range 70 – 90 mV; (ii) at about 50 mV, and (iii) in the range 20 – 30 mV. Large dips are observed near 40 and 60 mV. Locations of the features agree with the phonon DOS (PDOS) found in the inelastic neutron scattering experiments (dashed curves) [165]. Note however that the peak near 20 mV of our PC spectra is absent in the neutron data. This discrepancy might be due to the dispersion of the Fröhlich electron-phonon interaction $\alpha^2(\omega)$ [35] in MgB_2 or a full contribution $\alpha^2F(\omega)$ characterizing Cu electrode [166]. On the other hand, theoretical calculations [167] demonstrate that acoustic phonon modes below 30 meV might have anomalous dip in the phonon dispersion due either to acoustical plasmon [59] or inter-band transition [168] contributions. The anomaly, if any, may increase PDOS in this frequency region. The quantity $d^2I/dV^2(V)$ shown in Figure 25(b) obtained for another PC junction involves similar features to those of spectra depicted in Figure 25(a) and to the available data of other laboratories [169], [170].

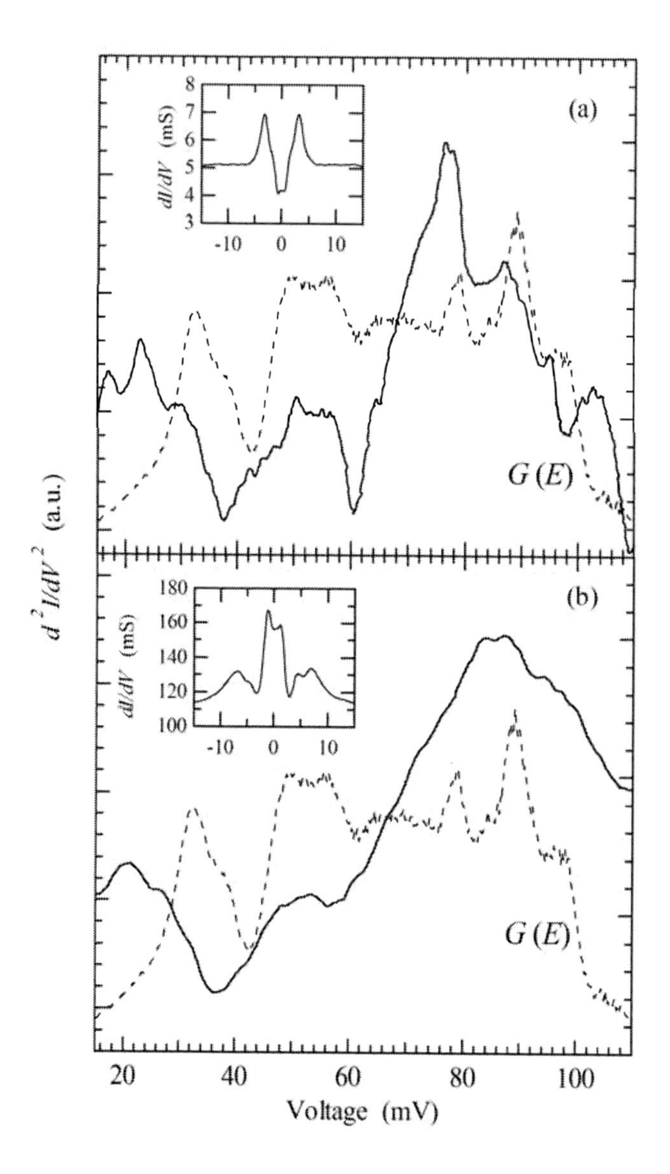

Figure 25. Second derivatives $d^2I/dV^2(V)$ of the PC current in the Cu-MgB_2 junction at 4.2 K. The dashed curves correspond to the inelastic neutron scattering data [165]. Insets demonstrate corresponding differential conductances dI/dV.

Peak structures between 70 – 90 mV in Figure 25 are enhanced conspicuously as compared with the rest of the spectral range. This energy region is close to that where boron atom vibration modes (E_{2g} or B_{1g}) are located [148], [165]. Therefore, boron modes should play an important role in the electron-phonon pairing in MgB_2. Indeed, theoretical calculations [165] predict strong coupling between the E_{2g} phonon mode near 75 meV and two-dimensional boron p_σ-band electrons.

Another example of d^2I/dV^2 point-contact spectra are depicted in Figure 26. This spectrum has peaks at the same positions as that shown in Figure 25. The peaks also correlate well with the PDOS maxima. However relative intensities of the peaks in Figures 25 and 26 are different. In particular, a significant increase of the 20 – 30 and 50 mV features and a reduction of 80 mV structures is observed for the PC spectra of Figure 26. The main gap value is estimated to be $\Delta \approx 4$ meV. This value is similar to the characteristics of the local low-T_c (~ 23 K) phase in Figure 15(c) and correlates with $\Delta = 3.5 \approx 5$ meV observed in Refs. [171], [172]. The d^2I/dV^2 spectrum of Figure 26 may correspond to the inherent properties of the low-T_c phase. This suggestion is supported by Raman scattering studies of MgB_2 thin films, which indicate that the E_{2g} mode frequency correlates with T_c [173]. It might reflect the weakening of the electron phonon interaction.

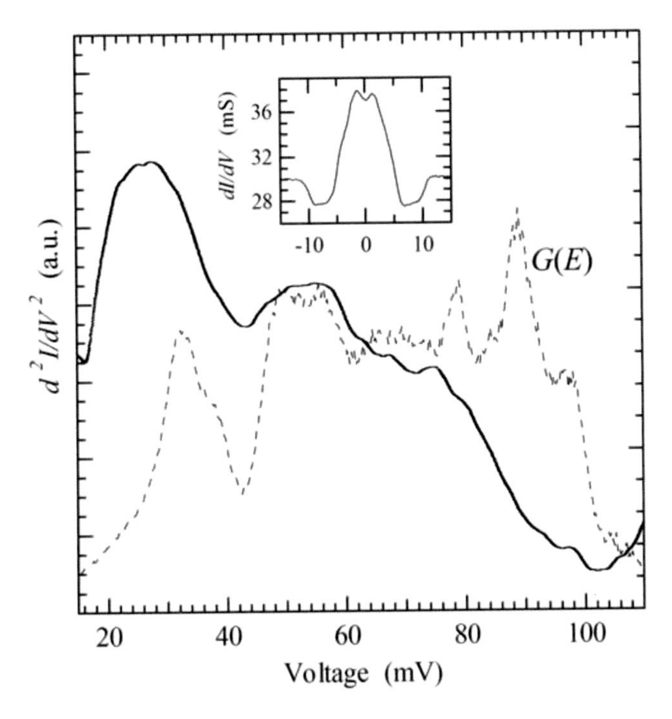

Figure 26. The same as in Figure 25 for another Cu-MgB_2 PC at 4.2 K. The dashed curve corresponds to inelastic neutron scattering data [165]. The corresponding $dI/dV(V)$ is shown in the inset.

3.4. STM spectra of MgB_2

Scanning tunneling microscopy (spectroscopy) has several advantages over more conventional tunneling technique. First of all, it allows building surface maps of the samples with a high (atomic-scale) spatial resolution [174], [175]. It is especially important, in particular, for high-T_c oxides with their small coherence lengths ξ [47], [108], [109], [175], [176] and intrinsic spatial inhomogeneities of normal and superconducting properties [175], [177], [178]. It is quite natural to apply STM for studying MgB_2 with its two-gap features suspicious of being the consequence of spatial inhomogeneities.

We have scanned several single-crystal facets to find the topmost flat *ab* plane normal to the *c* axis, which is expected from the layered hexagonal structure of MgB_2. Figure 27 shows a typical STM image of a clean single-crystal facet of MgB_2 in the superconducting state at 5 K. Characteristic hexagonal patterns of the spots can be seen with ridges between the spots. The image exhibits a slight corrugation, but no apparent modulations are observed. The sample bias voltage of 15 mV is well above the gap-edge voltage of $\Delta/e = 10$ mV. Therefore, the electron DOS at this energy is essentially the normal state one. Bright spots represent atomic positions, while the dark color corresponds to the depression region. The configuration and the average separation between the bright spots are consistent with those of Mg atoms [3], thereby confirming the observation of the Mg atomic layer on the *ab* plane. To our knowledge, this is the first observation of the surface atomic pattern of MgB_2 in the superconducting state, although room temperature patterns were obtained earlier [140].

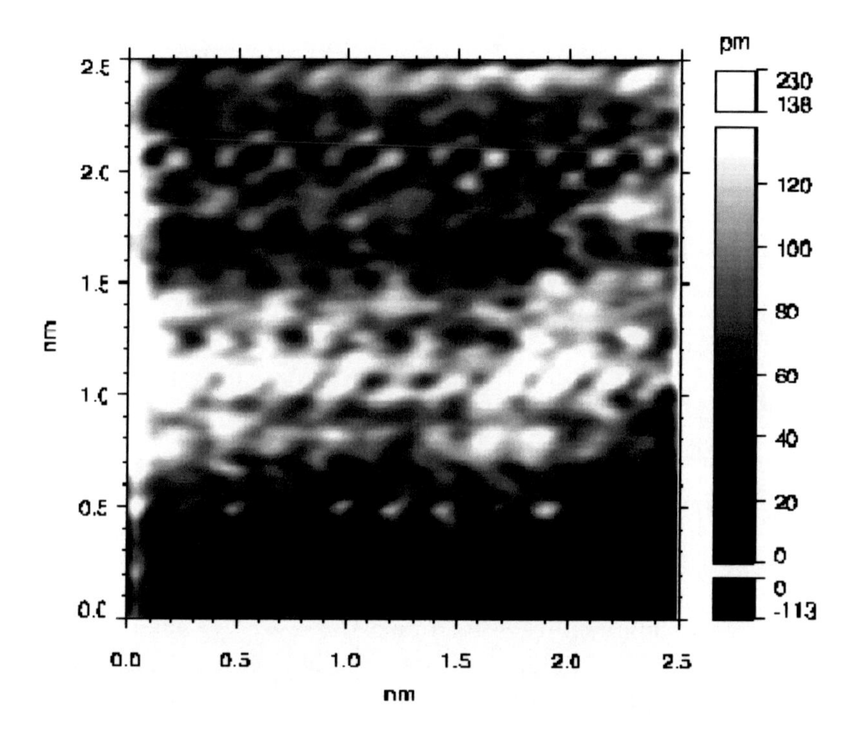

Figure 27. STM image of the MgB_2 surface at 5 K. The bright spots indicate Mg atoms. $V = 15$ mV, $I = 0.3$ nA.

STM images of two kinds of atomic arrangements are shown in Figure 28(a) and Figure 28(c), whereas Figure 28(a) is a magnified part of the scan presented in Figure 27. Figure 28(c) was measured at $V = 15$ mV and $I = 0.05$ nA. Both atomic patterns exhibit hexagonal structures with Figure 28(a) or without Figure 28(c) a centered atom. The appearance of either Figure 28(a) or Figure 28(c) depends on the cracked surface state. The distances between the nearest neighbor atoms for Figure 28(a) and Figure 28(c) are 0.3 - 0.35 nm and 0.15 - 0.2 nm, respectively in agreement with Mg and B lattice constants as is shown in Figure 28(b). Hence, these STM images should be attributed to Mg and B planes. We have also observed at 5 K complex lattice structures involving both Mg and B atoms similar to those of Ref. [140].

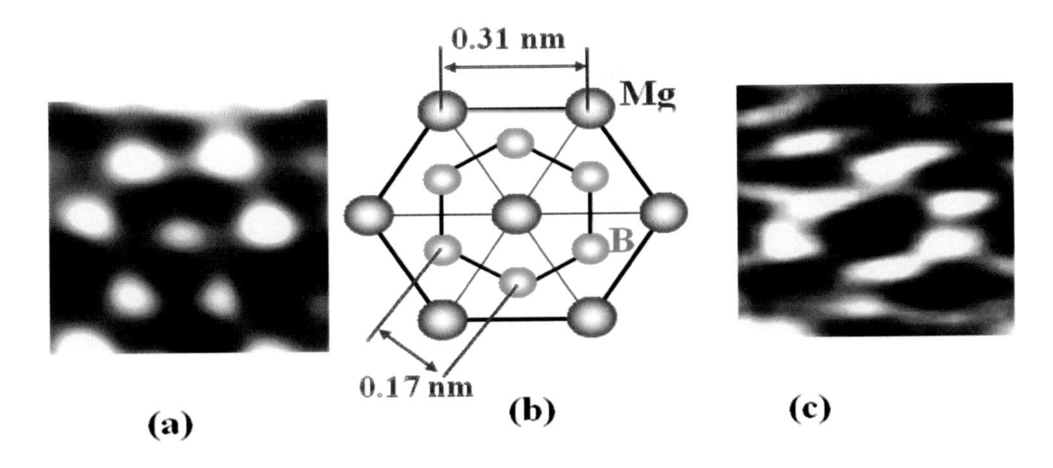

Figure 28. STM image of (a) Mg (V = 15 mV, I = 0.3 nA, 1 nm × 1nm cell) and (c) B (V = 15 mV, I = 0.05 nA, 0.5 nm × 0.5 nm cell) planes measured at 5 K. (b) Mg and B lattice structures.

We also measured superconducting energy gaps in the STS regime. Figure 29 shows the representative STS conductance. The bias voltage is reckoned from that of the tip. There exists slight asymmetry in dI/dV with respect to the bias polarity (a more pronounced asymmetric feature is seen in another spectrum presented in Figure 30), which is typical for our STS results. A similar conductance asymmetry (a higher dI/dV at negative sample voltages) characterizes STS data for high-T_c copper-oxide superconductors [179]. The gap structure is broadened as compared with the basic BCS model. As stems from Figure 29, dI/dV is better fitted by the correlated two-gap model [126], [127] than by the single-gap broadened BCS model, especially in the zero-bias region. This is consistent with the appearance of the second gap, which revealed itself in the BJTS and PCS results given above, although there is no conspicuous double-gap structure in the fitted and fitting curves. In terms of the correlated two-gap model the gap-peak broadening in Figure 30 can be due to the strong mutual quasiparticle scattering between two phases [127]. The gap value of 2Δ = 20 - 24 meV obtained by the fitting procedure is almost equal to the conductance peak separation being the largest gap value extracted from STS measurements [182].

Conductances obtained by STS and BJTS measurements are compared in Figure 30. The bias range of the BJTS is considered twice as much as that of the STS because the gap-edge peak separation for the former corresponds to $4\Delta/e$ appropriate to the SIS junction, while the peak separation for the latter is $2\Delta/e$ since in the STS one studies SIN junctions. It is obvious from Figure 30 that the gap value of the STS is very close to the BJTS result. On the other hand, dI/dV for the STS is almost two orders of magnitude lower than the BJTS conductance and is more broadened. These distinctions are probably due to the different junction geometries in the techniques concerned. Since the observed gap value 2Δ = 20-24 meV is the largest reproducible given by our measurements we believe it to be the predominant gap of MgB_2. Thus, the gap ratio $2\Delta/k_BT_c$ = 5-6 is inferred from both kinds of spectroscopies. This is the largest ratio among non-cuprate superconductors, which was earlier found in NMR measurements [184].

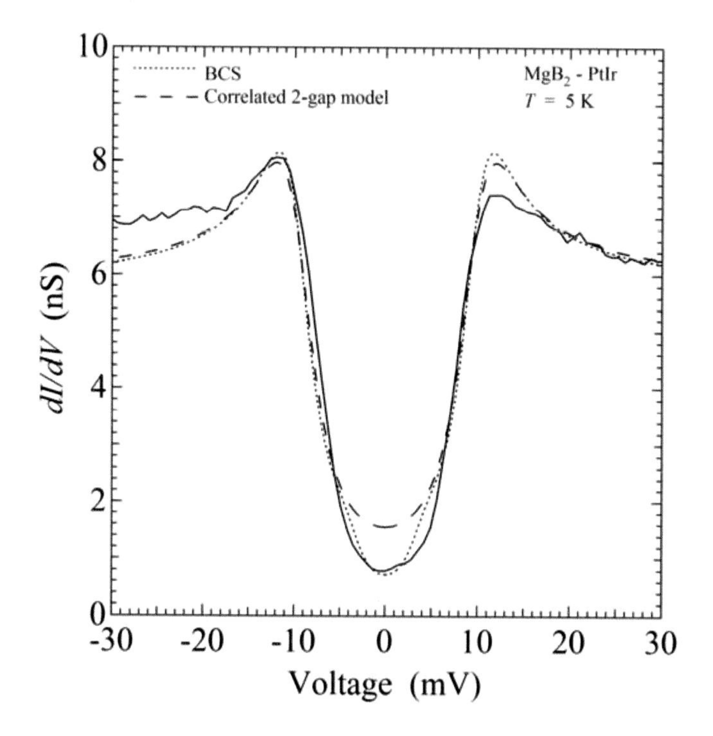

Figure 29. STS conductance of MgB_2.

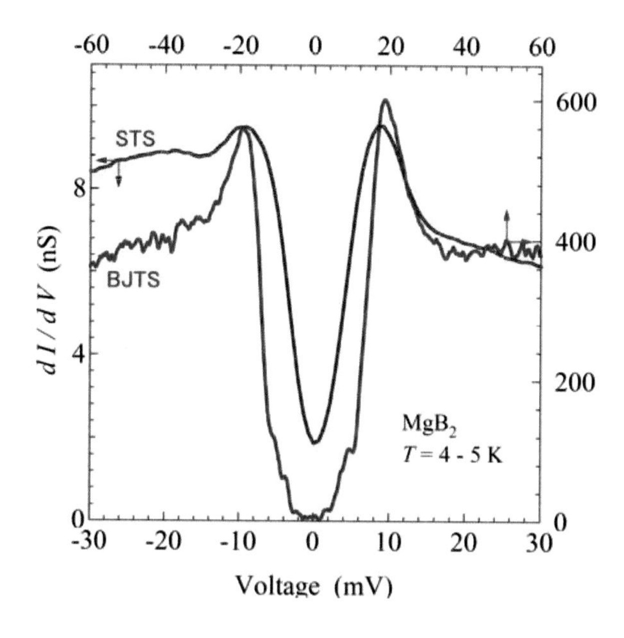

Figure 30. Comparison of MgB_2 conductances measured by STS and BJTS.

STS conductance at measured at $T = 5$ K with 6 nm intervals along a line segment of 200 nm on a single-crystal facet is shown in Figure 31. The gap-edge peak position corresponding to the gap of $2\Delta = 20$ meV is kept almost constant along the mapping line, although the overall $G(V)$ magnitude gradually varies. We also obtained a similar mapping along the line perpendicular to the first one. This stability indicates that the coherence peak location in the MgB_2 sample is constant at least within the domain of 104 – 105 nm^2. The identical broadening of the gap feature all over the swept domain demonstrated in Figure 31 suggests a small gap scatter, which could not be resolved even by precise STS measurements. Such a gap spread can be fitted well by the Gaussian distribution model and manifests itself also in thermodynamic properties, e. g., specific heat [120]. The relatively homogeneous spatial gap pattern of Figure 31 is in contrast to the gap inhomogeneity in cuprates, for which a large spread (~ ± 50 %) of the gap-edge peak positions is observed within a short range of ≈ 2-5 nm [117], [177], [179], [278]. However, it has been recently elucidated that the apparent gap inhomogeneity of high-T_c oxides is inherent to the so-called pseudogaps (in our opinion, CDW gaps) rather than to the superconducting gaps *per se* [86], [177], [180], [181], [278]. Recently, optical reflectivity measurements showed that the observed electron DOS inhomogeneity in cuprates, whatever the gapping nature, is a bulk phenomenon rather than a surface-related one [279]. We think that the same is true for MgB_2, where heat capacity studies revealed clear-cut multi-gap features [51], [118], [120], [239], [240].

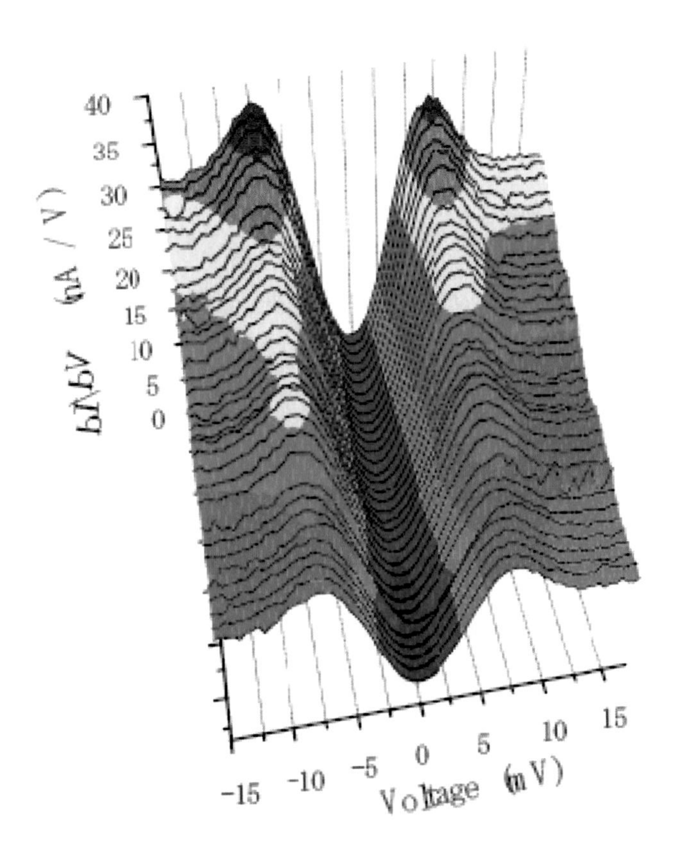

Figure 31. STS conductance mapping for MgB_2 at 5 K.

Sometimes our STS measurements revealed smaller gap patterns. The smallest of them is presented in Figure 32. The parent *I-V* curve is also shown there. The gap-edge peaks appear at $V \approx \pm 3$ mV against a richly structured higher-voltage background. Weak and smeared humps in $G(V)$ centered at around $V = \pm 9$ - 10 mV seem to be traces of the main-gap edges clearly reflected in Figures 29 – 31. Such a small apparent gap $2\Delta = 5$ - 6 meV did not frequently manifested itself in our studies although this is one of the main conductance features met in other STS and PC measurements and interpreted as the π-band gap inherent to the two-band superconductivity in MgB_2 [19], [182]. Dips at $V \approx \pm 4$ mV and shoulders at $V \approx \pm 1.5$ mV in Figure 32 can be associated with the proximity effect.

The variety of apparent gaps in tunnel and point-contact spectra of MgB_2 found here and elsewhere, being very impressive, is, nevertheless, not unique. For instance, one can mention totally different superconductor Rb_2CsC_{60}, polycrystalline samples of which with $T_c \approx 32.5$ K demonstrate wide energy gap histograms (3.3 ÷ 5.7 meV) in STM spectra [183].

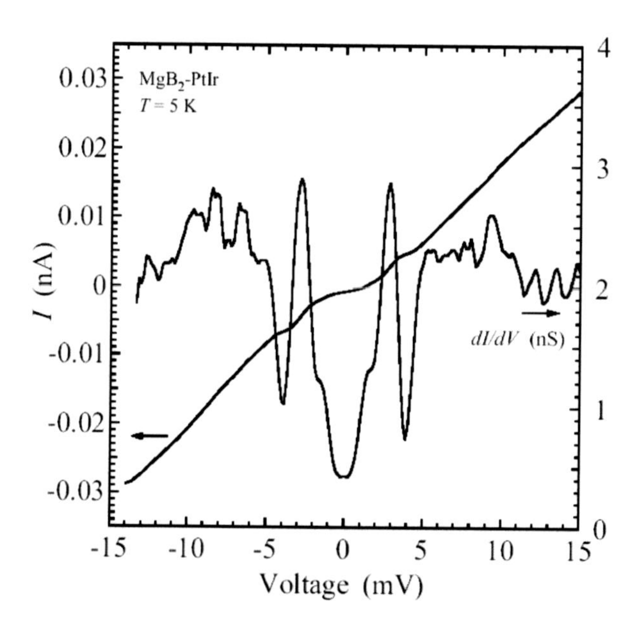

Figure 32. STS conductance of MgB_2 revealing the smallest gap found in our measurements.

3.5. Discussion of the MgB_2 multiple-gap features

3.5.1. Comparison with other studies

Superconducting gap values were determined in MgB_2 by different kinds of studies. Table 1 gives energy gap values inferred from experiments such as NMR [183], photoemission [16], [185], Raman spectroscopy [149], [150], [187], tunneling or PCS studies [19], [171], [172], [188]-[216]. Unfortunately for possible interpretations, the values as well as the very number of gap-like peculiarities vary from method to method and sample to sample.

Table 1. Superconducting gap values of MgB_2 reported by several experimental methods

Experiment	Energy gap values (meV)	References
NMR	$\Delta = 8.5$	[184]
ARPES	$\Delta_\pi = 1.5, \Delta_\sigma = 6.5, \Delta_{surface} = 6$	[16]
	$\Delta_\pi = 2.2, \Delta_\sigma = 5.5$	[185]
Raman Spectroscopy	$\Delta_{//c} = 1.9, \Delta_{//ab} = 6.5$	[149, 150]
	$\Delta_1 = 2.7, \Delta_2 = 6.2$	[186]
Break junction	$\Delta_S = 2\text{-}2.5, \Delta_M = 4.5 - 7.5, \Delta_L = 9 - 11$	This study
	$\Delta_1 = 2.5, \Delta_2 = 7.6$	[188]
Point contact	$\Delta = 1 - 12$	This study
	$\Delta_1 = 2.8, \Delta_2 = 7$	[189]
	$\Delta = 1.7 - 7$	[190]
	$\Delta = 3 - 4$	[191]
	$\Delta = 4.3 - 4.6$	[192]
	$\Delta_1 = 2.8, \Delta_2 = 9.8; \Delta_1 = 1.92, \Delta_2 = 3.45$	[193]
	$\Delta_1 = 2.3 - 2.6, \Delta_2 = 6.55 - 7.45$	[194]
	$\Delta_1 = 2 - 3, \Delta_2 = 6 - 8$	[195]
	$\Delta_{dirty} = 4, \Delta_{3D} = 2.6$	[196]
	$\Delta_1 = 2.7, \Delta_2 = 7.1$	[197]
	$\Delta_\pi = 2, \Delta_\sigma = 6 - 11$	[200]
STM/S	$\Delta = 2 - 10$	This study
	$\Delta = 5$	[172]
	$\Delta_1 = 1.9, \Delta_2 = 7.5, \Delta_{dirty} = 4$	[171]
	$\Delta = 2$	[201]
	$\Delta_{xy} \sim 5, \Delta_z \sim 8$	[202]
	$\Delta = 5 - 7$	[203]
	$\Delta_1 = 3.5, \Delta_2 = 7.5$	[204]
	$\Delta_\pi = 2.2$	[205]
	$\Delta_1 = 2.3, \Delta_2 = 7.1$	[206]
	$\Delta_1 = 2, \Delta_2 = 7.5$	[207]
	$\Delta_1 = 4.0, \Delta_2 = 9.5$	[208]
	$\Delta_1 = 2.2 - 2.8, \Delta_2 = 7.2$	[209]
	$\Delta = 8$	[210]
Planer (sandwich) junction	$\Delta_1 = 1.75, \Delta_2 = 8.2$	[211]
	$\Delta = 7.3$	[212]
	$\Delta = 2$	[213]
	$\Delta = 2.95$	[214]
	$\Delta = 2.2$	[215]
	$\Delta = 2.5$	[216]

A strong-coupling-value single gap was revealed by NMR measurements [184]. Since the inverse spin-lattice relaxation rate $1/T_1$ in metals is proportional to $[N(E_F)]^2$, it becomes singular just above T_c due to the electron spectrum gapping (see a full account in the frames of the BCS approach [217]). Therefore, a successful fitting of the NMR data with a single large gap Δ_L in Ref. [184] shows that only this gap makes a significant contribution into $1/T_1$. On the other hand, the (already conventional) two-band theory [46], [127], [218] predicts the domination of the DOS $N(\Delta_S, E_F)$ over its counterpart $N(\Delta_L, E_F)$, which should be true, in particular, for $1/T_1$. This prediction failed, which makes the validity of the indicated interpretation doubtful. On the contrary, NMR results are consistent with our conclusions drawn from the analysis of our tunnel measurements (BJTS, STS and PCS).

ARPES measurements showed that superconducting gap corresponding to different Fermi surface sheets are of different values although closing at the same critical temperature [16], [185], [186]. Specifically, superconducting gaps at $T = 0$ in σ and π bands are $\Delta_\sigma = 6.5$ or 5.5 meV and $\Delta_\pi = 1.5$ or 2.2 meV depending on the studied samples. In Ref. [16] a surface superconducting state with the gap of 6 meV was observed as well. This value is almost same as the bulk one $\Delta_\sigma = 6.5$ meV. In photoemission measurements [16], [185], [186] solid samples were exposed to photons with $h\nu = 20 \div 40$ eV. For such energies, as is well known [219], the absorption is very sensitive to the existence of surface states, which might prevent the detection of Δ_σ and observation of a well defined σ-band dispersion. We note that Δ_M values in the range 6 – 7.5 meV at $T = 4.2$ K found in various break junctions are consistent with ARPES surface gap values. Small gaps measured in the ARPES and attributed to the π band have the same magnitudes as those observed in a number of tunneling and PC studies.

The majority of groups measuring tunnel and point-contact spectra of MgB_2 (see Table 1) claim the material to possess intrinsic two-gap features being within the ranges $\Delta_1 = 2 - 3$ meV and $\Delta_2 = 6.5 - 7.5$ meV. Such a distinct two-gap structure is commonly accepted as a smoking gun of the intrinsic two-band superconductivity in MgB_2 [19], [20], [44]-[46], [49], [198], [199]. At the same time, other research groups report wide distributions of energy gap values [190], [193], [200] or single gap spectra [172], [192], [201], [210].

The existing controversy among available results of different groups deserves to be discussed in a more detail. To this end it is worthwhile reminding that our BJTS data presented above included from 2 to 5 dI/dV maxima for each V-polarity. Two break junction spectral extremes are shown in Figure 33. Figure 33(a) exhibits three gap-like maxima. Relative intensities of three peaks are different from typical features demonstrated in Figures 6 – 8, whereas the value of V_L ($\approx$ 36 mV) is larger than its analogue in Figure 6 although V_S for Figures 6 and 33(a) are similar. It is remarkable that polycrystalline samples including a lot of intrinsic defects show no sign of gap merging inherent to the two-gap model [44]-[46]], [49], [198], [199].

There are other typical MgB_2 tunneling spectra including large gaps $\Delta_L \approx 10 - 12$ meV such as shown in Figure 8. Those features are quite strong in spectra displayed in Figures 33(a) and 33(b). However in the case of Figure 33(b) there are extra interim maxima as well. Such 4- (or 5-) peak patterns testify that three or more gaps manifest themselves in the S_1IS_2 tunnel junction [21]. In the course of PC measurements we have also obtained three-gap-like spectra including $\Delta_L \approx 12$ meV, as can be seen from Figures 33(c) and 33(d). These spectra can be approximately fitted by the weighted sum of three BTK conductances (broken curves). Hence, the totality of tunnel and PC data for MgB_2 cannot be explained by simple two-band models being only a crude approximation to the rich variety of multiple-gap structures.

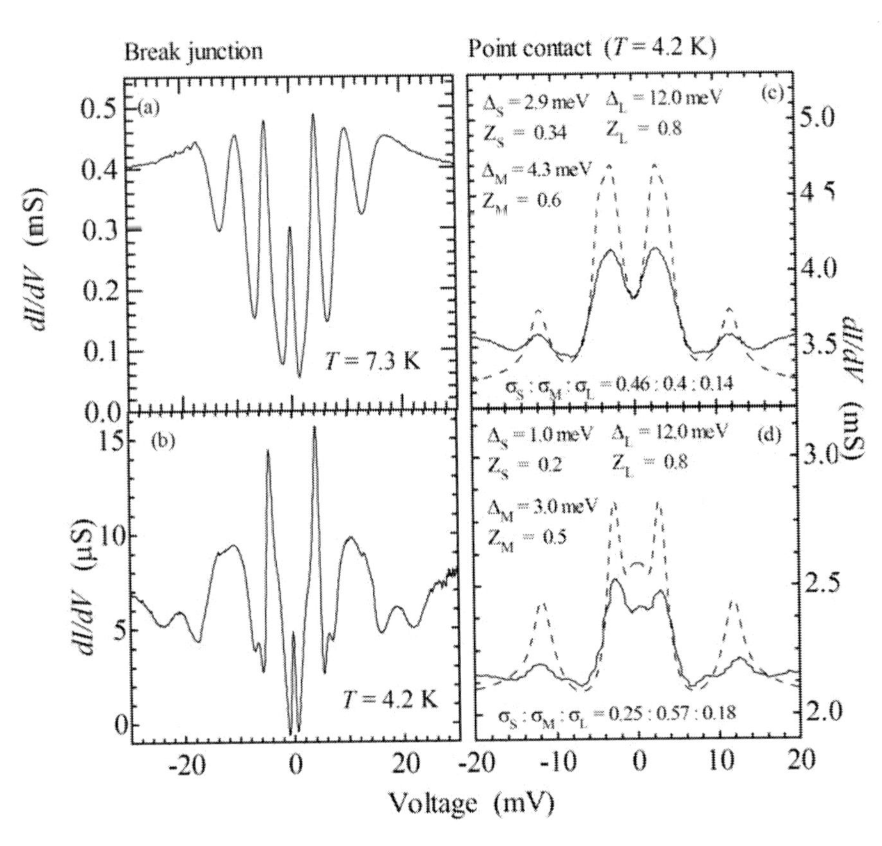

Figure 33. Examples of multiple-gap structures in BJTS ((a) and (b)) and PC spectra.

3.5.2. Chemical reactivity of MgB_2 as a factor affecting superconducting properties

The main benefit of an additional interband coupling in two-band superconductors, which enhances T_c in the framework of the two-gap model [47], [48], [104], [110], should disappear in the presence of strong nonmagnetic impurity scattering mixing electron states originating from different bands [4], [44], [49]. Accepting this prediction, which is simply a reformulation of the old good Anderson theorem [50], one should expect that any apparent two-gap structure would merge if measurements are carried out using polycrystalline or dirty samples [220], [221]. However, multiple gap structures were observed not only in high-purity wires or single crystals but also in polycrystalline pellets. Furthermore, two-gap superconductivity survived even in partially substituted samples ($Mg(B_{1-x}C_x)_2$, $Mg_{1-x}Al_xB_2$ etc) with strongly suppressed T_c (down to ~ 17 K) [222], [223]. The observed robustness of T_c is inconsistent with the two-band interpretation of MgB_2 superconductivity. Therefore, it seems natural to associate multiple-gap features in MgB_2 samples of a different quality with extrinsic factors, e. g., the proximity effect.

As for the actual sample quality, it is quite plausible to relate them first of all to Mg nonstoichiometry and oxygen contamination effects, which in their turn depend on the heat treatment temperature or annealing atmosphere during synthesis [224]. Phase separation into Mg-vacancy-rich and Mg-vacancy-poor regions is also possible as a consequence of the overall nonstoichiometry [226]. Oxygen impurity may lead to formation of MgO or phases of the coherent growth $Mg(B,O)_2$ [227 - 230]. Oxygen in a sample is believed to originate from flowing nominally high purity Ar or slightly oxidized metal containers (Mo, Ta etc) [227], [231]. At the temperature necessary to synthesize MgB_2 ($T \geq 973$ K), the yield of MgO or $Mg(B,O)_2$ increases as well [232]. Therefore, the presence of oxide inclusions can be hardly avoided even using the purest possible reagents.

Indeed, EDX (Energy Dispersive X-ray) spectra of cracked surfaces of our samples always indicate oxygen impurity. Since X-ray diffraction testifies that the MgB_2 phase dominates in the samples, it is natural to suggest oxide inclusions to concentrate in some specific local regions (e. g., at grain boundaries). This is really the case since EELS (Electron Energy Loss Spectroscopy) measurements [139] revealed oxide location in grain boundary neighborhoods of the thickness $\approx$ 150 Å, while STM studies [140] showed that grain boundary regions 50 to 200 Å wide demonstrate normal conductivity. Moreover, periodic alternation of B and O, which correspond to the MgB_2 crystal structure with a compositional modulation, was found in high-resolution transmission electron microscopy (HRTEM) studies [227]. Other HRTEM investigations of B_2O_3-doped MgB_2 samples revealed the nano-structure incorporated into MgB_2 grains as well as the existence of oxide phases [233]. Note that a degradation of the superconducting state triggered by the oxygen contamination, which may originate from oxide-based substrates or the inert gas with O_2 impurity, would be especially strong for thin films [234]-[238].

Hence, one should make a conclusion that such factors as unavoidable contaminations and structural defects may significantly alter superconducting characteristics of MgB_2 samples. Macroscopic transport properties are especially vulnerable to those extrinsic factors [4], [224], [225]. At the same time, inhomogeneities should affect microscopic properties leading to the data scatter presented in Table 1.

3.5.3. Modeling of multiple-gap structures

Our phenomenological approach to the apparent multiple-gap superconductivity of MgB_2 turned out to be successful not only for non-local tunnel phenomena but also for heat capacity as a typical bulk phenomenon. The corresponding numerical analysis [51], [118], [120] of the overall temperature dependence of the specific heat C near T_c and below is demonstrated in Figure 34. The experimental data of Refs. [239], [240] were fitted in the framework of the model for a mesoscopically disordered isotropic superconductor treated as a spatial ensemble of domains with continuously varying superconducting properties [117]. Domains were supposed to have sizes $L > \xi$, where ξ denotes the coherence length as before. Specific calculations were performed for a Gaussian distribution of actual domain superconducting gap values, partially intrinsic and partially induced by its neighbors. It was shown that the spatially averaged (observed) $\langle C(T) \rangle$ should be proportional to T^2 for low T, whereas the phase transition anomaly at T_c is substantially smeared even for small gap dispersions.

For narrow gap distributions there exists an intermediate T range, where the resulting curve $\langle C(T) \rangle$ can be relatively well approximated by an exponential BCS-like dependence with the proximity-induced gap smaller than the weak coupling value. An important point of our theory is that to mimic both low-T asymptotics ($\propto T^2$) and the transition region one should assume an existence of arbitrarily small gaps down to zero. The assumption correlates well with the occurrence of SIN behavior and Andreev reflection spectra in some nominally SIS break junctions discussed above (Figures 9(c) and 11). The existence of grain boundary regions of the width from 50 Å to 200 Å was revealed by EELS [139] and STM measurements [140]. EDX spectra suggest that certain degree of oxidation happens near grain boundaries. Thus, the appearance of the spatial ensemble of domains in MgB_2 samples assumed in the model [51], [118], [120] is supported both by the indicated experimental results and the chemical reactivity of this compound. Figure 34(b) shows a schematic diagram of a MgB_2 grain.

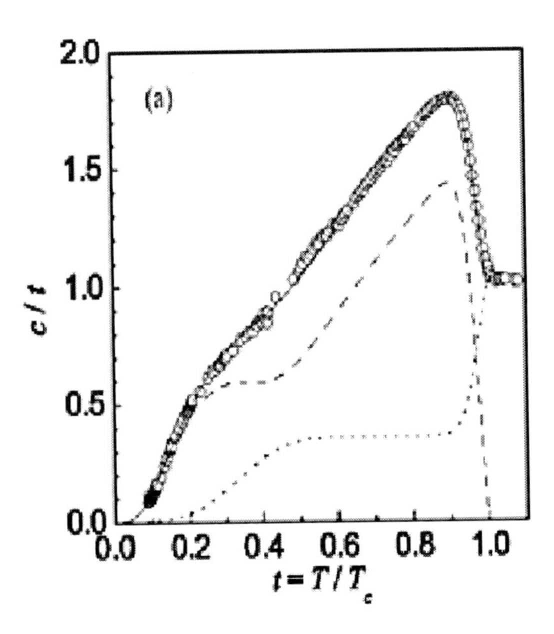

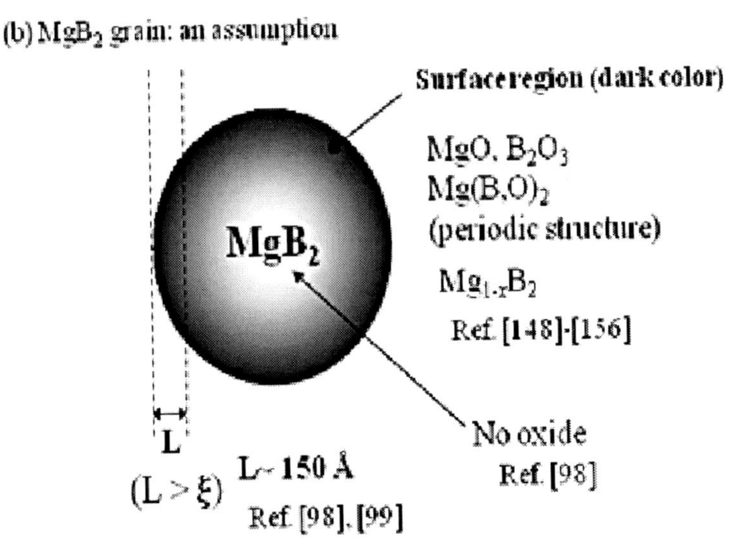

Figure 34. (a) Theoretical fitting of the experimental *T*-dependence of the heat capacity *C* measured in MgB_2 [239], [240]; (b) schematic drawing of a single MgB_2 grain.

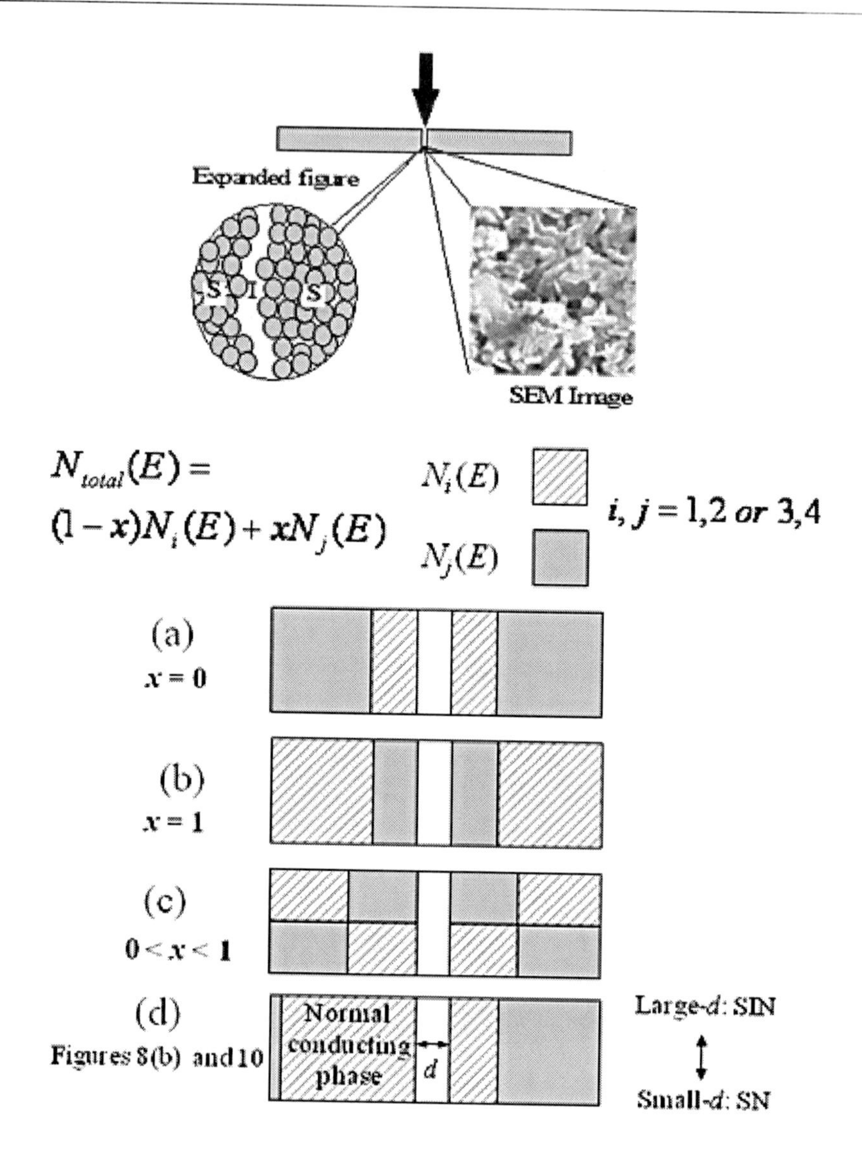

Figure 35. Schematic diagram of possible break-junction configurations: symmetrical S-I-S junction for the case of $x = 0$ (a); $x = 1$ (b); and $0 < x < 1$ (c) in $N_{total}(E)$ (Eq. 10), respectively. In the cases (a) and (b), either $N_i(E)$ or $N_j(E)$ contribute to the tunnel current in the junction. In the case (c), both $N_i(E)$ and $N_j(E)$ contribute to the tunnel current. In the models the combination of $\{i, j\} = \{1, 2\}$ can explain the conductance displayed in Figure 6, while $\{i, j\} = \{3, 4\}$ corresponds to the conductance shown in Figures 7 and 8. Figure 35(d) represents a possible SIN- or SN- junction formation in the MgB_2 break junction containing the normal-metal grain boundary. Depending on the thickness of insulating layer, SIN or SN junction spectra could be observed as shown in Figures 9(b) and 11.

As has been discussed above, DOSes in each superconducting MgB_2 area divided by an insulating layer could be expressed as $N_{total}(E) = (1-x)N_i(E) + xN_j(E)$ (Eq. 10). Figures 35(a), 35(b) and 35(c) describe symmetrical S-I-S junctions for the cases of $x = 0$, 1 and $0 < x < 1$ in $N_{total}(E)$, respectively. In the cases (a) and (b), either $N_i(E)$ or $N_j(E)$ contribute to the tunnel current in the junction. In the case of (c), both $N_i(E)$ and $N_j(E)$ contribute to the tunnel current. Quantities $N_i(E)$ and $N_j(E)$ are very sensitive to the proximity effect. In the adopted models, the combination of $\{i, j\} = \{1, 2\}$ can explain the features shown in Figure 6, while $\{i, j\} = \{3, 4\}$ can be used to reproduce results of Figures 7 and 8. Figure 35(d) represents a possible SIN- or SN- junction formation inside the MgB_2 break junction containing a normal-metal grain boundary. Depending on the thickness of the insulating interlayer, SIN- or SN-junction spectra could be observed, as shown in Figures 9(b) and 11. In order to explain a more general case of break junction spectra represented in Figure 33, one should elaborate a more sophisticated junction model. For example, an asymmetrical S_1-I-S_2 junction may be introduced.

3.5.4. Origin of the small gap Δ_S observed in MgB_2

Bearing in mind high chemical reactivity of MgB_2 it seems quite natural to suggest that the observed multiple-gap features reflect the existence of a certain phase separation (e. g., of the kind discussed for cuprates [272]) or at least an intrinsic spatially inhomogeneous nonstoichiometry [273]. The induced superconducting feature of $\Delta_M(T)$ can be interpreted in the framework of this model. At the same time, the energy gap Δ_S is different from any of the values Δ_M or Δ_L considered above. Furthermore, the magnetic field influence on BJTS reveals the appearance of two B_c vs. $\Delta(B = 0)$ features as is shown in Figure 21. The value Δ_S is similar to the gap Δ_π inferred from ARPES [16], [185] and *c*-axis polarized Raman scattering [150].

Theoretical explanations of the two-band scheme robustness for MgB_2 are based on calculations showing that the interband impurity-driven mixing between p_σ and p_π bands in MgB_2 is weak, because those bands are formed by different local orbitals and are orthogonal on the atomic scale. The layered structure and the compactness of the boron 2*s* and 2*p* orbitals make the disparity between p_σ and p_π bands unusually strong [218], [241]. To underline the alleged disparity, a superfluous analogy between boron honeycombs in MgB_2 and benzene-ring carbons is sometimes introduced [182]. One of the problems with this explanation of the T_c and gap-structure insensitivity to defects and impurities consists in the peculiarly unique position of MgB_2 among other superconducting borides. Why the same arguments shouldn't work for them as well as for other similar layered materials?

On the other hand, multiple-gap feature in MgB_2 could be also explained by the proximity effect combined with the phase separation, which was discussed above. Unfortunately, in most cases even the direct probing as in Refs. [222], [223] cannot discriminate between ordinary-space and momentum-space character of multiple-gap peculiarities. Theoretical analysis of ARPES spectra [242] and surface electronic structure calculations [243] predicts that Mg and B terminated surface (*c*-axis direction) form surface superconducting states. Furthermore, if σ band really dominates conductivity in dirty samples [241], it could be expected that dirty MgB_2 thin films with low RRR (= 1 ~ 2) and low T_c = 20 ~ 30 K reveal a large σ-band superconducting gap.

On the contrary, tunneling measurements studying such dirty samples reveal mainly the small gap feature ($2\Delta/k_BT_c < 3.53$), which is often interpreted as being of the π-band nature. Unfortunately, this experimental result does not fit the two-band model [49], [198], [199], [218], [241], according to which two gaps should merge into one entity for sufficiently strong inter-band mixing (the result being correct from the theoretical point of view!). Moreover, the **k**-space two-gap theory [199] predicts saturation of T_c for large degree of disordering, whereas heavy irradiation of MgB_2 revealed instead a linear reduction of T_c with resistivity quite similar to effects in A15 compounds [274].

It means that the basic two-band model [49], [198], [199], [218], [241], being interesting *per se*, can not be unequivocally applied to MgB_2. There is also no other theory able to adequately cover *all* features of superconductivity in this material and its derivatives. Hence, notwithstanding large efforts of a number of laboratories the real origin of the multiple-band behavior of MgB_2 is still not known. Nevertheless, in pure enough samples of superconductors with well separated sheets of their Fermi surfaces a true **k**-space two-gap or even three-gap [275] scenario might be realized.

4. Superconducting Gaps in Substances Related to MgB_2: NbB_2 and CaAlSi

4.1. BJTS data for NbB_2

It is quite natural that after the discovery of MgB_2 related superconducting compounds became the subject of a close scrutiny. NbB_2 is one of such materials [80], [244], [245]. A representative tunnel conductance curve for NbB_2 is shown in Figure 36. In Figure 36(a) SIS conductance of NbB_2 at 2 K is displayed together with the BCS-based fitting. Sharp gap-edge peaks and almost no-leakage conductance near zero bias are appropriate to this break junction. The observed gap-related positions determined by the peak-to-peak bias separation are $V_{p\text{-}p} = 4.8 - 5.1$ mV, which corresponds to $4\Delta/e$ of an SIS junction. Such tunnel spectra are quite reproducible except for the strength of the zero-bias peak, which perhaps depends on the interface quality. The experimental curve is fitted by the calculated one using the broadened BCS density of states (Eqs. (1) and (2)). One sees that tunnel $G(V)$ is well described by this expression contrary to the case of MgB_2. The gap parameters are estimated to be $\Delta(2\text{ K}) = 1.2 \pm 0.1$ meV ($\gamma \approx 0.13$ meV). Figure 36(b) represents BJTS data at 4.2 K. Larger γ values of ≈ 0.25 meV are necessary to fit those spectra.

BJTS with the bias separations $V_{p\text{-}p} = 3 - 3.6$ mV at 2 K are shown in Figure 37(a). These values $V_{p\text{-}p}$ are smaller than their counterparts in Figure 36. Dash-dotted curves correspond to the SIS conductance on the basis of the BCS model (Eqs. (1) and (2)). These fittings are quite consistent with the experiment except for the coherent peaks themselves. The observed pattern includes suppressed gap-edge peaks with comparably low conductance leakage near the zero-bias voltage. Such a behavior is described well by the correlated two-gap model as shown by dashed curves in Figure 37(a). The obtained fitting parameters $\Delta_1^{ph} \approx 0.1$ meV, $\Gamma_1/\Gamma_2 \approx 1$ and $\Delta_2 \approx 2$ meV are similar among different junctions. Since the best fittings correspond to the values $x < 0.2$ (the parameter x was defined in Eq. (10)), the spectra of Figure 37(a) are dominated by the feature $N_1(E)$ with a small gap Δ_1. The concomitant gap Δ_2 turns out to be larger than the typical gap value of Figure 36, while the small gap Δ_1 is mainly induced by the proximity term Γ_1 rather than by the intrinsic pairing Δ_1^{ph}. The observed moderate gap-edge peak and low-leakage conductance are well described by a large Γ_2 and comparable values of Γ_1 and Δ_1^{ph}. The approximate constancy of model parameters among different break junctions indicate that the gap structure is due to the proximity-induced superconducting phase with the almost zero intrinsic pairing parameter ($\Gamma_1 > \Delta_1^{ph} \approx 0$) and the larger gap with $\Delta_2^{ph} \approx 2$ meV.

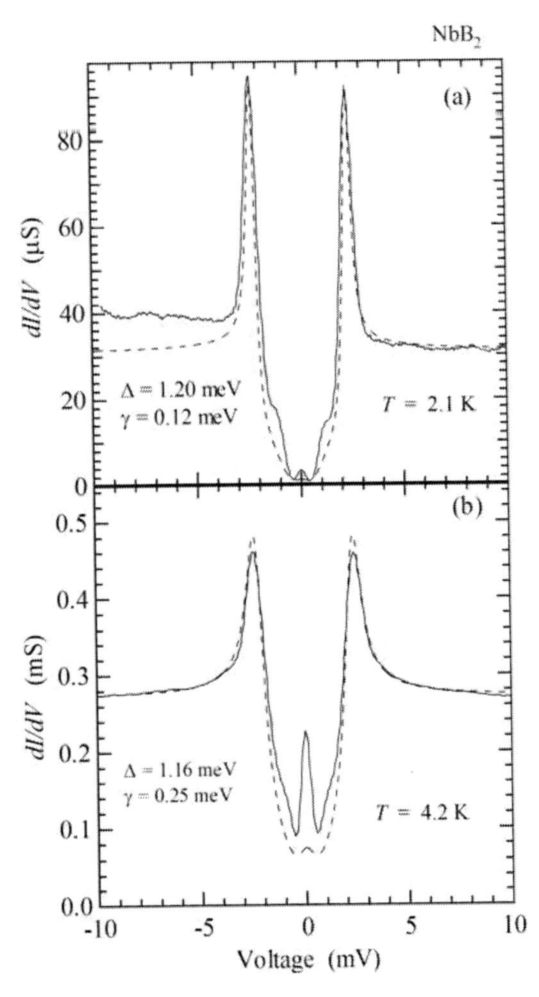

Figure 36. $G(V)$ for break junctions made of NbB_2 at 2.1 K (a) and 4.2 K (b). Dashed curves correspond to the fitting BCS-like $G(V)$.

Large gap structures with $V_{p\text{-}p} \approx 8 - 10$ mV were sometimes observed at 4.2 K in certain NbB_2 break junctions as is depicted in Figure 37(b). The tunnel spectrum demonstrates peak-to-peak bias separation $V_{p\text{-}p} \approx 8$ mV with sharp gap-edge structures. Accompanying broad humps at $V = \pm 1.7 - 1.8$ mV are probably due to the two-particle Schrieffer-Wilkins tunneling [21], [164], [266]. The zero-bias peak corresponds either to Josephson effect or the logarithmic feature existing in $G(V)$ for *non-equal energy gaps* on the l.h.s and r.h.s. of the junction and being due to thermally excited quasi-particles in an SIS′ junction [119], [267], [268].

Temperature variations of the tunnel conductance for NbB_2 are depicted in Figure 38. The spectrum shown in Figure 38(a) exhibits a typical gap feature similar to that of Figure 36, while the spectrum displayed in Figure 38(b) contains a T-variation of the small gap feature similar to that shown in Figure 37(a). Since the junctions concerned are rather stable and the gap-edge peaks are well defined up to highest temperatures when they are discernible, the T-dependent gap $2\Delta(T)$ is easily determined as $eV_{\text{p-p}}/2$. One sees from Figure 38(b) that the spectrum flattens above $T = 6.2$ K but is quite measurable for lower T. The gap-edge peaks are rather sharp, while the zero-bias conductance reveals a relatively large leakage. The small-gap structure can be interpreted as a phase with an induced gap.

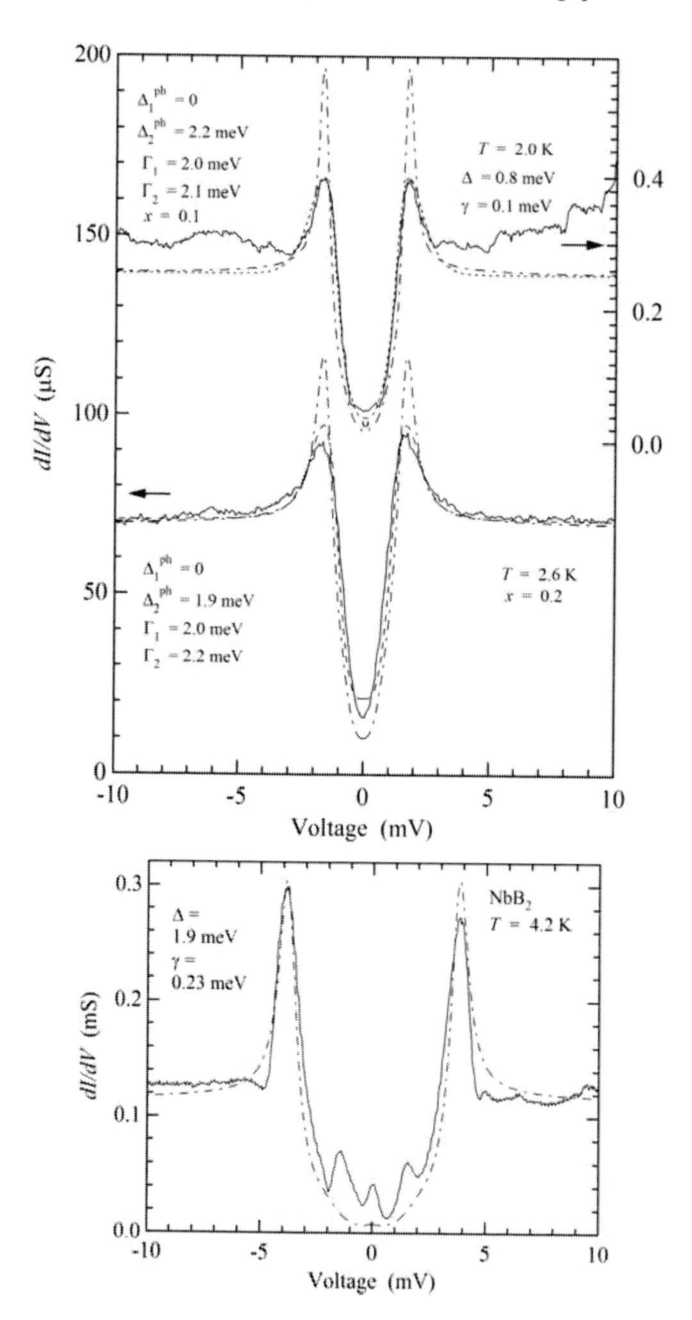

Figure 37. Examples of break-junction $G(V)$ for NbB_2 measured at about 2 K and revealing small gaps (a) as well as found at 4.2 K and demonstrating large gaps. Dashed curves in (a) correspond to the correlated two-gap fitting, while dot-dashed curves in (a) and (b) were obtained using the BCS theory.

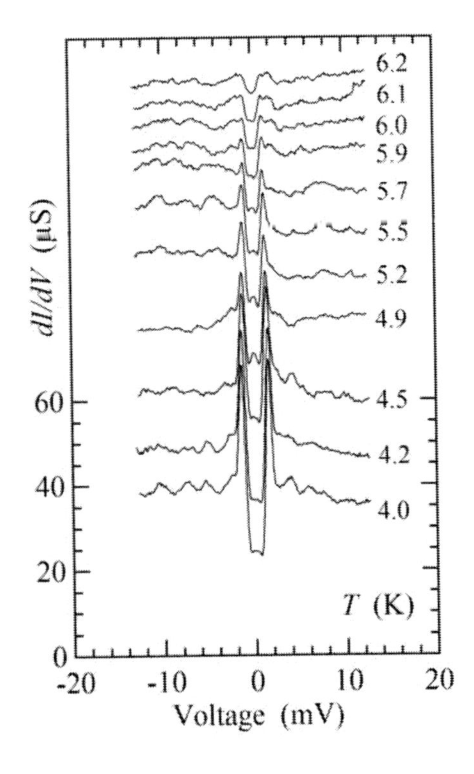

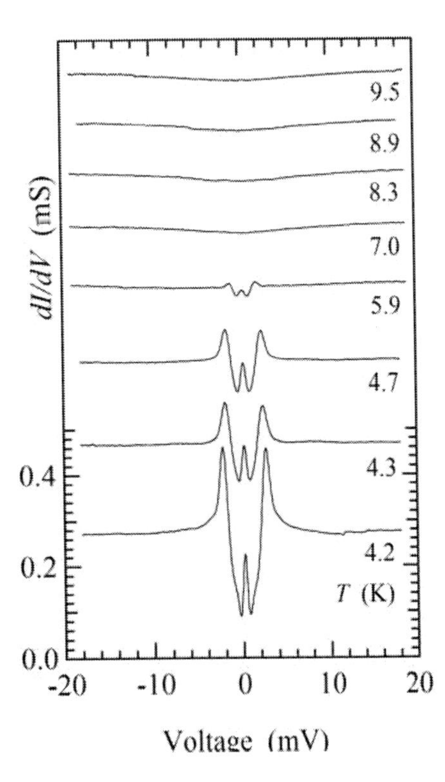

Figure 38. Temperature variations of tunnel break-junction $G(V)$ for NbB_2: (a) and (b) correspond to the typical gap and the small gap features, respectively. The conductance is shifted up for clarity without offsets between experimental and calculated curves.

T-dependences of the gap values and *NZBC* for the spectra of Figures 38(a) and 38(b) are shown in Figure 39. In Figure 39(a) the scaled BCS function $\Delta(T)$ should express the T-dependence of the small gap. The extrapolated values $\Delta(T = 0\text{ K})$ for Figures 38(a) and 38(b) are 1.2 – 1.3 meV and 0.7 – 0.8 meV, respectively. In both cases T_c is approximately 7 K. As comes about from Figure 39(b) the same value of T_c can be extrapolated from the *NZBC*. These values lead to the gap ratios $2\Delta(0)/k_BT_c = 4.2 - 4.5$ for the larger gap with $V_{p\text{-}p} = 5$ mV and $2\Delta(0)/k_BT_c = 2.4 - 2.6$ for the smaller gap with $V_{p\text{-}p} = 3$ mV. These values suggest that NbB_2 is a strong coupling BCS superconductor, the sample including another (induced) superconducting phase.

From the data presented above as well as other available evidence one should make a conclusion that at least three different phases coexist in NbB_2 samples. We mean a superconducting phase of NbB_2 with $T_c = 7$ K, the proximity-induced phase, and the superconducting phase with $\Delta(0) \approx 2$ meV. Since T_c of NbB_2 changes from 0 K to 9.1 K depending on the nominal composition $NbB_{2+\delta}$ [246], spatial inhomogeneity may be responsible for this phase separation. T-variation of $V_{p\text{-}p} = 5$ mV shows that the apparent T_c is about 7 K, which is consistent, e. g., with resistivity measurements. Some local patches in the samples may possess a relatively high T_c (≈ 9 K), while normal-conducting grains with smaller δ might be formed as well. Note that the quasi-stoichiometric NbB_2 is known as a normal conductor or low-T_c (≈ 3 K) superconductor [246]. Therefore, the proximity-induced and large-gap phases are unavoidable consequences of spatial inhomogeneity. If one assumes that the large-gap with $\Delta(0) \approx 2$ meV vanishes at $T_c = 9.1$ K, the corresponding gap to T_c ratio should be estimated as $2\Delta(0\ \mathrm{K})/k_B T_c \approx 5.2$.

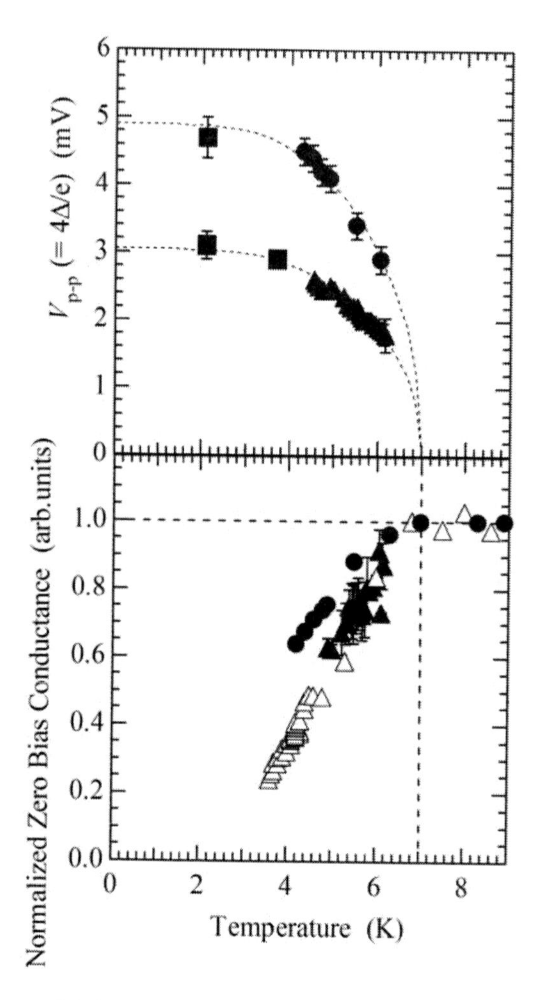

Figure 39. (a) Energy gaps $\Delta(T)$ from Figure 38 (a) (●) and Figure 38 (b) (▲). Symbols (■) correspond to other junctions. (b) Normalized zero-bias conductance (*NZBC*) inferred from data of Figure 38 (a) (●) and Figure 38 (b) (▲). Symbols (■) correspond to the T-dependence of *NZBC* for another tunnel spectrum.

4.2. BJTS data for CaAlSi

A ternary layered compound $Ca(Al_{0.5},Si_{0.5})_2$, isostructural with MgB_2 but having much lower $T_c \approx 7.7$ K than the latter, which has been discovered in Ref. [261] and possesses highly anisotropic superconducting properties [265], is of interest for our purposes, namely as a possible candidate for the multiple-gap superconductor. Representative tunnel conductances of CaAlSi found at 4.2 K are displayed in Figure 40. The observed gap values determined by the peak-to-peak bias separation are estimated to be $V_{p\text{-}p}$ (considered as $4\Delta/e$ assuming an SIS junction) = 4.2 – 4.6 mV at 4.2 K. These values are reproducible among different junctions. The experimental curves are compared with the calculated ones using Eqs. (1) and (2) [93]. The gap parameters are $\Delta(4.2\text{ K}) = 1.0 - 1.3$ meV ($\gamma \approx 0.16$ meV).

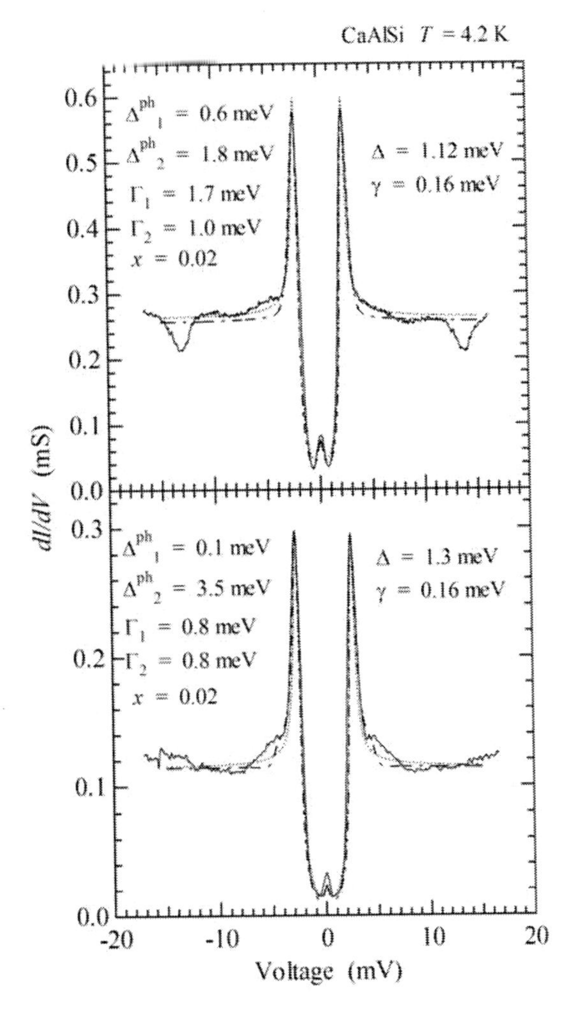

Figure 40. BJTS data for junctions involving CaAlSi at 4.2 K. Dashed and dash-dotted curves represent fittings by BCS and the correlated two-gap models, respectively.

One sees that the main observed gap structures in CaAlSi tunnel spectra are reasonably well described by the SIS conductance with the BCS density of states. However, shoulder structures are often found outside the gap-edge peaks. These structures are similar to the induced gap features of BJTS data in MgB_2, which have been fitted by the correlated two-gap model. Indeed, in Figure 40 the correlated two-gap model (Eqs. (5) – (10)) is demonstrated to satisfactorily reproduce the observed spectral features as is shown by dash-dotted curves. Still clear hump structures seen in Figure 40(b) can not be unambiguously reproduced by the correlated two-gap model. Therefore, the broadened dip-hump structures might have a different origin [21].

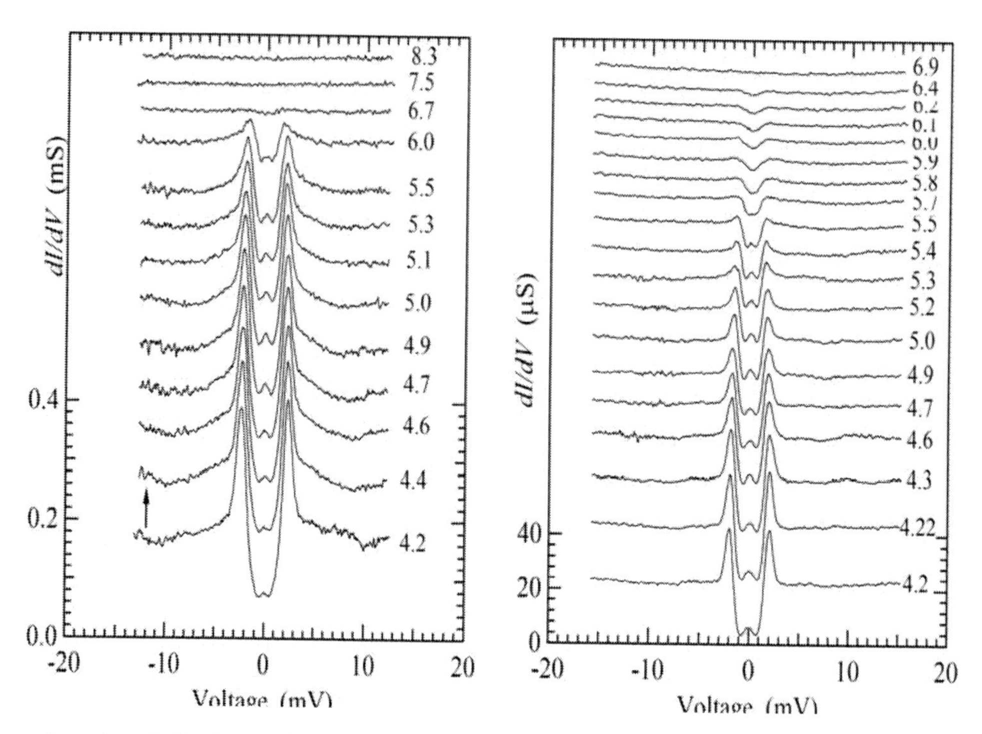

Figure 41. (a) and (b) T-variations of $G(V)$ for the same junctions as in Figures 40(a) and 40(b), respectively.

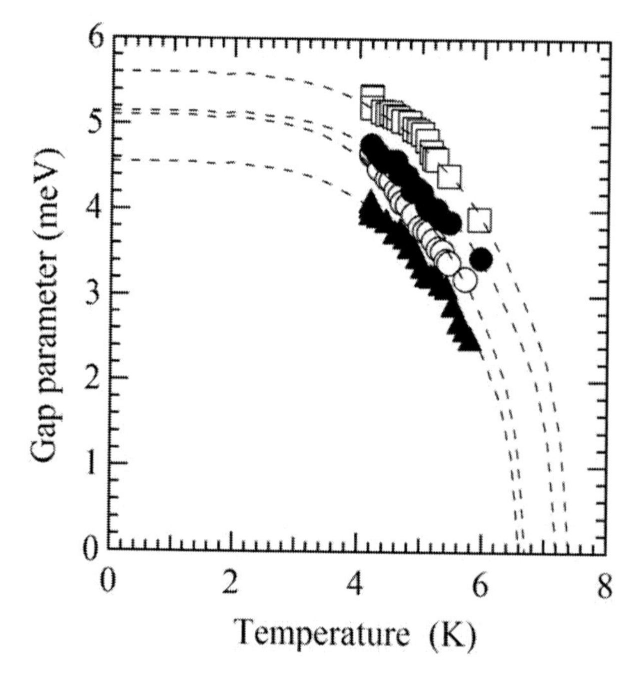

Figure 42. T-dependence of the energy gap $2\Delta(T)$ inferred from Figures 41(a) (●) and 41(b) (▲). Symbols (○) and (□) correspond to other junctions.

T-dependences of $G(V)$ in CaAlSi are shown in Figures 41(a) and 41(b). They describe tunnel spectra for the same junctions as in Figures 40(a) and 40(b), respectively. Since these junctions are quite stable and the gap-edge peaks are relatively well defined up to highest temperatures below T_c, the true T-dependent gap $2\Delta(T)$ may be identified with $eV_{p\text{-}p}/2$. This function is displayed in Figure 42. Using the scaled BCS fitting of $\Delta(T)$, we determine the extrapolated quantities $2\Delta(0\text{ K})$ and T_c as 2.2 – 2.6 meV and 6.6 – 7.3 K, respectively. The smaller is the zero-T gap, the lower is T_c. As a consequence, the ratio $2\Delta(0)/k_BT_c$ falls into the range 4.3 – 4.6, which means that CaAlSi is a strong-coupling s-wave superconductor. The ratio is slightly larger than that for an elemental superconductor Pb with similar T_c.

4.3. Discussion of superconductivity in compounds related to MgB_2

Since the critical temperature $T_c = 7 - 9$ K of NbB_2 is close to that of Nb ($T_c = 9.2$ K), it is instructive to compare tunnel spectral features of both materials, the more so as high-purity Nb was demonstrated long ago to exhibit two gaps in heat capacity measurements [247]. We have carried out tunneling measurements for the commercially available polycrystalline (3N) Nb. $G(V)$ for the corresponding break junction obtained at 4.2 K is shown in Figure 43(a). A sharp gap-edge structure is found at the bias positions of ± 3 mV. Other features are the Josephson peak as well as the inner fine structures at $V = \pm 1.5$ mV, which is due most probably to the two-particle tunneling [21], [164], [266]. The observed spectrum is well fitted by the SIS conductance in the framework of the BCS model (dashed curve). The fitting gap parameter is $\Delta(4.2\text{ K}) = 1.5$ meV. Such a value agrees with the conductance-peak positions being outside those of NbB_2. The gap to T_c ratio is $2\Delta(0)/k_BT_c \approx 3.8$ indicating a moderate strength of the electron-phonon coupling. This result agrees with the earlier tunnel data [248] being at the same time smaller than the results for NbB_2 (see Section 4.1).

Gapped DOSes calculated for typical NbB_2 and Nb junctions on the basis of the best fitting parameters are depicted in Figure 43(b). To make the comparison more descriptive we divided the results by T_c. The apparent discrepancy between spectral shapes might be due to samples' inhomogeneity or gap anisotropy (multiple nature). Since the gap position in the so normalized superconducting DOS of NbB_2 is larger than that of Nb, we believe that the tunneling spectra represented in Figure 36 reflect the intrinsic gap structure of the studied NbB_2 samples.

There have been quite not so many experiments directly measuring superconducting gaps in NbB_2 or CaAlSi. For NbB_2 NMR studies revealed the BCS weak coupling behavior with $2\Delta(0)/k_BT_c \approx 3.1$ [249]. This value is much smaller than ours. Since our ratio $2\Delta(0)/k_BT_c$ for MgB_2 inferred from tunneling measurements is in excellent agreement with the results of the same Ref. [249], the discrepancy still remains obscure (however, see below). On the other hand, photoemission [250] and heat capacity [251], [252] studies showed that CaAlSi is a strong coupling superconductor with $2\Delta(0)/k_BT_c = 4.1 - 4.4$. These ratios are very similar to our findings.

A plot $2\Delta(0)/k_BT_c$ versus T_c for MgB_2, NbB_2 and CaAlSi, which is based on our experimental results is shown in Figure 44. The gap ratio $2\Delta(0)/k_BT_c$ seems to increase with T_c. On the basis of this fact we might suppose that the discrepancy between the gap magnitudes in NbB_2 found by tunneling and NMR techniques could be attributed to the difference in T_c as well as stoichiometry of $NbB_{2+\delta}$ samples involved. Similar composition dependences were observed, e. g., in thin films of the A15 compound Nb_3Ge [176], [253].

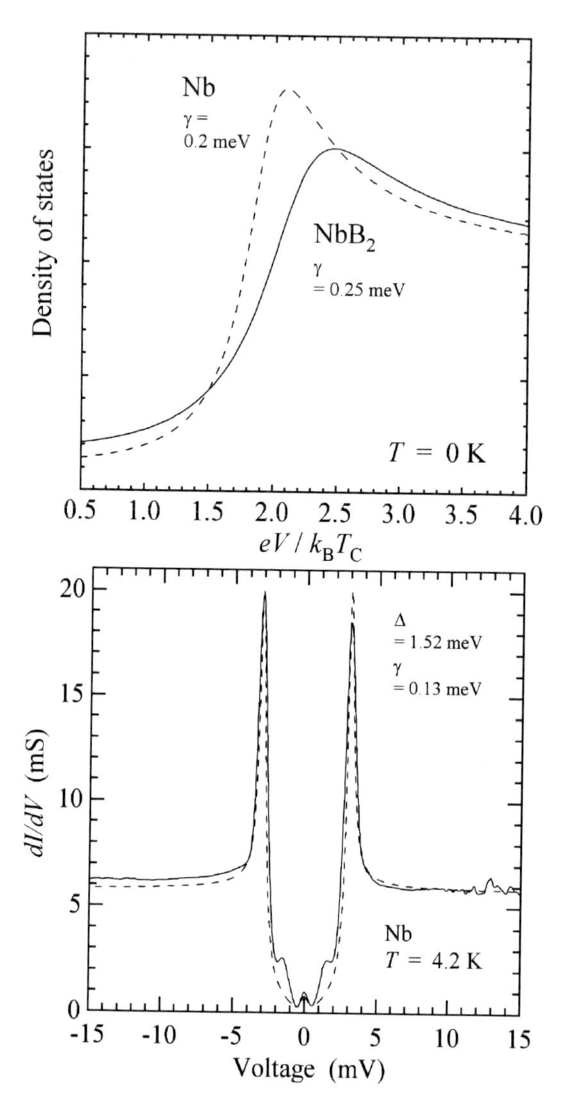

Figure 43. (a) $G(V)$ of Nb break junction at 4.2 K. The dashed curve corresponds to the calculated fitting by the BCS nodel; (b) dashed and solid curves represent calculated normalized BCS densities of states at 0 K for Nb and NbB_2, respectively.

The apparently strong coupling superconductivity in both MB_2 compounds (M = Nb, Mg) suggests that the origin of superconductivity is also the same there notwithstanding the large difference in T_c. At the same time, band structure calculations predict that the MgB_2 Fermi surface is mainly formed by the boron $2p$-orbital, while that of NbB_2 originates from niobium $4d$-orbital [254]. Nevertheless, the increase of boron $2p$-orbital contribution to the electron DOS at the Fermi level is predicted to occur in the superconducting $NbB_{2+\delta}$ with Nb vacancies [255], [256]. Therefore, both the ratio $2\Delta(0)/k_BT_c$ and T_c in MB_2 (M = Mg and Nb) depend on the boron $2p$-orbital DOS at the Fermi level.

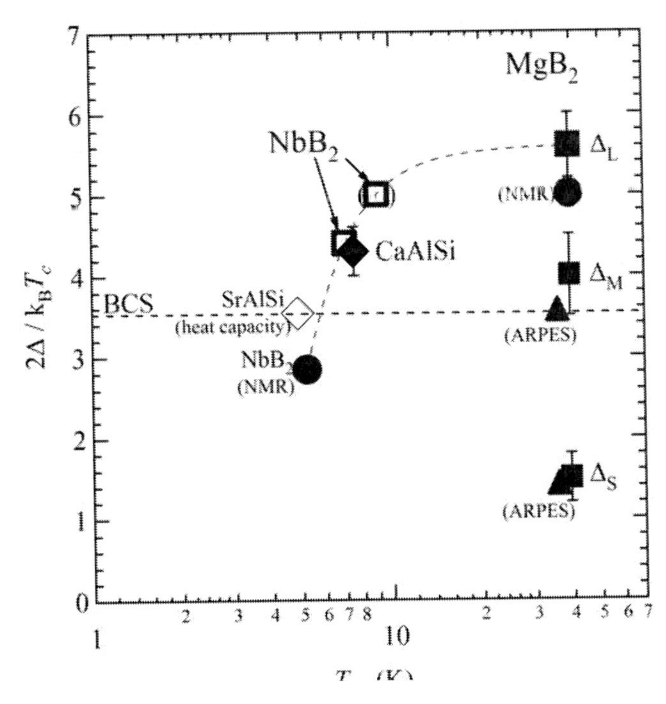

Figure 44. $2\Delta/k_BT_c$ vs T_c plot for MB_2 (M = Mg, Nb) and AAlSi (A = Ca, Sr). The data are our original or taken from NMR [249], ARPES [250] and specific heat data [251], [252]. The dashed curve is a guide to the eye.

Moreover, it turned out [246] that Nb-deficiency in $Nb_{1-x}B_2$, friendly to superconductivity, conspicuously affects not only $2\Delta(0)/k_BT_c$ and T_c but also the lattice parameter ratio of c/a. Namely, the values of $2\Delta(0)/k_BT_c$ and T_c increase with c/a. In MgB_2 the ratio c/a is larger than that of NbB_2, which correlates with the rise of T_c. This empirical correlation is not strange (see the probabilistic analysis of possible correlations between normal properties and superconductivity in various materials [258]), since a strong enough electron-phonon interaction should considerably affect the crystal lattice [259]. And indeed, a positive relationship between bond-stretching E_{2g} phonon softening and T_c was observed in MgB_2 [260]. On the basis of a calculated DOS in NbB_2 it was suggested [255], [256] that the large ratio $2\Delta(0)/k_BT_c$ in tunneling spectra of this compound is most probably due to the enhancement of the electron-phonon interaction (influencing the E_{2g} mode, crucial for superconductivity).

According to band calculations, the Fermi surface of CaAlSi is predominantly formed by Ca 3d states [257]. Indeed, ARPES measurements revealed the energy band dispersion [250], which is consistent with these calculations. Photoemission studies also demonstrated that superconducting gaps on the Fermi surface sheets around $M(L)$ and $\Gamma(A)$ points in the Brillouin zone have the same magnitude contrary to what is appropriate to MgB_2. The corresponding $2\Delta(0)/k_BT_c$ ratio is 4.2 ± 0.2. Similar strong coupling gap to T_c ratios 4 – 4.2 manifested themselves in heat capacity measurements [255], [256].

As has been mentioned above, our tunneling data show almost identical gap values Δ(4.2 K) = 1.0 – 1.3 meV well reproduced by the simple BCS model. Concomitant measurements for varying T lead to the same ratio $2\Delta(0)/k_BT_c = 4.4 \pm 0.2$ as in other experiments. It means that CaAlSi is a strong coupling superconductor with the almost isotropic gap function. The value of c/a in CaAlSi is similar to that of NbB_2 [246], [261]. However, the relationship between c/a and T_c in several silicides is opposite to the dependence found for MB_2 [246], [261], [262], as is shown in Figure 45. Namely, the ratio c/a grows when the values of $2\Delta(0)/k_BT_c$ and T_c decrease. The values of c/a in AAlSi (A = Ca, Sr and Ba) are reduced when the metal A changes from Ca to Ba. Band structure and phonon dispersion calculations [263], [264] disagree in their views concerning the role of electron-phonon interaction in superconductivity of silicides. In particular, in Ref. [264] superconductivity with $T_c \approx 8$ K in CaAlSi was suggested to be stimulated by the soft B_{1g} phonon mode, whereas the authors of Ref. [263] consider the phonon-exchange mechanism to be insufficient to result in such a strong coupling as in CaAlSi. Hence, further studies are required to elucidate the origin of Cooper pairing in the compounds concerned.

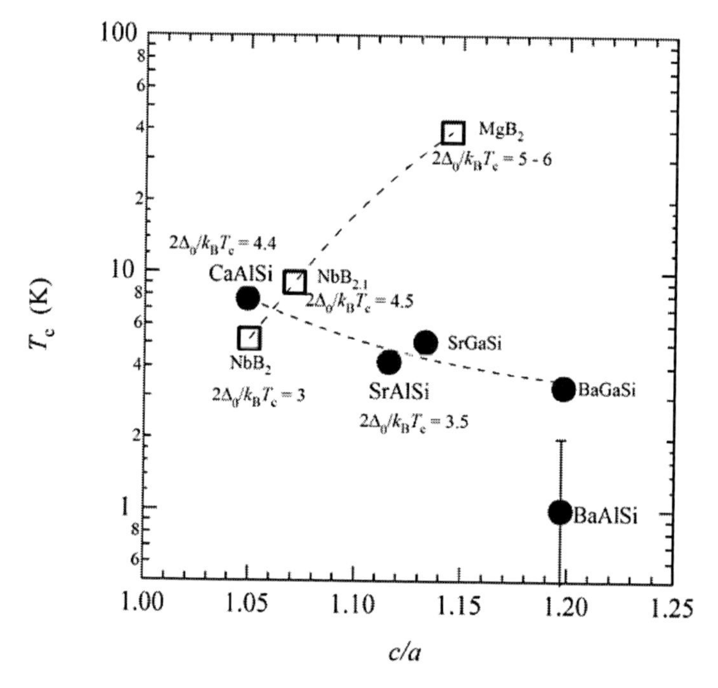

Figure 45. T_c vs c/a plot on the basis of data inferred from Refs. [246], [261], [262]. Dashed curves are guides to the eye.

5. Superconducting Gaps in Tunneling Studies of β-*M*NCl (*M* = Hf and Zr) Compounds

As has been already indicated in the Introduction, intercalated layered materials β-MNCl (M = Hf and Zr) are interesting primarily because of their high T_c. However, they proved to be [269] one more non-cuprate object with multiple gap features in addition to those discussed above. Therefore, it is reasonable to analyze their DOS-related properties here. Before presenting specific results we are going to make a few general remarks.

Almost all the $G(V)$ curves obtained in our measurements involve zero-bias peaks as well as the fine structures different from their gap-related counterparts and located in other bias regions. Zero-bias peaks may be signatures of the Josephson effect or a related weak-link behavior in constrictions appearing during formation of tunneling break junctions. The spectrum modulations and/or various fine structures are most probably due to the embedded additional series resistances corresponding to superconducting filaments in the main phase of the sample. The additional structures appear when those filaments are driven into the normal state by the critical current passing through narrow spatially confined regions. Alternatively, the unconventional features of dI/dV might reflect the excitation of lattice vibrations by the non-stationary (ac)-Josephson current, as was suggested, e. g., for high-T_c oxides [122], or the nonlinear self-detection of various harmonics in the Josephson junctions [124]. As for the sub-gap features, the n-particle tunneling [21], [164], [266] might be their origin. Here, to avoid less significant details, we focus on the predominant and most informative gap-related structures in dI/dV.

5.1. BJTS data for $Li_{0.48}(THF)_yHfNCl$ and $HfNCl_{0.7}$

Representative break-junction tunnel conductances dI/dV for $Li_{0.48}(THF)_xHfNCl$ are shown in Figure 46. In Figure 46(a), dI/dV reveals highly asymmetric broadened coherent gap peaks with the peak-to-peak separation $V_{p\text{-}p} \approx 24$ mV identified by us as $4\Delta(4.2\text{ K})/e$. The peak intensities are severely suppressed as compared to the SIS-type conductance calculated in the BCS model. Since the peaks should be located at $V = \pm 2\Delta/e$, the gap value is estimated to be $2\Delta \approx 11$–12 meV.

In Figure 46(a) an additional sub-gap structure and a small zero-bias peak are also seen. The broadened and V-shaped conductance background can be attributed to the reactive sample surface because we have obtained more conventional BCS-like conductance features for less-reactive samples. dI/dV shown in Figure 46(b) has slightly smaller Δ than that of Figure 46(a).

In Figure 46(c) another kind of the observed dependence $G(V)$ is presented. The overall multiple-gap structure consists of intensive inner peaks at ±4 mV ($V_{p\text{-}p}^{in} \approx 8$ mV) and outer ones of the comparable size near $V \approx \pm 11$ mV ($V_{p\text{-}p}^{out} \approx 22$ mV) with a concomitant strong zero-bias peak. The $V_{p\text{-}p}^{out}$ values are similar to those of Figures 46(a) and 46(b). One can additionally see smeared humps at biases $V = \pm 17$ mV larger than the main peak positions at $V = \pm V_{p\text{-}p}^{out}/2 \approx \pm 11$ mV.

Another type of observed $G(V)$ is demonstrated in the inset of Figure 46(c). Here a single-gap structure is found, well fitted by a simple BCS-like DOS. Hence, the inner gap-like features observed at $V \approx \pm 4$ mV in Figures 46(b) and 46(c) most probably correspond to another phase with a smaller gap rather than to an SIN junction formed instead of the nominal SIS one, not to say about the two-particle tunneling [21], [164], [266].

dI/dV for two different $HfNCl_{0.7}$ break junctions are displayed in Figure 47. The high-voltage conductance displayed in Figure 47(a) is relatively large, of the order of 0.3 mS, whereas that shown in Figure 47(b) is as low as ~ 10 μS. Well developed gap-edge structures at ± 8 mV and a zero-bias Josephson peak are clearly seen in Figure 47(a). A conventional (with a very small Γ) superconducting DOS with $\Delta \approx 3.9$ meV satisfactorily describes the data including the peak itself. Distinct dip-hump structures outside the gap region might be of intrinsic origin, corresponding to van Hove DOS singularities in layered systems [117], [118], [270], or be extrinsic, *i.e.* they might appear due to the current-driven suppression of superconductivity in small weak-link inter-grain contacts [271] or, possibly, at boundaries between segregated phases in the case of phase separation as in cuprates [272], [276]. The out-gap structure seems to be too large to be a standard strong-coupling peculiarity [34], [35] reflecting the inelastic boson-assisted electron tunneling.

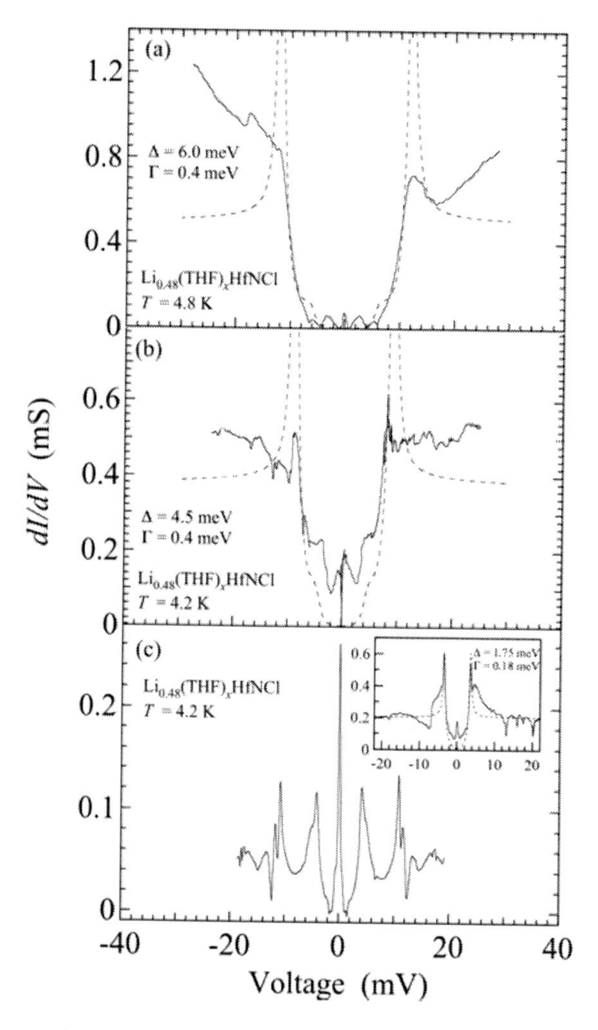

Figure 46. Tunneling conductances at T = 4.2 K for various break-junctions (a, b and c) involving $Li_{0.48}(THF)_xHfNCl$. Δ and Γ are the superconducting gap value and the gap broadening, respectively, for the DOS given by Eq. (2) with an accuracy of notations. The peak-to-peak distance $V_{p\text{-}p}$ is equal to $4\Delta(T)/e$. Dashed curves are calculated SIS conductances in the BCS model. The inset in (c) demonstrates $G(V)$ of the SIS type with the smallest achieved gap value $\Delta \approx 1.75$ meV.

In Figure 47(b) the smeared gap-edge peaks at $\pm V_{p\text{-}p}/2 \approx \pm 10 - 11$ mV (= $\pm 2\Delta/e$) are seen with a very low leakage and against a *V*-shaped background [277]. A BCS model can adequately reproduce the conductance only inside the gap. The additional small sub-gap peaks at approximately $\pm 2 - 3$ mV in Figure 47(a) and Figure 47(b) might be explained by Andreev quasiparticle reflection involving surface proximity effects [21].

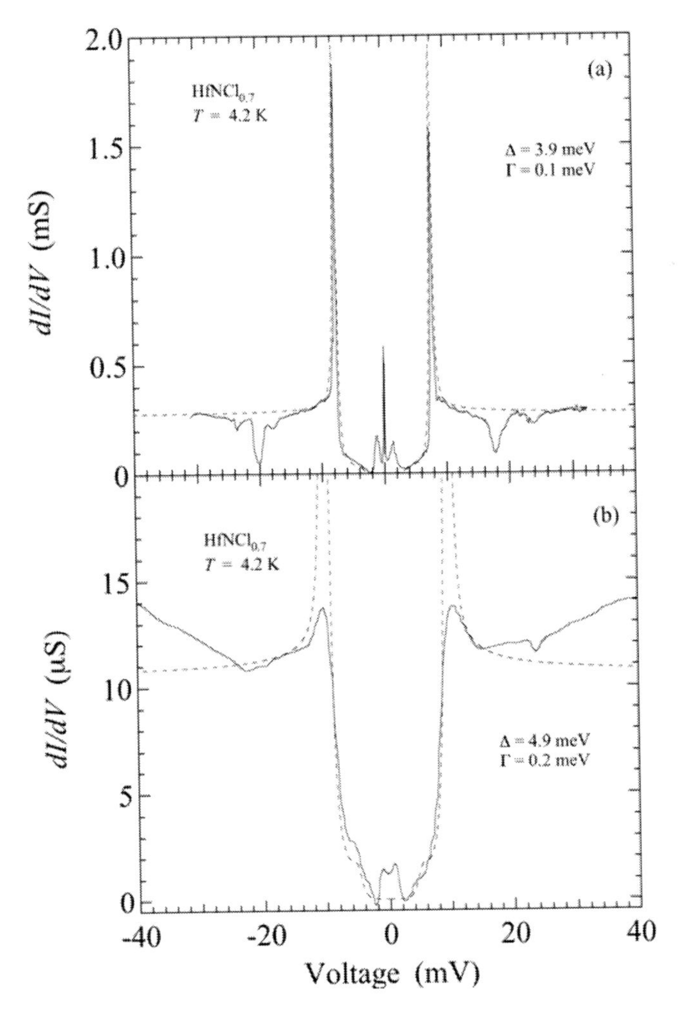

Figure 47. $G(V)$ at $T = 4.2$ K for different break junctions involving $HfNCl_{0.7}$. Dashed curves are fitting BCS conductances calculated for SIS junctions.

To get rid of the V-shaped background, tunneling conductances *dI*/*dV* presented in Figures 46(a) and 46(b) (for $Li_{0.48}(THF)_xHfNCl$), and Figure 47(b) (for $HfNCl_{0.7}$) were divided by $(V)^{1/2}$ and resulting normalized quantities were displayed in Figure 48 (a, b, and c, respectively). The gap-edge peaks are readily seen in all panels of Figure 48, and these curves became more similar to $G(V)$ predicted by the BCS theory than the raw data strongly influenced by the non-superconducting background. The positive-bias gap-edge peak in Figure 48(a) becomes more conspicuous as compared to that of Figure 46(a), whereas the reconstructed dI/dV still exhibits some asymmetry, appropriate to the raw data. In Figure 48(b), the normalization by $(V)^{1/2}$ makes the outer coherent peaks more clearly visible and the shoulders at $V \approx \pm$ 4 mV are transformed into discernible peaks. In Figure 48(c), shoulders appear outside the coherent peaks after the division by $(V)^{1/2}$, and the higher-bias region $|V| > 20$ mV becomes flat. Those shoulders might be a signature of the proximity-induced superconducting gap [21], [121]. The apparently $(V)^{1/2}$ dependence of the background presumably reflects scattering of conduction electrons on weakly disordered states in some regions caused by inhomogeneities left after carrier doping [21].

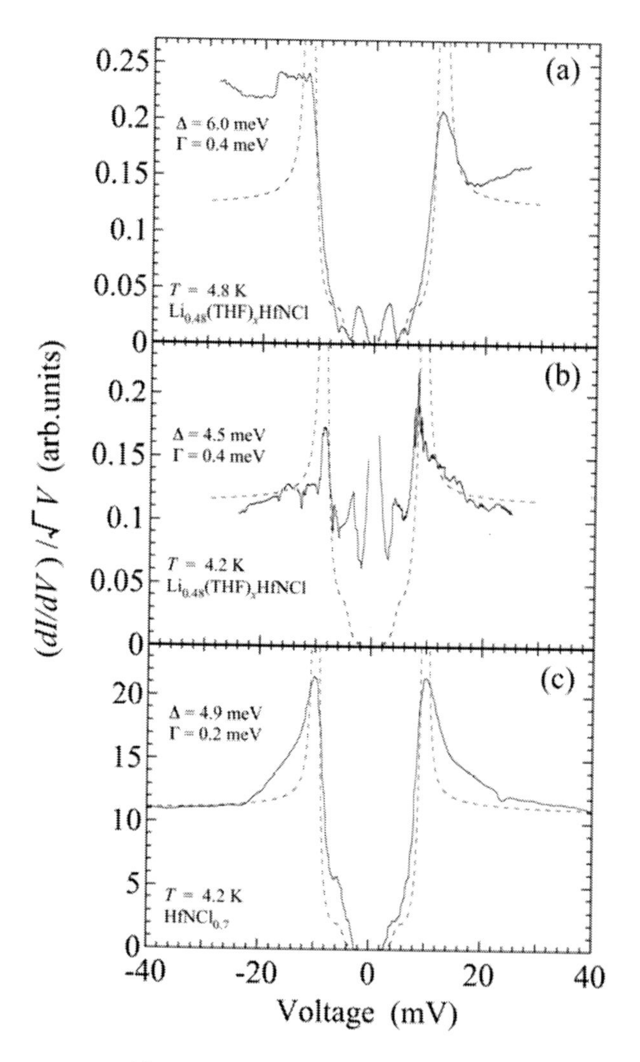

Figure 48. $G(V)$ divided by $(V)^{1/2}$. Here $G(V)$ corresponds to the data presented in Figures 46(a), (b) ($Li_{0.48}(THF)_xHfNCl$) and 47(b) ($HfNCl_{0.7}$) ((a), (b), and (c), respectively). The dashed curves are the calculated SIS conductances in the framework of the BCS theory.

In Figure 49 variations of $G(V)$ with T are shown in the gap region for $Li_{0.48}(THF)_xHfNCl$ (a) and $HfNCl_{0.7}$ (b). In both cases the gap structure is gradually wiped out with T to disappear at $T_c = 26$ K (a) and 24 K (b) for $Li_{0.48}(THF)_xHfNCl$ and $HfNCl_{0.7}$, respectively. In particular, one sees from Figure 49(b), that $G(V)$ in the neighborhood of the outer gap changes drastically in the range $10 < T < 12$ K but at the same time the inner gap remains almost unchanged, thereby suggesting that either the origin of the features concerned is different or, simply, the outer gap is more affected by spatial scatter. Such a disparity between different gaps was observed and explained for cuprates, where Cooper and charge-density-wave (CDW) pairings coexist [42], [86], [177], [180], [181], [268], [278]. With further increase of T above 12 K both gap structures broaden gradually and vanish at the bulk T_c.

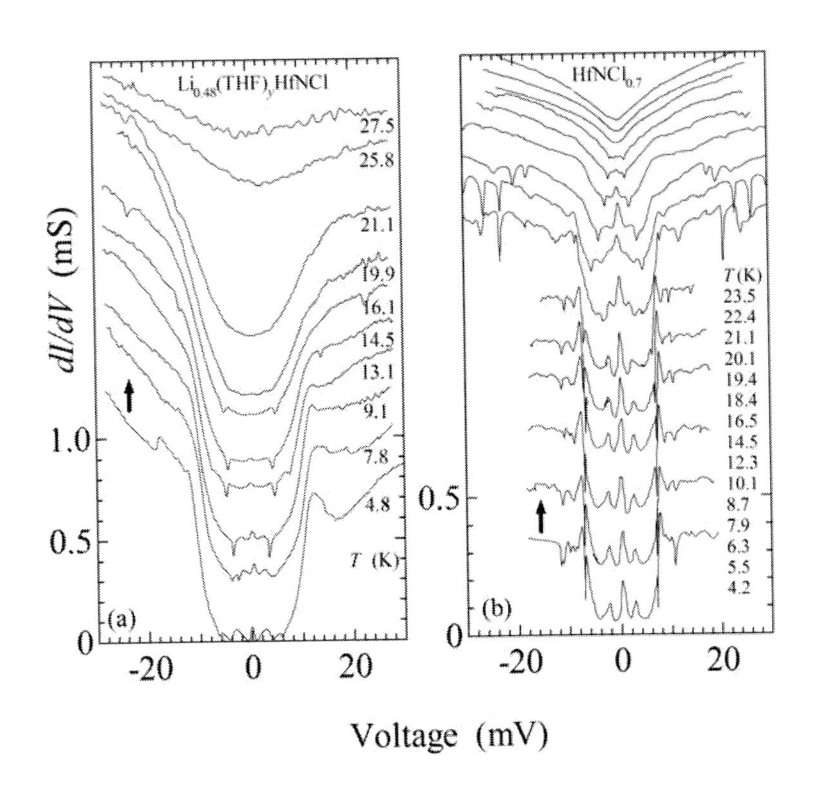

Figure 49. T variations of $G(V)$ for $Li_{0.48}(THF)_xHfNCl$ (a) and $HfNCl_{0.7}$ (b) break junctions, which were presented in Figures 46 (a) and 47 (a), respectively.

The multiple-gap structure of $G(V)$ for the $Li_{0.48}(THF)_xHfNCl$ break junction, which was shown above in Figure 46(c), is displayed in Figure 50 for different T. The gap structures are well defined at low T as shown in Figure 50(a). With increasing T, the magnitude of the outer-gap peaks, pronounced at low T, rapidly decreases so that the peaks evolve into broad humps at approximately 13 K, while the inner peak still remains conspicuous above this T. The zero-bias peak survives up to the bulk T_c. Figure 50(b) demonstrates vertically expanded spectra to see the detailed structure of $G(V)$ at $T > 15$ K. The humps at $V \approx \pm 12$ mV, i. e. outside the inner gap region, disappear at about 21 K, slightly below T_c. Subtle peaks near zero bias, which vanish at $T \approx 13$ K (see Figure 50(a)), can be attributed to the gaps, reduced by the proximity effect.

It might be well to point that the apparent V-shapes of dI/dV for some junctions does not have to be a consequence of the actual anisotropic nature of the superconducting order parameter leading to an incomplete Fermi surface gapping. Instead, such a behavior might reflect a contamination of the broken-contact surfaces, which has been shown, e.g., to degrade both the coherent peaks and the overall character of dI/dV for $Bi_2Sr_2CaCu_2O_{8+\delta}$ single crystals [280].

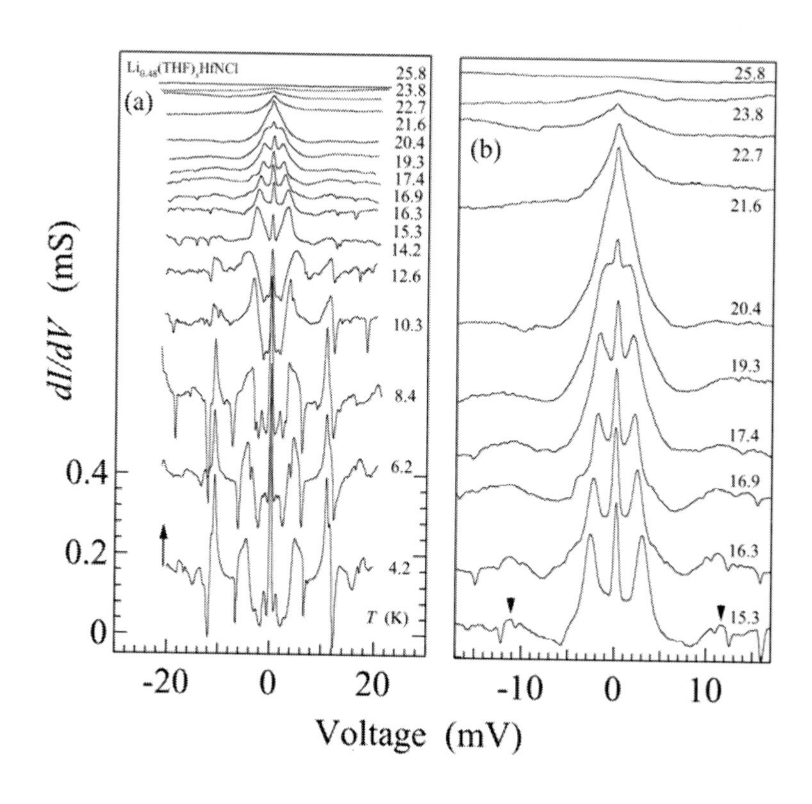

Figure 50. T variation of the double-peak structure for the sample $Li_{0.48}(THF)_xHfNCl$ presented above in Figure 46(c): (a) overall picture up to T_c, (b) enlarged fragment for T near T_c.

Dependences of gap values on T obtained from the data presented in Figure 50 for $Li_{0.48}(THF)_xHfNCl$ break junctions are demonstrated in Figure 51. Smearing of the gap features in $G(V)$ at higher T leads to uncertainties of gap values above 20 K, which is shown as large error bars. We note that the gap features with the amplitude $2\Delta(4.2\text{ K}) \approx 11 - 12$ meV ($V_{\text{p-p}} \approx 22 - 24$ mV) for different samples, shown in Figsures 49(a) and 50, exhibit similar T-dependences, which is well expressed by the scaled BCS curve with $T_c = 25.5$ K.

T-dependence of the inner gap for $Li_{0.48}(THF)_xHfNCl$ junctions with $V_{\text{p-p}}^{\text{in}} \approx 9 - 10$ mV shown in Figure 51 and derived from Figure 50 deviates downward from the BCS curve exhibiting almost linear decrease up to the temperatures near bulk T_c. This behavior seems to be a manifestation of the proximity induced gap [21], [121] and is similar to that observed in Pb-Sn-PbO-Pb junctions [143] and MgB_2 break junctions with their remarkable multiple-gap structures [281], [282]. The latter are usually considered [4] as an intrinsic effect, emerging from the interplay of two superconducting gaps each originating from a separate electron band [48], [104], [110], [111]. However, theoretical calculations show that the Fermi surface of $Li_{0.48}(THF)_xHfNCl$ involves a single band [23]. Therefore, possibility of the multiple-band superconductivity in $Li_{0.48}(THF)_xHfNCl$ seems to be ruled out, in particular, as an explanation of the smaller-gap feature. On the contrary, the origin of the proximity effect and the inner gap as its consequence may be local charge-carrier-density inhomogeneities [117], [118], [120]. As is readily seen from Figures 46(c) and 48 (for $V_{p\text{-}p}^{in}$), the small gap is quite reproducible. Therefore, the apparent inner-peak spectra in Figure 46 might be associated with the anisotropy in transport properties [26]. It should be noted that T-dependences of the multiple gap structures displayed in Figures 50 and 51 are similar to those of YNi_2B_2C where anisotropy of the superconducting gap is considered as the origin of such a pattern [284].

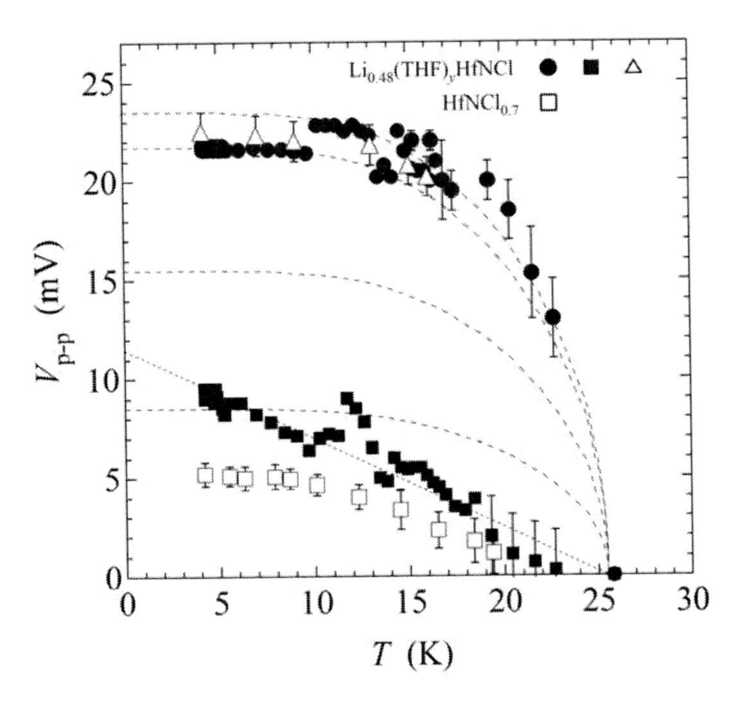

Figure 51. T-dependences of the peak-to-peak separations $V_{p\text{-}p}$ in $G(V)$ for break junctions made of $Li_{0.48}(THF)_xHfNCl$ and $HfNCl_{0.7}$. Closed symbols (● and ■) refer to $V_{p\text{-}p}^{out}$ and $V_{p\text{-}p}^{in}$, respectively, describe $Li_{0.48}(THF)_xHfNCl$ and are taken from Figure 50. The open symbols (Δ and □) denote peak-to-peak separations found for $Li_{0.48}(THF)_xHfNCl$ (Figure 49(a)) and $HfNCl_{0.7}$ (Figure 49(b)) break junctions, respectively. Symbols Δ correspond to predominant $V_{p\text{-}p}$, whereas □ describe the inner-gap $V_{p\text{-}p}^{in}$.

The indicative gap to T_c ratio $2\Delta(0)/k_BT_c$ for $Li_{0.48}(THF)_xHfNCl$ is of the same magnitude 5 – 5.6 [283] as that of the layered superconductor MgB_2 [281], [282]. As has been already pointed out above, these values are in fact larger than the weak-coupling BCS value $2\pi/\gamma \approx 3.52$ but much smaller than experimental values for high-T_c cuprates [144] and organic superconductors [285]. For $HfNCl_{0.7}$ samples the largest attainable gap value $2\Delta(0) \approx 11$ meV can be inferred from Figure 47(b), whereas $T_c \approx 23.5$ K, so that the ratio $2\Delta(0)/k_BT_c$ becomes 5.4. This is consistent with the results for $Li_{0.48}(THF)_xHfNCl$. The inner-gap small peaks at $V \approx \pm 2.5$ mV (at $T = 4.2$ K) taken from Figure 49(b) and describing $HfNCl_{0.7}$ break junctions tend to disappear near respective T_c as is shown in Figure 51 (open squares), although the low-temperature $V_{p\text{-}p}$ is half as that for $Li_{0.48}(THF)_xHfNCl$ samples (see Figures 46(b) and 46(c)). If we identify the smaller-gap features for $Li_{0.48}(THF)_xHfNCl$ break junctions with $V_{p\text{-}p} \approx 10$ mV $\approx 4\Delta/e$, the ratio $2\Delta(0)/k_BT_c$ becomes 2 – 2.3. This is substantially smaller than the

BCS value, which is consistent with our suggestion that those peaks have the proximity-induced nature.

$G(V)$ with the widest gap structures found by us are displayed in Figures 52 and 53 for $Li_{0.48}(THF)_xHfNCl$ and $HfNCl_{0.7}$ break junctions, respectively. Figure 52(a) exhibits the well-pronounced gap edges around $V \approx \pm 17$ mV and a zero-bias peak, the values corresponding to the outer peaks in Figure 46(c). This $G(V)$ is adequately reproduced by a single-junction SIS current determined by the convolution of BCS density of states (Eq. (1)), indicating that the largest gap-edge energy does not correspond to the intrinsically multiple junctions, as has been found, e.g., for oxide ceramics $BaPb_{1-x}Bi_xO_{3-\delta}$ [286]-[289] and stacks of tunnel junctions in layered cuprate mesas [290]-[293]. Conspicuous dip-hump structures are observed at $V \approx \pm 10 - 12$ mV *inside* the gap, the peak locations coinciding with those for the *predominant* coherent gap peaks at $V \approx \pm 11$–12 mV for those $Li_{0.48}(THF)_xHfNCl$ junctions, conductances of which are presented in Figure 46.

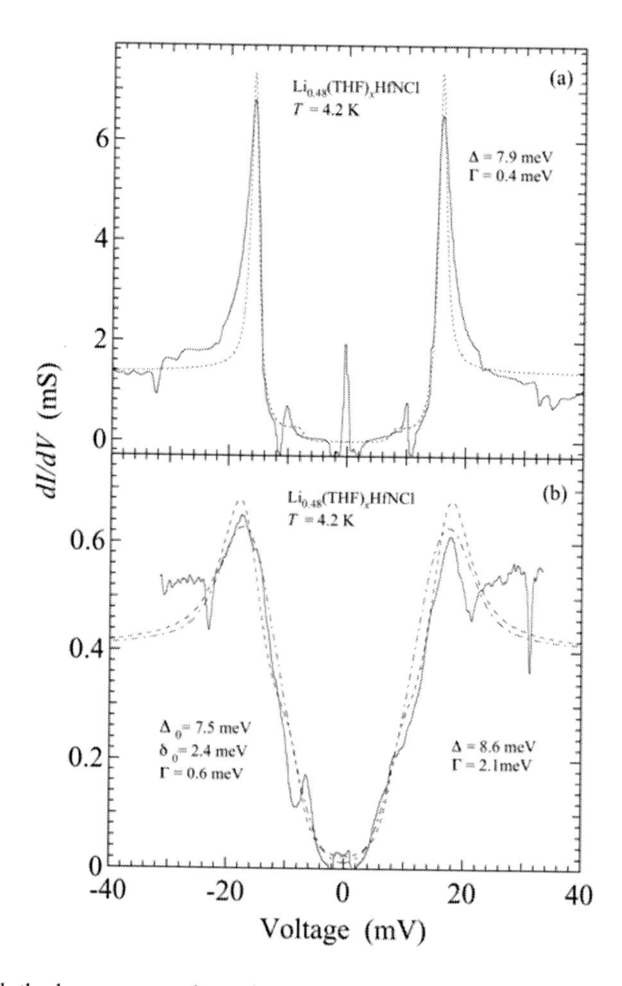

Figure 52. $G(V)$ with the largest gap values observed in $Li_{0.48}(THF)_xHfNCl$ break junctions. The dashed and dashed-dotted curves describe calculated SIS conductances with the smeared BCS density of states (Eq. (2)) and with the Gaussian gap distribution model (Eqs. (3) and (4)), respectively. The spectra (a) and (b) correspond to different break junctions.

The gap features of dI/dV measured for another junction and presented Figure 52(b) are substantially broadened, but their peak positions coincide with locations of fully developed BCS-like peaks of Figure 52(a). This gap structure can be reproduced using the Gaussian gap distribution model and/or smeared BCS density of states, the results of fitting calculations exhibited by dashed and dashed-dotted curves. For fittings to be successful both Γ (it corresponds to ε in Eq. (2)) and δ_0 in Eq. (4) should be about 0.25 Δ, which is much larger than $\Gamma \approx 0.05\ \Delta$ needed to fit experimental data in Figure 52(a). It means that, for certain junctions, dissipation and gap spread becomes essential, probably due to some kind of phase separation and proximity effect.

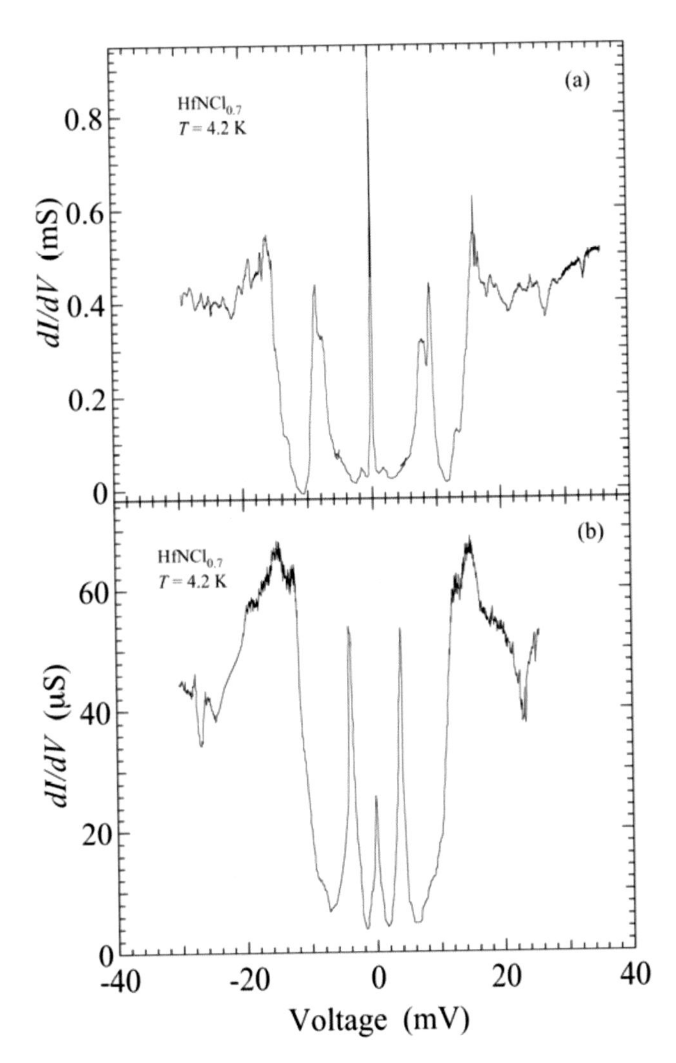

Figure 53. $G(V)$ with the apparent double-peak structure are presented in (a) and (b) for different $HfNCl_{0.7}$ break junctions.

We found a large outer gap with a distinct double-gap overall structure in $G(V)$ for $HfNCl_{0.7}$ break junctions, as is shown in Figure 53. The largest gap values are similar to those of $Li_{0.48}(THF)_xHfNCl$ depicted in Figure 52. Such an apparent double-gap structure has been already seen in Figure 46(c), although the gap energies are different for various junctions. The outer peak-to-peak distance $V_{p\text{-}p}^{out} \approx 32$ mV in Figure 53(a) is of the same order of magnitude as that in Figure 53(b), while the inner-gap parameter $V_{p\text{-}p}^{in} \approx 17$ mV in Figure 53(a) appreciably exceeds $V_{p\text{-}p}^{in} = 8 - 10$ mV from Figure 53(b). Two values 10 and 17 mV for $V_{p\text{-}p}^{in}$ are the same as those given in Figures 46 and 47. In particular, the multiple-gap values for $Li_{0.48}(THF)_xHfNCl$ in Figure 46(c), $V_{p\text{-}p} =$ 8-10 mV, 20-22 mV, 34 mV, are quite similar to those for $HfNCl_{0.7}$ in Figures 53(a) and 53(b). Therefore, the spectrum of experimental $V_{p\text{-}p}$ involves 8 – 10 mV, 18 – 24 mV, and 32 – 35 mV. To further confirm the existence of the largest $V_{p\text{-}p} = 32 - 35$ mV, we have carried out complementary STS measurements having well-defined vacuum tunneling barrier.

STM measurements were also carried out on $HfNCl_{0.7}$. The STM image taken at 5 K on the cleaved *ab* surface of the $HfNCl_{0.7}$ sample is shown in Figure 54 [294]. The sample bias was $V = 0.2$ V and the tunnel current was $I = 0.4$ nA. A triangular arrangement of bright spots is clearly visible against the weakly inhomogeneous background. These bright spots correspond to a higher altitude of the scanning tip. The total difference in altitude was 0.2 nm. We could always obtain such quality of the surface, which means that the cleaved surface of $HfNCl_{0.7}$ is clean and stable under the UHV condition in strong contrast to the cleavage in the ambient atmosphere. From the two-dimensional fast Fourier-transformation (FFT) analysis, the separation of the nearest-neighbor spots was found to be 0.369 nm, which corresponds to the lattice parameter *a* deduced from X-ray diffraction measurements. By examining the atomic arrangement in the conducting double honeycomb metal-nitrogen network on the *ab*-plane of $HfNCl_{0.7}$, we attributed the bright spots to metallic Hf atoms.

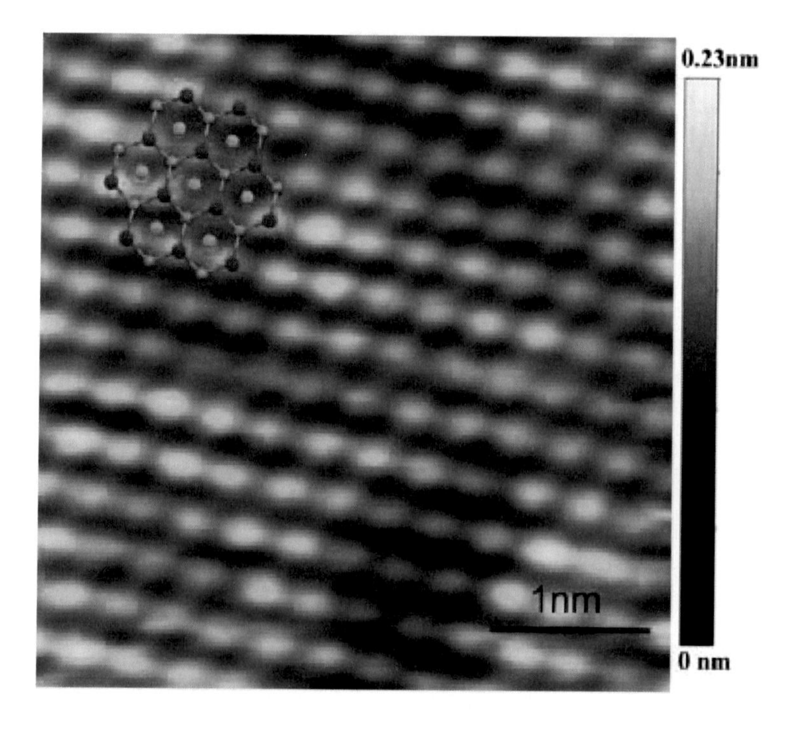

Figure 54. STM image of β-$HfNCl_{1-x}$ ($V = 0.2$V, $I = 0.4$ nA, $T = 5$ K).

To investigate the local superconducting gap related features on the nanometer scale we carried out STS and STM measurements at 5 K simultaneously. The conductance, $G(V) = dI/dV$, map within the area of at least 10 nm × 10 nm in the *ab* plane reveals fairly uniform magnitudes exhibiting almost constant gap-edge peak positions. In Figure 55 such profiles of the STS conductance were displayed along the distance of 10 nm. Relatively homogeneous gap structures were found with weak variations of $G(V)$ magnitudes. In general, $G(V)$ are weakly asymmetric, i.e., the negative-bias (corresponding to the electron tunneling from the sample into the tip) backgrounds are somewhat more pronounced than their positive-bias counterparts, and the junction exhibits substantial leakage at zero bias. The gap-edge peaks at $V \approx \pm$ 10 mV are discernible, although smeared.

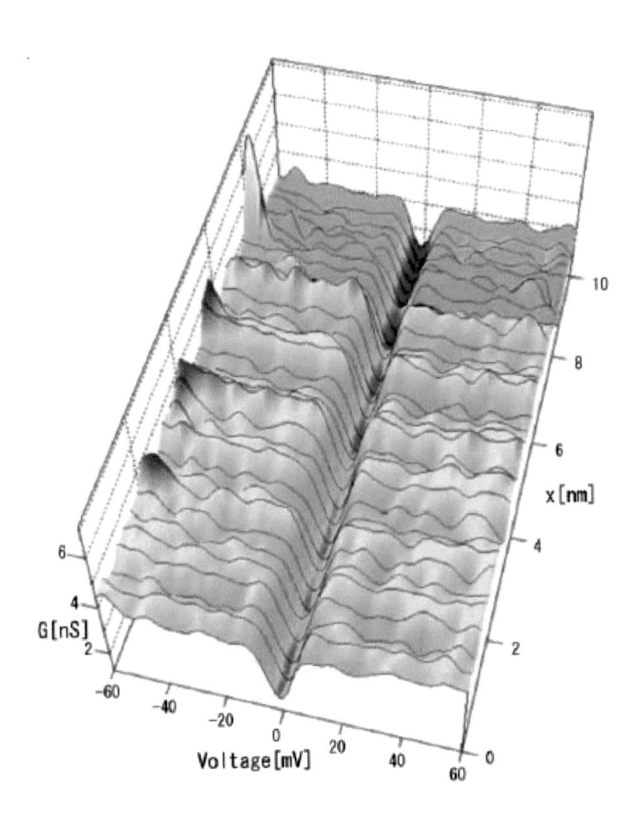

Figure 55. Spatial (x) scan of conductance profiles for β-$HfNCl_{1-x}$ at 5 K.

The gap value of 2Δ = 20 meV in $HfNCl_{0.7}$ revealed by the STS measurements presented here is in accord with that obtained in our previous break-junction tunneling experiments (see Figure 56). The break-junction SIS conductance for $Li_{0.48}(THF)_xHfNCl$ duplicated from Figure 52(a) is displayed in Figure 56 for comparison. We note that STS data presented in Figures 54 – 56 were taken on $HfNCl_{0.7}$ rather than on the related material $Li_{0.48}$ $(THF)_xHfNCl$, which is extremely reactive. Nevertheless, it is worthwhile to put data for both compounds in the same Figure 56, because T_c's of $Li_{0.48}(THF)_xHfNCl$ and $HfNCl_{0.7}$ are quite similar (25 K and 24K) and other electronic properties are basically the same in spite of the chemical differences. The peak positions of both curves almost coincide after compensatory rescaling between SIN and SIS junctions, as is clearly demonstrated in Figure 56. Hence, the extremely large gap features found in break junction (SIS) measurements are confirmed by STM and can be unambiguously identified with the revealed intrinsic superconducting gaps. Thus, large gap values $2\Delta \approx 16 - 20$ meV ($V_{p\text{-}p} = 32 - 40$ mV) are *inherent* to both $Li_{0.48}(THF)_xHfNCl$ and $HfNCl_{0.7}$ compounds.

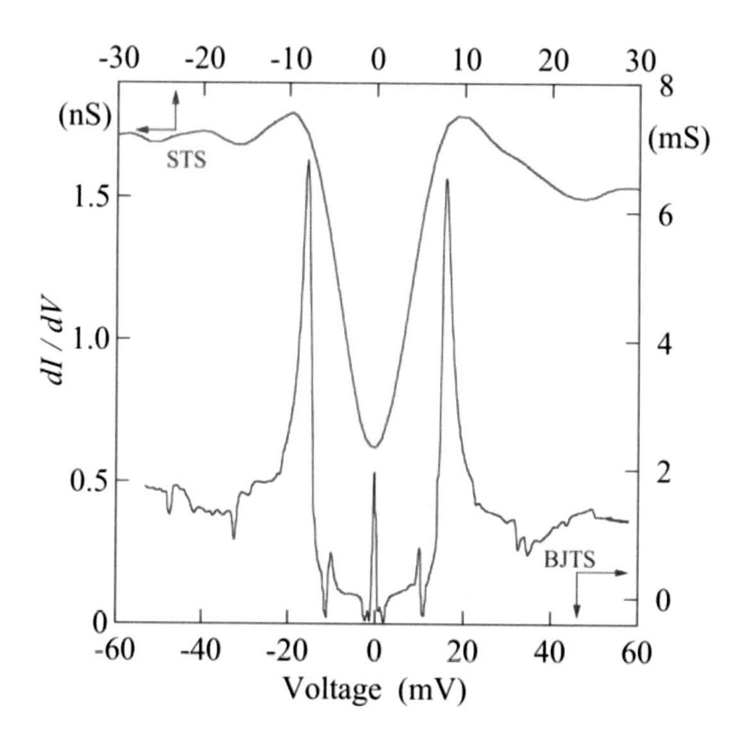

Figure 56. $G(V)$ for SIN junction (STS) between the PtIr tip and $HfNCl_{0.7}$ sample at $T = 5$ K in comparison with that of $Li_{0.48}(THF)_xHfNCl$ SIS break junction (BJTS) at 4 K borrowed from Figure 52(a). The bias scale for the SIN junction is twice enlarged to match with the scale for the SIS junction.

Figure 57 shows thermal evolution of the break-junction spectrum for another junction involving $HfNCl_{0.7}$ at high T above 17 K, when the value of $V_{p\text{-}p}(T = 17\text{ K})$ becomes 25 mV. Since this voltage exceeds the representative value of $4\Delta/e = V_{p\text{-}p} = 16 - 20$ mV at 4.2 K, the corresponding junction is probably considered as an example of the systems with the largest gap features appropriate, e. g., to the data presented in Figure 53. The resulting T-dependence of $4\Delta/e = V_{p\text{-}p}$ together with its value at 4.2 K, taken from Figure 53, is plotted in the inset of Figure 57. One sees that the observed largest gap value at 4.2 K in Figure 53 is consistent with the extrapolated BCS curve using the gap at 17 K from the main frame of Figure 57 and $T_c = 24$ K. Hence, $G(V)$ displayed in Figure 57 describes the high-T continuation of $G(V)$ with low-T peak-to-peak distance $V_{p\text{-}p} = 4\Delta/e \sim 32 - 35$ mV. From this $V_{p\text{-}p}(T)$, the gap to T_c ratio $2\Delta(0)/k_BT_c$ is estimated as 7.6 – 8.6. This value is quite similar to the ratio appropriate to $Li_{0.48}(THF)_xHfNCl$ as stems from Figure 52.

Among the gap energies found for $HfNCl_{0.7}$ in our break-junction studies, the largest gap is typical for STS measurements. The resultant large ratio $2\Delta(0)/k_BT_c$ has been also previously discovered in $ZrNCl_{0.7}$ [277]. Such values are comparable to ratios for high-T_c cuprates [144], [145] or organic superconductors [285], although the origin of the similarity for different classes of materials is not yet clear.

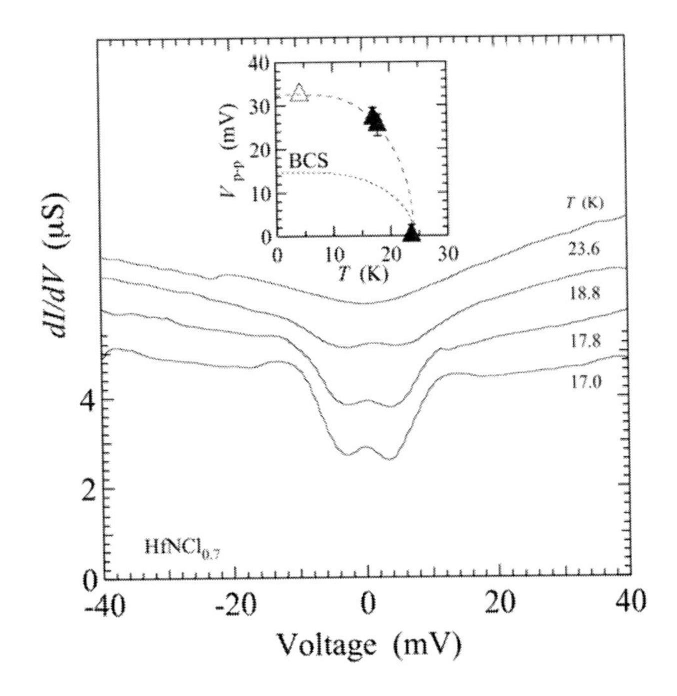

Figure 57. $G(V)$ for various T for the largest superconducting gap in $HfNCl_{0.7}$. Inset shows the T-dependence of $V_{p\text{-}p}$ together with the low-T value $V_{p\text{-}p}^{out}$ (open triangle) from Figure 53.

5.2. BJTS data for $ZrNCl_{0.7}$

Energy gap structures of $ZrNCl_{0.7}$ are shown in Figure 58. The spectra exhibit well developed gap-edge peaks at the bias of $\pm 8 - 10$ mV, (the peak-to-peak separation $V_{p\text{-}p} \approx 16 \sim 20$ mV). Zero-bias peak structures, which we attribute to the dc Josephson current, and sub-gap peak structures ($\pm$ 4 – 5 mV) with $V_{p\text{-}p} = 8 - 10$ mV are also observed. Such patterns are typical for SIS junctions. Sharp gap-edge peaks for Figures 58(a) and 58(b) and hump structures near $\pm$ 20 mV outside the gap-edge peaks for Figure 58(b) are clearly distinguished. The dashed curves are the calculated SIS conductances using the BCS density of states with a small broadening parameter [295]. The calculation well reproduces the peak positions and the low leakage structure near zero-bias conductance. The gap parameters Δ at 4.2 K are estimated to be $\Delta = 4.0$ meV and $\Delta = 4.9$ meV for Figures 58(a) and 58(b), respectively.

A smaller gap of $V_{p\text{-}p} = 7 \sim 10$ mV was observed at 4.2 K in addition to larger ones indicated above. Figure 59 reveals two examples of such spectra. In Figure 59(a) one can see relatively sharp gap-edge peaks at voltage $V \approx \pm 4.8$ mV with no leakage conductance near zero bias. Figure 59(b) involves gap structures at $V \approx \pm 3.5$ mV together with peculiar sharp hump structures at $V \approx \pm 9$ mV, which corresponds to the main gap peak $V_{p\text{-}p} \approx 20$ mV of Figure 58. The fitting BCS dependences for SIS tunneling are shown by dashed curves. The calculations reproduce fairly well the gap peak and the conductance bottom with $\Delta \approx 2.3$ and 1.6 meV for Figures 59(a) and 59(b), respectively. Smaller hump features cannot be reproduced by such a simple fitting function.

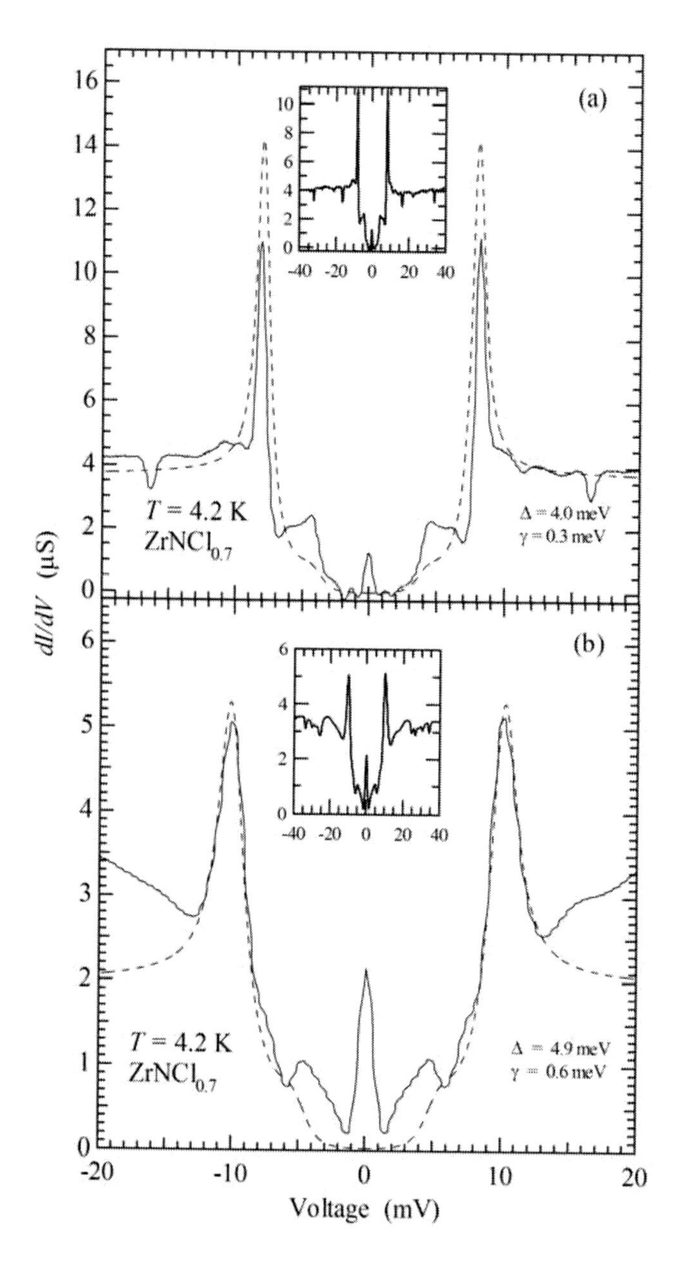

Figure 58. $G(V)$ measured at 4.2 K for different break junctions (a) and (b) made of $ZrNCl_{0.7}$. The dashed curves are calculated in the framework of the BCS model. Insets show $G(V)$ for high bias regions.

A remarkable anti-correlation takes place between dI/dV values and gap widths in $ZrNCl_{0.7}$ as is readily seen in Figures 58 and 59. Specifically, for junctions represented in Figure 58 the gaps are large, whereas the conductance is relatively small of about 1 μS. On the contrary, junctions presented in Figure 59 exhibit smaller gaps and dI/dV of the order 100 μS. The maximal ratios between largest gaps of Figure 58 and smallest gaps of Figure 59 are about 2 – 3. It is plausible that all these discrepancies reflect locally varying charge carrier concentration or gap anisotropy, which directly manifests itself as a consequence of spatially varying junction plane orientations. Distributions in $V_{p\text{-}p}$ are also inherent to $Li_{0.48}(THF)_xHfNCl$, as has been shown above, thereby being a common feature of the layered nitride superconductors.

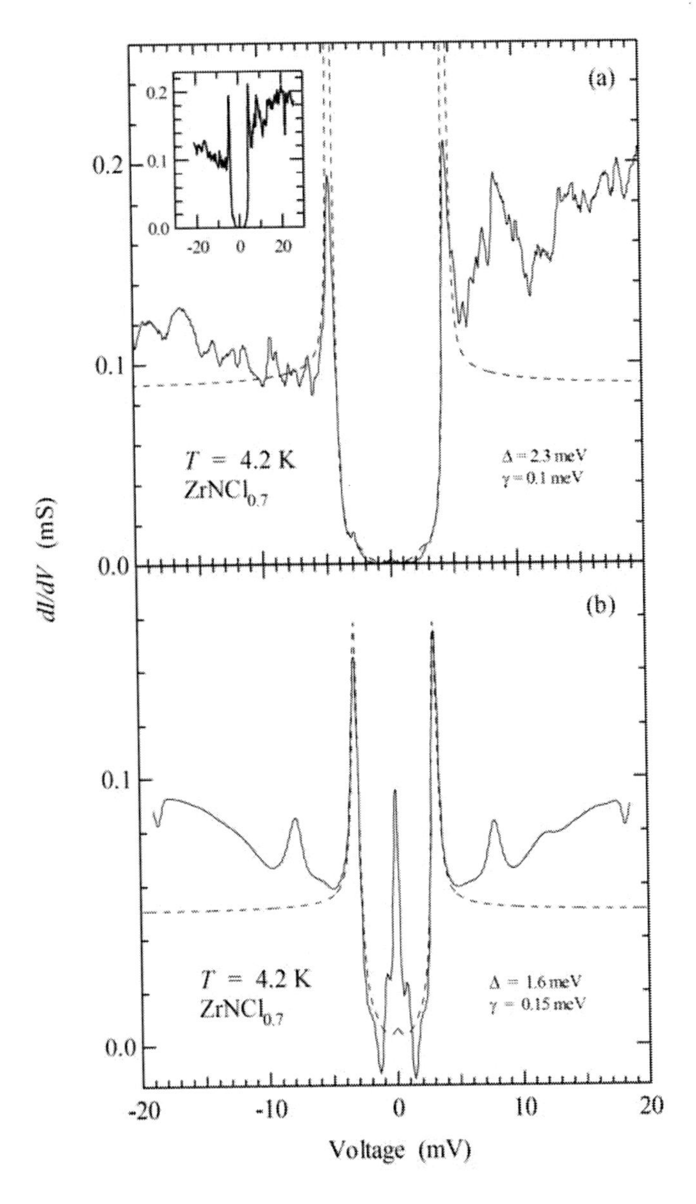

Figure 59. (a) and (b) the same as in Figure 58 but for other break junctions.

T-dependences of $ZrNCl_{0.7}$ tunnel spectra are demonstrated in Figure 60 for the same junctions as in Figure 58. Upon warming the gap structures gradually broaden and disappear. The gap depletion merges with the background conductance at $T \approx 14$ K, which coincides with the bulk T_c of this compound [29]. The dashed curves at each T correspond to the calculated SIS conductances in the framework of the BCS theory. As is seen the fitting is quite successful.

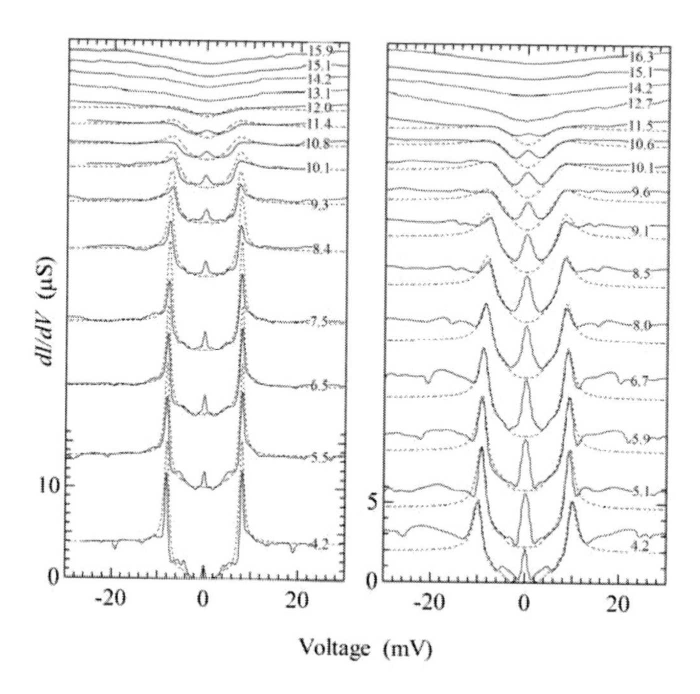

Figure 60. T variations of $G(V)$ for $ZrNCl_{0.7}$. The data (a) and (b) correspond to Figures 58(a) and 58(b), respectively. Dashed curves are SIS conductances calculated on the basis of the BCS model. $G(V)$ curves are shifted up for clarity without any relative offset between experimental and calculated curves.

Thermal evolution of small-gap tunnel spectra is demonstrated in Figure 61. Figure 61(a) corresponds to the junction described earlier in Figure 59(b). Although the gap structure with $V_{p\text{-}p} \approx 7$ mV and the Josephson peak persist up to T_c, the hump structures at $\pm$ 9 mV can hardly be distinguished against the background above $T \approx 10$ K. The BCS-based fitting of the SIS conductance cannot reproduce in full the complex T-dependent gap structure. In Figure 61(b), which represents another junction, a clear-cut Josephson peak can be seen at any T up to about T_c. Both gap edges and zero-bias peaks broaden with T, vanishing at $T \approx 14$ K. Moderate hump peculiarities near $V = \pm 10$ mV can be hardly recognized above $T \approx 10$ K.

Gap values obtained from data of Figures 60 and 61 are plotted in Figure 62. A solid curve is the standard BCS T-dependence of the gap, while dashed curves are scaled BCS ones in the sense of the strong-coupling "α-model" [142]. One sees that for spectra with smaller gaps the T-dependences are almost identical with the conventional BCS curve. Experimental data for large-gap spectra are well described by the scaled BCS theory. T_c is approximately 14 K. The value of $4\Delta(0)/e$ is in the range 16 – 20 mV. The estimated gap to T_c ratio is

$2\Delta(0)/k_B T_c = 6.6 - 8.4$. This is more than twice as large as the weak-coupling BCS value $\approx$ 3.52. As we have already indicated, such anomalously large values are nevertheless similar to those found in cuprate [144], [145] as well as organic superconductors [285]. For the smaller gaps $\Delta(0) = 1.6 - 2.3$ meV, the ratios $2\Delta(0)/k_B T_c = 2.7 - 3.8$ approximately agree with the BCS value.

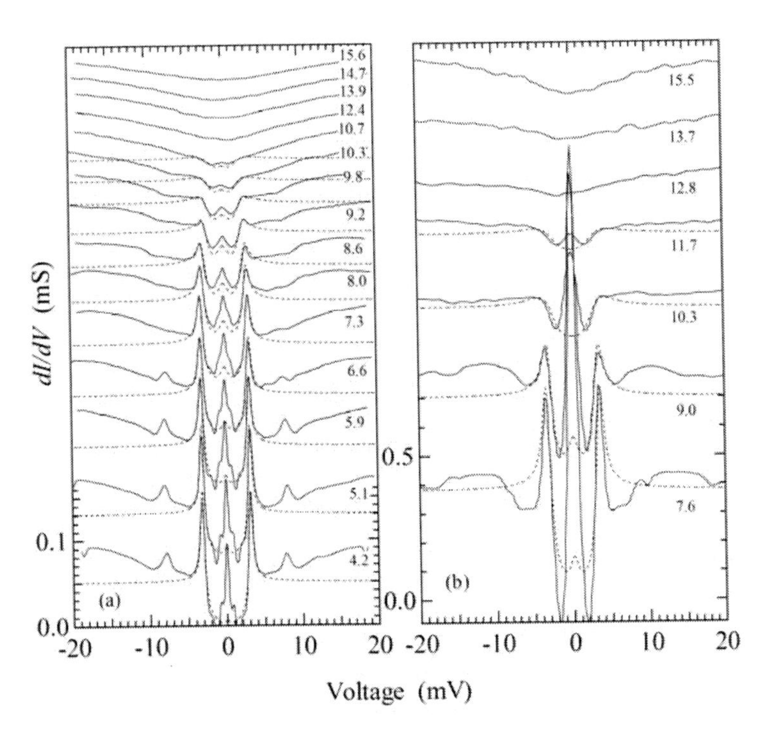

Figure 61. The same as in Figure 60 but for other break junctions. Figure (a) corresponds to the junction presented in Figure 59(b), while Figure (b) demonstrates properties of one more junction.

As can be seen from Figure 59, in $ZrNCl_{0.7}$, V-shaped conductance behavior outside the gapped regions of $G(V)$ is often observed. It seems that this dependence is connected to the overall semiconducting resistivity of the undoped β-$ZrNCl_{0.7}$. Moreover, it is known from the previous study [282] that the background tunnel conductance in β-$ZrNCl_{0.7}$ flattens when divided by $V^{1/2}$. This can be seen from the main frame of Figure 63, whereas the original V-shaped structure is demonstrated in the inset of Figure 63. The introduced $V^{1/2}$ dependence presumably reflects scattering of conduction electrons on weakly disordered states in some regions, caused by inhomogeneities left after carrier doping [21], [269]. The observation of V-shaped background suggests that the semiconducting electronic conduction preserves even in the superconducting phase of $ZrNCl_{0.7}$. It should be noted that the V-shaped background conductance has been observed in high-T_c cuprates as well [144, 295]. Such a peculiar form of the conductance is sometimes discussed in terms of the pseudogap states above T_c. Recent STS measurements in the organic superconductor κ-$(BEDT\text{-}TTF)_2Cu(NCS)_2$ also revealed both the unusually large gap to T_c value ($2\Delta(0)/k_B T_c = 6 - 9$) and the existence of normal-state gaps [285]. Therefore, detailed measurements of $G(V)$ in β-ZrNCl above T_c is needed to clarify the origin of the anomaly concerned.

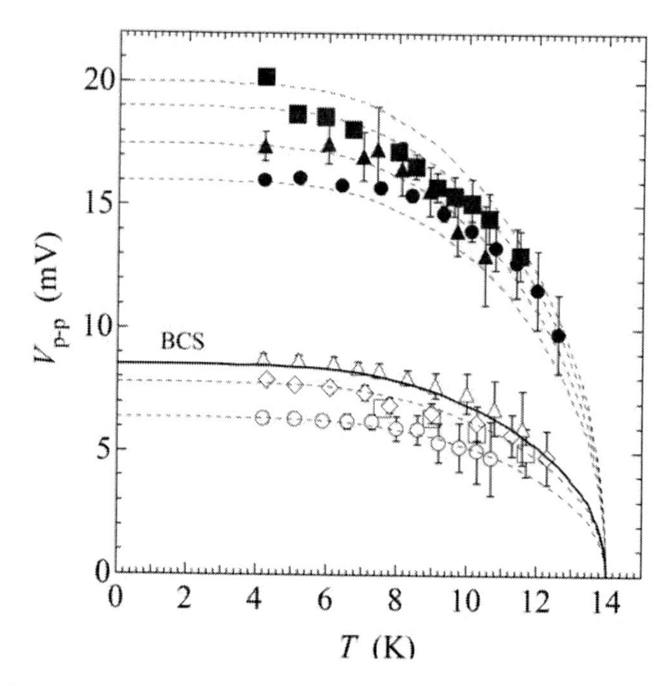

Figure 62. T-dependences of V_{p-p} (= 4Δ/e) for the data of Figures 60(a) (●) and 58(b) (■) – large gaps; and Figures 61(a) (□) and 61(b) (○) – smaller gaps. Other symbols (▲, Δ and ◊) denote junctions, spectra of which are not shown in the paper.

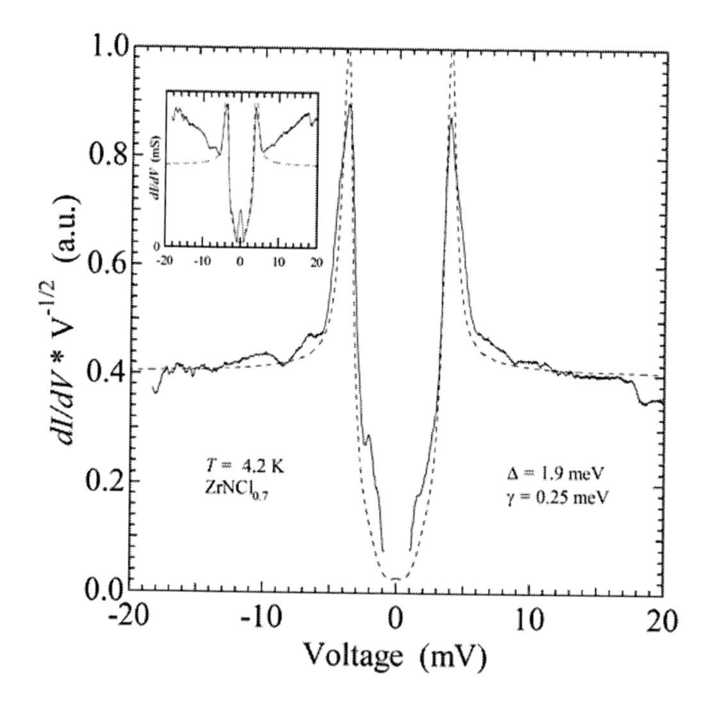

Figure 63. $G(V)$ divided by $(V)^{1/2}$ for a $ZrNCl_{0.7}$ break junction. The inset corresponds to the raw data. The dashed curves are the calculated SIS conductances in the framework of the BCS theory.

5.3. Discussion of the results for β-*M*NCl materials

Among the gap energies found for $HfNCl_{0.7}$ in our break-junction measurements the largest gap is a typical one inherent to STS measurements. The resultant anomalously large gap to T_c ratio was also discovered in $ZrNCl_{0.7}$. As has been mentioned, such values are observed in high-T_c cuprates [144], [145], [295] and organic superconductors [285] as well. The similarity of deviations from the conventional BCS behavior for apparently different classes of materials reflects most probably the strong electron-phonon coupling. Nevertheless, the extremely large values of $2\Delta(0)/k_BT_c > 5$ in the layered compounds $Li_{0.48}(THF)_xHfNCl$, $HfNCl_{0.7}$ and $ZrNCl_{0.7}$ cannot be quantitatively explained even taking into account the strong-coupling renormalization.

Indeed, studies of possible restrictions on this dimensionless parameter (a key one according to the α-theory of Ref. [142]) have been carried out on the basis of the semi-microscopic Eliashberg gap equations [296], [297] starting from works of Geilikman *et al.* [298]. As a result, a number of approximate expressions appeared which contained characteristics of the materials phonon spectra [299], [300]. In particular, a semi-empirical formula Eq.(14) for $2\Delta(0)/k_BT_c$ can be used for estimations and insight [146], [298]-[300].

We emphasize that large values $2\Delta(0)/k_BT_c > 5$ for superconducting order parameters found in our measurements are typical for observed CDW or spin-density-wave (SDW) energy gaps, although in the mean-field approximation they are described by the *same* BCS gap equation as weak-coupling superconductors [301]. For instance, SDWs in underdoped $SrFe_2As_2$ reveal the ratio $2\Delta_{SDW}(0)/k_BT_{SDW} > 7.2 \pm 1$ [302].

Recent Raman-scattering studies carried out under pressure led to the conclusion that low-energy phonons and actual huge electron-phonon coupling constant $\lambda > 3$ constitute a background for superconductivity in *M*NCl ($ZrNCl_{0.7}$ and $Li_{0.48}(THF)_xHfNCl$) [78]. Unfortunately, so far our tunnel measurements did not reveal such phonons.

On the other hand, there is some evidence that optical high-frequency phonons might be the driving force of superconductivity in halides. For instance, Raman investigation of Na-intercalated HfNCl showed that the nitrogen vibration modes with energies about 75 meV are broadened by electron doping [303]. NMR experiments show that the same group of phonons may explain a small isotope effect in $Li_{0.48}(THF)_xHfNCl$ [53]. Moreover, neutron scattering measurements of $Li_{1.16}ZrNCl$ samples exhibited phonon softening around 20 meV and 80 meV mainly corresponding to vibrations of Zr and N ions [304].

As for the theoretical picture of superconductivity in these layered materials, it is far from being full and self-consistent. On the one hand, vibrations of light N ions are expected to have large amplitudes, which implies their large anharmonicity and strong involvement in the electron-phonon interaction [305]. On the other hand, a recent theoretical study of lattice dynamics and electron-phonon interaction in Li_xZrNCl reveal a small coupling constant $\lambda \approx 0.5$ [306]. This might indicate, e.g., the necessity to go beyond the Eliashberg theory [296]-[300]. In particular, the actual Cooper pairing in these and other layered compounds with large T_c might be (i) unconventional with different order parameter symmetry; (ii) induced by other gluing bosons; and (iii) dominated by certain specific peculiarities of layered structures [307]-[311].

One should note that the observed strong-coupling features are similar to those appropriate to the largest gap of MgB_2. Therefore, the very existence of the layered structures with the honeycomb lattice involving light-mass ions might be crucial for superconductivity in both systems.

At the same time, the appearance of ratios $2\Delta(0)/k_BT_c \approx 2 - 2.3$ for the smallest gaps in $Li_{0.48}(THF)_xHfNCl$ and $HfNCl_{0.7}$ (see Figure 62) as well as in $ZrNCl_{0.7}$ indicates that the coexistence of anomalously large and/or small gaps seems to be a common characteristics for $HfNCl_{0.7}$ and $ZrNCl_{0.7}$ compounds. Peculiarities of the electron-phonon interaction arising from the quasi-two-dimensional character of the material and a strong momentum dependence of the interaction vertex may play an important role in this connection. In particular, one should mention a hypothetic purely electron, e.g., dynamically screened Coulomb mechanism of superconductivity [312]-[314]. Crucial role of the *intra-layer* pairing interaction in nitrides can be deduced from heat capacity measurements of intercalated material Li_xM_yZrNCl (M = DMF or DMSO) with varying inter-layer distances d [315]. Namely, after some increase with the growing separation between layers, T_c became a weakly decreasing function of this parameter showing no steep reduction even for $d \geq 20$ Å.

One can notice that the gap value and T_c for $Li_{0.48}(THF)_xHfNCl$ are slightly larger than those of $HfNCl_{0.7}$. This difference could be related to the increase of the inter-layer spacing d by the cointercalation. For $Li_{0.48}(THF)_xHfNCl$, THF molecules are cointercalated with Li between Cl layers. The spacing d increases from 0.923 nm for β-HfNCl to 1.87 nm for $Li_{0.48}(THF)_xHfNCl$, while the d value in $HfNCl_{0.7}$ is kept almost unaltered during de-intercalation process [29]. Since variation of the lattice constant should affect both the phonon dispersion and electron anisotropy, this might cause the change of gaps and T_c. For instance, a slight shift of T_c ($\approx$ 2 K) has been observed after the co-intercalation of propylene carbonate (PC) when d increased for $Na_{0.28}(PC)_{0.55}HfNCl$ in comparison to that for $Na_{0.28}HfNCl$ [316].

Modification of the electron-phonon interaction must inevitably affect superconducting properties. So it is no wonder that Raman scattering experiments showed different doping levels result in different linewidths of the superconductivity-related A_{1g} optical mode [303]. However, similar to the case of $ZrNCl_{0.7}$, detailed measurements demonstrated that high-energy phonons at 77 meV, which are theoretically predicted to be the most strongly interacting with conduction electrons, exhibit weaker electron-phonon interaction when T_c increases following the reduction of the doping level [223]. This seems to be in apparent and unexplained contradiction with the (quite plausible!) assumption that the electron-phonon interaction dominates the Cooper pairing in this compound.

Let us discuss now peculiar large and multiple-gap structures of dI/dV for MNCl compounds. SIS junctions of $HfNCl_{0.7}$ reveal three characteristic gap values, namely, $V_{p\text{-}p}$ = 8 – 10 mV, 18 – 24 mV and 32 – 35 mV, while two peak-to-peak values $V_{p\text{-}p}$ = 7 – 10 mV and 16 – 20 mV are observed in $ZrNCl_{0.7}$. Note, that the former two gap values $V_{p\text{-}p}$ = 8 – 10 mV and 18 – 24 mV in HfNCl are similar to $V_{p\text{-}p}$ = 7 – 10 mV and 16 – 20 mV in $ZrNCl_{0.7}$, respectively. The ratio between the largest 32 – 35 mV and middle 18 – 24 mV values in $HfNCl_{0.7}$ is roughly similar to that between two gap values in $ZrNCl_{0.7}$. Gap varieties most probably reflect a strongly anisotropic electron band structures in those layered crystals, as has been shown by band calculations [23]-[25] and confirmed by X-ray absorption [28] and μSR [318] experiments.

Magnetization measurements also revealed anisotropic features. Specifically, upper critical fields in both $HfNCl_{0.7}$ and $ZrNCl_{0.7}$ compounds [26, 30] with the anisotropy parameter $\gamma_H = H_{c2//ab}/H_{c2//c} = \xi_c/\xi_{ab} > 1$ (as has been defined above, ξ: is the superconducting coherence length) were found, suggesting the superconducting gap anisotropy. In Ref. [23] it was predicted that the Fermi surface of electron doped HfNCl is formed by an only one band originating from Hf-N hybridized orbital, while for electron doped ZrNCl three bands are important, originating from Zr-N and Zr-Zr orbitals. Therefore, for the same doping level x = 0.3, carrier density of the single Hf-N band should be higher than that of the Zr-N one. Since the MN (M = Hf or Zr) orbital involves a light-mass element (N), it is natural to expect that the high-energy mode(s) coupled with doped electrons are responsible for superconductivity. For this reason, it is possible to consider that relatively large charge carrier concentration in the Hf-N band of HfNCl may raise T_c as compared to that of β-ZrNCl.

From specific heat measurements the strong-coupling gap to T_c ratio $2\Delta(0)/k_BT_c = 4.6 - 5.2$ was found for $ZrNCl_{0.7}$ [52]. This ratio is similar to our ratio $2\Delta(0)/k_BT_c = 5 - 5.6$ for the predominant middle-size gap feature or the averaged value between the smallest and largest gap to T_c ratios (≈ 2 and ≈ 8). Specific heat data represent a bulk property, while tunneling spectroscopy probes local interface electronic states. Making an assumption indicated above local and bulk measurements can be reconciled. Actually, similar considerations work well while describing multiple gap structures in MgB_2 and the gap distributions in high-T_c cuprates [179].

CONCLUSION

We used the tunneling spectroscopy technique to study superconducting gapping in recently discovered novel superconductors MgB_2, $Li_{0.48}(THF)_{0.3}HfNCl$ and related layered substances with an emphasize on the multiple-gap manifestations. One should take into consideration that the very multi-layer structure of cuprates, pnictides ($T_c \leq 56$ K in $Gd_{1-x}Th_xFeAsO$ [319]) or MgB_2 points to the fact that these compounds are in essence naturally intercalated materials [7], whereas $Li_{0.48}(THF)_{0.3}HfNCl$ [5]-[7], intercalated graphite or other carbon-based systems [320, 321] as well as certain dichalcogenides [322] are artificially intercalated objects with arising (relatively) high-T_c superconductivity.

Probing MgB_2 samples (polycrystalline pellets and high purity wire segments) by BJTS-, STS- and PCS- methods revealed wide distributions of gap values. The latter most probably reflects the proximity-induced superconductivity in a spatially inhomogeneous environment. The measured spectra were described on the basis of the correlated two-gap model. The apparent spectra contain three groups of gaps, namely, $\Delta_S = 2.2 \pm 0.3$ meV, $\Delta_M = 6.0 \pm 1.5$ meV, and Δ_L = 10-12 meV at 4.2 K. All these features survive almost up to T_c. The middle gap Δ_M with the non-BCS-like behavior $\Delta(T)$ is most likely governed by the proximity effect. The values and T dependences of Δ_S and Δ_L are well reproducible as compared to Δ_M. The largest gap Δ_L indicates the unusually strong-coupling superconductivity with the ratio $2\Delta/k_BT_c = 5 - 6$.

Under the influence of the external magnetic field B, the gap-closing field B_c for Δ_L and Δ_M obeys the relationship $B_c = a\Delta(B = 0)^2$ coming from Ginzburg-Landau theory, contrary to what is appropriate for Δ_S. The totality of data testifies that the largest gap Δ_L induces Δ_M by the proximity effect. Moreover, $B_c \approx 7 - 13$ T for Δ_S and $B_c \approx 18 - 20$ T for Δ_L are approximately the same as the upper critical field at 0 K along the c-axis and ab-plane directions, respectively.

A rich diversity of facts thoroughly described above shows that the widely applied schemes of the two-band superconductivity are at least a crude approximation in their attempts to explain actual multiple-gap spectra. A model of the disordered superconductor with a spatial ensemble of relatively thick mesoscopic domains, which is presumably a consequence of the existing contaminated oxide phases in MgB_2, can serve as a satisfactory phenomenological approach to explain both multiple-gap superconducting features and normal-state properties.

As to the microscopic background, our measurements of the second current derivative $d^2I/dV^2(V)$ for Cu-MgB_2 point contacts indicate that electron-phonon interaction with boron in-plane modes plays a decisive role in the superconductivity of MgB_2.

BJTS measurements demonstrated that BCS-like gapped patterns with typical ratios $2\Delta(0)/k_BT_c = 4.2 - 4.5$ take place for compounds NbB_2 and CaAlSi related to MgB_2. In the case of NbB_2 more or less conventional gap structures, which were well reproduced by the correlated two-gap model, were observed for the majority of measurements. Small observed gaps most likely originate from the mesoscopical inhomogeneity of the non-stoichiometric NbB_2.

Irregular multiple-gap structures were also found by means of BJTS in superconducting layered compounds β-HfNCl (specifically, $Li_{0.48}(THF)_x$HfNCl and $HfNCl_{0.7}$) and $ZrNCl_{0.7}$. The technique concerned turned out to be most effective to elucidate the electronic properties of strongly reactive materials like those indicated above. The BCS-like gap structures were revealed corresponding to the apparent s-wave symmetry of the superconducting gap.

For $Li_{0.48}(THF)_x$HfNCl with the highest $T_c = 25.5$ K the predominant gap is $2\Delta(4.2\text{ K}) = 11 - 12$ meV, which corresponds to $2\Delta(0)/k_BT_c = 5 - 5.6$. Besides, we observed a much larger gap $2\Delta(4.2\text{ K}) = 17 - 20$ meV and a smaller gap $2\Delta(4.2\text{ K}) = 4 - 5$ meV. The highest gap to T_c ratio is estimated as $2\Delta(0)/k_BT_c = 7.5$ - 8.6, which agrees well with our complementary STM measurements. The results are reproducible and the observed stable features seem to be intrinsic. The latter ratios are 2-3 times larger than the BCS weak-coupling values, whatever the superconducting order-parameter symmetry [40], [155], and are quite similar to those appropriate to $ZrNCl_{0.7}$ as well as to high-T_c copper-oxide and organic superconductors. Obviously, such anomalously huge energy gaps could be hardly explained even by an extremely strong electron-boson coupling, so that their origin remains to be elucidated if we remain in the framework of the conventional Cooper-pairing concept. On the other hand, forms and T dependences of tunnel dI/dV for these compounds are in accord with the original BCS picture based on the conventional loose Cooper-pair picture rather than on the Bose-Einstein-condensate (bipolaron) scenario [323], [324].

Several common factors were found in both MgB_2 and β-MNCl materials in addition to their multiple-gap character of the electron spectra. One of them is their unconventionally strong coupling with $2\Delta(0)/k_BT_c > 5$ found in our tunneling data. Another common point could be associated with the two-dimensionality of the electron conduction bands for crystal lattices containing hexagonal networks of boron or nitrogen atoms. Corresponding optical phonon modes are high energy ones due to light masses of those atoms, which could be an important factor to induce a strong electron-phonon interaction.

It should be emphasized that all our considerations were based either on explicit or tacit assumptions of electron-phonon interaction being the driving force of superconductivity and Cooper pairing being the underlying microscopic mechanism of the latter [155]. For materials studied here this viewpoint is more or less shared even by advocates of more sophisticated approaches (see, e. g., Refs. [11], [325]). As regards cuprates, pnictides, organic superconductors and other (sometimes called "exotic") superconductors [326]-[329], they are most often considered as "class-2" (the term borrowed from Ref. [325]) materials, which become superconductors of different type under the influence of a different agent (agents?) [33], [62], [330]-[332]. The main argument is the (explicit or implicit) notorious 40 K limit [325], [333] for phonon-induced superconductivity shown (see, e. g., [334]) to be an artifact of wrong theoretical reasoning. Nevertheless, one should never reject a chance of important unexpected findings even in the (at the first glance) conventional "class-1" superconductors. Multiple-gapness is a hint about such a promising possibility.

As for observations of multiple-gap features in other superconducting compounds than those studied here, they have a long history and are quite common, although not always recognized. For instance, we can mention high-purity metals [105], [106], pnictides $Ba_{1-x}K_xFe_2As_2$ ($T_c \approx$ 32 K) [335], $Ba_{0.55}K_{0.45}Fe_2As_2$ ($T_c \approx$ 27 K) [336], LiFeS ($T_c \approx$ 15.3 K) [337], $FeSe_{1-x}$ ($T_c \approx$ 13 K) [338], and $SmFeAsO_{1-x}F_x$ ($T_c \approx$ 42 K and $T_c \approx$ 52 K) [339], Cu_xTiSe_2 ($T_c \approx$ 3.5 K) [340], ZrB_{12} ($T_c \approx$ 6 K) [341], borocarbides YNi_2B_2C ($T_c \approx$ 13.8 K) [342], $LuNi_2B_2C$ ($T_c \approx$ 16.9 K) [343], and $ErNi_2B_2C$ ($T_c \approx$ 11 K) [344], A15 compounds V_3Si ($T_c \approx$ 12.5 K and 16.5 K) [345] and Nb_3Sn ($T_c \approx$ 18 K) [346], [347] (with a large spread of gap values), dichalcogenides NbS_2 ($T_c \approx$ 6.1 K) [348] and $NbSe_2$ ($T_c \approx$ 7 K) [349], ($T_c \approx$ 7.2 K) [350], [351], ($T_c \approx$ 6.7 K) [352], oxide $Na_{0.35}CoO_2{\cdot}1.3H_2O$ ($T_c \approx$ 4.5 K) [353], ternary-iron silicides $Lu_2Fe_3Si_5$ ($T_c \approx$ 6.1 K) [354]-[356] and $Sc_2Fe_3Si_5$ ($T_c \approx$ 4.5 K) [356], iridium silicide $Sc_5Ir_4Si_{10}$ ($T_c \approx$ 8.4 K) [255], as well as $Mg_{10}Ir_{19}B_{16}$ ($T_c \approx$ 5.7 K) [357]. Since the listed materials are numerous and possess varying normal-state and superconducting properties, it is reasonable to consider multiple-gapness as a manifestation of samples' intrinsic or extrinsic inhomogeneity rather than a unique microscopic feature surviving extreme conditions of impurity electron scattering and other disordering effects. It might also that some of these materials are ordinary single-gap superconductors, the deviations from the conventional behavior being fitting-procedure artifacts. Anyway, BJ technique would be of help in all ambiguous cases because of its probing freshly prepared surfaces.

To summarize, our studies show that multiple-gap superconductors are not unique and possess many interesting properties. At the same time, the origin of this behavior is still not clear and should be elucidated in order to successfully look for new high-T_c and technologically promising superconductors.

ACKNOWLEDGMENTS

This work was supported by a Grant-in-Aid for Scientific Research (Nos. 19540370, 19105006, 19014016) of Japan Society of Promotion of Science. AMG highly appreciates FY2007 fellowship no. S-07042 granted by the Japan Society for the Promotion of Science and the grants given in the framework of the 2009 and 2010 Visitors Programs of the Max Planck Institute for the Physics of Complex Systems (Dresden, Germany) as well as the support by the Project N 23 of the 2009–2011 Scientific Cooperation Agreement between Poland and Ukraine. We thank the Cryogenic Center, Hiroshima University for supplying liquid helium. Superconductivity of complex systems was discussed with many colleagues from a number of laboratories in different countries. We are grateful to all of them.

REFERENCES

[1] J. G. Bednorz, K. A. Müller, Possible High T_c superconductivity in the Ba-La-Cu-O System, *Zeitschrift für Physik, B* vol. **64**, no. 2, pp. 189-193, 1986.

[2] A. Schilling, M. Cantoni, J. D. Guo, H. R. Ott, Superconductivity above 130 K in the Hg-Ba-Ca-Cu-O system, *Nature*, vol. **363**, no. 6427, pp. 347-349, 1993.

[3] J. Nagamatsu, N. Nakagawa, T. Muranaka, Y. Zenitani and J. Akimitsu, Superconductivity at 39K in magnesium diboride, *Nature*, vol. **410**, no. 6824, pp. 63-64, 2001.

[4] X. X. Xi, Two-band superconductor magnesium diboride, *Reports on Progress in Physics*, vol. **71**, no. 11, Article ID 116501 (26 pages), 2008.

[5] S. Yamanaka, K. Hotehama and H. Kawaji, Superconductivity at 25.5 K in electron-doped layered hafnium nitride, *Nature*, vol. **392**, no. 6676, pp. 580-582, 1998.

[6] S. Yamanaka, High-T_c superconductivity in electron-doped layer structured nitrides, *Annual Review of Material Science*, vol. **30**, pp. 53-82, 2000.

[7] S. Yamanaka, Intercalation and superconductivity in ternary layer structured metal nitride halides (*MNX*: M = Ti, Zr, Hf; X = Cl, Br, I), *Journal of Materials Chemistry*, vol. **20**, no. 15, pp. 2922-2933, 2010.

[8] G. Amano, S. Akutagawa, T. Muranaka, Y. Zenitani, J. Akimitsu, Superconductivity at 18 K in Yttrium Sesquicarbide System Y_2C_3, *Journal of the Physical Society of Japan*, vol. **73**, no.3, pp. 530-532, 2004.

[9] Y. Kamihara, T. Watanabe, M. Hirano, H. Hosono, Iron-Based Layered Superconductor $La[O_{1-x}F_x]FeAs$ (x = 0.05 – 0.12) with T_c = 26 K, *Journal of the American chemical Society*, vol. **130**, no. 11, pp. 3296-3297, 2008.

[10] H. Takahashi, K. Igawa, K. Arii, Y. Kamihara, M. Hirano, H. Hosono, Superconductivity at 43 K in an iron-based layered compound $LaO_{1-x}F_xFeAs$, *Nature*, vol. **453**, no. 7193, pp. 376-378, 2008.

[11] W. E. Pickett, The next breakthrough in phonon-mediated superconductivity, *Physica C*, vol. **468**, no. 2, pp. 126-135, 2008.

[12] Z. A. Ren, J. Yang, W. Lu, X.-L. Shen, Z. Cai Li, G.-C. Che, X.-L. Dong L.-L. Sun, F. Zhou, Z.-X. Zhao, Superconductivity in the iron-based F-doped layered quaternary compound $Nd[O_{1-x}F_x]FeAs$, *Europhysics Letters*, vol. **82**, no. 5, Article ID 57002 (2 pages), 2008.

[13] R. J. Cava, Genie in a bottle, *Nature*, vol. **410**, no. 6824, pp. 23-24, 2001.

[14] J. Kortus, I. I. Mazin, K. D. Belashchenko, V. P. Antropov, L. L. Boyer, Superconductivity of Metallic Boron in MgB_2, *Physical Review Letters*, vol. **86**, no. 20, pp. 4656-4659, 2001.

[15] H. Uchiyama, K. M. Shen, S. Lee, A. Damascelli, D. H. Lu, D.L. Feng, Z.-X. Shen, S. Tajima, Electronic Structure of MgB_2 from Angle-Resolved Photoemission Spectroscopy, *Physical Review Letters*, vol. **88**, no. 15, Article ID 157002, 2002.

[16] S. Souma, Y. Machida, T. Sato, T. Takahashi, H. Matsui, S.-C. Wang, H. Ding, A. Kaminski, J. C. Campuzano, S. Sasaki and K. Kadowaki, The origin of multiple superconducting gaps in MgB_2, *Nature*, vol. **423**, no. 6935, pp. 65-67, 2003.

[17] R. Osborn, E. A. Goremychkin, A. I. Kolesnikov and D. G. Hinks, Phonon Density of States in MgB_2, *Physical Review Letters*, vol. **87**, no.1, Article ID017005, 2001.

[18] Y. Kong, O. V. Dolgov, O. Jepsen, O. K. Andersen, Electron-phonon interaction in the normal and superconducting states of MgB_2, *Physical Review B*, vol. **64**, no. 2, ID 020501, 2001.

[19] D. Daghero, R. S. Gonnelli, Probing multiband superconductivity by point-contact spectroscopy, *Superconducting Science and Technology*, vol. **23**, no. 4, ID 043001, 2010.

[20] I. N. Askerzade, Study of layered superconductors in the framework of electron-phonon coupling theory, *Uspekhi Fizicheskikh Nauk*, vol. **179**, no. 10, pp. 1033-1045, 2009.

[21] E. L. Wolf, *Principles of Electron Tunneling Spectroscopy*, Oxford University Press, New York and Claredon Press, Oxford, 1985.

[22] J. Müller, A15-type superconductors, *Reports on Progress in Physics*, vol. **43**, no. 5, pp. 643-687, 1980.

[23] C. Felser, R. Seshadri, "Electronic structures and instabilities of ZrNCl and HfNCl: implications for superconductivity in the doped compounds, *Journal of Materials Chemistry*, vol. **9**, no. 2, pp. 459-464, 1999.

[24] I. Hase, Y. Nishihara, Electronic structure of superconducting layered zirconium and hafnium nitride, *Physical Review B*, vol. **60**, no. 3, pp. 1573-1581, 1999.

[25] I. Hase, Y. Nishihara, Electronic band structure of ZrNCl and HfNCl, *Physica B*, vol. **281&282**, pp. 788-789, 2000.

[26] H. Tou, Y. Maniwa, K. Ito, S. Yamanaka, NMR studies of layered superconductor $Li_{0.48}(THF)_y HfNCl$, *Physica C*, vol. **341-348**, no. 4, pp. 2139-2140, 2000.

[27] Y. J. Uemura, Y. Fudamoto, I. M. Gat, M. I. Larkin, G. M. Luke, J. Merrin, K. M. Kojima, K. Itoh, S. Yamanaka, R. H. Heffner, D. E. MacLaughlin, μSR studies of intercalated HfNCl superconductor, *Physica B*, vol. **289-290**, pp. 389-392, 2000.

[28] T. Yokoya, Y. Ishiwata, S. Shin, S. Shamoto, K. Iizawa, T. Kajitani, I. Hase and T. Takahashi, Changes of electronic structure across the insulator-to-metal transition of quasi-two-dimensional Na-intercalated β-HfNCl studied by photoemission and X-ray absorption, *Physical Review B*, vol. **64**, no. 15, Article ID153107, 2001.

[29] L. Zhu and S. Yamanaka, Preparation and Superconductivity of Chlorine-Deintercalated Crystals β -$MNCl_{1-x}$ (M = Zr, Hf), *Chemistry of Materials*, vol. **15**, no. 9, pp. 1897-1902, 2003.

[30] H. Tou, Y. J. Tanaka, M. Sera, Y. Taguchi, T. Sasaki, Y. Iwasa, L. Zhu, S. Yamanaka, Upper critical field in the electron-doped layered superconductor $ZrNCl_{0.7}$: Magnetoresistance studies, *Physical Review B*, vol. 72, no. 2, Article ID 020501(R), 2005.

[31] M. Karppinen, H. Yamauchi, Control of the charge inhomogeneity and high-T_c superconducting properties in the homologous series of multi-layered copper oxides, *Materials Science and Engineering R*, vol. **26**, no. 3, pp. 51-96, 1999.

[32] R. J. Cava, Oxide superconductors", Journal of the American Ceramic Society, vol. **83**, no. 1, pp. 5-28, 2000.

[33] D. Manske, *Theory of Unconventional Superconductors. Cooper-Pairing Mediated by Spin Excitations*, Springer, New York, USA, 2004.

[34] E. G. Maksimov, M. L. Kulić and O. V. Dolgov, Bosonic spectral function and the electron-phonon interaction in HTSC cuprates, *Advances in Condensed Matter Physics*, vol. **2010**, ID 423725, 2010.

[35] F. Marsiglio, J. P. Carbotte, Electron–phonon superconductivity, Ch. 3, pp. 73-162, In the Book: *Superconductivity. Vol. 1: Conventional and Unconventional Superconductors*, Edited by K. H. Bennemann and J. B. Ketterson, Springer, Berlin, Germany, 2008.

[36] J. Bardeen, L. N. Cooper, J. R. Schrieffer, Theory of Superconductivity, *Physical Review*, vol. **108**, no. 5, pp. 1175-1204, 1957.

[37] D. Kondepudi, I. Prigogine, *Modern Thermodynamics. From Heat Engines to Dissipative Structures*, John Wiley and Sons, Chichister, USA, 1999.

[38] M. de Llano, J. F. Annett, Generalized Cooper pairing in superconductors, International *Journal of Modern Physics B*, vol. **21**, no. 21, pp. 3657-3686, 2007.

[39] I. Giaever, K. Megerle, Study of Superconductors by Electron Tunneling, *Physical Review*, vol. **122**, no. 4, pp. 1101-1111, 1961.

[40] H. Won, K. Maki, *d*-wave superconductor as a model of high-T_c superconductors, *Physical Review B*, vol. **49**, no. 2, pp. 1397-1402, 1994.

[41] V. P. Mineev, K. V. Samokhin, *Intoduction to Unconventional Superconductivity*, Gordon and Breach Science Publishers, Amsterdam, The Netherlands, 1999.

[42] A. M. Gabovich, A. I. Voitenko, Model for the coexistence of d-wave superconducting and charge-density-wave order parameters in high-temperature cuprate superconductors, *Physical Review B*, vol. **80**, no. 22, ID 224501, 2009.

[43] J. Akimitsu, T. Muranaka, Superconductivity in MgB_2, *Physica C*, vol. **388-389**, pp. 98-102, 2003.

[44] A.Y. Liu, I. I. Mazin, J. Kortus, Beyond Eliashberg superconductivity in MgB_2: Anharmonicity, two-phonon scattering, and multiple gaps, *Physical Review Letters*, vol. **87**, no. 8, Article ID 087005, 2001.

[45] H. J. Choi, D. Roundy, H. Sun, M. L. Cohen, S. G. Louie, The origin of the anomalous superconducting properties of MgB_2, *Nature*, vol. **418**, no. 6899, pp. 758-760, 2002.

[46] A. A. Golubov, J. Kortus, O. V Dolgov, O. Jepsen, Y. Kong, O. K. Andersen, B. J. Gibson, K. Ahn, R. K. Kremer, Specific heat of MgB_2 in a one- and a two-band model from first-principles calculations, *Journal of Physics: Condensed Matter*, vol. **14**, no. 6, pp. 1353-1360, 2002.

[47] H. Suhl, B. T. Matthias, L. R. Walker, Bardeen-Cooper-Schrieffer theory of superconductivity in the case of overlapping bands, *Phys. Rev. Lett.*, vol. **3**, no. 12, pp. 552-554, 1959.

[48] V. A. Moskalenko, Superconductivity of metals taking into account the overlap of the energy bands, *Fizika Metallov I Metallovedenie*, vol. **8**, no. 4, pp. 503-513, 1959 (in Russian).

[49] A. A. Golubov, I. I. Mazin, Effect of magnetic and nonmagnetic impurities on highly anisotropic superconductivity, *Physical Review B*, vol. **55**, no. 22, pp. 15146-15152, 1997.

[50] P. W. Anderson, Theory of dirty superconductors, *Journal of Physics and Chemistry of Solids*, vol. **11**, no. 1-2, pp. 26-30, 1959.

[51] A. M. Gabovich, M. S. Li, M. Pękała, H. Szymczak, A. I. Voitenko, Heat capacity of mesoscopically disordered superconductors with emphasis on MgB_2, *Journal of Physics: Condensed Matter*, vol. **14**, no. 41, pp. 9621-9629, 2002.

[52] Y. Taguchi, M. Hisakabe, Y. Iwasa, Specific Heat Measurement of the Layered Nitride Superconductor Li_xZrNCl, *Physical Review Letters*, vol. **94**, no. 21, Article ID217002, 2005.

[53] H. Tou, Y. Maniwa, S. Yamanaka, Superconducting characteristics in electron-doped layered hafnium nitride: ^{15}N isotope effect studies, *Physical Review B*, vol. **67**, no. 10, Article ID 100509(R), 2003.

[54] Y. Taguchi, T. Kawabata, T. Takano, A. Kitora, K. Kato, M. Takata, Y. Iwasa, Isotope effect in Li_xZrNCl superconductors, *Physical Review B*, vol. **76**, no. 6, Article ID064508, 2007.

[55] A. M. Gabovich, D. P. Moiseev, Isotope effect in jellium and Brout models, *Fizika Tverdogo Tela*, vol. **23**, no. 5, pp. 1511-1514, 1981.

[56] H. Stern, Trends in superconductivity in the periodic table, *Physical Review B*, vol. **8**, no. 11, pp. 5109-5121, 1973.

[57] H. Stern, Superconductivity in transition metals and compounds, *Physical Review B*, vol. **12**, no. 3, pp. 951-960, 1975.

[58] M. Tachiki and S. Takahashi, Charge-fluctuation mechanism of high-T_c superconductivity and the isotope effect in oxide superconductors, *Physical Review B*, vol. **39**, no. 1, pp. 293-299, 1989.

[59] J. Ruvalds, Are there acoustic plasmons? *Advances in Physics*, vol. **30**, no. 5, pp. 677-695, 1981.

[60] R. Englman, B. Halperin, M. Weger, Jahn-Teller (reverse sign) mechanism for superconductive pairing, *Physica C*, vol. **169**, no. 3-4, pp. 314-324, 1990.

[61] N. P. Netesova, Plasma oscillation and isotope effect, *Physica C*, vol. **460-462**, pp. 918-919, 2007.

[62] A. Bussmann-Holder, H. Keller, Unconventional isotope effects, multi-component superconductivity and polaron formation in high temperature cuprate superconductors, *Journal of Physics: Conference Series*, vol. **108**, no. 1, ID 012019, 2008.

[63] B. M. Klein, R. E. Cohen, Anharmonicity and the inverse isotope effect in the palladium-hydrogen system, *Physical Review B*, vol. **45**, no. 21, pp. 12405-12414, 1992.

[64] E. Matsushita, Electron theory of inverse isotope effect in superconducting PdH_x system, *Solid State Communications*, vol. **38**, no. 5, pp. 419-421, 1981.

[65] S. Flach, P. Härtwich, J. Schreiber, On the influence of structural instabilities on superconductivity in the framework of the electron-phonon mechanism, *Solid State Communications*, vol. **75**, no. 8, pp. 647-650, 1990.

[66] R. Szczęsniak, M. Mierzejewski, J. Zielinski, P. Entel, Modification of the isotope effect by the van Hove singularity of electrons on a two-dimensional lattice, *Solid State Communications*, vol. **117**, no. 6, pp. 369-371, 2001.

[67] R. Grassme, P. Seidel, The BCS gap equation within the van Hove scenario of high-T_c superconductivity, *Journal of Superconductivity*, vol. **9**, no. 6, pp. 619-624, 1996.

[68] V. M. Loktev, V. M. Turkowski, On the theory of isotope-effect in the *d*-wave superconductors, *Journal of Low Temperature Physics*, vol. **143**, no. 3-4, pp. 115-130, 2006.

[69] T. W. Ebbesen, J. S. Tsai, K. Tanigaki, J. Tabuchi, Y. Shimakawa, Y. Kubo, I. Hirosawa, J. Mizuki, Isotope effect on superconductivity in Rb_3C_{60}, *Nature*, vol. **355**, no. 6361, pp. 620-622, 1992.

[70] D. G. Hinks, H. Claus, J. D. Jorgensen, The complex nature of superconductivity in MgB_2 as revealed by the reduced total isotope effect, *Nature*, vol. **411**, no. 6836, pp. 457-460, 2001.

[71] P. Konsin, N. Kristoffel, T. Örd, On the composition-dependent isotope effect of HTSC in the two-band model, *Annalen der Physik*, vol. **505**, no. 3, pp. 279-283, 1993.

[72] G. Gusman, Dirty superconductors and isotope effect, *Journal of Physics and Chemistry of Solids*, vol. **28**, no. 1, pp. 2327-2333, 1967.

[73] P. M. Shirage, K. Kihou, K. Miyazawa, C. H. Lee, H. Kito, Y. Yoshida, H. Eisaki, Y. Tanaka, A. Iyo, Iron isotope effect on T_c in optimally-doped $(Ba,K)Fe_2As_2$ (T_c = 38 K) and $SmFeAsO_{1-y}$ (T_c = 54 K) superconductors, *Physica C*, vol. **470**, no. 20, pp. 986-988, 2010.

[74] J. Appel, Role of thermal phonons in high-temperature superconductivity, *Physical Review Letters*, vol. **21**, no. 16, pp. 1164-1167, 1968.

[75] A. M. Gabovich, A. I. Voitenko, Temperature-dependent inelastic electron scattering and superconducting state properties, *Physics Letters A*, vol. **190**, no. 2, pp. 191-195, 1994.

[76] K. Hanzawa, Isotope effects in the cuprates, *Journal of the Physical Society of Japan*, vol. **64**, no. 12, pp. 4856-4866, 1995.

[77] W. L. McMillan, Transition temperature of strong coupled superconductors, *Physical Review*, vol. **167**, no. 2, pp. 331-344, 1968.

[78] Y. Taguchi, M. Hisakabe, Y. Ohishi, S. Yamanaka, Y. Iwasa, High-pressure study of layered nitride superconductors, *Physical Review B*, vol. **70**, no. 10, Article ID104506, 2004.

[79] R. A. Ribeiro, S. L. Bud'ko, C. Petrovic, P. C. Canfield, Effects of stoichiometry, purity, etching and distillng on resistance of MgB_2 pellets and wire segments, *Physica C*, vol. 382, no. 2-3, pp. 194-202, 2002.

[80] H. Takagiwa, S. Kuroiwa, M. Yamazawa, J. Akimitsu, K. Ohishi, A. Koda, W. Higemoto, R. Kadono, Correlation between superconducting carrier density and transition temperature in NbB_{2+x}, *Journal of the Physical Society of Japan*, vol. **74**, no. 5, pp. 1386-1389, 2005.

[81] S. Yamanaka, T. Otsuki, T. Ide, H. Fukuoka, R. Kumashiro, T. Rachi, K. Tanigaki, F. Z. Guo, K. Kobayashi, Missing superconductivity in BaAlSi with the AlB_2 type structure, *Physica C*, vol. **451**, no. 1, pp. 19-23, 2007.

[82] T. Ekino, T. Takabatake, H. Tanaka, H. Fujii, Tunneling Evidence for the Quasiparticle Gap in Kondo Semiconductors CeNiSn and CeRhSb, *Physical Review Letters*, vol. **75**, no. 23, pp. 4262-4265, 1995.

[83] T. Ekino, H. Fujii, M. Kosugi, Y. Zenitani, J. Akimitsu, Tunneling spectroscopy of the superconducting energy gap in RNi_2B_2C (R = Y and Lu), *Physical Review B*, vol. 53, no. 9, pp. 5640-5649, 1996.

[84] J. T. Okada, T. Ekino, T. Takasaki, Y. Yokoyama, Y. Watanabe, H. Fujii, S. Nanao, Observation of a pseudogap in α-AlMnSi by break-junction tunneling spectroscopy, *Journal of Non-Crystalline Solids*, vol. **312-314**, pp. 513-516, 2002.

[85] J. T. Okada, T. Ekino, Y. Yokoyama, T. Takasaki, Y. Watanabe, S. Nanao, Evidence of the Fine Pseudogap at the Fermi Level in Al-based Quasicrystal, *Journal of the Physical Society of Japan*, vol. **76**, no. 3, Article ID 033707 (4pages), 2007.

[86] T. Ekino, A. M. Gabovich, M. S. Li, M. Pękała, H. Szymczak, A. I. Voitenko, Temperature-dependent pseudogap-like features in tunnel spectra of high-T_c cuprates as a manifestation of charge-density waves, *Journal of Physics: Condensed Matter*, vol. **20**, no. 42, Article ID 425218 (15 pages), 2008.

[87] A. F. Andreev, Thermal conductivity of intermediate state of superconductors, Zhurnal *Eksperimental'noi i Teoreticheskoi Fiziki*, vol. **46**, no. 5, pp. 1823-1826, 1964.

[88] I. K. Yanson, I. O. Kulik, A. G. Batrak, Point-contact spectroscopy of electron-phonon interaction in normal-metal single crystals, *Journal of Low Temperature Physics*, vol. **42**, no. 5/6, pp. 527-556, 1981.

[89] I. K. Yanson, V. V. Fisun, N. L. Bobrov, Yu. G. Naidyuk, W. N. Kang, E.-M. Choi, H.-J. Kim, S.-I. Lee, Phonon structure in I-V characteristic of MgB_2 point-contacts, *Physical Review B*, vol. **67**, no. 2, ID 024517, 2003.

[90] Yu. G. Naidyuk, I. K. Yanson, *Point-Contact Spectroscopy*, Springer, New York, USA, 2005.

[91] J. R. Waldram, The Josephson effects in weakly coupled superconductors, *Reports on Progress in Physics*, vol. **39**, no. 8, pp. 751-821, 1976.

[92] M. Ledvij, R. A. Klemm, Dependence of the Josephson coupling of unconventional superconductors on the properties of the tunneling barrier, *Physical Review B*, vol. **51**, no. 5, pp. 3269-3272, 1995.

[93] R. C. Dynes, V. Narayanamurti, J. P. Garno, Direct Measurement of Quasiparticle-Lifetime Broadening in a Strong-Coupled Superconductor, *Physical Review Letters*, vol. **41**, no. 21, pp. 1509-1512, 1978.

[94] R. C. Dynes, J. P. Garno, G. B. Hertel, T. P. Orlando, Tunneling study of superconductivity near the metal-insulator transition, *Physical Review Letters*, vol. **53**, no. 25, pp. 2437-2440, 1984.

[95] F. Giazotto, T. T. Heikkilä, A. Luukanen, A. M. Savin, J. P. Pekola, Opportunities for mesoscopics in thermometry and refrigeration: Physics and applications, *Reviews of Modern Physics*, vol. **78**, no. 1, pp. 217-274, 2006.

[96] J. P. Pekola, V. F. Maisi, S. Kafanov, N. Chekurov, A. Kemppinen, Yu. A. Pashkin, O.-P. Saira, M. Möttönen, J. S. Tsai, Environment-assisted tunneling as an origin of the Dynes density of states, *Physical Review Letters*, vol. **105**, no. 2, ID 026803, 2010.

[97] W. A. Harrison, Tunneling from an Independent-Particle Point of View, *Physical Review*, vol. **123**, no. 1, pp. 85-89, 1961.

[98] M. H. Cohen, L. M. Falicov, J. C. Phillips, Superconductive Tunneling, *Physical Review Letters*, vol. 8, no. 8, pp. 316-318, 1962.

[99] D. H. Douglass, Jr., L. M. Falicov, *Progress in Low Temperature Physics*, vol. IV, p. 97, The superconductive Energy Gap, Edited by C. J. Gorter, North Holland Publishing Co., Amsterdam, Netherlands, 1964.

[100] G. D. Mahan, *Many-Particle Physics (Physics of Solids and Liquids)* (Third edition), Plenum Press, New York and London, 2000.

[101] J. Bardeen, Tunneling from a Many-Particle Point of View, *Physical Review Letters*, vol. **6**, no. 2, pp. 57-59, 1961.

[102] V. Ambegaokar, A. Baratoff, Tunneling Between Superconductors, *Physical Review Letters*, vol. **10**, no. 11, pp. 486-489, 1963.

[103] I. O. Kulik, I. K. Yanson, *Josephson Effect in Superconducting Tunneling Structures*, (Coronet, New York 1971).

[104] N. Kristoffel, P. Konsin, T. Örd, The sign of interband interaction and stability of superconducting ground state in the multiband model, *La Rivista del Nuovo Cimento*, vol. **17**, no. 9, pp. 1-41, 1994.

[105] L. Y. L Shen, N. M. Senozan, N. E. Phillips, Evidence for Two Energy Gaps in High-Purity Superconducting Nb, Ta, and V, *Physical Review Letters*, vol. **14**, no. 25, pp. 1025-1027, 1965.

[106] S. I. Vedeneev, A. I. Golovashkin, G. P. Motulevich, Investigations of the electron-phonon interaction in superconducting alloy Nb_3Sn, *Trudy FIAN SSSR*, vol. **86**, pp. 140-160, 1975.

[107] J. R. Kirtley, R. M. Feenstra, A. P. Fein, S. I. Raider, W. J. Gallagher, R. Sandstrom, T. Dinger, M. W. Shafer, R. Koch, R. Laibowitz, B. Bumble, Studies of superconductors using a low-temperature, high-field scanning tunneling microscope, *Journal of Vacuum Science and Technology A: Vacuum. Surfaces, and Films*, vol. **6**, no. 2, pp. 259-262, 1988.

[108] J. R. Kirtley, Tunneling measurements of the energy gap in high-T_c superconductors, *International Journal of Modern Physics B*, vol. **4**, no. 2, pp. 201-237, 1990.

[109] B. Barbiellini, Ø. Fischer, M. Peter, Ch. Renner, M. Weger, Gap distribution of the tunneling spectra in $Bi_2Sr_2CaCu_2O_x$ and some other superconductors, *Physica C*, vol. **220**, no.1, pp. 55-60, 1994.

[110] A. M. Gabovich, E. A. Pashitskii, Superconductivity of many-valley semiconductors, *Ukrainskii Fizicheskii Zhurnal*, vol. **20**, no. 11, pp. 1814-1823, 1975 (in Russian).

[111] A. Bussmann-Holder, L. Genzel, A. Simon, R. R. Bishop, Gap distribution and multiple-coupling in high T_c's, *Zeitschrift fur Physik B Condensed Matter*, vol. **92**, no. 2, pp. 149-154, 1993.

[112] V. L. Pokrovsky, M. S. Ryvkin, On thermodynamics of anisotropic superconductors, *Zhurnal Eksperimental'noi i Teoreticheskoi Fiziki*, vol. **43**, no. 1, pp. 92-104, 1962.

[113] D. W. Youngner, R. A. Klemn, Theory of the upper critical field in anisotropic superconductors, *Physical Review B*, vol. **21**, no. 9, pp. 3890-3896, 1980.

[114] S. V. Pokrovsky, V. L. Pokrovsky, Density of states and order parameter in dirty anisotropic superconductors, *Physical Review B*, vol. **54**, no. 18, pp. 13275-13287, 1996.

[115] M. B. Walker, Fermi-liquid theory for anisotropic superconductors, *Physical Review B*, vol. **64**, no. 13, Article ID134515, 2001.

[116] A. I. Posazhennikova, T. Dahm, K. Maki, Anisotropic *s*-wave superconductivity: comparison with experiments on MgB_2 single crystals, *Europhysics Letters*, vol. **60**, no. 1, pp. 134-140, 2002.

[117] A. M. Gabovich, A. I. Voitenko, Influence of the order parameter nonhomogeneities on low-temperature properties of superconductors, *Physical Review B*, vol. **60**, no. 10, pp. 7465-7472, 1999.

[118] A. M. Gabovich, Mai Suan Li, M. Pękała, H. Szymczak, A. I. Voitenko, Heat capacity of mesoscopically inhomogeneous superconductors: Theory and applications to MgB_2, *Physica C*, vol. **405**, no. 3-4, pp. 187-211, 2004.

[119] A. I. Larkin, Yu. N. Ovchinnikov, Tunnel effect between superconductors in an alternating field, *Zhurnal Eksperimental'noi i Teoreticheskoi Fiziki*, vol. **51**, no. 5, pp. 1535-1543, 1966.

[120] T. Ekino, A. M. Gabovich, Mai Suan Li, T. Takasaki, A. I. Voitenko, J. Akimitsu, H. Fujii, T. Muranaka, M. Pękała, H. Szymczak, Spatially heterogeneous character of superconductivity in MgB_2 as revealed by local probe and bulk measurements, *Physica C*, vol. **426-431,** pp. 230-233, 2005.

[121] Z. Long, M. D. Stewart, Jr., T. Kouh, J. M. Valles, Jr., Subgap Density of States in Superconductor-Normal Metal Bilayers in the Cooper Limit, *Physical Review Letters*, vol. **93**, no. 25, Article ID257001, 2004.

[122] Ya. G. Ponomarev, E. B. Tsokur, M. V. Sudakova, S. N. Tchesnokov, M. E. Shabalin, M. A. Lorenz, M. A. Hein, G. Müller, H. Piel, B. A. Aminov, Evidence for strong electron-phonon interaction from inelastic tunneling of Cooper pairs in *c*-direction in $Bi_2Sr_2CaCu_2O_8$ break junctions, *Solid State Communications*, Vol. **111**, no. 9, pp. 513-518, 1999.

[123] K. Masuda, S. Kurihara, Coherence effect in a two-band superconductor: application to iron pnictides, *Journal of the Physical Society of Japan*, vol. **79**, no. 7, ID 074710, 2010.

[124] N. R. Werthamer, Nonlinear Self-Coupling of Josephson Radiation in Superconducting Tunnel Junctions, *Physical Review*, vol. **147**, no. 1, pp. 255-263, 1966.

[125] V. Z. Kresin, H. Morawitz, S. A. Wolf, *Mechanisms of Conventional and High-T_c Superconductivity*, Oxford University Press, New York, 1993.

[126] W. L. McMillan, Tunneling Model of the Superconducting Proximity Effect, *Physical Review*, vol. **175**, no. 2, pp. 537-542, 1968.

[127] N. Schopohl, K. Scharnberg, Tunneling density of states for the two-band model of superconductivity, *Solid State Communications*, vol. **22**, no. 6, pp. 371-374, 1977.

[128] C. Noce, L. Maritato, Microscopic equivalence between the two-band model and McMillan proximity-effect theory, *Physical Review B*, vol. **40**, no.1, pp. 734-736, 1989.

[129] Y. Noat, T. Cren, F. Debontridder, D. Roditchev, W. Sacks, P. Toulemonde, A. San Miguel, Signatures of multigap superconductivity in tunneling spectroscopy, *Physical Review B*, vol. **82**, no. 1, ID 014531, 2010.

[130] G. E. Blonder, M. Tinkham, T. M. Klapwijk, Transition from metallic to tunneling regimes in superconducting microconstrictions: Excess current, charge imbalance, and supercurrent conversion, *Physical Review B*, vol. **25**, no. 7, pp. 4515-4532, 1982.

[131] J. Y. T. Wei, N-C. Yeh, D. F. Garrigus, M. Strasik, Directional tunneling and Andreev reflection on $YBa_2Cu_3O_{7-\delta}$ single crystals: Predominance of *d*-wave pairing symmetry verified with the generalized Blonder, Tinkham, and Klapwijk theory, *Physical Review Letters*, vol. **81**, no. 12, pp. 2542-2545, 1998.

[132] C. J. Pethick, H. Smith, Relaxation and collective motion in superconductors: a two-fluid description, *Annals of Physics*, vol. **119**, no. 1, pp. 133-169, 1979.

[133] I. O. Kulik, Spatial quantization and proximity effect in $S-N-S$ junctions, *Zhurnal Eksperimental'noi i Teoreticheskoi Fiziki*, vol. **57**, no. 5, pp. 1745-1759, 1969.

[134] S. Kashiwaya and Y. Tanaka, Tunnelling effects on surface bound states in unconventional superconductors, *Reports on Progress in Physics*, vol. **63**, no. 10, pp. 1641-1724, 2000.

[135] A. Furusaki, Josephson current carried by Andreev levels in superconducting quantum point contacts, *Superlattices and Microstructures*, vol. **25**, no. 5-6, pp. 809-818, 1999.

[136] C. C. Tsuei, J. R. Kirtley, Pairing symmetry in cuprate superconductors, *Reviews of Modern Physics*, vol. **72**, no. 4, pp. 969-1016, 2000.

[137] G. Deutscher, Andreev-Saint-James reflections: A probe of cuprate superconductors, *Reviews of Modern Physics*, vol. **77**, no. 1, pp. 109-135, 2005.

[138] G. Logvenov, A. Gozar, V. Y. Butko, A. T. Bollinger, N. Bozovic, Z. Radovic, I. Bozovic, Comprehensive study of high-T_c interface superconductivity, *Journal of Physics and Chemistry of Solids*, vol. **71**, no. 8, pp. 1098-1104, 2010.

[139] R. F. Klie, J. C. Idrobo, N. D. Browning, A. Serquis, Y. T. Zhu, X. Z. Liao, F. M. Mueller, Observation of coherent oxide precipitates in polycrystalline MgB_2, *Applied Physics Letters*, vol. **80**, no. 21, pp. 3970-3972, 2002.

[140] H. Narayan, S. B. Samanta, A. Gupta, A. V. Narlikar, R. Kishore, K. N. Sood, D. Kanjilal, T. Muranaka, J. Akimitsu, SEM, STM/STS and heavy ion irradiation studies on magnesium diboride superconductor, *Physica C*, vol. **377**, no. 1-2, pp. 1-6, 2002.

[141] P. G. de Gennes, D. Saint-James, Elementary excitations in the vicinity of a normal metal-superconducting metal contact, *Physics Letters*, vol. **4**, no. 2, pp. 151-152, 1963.

[142] H. Padamsee, J. E. Neighbor, C. A. Shiffman, Quasiparticle phenomenology for thermodynamics of strong-coupling superconductors, *Journal of Low Temperature Physics*, vol. **12**, no. 3-4, pp. 387-411, 1973.

[143] A. Gilabert, J. P. Romagnan, E. Guyon, Determination of the energy gap of a superconducting tin-lead sandwich by electron tunneling, *Solid State Communication*, vol. **9**, no. 15, pp. 1295-1297, 1971.

[144] T. Ekino, S. Hashimoto, T. Takasaki, H. Fujii, Tunneling spectroscopy of the normal-state gap in $(Bi, Pb)_2Sr_2Ca_2Cu_3O_{10+\delta}$, *Physical Review B*, vol. **64**, no. 9, Article ID092510, 2001.

[145] A. Damascelli, Z. Hussain, Z.-X. Shen, Angle-resolved photoemission studies of the cuprate superconductors, *Reviews of Modern Physics*, vol. **75**, no. 2, pp. 473-541, 2003.

[146] F. Marsiglio, J. P. Carbotte, Strong-coupling corrections to Bardeen-Cooper-Schrieffer ratios, *Physical Review B*, vol. **33**, no. 9, pp. 6141-6146, 1986.

[147] T. Yildirim, O. Gulseren, J. W. Lynn, C. M. Brown, T. J. Udovic, Q. Huang, N. Rogado, K. A. Regan, M. A. Hayward, J. S. Slusky, T. He, M. K. Haas, P. Khalifah, K. Inumaru, R. J. Cava, Giant Anharmonicity and Nonlinear Electron-Phonon Coupling in MgB_2: A Combined First-Principles Calculation and Neutron Scattering Study, *Physical Review Letters*, vol. **87**, no. 3, Article ID037001, 2001.

[148] K.-P. Bohnen, R. Heid, B. Renker, Phonon Dispersion and Electron-Phonon Coupling in MgB_2 and AlB_2, *Physical Review Letters*, vol. **86**, no. 25, pp. 5771-5774, 2001.

[149] J. W. Quilty, S. Lee, A. Yamamoto, S. Tajima, Superconducting Gap in MgB_2: Electronic Raman Scattering Measurements of Single Crystals, *Physical Review Letters*, vol. **88**, no. 8, Article ID087001, 2002.

[150] J. W. Quilty, S. Lee, S. Tajima, A. Yamamoto, *c*-Axis Raman Scattering Spectra of MgB_2: Observation of a Dirty-Limit Gap in the π Bands, *Physical Review Letters*, vol. **90**, no. 20, Article ID207006, 2003.

[151] J. Hlinka, I. Gregora, J. Pokorný, A. Plecenik, P. Kúš, L. Satrapinsky, S. Beňačka, Phonons in MgB_2 by polarized Raman scattering on single crystals, *Physical Review B*, vol. **64**, no. 14, Article ID140503, 2001.

[152] Yu. G. Naidyuk, H. V. Loehneysen and I. K. Yanson, Temperature and magnetic-field dependence of the superconducting order parameter in Zn studied by point-contact spectroscopy, *Physical Review B*, vol. **54**, no. 22, Article ID16077, 1996.

[153] T. Ekino, Y. Sezaki, H. Fujii, Tunneling Spectroscopy of $Bi_2Sr_2CuO_{6+\delta}$, *Advances in Superconductivity X*, Springer-Verlag Tokyo, pp. 183-186, 1998.

[154] A. A. Abrikosov, On the Magnetic Properties of Superconductors of the Second Group, *Zhurnal Ekspermental'noi i Teoreticheskoi Fiziki*, vol. 32, no. 6, pp. 1442-1452, 1957.

[155] A. A. Abrikosov, *Fundamentals of the Theory of Metals*, Elsevier Science Publishers B.V., Amsterdam, 1988.

[156] X. Zhang, Y. S. Oh, Y. Liu, L. Yan, S. R. Saha, N. P. Butch, K. Kirshenbaum, K. H. Kim, J. Paglione, R. L. Greene, I. Takeuchi, Evidence of a universal and isotropic $2\Delta/k_BT_c$ ratio in 122-type iron pnictide superconductors over a wide doping range, *Physical Review B*, vol. **82**, no. 2, ID 020515, 2010.

[157] D. Saint-James, G. Sarma, E. J. Thomas, *Type II Superconductivity*, Pergamon Press, Oxford, United Kingdom, 1969.

[158] M. Eisterer, Magnetic properties and critical currents of MgB_2, *Superconducting Science and Technology*, vol. **20**, no. 12, pp. R47-R73, 2007.

[159] A. E. Koshelev, A. A. Golubov, Why magnesium diboride is not described by anisotropic Ginzburg-Landau theory, *Physical Review Letters*, vol. **92**, no. 10, ID 107008, 2004.

[160] M. Eisterer, H. W. Weber, The influence of weak texture on the critical currents in polycrystalline MgB_2, *Superconducting Science and Technology*, vol. **23**, no. 3, ID 034006, 2010.

[161] M. Xu, H. Kitazawa, Y. Takano, J. Ye, K. Nishida, H. Abe, A. Matsushita, N. Tsuji, G. Kido, Anisotropy of superconductivity from MgB_2 single crystals, *Applied Physics Letters*, vol. **79**, no. 17, pp. 2779-2781, 2001.

[162] Yu. Eltsev, S. Lee, K. Nakao, N. Chikumoto, S. Tajima, N. Koshizuka, M. Murakami, Anisotropic superconducting properties of MgB_2 single crystals probed by in-plane electrical transport measurements, *Physical Review B*, vol. **65**, no. 14, Article ID140501, 2002.

[163] M. Angst, R. Puzniak, A. Wisniewski, J. Jun, S. M. Kazakov, J. Karpinski, J. Roos, H. Keller, Temperature and Field Dependence of the Anisotropy of MgB_2, *Physical Review Letters*, vol. **88** no. 16, Article ID167004, 2002.

[164] Yu. N. Ovchinnikov, R. Cristiano, C. Nappi, A. Barone, Two-particle tunneling current in Josephson junctions, *Journal of Low Temperature Physics*, vol. **99**, no. 1-2, pp. 81-105, 1995.

[165] T. Muranaka, J. Akimitsu, Superconductivity in MgB_2, *Solid Physics-Topics*, vol. **36**, no. 11, pp. 55, 2002 (in Japanese).

[166] A. G. M. Jansen, F. M. Mueller, P. Wyder, Direct measurements of electron-phonon coupling $\alpha^2F(\omega)$ using point contacts: noble metals, *Physical Review B*, vol. **16**, no. 4, pp. 1325-1328, 1977.

[167] R. F. Wood, Bo E. Sernelius, A. L. Chernyshev, Acoustic-phonon anomaly in MgB_2, *Physical Review B*, vol. **66**, no. 1, Article ID014513, 2002.

[168] W. Ku, W. E. Pickett, R. T. Scalettar, A. G. Eguiluz, *Ab initio* investigation of collective charge excitations in MgB_2, *Physical Review Letters*, vol. **88**, no. 5, ID 057001, 2002.

[169] I. K. Yanson, Yu. G. Naidyuk, Advances in point-contact spectroscopy: tow-band superconductor MgB_2 (Review), *Low Temperature Physics*, vol. **30**, no. 4, pp. 261-274, 2004.

[170] P. Szabó, P. Samuely, J. Kačmarčík, T. Klein, J. Marcus, A. G. M. Jansen, Point-contact spectroscopy of MgB_2 in high magnetic fields, *Physica C*, vol. **388-389**, pp. 145-146, 2003.

[171] P. Martinez-Samper, J. G. Rodrigo, G. Rubio-Bollinger, H. Suderow, S. Vieira, S. Lee, S. Tajima, Scanning tunneling spectroscopy in MgB_2, *Physica C*, vol. **385**, no. 1-2, pp. 233-243, 2003.

[172] G. Karapetrov, M. Iavarone, W. K. Kwok, G. W. Crabtree, D. G. Hinks, Scanning Tunneling Spectroscopy in MgB_2, *Physical Review Letters*, vol. **86**, no. 19, pp. 4374-4377, 2001.

[173] K. A. Yates, G. Burnell, N. A. Stelmashenko, D.-J. Kang, H. N. Lee, B. Oh, M. G. Blamire, Disorder-induced collapse of the electron-phonon coupling in MgB_2 observed by Raman spectroscopy, *Physical Review B*, vol. **68**, no. 22, Article ID220512(R), 2003.

[174] G. Binnig, H. Rohrer, Scanning tunneling microscopy – from birth to adolescence, *Reviews of Modern Physics*, vol. **59**, no. 3, pp. 615-625, 1987.

[175] Ø. Fischer, M. Kugler, I. Maggio-Aprile, C. Berthod, Scanning tunneling spectroscopy of high-temperature superconductors, *Reviews of Modern Physics*, vol. **79**, no. 1, pp. 353-419, 2007.

[176] K. I. Wysokiñski, A. Ciechan, J. Krzyszczak, Theoretical analysis of the short coherence length superconductors, *Acta Physica Polonica A*, vol. **118**, no. 2, pp. 232-237, 2010.

[177] A. M. Gabovich, A. I. Voitenko, T. Ekino, Mai Suan Li, H. Szymczak, M. Pękała, Competition of superconductivity and charge density waves in cuprates: Recent evidence and interpretation, *Advances in Condensed Matter Physics*, vol. **2010**, ID 681070, 2010.

[178] A. Sugimoto, S. Kashiwaya, H. Eisaki, H. Kashiwaya, H. Tsuchiura, Y. Tanaka, K. Fujita, S. Uchida, Enhancement of electronic inhomogeneities due to out-of-plane disorder in $Bi_2Sr_2CuO_{6+\delta}$ superconductors observed by scanning tunneling spectroscopy, *Physical Review B*, vol. **74**, no. 9, ID 094503, 2006.

[179] K. M. Lang, V. Madhavan, J. E. Hoffman, E. W. Hudson, H. Eisaki, S. Uchida, J. C. Davis, Imaging the granular structure of high-T_c superconductivity in underdoped $Bi_2Sr_2CaCu_2O_{8+\delta}$, *Nature*, **415**, no. 6870, pp. 412-416, 2002.

[180] M. C. Boyer, W. D. Wise, K. Chatterjee, M. Yi, T. Kondo, T. Takeuchi, H. Ikuta, E. W. Hudson, Imaging the two gaps of the high-temperature superconductor $Bi_2Sr_2CuO_{6+x}$, *Nature Physics*, vol. **3**, no. 11, pp. 802-806, 2007.

[181] T. Kurosawa, T. Yoneyama, Y. Takano, M. Hagiwara, R. Inoue, N. Hagiwara, K. Kurusu, K. Takeyama, N. Momono, M. Oda, M. Ido, Large pseudogap and nodal superconducting gap in $Bi_2Sr_{2-x}La_xCuO_{6+\delta}$ and $Bi_2Sr_2CaCu_2O_{8+\delta}$: Scanning tunneling microscopy and spectroscopy, *Physical Review B*, vol. **81**, no. 9, ID 094519, 2010.

[182] P. C. Canfield, G. W. Crabtree, Magnesium diboride: better late than never, *Physics Today*, vol. **56**, no. 3, pp. 34-40, 2003.

[183] P. Jess, U. Hubler, H. P. Lang, H.-J. Güntherodt, H. Werner, R. Schlögl, K. Lüders, Energy gap determination on polycrystalline Rb_2CsC_{60} by scanning tunneling spectroscopy, *Journal of Physics and Chemistry of Solids*, vol. **58**, no. 1, pp. 1803-1805, 1997.

[184] H. Kotegawa, K. Ishida, Y. Kitaoka, T. Muranaka, J. Akimitsu, Evidence for Strong-Coupling *s*-Wave Superconductivitiy in MgB_2: ^{11}B NMR Study, *Physical Review Letters*, vol. **87**, no. 12, Article ID127001, 2001.

[185] S. Tsuda, T. Yokoya, Y. Takano, H. Kito, A. Matsushita, F. Yin, J. Itoh, H. Harima, S. Shin, Definitive Experimental Evidence for Two-Band Superconductivity in MgB_2, *Physical Review Letters*, vol. **91**, no. 12, Article ID127001, 2003.

[186] S. Tsuda, T. Yokoya, Y. Takano. H. Kito, A. Matsushita, F. Yin, J. Itoh, H. Harima, S. Shin, Definitive experimental evidence for two-band superconductivity in MgB_2, *Physical Review Letters*, vol. **91**, no. 12, ID 127001, 2003.

[187] X. K. Chen, M. J. Konstantinovic, J. C. Irwin, D. D. Lawrie, J. P. Franck, Evidence for two superconducting gaps in MgB_2, *Physical Review Letters*, vol. **87**, no. 15, Article ID157002, 2001.

[188] H. Schmidt, J. F. Zasadzinski, K. E. Gray, D. G. Hinks, Break-junction tunneling on MgB_2, *Physica C*, vol. **385**, no. 1-2, pp. 221-232, 2003.

[189] P. Szabó, P. Samuely, J. Kačmarčík, T. Klein, J. Marcus, D. Fruchart, S. Miraglia, C. Marcenat, A. G. M. Jansen, Evidence for Two Superconducting Energy Gaps in MgB_2 by Point-Contact Spectroscopy, *Physical Review Letters*, vol. **87**, no. 13, Article ID137005, 2001.

[190] F. Laube, G. Goll, J. Hagel, H. V. Lohneysen, D. Ernst, T. Wolf, Superconducting energy gap distribution of MgB_2 investigated by point-contact spectroscopy, *Europhysics Letters*, vol. **56**, no. 2, pp. 296-301, 2001.

[191] A. Kohen, G. Deutscher, Symmetry and temperature dependence of the order parameter in MgB_2 from point contact measurements, *Physical Review B*, vol. **64**, no. 6, Article ID060506 (R), 2001.

[192] H. Schmidt, J. F. Zasadzinski, K. E. Gray, D. G. Hinks, Energy gap from tunneling and metallic contacts onto MgB_2: Possible evidence for a weakened surface layer, *Physical Review B*, vol. **63**, no. 22, Article ID220504 (R), 2001.

[193] Zhuang-Zhi Li, Hong-Jie Tao, Yi Xuan, Zhi-An Ren, Guang-Can Che, Bai-Ru Zhao, Andreev reflection spectroscopy evidence for multiple gaps in MgB_2, *Physical Review B*, vol. **66**, no. 6, Article ID064513, 2002.

[194] Yu. G. Naidyuk, I. K. Yanson, L. V. Tyutrina, N. L. Bobrov, P. N. Chubov, W. N. Kang, Hyeong-Jin Kim, Eun-Mi Choi, Sung-Ik Lee, Superconducting energy gap distribution in *c*-axis oriented MgB_2 thin film from point contact study, *Pis'ma v Zhurnal Eksperimental'noi i Teoreticheskoi Fiziki*, vol. **75**, no. 5, pp. 283-286, 2002.

[195] Suyoun Lee, Z. G. Khim, Yonuk Chong, S. H. Moon, H. N. Lee, H. G. Kim, B. Oh, Eun Jip Choi, Meaurement of the superconducting gap of MgB_2 by point contact spectroscopy, *Physica C*, vol. **377**, no. 3, pp. 202-207, 2002.

[196] A. Plecenik, Š. Beňačka, P. Kúš, M. Grajcar, Superconducting gap parameters of MgB_2 obtained on MgB_2/Ag and MgB_2/In junctions, *Physica C*, vol. **368**, no. 1-4, pp. 251-254, 2002.

[197] R. S. Gonnelli, D. Daghero, G. A. Ummarino, V. A. Stepanov, J. Jun, S. M. Kazakov, J. Karpinski, Direct Evidence for Two-Band Superconductivity in MgB_2 Single Crystals from Directional Point-Contact Spectroscopy in Magnetic Fields, *Physical Review Letters*, vol. **89** , no. 24, Article ID247004, 2002.

[198] A. Brinkman, A. A. Golubov, H. Rogalla, O. V. Dolgov, J. Kortus, Y. Kong, O. Jepsen, O. K. Andersen, Multiband model for tunneling in MgB_2 junctions, *Physical Review B*, vol. **65**, no. 18, ID 180517, 2002.

[199] O. V. Dolgov, R. K. Kremer, J. Kortus, A. A. Golubov, S. V. Shulga Thermodynamics of two-band superconductors: The case of MgB_2, *Physical Review B*, vol. **72**, no. 2, ID 024504, 2005.

[200] Ya. G. Ponomarev, S. A. Kuzmichev, M. G. Mikheev, M. V. Sudakova, S. N. Tchesnokov, N. Z. Timergaleev, A. V. Yarigin, E. G. Maksimov, S. I. Krasnovobodtsev, A. V. Varlashkin, M. A. Hein, G. Müller, H. Piel, L. G. Sevastyanova, O. V. Kravchenko, K. P. Burdina, B. M. Bulychev, Evidence for a two-band behavior of MgB_2 from point contact and tunneling spectroscopy, *Solid State Communications*, vol. **129**, no. 2, pp. 85-89, 2004.

[201] G. Rubio-Bollinger, H. Suderow, S. Vieira, Tunneling Spectroscopy in Small Grains of Superconducting MgB_2, *Physical Review Letters*, vol. **86**, no. 24, pp. 5582-5584, 2001.

[202] P. Seneor, C.-T. Chen, N.-C. Yeh, R. P. Vasquez, L. D. Bell, C. U. Jung, Min-Seok Park, H.-J. Kim, W. N. Kang, S.-I. Lee, Spectroscopic evidence for anisotropic *s*-wave pairing symmetry in MgB_2, *Physical Review B*, vol. **65**, no. 1, Article ID012505, 2001.

[203] Amos Sharoni, Israel Felner, Oded Millo, Tunneling spectroscopy and magnetization measurements of the superconducting properties of MgB_2, *Physical Review B*, vol. **63**, no. 22, Article ID220508(R), 2001.

[204] F. Giubileo, D. Roditchev, W. Sacks, R. Lamy, D. X. Thanh, J. Klein, S. Miraglia, D. Fruchart, J. Marcus, Ph. Monod, Two-Gap State Density in MgB_2: A True Bulk Property Or A Proximity Effect? *Physical Review Letters*, vol. **87**, no. 17, Article ID177008, 2001.

[205] M. R. Eskildsen, M. Kugler, S. Tanaka, J. Jun, S. M. Kazakov, J. Karpinski, Ø. Fischer, Vortex Imaging in the *p* Band of Magnesium Diboride, *Physical Review Letters*, vol. **89**, no. 18, Article ID187003, 2002.

[206] M. Iavarone, G. Karapetrov, A. E. Koshelev, W. K. Kwok, G. W. Crabtree, D. G. Hinks, W. N. Kang, E.-M. Choi, H. J. Kim, H.-J. Kim, S. I. Lee, Two-Band Superconductivity in MgB_2, *Physical Review Letters*, vol. **89**, no. 18, Article ID187002, 2002.

[207] H. Suderow, M. Crespo, P. Martinez-Samper, J. G. Rodrigo, G. Rubio-Bollinger, S. Vieira, N. Luchier, J. P. Brison, P. C. Canfield, Scanning tunneling microscopy and spectroscopy at very low temperatures, *Physica C*, vol. **369**, no. 1-4, pp. 106-112, 2002.

[208] M. Xu, Z. Xiao, Z. Wang, Y. Takano, T. Hatano, K. Sagisaka, M. Kitahara, D. Fujita, Local density of electronic states in MgB_2 studied by low temperature STM and STS: direct evidence for a multiple-gap superconductor, *Surface Science*, vol. **541**, no. 1-3, pp. 14-20, 2003.

[209] T. W. Heitmann, S. D. Bu, D. M. Kim, J. H. Choi, J. Giencke, C. B. Eom, K. A. Regan, N. Rogado, M. A. Hayward, T. He, J. S. Slusky, P. Khalifah, M. Haas, R. J. Cava, D. C. Larbalestier, M. S. Rzchowski, MgB_2 energy gap determination by scanning tunneling spectroscopy, *Supercoductor Science and Technology*, vol. **17**, no. 2, pp. 237-242, 2004.

[210] Y. Bugoslavsky, G. K. Perkins, X. Qi, L. F. Cohen, A. D. Caplin, Vortex dynamics in superconducting MgB_2 and prospects for applications, *Nature*, **410**, no. 6828, pp. 563-565, 2001.

[211] M. H. Badr, M. Freamat, Yu. Suchko, K.-W. Ng, Temperature and field dependence of the energy gap of MgB_2/Pb planar junctions, *Physical Review B*, vol. **65**, no. 18, Article ID184516, 2002.

[212] D. K. Aswal, S. Sen, S. C. Cadkari, A. Singh, S. K. Gupta, L. C. Gupta, A. Bajpai, A. K. Nigam, Andreev reflections on a MgB_2 superconductor, *Physical Review B*, vol. **66**, no. 1, Article ID012513, 2002.

[213] Y. Zhang, D. Kinion, J. Chen, J. Clarke, D. G. Hinks, G. W. Crabtree, MgB_2 tunnel junctions and 19 K low-noise dc superconducting quantum interference devices, *Applied Physics Letters*, vol. **79**, no. 24, pp. 3995-3997, 2001.

[214] A. Saito, A. Kawakami, H. Shimakage, H. Terai, Z. Wang, Josephson tunneling properties in MgB_2/AlN/NbN tunnel junctions, *Journal of Applied Physics*, vol. **92**, no. 12, pp. 7369-7372, 2002.

[215] G. Carapella, N. Martucciello, G. Costabile, C. Ferdeghini, V. Ferrando, G. Grassano, Josephson effect in Nb/Al_2O_3/Al/MgB_2 large-area thin film heterostructures, *Applied Physics Letters*, vol. **80**, no. 16, pp. 2949-2951, 2002.

[216] K. Ueda, M. Naito, Tunnel junctions on as-grown MgB_2 films, *Physica C*, vol. **408-410**, pp. 134-135, 2004.

[217] J. Winter, *Magnetic Resonance in Metals*, Clarendon Press, Oxford, UK, 1971.

[218] I. I. Mazin, V. P. Antropov, Electronic structure, electron-phonon coupling, and multiband effects in MgB_2, *Physica C*, vol. **385**, no. 1-2, pp .49-65, 2003.

[219] *X-ray photoemission spectroscopy*, Edited by the Surface Science Society of Japan, Maruzen Co. Ltd. 2002, (in Japanese).

[220] T. Masui, S. Lee, A. Yamamoto, H. Uchiyama, S. Tajima, Carbon-substitution effect on superconducting properties in MgB_2 single crystals, *Physica C*, vol. **412 – 414**, no. 1, pp. 303-306, 2004.

[221] H. D. Yang, H. L. Liu, J.-Y. Lin, M. X. Kuo, P. L. Ho, J. M. Chen, C. U. Jung, M. S. Park, S. I. Lee, Superconducting three-dimensional networks in a magnetic field: Frustrated systems, *Physical Review B*, vol. **69**, no. 9, Article ID092505, 2004.

[222] Z. Holánová, P. Szabó, J. Kačmarčík, P. Samuely, R. A. Ribeiro, S. L. Bud'ko, P. C. Canfield, Energy gaps in carbon-substituted MgB_2, *Physica C*, vol. **408-410**, pp. 610-611, 2004.

[223] Ya. G. Ponomarev, S. A. Kuzmichev, N. M. Kadomtseva, M. G. Mikheev, M. V. Sudakova, S. N. Chesnokov, E. G. Maksimov, S. I. Krasnosvovodtsev, L. G. Sevast'yanova, K. P. Burdina, B. M. Bulychev, Investigation of a superconducting $Mg_{1-x}Al_xB_2$ system by tunneling and microjunction (Andreev) spectroscopies, *Pis'ma v Zhurnal Eksperimental'noi i Teoreticheskoi Fiziki*, vol. **79**, no. 10, pp. 597-601, 2004.

[224] V. A. Drozd, A. M. Gabovich, P. Gierłowski, M. Pękała, H. Szymczak, Transport properties of bulk and thin-film MgB_2 superconductors: effects of preparation conditions, *Physica C*, vol. **402**, no. 4, pp. 325-334, 2004.

[225] M. Eisterer, J. Emhofer, S. Sorta, M. Zehetmayer, and H. W. Weber, Connectivity and critical currents in polycrystalline MgB_2, *Superconducting Science and Technology*, vol. **22**, no. 3, ID 034016, 2009.

[226] P. A. Sharma, N. Hur, Y. Horibe, C. H. Chen, B. G. Kim, S. Guha, M. Z. Cieplak, S. W. Cheong, Percolative Superconductivitiy in $Mg_{1-x}B_2$, *Physical Review Letters*, vol. **89**, no. 16, Article ID167003, 2002.

[227] X. Z. Liao, A. C. Serquis, Y. T. Zhu, J. Y. Huang, D. E. Peterson and F. M. Mueller, H. F. Xu, Controlling flux pinning precipitates during MgB_2 synthesis, *Applied Physics Letters*, vol. **80**, no. 23, pp. 4398-4400, 2002.

[228] R. P. Vasquez, C. U. Jung, M.-S. Park, H.-J. Kim, J. Y. Kim, S.-I. Lee, X-ray photoemission study of MgB_2, *Physical Review B*, vol. **64**, no. 5, Article ID052510, 2001.

[229] Y. Zhu, L. Wu, V. Volkov, Q. Li, G. Gu, A. R. Moodenbaugh, M. Malac, M. Suenaga, J. Tranquada, Microstructure and structural defects in MgB_2 superconductor, *Physica C*, vol. **356**, no. 4, pp. 239-253, 2001.

[230] D. Eyidi, O. Eibl, T. Wenzel, K. G. Nickel, M. Giovannini, A. Saccone, Phase analysis of superconducting polycrystalline MgB_2, *Micron*, vol. **34**, no. 2, pp. 85-96, 2003.

[231] Sergy Lee, Crystal growth of MgB_2, *Physica C*, vol. **385**, no. 1-2, pp. 31-41, 2003.

[232] W. Weimin, F. Zhengyi, W. Hao, Y. Runzhang, Chemistry reaction processes during combustion synthesis of B_2O_3-TiO_2-Mg system, *Journal of Material Processing Technology*, vol. **128**, no. 1-3, pp. 162-168, 2002.

[233] S. Ueda, J. Shimoyama, A. Yamamoto, Y. Katsura, I. Iwayama,S. Horii, K. Kishio, Flux pinning properties of impurity doped MgB_2 bulks synthesized by diffusion method, *Physica C*, vol. **426-431**, no. 2, pp. 1225-1230, 2005.

[234] C. B. Eom, M. K. Lee, J. H. Choi, L. J. Belenky, X. Song, L. D. Cooley, M. T. Naus, S. Patnaik, J. Jiang, M. Rikel, A. Polyanskii, A. Gurevich, X. Y. Cai, S. D. Bu, S. E. Babcock, E. E. Hellstrom, D. C. Larbalestier, N. Rogado, K. A. Regan, M. A. Hayward, T. He, J. S. Slusky, K. Inumaru, M. K. Haas, R. J. Cava, High critical current density and enhanced irreversibility field in superconducting MgB_2 thin films, *Nature*, vol. **411**, no. 6837, pp. 558-560, 2001.

[235] D. H. A. Blank, H. Hilgenkamp, A. Brinkman, D. Mijatovic, G. Rijnders, H. Rogalla, Superconducting Mg-B thin films by pulsed-laser deposition in an *in situ* two-step process using multicomponent targets, *Applied Physics Letters*, vol. **79**, no. 3, pp. 394-396, 2001.

[236] K. Ueda, N. Saito, *In situ* growth of superconducting MgB_2 thin films by molecular-beam epitaxy, *Journal of Applied Physics*, vol. **93**, no. 4, pp. 2113-2120, 2003.

[237] A. Saito, H. Shimakage, A. Kawakami, Z. Wang, K. Kuroda, H. Abe, M. Naito, W. J. Moon, K. Kaneko, M. Mukaida, S. Ohshima, XRD and TEM studies of as-grown MgB_2 thin films deposited on *r*- and *c*-plane sapphire substrates, *Physica C*, vol. **412-414**, Part 2, pp. 1366-1370, 2004.

[238] S. R. Shinde, S. B. Ogale, R. L. Greene, T. Venkatesan, P. C. Canfield, S. L. Bud'ko, G. Lapertot, C. Petrovic, Superconducting MgB_2 thin films by pulsed laser deposition, *Applied Physics Letters*, vol. **79**, no. 2, pp. 227-229, 2001.

[239] R. A. Fisher, G. Li, J. C. Lashley, F. Bouquet, N. E. Phillips, D. G. Hinks, J. D. Jorgensen, G. W. Crabtree, Specific heat of $Mg_{11}B_2$, *Physica C*, vol. **385**, no. 1-2, pp. 180-191, 2003.

[240] F. Bouquet, Y. Wang, I. Sheikin, P. Toulemonde, M. Eisterer, H. W. Weber, S. Lee, S. Tajima, A. Junod, Unusual effects of anisotropy on the specific heat of ceramic and single crystal MgB_2, *Physica C*, vol. **385**, no. 1-2, pp. 192-204.

[241] I. I. Mazin, O. K. Anderson, O. Jepsen, O. V. Dolgov, J. Kortus, A. A. Golubov, A. B. Kuz'menko and D. van der Marel, Superconductivity in MgB_2: Clean or Dirty? *Physical Review Letters*, vol. **89**, no. 10, Article ID107002, 2002.

[242] V. D. P. Servedio, S.-L. Drechsler, T. Mishonov, Surface states and their possible role in the superconductivity of MgB_2, *Physical Review B*, vol. **66**, no. 14, Article ID140502(R), 2002.

[243] V. M. Silkin, E. V. Chulkov, P. M. Echenique, Surface and image-potential states on MgB_2(0001) surfaces, *Physical Review B*, vol. **64**, no. 17, Article ID172512, 2001.

[244] H. Takeya, K. Togano, Y. S. Sung, T. Mochiku, K. Hirata, Metastable superconductivity in niobium diborides, *Physica C*, vol. **408-410**, pp. 144-145, 2004.

[245] S. Kuroiwa, Y. Tomita, A. Sugimoto, T. Ekino, J. Akimitsu, Specific heat and tunneling spectroscopy study of NbB_2 with maximum T_c ~ 10 K, *Journal of the Physical Society of Japan*, vol. **76**, no. 9, ID 094705, 2007.

[246] A. Yamamoto, C. Takao, T. Masui, M. Izumi, S. Tajima, High-pressure synthesis of superconducting $Nb_{1-x}B_2$ (x = 0 - 0.48) with the maximum T_c = 9.2 K, *Physica C*, vol. **383**, no. 3, pp. 197-206, 2002.

[247] L. Y. L. Shen, N. M. Senozan, N. E. Phillips, Evidence for two energy gaps in high-purity superconducting Nb, Ta, and V, *Physical Review Letters*, vol. **14**, no. 25, pp. 1025-1026, 1965.

[248] J. Bostock, V. Diadiuk, W. N. Cheung, K. H. Lo, R. M. Rose, M. L. A. MacVicar, Does Strong Coupling Theory Describe Superconducting Nb? *Physical Review Letters*, vol. **36**, no. 11, pp. 603-606, 1976.

[249] H. Kotegawa, K. Ishida, Y. Kitaoka, T. Muranaka, N. Nakagawa, H. Takagiwa, J. Akimitsu, Evidence for strong-coupling *s*-wave superconductivity in MgB_2: 11B-NMR study of MgB_2 and the related materials, *Physica C*, vol. **378-381**, no. 1, pp. 25-32, 2002.

[250] S. Tsuda, T. Yokoya, S. Shin, M. Imai, I. Hase, Identical superconducting gap on different Fermi surfaces of $Ca(Al_{0.5}Si_{0.5})_2$ with the AlB_2 structure, *Physical Review B*, vol. **69**, no. 10, Article ID100506(R), 2004.

[251] B. Lorenz, J. Cmaidalka, R. L. Meng, C. W. Chu, Thermodynamic properties and pressure effect on the superconductivitiy in CaAlSi and SrAlSi, *Physical Review B*, vol. **68**, no. 1, Article ID014512, 2003.

[252] B. Lorenz, J. Cmaidalka, R. L. Meng, Y. Y. Xue, C. W. Chu, Thermodynamic and superconducting properties of the C32 intermetallic compounds CaAlSi and SrAlSi, *Physica C*, vol. **408-410**, pp. 171-172, 2004.

[253] K. E. Kihlstrom, T. H. Geballe, Tunneling $\alpha^2F(\omega)$ as a function of composition in A15 NbGe, *Physical Review B*, vol. **24**, no. 7, pp. 4101-4104, 1981.

[254] I. R. Shein, A. L. Ivanovskii, Band structure of ZrB_2, VB_2, NbB_2, and TaB_2 hexagonal diborides: Comparison with superconducting MgB_2, *Phys. Solid. State*, vol. **44**, no. 10, pp. 1833-1839, 2002.

[255] A. L. Ivanovskii, Band structure and properties of the superconducting MgB_2 and related compounds (Review), *Fizika Tverdogo Tela*, vol. **45**, no. 10, pp. 1742-1769, 2003.

[256] P. J. T. Joseph, P. P. Singh, Theoretical study of electronic structure and superconductivity in $Nb_{1-x}B_2$ alloys, *Physica C*, vol. **391**, no. 2, pp. 125-130, 2003.

[257] I. R. Shein, V. V. Ivanovskaya, N. I. Medvedeva, A. L. Ivanovskii, Electronic properties of new $Ca(Al_xSi_{1-x})_2$ and $Sr(Ga_xSi_{1-x})_2$ superconductors in crystalline and nanotubular states, *JETP Letters*, vol. **76**, no. 3, pp. 189-193, 2002.

[258] J. E. Hirsch, Correlations between normal-state properties and superconductivity, *Physical Review B*, vol. **55**, no. 14, pp. 9007-9024, 1997.

[259] R. A. Cowley, Structural phase transitions. I. Landau theory? *Advances in Physics*, vol. **29**, no. 1, pp. 1-110, 1980.

[260] A. V. Pogrebnyakov, J. M. Redwing, S. Raghavan, V. Vaithyanathan, D. G. Schlom, S. Y. Xu, Q. Li, D. A. Tenne, A. Soukiassian, X. X. Xi, M. D. Johannes, D. Kasinathan, W. E. Pickett, J. S. Wu, J. C. H. Spence, Enhancement of the superconducting transition temperature of MgB_2 by a strain-induced bond-strctching mode softening, *Physical Review Letters*, vol. **93**, no. 14, ID 147006, 2004.

[261] M. Imai, K. Nishida, T. Kimura, H. Abe, Superconductivity of $Ca(Al_{0.5},Si_{0.5})_2$, a ternary silicide with the AlB_2-type structure, *Applied Physics Letters*, vol. **80**, no. 6, pp. 1019-1021, 2002.

[262] M. Imai, K. Nishida, T. Kimura, H. Kitazawa, H. Abe, H. Kito, K. Yoshii, Superconductivity of $M_I(M_{II0.5},Si_{0.5})_2$ (M_I = Sr and Ba, M_{II} = Al and Ga), ternary silicides with the AlB_2-type structure, *Physica C*, vol. **382**, no. 4, pp. 361-366, 2002.

[263] I. I. Mazin, D. A. Papaconstantopoulos, Electronic structure and superconductivity of CaAlSi and SrAlSi, *Physical Review B*, vol. **69**, no. 18, Article ID180512(R), 2004.

[264] G. Q. Huang, L. F. Chen, M. Liu, D. Y. Xing, Electronic structure and electron-phonon interaction in the ternary silicides MAlSi (M = Ca, Sr, and Ba), *Physical Review B*, vol. **69**, no. 6, Article ID064509, 2004.

[265] T. Tamegai, K. Uozato, A. K. Ghosh, M. Tokunaga, Anisotropic superconducting properties of CaAlSi and CaGaSi, *International Journal of Modern Physics B*, vol. **19**, no. 3, pp. 369-374, 2005.

[266] J. R. Schrieffer, J. W. Wilkins, Two-Particle Tunneling Processes Between Superconductors, *Physical Review Letters*, vol. **10**, no. 1, pp. 17-20, 1963.

[267] S. Shapiro, P. H. Smith, J. Nicol, J. L. Miles, P. F. Strong, Superconductivity and electron tunneling, *IBM Journal of Research and Development*, vol. **6**, no. 1, pp. 34-43, 1962.

[268] A. M. Gabovich, A. I. Voitenko, Nonstationary Josephson effect for superconductors with charge-density waves, *Physical Review B*, vol. **55**, no. 2, pp. 1081-1099, 1997.

[269] T. Takasaki, T. Ekino, A. Sugimoto, K. Shohara, S. Yamanaka, A. M. Gabovich, Tunneling spectroscopy of layered superconductors: intercalated $Li_{0.48}(C_4H_8O)_x$HfNCl and de-intercalated $HfNCl_{0.7}$, *European Physical Journal B*, vol. **73**, no. 4, pp. 471-482, 2010.

[270] J. Bok, J. Bouvier, Tunneling in anisotropic gap superconductors, *Physica C*, vol. **274**, no. 1-2, pp. 1-8, 1997.

[271] K. K. Likharev, Superconducting weak links, *Reviews of Modern Physics*, vol. **51**, no. 1, pp. 101-159, 1979.

[272] E. L. Nagaev, Phase separation in high-temperature superconductors and related magnetic systems, *Uspekhi Fizicheskikh Nauk*, vol. **165**, no. 5, pp. 529-554, 1995.

[273] R. Collongues, *La Non-Stoechiometrie*, Masson, Paris, France, 1971.

[274] M. Putti, R. Vaglio, J. M. Rowell, Radiation effects on MgB_2: a review and a comparison with A15 superconductors, *Superconducting Science and Technology*, vol. **21**, no. 4, ID 043001, 2008.

[275] Y. Tanaka, T. Yanagisawa, Chiral state in three-gap superconductors, *Solid State Communications*, vol. **150**, no. 41-42, pp. 1980-1982, 2010.

[276] E. V. L. de Mello, E. S. Caixeiro, Effects of phase separation in the cuprate superconductors, *Physical Review B*, vol. **70**, no. 22, Article ID224517, 2004.

[277] T. Takasaki, T. Ekino, H. Fujii, S. Yamanaka, Tunneling Spectroscopy of Deintercalated Layered Nitride Superconductor $ZrNCl_{0.7}$, *Journal of the Physical Society of Japan*, vol. **74**, no. 9, pp. 2586-2591, 2005.

[278] T. Ekino, A. M. Gabovich, Mai Suan Li, M Pękała, H. Szymczak, A. I. Voitenko, Analysis of the pseudogap-related structure in tunneling spectra of superconducting $Bi_2Sr_2CaCu_2O_{8+\delta}$ revealed by the break-junction technique, *Physical Review B*, vol. **76**, no. 18, Article ID180503, 2007.

[279] M. Ishikado, K. M. Kojima, S. Uchida, Electronic inhomogeneity and optical response of Bi2212, *Physica C*, vol. **470**, no. 20, pp. 1045-1047, 2010.

[280] Ch. Renner, Ø. Fischer, Vacuum tunneling spectroscopy and asymmetric density of states of $Bi_2Sr_2CaCu_2O_{8+\delta}$, *Physical Review B*, vol. **51**, no. 14, pp. 9208-9218, 1995.

[281] T. Ekino, T. Takasaki, T. Muranaka, J. Akimitsu, H. Fujii, Tunneling spectroscopy of the superconducting gap in MgB_2, *Phys. Rev. B*, vol. **67**, pp.094504, 2003.

[282] T. Takasaki, T. Ekino, T. Muranaka, T. Ichikawa, H. Fujii, J. Akimitsu, Multiple-Gap Features from Break-Junction Tunneling in the Superconducting MgB_2, *Journal of the Physical Society of Japan*, vol. **73**, no.7, pp. 1902-1913, 2004.

[283] T. Ekino, T. Takasaki, H. Fujii, S. Yamanaka, Tunneling spectroscopy of the electron-doped layered superconductor $Li_{0.48}(THF)_{0.3}HfNCl$, *Physica C*, vol. **388-389**, pp. 573-574, 2003.

[284] P. Raychaudhuri, D. Jaiswal-Nagar, G. Sheet, S. Rarnakrishnan, H. Takeya, Evidence of Gap Anisotropy in Superconducting YNi_2B_2C Using Directional Point-Contact Spectroscopy, *Physical Review Letters*, vol. **93**, no. 15, Article ID156802, 2004.

[285] T. Arai, K. Ichimura, K. Nomura, S. Takasaki, J. Yamada, S. Nakatsuji, H. Anzai, Tunneling spectroscopy on the organic superconductor κ-$(BEDT\text{-}TTF)_2Cu(NCS)_2$ using STM, *Physical Review B*, vol. **63**, no. 10, Article ID104518, 2001.

[286] N. A. Belous, A. M. Gabovich, I. V. Lezhnenko, D. P. Moiseev, V. M. Postnikov, S. K. Uvarova, Josephson effects in bulk solid state samples of superconducting ceramics $BaPb_{1-x}Bi_xO_3$, *Physics Letters A*, vol. **92**, no. 9, pp. 455-456, 1982.

[287] N. A. Belous, A. E. Chernyakhovskii, A. M. Gabovich, D. P. Moiseev, V. M. Postnikov, Multiple effects in the disordered Josephson medium $BaPb_{1-x}Bi_xO_3$, *Journal of Physics C: Solid State Physics*, vol. **21**, no. 6, pp. L153-L159, 1988.

[288] Y. Enomoto, M. Suzuki, T. Murakami, T. Inukai, T. Inamura, Observation of Grain Boundary Josephson Current in $BaPb_{0.7}Bi_{0.3}O_3$ Films, *Japanese Journal of Applied Physics*, vol. **20**, no. 9, pp. L661-L664, 1981.

[289] M. Suzuki, Y. Enomoto, T. Murakami, Study on grain boundary Josephson junctions in $BaPb_{1-x}Bi_xO_3$ thin films, *Journal of Applied Physics*, vol. **56**, no. 7, pp. 2083-2092, 1984.

[290] V. M. Krasnov, A. Yurgens, D. Winkler, P. Delsing, T. Claeson, Evidence for Coexistence of the Superconducting Gap and the Pseudogap in Bi-2212 from Intrinsic

Tunneling Spectroscopy, *Physical Review Letters*, vol. **84**, no. 25, pp. 5860-5863, 2000.

[291] A. Yurgens, Intrinsic Josephson junctions: recent developments, *Superconductor Science and Technology*, vol. **13**, no. 8, pp. R85-R100, 2000.

[292] A. Yurgens, M. Torstensson, L. X. You, T. Bauch, D. Winkler, I. Kakeya, K. Kadowaki, Small-number arrays of intrinsic Josephson junctions, *Physica C*, vol. **468**, no. 7-10, pp. 674-678, 2008.

[293] M. Suzuki, T. Watanabe, Discriminating the Superconducting Gap from the Pseudogap in $Bi_2Sr_2CaCu_2O_{8+\delta}$ by interlayer Tunneling Spectroscopy, *Physical Review Letters*, vol. **85**, no. 22, pp. 4787-4790, 2000.

[294] T. Ekino, A. Sugimoto, K. Shohara, S. Yamanaka, A. M. Gabovich, STM/STS measurements of the layered superconductor β-$HfNCl_{1-x}$, *Physica C*, vol. 470, Suppl. 1, pp. S725-S727, 2010.

[295] T. Ekino, Y. Sezaki, H. Fujii, Features of the energy gap above T_c in $Bi_2Sr_2CaCu_2O_{8+\delta}$ as seen by break-junction tunneling, *Physical Review B*, vol. **60**, no. 9, pp. 6916-6922, 1999.

[296] G. M. Eliashberg, Interaction of electrons with lattice vibrations in a superconductor, *Zhurnal Eksperimental'noi i Teoreticheskoi Fiziki*, vol. 38, no. 3, pp. 966-976, 1960.

[297] G. M. Eliashberg, Temperature Green's functions of electrons in a superconductor, *Zhurnal Eksperimental'noi i Teoreticheskoi Fiziki*, vol. 39, no. 5, pp. 1437-1441, 1960.

[298] B. T. Geilikman, V. Z. Kresin, N. F. Masharov, Transition temperature and energy gap for superconductors with strong coupling, *Journal of Low Temperature Physics*, vol. **18**, no. 3/4, pp. 241-271, 1975.

[299] P. B. Allen, B. Mitrović, Theory of Superconducting T_c., pp. 1-92, In the Book: *Solid State Physics*, Edited by F. Seitz, D. Turnbull, and H. Ehrenreich, Academic, New York, USA, 1982.

[300] J. P. Carbotte, Properties of boson-exchange superconductors, *Reviews of Modern Physics*, vol. **62**, no. 4, pp. 1027-1157, 1990.

[301] Yu. V. Kopaev, About the interplay theory between electron and structural transformations and superconductivity, *Trudy Fizicheskogo Instituta Akademii Nauk SSSR*, vol. **86**, pp. 3-100, 1975.

[302] L. Stojchevska, P. Kusar, T. Mertelj, V. V. Kabanov, X. Lin, G. H. Cao, Z. A. Xu, D. Mihailovic, Electron-phonon coupling and the charge gap of spin-density wave iron-pnictide materials from quasiparticle relaxation dynamics, *Physical Review B*, vol. **82**, no. 1, ID 012505, 2010.

[303] A. Cros, A. Cantarero, D. Beltran-Porter, J. Oro-Sole, A. Fuertes, Lattice dynamics of superconducting zirconium and hafnium nitride halides, *Physical Review B*, vol. **67**, no. 10, Article ID104502, 2003.

[304] P. Adelmann, B. Renker, H. Schober, M. Braden, F. Fernandez-Diaz, Lattice Dynamics of Li-ZrNCl: An Electron doped Layered Superconductor, *Journal of Low Temperature Physics*, vol. **117**, no. 3/4, pp. 449-453, 1999.

[305] N. Tsuda, Private communications.

[306] R. Heid, K.–P. Bohnen, Ab initio lattice dynamics and electron-phonon coupling in Li_xZrNCl, *Physical Review B*, vol. **72**, no. 13, Article ID134527, 2005.

[307] M. V. Sadovskii, High-temperature superconductivity in layered iron compounds, *Uspekhi Fizicheskikh Nauk*, vol. **178**, no. 12, pp. 1243-1271, 2008.

[308] A. L. Ivanovskii, New high-temperature superconductors based on rare earth and transition metal oxyarsenides and related phases: synthesis, properties and simulation, *Uspekhi Fizicheskikh Nauk*, vol. **178**, no. 12, pp. 1273-1306, 2008.

[309] Yu. A. Izyumov, E. Z. Kurmaev, FeAs systems: a new class of high-temperature superconductors, *Uspekhi Fizicheskikh Nauk*, vol. **178**, no. 12, pp. 1307-1334, 2008.

[310] K. Kuroki, S. Onari, R. Arita, H. Usui, Y. Tanaka, H. Kontani, H. Aoki, Unconventional Pairing Originating from the Disconnected Fermi Surfaces of Superconducting $LaFeAsO_{1-x}F_x$, *Physical Review Letters*, vol. **101**, no. 8, Article ID087004, 2008.

[311] K. Kuroki, Unconventional pairing in doped band insulators on a honeycomb lattice: the role of the disconnected Fermi surface and a possible application ot superconducting β-*M*NCl (M = Hf, Zr), *Science and Technology of Advanced Materials*, vol. **9**, no. 4, Article ID 044202 (7 pages), 2008.

[312] E. A Pashitskii, Yu. A Romanov, Plasma waves and superconductivity in quantizing semiconducting (semimetallic) films and layered structures, *Ukrainskii Fizichskii Zhurnal*, vol. **15**, no. 10, pp. 1594-1606, 1970 (in Russian).

[313] Y. Takada, Plasmon Mechanism of Superconductivity in Two- and Three-Dimensional Electron Systems, *Journal of the Physical Society of Japan*, vol. **45**, no. 3, pp. 786-794, 1978.

[314] A. Bill, H. Morawitz, V. Z. Kresin, Dynamical screening and superconducting state in intercalated layered metallochloronitrides, *Physical Review B*, vol. **66**, no. 10, Article ID100501(R), 2002.

[315] Y. Kasahara, T. Kishiume, K. Kobayashi, Y. Taguchi, Y. Iwasa, Superconductivity in molecule-intercalated Li_xZrNCl with variable interlayer spacing, *Physical Review B*, vol. **82**, no. 5, ID 054504, 2010.

[316] S. Yamanaka, K. Hotehama, T. Koiwasaki, H. Kawaji, H. Fukuoka, S. Shamoto, T. Kawajiri, Substitution and cointercalation effects on superconducting electron-doped layer structured metal nitride halides, *Physica C*, vol. **341-348**, no. 2, pp. 699-702, 2000.

[317] A. Kitora, Y. Taguchi, Y. Iwasa, Probing Electron-Phonon interaction in Li_xZrNCl Superconductors by Raman Scattering, *Journal of the Physical Society of Japan*, vol. **76**, no. 2, Article ID 023706, 2007.

[318] T. Ito, Y. Fudamoto, A. Fukaya, I. M. Gat-Malureanu, M. I. Larkin, P. L. Russo, A. Savici, Y. J. Uemura, K. Groves, R. Breslow, K. Hotehama, S. Yamanaka, P. Kyriakou, M. Rovers, G. M. Luke, K. M. Kojima, Two-dimensional nature of superconductivity in the intercalated layered systems Li_xHfNCl and Li_xZrNCl: Muon spin relaxation and magnetization measurements, *Physical Review B*, vol. **69**, no. 13, Article ID134522, 2004.

[319] C. Wang, L. Li, S. Chi, Z. Zhu, Z. Ren, Y. Li, Y. Wang, X. Lin, Y. Luo, S. Jiang, X. Xu, G. Cao, Z. Xu, Thorium-doping–induced superconductivity up to 56K in $Gd_{1-x}Th_xFeAsO$, *Europhysics Letters*, vol. **83**, no. 6, ID 67006, 2002.

[320] N. B. Hannay, T. H. Geballe, B. T. Matthias, K. Andres, P. Schmidt, D. MacNair, Superconductivity in graphitic compounds, *Physical Review Letters*, vol. **14**, no. 7, pp. 225-226, 1965.

[321] Y. Kopelevich, P. Esquinazi, Ferromagnetism and superconductivity in carbon-based systems, *Journal of Low Temperature Physics*, vol. **146**, no. 5-6, pp. 629-639, 2007.

[322] L. N. Bulaevskii, Superconductivity and electronic properties of layered compounds, *Uspekhi Fizicheskikh Nauk*, vol. **116**, no. 3, pp. 449-483, 1975.

[323] A. S. Alexandrov, Bose-Einstein condensation of strongly correlated electrons and phonons in cuprate superconductors, *Journal of Physics: Condensed Matter*, vol. **19**, no. 12, Article ID 125216 (23 pages), 2007.

[324] N. M. Plakida, *High-Temperature Cuprate Superconductors. Experiment, Theory, and Applications,* Springer, Berlin, 2010.

[325] M. L. Cohen, "Predicting and explaining T_c and other properties of BCS superconductors, *Modern Physics Letters B*, vol. **24**, no. 28, pp. 2755-2768, 2010.

[326] *Superconductivity in Magnetic and Exotic Materials*, Edited by T. Matsubara and A. Kotani, Springer, Berlin, 1984.

[327] B. Brandow, Characteristic features of the exotic superconductors, *Physics Reports*, vol. **296**, no. 1, pp. 1-63, 1998.

[328] B. Brandow, Explanation of the exotic superconductors by a valence-fluctuation pairing mechanism, *Philosophical Magazine B*, vol. **80**, no. 6, pp. 1229-1297, 2000.

[329] G. Deutscher, *New Superconductors: From Granular to High-T_c*, World Scientific, Singapore, 2006.

[330] J. Krzyszczak, T. Domañski, K. I. Wysokiñski, R. Micnas, S. Robaszkiewicz, Real space inhomogeneities in high temperature superconductors: the perspective of the two-component model, *Journal of Physics: Condensed Matters*, vol. **22**, no. 25, ID 255702, 2010.

[331] S. Deng, A. Simon, J. Köhler, Pairing mechanisms viewed from physics and chemistry, In Book: *Superconductivity in Complex Systems. Series: Structure and Bonding*, edited by K. A. Müller and A. Bussmann-Holder (Springer, Berlin), vol. **114**, pp. 103-141, 2005.

[332] P. W. Anderson, Is there glue in cuprate superconductors? *Science*, vol. **316**, no. 5832, pp. 1705-1707, 2007.

[333] M. L. Cohen, P. W. Anderson, Comments on the maximum superconducting transition temperature, In Book: *Superconductivity in d- and f-band Metals*, edited by H. C. Wolfe and D. H. Douglass (American Institute of Physics, New York), Conference Proceedings, vol. **4**, pp. 17-27, 1972.

[334] E. G. Maksimov and O. V. Dolgov, A note on the possible mechanisms of high-temperature superconductivity, *Uspekhi Fizicheskikh Nauk*, vol. **177**, no. 9, pp. 983-988, 2007.

[335] R. Khasanov, D. V. Evtushinsky, A. Amato, H.-H. Klauss, H. Lütkens, Ch. Niedermayer, B. Büchner, G. L. Sun, C. T. Lin, J. T. Park, D. S. Inosov, V. Hinkov, Two-gap superconductivity in $Ba_{1-x}K_xFe_2As_2$: A complementary study of the magnetic penetration depth by muon-spin rotation and angle-resolved photoemissions, *Physical Review Letters*, vol. **102**, no. 18, ID 187005, 2009.

[336] P. Szabó, Z. Pribulová, G. Pristáš, S. L. Bud'ko, P. C. Canfield, P. Samuely, Evidence for two-gap superconductivity in $Ba_{0.55}K_{0.45}Fe_2As_2$ from directional point-contact Andreev-reflection spectroscopy, *Physical Review B*, vol. **79**, no. 1, ID 012503, 2009.

[337] K. Sasmal, B. Lv, Z. Tang, F. Y. Wei, Y. Y. Xue, A. M. Guloy, C. W. Chu, Lower critical field, anisotropy, and two-gap features of LiFeAs, *Physical Review B*, vol. **81**, no. 14, ID 144512, 2010.

[338] R. Khasanov, M. Bendele, A. Amato, K. Conder, H. Keller, H.-H. Klauss, H. Luetkens, E. Pomjakushina, Evolution of two-gap behavior of the superconductor $FeSe_{1-x}$, *Physical Review Letters*, vol. **104**, no. 8, ID 087004, 2010.

[339] D. Daghero, M. Tortello, R. S. Gonnelli, V. A. Stepanov, N. D. Zhigadlo, J. Karpinski, Evidence for two-gap nodeless superconductivity in $SmFeAsO_{1-x}F_x$ from point-contact Andreev-reflection spectroscopy, *Physical Review B*, vol. **80**, no. 6, ID 060502, 2009.

[340] M. Zaberchik, K. Chashka, L. Patlgan, A. Maniv, C. Baines, P. King, A. Kanigel, Possible evidence of a two-gap structure for the Cu_xTiSe_2 superconductors, *Physical Review B*, vol. **81**, no. 22, ID 220505, 2010.

[341] V. A. Gasparov, N. S. Sidorov, I. I. Zver'kova, Two-gap superconductivity in ZrB_{12}: Temperature dependence of critical magnetic fields in single crystals, *Physical Review B*, vol. **73**, no. 9, ID 094510, 2006.

[342] C. L. Huang, J.-Y. Lin, C. P. Sun, T. K. Lee, J. D. Kim, E. M. Choi, S. I. Lee, H. D. Yang, Comparative analysis of specific heat of YNi_2B_2C using nodal and two-gap models, *Physical Review B*, vol. **73**, no. 1, ID 012502, 2006.

[343] N. L. Bobrov, S. I. Beloborod'ko, L. V. Tyutrina, V. N. Chernobay, I. K. Yanson, D. G. Naugle, K. D. D. Rathnayaka, Investigation of the superconducting energy gap in the compound $LuNi_2B_2C$ by the method of point contact spectroscopy: two-gap approximation, *Fizika Nizkikh Temperatur*, vol. **32**, no. 4-5, pp. 641–650, 2006.

[344] N. L. Bobrov, V. N. Chernobay, Yu. G. Naidyuk, L. V. Tyutrina, I. K. Yanson, D. G. Naugle, K. D. D. Rathnayaka, Observation of anisotropic effect of antiferromagnetic ordering on the superconducting gap in $ErNi_2B_2C$, *Fizika Nizkikh Temperatur*, vol. **36**, no. 10-11, pp. 1228-1243, 2010.

[345] Yu. A. Nefyodov, A. M. Shuvaev, M. R. Trunin, Microwave response of V_3Si single crystals: Evidence for two-gap superconductivity, *Europhysics Letters*, vol. **72**, no. 4, pp. 638-644, 2005.

[346] V. Guritanu, W. Goldacker, F. Bouquet, Y. Wang, R. Lortz, G. Goll, A. Junod, Specific heat of Nb_3Sn: The case for a second energy gap, *Physical Review B*, vol. **70**, no. 18, ID 184526, 2004.

[347] M. Marz, G. Goll, W. Goldacker, R. Lortz, Second superconducting energy gap of Nb_3Sn observed by breakjunction point-contact spectroscopy, *Physical Review B*, vol. **82**, no. 2, ID 024507, 2010.

[348] J. Kačmarčík, Z. Pribulová, C. Marcenat, T. Klein, P. Rodière, L. Cario, P. Samuely, Specific heat measurements of a superconducting NbS_2 single crystal in an external magnetic field: Energy gap structure, *Physical Review B*, vol. **82**, no. 1, ID 014518, 2010.

[349] E. Boaknin, M. A. Tanatar, J. Paglione, D. Hawthorn, F. Ronning, R. W. Hill, M. Sutherland, L. Taillefer, J. Sonier, S. M. Hayden, J. W. Brill, Heat conduction in the vortex state of $NbSe_2$: Evidence for multiband superconductivity, *Physical Review Letters*, vol. **90**, no. 11, ID 117003, 2003.

[350] J. G. Rodrigo, S. Vieira, STM study of multiband superconductivity in $NbSe_2$ using a superconducting tip, *Physica C*, vol. **404**, no. 1-4, pp. 306-310, 2004.

[351] M. Zehetmayer, H. W. Weber, Experimental evidence for a two-band superconducting state of $NbSe_2$ single crystals, *Physical Review B*, vol. **82**, no. 1, ID 014524, 2010.

[352] C. L. Huang, J.-Y. Lin, Y. T. Chang, C. P. Sun, H. Y. Shen, C. C. Chou, H. Berger, T. K. Lee, H. D. Yang, Experimental evidence for a two-gap structure of superconducting $NbSe_2$: A specific-heat study in external magnetic fields, *Physical Review B*, vol. **76**, no. 21, ID 212504, 2007.

[353] N. E. Phillips, N. Oeschler, R. A. Fisher, J. E. Gordon, M.-L. Foo, R. J. Cava, Specific-heat of $Na_{0.35}CoO_2{\cdot}1.3H_2O$: Effects of sample age and pair breaking on two-gap superconductivity, *Physica C*, vol. **460-462**, pp. 473-474, 2007.

[354] Y. Nakajima, T. Nakagawa, T. Tamegai, H. Harima, Specific-heat evidence for two-gap superconductivity in the ternary-iron silicide $Lu_2Fe_3Si_5$, *Physical Review Letters*, vol. **100**, no. 15, ID 157001, 2008.

[355] R. T. Gordon, M. D. Vannette, C. Martin, Y. Nakajima, T. Tamegai, R. Prozorov, Two-gap superconductivity seen in penetration-depth measurements of $Lu_2Fe_3Si_5$ single crystals, *Physical Review B*, vol. **78**, no. 2, ID 024514, 2008.

[356] T. Tamegai, Y. Nakajima, T. Nakagawa, G. Li, and H. Harima, Two-gap superconductivity in $R_2Fe_3Si_5$ (R = Lu, Sc) and $Sc_5Ir_4Si_{10}$, *Science and Technology of Advanced Materials*, vol. **9**, no. 4, ID 044206, 2008.

[357] I. Bonalde, R. L. Ribeiro, W. Brämer-Escamilla, G. Mu, H. H. Wen, Possible two-gap behavior in noncentrosymmetric superconductor $Mg_{10}Ir_{19}B_{16}$: A penetration depth study, *Physical Review B*, vol. **79**, no. 5, ID 052506, 2009.

In: Superconductivity
Editor: Vladimir Rem Romanovskiĭ

ISBN 978-1-61324-843-0

Chapter 2

Macroscopic Dynamics and Stability of Thermo-Electrodynamics States Induced in High-Temperature Superconductors by Applied Current

V. R. Romanovskii

National Research Center Kurchatov Institute
NBIK- center, Moscow, Russia

Abstract

Macroscopic peculiarities of applied current penetration are theoretically investigated to understand the basic physical trends, which are characteristic for the stable formation of the thermo-electrodynamics states of high-T_c superconductors without stabilizing matrix placed in DC external magnetic field. The performed analysis shows that the definition of the current stability conditions of high-T_c superconductors must take into consideration the development of the interconnected thermal and electrical states when the temperature of superconductor may stably rise before instability from the temperature of coolant to its critical temperature. As a result, the thermal degradation effect of the current-carrying capacity of superconductor exists. The mechanisms of this effect are discussed. The boundary values of the electric field and the current above which the charged current is unstable are defined taking into account the size effect, cooling condition peculiarities, non-linear temperature dependences of the critical current density of superconductor and *n*-value of its *E-J* relation. It is proved that the allowable stable values of electric field and current can be both below and above those determined by a priori chosen critical current and electric field of the superconductor. The violation features of the steady current distribution in high-T_c superconductors cooled by liquid refrigerant (helium, hydrogen and nitrogen) are studied. The necessary criteria allowing one to determine the influence of the properties of superconductor and coolant on the current instability mechanism are written.

The peculiarity formation of possible resistive states of high-T_c superconductor are investigated in the static approximation considering the temperature-decreasing dependence of the power exponent (n-value) of its voltage-current characteristics.

1. Introduction

The stability investigations of current charging into low-T_c and high-T_c superconducting materials are one of main problems of applied superconductivity [1], [2]. The limiting current-carrying capacity of low-T_c superconductors, which have steep voltage-current characteristics, is described with a good degree of accuracy by their critical current [1], [2]. However, the finite voltage induced in high-T_c superconductors by the current charging appears long before the current instability onset. As known, this is associated with a broad shape of the voltage-current characteristics of high-T_c superconductors. Nevertheless, the voltage-current characteristics are widely used to determine the limiting current-carrying capacity of high-T_c superconductors as their basic property.

Typically, the voltage-current characteristics of high-T_c superconductors are investigated in the low electric field range close to 10^{-6} V/cm, which are characterized by small overheating of superconductors (ΔT~0.01-0.1K). But even in these cases, superconductors may be in a resistive state [3]. In this case, the voltage-current characteristic of superconductor is essentially non-linear, since the vortex lattice is gradually coming to the motion.

Resistive states of high-T_c superconductors exist in a wide current range. As it is proved experimentally and theoretically [4]-[8], they can be stable at higher currents and overheating than the corresponding values following from the fixed electric field criterion, for example, 10^{-6} V/cm. These resistive states significantly expand the range of permissible stable operating mode of high-T_c superconducting materials [9].

In the macroscopic approximation, the formulation of the current stability conditions must be based on the analysis of violation of thermal equilibrium of their electrodynamics states, as it was formulated for the first time in [10] for fully penetrated current states of low-T_c superconductors. The current-carrying capacities of high-T_c superconducting composites (superconductor + matrix) have been experimentally and theoretically investigated in various operating modes (see, for example, [4]-[8], [11]-[20] and references cited therein). However, the basic thermal formation features of stable and unstable current states of the high-T_c superconductors without a stabilizing matrix have still not been discussed. This study is important for both the characterization of the performances of high-T_c superconductors and the understanding of the macroscopic mechanisms limiting their current-carrying capacity taking into consideration their huge flux-creep states.

The heating issues of current-voltage characteristics of high-T_c superconductors are also important for the c-axis intrinsic Josephson junction geometries. There is a big discussion on the pseudogap in high-T_c superconductors inferred from the V-I curves during the intrinsic Josephson junction tunneling spectroscopy [19], [20]. The intrinsic tunneling spectroscopy of bismuth cuprate superconductors has received considerable attention. This technique utilizes intrinsic Josephson junctions, which are unique in probing bulk electronic properties, among other important methods such as scanning tunneling spectroscopy and the break-junction technique. However, intrinsic tunneling spectra can be seriously affected by self-heating due to poor thermal conductivity of the cuprate materials and relatively large current densities required to reach the gap voltage. Therefore, experimentally, intrinsic tunneling spectra are usually different from those obtained in scanning tunneling spectroscopy and break-junction experiments. As a result, it is important to eliminate the heating in order to obtain the genuine intrinsic tunneling spectra.

In general, thermo-electrodynamics analysis must be based on the solution of the multidimensional Fourier and Maxwell equations that allow one to take into account the properties of the superconductor, conditions of its cooling, etc. However, such numerical calculations are cumbersome and time-consuming because of the complexity of computation models, which, as a rule, are based on the use of the finite element method (see, for example, [14], [21], [22]). Therefore, in this approximation, the analyses of the basic physical peculiarities and specific formation of stable and unstable thermo-electrodynamics states are difficult. Simpler models are preferable for these investigations. They allow one to find an analytical solution of practically important problems, which enables to formulate appropriate stability criteria.

In this paper, the role of the heat transfer mechanisms in the formation of the macroscopic electrodynamics states of high-T_c superconductor superconductors without stabilizing matrix during current charging is discussed in detail. The proposed results were obtained from the solution of the zero-dimensional and one-dimensional static and transient equations describing the formation of the thermal and electrodynamics states under DC external magnetic field. These approximations permit:

- to study the non-isothermal voltage-current characteristic of a high-T_c superconductor;
- to discuss the thermal peculiarities of the formation of the voltage-current characteristic of high-T_c superconductor accounting for non-uniform temperature distribution in its cross section and to get the appropriate current instability conditions defining the maximum stable value of the charged current;
- to find some thermal features of violation of the stable current mode of high-T_c superconductors under different cooling regimes, and to formulate the criteria defining the possible mechanisms of current instability in high-T_c;
- to investigate the features of the resistive state formation of high-T_c superconductors when a power law exponent of the *E-J* relation decreases with increasing temperature.

As a whole, the study made allows to formulate the basic thermal features of the onset of current instability in high-T_c superconductor having no stabilizing matrix.

2. Thermo-Electrodynamics Models

Let us consider a superconductor with a slab geometry ($-a<x<a$, $-\infty<y<\infty$, $-b<z<b$, $b>>a$) placed in a constant external magnetic field parallel to its surface in the z-direction that is penetrated over its cross section ($S = 4ab$). Suppose that the applied current is charged in the y-direction increasing linearly from zero at the constant sweep rate dI/dt and its self magnetic field is negligibly lower than the external magnetic field [2]. Let us describe the voltage-current characteristic of the superconductor by a power law and approximate the dependence of the critical current on the temperature by the linear relationship. Assume also that the superconductor has the transverse size in the x-direction, which does not lead to the magnetic instability. As it was shown in [23], [24], the flux penetration phenomena during creep are characterized by a finite velocity. Therefore, taking into consideration the existence of the moving boundary of the current penetration region, the transient equations describing the evolution of the temperature and electric field inside the superconducting slab is independent of z and y coordinates and may be written as follows

$$C(T)\frac{\partial T}{\partial t}=\frac{\partial}{\partial x}\left(\lambda(T)\frac{\partial T}{\partial x}\right)+\begin{cases}0, & 0<x<x_p\\ EJ, & x_p<x<a\end{cases}, \tag{1}$$

$$\mu_0\frac{\partial J}{\partial t}=\frac{\partial^2 E}{\partial x^2},\quad t>0,\quad 0\le x_p<x<a. \tag{2}$$

Here, the electric field $E(x,t)$, current density $J(x,t)$ and critical current density $J_c(T,B)$ conform the following relationships

$$E=E_c\left[\frac{J}{J_c(T,B)}\right]^n, \tag{3}$$

$$J_c(T,B)=J_{c0}(B)\frac{T_{cB}(B)-T}{T_{cB}(B)-T_0}. \tag{4}$$

For the problem under consideration, the initial and boundary thermo-electrodynamics conditions are given by

$$T(x,0)=T_0,\quad E(x,0)=0, \tag{5}$$

$$\frac{\partial T}{\partial x}(0,t)=0,\ \lambda\frac{\partial T}{\partial x}(a,t)+h\left[T(a,t)-T_0\right]=0, \tag{6}$$

$$\begin{aligned}&E(x_p,t)=0,\quad x_p>0,\\ &\frac{\partial E}{\partial x}(0,t)=0,\quad x_p=0,\\ &\frac{\partial E}{\partial x}(a,t)=\frac{\mu_0}{4b}\frac{dI}{dt}.\end{aligned} \tag{7}$$

Here, C and λ are the specific heat capacity and thermal conductivity of the superconductor, respectively; h is the heat transfer coefficient; T_0 is the cooling bath temperature; n is the power law exponent of the E-J relation; E_c is the voltage criterion defining the critical current density of the superconductor; J_{c0} and T_{cB} are the known constants at the given external magnetic field B; x_p is the moving boundary of the current penetration area following from the integral relation

$$4b\int_{x_p}^{a} J(x,t)\,dx = \frac{dI}{dt}t \tag{8}$$

defining the conservation law of charged current [23], [24].

The approach (4) describing linear temperature dependence of the critical current of superconductor is reasonable for operating states when the temperature variation of a superconductor is not noticeable. However, the huge flux creep of high-T_c superconductors leads to the strong nonlinear temperature dependence of their critical currents in the high magnetic field. Therefore, the relationship (4) is not universal. In these cases, the various approximations are used [25]-[31] to describe the temperature-dependent critical current of high-T_c superconductors. In particular, this quantity may be described by equation

$$J_c(T,B) = J_0\left(1-\frac{T}{T_c}\right)^{\gamma}\left[(1-\chi)\frac{B_0}{B_0+B} + \chi\exp\left(-\frac{\beta B}{B_{c0}\exp(-\alpha T/T_c)}\right)\right] \tag{9}$$

proposed in [31]. It is good approximation for Bi-based superconductors. Figures 1 and 2 show the possible temperature variation of the critical current $I_c = \eta J_c S$ scaled for the $Bi_2Sr_2CaCu_2O_8$ (Bi2212) composite (η is the volume fraction of a superconductor in a composite). The calculations presented in Figure 1 were made at T = 4.2 K using the following constants

$$\begin{gathered} T_c = 87.1\text{ K},\quad B_{c0} = 465.5\text{ T}, \\ \alpha = 10.33,\quad \beta = 6.76,\quad \gamma = 1.73,\quad \chi = 0.27, \\ B_0 = 1\text{T},\quad J_0 = 5.9\times10^4\text{ A/cm}^2 \end{gathered} \tag{10}$$

which were reported for the Ag/Bi2212 superconductor in [31] and were adopted according to the measured data published in [32] for Bi2212 composite (S = 0.01862 cm^2, η = 0.2). The curves depicted in Figure 2 were calculated at

$$\begin{gathered} T_c = 87.1\text{ K},\quad B_{c0} = 465.5\text{ T}, \\ \alpha = 10.33,\quad \beta = 5,\quad \gamma = 1.73,\quad \chi = 0.2, \\ B_0 = 0.075\text{ T},\quad J_0 = 1.1\times10^6\text{ A/cm}^2 \end{gathered} \tag{11}$$

in accordance with [33] for the composite based on Bi2212 (S = 0.012 cm^2, η = 0.263).

The diffusion one-dimensional model defined by equations (1) – (8) may be simplified in the cases when the temperature distribution inside the superconductor is practically uniform. Under the considered slab geometry, the uniform temperature distribution exists when the condition $ha/\lambda \ll 1$ takes place. (Below this condition will be strictly proved). Then integrating equation (1) with respect to x from 0 to a and considering the boundary conditions (6), it is easy to get the following transient heat balance equation

$$C(T)\frac{dT}{dt} = -\frac{h}{a}(T - T_0) + E(t)J(t). \tag{12}$$

This equation with the relations (3) and (4) describes the uniform time variation of the temperature and electric field as a function of the induced current $I(t) = J(t)S = dI/dt \times t$.

The limiting transition at $dI/dt \to 0$, applying to the model described by equation (12), leads to the static zero-dimensional heat balance equation [6]-[8]. In this case, i. e. at $ha/\lambda <<$ 1, uniform static thermal state of the superconducting slab is the solution of the following equation

$$EJ = h(T - T_0)/a. \tag{13}$$

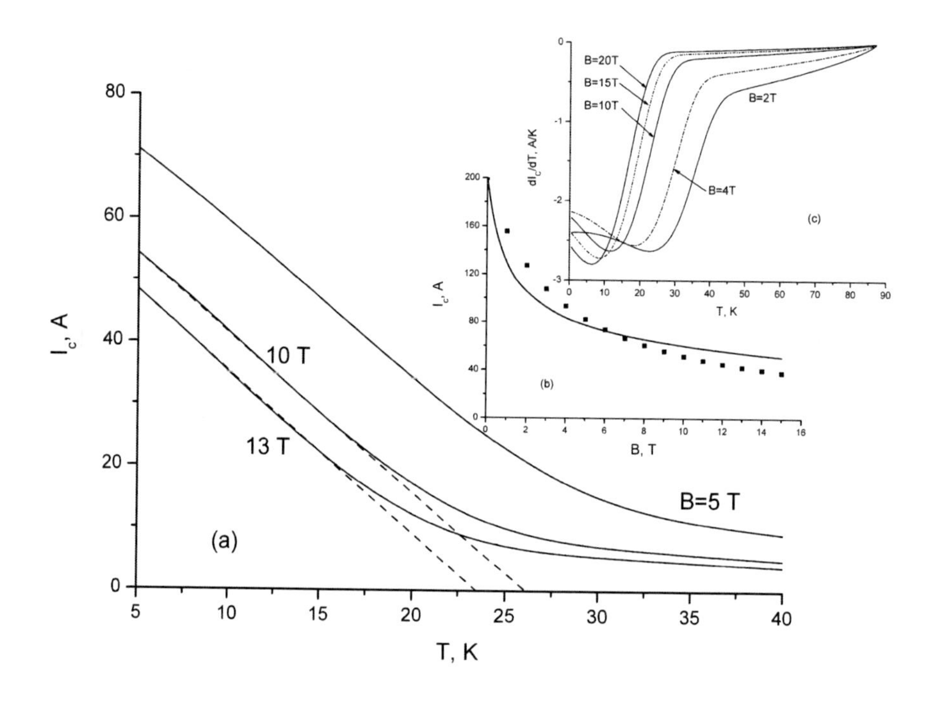

Figure 1. Critical current of Ag/Bi2212 versus temperature (a) and magnetic induction (b): (▪) – after [29], (——) – calculations according to (9), (- - - -) – linear approximation; (c) – temperature dependence of dI_c/dT.

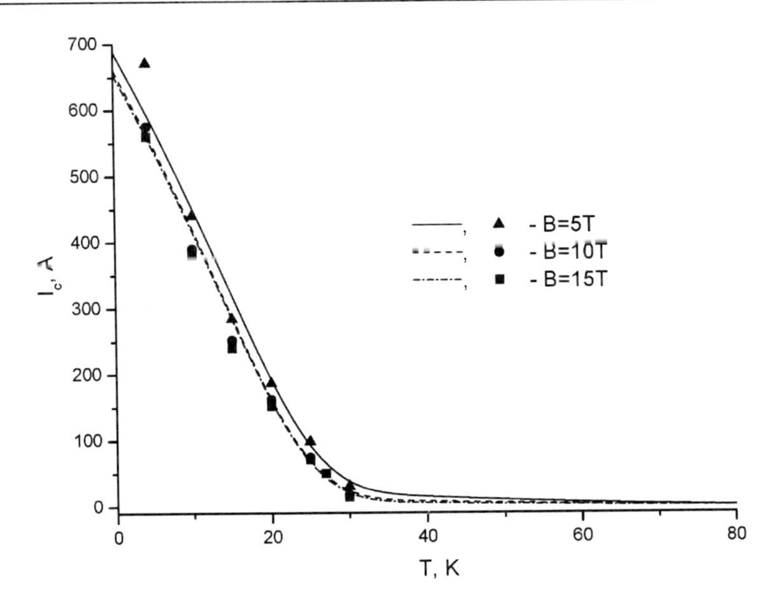

Figure 2. Critical current of Ag/Bi2212 versus temperature: markers – after [33], (——), (- - - -), (– · – · – · –) – calculations according to (9).

It is easy to find that the temperature-voltage and current-voltage characteristics of a superconducting slab in the framework of the static zero-dimensional approximation described by relations (3), (4), (13) are given by

$$T = T_0 + \frac{aJ_{c0}E}{h\left(1+\dfrac{J_{c0}}{E_c^{1/n}}\dfrac{a}{h\left(T_{cB}-T_0\right)}E^{\frac{n+1}{n}}\right)}\left(\frac{E}{E_c}\right)^{1/n},$$

$$J = \frac{J_{c0}}{1+\dfrac{J_{c0}}{E_c^{1/n}}\dfrac{a}{h\left(T_{cB}-T_0\right)}E^{\frac{n+1}{n}}}\left(\frac{E}{E_c}\right)^{1/n}. \tag{14}$$

Besides, the development of the incomplete penetration modes may be also estimated using the scaling model proposed in [23], if the temperature of the superconductor during this mode is close to the cooling bath temperature. Then the electric field distribution inside the superconducting slab and the evolution of the moving current penetration boundary may be described by the following approximate formulae

$$E(x,t)=\begin{cases}0, & 0\le x\le x_p\\ \dfrac{\mu_0}{4b}\dfrac{dI}{dt}\left(x-x_p\right), & x_p\le x\le a\end{cases}$$

$$x_p(t)=a-\left(\frac{n+1}{n}t\right)^{\frac{n}{n+1}}\left(\frac{\mu_0}{4b}\frac{dI}{dt}\right)^{\frac{n-1}{n+1}}\left[\frac{E_c}{\mu_0^n J_{c0}^n}\right]^{\frac{1}{n+1}}, \qquad (15)$$

$$t\le t_f=\frac{n}{n+1}\frac{\mu_0 J_{c0}}{E_c^{1/n}}a^{(n+1)/n}\left(\frac{\mu_0}{4b}\frac{dI}{dt}\right)^{(1-n)/n}$$

in the weak flux-creep regimes ($n \ge 10$) and the temperature dependence (4).

The models formulated allow one to investigate the role of the thermal mechanisms in the macroscopic formation of the electrodynamics states of high-T_c superconductors during current charging and to find the corresponding stability criteria. Note that below-made analysis was performed in the frameworks both of simple approximations based on the postulation, which is assumed that n=const and J_c has the linear temperature dependence, and under more general assumptions when the quantities n and J_c versus temperature are non-linear functions. It allowed to formulate the basic features of the thermo-electrodynamics state formation. They must be taken into account when the operating stability conditions of the superconductors are investigated.

3. Potential Thermo-Electrodynamic States of High-T_C Superconductors

To appreciate the temperature influence on the electric field dynamics inside superconductor, let us reduced the set of equations (1) – (3), eliminating the current density. Omitting the intermediate algebra, it is easy to get the following equation

$$\frac{\mu_0 J_c}{nE}\left(\frac{E}{E_c}\right)^{1/n}\frac{\partial E}{\partial t}=\frac{\partial^2 E}{\partial x^2}-\frac{\mu_0}{C(T)}\frac{dJ_c}{dT}\left(\frac{E}{E_c}\right)^{1/n}\left[EJ_c\left(\frac{E}{E_c}\right)^{1/n}+\frac{\partial}{\partial x}\left(\lambda\frac{\partial T}{\partial x}\right)\right]$$

which is valid in the region $0 \le x_p(t) < x < a$. Note that the existence only the temperature dependence of J_c was assumed during this transformation.

The term in left part and the first term in right part of this equation describe the electric field evolution in the isothermal approximation. In this case, the inequality $\partial^2 E/\partial x^2 > 0$ always takes place in accordance with equation (2) when $\partial J_c/\partial B$ is negligible. Besides, as it follows from (15), the isothermal sweep dynamics of the electric field on the surface of superconductor satisfy the estimator $E(a,t) \sim t^{n/(n+1)}$. Therefore, during these regimes the next inequalities $\partial E(a,t)/\partial t > 0$ and $\partial^2 E(a,t)/\partial t^2 < 0$ will exist when electric field is small ($E << E_c$) and temperature distribution is uniform. However, the temperature rise of the superconductor changes these regularities. This effect depends on the third term in right part of equation written. Using equation (3), it is easy to prove that the finite values dJ_c/dT, which are negative for practicable superconductors, will lead to other relations when the magnetic field inside superconductor and induced temperature become noticeably differ from their initial values. Namely, one can obtain $\partial E(a,t)/\partial t > 0$ and $\partial^2 E(a,t)/\partial t^2 > 0$. In other words, the dynamics of the electric field becomes more intensive when the critical current density decreases with temperature. Physically, this behavior takes place because the flux diffusion is accompanied with the energy dissipation stored by the applied current. Indeed, the temperature rise increases the electric field and changes the density of the applied current. Then the magnetic flux penetrates more deeply into the superconductor. In turn, this will lead to new dissipation producing the corresponding rise of temperature and so on. As a result, the sharp rise of the electric field and temperature may occur when the current penetration front exceeds some boundary that is known as the stability boundary [1], [2].

To demonstrate this feature stricter, let us rewrite the latter equation in the form

$$\frac{1}{D_m}\frac{\partial E}{\partial t} = \frac{\partial^2 E}{\partial x^2} + \gamma E + q,$$

Where

$$D_m = \frac{nE}{\mu_0 J_c(T)}\left(\frac{E_c}{E}\right)^{1/n},$$

$$\gamma = \frac{\beta(T)}{a^2}\left(\frac{E}{E_c}\right)^{2/n}, \quad q = \Lambda(T)\tau(T)\left(\frac{E}{E_c}\right)^{1/n}$$

Here, D_m is the magnetic diffusion coefficient,

$$\beta(T) = \frac{\mu_0 J_c(T) a^2}{C(T)}\left|\frac{\partial J_c}{\partial T}\right|$$

is the magnetic stability parameter [1], [2],

$$\Lambda(T) = \mu_0 \frac{\lambda(T)}{C(T)}\left|\frac{\partial J_c}{\partial T}\right|,$$

$$\tau(T) = \frac{1}{\lambda(T)}\frac{d\lambda}{dT}\left(\frac{\partial T}{\partial x}\right)^2 + \frac{\partial^2 T}{\partial x^2}.$$

This is diffusion type equation describing the fission-chain-reaction phenomena, in which the quantity γ is the like-fission coefficient and q is the volume source. Since the fission coefficient is positive due to the negative value of dJ_c/dT, then the electric field evolution induced by applied current has the fission-chain-reaction nature. As known, such processes become an irreversible when γ exceeds some value γ_q in accordance with the fission quality obeying the time-dependent law depending on $exp(\gamma - \gamma_q)t$-terms. Under this condition, the electric field evolution in superconductor has the instability nature leading to the avalanche modes. The latter happens as the spontaneous rise of the electric field and temperature. However, there exist the peculiarities that underlie the basis of the instability phenomena in superconductors.

First, the quantity D_m increases with increasing n-value. Therefore, the instability develops more intensively when the voltage-current characteristic of superconductor is steeper. As a result, instability has sharp character in low-T_c superconductors and gets more smoothed nature in high-T_c superconductors. Second, the quantity γ has essential temperature dependence. First of all, it varies in inverse proportion to the heat capacity of a superconductor. So, the fission-chain evolution of the induced electric field in high-T_c superconductors may decay after a certain temperature increase. In other words, high-T_c superconductors have the thermal self-stabilization modes even after instability onset. So, the swept electric field may come close to the steady modes without the transition of the high-T_c superconductor to the normal state due to its large temperature margin. There exist other reasons of thermal effect on the electric field evolution. They are due to the specified temperature variation of the critical current density and the value dJ_c/dT. Indeed, Figure 1 shows that not only J_c may be small but the dJ_c/dT-values become very small in accordance with the huge flux-creep degradation in high temperature range, which, however, is not close to the critical temperature. Thereby, the temperature effect on the electric field essentially depends on the quantity dJ_c/dT after relatively high temperature rise of superconductor. Besides, the effect of the temperature on the current diffusion depends also on the value and sign of $\partial^2 T/\partial x^2$-term. The latter may be both positive (generation mode) and negative (absorption mode) over the cross-section of a superconductor. This depends on the cooling conditions, cross section of superconductor, its thermal conductivity and critical parameters. Therefore, if this quantity is positive, the diffusion character of the electric field is more unstable. On the contrary, the temperature distribution will render stabilizing support on the electric field diffusion at $\partial^2 T/\partial x^2 < 0$.

Thus, there exist the thermal effects according to which the electrodynamics states of high-T_c superconductors become more stable in the high temperature range, which is not close to the critical one. In other words, the possible temperature rise of high-T_c superconductors may lead to their stable states. For example, in the case of $Bi_2Sr_2CaCu_2O_8$ superconductor under consideration, similar states may exist in the temperature range close to about 30 K at high applied magnetic fields, as follows from Figure 1. As a whole, the existence of the corresponding mode follows from the joint non-trivial temperature variation of the properties of high-T_c superconductor, which are $C(T)$, $\lambda(T)$, $J_c(T)$ and dJ_c/dT. The importance of this conclusion should be emphasized. To understand the stability mechanism of high-T_c superconductors, one usually allows for only the heat capacity of high-T_c superconductor believing that it just decreases the probability of the instability onset. In the meantime, it appears that the probability of the instability onset in high-T_c superconductor not only decreases with increasing temperature but instability may be absent due to the temperature dependences of above-mentioned properties of high-T_c superconductors. As a result of this temperature effect, the fission-chain-reaction character of the electromagnetic state evolution in high-T_c superconductor is absent in the high temperature range and its swept modes may become stable.

4. Dependence of the Voltage-Current Characteristic of High-T_C Superconductor on Heat Capacity and Features of Stability Conditions of Charged Current

As it is well known, the critical properties of Bi-based superconductors essentially exceed critical properties of low-T_c superconductors in high magnetic fields. This opens up many kinds of applications of these superconductors. For example, a conduction-cooled Bi2212-magnet is one of possible new generations of a high field magnet system [34]-[36]. In this connection, let us investigate the physical features of stable and unstable formation of thermo-electrodynamics states in the current charging into a Bi2212-superconductor under non-intensive cooling conditions, which take place in conduction-cooled magnets [36].

The time variations of the current penetration depth x_p and electric field on the surface of the slab are shown in Figure 3. The results presented were based on the numerical solution of the problem defined by equations (1) – (8). For convenience of the performed analysis, it was done using the current normalized by the slab width ($I^* = 0.5I/b$). In this case, normalized current sweep rate $dI^*/dt = 0.5b^{-1}dI/dt$ was single variable quantity. The simulation was made for a $Bi_2Sr_2CaCu_2O_8$ superconducting slab ($a = 5 \times 10^{-2}$ cm) initially cooled at $B = 10$ T, $T_0 = 4.2$ K for two values of dI^*/dt. The following constants $h = 10^{-3}$ W/(cm^2 ×K), $E_c = 10^{-6}$ V/cm, $n = 10$ were used. According to the linear fitting presented in Figure 1, the critical parameters of the superconductor in the framework of approximation (4) are equal to

$$T_{cB} = 26.1 \text{ K}, \quad J_{c0} = 1.52 \times 10^4 \text{ A/cm}^2 . \tag{16}$$

The specific heat capacity and thermal conductivity of the superconductor were defined as follows

$$C\left[\frac{J}{cm^3 \cdot K}\right] = 10^{-6} \times \begin{cases} 58.5T + 22T^3, & T \le 10K \\ -10.54\times10^4 + 1.28\times10^4 T, & T > 10K \end{cases}, \tag{17}$$

$$\lambda(T)\left[\frac{W}{cm \cdot K}\right] = \left(-1.234\times10^{-5} + 1.654\times10^{-4}T \right.$$
$$\left. +4.608\times10^{-6}T^2 - 1.127\times10^{-7}T^3 + 6.061\times10^{-10}T^4\right) \tag{18}$$

according to [37], [38], respectively.

As it follows from Figure 4(a), the temperature variation of the superconductor under consideration has a minor effect on the dynamics of electrodynamics states in the stage of partially penetrated mode. Therefore, it is easy to find the electric field on the surface of the superconductor when its cross section is completely filled with current. According to the scaling approximation (15), this value is given by $E_f = \frac{a\mu_0}{4b}\frac{dI}{dt}$.

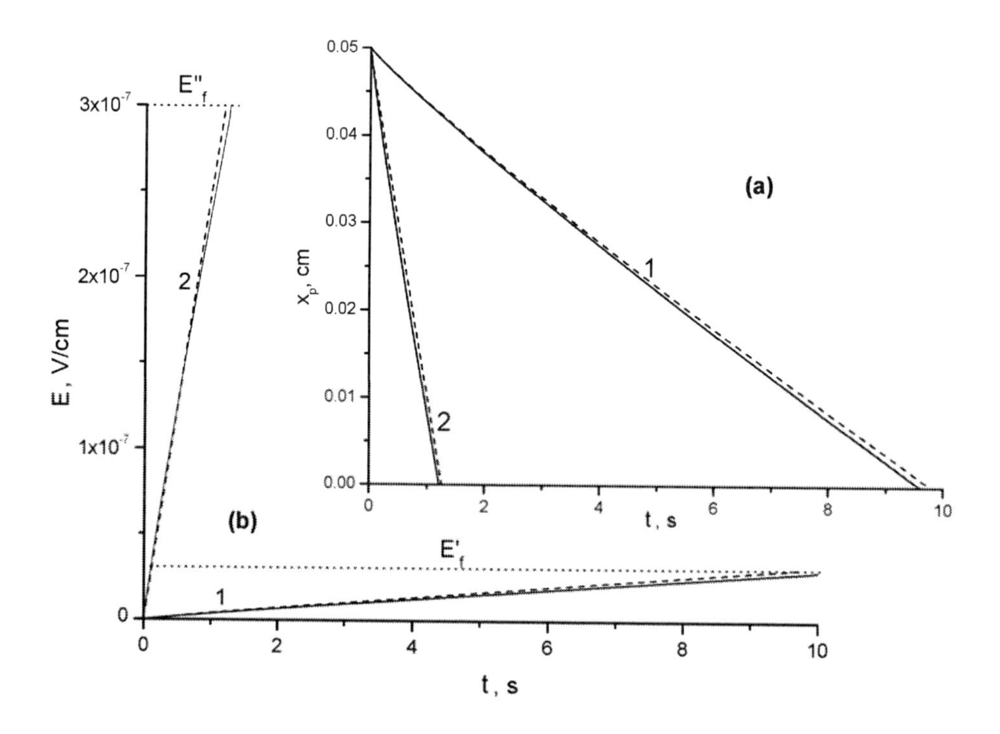

Figure 3. Time variation of the current penetration depth (a) and electric field on the surface of the slab (b) calculated using model (1)–(8) (solid lines) and model (15) (dashed lines): 1 – dI^*/dt = 10^2A/(s × cm); 2 – dI^*/dt = 10^3A/(s × cm). Values E_f and E'_f are the electric field on the superconductor's surface calculated in the scaling approximation for given values of dI^*/dt.

The corresponding values are shown in Figure 3 for each current charging rate. It is seen that the scaling approximation describes the partially penetrated states with satisfactory accuracy. Using it, let us estimate the influence of the current charging rate on the non-uniform redistribution character of the electric field inside superconductor during initial stage of the fully penetrated mode ($I > I_f = t_f dI/dt$). Curves 2, 2', 3 and 3' in Figure 4 depict the evolution of electric field and temperature in the superconductor during both partially and fully penetrated states on the surface of a slab (2, 3, 4) and in the center (2', 3'). Curve 1 corresponds to a static uniform distribution induced in the slab by infinitely slow current charging and follows from approximation (14). Comparing curves 1-3 in the sub-critical electric fields ($E < E_c$), it is easy to understand that the electrodynamics state formation will be practically uniform in the fully penetrated mode when $E_f << E_c$, i.e. at $dI/dt << 4bE_c/(a\mu_0)$. As a whole, the curves in Figure 4 indicate the existence of the characteristic dynamics of the fully penetrated states that are proper for the stable and unstable current charging modes of high-T_c superconductors and depend on dI/dt.

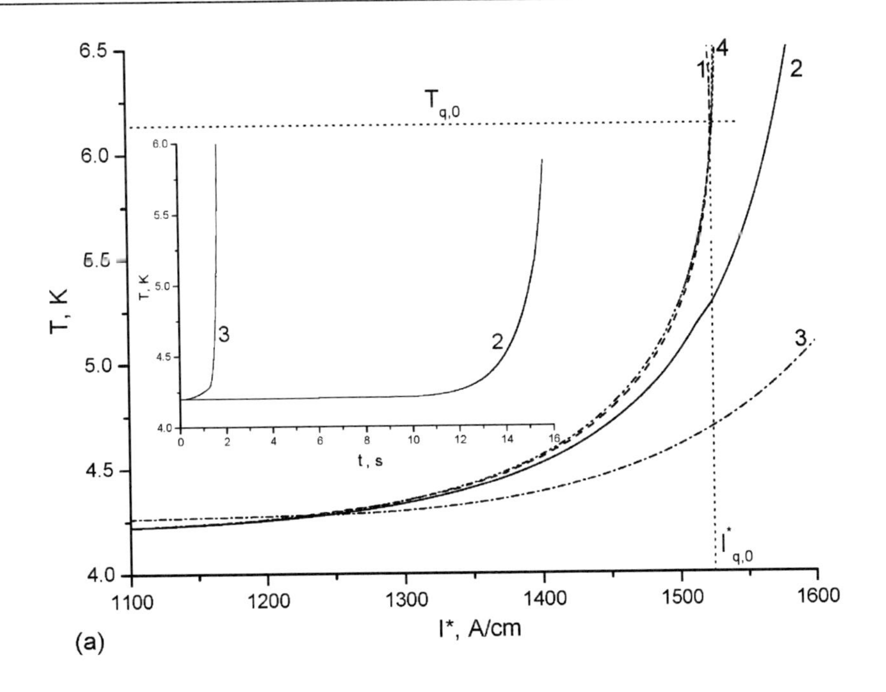

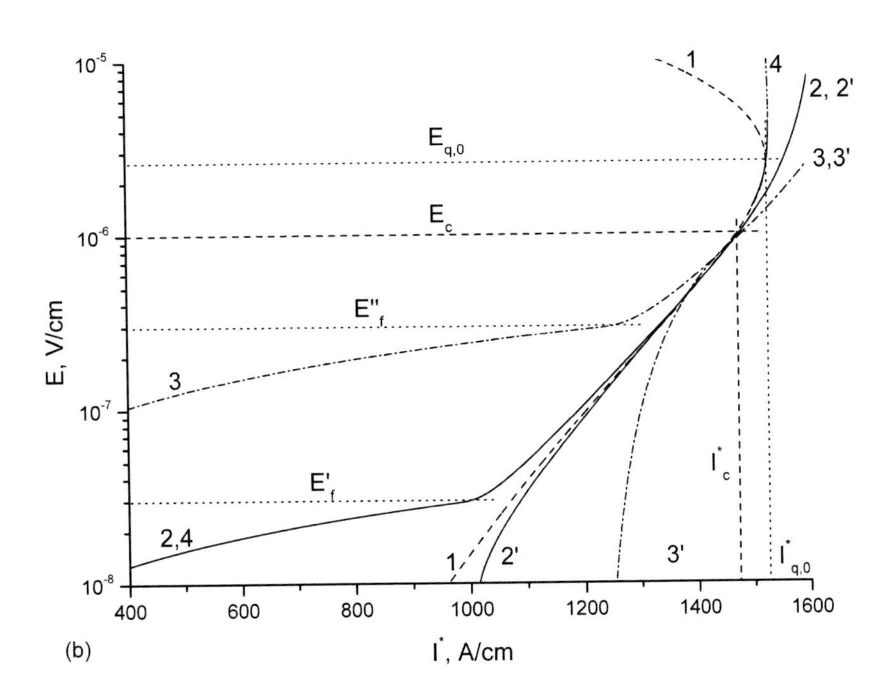

Figure 4. Sweep rate dependence of the temperature-current (a) and voltage-current (b) characteristics of a high-T_c superconductor: 1 – $dI^*/dt \to 0$; 2, 2' – $dI^*/dt = 10^2$ A/(s × cm); 3, 3' – $dI^*/dt = 10^3$ A/(s × cm), 4 – $dI^*/dt = 10^2$ A/(s × cm), $C = C(T_0)$.

First, there exists the transient period after which the fully penetrated electric fields on the surface of the slab and in its centre become practically equal. To estimate the influence of the sweep rate on the time during which the non-uniform distribution of the electric field exists let us use the estimator

$$\partial E/\partial t \sim \left(E/a^2\right)\partial E/\partial J \tag{19}$$

following from equation (2). It shows that the redistribution of the electric field in the cross-section of the slab becomes more intensive with increasing differential resistivity of a superconductor $\partial E/\partial J$ and value of the induced electric field. Figure 5 demonstrates the corresponding variation of the $\partial E/\partial J$. Taking into consideration the results presented in Figure 5 and evident fact that induced electric field increases with increasing dI/dt (Figure 4), it is easy to understand that the redistribution time window between non-uniform and practically uniform fully penetrated states will decrease with increasing sweep rate, as it is depicted in Figure 4.

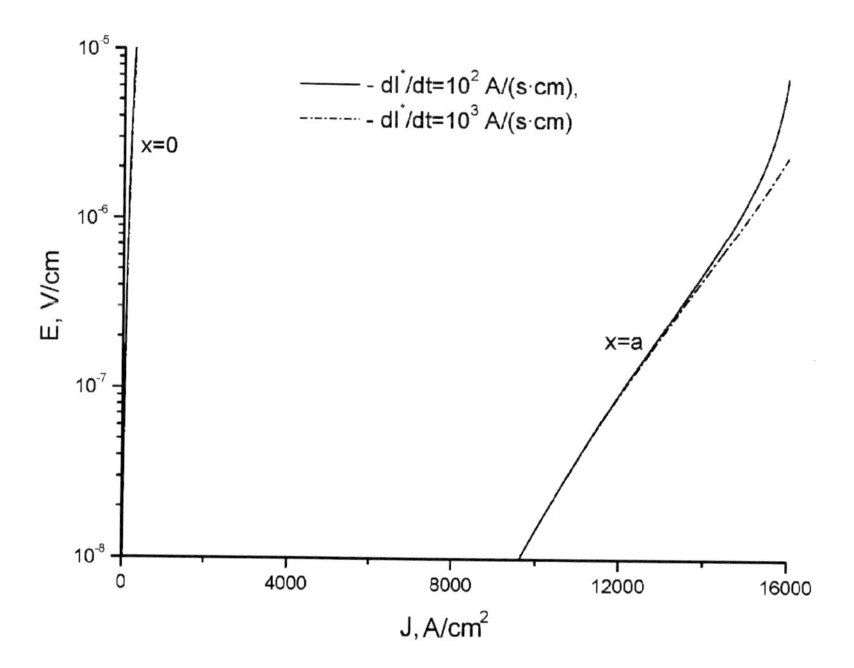

Figure 5. Electric field in the centre and on the surface of a slab as a function of the current density.

Second, Figure 4 also reveals that in the over-critical electric field region ($E > E_c$) the fully penetrated dependences $E(I^*, dI^*/dt)$ and $T(I^*, dI^*/dt)$ that follows from 1D unsteady model (1) – (8) do not identically with the corresponding static dependences defined by static model (13, 14). Moreover, the transient differential resistivity of the superconductor not only decreases with increasing dI^*/dt but has just positive values. Consequently, the transient voltage-current characteristic of a high-T_c superconductor does not allow to find the boundary of the current instability onset in the continuous current charging experiments. Really, in the static zero-dimensional approximation, the limiting current-carrying capacity is defined by the condition

$$\partial E/\partial J \to \infty . \tag{20}$$

This condition was formulated for the first time in [10] for low-temperature superconductors cooled by a coolant with the constant coefficient of heat transfer. This cooling mode is characteristic of the conduction-cooled or gas-cooled superconductors. The condition (20) exists because differential resistivity of a superconductor can have infinitely large value, if the crisis of the boiling mode is absent. It corresponds to the boundary between stable ($\partial E/\partial J > 0$, $\partial T/\partial J > 0$) and unstable ($\partial E/\partial J < 0$, $\partial T/\partial J < 0$) branches on the voltage-current and temperature current characteristics when they are calculated in a static approximation assuming uniform distribution of temperature and current in the cross-section of superconductor. As a result, there are the so-called quench parameters above which no stable states can be found. The corresponding quench quantities of the electric field $E_{q,0}$ and current per width $I^*_{q,0}$ defined according to the (14) and (20) are depicted in Figure 4 by the dotted lines. They are equal to

$$E_{q,0} = \left[\frac{E_c^{1/n}}{nJ_{c0}}\frac{h}{a}\left(T_{cB} - T_0\right)\right]^{\frac{n}{n+1}}, \qquad I^*_{q,0} = \frac{2an}{n+1}\left[\frac{J^n_{c0}}{nE_c}\frac{h}{a}\left(T_{cB} - T_0\right)\right]^{\frac{1}{n+1}}. \tag{21}$$

Note that the calculated boundary of stability is located in the over-critical electric field region for the superconductor under consideration. The conditions describing the existence of the stable over-critical static states are formulated below.

As follows from Figure 4, the condition (20) is not observed in the unsteady thermo-current states as the transient voltage-current characteristic of a high-T_c superconductor has only positive slope. To explain this peculiarity, let us use the transient zero-dimensional approximation. Assuming that the critical current of a superconductor does not depend on the magnetic field, it is easy to get

$$\frac{dE}{dJ} = \frac{1 + \dfrac{dT}{dJ}\left|\dfrac{dJ_c}{dT}\right|\left(\dfrac{E}{E_c}\right)^{1/n}}{\dfrac{J_c(T)}{nE}\left(\dfrac{E}{E_c}\right)^{1/n}} \tag{22}$$

using the voltage-current characteristic of superconductor described by equation (3). To find the term dT/dJ, let us utilize equation (12) taking into consideration that

$$\frac{dT}{dt} = \frac{1}{S}\frac{dT}{dJ}\frac{dI}{dt}.$$

Then unsteady dependence dT/dJ is given by

$$\frac{dT}{dJ} = \frac{EJ - \left(T - T_0\right)h/a}{C(T)\, {}^{dI}\!/_{dt}}S. \tag{23}$$

Since $dT/dt > 0$ during current charging then $dT/dJ > 0$. Therefore, as follows from equations (22) and (23), the differential resistivity of a superconductor is always positive in the monotonously increasing current charging and the slope of the unsteady $E(I^*, dI^*/dt)$ and $T(I^*, dI^*/dt)$ dependences will become smaller when the heat capacity of a superconductor is higher, i.e., will decrease with increasing temperature of a superconductor during both stable and unstable states. At the same time, as follows from Figure 4(a), there exists stable temperature rise of a superconductor before current instability. This unavoidable temperature variation of a superconductor will change its heat capacity and the discussed effect of heat capacity of superconductor on its E-J relation is unavoidable during current charging modes.

To illustrate the role of the temperature dependence of heat capacity of the superconductor in the formation of its thermo-electrodynamics states, Figure 4 also presents the dependencies $E(I^*)$ and $T(I^*)$ obtained under the assumption that the heat capacity of the superconductor calculated at the temperature of the refrigerant does not change in the temperature (curve 4). Besides, Figure 4(a) clearly depicts that the temperature of the superconductor before the current instability onset is not equal to the temperature of the refrigerant. In particular, allowable temperature rise of a superconductor before current instability onset depends on the current charging rate: the higher the dI/dt, the higher the stable overheating of a superconductor, as shown in the inset of Figure 4(a), and, hence, the effect on the heat capacity of a superconductor on the voltage-current and temperature-current characteristics of a superconductor becomes more noticeable. The latter, as it was already discussed, will grow only with a positive slope in the continuous current charging. These typical dependences are indicated in Figure 4.

The results obtained demonstrate the reason according to which the unsteady voltage-current characteristics of high-T_c superconductor do not allow to find the boundary of the current instability onset. To avoid this complexity, the current charging with break (method of the fixed current) is used in the experiments [4], [5], [11]-[13], [15], [18]. According to this method, there are two final states: temperature and voltage either are stabilized or begin to grow spontaneously. In the first case, the stable current distribution is a direct consequence of the existence of stable steady states that underlie the formation of a stable branch of the voltage-current characteristic of a superconductor. But if the charged current exceeds the corresponding quench current $I^*_{q,0}$ then the formed states are unsteady even in spite of the fact that the corresponding values of the electric field and the temperature on the non-stationary voltage-current and temperature-current characteristics of a superconductor will be below than the boundary values $E_{q,0}$ and $T_{q,0}$. The latter in the static approximation are described by (14) and equal to

$$T_{q,0} = T_0 + E_{q,0} J_{q,0} a / h . \tag{24}$$

Physically, the method of the fixed current is based on the existence of the static stable thermo-current states that precedes instability onset. Therefore, all currents in Figure 4 exceeding the relevant value I^*_q correspond to unstable currents. Note that this method of determination of the boundary of the current instability, for the first time was proposed in [39], [40] analyzing the conditions of the current instability of the low-T_c superconductors.

To prove these peculiarities, the evolution of the temperature of the superconductor at different current charging rate is shown in Figures 6 and 7. The calculations were carried out both in the continuous current charging and during current charging to a fixed value I^*_0 after which dI/dt is equal to zero. Here, the corresponding value of $T_{q,0}$ that is the same as in Figure 4(a) is also given. Figure 6 also presents the results of calculations performed at a constant value of heat capacity of a superconductor and Figure 7 demonstrates the existence of almost uniform temperature distribution in the cross section of the superconductor during the completely current charging process. This thermal effect is observed due to the condition $ha/\lambda << 1$, which is performed for given initial parameters. Figures 6 and 7 clearly depicts that the static value $T_{q,0}$ is the temperature boundary between stable and unstable states during fully penetrated modes and when $ha/\lambda << 1$ regardless of the current charging rate and the temperature dependence of heat capacity of a superconductor. Therefore, if the temperature of the superconductor does not exceed the value $T_{q,0}$ in the transition process, which occurs after fixing the charged current, then this current is stable. Otherwise, the charged current is unstable and the current instability is accompanied by spontaneous heating of the superconductor. This conclusion defines thermal reason of a current instability onset.

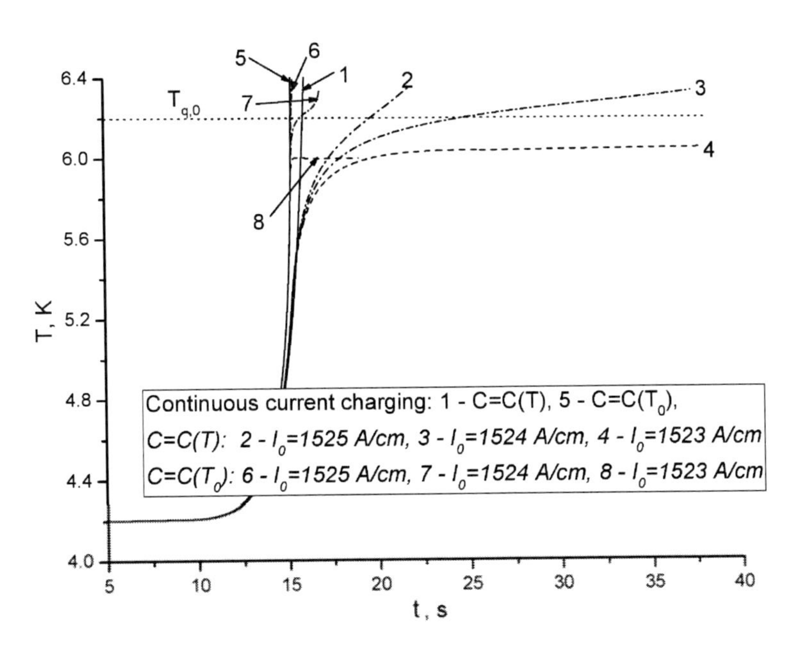

Figure 6. Time variation of the surface temperature of superconductor during continuous charging and current charging with break at various temperature dependences of heat capacity.

Note the feature that is characteristic of modern high-T_c superconductors. The results, presented above and discussed below, show that the current instability occurs during fully penetrated modes. In this case, the instability onset does not depend on the current charging rate. At the same time, the current instability onset in the low-T_c superconductors can also occur in the partially penetrated modes [2], [39], [40]. This is due to differences in critical properties of low-temperature and high-temperature superconductors. It is not difficult to write the condition under which the current instability onset in high-T_c superconductors will occur in the fully penetrated mode. Obviously, this would be in the case when $E_{q,0} > E_f$, i.e. at

$$\left[\frac{E_c^{1/n}}{nJ_{c0}}\frac{h}{a}\left(T_{cB}-T_0\right)\right]^{\frac{n}{n+1}} > \frac{a\mu_0}{4b}\frac{dI}{dt}. \quad (25)$$

If this condition is not valid then the charged current will depend on the sweep rate [2], [39], [40]. In this case, the permissible overheating of the superconductor before the instability may essentially differ from $T_{q,0}$ and increases with increasing dI/dt [39], [40]. The condition (25) shows that, in general, the current instability onset depends on dJ_c/dT as $dJ_c/dT \sim J_{c0}/(T_{cB} - T_0)$ both in the partially and fully penetrated modes. Therefore, the probability of the current instability onset in the partially penetrated mode will increase with increasing critical current density of high-T_c superconductors. In other words, there may be such current charging modes when the current-carrying properties of the high-T_c superconductor with high critical current density will be not fully used due to the instability of the charged currents that may be in the partially penetrated mode. In addition, the partially or fully penetrated regimes of instability onset depend on the n-value of voltage-current characteristic of a superconductor. In particular, the probability of unstable states in the partially penetrated mode decreases with decreasing n. That is why their existence is not observed in the modern high-T_c superconductors.

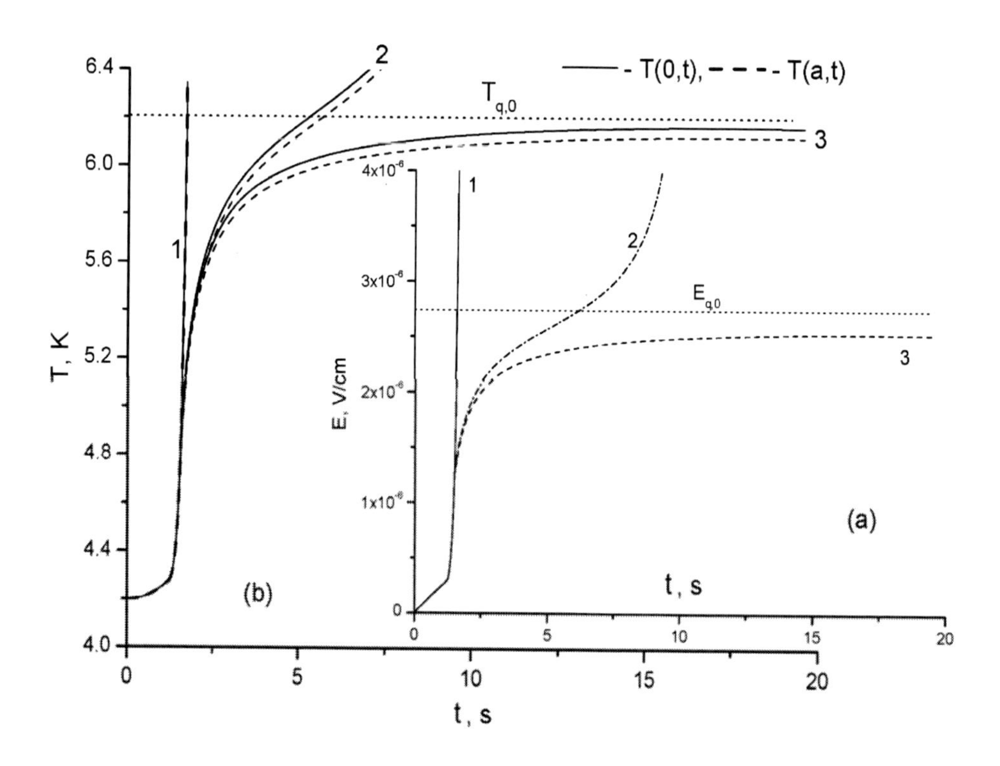

Figure 7. Time variation of the electric field (a) on the surface of superconductor and temperature (b) during continuous charging at $dI^*/dt = 10^3$A/(s × cm) (curve 1) and current charging with break: curve 2 – I^*_0= 1524 A/cm; curve 3 – I^*_0= 1523 A/cm.

Besides the conditions (20) and general method of the fixed current, one can use another method of the determination of current instability onset if the temperature distribution in the cross-section of the superconductor is uniform. The boundary of stable current mode may be defined analyzing the removal efficiency of the Joule heating into the refrigerant [10, 18]. Accordingly, the stable regimes must satisfy the following relations

$$G(T) = W(T), \quad \partial G(T)/\partial T = \partial W(T)/\partial T \tag{26}$$

in the stationary homogeneous states. Here, $G = EJa$ is the power of the Joule heating, $W = h(T - T_0)$ is the power of the heat removal. To illustrate the legitimacy of the conditions (26), the typical change in temperature of G (curves 1 – 3) and W (curve 4) is presented in Figure 8. These traces are observed in experiments [32]. The calculations were carried out under different conditions of current charging: continuous (curve 1) and with interruption (curves 2 and 3). Here, the curve 2 corresponds to the unstable current state and the curve 3 describes stable state. These dependencies were obtained by numerical solution of the problem (1) – (8). The results depicted show that the existence of the stable states, which satisfy the condition (26), are possible if the Joule heating does not lead to an increase in temperature of the superconductor above a certain limiting value after the termination of the current charging. To find this temperature, an appropriate solution of the stationary heat diffusion equation must be defined. In the case under consideration, which is characterized by almost uniform thermo-electrodynamics states, the corresponding boundary value of $G_{q,0} = E_{q,0}J_{q,0}a$ is related to the temperature $T_{q,0}$, which follows from relations (13), (21) and (24). Accordingly, the corresponding interrelation between them has the form

$$T_{q,0} = T_0 + \frac{1}{h} G_{q,0},$$

where

$$G_{q,0} = \frac{h}{n+1}\left(T_{cB} - T_0\right).$$

Therefore, $T_{q,0}$, shown both in Figure 4 and Figures 6 – 8, is the same. As a result, if the amount of the generated Joule heating in the superconductor exceeds a value $G_{q,0}$ in the continuous current charging, then these states will be obviously unstable. So, the determination of the boundary of the stable current states can be done in the framework of various equivalent models based, firstly, on an analysis of the changes in the differential resistance of a superconductor taking into account condition (20), and, secondly, on the analysis of admissible heat release, which occurs in a superconductor in the current charging considering the condition (26).

Thus, the heat capacity of high-T_c superconductor plays essential role in the formation of its voltage-current and temperature-current characteristics. It appears that it is impossible to see in an experiment a steep transition from the stable state to the unstable one even at helium temperature level.

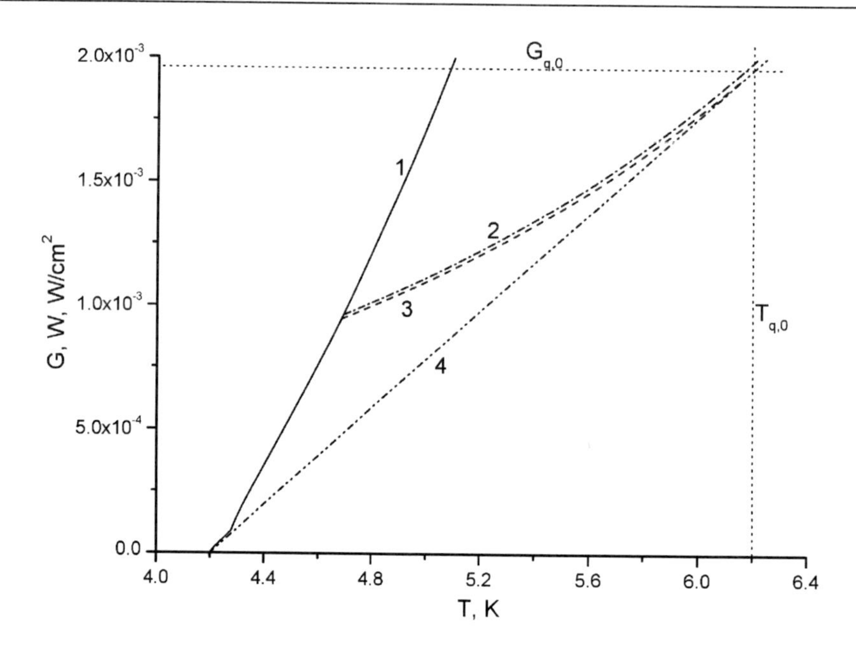

Figure 8. Heat generation (curves 1, 2, 3) and cooling power (curve 4) versus temperature at dI^*/dt = 10^3A/(s×cm): continuous current charging – 1, current charging with break: 2 – I^*_0= 1524 A/cm, 3 – I^*_0 = 1523 A/cm.

Strictly speaking, this is possible only at very low current charging rates. This is due to the allowable increase in temperature of the high-T_c superconductors before the current instability onset. Note that the heat capacity of low-T_c superconductors also influences on the formation of their voltage-current and temperature-current characteristics. However, its role is insignificant because of the small temperature margin. Curves 4 in Figure 4 clearly show the typical type of voltage-current and temperature-current characteristics of superconductors with low heat capacity.

The role of the heat capacity of a superconductor on the formation of both the stable and unstable states will increase when the operating temperature increases. Let us discuss this peculiarity in more detail. Figure 9 shows the influence of the current charging rate on the voltage-current and the temperature-current characteristics of the superconducting slab under consideration cooled at operating temperature T_0 = 20 K. As above, the numerical simulation was performed under the assumption of the non-intensive cooling condition (h = 10^{-3} W/(cm^2×K) at B = 10 T. This allows to use the unsteady and static zero-dimensional models (12) and (13). In this case, the critical current density is described by equation (9) with parameters (10) in the voltage-current characteristic (3). Accordingly, solid lines 2-4 describe the unsteady thermo-electrodynamics states of the superconductor and dashed lines 1 show the static voltage-current and temperature-current characteristics. Dash-dotted curve 5 corresponds to the isothermal voltage-current characteristic calculated at T = 20 K.

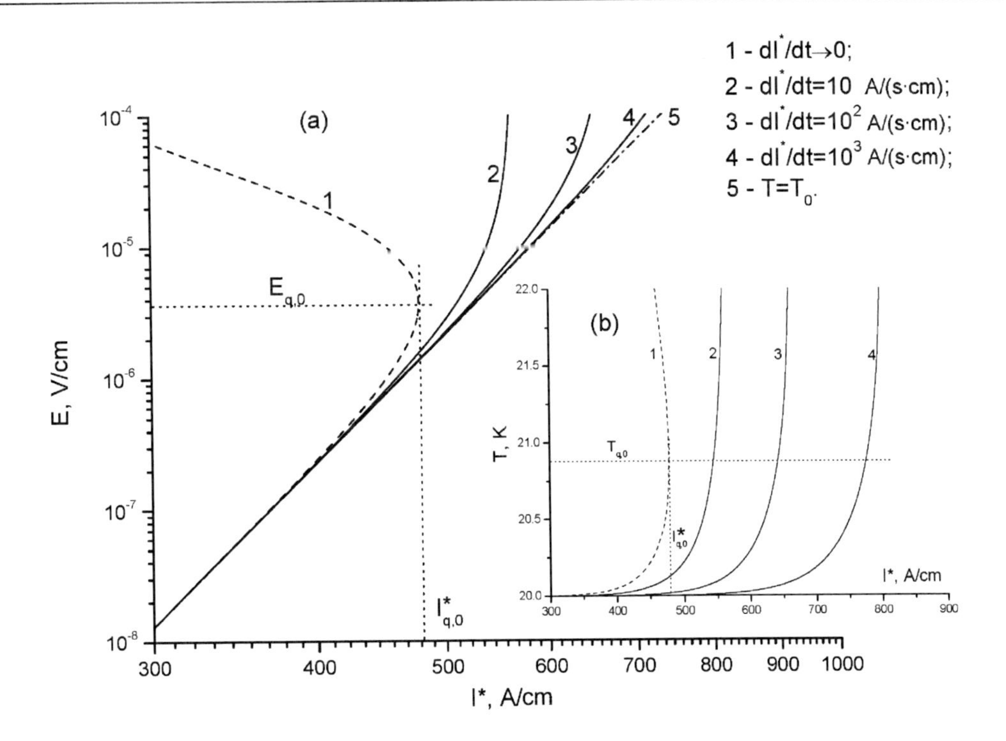

Figure 9. Voltage-current (a) and temperature-current (b) characteristics of Bi2212 at $T_0 = 20$K and various current charging rate.

Figure 9 clearly shows that the slope of the non-stationary voltage-current and temperature-current characteristics not only decreases with increasing *dI/dt* because of the heat capacity of the superconductor increase in temperature of the coolant and the superconductor. Comparing the temperature-current characteristics shown in Figure 4(a) and Figure 9(b), it is easy to see the following operating temperature effect: the higher the temperature of the refrigerant, the more close the non-isothermal dependences $E(J)$ and $T(J)$ to the corresponding isothermal dependences in the relatively high current charging rate (curve 5). This peculiarity follows from the relationships (22) and (23) at the limiting transition $dT/dJ \to 0$ or $C(T)dI/dt \to \infty$. It allows to make unexpected conclusion that may be important for practical applications: an isothermal voltage-current characteristics in over-critical electrical field range can be obtained charging the current into the non-intensive cooled superconductor at high value of the current charging rate. It is obvious that in this case condition ha/□<< 1 must take place. (Features of these states at ha/□>>1, for example, an intensive cooling conditions, will be discussed .

Note another feature of the influence of the heat capacity on the processes in the high-temperature superconductors, which must be taken into account in the theoretical and the experimental determination of the boundary of the stable current regimes. Figure 6 shows that the heat capacity of a superconductor increasing with temperature leads to a corresponding increase in the transition process that occurs in a superconductor after charging a given current I^*_0. This feature occurs as in the formation of stable states, and in the violation of a stable distribution charged current. Consequently, the higher the temperature of the coolant, the higher the duration of the transition process. It will increase not only in accordance with the increase in temperature of the refrigerant, but also appropriate to the steady increase in temperature of superconductor preceding the instability onset.

Thus, in the continuous current charging modes, non-stationary voltage-current and temperature-current characteristics of high-T_c superconductors during fully penetrated states have only one branch with the positive slope, which depends on the current sweep rate: the slope decreases when the value dI/dt increases. It exists both in the stable and unstable thermo-current states. This peculiarity is due to the finite temperature rise of a high-T_c superconductor both before and after instability onset that increases the heat capacity of a superconductor. Its role is the most significant in the high electric fields, where it can be observed a noticeable thermal capacity effect even at helium temperature level of the refrigerant. That is why the transient voltage-current characteristics of high-T_c superconductors do not permit to find the current instability conditions and the possible stable increase in temperature of high-T_c superconductors should be taken into account for correct investigation of their critical currents.

These peculiarities justify the experimental method of the determination of the boundaries of stable current state of high-T_c superconductors. It is based on an analysis taking place in the thermo-electrodynamics processes after the charging of a given current. In this case, the current state is stable if the values of the electric field, temperature and Joule heating in the end of the transition process does not exceed the corresponding static boundary values, namely, $E_{q,0}$, $T_{q,0}$, $G_{q,0}$ if zero-dimensional states exist. The later determine allowable maximum values during uniform modes.

5. Voltage-Current Characteristic and Current Instability Conditions of a High-T_C Superconductor in the Non-Uniform Temperature Distribution

As it was shown above, the current after which unstable states exist can be determined by the analysis of the static voltage-current characteristic of a superconductor and can be obtained from the conditions (20) or (26). However, these criteria were formulated under the assumption that the distributions of the temperature, the electric field E and the current density J inside the superconductor would be uniform. As follows from the heat conduction theory [41], [42], uniform thermal mode take place at $ha/\lambda \ll 1$. Under this condition, the uniform temperature distribution in the cross section of superconductor is the most probable in the cases of low thickness, non-intensive cooling conditions or high thermal conductivity. For example, operating regimes with uniform temperature distribution may be observed during current charging into a superconducting composite with a high conductivity matrix [6]-[8]. At the same time, the thermal conductivity of superconducting materials having no stabilizing matrix is low. Therefore, the temperature distribution inside the superconductor is not, strictly speaking, uniform [1], [2]. In addition, the criterion of the spatial heterogeneity of the thermal states, i.e. condition $ha/\lambda \gg 1$, does not describe the influence of the critical properties of superconductor on the spatial features of the development of stable thermo-electrodynamics states. Therefore, let us investigate the formation peculiarities of the voltage-current characteristic of high-T_c superconductor having no stabilizing matrix accounting for non-uniform temperature distribution in its cross section and formulate the appropriate current instability conditions.

To discuss the basic physical features, let us use, first of all, the one-dimensional diffusion model described by equations (1) – (8) considering the non-isothermal penetration of applied current into examined $Bi_2Sr_2CaCu_2O_8$ superconductor at B = 10 T and T_0 = 4.2 K under parameters (16) – (18) assuming that intensive cooling condition (h = 0.1 W/(cm^2 ×K)) takes place. As above, the performed analysis will be done using the current and current sweep rate normalized by the slab width ($I^* = 0.5I/b$, $dI^*/dt = 0.5b^{-1}dI/dt$).

Figure 10 shows the changes in time of the electric field and the temperature, which will occur on the surface and in the center of the superconducting slab at $dI^*/dt = 10^2$ A/(s ×cm). Curves 3, 3', 4, 4', 5, 5' in Figure 11 show the influence of the current sweep rate on the formation of voltage-current and temperature-current characteristics of the superconductor examined. To understand the peculiarities of the non-isothermal development of electrodynamics states in the superconductor that take place both in stable and unstable regimes, the isothermal voltage-current characteristic of a superconductor (curve 1) calculated according to the formulae (3) and (4) at T = 4.2 K is also presented in Figure 11. Besides, the curve 2 describes the non-isothermal static voltage-current characteristic determined under the assumption of the uniform temperature and current distributions that takes place over the whole range of the transport current charging. In this case, the static thermo-electrodynamics states of the superconductor are described by equations (3), (4) and (13) in which the density of the transport current is equal to $J = I^*/(2a)$. Figure 11(a) also depicts the electric field $E_{q,0}$ and the current $I^*_{q,0}$ defining the stability boundary of spatially homogeneous states, which follow from the relationships (21).

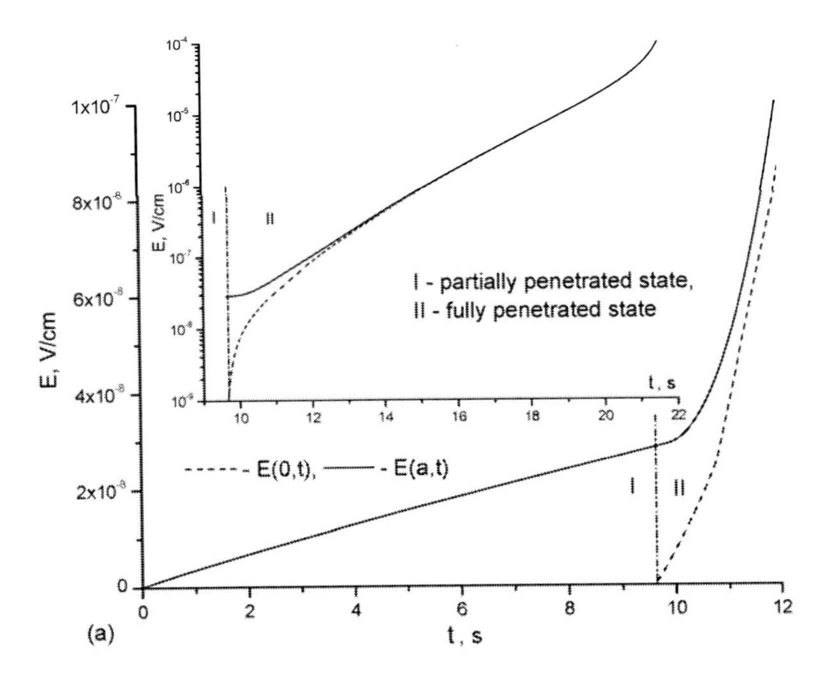

Figure 10. (Continued).

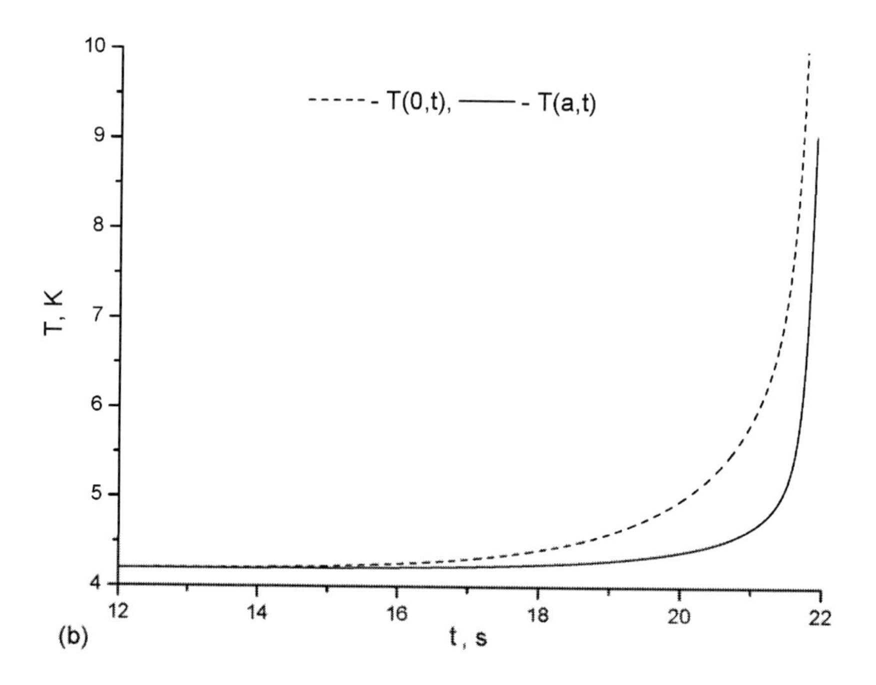

Figure 10. Time variation of the electric field (a) and the temperature (b) on the surface (——) and in the centre (- - - -) of intensive cooled Bi2212 superconductor in the continuous current charging.

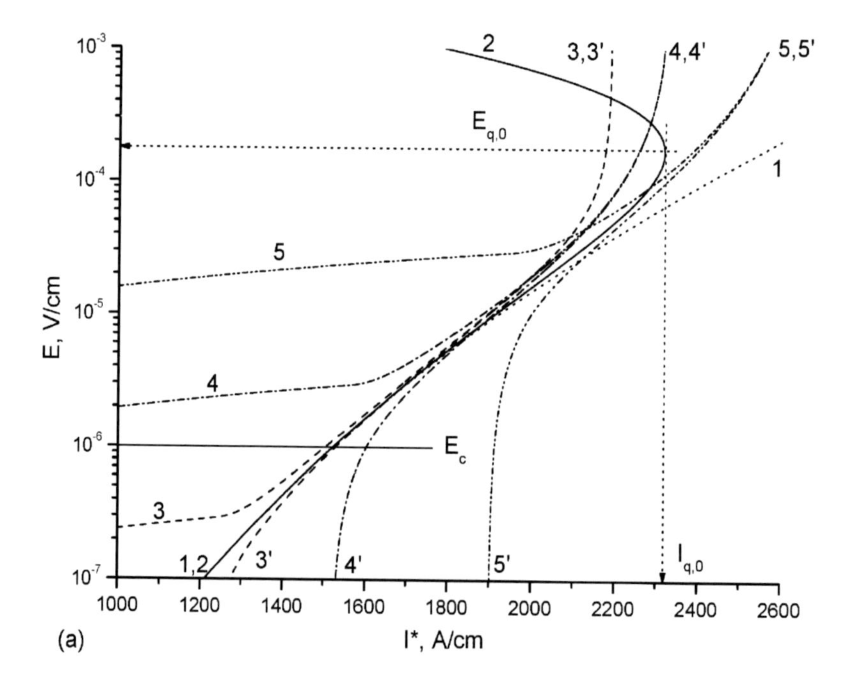

Figure 11. (Continued).

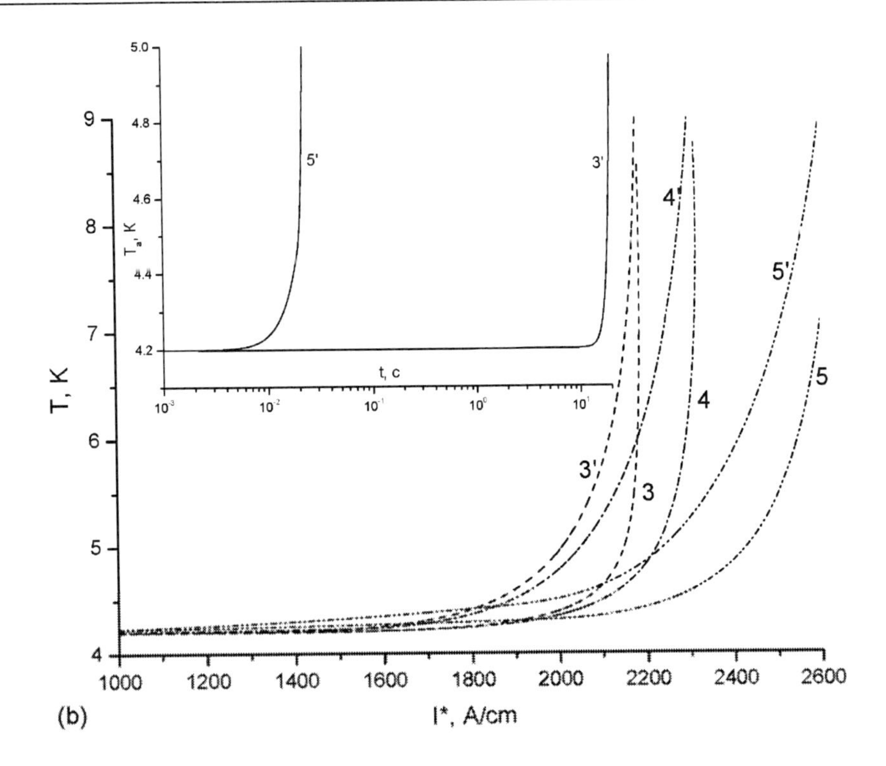

Figure 11. Dependences electric field (a) and temperature (b) on the surface (curves 3, 4, 5) and in the center (curves 3', 4', 5') of Bi2212 in the various operating modes: 1 – $T = 4.2$ K, 2 – $dI^*/dt \to 0$, 3, 3' – $dI^*/dt = 10^3$ A/(s × cm), 4, 4' – $dI^*/dt = 10^4$ A/(s × cm), 5, 5' – $dI^*/dt = 10^5$ A/(s × cm).

Figures 10 and 11 illustrate the characteristic size peculiarities of the temperature and electric field formation in the superconducting slab in the continuous current charging that has to be taken into account when the current stability conditions are determined.

First, the temperature distribution in a superconductor ceases to be uniform even at a relatively low current charging rate while the distribution of the electric field becomes almost uniform after some transient period. This trend is kept at higher current charging rates, as can be seen from Figure 11. Nevertheless, it should be noted that the uniform distribution of the electric field in a superconductor would occur in the fully penetrated mode when the current charging rate satisfies the condition $dI/dt << 4bE_c/(\mu_0 a)$, as it was proved above. Therefore, the temperature distribution may be non-uniform while the electric field in the slab may be practically uniform at $dI/dt << 4bE_c/(\mu_0 a)$ in the fully penetrated mode.

Second, during intensive cooling condition the unsteady voltage-current characteristics of the high-T_c superconductor (curves 3, 3'- 5, 5' in Figure 11) quantitatively and qualitatively differ from the dependences $E(I^*)$ calculated not only in the isothermal approximation (curve 1) but also in terms of the static non-isothermal zero-dimensional model (curve 2) described by equations (3), (4) and (13). In the high electric field ($E > E_c$), this feature is observed because of the thermal influence of the heat capacity of the superconductor on the non-stationary thermal and electrodynamics states in the continuous current charging, as it was discussed above. Therefore, the slope of curves 3, 3'- 5, 5' is always positive and decreases with the increasing current charging rate due to the corresponding increase in the superconductor's heat capacity with temperature, which will occur in the continuous current charging. Due to this peculiarity, the non-stationary voltage-current characteristic of superconductor during fully penetrated regime should be below the corresponding static voltage-current characteristic before the current instability, if the formation of the electric field in the superconductor depends only on its heat capacity. However, Figure 11 depicts that the non-stationary voltage-current characteristics of a superconductor (curves 3, 3' and 4, 4') may be higher at a relatively low current charging rate than the static voltage-current characteristic calculated according to the non-isothermal zero-dimensional model in the area of the electric fields preceding the current instability. As a result, this rise of the electric field will change the range of the stable currents calculated in the framework of the non-uniform distribution of temperature in comparison with a similar range obtained by the zero-dimensional model. This formation feature of the voltage-current characteristic of the superconductor taking place at low values of dI/dt cannot be explained by the influence of superconductor's heat capacity on the electrodynamics processes occurring in superconductors.

In order to understand the physical reason of this peculiarity, let us find the static temperature distribution in the slab under consideration, which takes place in the fully penetrated mode at $dI/dt \to 0$ but considering the thermal heterogeneity of the operating mode of the superconductor. Accordingly, let us find the solution of the following system

$$\frac{\partial}{\partial x}\left(\lambda(T)\frac{\partial T}{\partial x}\right)+EJ=0, \tag{27}$$

$$\frac{\partial^2 E}{\partial x^2}=0 \tag{28}$$

under the boundary conditions

$$\frac{\partial T}{\partial x}(0,t)=0, \quad \lambda\frac{\partial T}{\partial x}(a,t)+h\left[T(a,t)-T_0\right]=0, \tag{29}$$

$$\frac{\partial E}{\partial x}(0,t)=0, \quad \frac{\partial E}{\partial x}(a,t)=0 \tag{30}$$

taking into account the relationship (3) and (4).

As follows from (28) and (30) E = const (at $dI/dt \to 0$). This limiting case proves the above-discussed feature according to which the distribution of the electric field induced by the charged current at a small rate is practically uniform.

In general, the solution of the problem described by equation (27) and condition (29) requires the use of numerical methods. However, assuming that the thermal conductivity coefficient is constant ($\lambda = \lambda_0$ = const) it can be solved analytically. Indeed, let us introduce the dimensionless variables

$$\theta = \frac{T_{cB} - T}{T_{cB} - T_0}, \quad X = \frac{x}{a}$$

Then the problem (3), (4), (27) and (29) can be rewritten as follows

$$\frac{d^2\theta}{dX^2} \quad \gamma\theta = 0,$$
$$\frac{d\theta}{dX}(0) = 0, \quad \frac{d\theta}{dX}(1) + H\theta(1) = 0 \tag{31}$$

Here, $H = ha/\lambda_0$ is the dimensionless thermal resistance of superconductor,

$$\gamma = \frac{J_{c0} E_c a^2}{\lambda_0 (T_{cB} - T_0)} \left(\frac{E}{E_c} \right)^{(n+1)/n}$$

is the dimensionless parameter that, as it will be proved strictly below, describes the influence of the material properties of the superconductor on the heterogeneous character of the thermo-electrodynamics state formation. This parameter can be rewritten in the form of

$$\gamma = \frac{JEa}{\lambda_0 (T_{cB} - T)/a}.$$

Therefore, the physical meaning of the parameter γ is equal to the ratio of the power of the Joule heating in the superconductor to the power of the heat flux transferred by thermal conductivity. Thus, it describes directly the influence of the conductive heat transfer mechanism on the electrodynamics state formation in the superconductor. In particular, it is not difficult to understand that boundary problem (31) leads to a model with a uniform temperature distribution in the superconductor at $\gamma << 1$, for example, in the effective conductive heat transfer.

The analytical solution of the problem (31) is given by

$$\theta(X) = \theta_0 \cosh\left(\sqrt{\gamma} X\right), \quad \theta_0 = \frac{H}{\sqrt{\gamma} \sinh\sqrt{\gamma} + H \cosh\sqrt{\gamma}}$$

Therefore, in the dimensional form, the static distribution of temperature in the superconducting slab with a power voltage-current characteristic is described by the expression

$$T(x) = T_{cB} - (T_{cB} - T_0)\theta_0 \cosh\left(\sqrt{\gamma}\frac{x}{a}\right) \tag{32}$$

during fully penetrated mode. In this case, the relationship between current and electric field determined as $I^*(E) = 2\int_0^a J(E)dx$ according to the symmetry of the examined problem is written in the form of

$$I^* = 2J_{c0}a\left(\frac{E}{E_c}\right)^{1/n} \frac{H}{\gamma + H\sqrt{\gamma}/\tanh\sqrt{\gamma}}. \tag{33}$$

Figure 12 shows the static voltage-current characteristics determined in the assumption of the uniform and non-uniform distribution of temperature in the slab. In the first case, curve 1 was defined in accordance with formula

$$I^* = 2J_{c0}a\left(\frac{E}{E_c}\right)^{1/n} \frac{H}{\gamma + H} \tag{34}$$

as follows from the zero-dimensional model defined by relations (3), (4) and (13). For heterogeneous states, the performed calculations were based on the formula (33) under assumption that the thermal conductivity of the superconductor is defined as $\lambda_0 = \lambda(T^*)$ at different T^* according to (18) (curves 2 – 5), as well as on the numerical solution of the problem (3), (4), (27), (29) taking into account the dependence (18) (curve 6). In addition, the change in curve 6 near the boundary of stability is shown in more detail in the inset of Figure 12 where the appropriate value of the instability current I_q is also depicted.

These results show clearly that during theoretical analysis of current instability conditions the non-uniform temperature distribution in the cross-section of superconductor leads to a reduction in the stable currents relative to the corresponding values determined in the framework of the zero-dimensional model. To understand the physical meaning of this feature, let us find the average temperature of the superconductor both in the case of one-dimensional and zero-dimensional approximations. In accordance with one-dimensional model and formula (32) its value is equal to

$$T_v = \frac{1}{a}\int_0^a T(x)dx = T_0 + (T_{cB} - T_0)\left(1 - \frac{H}{\gamma + H\sqrt{\gamma}/\tanh\sqrt{\gamma}}\right). \tag{35}$$

At the same time, according to the zero-dimensional model, the temperature of superconductor is defined as

$$T_{v,0} = T_0 + (T_{cB} - T_0)\left(1 - \frac{H}{\gamma + H}\right) \tag{36}$$

in terms of the variables used. Comparing the values of T_v and $T_{v,0}$ it is easy to see that $T_v > T_{v,0}$ because the condition $\sqrt{\gamma} > \tanh\sqrt{\gamma}$ is always observed. So, the average temperature of superconductor at any finite γ always exceeds the corresponding values calculated on the assumption of existence of homogeneous states. The difference the higher, the higher the parameter γ due to the corresponding increase in the term $\sqrt{\gamma}/\tanh\sqrt{\gamma}$. The latter has the same influence on the voltage-current characteristic of a superconductor. This follows from the comparison of voltage-current characteristics described by formulas (33) and (34), which were obtained under the one- and zero-dimensional stationary approximations, respectively. In other words, there is a thermal size effect, which is based on spatial peculiarities of the temperature distribution in the superconductor and depends on the value of γ. In accordance with this effect, the current stability conditions will be modified.

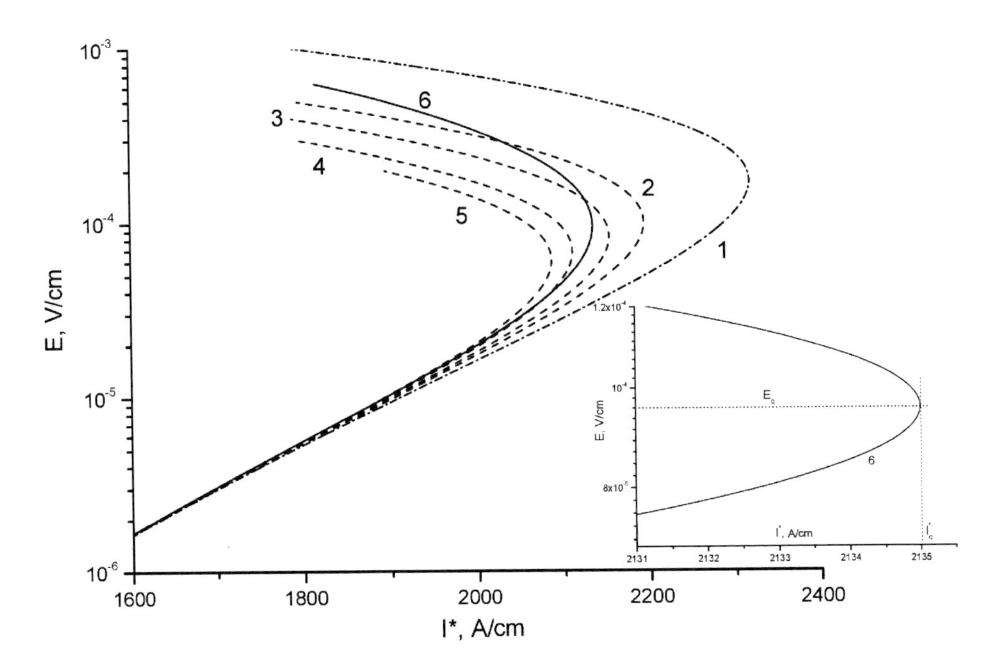

Figure 12. Influence of the thermal conductivity coefficient of superconductor on the voltage-current characteristic: 1 – $H\rightarrow 0$, 2 – $\lambda_0 = \lambda(10\text{ K})$, 3 – $\lambda_0 = \lambda(7\text{ K})$, 4 – $\lambda_0 = \lambda(5\text{ K})$, 5 – $\lambda_0 = \lambda(4.2\text{ K})$, 6 – $\lambda(T)$. Here, $\lambda(T)$ is defined by (18).

Parameter γ also allows one to estimate the legitimacy of the usage of the zero-dimensional model when the critical properties of the superconductor change. In particular, it is not difficult to understand that the influence of the non-uniform distribution of temperature on the electrodynamics state of superconductor will be more noticeable when its critical current density is higher or the so-called temperature margin of superconductor, which is equal to $(T_{cB} - T_0)$, is lower with other parameters being equal. Their influence is the more essential, the more intensive the cooling, as it can be seen from (33) and (35).

Note that in the cases discussed, the change in the stable current range is associated with the influence of the temperature-dependent thermal conductivity coefficient on the electrodynamics processes in a superconductor. As a result, difference between the voltage-current characteristics calculated according to the size effect and without it, becomes more noticeable when the thermal conductivity coefficient of the superconductor is reduced. Therefore, the influence of spatial heterogeneity of temperature distribution in the superconductor on the formation of its voltage-current characteristic is more noticeable, if the operating temperature of coolant, which affects the thermal conductivity coefficient, is lower with other parameters being equal.

The results presented in Figure 12, also prove strongly that the boundary of the current instability mation follows from the condition

$$\partial E / \partial I \to \infty . \tag{37}$$

in the static non-uniform approxi. Its formal difference from the condition (20) shows that the current instability condition depends on the character of the current redistribution in the superconducting slab when the temperature distribution in the superconductor is non-uniform. To verify this feature and to show the contravention of the condition (20) more strictly, Figure 13 depicts the dependences $E(J)$, which take place on the surface of the superconducting slab and in its center during current charging. The simulation was made using numerical solution of the problem (1) – (8) for different current charging regimes. Curves 1 and 1' describe the growth of the electric field in the continuous current charging. Curves 2, 2' and 3, 3' were calculated in the current charging with a break when the value dI^*/dt was taken zero if the current is equal to some value I^*_0. According to the results shown in the inset of Figure 12, curves 3 and 3' describe stable dynamics of the electric field, and curves 2 and 2' correspond to unstable states. Figure 13 shows that the current is redistributed from the central part of the superconducting slab to its external part in the charging both the stable (I^*_0 = 2130 A/cm) and unstable (I^*_0 = 2140 A/cm) currents even in spite of the fact that the development of the transition process depends on the current charging rate. This result reveals that the currents in the internal part of superconductor are always more unstable in comparison with the ones flowing on its surface due to the corresponding non-uniform temperature distribution. As a result, the condition (20) loses its physical meaning. Consequently, to determine in a static approximation the boundary of stable currents taking into account the heterogeneity of the thermal states, it is necessary to use the condition (37).

The non-uniform distribution of the temperature in the superconductor also changes the formulation of the stability criterion allowing one to find a stable boundary based on analysis of the energy balance between the cooling power and heat release during current charging. Accordingly, the relations (26) describe the stability boundary at the uniform thermo-electrodynamics states. In order to formulate this condition taking into account the non-uniform distribution of temperature, let us integrate the heat equation (27) with respect to x from 0 to a. Then taking into account the boundary conditions at x = 0 and x = a it is not difficult to obtain the following relationship

$$\int_0^a EJdx = h\left(T - T_0\right)\Big|_{x=a} . \tag{38}$$

This relation describes the first equality of the conditions (26) in the integral form. Obviously, the second equality is modified according to the given formulation of the quantities G and W.

Figure 14 verifies the determination of these quantities for spatially non-uniform thermo-electrodynamics states. Namely, it demonstrates the changes of the quantities $G = \int_0^a EJdx$ (curves 2, 2', 2", 3, 3', 3") and $W = h(T - T_0)\big|_{x=a}$ (curve 1) as a function of the temperature during current charging with different charging rate.

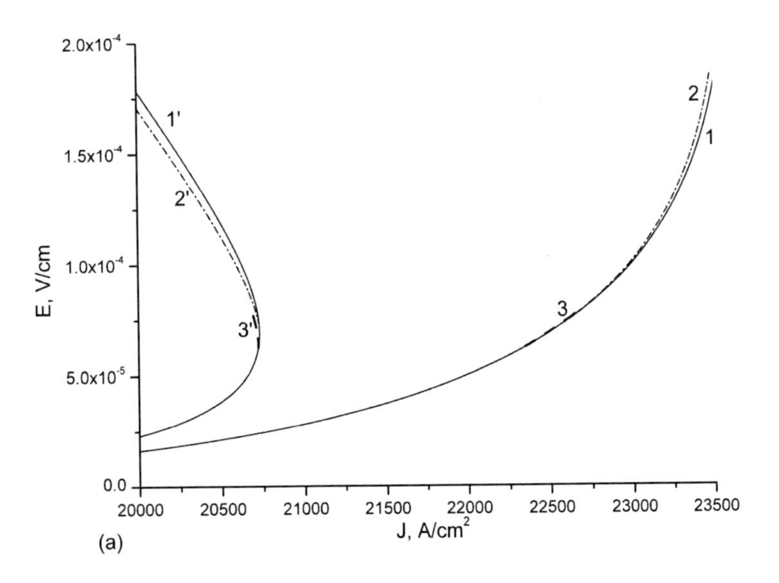

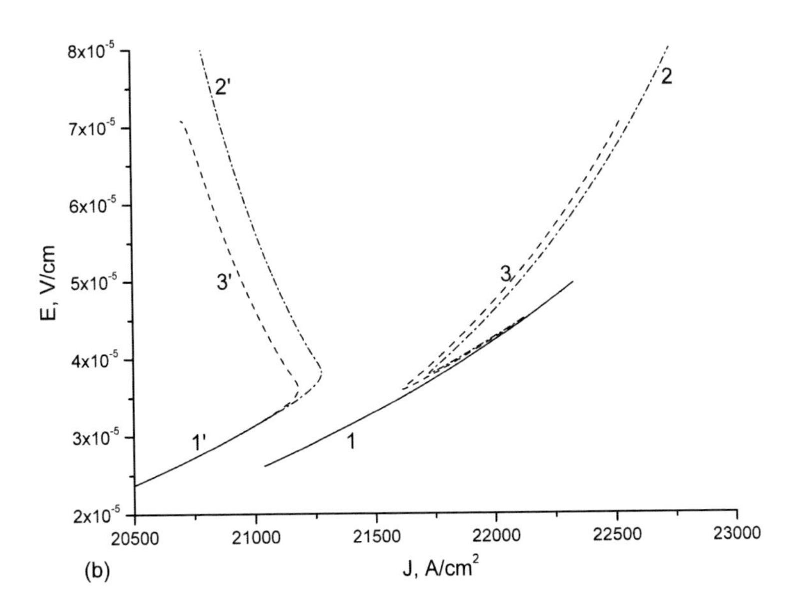

Figure 13. Electric field on the surface (1, 2, 3) and in the centre (1', 2', 3') of superconductor as a function of current density at $dI^*/dt = 10^2$A/(s × cm) (a), $dI^*/dt = 10^5$A/(s × cm) (b): 1, 1' – continuous current charging, 2, 2' – I^*_0 = 2140 A/cm, 3, 3' – I^*_0 = 2130 A/cm.

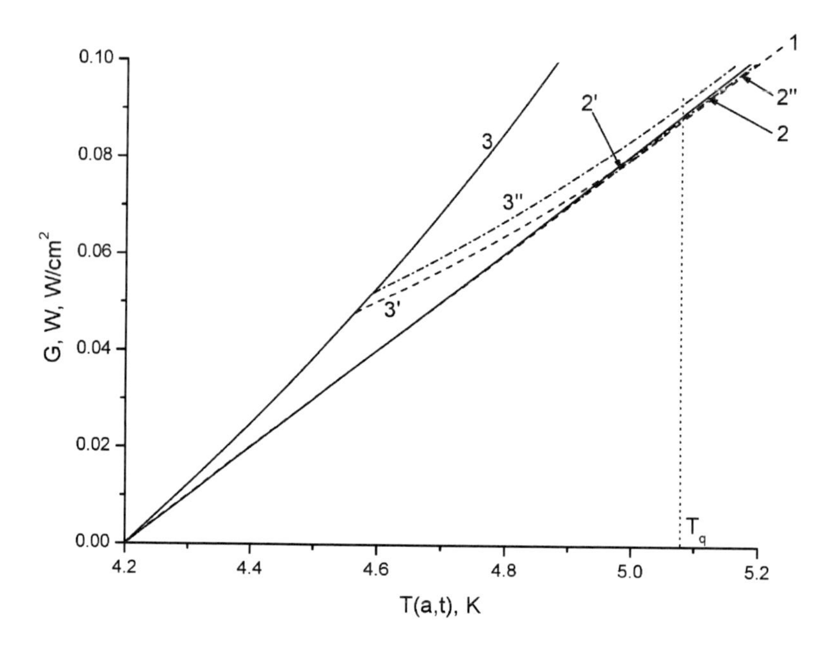

Figure 14. Temperature dependence of the surface heat removal (1) and Joule heating at dI^*/dt = 100A/(s × cm) (2, 2', 2") and $dI^*/dt = 10^5$ A/(s × cm) (3, 3', 3"): 2, 3 – continuous current charging, 2', 3' – I^*_0= 2130 A/cm, 2", 3" – I^*_0= 2140 A/cm.

The simulation was based on the numerical solution of the problem defined by equations (1) – (8). Two kinds of current charging mode were considered: continuous current charging (curves 2 and 3) and charging with a break (curves 2', 2", 3', 3"). Figure 14 clearly proves that the stable current distribution in a superconductor depends on the thermal state of its surface. Therefore, in the experimental determination of the boundary of the instability onset using method of fixed current it is sufficiently to control only surface temperature of superconductor (Figure 15).

Using the voltage-current characteristic (33) and the condition (37), it is easy to find the electric field E_q and current per width of the slab I^*_q preceding the current instability onset in the non-uniform distribution of temperature in the slab. Accordingly, in terms of the dimensionless variables used, the value E_q is determined by the solution of equation

$$H = \frac{2n}{n+1}\frac{\sqrt{\gamma_q}\tanh\sqrt{\gamma_q}}{\dfrac{2\sqrt{\gamma_q}}{\sinh\left(2\sqrt{\gamma_q}\right)} - \dfrac{n-1}{n+1}}. \tag{39}$$

Then the value of I^*_q is equal to

$$I_q^* = 2J_{c0}a\left(\frac{E_q}{E_c}\right)^{1/n}\frac{H}{\gamma_q + H\sqrt{\gamma_q}/\tanh\sqrt{\gamma_q}}. \tag{40}$$

Here,

$$\gamma_q = \frac{J_{c0}E_c a^2}{\lambda_0 (T_{cB} - T_0)} \left(\frac{E_q}{E_c} \right)^{1+\frac{1}{n}}$$

The quantity γ_q as a function of thermal resistance H is presented in Figure 16 for various values of the exponent of the *E-J* curve. The calculations were made both for the case of spatially homogeneous thermo-electrodynamics states (curves 1, 2, 3) and taking into account their heterogeneity (curves 1', 2', 3'). The dependence $\gamma_q(H)$ in the framework of the zero-dimensional approximation is calculated by the formula

$$\gamma_q = H / n \tag{41}$$

which follows from (21) or can be obtained from (39) under the assumption that the $\gamma_q << 1$.

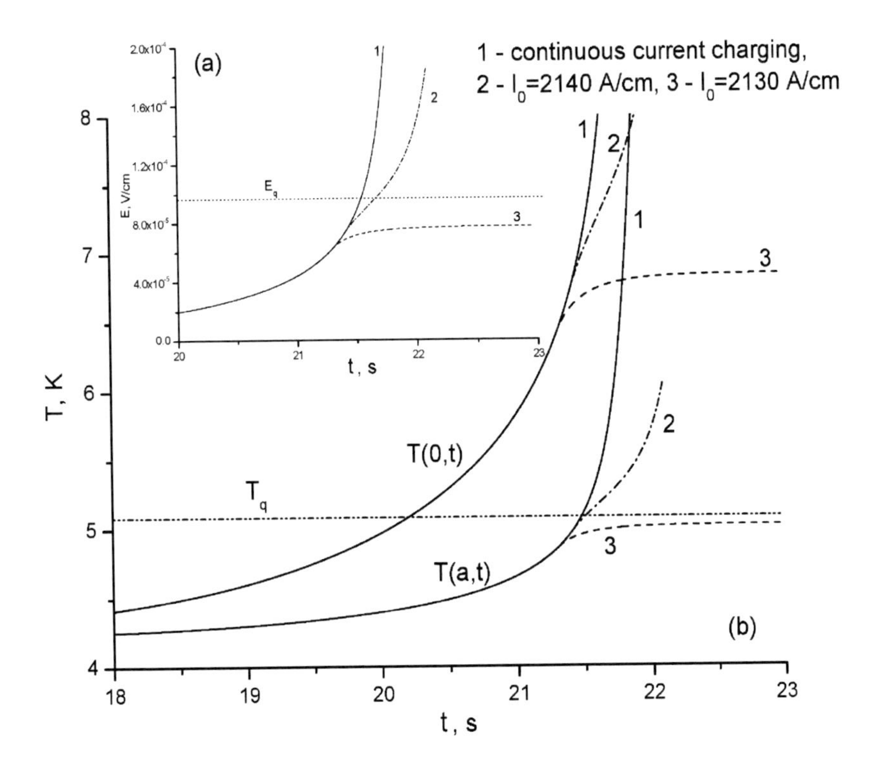

Figure 15. Time variation of the electric field (a) and the temperature (b) on the surface and in the centre of intensive cooled superconductor at $dI^*/dt = 100 \text{A}/(\text{s} \times \text{cm})$.

Figure 16 shows that the value γ_q calculated for thermally heterogeneous states does not exceed unit within a wide range of the thermal resistance variations. This conclusion allows one to find the approximate solution of equation (39) describing the boundary of stable values of electric field. After some transformation, it is easy to get

$$\gamma_q \approx \frac{H}{n+H(n-1)/3}. \quad (42)$$

Approximate values γ_q calculated according to (42) are also shown in Figure 16 (curves 1", 2", 3"). It is seen that the formula (42) allows one to find the quantity γ_q with a satisfactory accuracy. In the dimensional variables, the equality (42) is written as follows

$$E_q \approx E_c\left[\frac{1}{n+(n-1)H/3}\frac{h(T_{cB}-T_0)}{J_{c0}E_c a}\right]^{n/(n+1)}. \quad (43)$$

In this case, current of instability is equal to

$$I_q \approx I_c\frac{\left[\frac{1}{n+H(n-1)/3}\frac{h(T_{cB}-T_0)}{J_{c0}E_c a}\right]^{1/(n+1)}}{\frac{1}{n+H(n-1)/3}+\sqrt{\frac{H}{n+H(n-1)/3}}/\tanh\sqrt{\frac{H}{n+H(n-1)/3}}}. \quad (44)$$

Here, I_c is the critical current of a superconductor at $T = T_0$ ($I_c = J_{c0}S$).

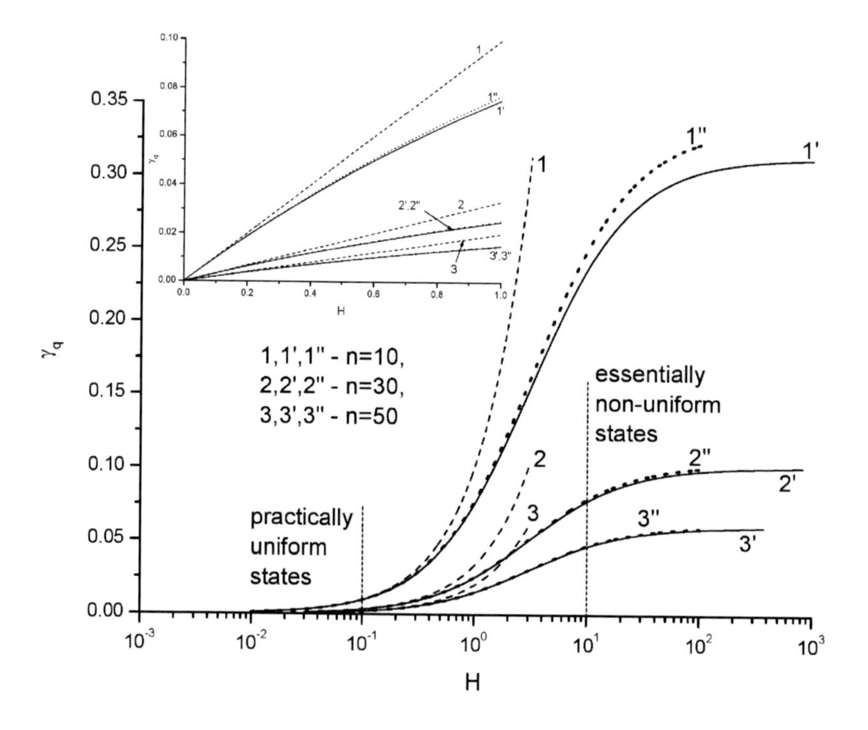

Figure 16. Influence of thermal resistance of the superconducting slab on the quantity γ_q calculated in the framework of zero-dimensional (1, 2, 3) and one-dimensional models. Here, the curves (1', 2', 3') and (1", 2", 3") are obtained according to (21) and (23), respectively.

Comparing this value of quench electric field with the value described by the formula (21) following from the zero-dimensional model, it is not difficult to find difference between the estimated boundary values of the electric field calculated in the framework of the zero-dimensional and one-dimensional approximations. The corresponding relationship between $E_{q,0}$ and E_q can be written as

$$\frac{E_{q,0}}{E_q} = \left(1+\frac{n-1}{3n}H\right)^{n/(n+1)}. \tag{45}$$

In this case, difference between the corresponding values of the current of instability is defined as

$$\frac{I^*_{q,0}}{I^*_q} = \frac{n}{n+1}\left(\frac{1}{n+H(n-1)/3} + \frac{\sqrt{\dfrac{H}{n+H(n-1)/3}}}{\tanh\sqrt{\dfrac{H}{n+H(n-1)/3}}}\right)\left(1+\frac{n-1}{3n}H\right)^{1/(n+1)}. \tag{46}$$

These expressions allow one to estimate quantitatively the influence of the spatial heterogeneity of thermo-electrodynamics states on the current stability conditions. In particular, the formulae (45) and (46) show that difference between the boundary values of electric field and current, which follow from the zero-dimensional and one-dimensional calculations, always takes place for any finite value of γ because of the above-mentioned features of the temperature changes in cross-section of a superconductor. Moreover, difference increases with increasing γ. As a whole, according to the results presented in Figure 16, it is necessary to take into account the following definition peculiarities of current instability conditions of high-temperature superconductors in their theoretical analysis.

First, as expected, the thermal size effect is not observed at $H \ll 1$. However, it is a rough estimate. According to Figure 16, variation of H will not be accompanied by a noticeable influence of the spatial heterogeneity on the thermo-electrodynamics states at $H < 0.1$. In this case, the boundary values of the electric field and the current are described by the relations (21) with a good degree of accuracy.

Second, the one-dimensional approximation correctly describing the current stability conditions at intensive cooling conditions ($H > 10$) allows one to find the limiting value $\gamma_{q,i}$ at $H \to \infty$. According to (39), it is the solution of equation

$$\frac{2\sqrt{\gamma_{q,i}}}{\sinh\left(2\sqrt{\gamma_{q,i}}\right)} = \frac{n-1}{n+1}.$$

It is essential that this value $\gamma_{q,i}$ cannot be found in terms of the zero-dimensional model because the spatially heterogeneous character of the thermo-electrodynamics processes is critically dependent on its thermal resistance at $H > 10$ (Figure 16). In these states, the increase in the limiting values of the electric field and current becomes significantly non-linear with increasing H and the following estimate:

$$I_q^* \approx I_{q,0}^* \frac{n+1}{n} \frac{\tanh\sqrt{\gamma_q}}{\sqrt{\gamma_q}\left(1+\frac{n-1}{3n}H\right)^{1/(n+1)}}$$

is found according to (46). This relation shows that the current of instability will not increase proportionally to an increase in the thickness or the critical current density of superconductor in the intensive cooling conditions unlike the corresponding proportional change of the critical current of the superconductor with increasing quantities a and J_{c0}. In other words, the thermal size effect will lead to the thermal degradation of the current-carrying capacity of superconductor. In this case, the increase in the quantities a and J_{c0} leading to increase in the critical current of the superconductor is not accompanied by a corresponding increase in the currents of instability. Let us discuss in detail the thermal mechanisms leading to the non-uniform temperature distribution in the cross-section of a superconductor, which may cause the thermal degradation of the current-carrying capacity.

6. Thermal Degradation of the Limiting Charged Currents

Rewrite the relationship (33) in the form of

$$I = I_c \frac{(E/E_c)^{1/n}}{\sqrt{(E/E_\lambda)(E/E_c)^{1/n}} / \tanh\sqrt{(E/E_\lambda)(E/E_c)^{1/n}} + (E/E_h)(E/E_c)^{1/n}}. \quad (47)$$

Here,

$$E_h = \frac{h(T_{cB}-T_0)}{J_{c0}a}, \quad E_\lambda = \frac{\lambda_0(T_{cB}-T_0)}{J_{c0}a^2}.$$

This equation of the current-voltage characteristic of the superconductor shows that there exist two characteristic values E_h and E_λ above which their intensity of growth changes during current charging. The quantities E_h and E_λ are equal to the ratio of the power of the heat flux into the coolant and the heat flux transferred by thermal conductivity to the power of the Joule heating, respectively. They take into account the influence of the convective and conductive heat transfer mechanisms on the electrodynamics processes in the superconductor and relate as

$$\frac{E_h}{E_\lambda} = \frac{ha}{\lambda_0} = H. \quad (48)$$

For this reason, the voltage-current characteristic of the superconductor depends not only on its critical properties but also on its thermal resistance H. Thus, the growth of the voltage-current characteristic of the high-T_c superconductor will depend on the character of the temperature variation, as it was shown above. According to (32), the temperature of superconductor rewritten in the form of

$$T(x) = T_0$$

$$+\left(T_{cB} - T_0\right)\left[1 - \frac{1}{1 + \dfrac{E}{E_h}\left(\dfrac{E}{E_c}\right)^{1/n} \dfrac{\tanh\sqrt{(E/E_\lambda)(E/E_c)^{1/n}}}{\sqrt{(E/E_\lambda)(E/E_c)^{1/n}}}} \frac{\cosh\left(\sqrt{\dfrac{E}{E_\lambda}\left(\dfrac{E}{E_c}\right)^{1/n}}\dfrac{x}{a}\right)}{\cosh\left(\sqrt{\dfrac{E}{E_\lambda}\left(\dfrac{E}{E_c}\right)^{1/n}}\right)}\right]$$

increases with the reduction in values E_h and E_λ. As a result, this formula allows to formulate the thermal nature of the characteristic quantities E_h and E_λ considering the limiting cases at $E_\lambda \to \infty$, i.e. at $\lambda_0 \to \infty$, and at $E_h \to \infty$, i.e. at $h \to \infty$.

Indeed, it is easy to find that the temperature distribution in the superconductor becomes uniform, when $E_\lambda \to \infty$. In this case, the temperature and current can be expressed as

$$T = T_0 + \left(T_{cB} - T_0\right)\left[1 - \frac{1}{1 + (E/E_h)(E/E_c)^{1/n}}\right], \tag{49}$$

$$I = I_{c0} \frac{(E/E_c)^{1/n}}{1 + (E/E_h)(E/E_c)^{1/n}}. \tag{50}$$

Therefore, the quantity E_λ is the characteristic electric field exceeding which there will be a non-uniform temperature distribution in the slab. It exists because of the power of the conductivity heat flux is finite due to the finite value of λ_0. Further, as follows from (49) and (50), the non-isothermal voltage-current characteristic of superconductor in the uniform current distribution becomes isothermal ($T = T_0$) at $E_h \to \infty$. Consequently, the quantity E_h is the characteristic electric field allowing one to estimate the influence of the external cooling intensity on the character of growth of its voltage-current characteristics taking into account the thermal deviation of electrodynamics states from those that exist under the ideal cooling conditions ($h \to \infty$, $\lambda_0 \to \infty$).

Obviously, ideal conditions $E_h \to \infty$ and $E_\lambda \to \infty$ should be kept in the experiments in which the critical current of superconductor is measured. However, the quantities E_h and E_λ are finite. Therefore, the contribution of the convective and conductive mechanisms in the formation of the voltage-current characteristic of the high-T_c superconductor will depend on the difference between E_h and E_λ. First, if $E_h/E_\lambda \ll 1$, i.e. at $H \ll 1$, then according to (47) – (50) the growth of the voltage-current characteristic of the superconductor will be practically uniform and isothermal in the range of electric fields $E \ll E_h$. Second, if $E_h \ll E \ll E_\lambda$, then the non-isothermal formation of the electrodynamics states will be spatially homogeneous. Third, the voltage-current characteristic of the superconductor will be influenced by the non-uniform distribution of temperature in its cross-sectional area when $E \gg E_\lambda$. Therefore, if $E \gg E_h \gg E_\lambda$, i.e. at $H \gg 1$, then non-isothermal electrodynamics processes occurring in the superconductor during current charging depends on the spatial heterogeneity of the temperature. In this case, the difference between non-isothermal voltage-current characteristics of the superconductor described by the equalities (47) and (50), which are defined in the framework of one-dimensional and zero-dimensional approximations, respectively, increases with increasing its thermal resistance (Figure 16). Therefore, the analysis of the formation of the thermo-electrodynamics states of superconductor should be done allowing for the non-uniform distribution of temperature at $E_h >> E_\lambda$, i.e. taking into account the thermal size effect.

The comparison between the critical current of the superconductor with the corresponding values of the current of instability calculated in accordance with the zero-dimensional and one-dimensional approximations is shown in Figure 17(a). These quantities are normalized to the width of the slab ($I^*_c = 0.5I_c/b$, $I^*_{q,0} = 0.5I_{q,0}/b$, $I^*_q = 0.5I_q/b$). Figure 17(b) depicts the corresponding curves in the range of small J_{c0}. Figure 17(c) presents the electric field preceding the onset of instability as a function of the critical current density. This curve was defined by the solution of equation (39). The calculations were made for the examined $Bi_2Sr_2CaCu_2O_8$ superconductor at $\lambda_0 = 0.92 \times 10^{-3}$ W/(cm×K).

The results presented clearly demonstrate the existence of the features of the variation in the current stability boundary of high-T_c superconductors. First, the zero-dimensional approximation leads to the overestimated boundary values of the current (also of the electric field) comparing with the corresponding values determined in the framework of the one-dimensional model. As discussed above, this feature is determined by the relevant difference between values T_v and $T_{v,0}$. Second, it is apparent that the values of E_q and I_q may be both lower and higher than the critical quantities E_c and I_c. Therefore, the stability areas may be both sub-critical and over-critical. In particular, the current instabilities may happen at $E_q < E_c$ and $I_q < I_c$ (sub-critical stability area). As follows from Figure 17(a), the higher the value J_{c0}, the more the difference between the values of I_q and I_c and also the difference between the values of I^*_q and $I^*_{q,0}$ calculated in terms of the one-dimensional and zero-dimensional models. However, the given peculiarities are modified with the decrease in the quantity J_{c0}, as follows from Figures 17(b) and 17(c). First, one can observe the over-critical regimes of stability when the allowed values of electric field and current exceed simultaneously the relevant values of E_c and I_c. Second, Figure 17 also shows possible existence of the intermediate stability area when the sub-critical stable currents ($I_q < I_c$) exist at the over-critical values of the electric field ($E_q > E_c$).

In order to find the boundary of the stability areas discussed, let us use the relations (41) and (42). Then, the boundary between the sub-critical and over-critical values of the electric field follows from the equality $E_q = E_c$. In this case, the current of the instability is obviously lower than the critical current of the superconductor. Accordingly, the condition of existence of the sub-critical stability area ($E_q < E_c$, $I_q < I_c$) is described by the inequality

$$\frac{E_h}{E_c} < n + H(n-1)/3 .$$

In the thermally uniform states ($H << 1$), this condition rewritten as

$$E_h / E_c < n$$

If these conditions are violated, the electric field induced in the superconductor before the onset of instability will exceed the value of E_c. In this case, as noted above, the current of the instability may be lower than I_c. Therefore, the equality $I_q = I_c$ allows to find the over-critical stable states ($E_q > E_c$, $I_q > I_c$). The latter will occur at

$$\frac{E_h}{E_c} > 1+\sqrt{H\left[n+H(n-1)/3\right]}/\tanh\sqrt{\frac{H}{n+H(n-1)/3}}$$

in the framework of the non-uniform model or if inequality

$$\frac{E_h}{E_c} > (n+1)\left(\frac{n+1}{n}\right)^n$$

takes place during zero-dimensional approximation. Then the intermediate current stability area, i.e. over-critical electric field and sub-critical currents ($E_q > E_c$, $I_q < I_c$), will take place at

$$1+\sqrt{H\left[n+H(n-1)/3\right]}/\tanh\sqrt{\frac{H}{n+H(n-1)/3}} > \frac{E_h}{E_c} > n+H(n-1)/3$$

or

$$(n+1)\left(\frac{n+1}{n}\right)^n > \frac{E_h}{E_c} > n$$

when $H << 1$.

The obtained criteria indicate that the existence of sub-critical or over-critical areas of stability in both thermally homogeneous and thermally heterogeneous states depends on the dimensionless parameter

$$\alpha = \frac{J_{c0}E_c a}{h(T_{cB}-T_0)} = \frac{E_c}{E_h}$$

It has a similar physical meaning as the known Stekly parameter [1], [2] and can be defined as the parameter of thermal stabilization of superconductor. In other words, the current stability boundary of high-T_c superconductors having no matrix depends, fist of all, on the efficiency of the Joule heating transferred into the coolant. In particular, the area of over-critical states together with the discussed influence of the critical current density of superconductor will increase with decrease in the thickness of the superconductor, with increase in its critical temperature or decrease in the temperature of the coolant. On the contrary, the sub-critical states are the most probable in massive high-T_c superconductors with high critical currents or at a low temperature margin when the thermal heterogeneity of the electrodynamics states of high-T_c superconductors effects significantly on the stability conditions of charged current.

As follows from the data presented in Figure 17, there exists the degradation mechanism of the current-carrying capacity of the superconductor. Namely, the allowable values of charged currents may be not only lower than a priori defined value of I_c, but also do not increase proportionally to the increase in this value. This effect is because the penetration of the charged current depends on the corresponding change in the superconductor's temperature, which inevitably increases with increasing current. Figure 18 depicts the dependence of degradation parameters $\Delta = (I_c - I_q)/I_q$ and $\Delta_0 = (I_c - I_{q,0})/I_{q,0}$ on the dimensionless values of thermal resistance H and the quantity $\varepsilon_2 = (E_\lambda / E_c)^{n/(n+1)}$, which were calculated at $n = 10$ according to the one-dimensional and zero-dimensional approximations, respectively. These results make it possible to determine the value of H and ε_2 at which the thermal degradation is most noticeable.

As it is expected, a significant reduction in the instability current relative to the critical current of the superconductor will be observed at $H << 1$, i.e. under the non-intensive cooling conditions. In this case (more exactly at $H < 0.1$), the values of Δ and Δ_0 do not practically differ from each other, and the size effect is insignificant. Therefore, the degradation parameter can be estimated in the spatially uniform approximation. Accordingly, it is easy to find

$$\Delta \approx \Delta_0 = \frac{n+1}{n}\left(\frac{n}{\alpha}\right)^{1/(n+1)} - 1$$

This expression shows that the degradation of current-carrying capacity of the high-T_c superconductors may be essential during non-intensive cooling conditions due to the finite value of the exponent n of their voltage-current characteristics. However, at $H > 10$, for example, in the intensive cooling, this estimation is too rough, because, as disused above, in this case the calculations of the stability conditions must take into account the thermal heterogeneity of the electrodynamics states induced in high-T_c superconductors by current charging. In these cases, the thermal degradation depends on the efficiency of the conductive transfer of the Joule heating and the critical properties of the superconductor.
The influence of the latter on the thermal degradation of current-carrying capacity can be determined analyzing the influence of value ε_2. As follows from Figure 18(b), the reduction in the stability currents in comparison with the critical current is the most noticeable at $\varepsilon_2 << 1$. In this case, the degradation of the current-carrying capacity will exceed 20% even at intensive cooling ($H = 10$). At the same time, the effect of degradation might be not so significant at $\varepsilon_2 >> 1$ because in this case it would depend, fist of all, on the heat removal to the coolant. Thereby, the characteristic value $\varepsilon_2 = 1$ defines the boundary between the areas of high or low thermal degradation of current-carrying capacity of the superconductor under the intensive cooling conditions.

Thus, the theoretical analysis taking into account the thermal heterogeneity of the electrodynamics processes in the cross section of high-T_c superconductor leads to the increased values of electric field induced before the current instability. This is due to the increase in its average temperature, which is inevitable before the instability onset. As a result, the estimated range of currents steadily flowing in the superconductor decreases. The proposed analysis show that the non-uniform temperature distribution in the superconductor depends not only on its thermal resistance but also on its critical parameters. Their influence will increase with increasing heat transfer coefficient.

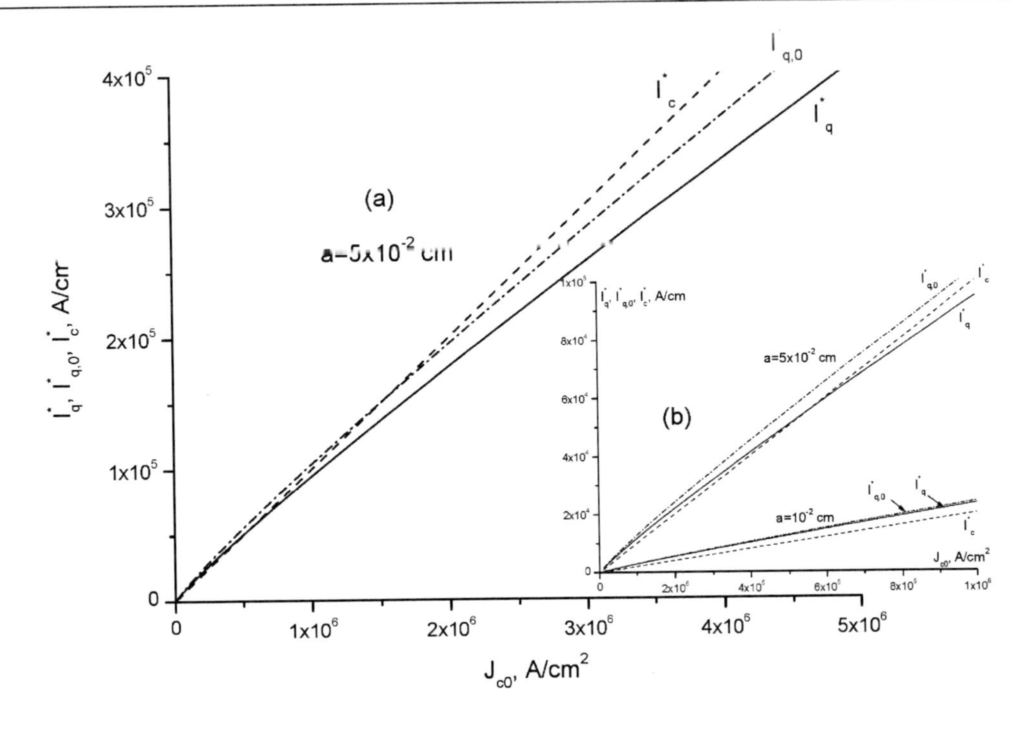

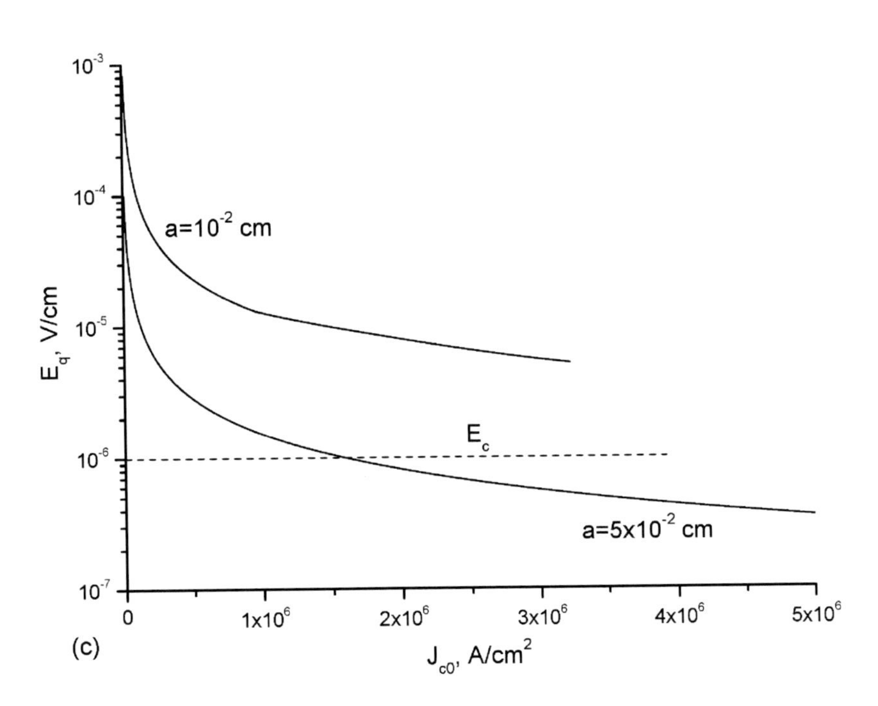

Figure 17. Influence of the critical current density of the superconductor on the allowable values of charged current (a, b) and electric field (c) before current instability onset.

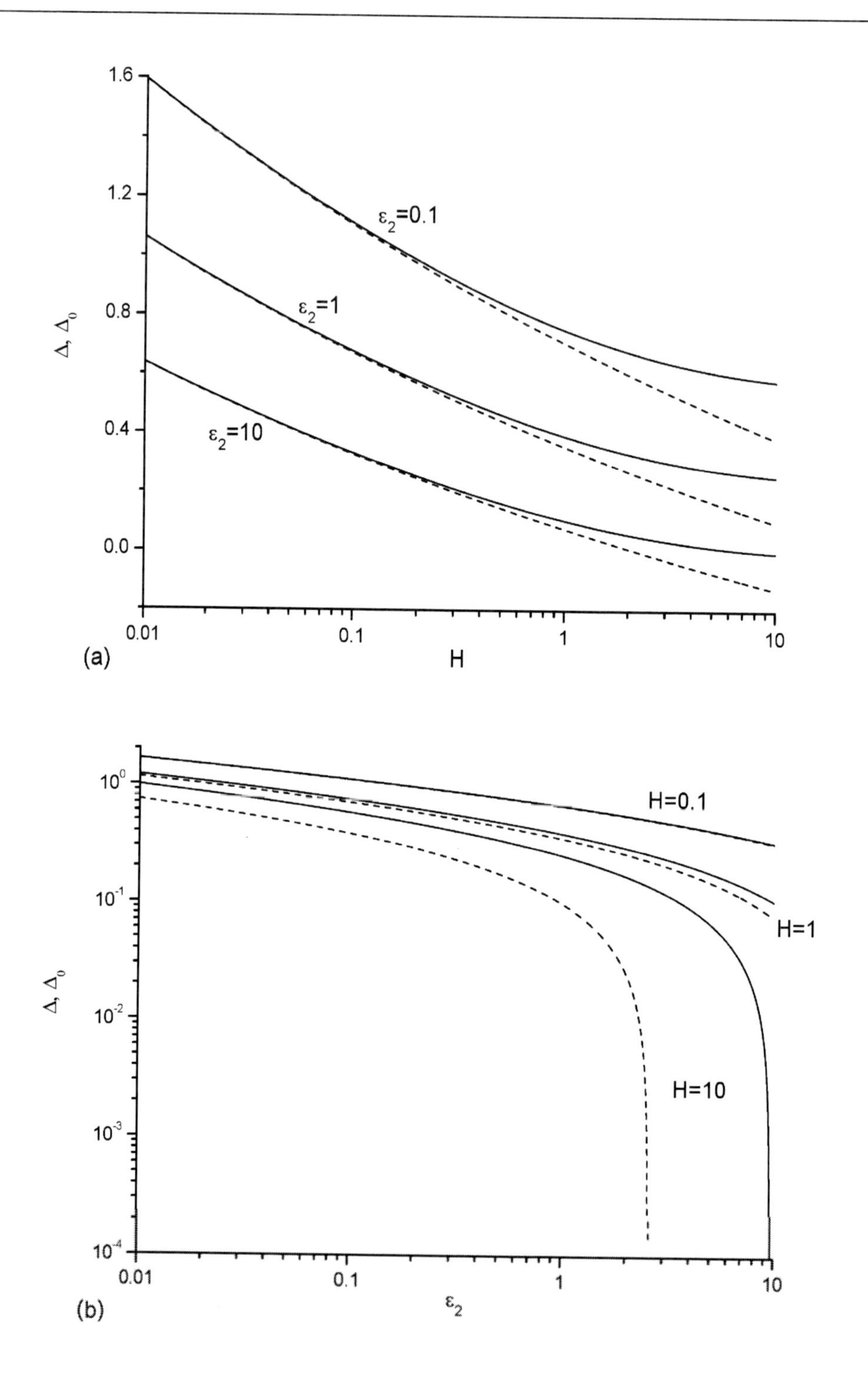

Figure 18. Dependence of the degradation parameters determined in the zero- and one-dimensional approximations on the thermal resistance of superconductor (a) and characteristic value of electric field ε_2 (b): (——) – Δ, (- - - -) – Δ_0.

Besides, there exist two typical values of the electric field, which affect the intensity of thermal growth of the voltage-current characteristics of the high-T_c superconductor and the stable range of charged current. They depend on the cooling condition, the thickness of the superconductor, its thermal conductivity coefficient and the critical properties, defining the role of the convective and conductive thermal mechanisms in the current state formation of high-T_c superconductors. In accordance with the thermal peculiarities of the stable electrodynamics state formation of the superconductor, the conditions of current stability change. The criteria of current instability onset, which were written taking into account the non-uniform temperature distribution in the cross-sectional area of superconductor, show that the allowable values of electric field and current can be sub-critical or over-critical relative to a priori chosen critical parameters of the superconductor E_c and J_c. As a result, the thermal degradation effect of current-carrying capacity of the superconductor exists. In this case, the valid values of charged currents do not increase with the proportional increase in cross-sectional area of the superconductor and its critical current density. The effect of the thermal degradation should be taken into account when the range of permissible current charged into the superconductor with a high critical current density is determined.

SUPERCONDUCTORS COOLED BY LIQUID COOLANT

The results above-discussed have allowed to formulate the physical features of the formation of stable thermo-electrodynamics states of high-T_c superconductors cooled by a cooler with constant heat transfer coefficient, i.e. without the infringement of the heat exchange conditions with a coolant on a surface of a superconductor. Let us discuss formation features of steady current modes in the superconductors cooled by liquid coolants (helium, hydrogen or nitrogen) when the nucleate and film heat transfer regimes exist. It allows to formulate the criteria allowing to estimate possible mechanisms of current instability onset.

As above, let us consider the problem of the determination of the limiting current stably flowing in the superconducting slab with a half-thickness a and half-width b placed in a constant external magnetic field that parallel to the slab surface and has penetrated over the cross-section of the slab. To simplify the analysis, let us, first of all, assume that the current is charged with an infinitely low rate and its self magnetic field is negligibly lower than the external magnetic field, the temperature T, the electric field E and the current density J are uniformly distributed inside the slab. Suppose that the surface of the superconductor is cooled by liquid coolant having the operating temperature T_0. As above, let us use a power law to describe the voltage-current characteristic of the superconductor. Then the static uniform distribution of the temperature and electric field for a given value of the current density may be defined solving the following system of equations

$$E = E_c\left[J / J_c(T)\right]^{n(T)}, \tag{51}$$

$$EJ = q(T)/a, \tag{52}$$

where $q(T)$ is the heat flux to the coolant. Below the liquid helium, hydrogen and nitrogen are considered as coolants, which have the nucleate and film boiling modes. Accordingly, the values of $q(T)$ during these boiling regimes were fitted by the expression

$$q(T)\left[W/cm^2\right]=\begin{cases} h_1\left(T-T_0\right)^{\nu_1}, & T\le T_0+\Delta T_{cr} \\ h_2\left(T-T_0\right)^{\nu_2}, & T>T_0+\Delta T_{cr} \end{cases}. \qquad (53)$$

Here, ΔT_{cr} is the critical overheating depending on the coolant, beyond which the transition from the nucleate boiling regime to the film boiling one occurs. Values h_1, h_2, ν_1, ν_2, T_0 and ΔT_{cr} estimated according to [43] are given in Table 1 for each type of coolant.

Table 1. Coolant properties

Coolant	T_0, K	ΔT_{cr}, K	h_1	h_2	ν_1	ν_2
Liquid helium	4.2	0.6	2.15	0.06	1.5	0.82
Liquid hydrogen	20	3	0.66	0.024	2.6	1.1
Liquid nitrogen	77.3	10	0.04	0.036	2.4	0.76

Note that unlike considered above models, the temperature dependence of n-value as well as the nonlinear nature of $J_c(T)$ are taken into account below.

Figure 19 depicts the static temperature-current and voltage-current characteristics that take place in the $Bi_2Sr_2CaCu_2O_8$ slab ($a = 10^{-2}$ cm) placed in the external magnetic field B=10 T and cooled by liquid helium. Two types of simulation were made. The calculations were performed both for the nucleate boiling regime and for the cooling conditions that could exist if only the film boiling mode takes place. The temperature boundary between these regimes is indicated by the dotted horizontal curve. The temperature dependences $J_c(T)$ and $n(T)$ were described as follows. The critical current density of the $Bi_2Sr_2CaCu_2O_8$ was determined by the formulae (9) and (11). The n-value is described by the expression

$$n(T)=n_0\frac{T_{cn}-T}{T_{cn}-T_0}+n_c\frac{T-T_0}{T_{cn}-T_0}. \qquad (54)$$

Certainly, the $n(T)$-relationship may have a more complicated form. However, the existing experiments describing the $n(T)$-dependence are few and, therefore, the exact theory describing the $n(T)$-dependences has still not been fully developed. At the same time, the formula (54) has certain advantages. First, it allows one to get temperature-independent value of n assuming that $n_0 = n_c$. In Figure 19 the solid curves 1 and 1' correspond to this approximation calculated at n_0=10. Second, the formula (54) allows also to estimate the influence of n-value decreasing with the temperature on the current instability conditions using simple temperature-linear approximation. This approximation is shown by the dashed curves 2 and 2' in Figure 19. They were obtained at n_0 = 10, n_c = 1, T_{cn} = 87.1 K. These temperature dependences of $n(T)$ are given in the inset of Figure 19(a).

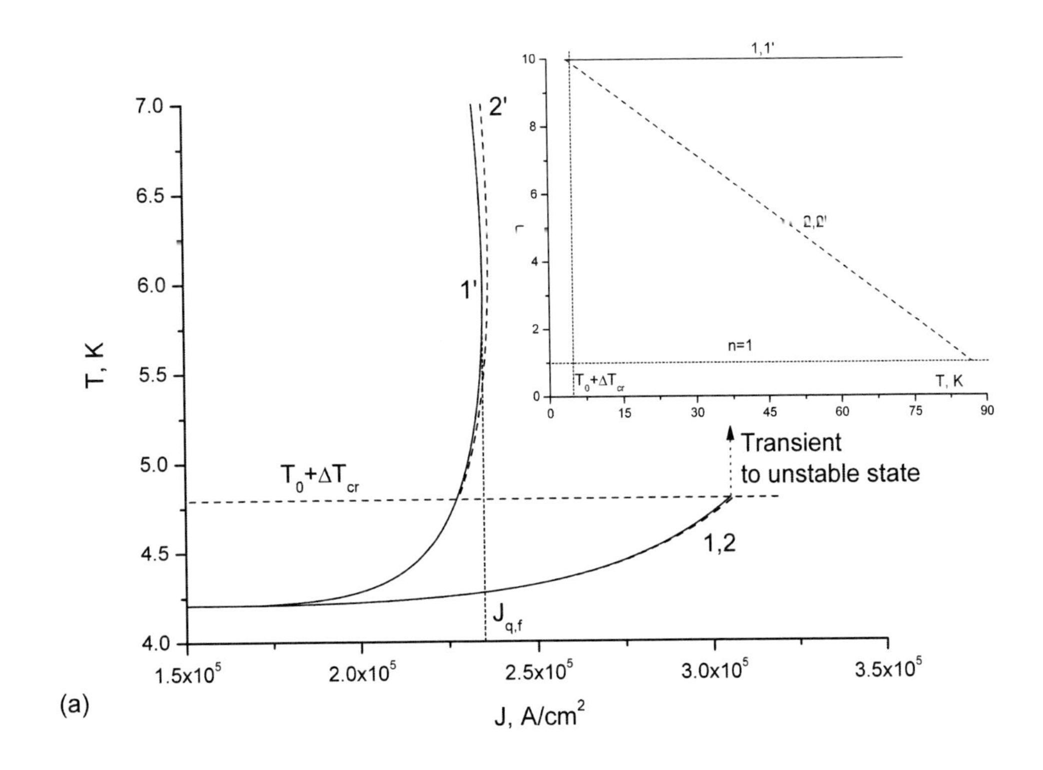

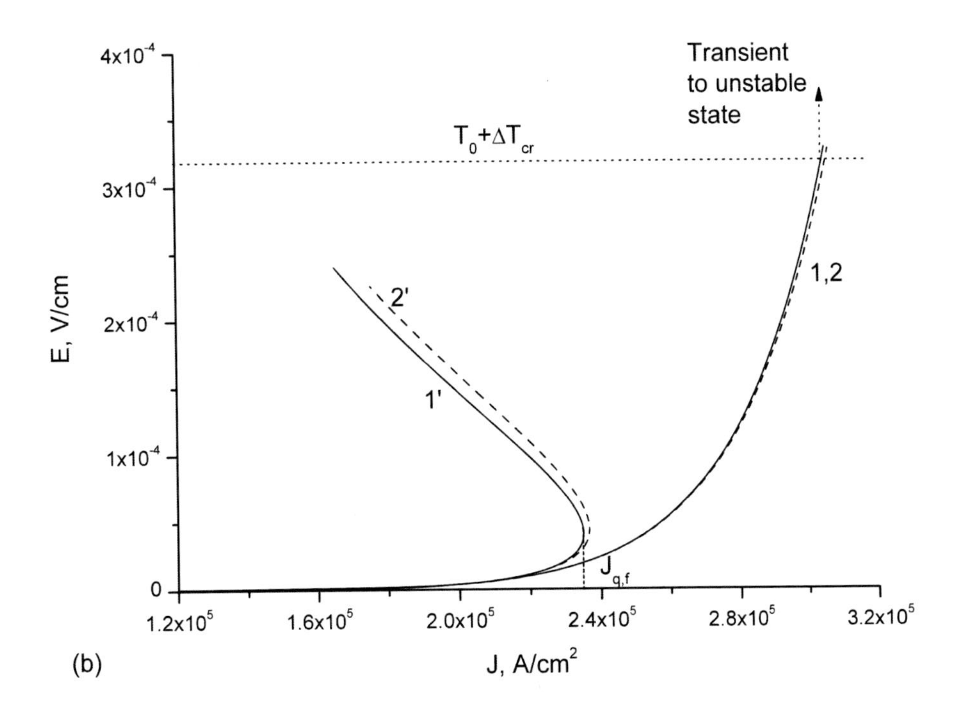

Figure 19. Temperature-current (a) and voltage-current (b) characteristics of the $Bi_2Sr_2CaCu_2O_8$ cooled by liquid helium: 1, 2 – the nucleate boiling regime, 1', 2' – the film boiling regime.

The results presented in Figure 19 show that the current instability in the examined superconductor is a consequence of its trivial overheating exceeding the quantity ΔT_{cr}. This overheating initiates the thermal transition from the nucleate to the film boiling modes. In this case, the slopes of curves $E(J)$ and $T(J)$ are positive ($\partial E/\partial J > 0$, $\partial T/\partial J > 0$) before the current instability onset. Therefore, in terms of the formation of the temperature-current and voltage-current characteristics of the superconductor, the current instability starts at their steady growth. At the same time, as it is discussed, the current instability in the superconductors can also occur if the condition (20) takes place. Hence, the condition (20) raises the question about the possibility of the current instability in superconducting materials cooled by liquid coolant, which can occur before the transition from the nucleate boiling to the film boiling regime according to the condition (20). In order to answer it, let us find the temperature of the superconductor before the current instability in the framework of the model (51) - (53), using the condition (20). Note that to take directly the advantage of this condition is mathematically difficult because of non-linearity of the system (51) - (53). At the same time, the problem is simplified, if the condition (20) will be rewritten in the form of the inequality

$$\partial J / \partial T < 0. \tag{55}$$

It is a consequence of the fact that the condition (20), as noted above, divides the temperature-current and voltage-current characteristics of a superconductor into two areas: stable, which satisfies the conditions $\partial E/\partial J>0$ or $\partial T/\partial J>0$, and unstable, when $\partial E/\partial J<0$ or $\partial T/\partial J<0$.

Excepting the electric field from equations (51) and (52), it is easy to get a relation between the current density and temperature. The latter may be written as

$$J(T) = \left[W(T) J_c^n(T) / E_c\right]^{1/(n+1)}, \tag{56}$$

where

$$W(T) = q(T)/a.$$

Then considering that the critical current density of the superconductor and its n-value decrease with temperature, the current distribution in the superconductor is unstable, if, according to the condition (55) and relation (56), the inequality

$$\frac{n}{J_c}\left|\frac{dJ_c}{dT}\right| + \left|\frac{dn}{dT}\right| \ln\frac{J_c}{J} > \frac{1}{W}\frac{dW}{dT} \tag{57}$$

is realized. Accordingly, the boundary of stable states is determined by solution of equation

$$\frac{n}{J_c}\left|\frac{dJ_c}{dT}\right| + \frac{1}{n+1}\left|\frac{dn}{dT}\right| \ln\frac{E_c J_c}{W} = \frac{1}{W}\frac{dW}{dT} \tag{58}$$

in accordance with the expression (56).

Criteria (57) or (58) allow one to determine directly the temperature of superconductor T_q before the current instability, which is a result of the condition (20). In particular, let us find this value assuming for simplicity that the critical current density is described by the linear relationship (4). Then, according to (58), the value of T_q follows from the solution of equation

$$\frac{1}{n(T_q)+1}\left|\frac{dn}{dT}\right|\Bigg|_{T=T_q} \ln\frac{aE_c J_c(T_q)}{h_1(T_q-T_0)^{\nu_1}} = \frac{\nu_1}{T_q-T_0} - \frac{n}{T_{cB}-T_q}.$$

In the simplest approximation $n = const$, the quantity T_q equals

$$T_q = T_0 + \frac{T_{cB}-T_0}{1+n/\nu_1}. \tag{59}$$

Let us use this simplified solution and find the condition under which the current instability occurs because of the conditions (20) rather than the thermal transition from the nucleate to the film boiling regimes. It is clear that it takes place at $T_0+\Delta T_{cr} > T_q$, i.e. at

$$\frac{n}{\nu_1} > \frac{T_{cB}-T_0}{\Delta T_{cr}} - 1. \tag{60}$$

Thus, the parameters of the superconductor and the liquid coolant are connected by the non-trivial relationship leading to different mechanisms of the current instability onset. In particular, they depend on difference in the rise rates of the voltage-current characteristic of superconductor (n-value) and the heat flux to the coolant during nucleate boiling regime (ν_1-value). As a whole, the condition (60) indicates the following characteristic peculiarities of the current instability onset in the high-T_c superconductor cooled by liquid coolant.

First, as follows from the condition (60), a characteristic n-value exists. It equals

$$n_\nu = \left(\frac{T_{cB}-T_0}{\Delta T_{cr}} - 1\right)\nu_1.$$

If $n > n_\nu$, then the current instability during nucleate boiling mode will occur according to the condition (20). As a result, the current instability will take place before the violation of the nucleate boiling mode at the lower n-value, if the operating temperature T_0 or the critical overheating ΔT_{cr} are higher while the quantity T_{cB} is smaller. As a consequence, in the limiting case, satisfying the condition

$$\left(\frac{T_{cB}-T_0}{\Delta T_{cr}} - 1\right)\nu_1 < 1$$

the current instability will always occur before the crisis of the coolant boiling for all $n > 1$.

Table 2. Potential temperature margin versus n-value

n-value	10	20	30	40	50
ΔT, liquid helium	4.6	8.6	12.6	16.6	20.6
ΔT, liquid hydrogen	14.54	26.08	37.62	49.15	60.69
ΔT, liquid nitrogen	51.67	93.33	135	177	218

Second, the electrodynamics mechanism of the current instability, i.e. the criterion (20), is the most probable when the temperature margin of the superconductor that equals $\Delta T = T_{cB} - T_0$ is higher at the fixed value of the quantity ΔT_{cr}. Let us use the condition (60) and estimate the potential values of ΔT at which the current instability follows from the conditions (20) rather than the boiling crisis. Relevant values of ΔT as a function of n-value are presented in Table 2 for different types of coolant. They were obtained according to the data given in Table 1.

Table 2 shows that the current instability may satisfy the condition (20) even when the high-T_c superconductor is cooled by liquid helium. However, these unstable regimes can be observed when the high-T_c superconductor will be placed in very high external magnetic field. At the same time, in the cases of high-T_c superconductor cooled by liquid nitrogen the electromagnetic mechanism of the destruction of the stable current distribution described by the condition (20) will be realized in a wide range of variations of the external magnetic field. Moreover, as it will be discussed below, the values of $n(T)$ should be close to unit to start the current instability because of the boiling crisis of liquid nitrogen.

To illustrate the peculiarities discussed, some temperature-current and voltage-current characteristics are presented in Figures 20 and 21. They were calculated for the $Bi_2Sr_2Ca_2Cu_3O_8$ slab ($a = 10^{-2}$ cm), which is in the external magnetic field $B = 1$ T, and for the $YBa_2Cu_3O_7$ film ($a = 10^{-5}$ cm) placed in the external magnetic field $B = 5$ T. The calculations were performed both for the nucleate and for the film boiling modes when the $Bi_2Sr_2Ca_2Cu_3O_8$ slab is cooled by liquid hydrogen and the $YBa_2Cu_3O_7$ film is cooled by liquid nitrogen. The temperature dependence of $n(T)$ was determined by the formula (54). In this case, the n-values for the $Bi_2Sr_2Ca_2Cu_3O_8$ are assumed to be equal to $n_0 = n_c = 30$ (curves 1 and 1') and $n_0 = 30$, $n_c = 1$ and $T_{cn} = 62.72$ K (curves 2 and 2'). For the $YBa_2Cu_3O_7$ the following values $n_0 = n_c = 10$ (curves 1 and 1') and $n_0 = 10$, $n_c = 1$ and $T_{cn} = 90$ K (curves 2 and 2') were set. The relevant dependences of $n(T)$ are shown in the insets of Figures 20(a) and 21(a). The temperature dependences of the critical current density of the superconductors under consideration were set as follows. The critical current density of the $Bi_2Sr_2Ca_2Cu_3O_8$ is determined by the formula (4) for which the values J_{c0} and T_{cB} were adopted from the linear interpolation of the experimental data given in [29]. Therefore, they were equal to $J_{c0} = 8.42 \times 10^4$ A/cm^2 and $T_{cB} = 62.72$ K at B = 1 T, $T_0 = 20$ K, $E_c = 10^{-6}$ V/cm. For the $YBa_2Cu_3O_7$ the critical current density is calculated from the formula

$$J_c(T,B) = J_0\left(1 - T/T_c\right)^{1.5} B^{-0.62}\left(1 - B_{irr}\right)^{2.27}, \quad B_{irr} = 376.235\exp(-0.048T)$$

at $E_c = 2 \times 10^{-6}$ V/cm, $J_0 = 2.8 \times 10^6$ A/cm^2, $T_c = 90$ K in accordance with the results presented in [44], [45].

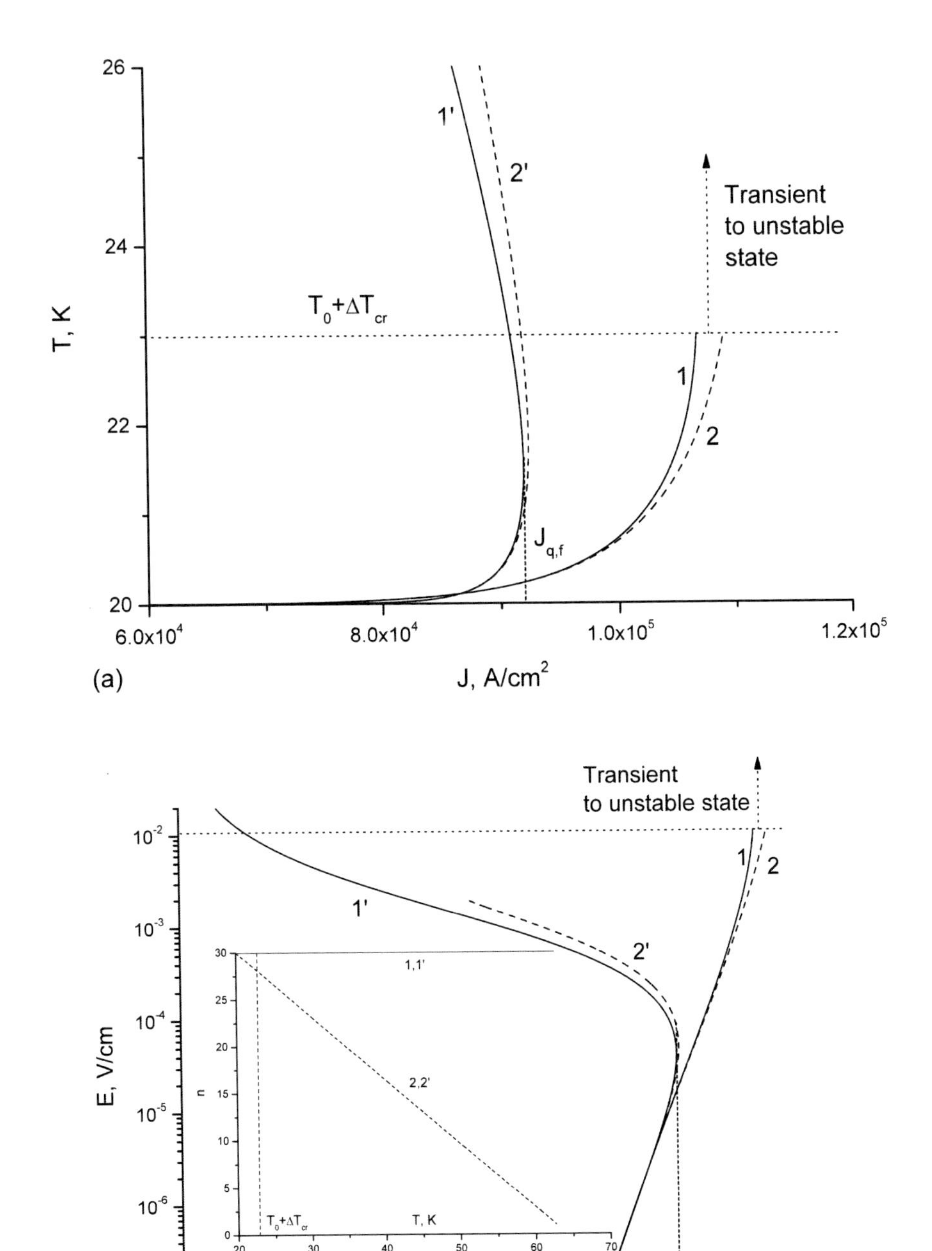

Figure 20. Temperature-current (a) and voltage-current (b) characteristics of the $Bi_2Sr_2Ca_2Cu_3O_8$ cooled by liquid hydrogen: 1, 2 – the nucleate boiling regime, 1', 2' – the film boiling regime.

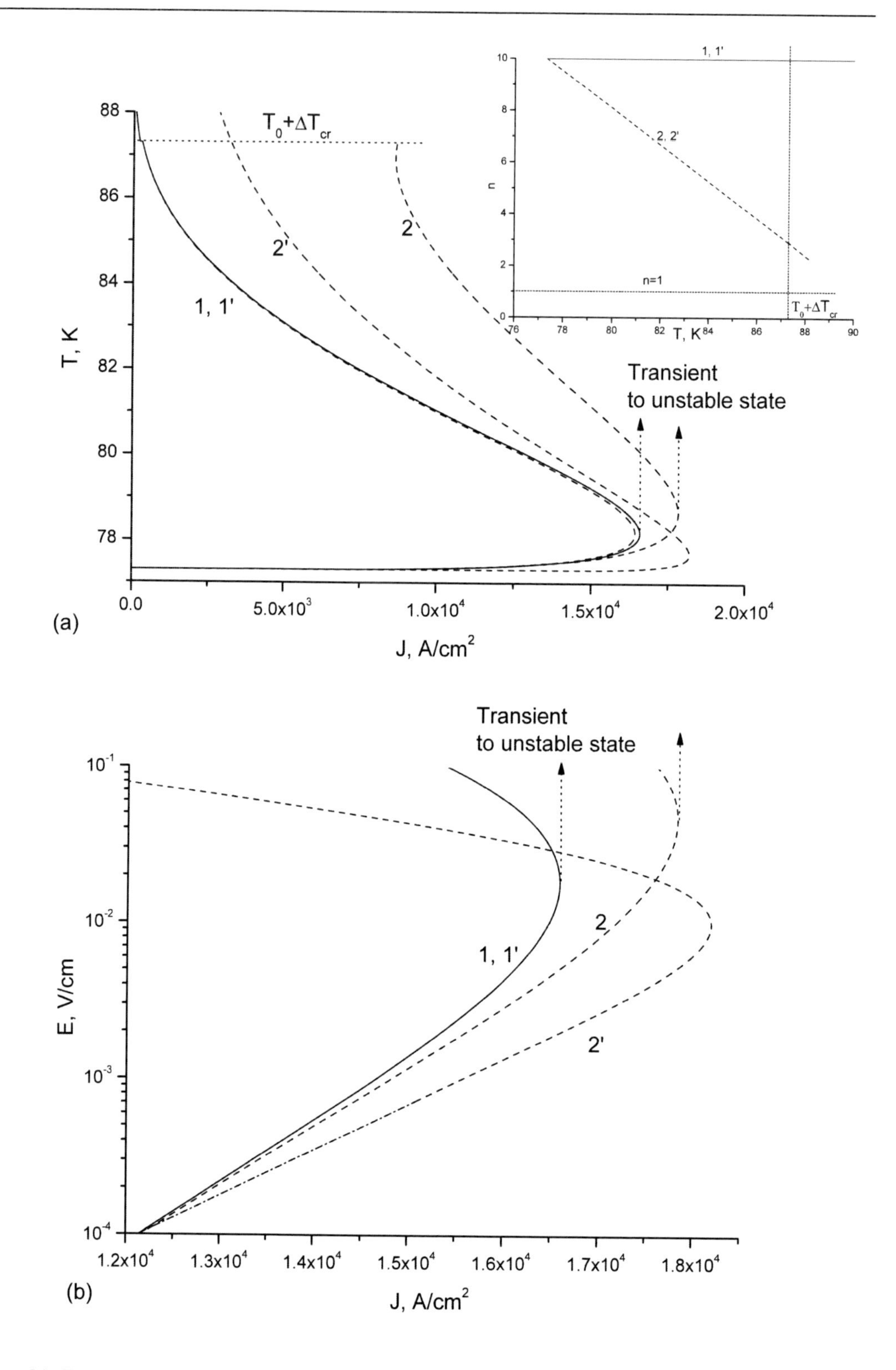

Figure 21. Temperature-current (a) and voltage-current (b) characteristics of the $YBa_2Cu_3O_7$ cooled by liquid nitrogen: 1, 2 – the nucleate boiling regime, 1', 2' – the film boiling regime.

As can be seen from Figure 20, the current instability in the $Bi_2Sr_2Ca_2Cu_3O_8$ slab occurs when the nucleate boiling mode of the cooling is destabilized. Nevertheless, the value of T_q = 23.4 K calculated for the superconductor under consideration using the formula (59) at $n = 30$ is close to the corresponding value of ΔT_{cr} that equals 23 K. So, the curves 1 depicted the dependences $E(J)$ and $T(J)$ in Figure 20 have sharply increasing character before the current instability unlike similar curves 1 shown in Figure 19. Consequently, the mechanism of the current instability in the $Bi_2Sr_2Ca_2Cu_3O_8$ slab may be modified. As a result, the violation of its current stability may be described by the condition (20), for example, due to the corresponding decrease in the quantity T_{cB}.

The current instability in the $YBa_2Cu_3O_7$ film under consideration occurs before a violation of the nucleate boiling regime (Figure 21) according to both the criteria (57) in the case of constant n-value and (60) when it decreases with temperature. Let us estimate the value of n in the approximation $n = const$ when the violation of the stable state of current charged into the high-T_c superconductors cooled by liquid nitrogen will occur due to the unlimited growth of their differential resistivity. Assuming T_{cB}-$T_0 \sim$ 20 K, it is easy to find from (60) that the current instability in the high-T_c superconductors cooled by liquid nitrogen will occur before the onset of the boiling crisis of the coolant at $n > 2.4$. In the meantime, it is known [29], [30] that the values of n of high-T_c superconductors cooled by liquid nitrogen can become close to unit with increasing temperature of the superconductor. Therefore, as follows from the comparison of the results presented in Figures 19 – 21, it is necessary to take into account the temperature-dependent nature of the values n only in the cases when liquid nitrogen is used. This is because its value ΔT_{cr} is higher than the corresponding values of liquid helium or hydrogen. So, the more general criteria (57) or (59) should be used to determine the possible mechanism of the current instability when a noticeable increase in the allowable temperature of a superconductor occurs.

The results and criteria discussed above were obtained under the assumption that there exists a spatial homogeneity of the thermo-electrodynamics states during current charging, i.e. at $ha/\lambda \ll 1$. If this condition is valid, the current stability problem should be solved taking into account the size effect, i.e. considering the heterogeneous distribution of the temperature in the cross section of the superconductor. Let us discuss its possible influence on the current instability conditions of the above-discussed bismuth-based superconductors. (It is easy to find that the size effect is insignificant for the $YBa_2Cu_3O_7$ film considered because of its small thickness).

Let us determine the limiting current stably flowing in the superconductor using the solution of the one-dimensional stationary equation (27) with the boundary conditions

$$\frac{\partial T}{\partial x}(0,t)=0,\ \lambda\frac{\partial T}{\partial x}(a,t)=-q(T).$$

In this case, describing the voltage-current characteristic of superconductor and its temperature dependence on the critical current density and n-value as it was made above, we will use the relationship (18) for the $Bi_2Sr_2CaCu_2O_8$ and

$$\lambda(T) = 2\times10^{-4}T\left[\frac{W}{cm\times K}\right]$$

for the $Bi_2Sr_2Ca_2Cu_3O_8$ according to [46] to define thermal conductivity.

Figures 22 and 23 depict the temperature-current characteristics of the $Bi_2Sr_2CaCu_2O_8$ and the $Bi_2Sr_2Ca_2Cu_3O_8$ numerically calculated at $n_0 = n_c = 10$ and $n_0 = n_c = 30$, respectively. These curves were determined for the nucleate boiling regime both excluding the size effect (curve 1) and taking it into account (curves 2, 3).

Figure 22 shows that in the case of the $Bi_2Sr_2CaCu_2O_8$ examined the current instability occurs because of the violation of the nucleate boiling regime even in spite of the significantly heterogeneous temperature distribution in the cross section of the superconductor. However, the attention should be paid to the difference in the growth character of temperature-current curves calculated in the framework of zero and one-dimensional approximations. It is seen that the slope of the curve 1 obtained in terms of the zero-dimensional model is less smooth than the slope of the curves 2 and 3, which were calculated taking into account the heterogeneous distribution of temperature in the cross section of the superconductor. This feature is a consequence of the size effect discussed: the formation of a spatially heterogeneous temperature field is accompanied by its sharper rise in comparison with the corresponding change in the uniform temperature field during current charging. This feature becomes more evident before the onset of the boiling crisis when the curves 2 and 3, unlike the curve 1, start to grow quickly. Obviously, this difference between the zero-dimensional and one-dimensional approximations also takes place in the formation of the electric field. Therefore, the current instability will be a result of the electromagnetic mechanism described by the condition (20) rather than by the thermal mechanism based on the transition from the nucleate to film boiling modes when the subsequent increase in the transversal size of the given superconductor will be made. Such an influence of the size effect on the current stability conditions also depends on the type of coolant. This can be seen from Figure 23. Thereby, if, in the framework of the zero-dimensional model, the current instability in $Bi_2Sr_2Ca_2Cu_3O_8$ cooled by liquid hydrogen is the result of its thermal overheating exceeding the quantity $T_0 + \Delta T_{cr}$, then according to the one-dimensional approximation the current instability occurs before the violation of the nucleate boiling regime. In this case, the temperature-current characteristics of the $Bi_2Sr_2Ca_2Cu_3O_8$ calculated both on the surface of the superconductor and inside its central part have two character branches: stable ($\partial T/\partial J > 0$) and unstable ($\partial T/\partial J < 0$). Therefore, an adequate description of the current instability mechanisms in the high-T_c superconductors cooled by liquid coolant at the intermediate operating temperatures also depends on the spatial homogeneity of the temperature field induced in the superconductor due to current charging. The results presented in Figures 22 and 23 also show that the size effect leads to a steady reduction in the stable current range of the liquid-cooled superconductors. The discussion of this peculiarity was made above.

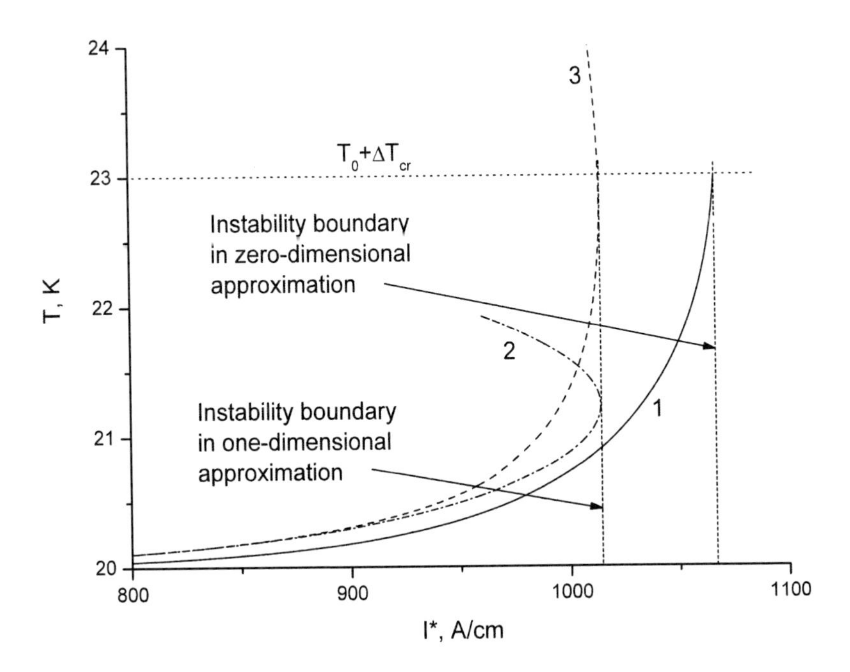

Figure 22. Temperature of the $Bi_2Sr_2CaCu_2O_8$ ($a = 10^{-2}$ cm) cooled by liquid helium versus normalized current ($I^* = J \times a$): 1 – zero-dimensional approximation, 2, 3 – one-dimensional approximation (2 – $T(a)$, 3 – $T(0)$).

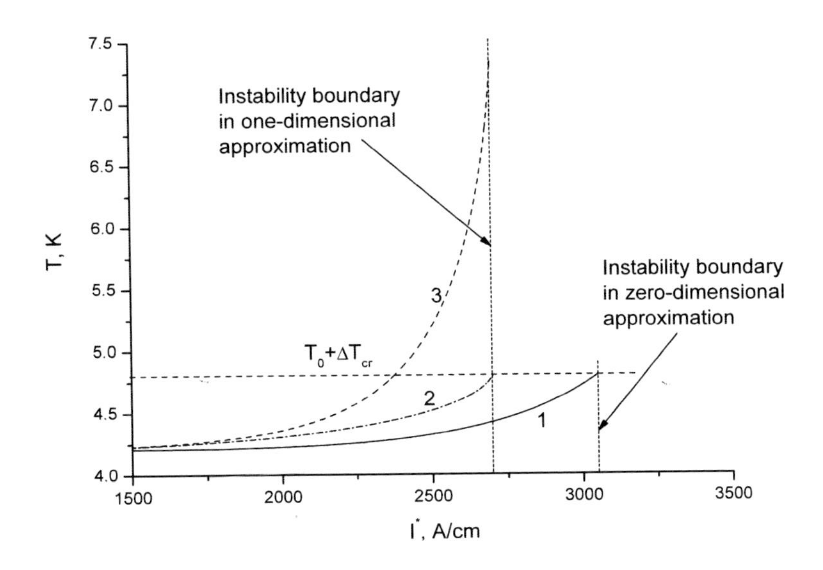

Figure 23. Temperature of the $Bi_2Sr_2Ca_2Cu_3O_8$ ($a = 10^{-2}$ cm) cooled by liquid hydrogen versus normalized current ($I^* = J \times a$): 1 – zero-dimensional approximation, 2, 3 – one-dimensional approximation. Here, 2 – $T(a)$, 3 – $T(0)$.

Thus, the current instability mechanisms in high-T_c superconductors cooled by the liquid coolant have intrinsic peculiarity defined by the allowable stable variation of their temperature before the onset of instability. There exists a non-trivial correlation between properties of superconductor and coolant leading to different mechanisms of the current instability onset. The latter also depends on the character of the temperature distribution in the cross section of the superconductor. As a result, the instability onset may be caused by both the trivial overheating of the superconductor's surface above the temperature transition from the nucleate boiling regime to the film boiling one and by the appearance of an unstable branch in the voltage-current characteristic even under the nucleate boiling mode of cooling. The criteria formulated allow one to define each of the mentioned mechanisms. They show that the probability of the thermal mechanism of current instability, which is the result of the boiling crisis, increases with decreasing n-value. Besides, this thermal mechanism of the current instability onset is the most typical instability mechanism for high-T_c superconductors cooled by a liquid coolant with low operating temperature, especially when they are cooled by liquid helium. At the same time, the boiling crisis will lead to unstable current states of high temperature superconductors cooled by liquid nitrogen only when the n-value is reduced to possible values closed to unit. Accordingly, the stability analysis of the thermo-electrodynamics states of superconductors cooled by liquid nitrogen should take into account the decrease in the temperature of the n-value.

These peculiarities make the current instability onset in the high-T_c superconductors cooled by liquid hydrogen or nitrogen different compared with the low-temperature superconductors cooled by liquid helium. Therefore, the study of the current stability conditions of high-T_c superconductors must consider this peculiarity, particularly, when liquid nitrogen is used.

8. Possible Multi-Stable Static Resistive States of High-T_C Superconductors with Temperature-Decreasing Power Exponent of their Voltage-Current Characteristic

The results discussed in the previous paragraph show that it is a necessity to account of temperature dependence of n-value in the cases when allowable overheating of a superconductor changes in a wide range. To investigate the effect of n-value, let us consider a simple problem regarding stable current distribution in an infinitely long $Bi_2Sr_2Ca_2Cu_3O_8$ with slab geometry. Assume that the external magnetic field B is constant; its value much more than self magnetic field and longitudinal variation is negligible. Suppose also that the power law describes the voltage-current characteristic of a superconductor. To evaluate the temperature and electric field without excessive computations, let us assume that their distributions over the cross-section of a superconductor are uniform. Under these assumptions, the following set of equations (13) and (51) describes the static distribution of the temperature and electric field in a superconductor. In the analysis, we take the superconductor to be $Bi_2Sr_2Ca_2Cu_3O_8$ at $B = 10$ T with the following parameters $E_c = 10^{-6}$ V/cm, $a = 10^{-2}$ cm, $h = 10^{-3}$ W/cm^{2}K. The critical current density and n-value of Bi2223 versus temperature and magnetic field are described in the form proposed in [29] during calculations. Figure 24 shows the corresponding dependences below used.

The problem formulated allows one to study the change of temperature and electric field inside the superconductor in a static approximation, which will occur at an infinitely slow increase in the transport current. In particular, it is not difficult to find the quantities $\partial E/\partial J$ and $\partial T/\partial J$ determining the steepness and the monotony of growth of electric field and temperature of superconductor in the resistive states. Omitting the intermediate transformations and taking the temperature-decreasing quantities J_c and n into consideration, one can find

$$\frac{\partial E}{\partial J}=\frac{\left(\frac{E_c}{E}\right)^{1/n(T)}+\frac{Ea}{h}\left[\left|\frac{dJ_c}{dT}\right|-\frac{J_c(T)}{n(T)^2}\left|\frac{dn}{dT}\right|\ln\left(\frac{E}{E_c}\right)\right]}{\frac{J_c(T)}{n(T)E}-\frac{Ja}{h}\left[\left|\frac{dJ_c}{dT}\right|-\frac{J_c(T)}{n(T)^2}\left|\frac{dn}{dT}\right|\ln\left(\frac{E}{E_c}\right)\right]}, \tag{61}$$

$$\frac{\partial T}{\partial J}=\frac{a}{\frac{h}{E}-\frac{aJn(T)}{J_c(T)}\left[\left|\frac{dJ_c}{dT}\right|-\frac{J_c(T)}{n(T)^2}\left|\frac{dn}{dT}\right|\ln\left(\frac{E}{E_c}\right)\right]}. \tag{62}$$

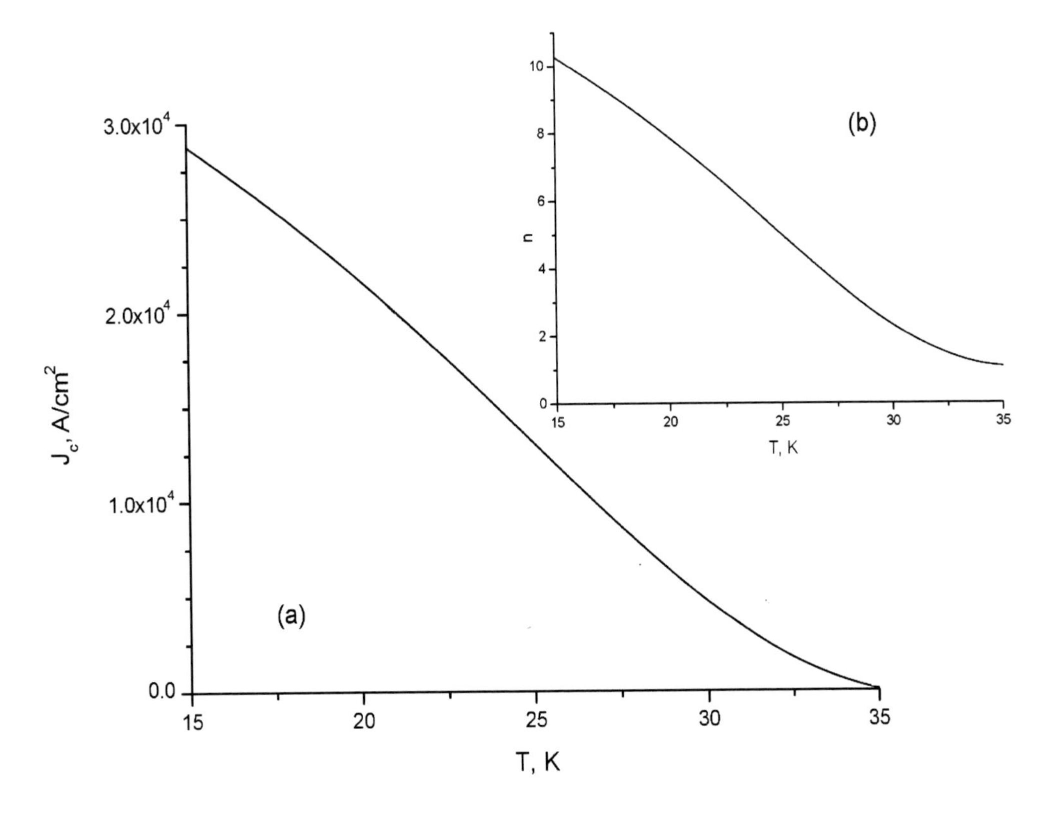

Figure 24. Critical current density (a) and n-value (b) of Bi2223 versus temperature after [30].

These expressions demonstrate the basic physical characteristics that underlie the formation of resistive states of the superconductor with n-value depending on the temperature. First of all, formulae (61) and (62) show that the stable rise of the electric field and temperature depends on the temperature variation of the quantities J_c, dJ_c/dT, n and dn/dT taking place in the various values of induced electric field. As a result, their effect on the $\partial E/\partial J$ and $\partial T/\partial J$ dependences is ambiguous. As a result, the decrease of $n(T)$ plays unstable role in the formation of the sub-critical regimes ($E < E_c$). However, the temperature-decreasing character of $n(T)$ may occur stable influence on the thermo-electric regimes at $E > E_c$. In other words, the voltage-current and temperature-current characteristics of high-T_c superconductor may have an additional stable branch in the over-critical electric fields. Note that possible existence of the multi-stable branches during formation of the voltage-current and temperature-current characteristics of superconducting composite was early discussed in [47] taking into consideration the stable jump-current-redistribution between the superconductor and matrix. In that case, the existence of the multi-stable branches was a result of the current sharing mechanism. In the case under consideration, the multi-stable states may be observed in superconductor without stabilizing matrix due to temperature decreasing dependence of n-value. This feature can lead to significant functional differences between the voltage-current characteristics of a superconductor calculated both at n-value decreasing with temperature and at a constant value of n. Indeed, in the latter case ($n=const$) the expression (61) and (62) are rewritten as follows

$$\frac{\partial E}{\partial J}=\frac{\left(\frac{E_c}{E}\right)^{1/n}+\frac{Ea}{h}\left|\frac{dJ_c}{dT}\right|}{\frac{J_c(T)}{nE}-\frac{aJ}{h}\left|\frac{dJ_c}{dT}\right|},\qquad \frac{\partial T}{\partial J}=\frac{a}{\frac{h}{E}-\frac{aJn}{J_c(T)}\left|\frac{dJ_c}{dT}\right|}. \tag{63}$$

As it was shown above, these formulae depict that the voltage-current and temperature-current characteristics of these superconductors will have only one stable and unstable branches satisfying the conditions $\partial E/\partial J > 0$, $\partial T/\partial J > 0$ and $\partial E/\partial J < 0$, $\partial T/\partial J < 0$, respectively. Formally, the existence of these conditions is associated with the change of signs in denominator of expressions describing the values of $\partial T/\partial J$ and $\partial E/\partial J$ due to the reduction with the temperature the values of $J_c(T)$. In this case, the electric field increases monotonically and stably up to the boundary values known as the quench electric field separating the stable state from unstable one. Therefore, in the static approximation, the quench electric field and the corresponding values of the quench current or the allowable temperature rise of superconductor are determined from the condition (20). Accordingly, the mono-stable resistive states are unstable if the applied current exceeds the quench current.

In Figure 25, the voltage-current and the temperature-current characteristics of $Bi_2Sr_2Ca_2Cu_3O_8$ are depicted. They were derived under the assumption of constant value of n (dashed curve 5) and its reduction with temperature (solid curves 1-4).

The results presented show that the dependences $E(J)$ and $T(J)$ can have both mono-stable and multi-stable branches when the values of $n(T)$ decrease with temperature. The first of them are observed in the high temperature range of the coolant, close to the critical temperature of superconductor. At the same time, at low coolant temperature, the resistive states of Bi2223-superconductor can be transformed into the multi-stable ones. As a result, the existence of the following states is possible (curve 3). At the initial stage of current charging, which exists in the current density range changing from 0 to J_1, the electric field and temperature of superconductor steadily increase with increasing current density since the conditions $\partial E/\partial J > 0$ and $\partial T/\partial J > 0$ take place. The value of current density J_1 of the first stable part of the voltage-current characteristics is defined by the condition $\partial E / \partial J \to \infty$ or $\partial T / \partial J \to \infty$ at $J \to J_1$ In the current density range from J_2 to J_1, where J_2 also follows from the conditions $\partial E / \partial J \to \infty$ or $\partial T / \partial J \to \infty$ at $J \to J_2$, an unstable branch exists due to the conditions $\partial E/\partial J < 0$ and $\partial T/\partial J < 0$. In the same current range, but at higher electric field, the stability conditions $\partial E/\partial J > 0$ and $\partial T/\partial J > 0$ are possible and, therefore, second stable region appears. As a result, the stable resistive mode of Bi2223-superconductor under consideration will move to a new stable regime at currents above the value J_1. Its boundaries vary in the current density range from J_2 to J_3. If applied current exceeds the value of J_3 then the thermo-electrodynamics states of Bi2223-superconductor under consideration will be completely unstable, because subsequent branches in the voltage-current and temperature-current characteristics have only negative slope. Thus, the value of J_3 defines limiting current-carrying capacity of the superconductor determined in accordance with the temperature dependence of n-value at a given coolant temperature.

From a formal point of view, the possible existence of the multi-stable modes is a consequence of the multiple-valued solutions of equation

$$\frac{h}{aJE} = \frac{n(T)}{J_c(T)}\left|\frac{dJ_c}{dT}\right| - \frac{1}{n(T)}\left|\frac{dn}{dT}\right| \ln\left(\frac{E}{E_c}\right) \tag{64}$$

which follows from (61) or (62) at $\partial E / \partial J \to \infty$ or $\partial T / \partial J \to \infty$. Accordingly, if this equation rewritten, for example, in the temperature terms defining the range of allowable overheating of superconductor, has one root, then the resistive modes of superconductor is mono-stable. In the cases when the number of roots will be equal to three, then there will be multi-stable states. There is also a state when the roots of equation (64) are absent. According to (64), this takes place at

$$\frac{1}{J_c(T)}\left|\frac{dJ_c}{dT}\right| < \frac{1}{n^2(T)}\left|\frac{dn}{dT}\right| \ln\left(\frac{E}{E_c}\right). \tag{65}$$

As can be seen from (62), this criterion corresponds to current mode during which the temperature of superconductor at current charging will steadily increase until the critical temperature, i.e. there will be the condition $\partial T/\partial J > 0$ at $T_0 < T < T_{cB}$. In general, as noted above and follows from equation (65), the stable formation of additional resistive states depends on the interrelated changes with the temperature of the values J_c, dJ_c/dT, n and dn/dT. In Figure 25, this feature is illustrated by curves 4 and 5 calculated under various approximations of $n(T)$.

Therefore, the correct description of the quantities J_c and n as a function of temperature and, particularly, near the critical temperature plays an important role in the theoretical analysis of the electrodynamics states of high-T_c superconductors.

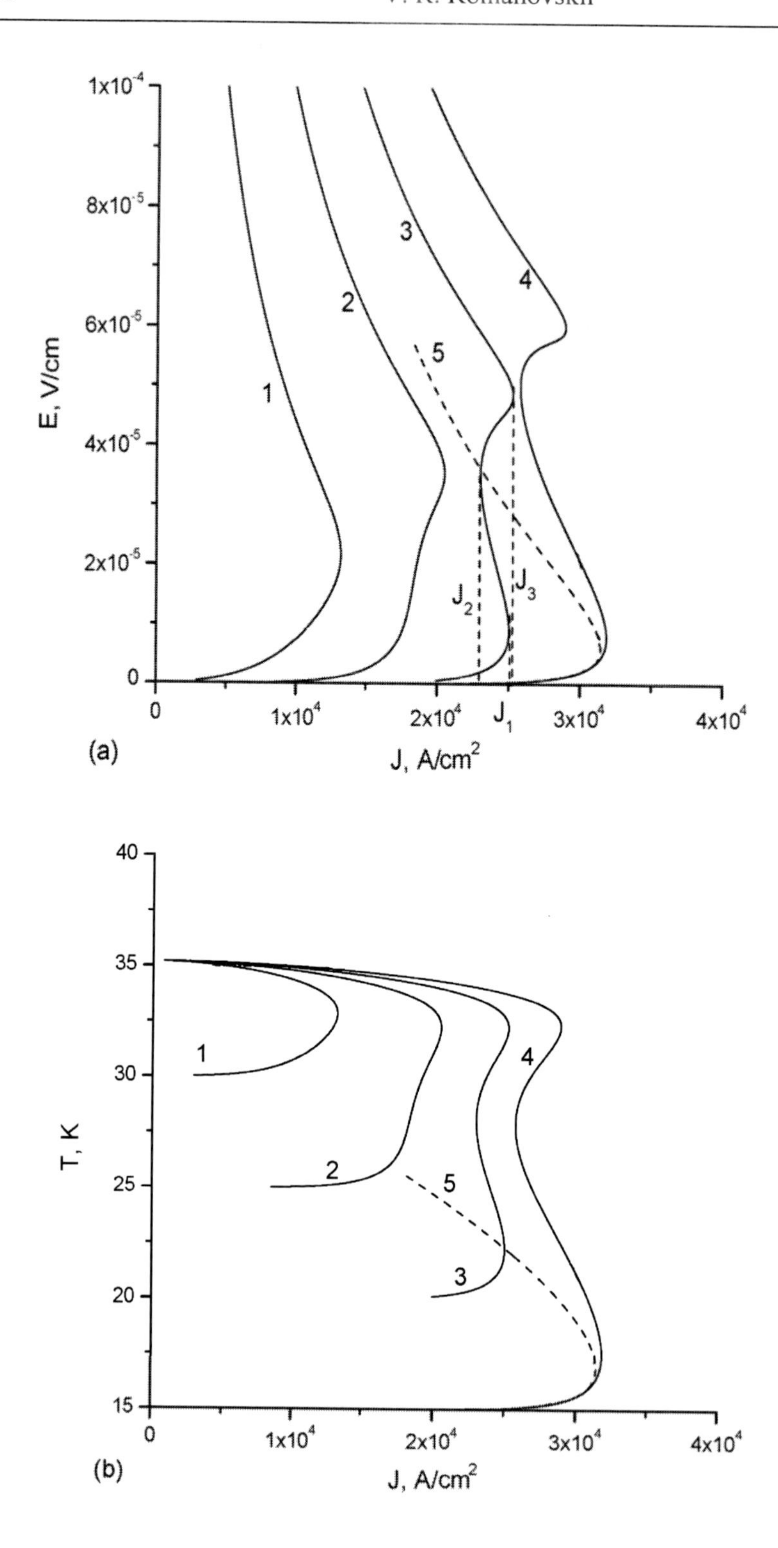

Figure 25. Possible voltage-current (a) and temperature-current (b) characteristics of Bi2223 at $B = 10$ T and different operating temperatures. Non-linear approximation of $n(T)$: 1 – $T_0 = 30$ K, 2 – $T_0 = 25$ K, 3 – $T_0 = 20$ K, 4 – $T_0 = 15$ K. Linear approximation n: ($n = 10$), 5 – $T_0 = 15$ K.

9. Physical Peculiarities of Dissipative Phenomena in High-T_C Superconductors Due to Self-Field Losses

Knowledge of the AC losses in high-T_c superconductors is crucial for designing commercial superconducting devices. They are due to the changing magnetic flux within the superconductor. The flux variation may be due to the applied magnetic field or due to the self-field of a charged current or combination of them. As the reduction of AC losses is one of the main research targets concerning the application of superconducting materials, then their accurate investigation is an important issue in superconductivity.

The first model describing AC properties has been proposed by Bean [1], to explain the observed irreversible isothermal magnetization of hard type-II superconductors. It is so-called critical state model (CSM). It is still most often employed for simulating the magnetization of type-II superconductors for the geometry of a slab and a cylinder in a parallel external field including field dependent critical current density. In these cases, the $E(J)$ dependence of the superconducting material is characterized by a step-like function. Accordingly, the current density of superconductor cannot exceed some critical value and, until this threshold is reached, the electric field is zero. This approximation allows one to describe analytically the AC losses in superconductors having simple geometry [1], [2]. Efforts were continuously on for solving the CSM for samples with non-zero demagnetization factor, in particular, infinite cylinders and thin strips in a transverse field [48]-[53]. In these cases, the self-field losses have been derived for the first time by Norris [54] in the frame of the corresponding modification of the critical state model.

Using these analytical models detailed investigations have been made over wide ranges of applied currents and frequencies in mono- and multifilamentary superconductors [55]-[58]. However, systematic studies have revealed a significant deviation between the measured AC losses and that predicted by the CSM. There are many reasons based on these features. The main of them is as follows. The CSM assumes that a homogenous current distribution in the fully penetrated states cannot be changed during current charging. It is a result of the step-like $E(J)$ dependence used by the CSM. Besides, the CSM does not allow investigate loss dependence as a function current charging rate [2]. At the same time, smooth character of the $E(J)$ dependence of real superconductors (both low- and high temperature) leads to the stable existence of the corresponding fully penetrated states even in the low-T_c superconductors [10]. The latter may be characterized by finite temperature rise at stable modes, as it was shown above. Thus, the parameters determining the AC losses in high-T_c superconductors are still to be conclusively established. So, let us discuss basic physical peculiarities based on the dissipative phenomena in high-T_c superconductors.

As follows from the above-discussed results, there exist three thermo-electrodynamics modes during current charging: partially penetrated mode with negligible overheating, fully penetrated mode with inconsiderable but finite overheating and fully penetrated mode with noticeable overheating. According to the results presented in Figures 4 and 11, the value E_c allows one to find the characteristic temperature that a superconductor will has in the fully penetrated stable state at $dI/dt \to 0$. This value is maximum at $x = 0$ in accordance with (32) and equals

$$T(0)\big|_{E=E_c} = T_{\max,c} = T_{cB} - (T_{cB} - T_0)\frac{H}{\sqrt{\gamma_c}\sinh\sqrt{\gamma_c} + H\cosh\sqrt{\gamma_c}},$$

$$\gamma_c = \gamma(E_c) = \frac{E_c}{E_\lambda}$$

Therefore, the overheating $\Delta T_c = T_{\max,c} - T_0$ at $E = E_c$ is equal to

$$\Delta T_c = (T_{cB} - T_0)\left[1 - \frac{1}{(\sqrt{\gamma_c}/H)\sinh\sqrt{\gamma_c} + \cosh\sqrt{\gamma_c}}\right]$$

Since γ_c does not exceed unit (Figure 16) then after some algebra the approximate value of ΔT_c may be written as follows

$$\Delta T_c \approx (T_{cB} - T_0)\Big/\left(1 + \frac{2H}{2+H}\frac{E_\lambda}{E_c}\right)$$

This value may be calculated as

$$\Delta T_c \approx \frac{T_{cB} - T_0}{1 + \dfrac{E_h}{E_c}} = \frac{T_{cB} - T_0}{1 + \dfrac{h(T_{cB} - T_0)}{aJ_{c0}E_c}}$$

when $H \ll 1$.

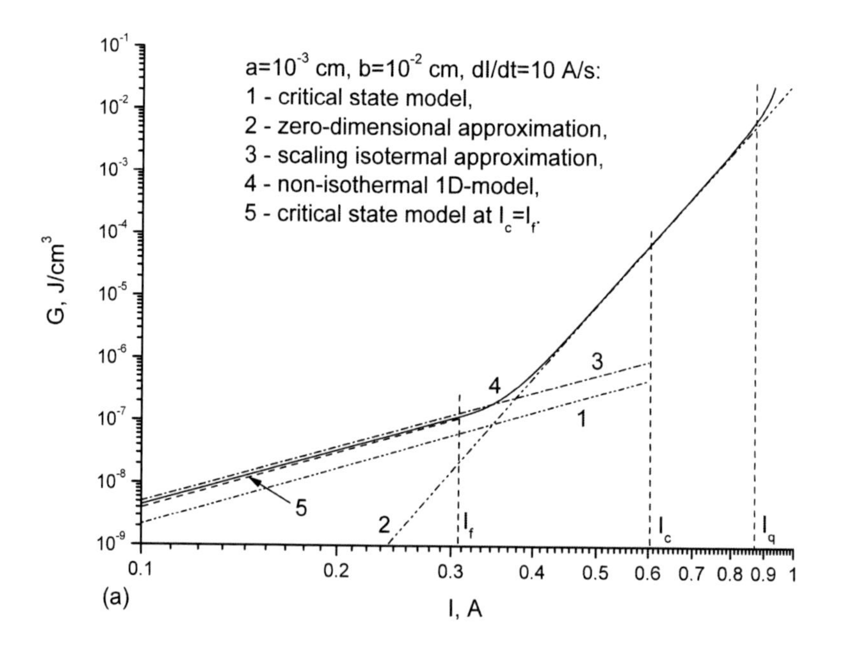

Figure 26. (Continued).

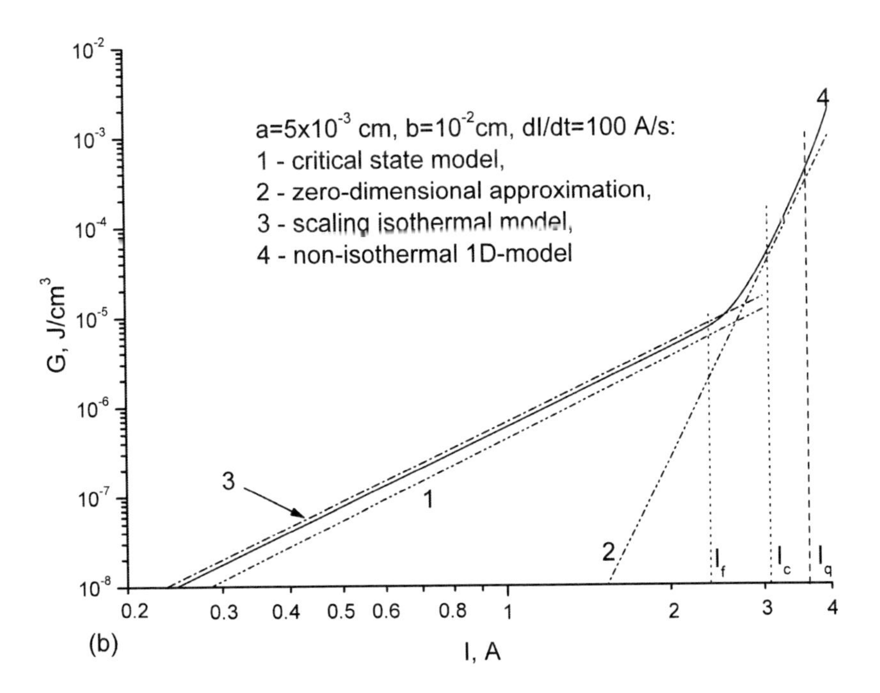

Figure 26. Self-field losses as a function of applied current during partially and fully penetrated regimes.

Thereby, the overheating of a superconductor is inconsiderable but finite ($0 < \Delta T_c < T_{cB} - T_0$) at $E = E_c$ and $dI/dt \to 0$, if

$$\frac{E_\lambda}{E_c} >> \frac{2+H}{2H} \tag{66}$$

or

$$\frac{h\left(T_{cB} - T_0\right)}{aJ_{c0}E_c} >> 1 \tag{67}$$

at $H << 1$. The latter estimator also follows from (14) and corresponds to the practically isothermal voltage-current characteristic of superconductor ($\Delta T_c = aJ_{c0}E_c/h$).

These estimates allow one to find exactly the boundary between isothermal and non-isothermal states and, as a result, to use isothermal models during AC calculations. To illustrate this conclusion, Figure 26 shows the effect of the stable temperature rise on the self-field losses

$$G = \frac{1}{a}\int_{x_p}^{a} dx \int_0^t EJdt$$

as a function of applied current ($I(t) = J(t)S = dI/dt \times t$) in the conduction-cooled high-T_c superconductor ($h = 10^{-3}$ W/cm^{2}K) that has the critical properties describing by (4) and (16). The simulation was done for the superconducting slab with different cross section at B = 10 T, $T_0 = 4.2$ K, $E_c = 10^{-6}$ V/cm, $n = 10$ and various current charging rates. Different models were used during calculations. Curve 1 corresponds to the CSM-calculations ($T = T_0$). In this case, the following formula was used

$$G_{csm}(t) = \frac{\mu_0 I_c^2}{24S}\frac{a}{b}\left(\frac{I(t)}{I_c}\right)^3, \quad I_c = J_{c0}S. \tag{68}$$

according to [1]. Curve 2 was obtained using isothermal zero-dimensional model that follows from equations (3) and (4) at $T = T_0$. Correspondingly, the self-field losses are described by formula

$$G_f(t) = \frac{I_c^2}{S(n+2)}\left(\frac{I(t)}{I_c}\right)^{n+2}\frac{E_c}{dI/dt}, \quad I_c = J_{c0}S. \tag{69}$$

in the fully penetrated case. Curve 3 illustrates the self-field losses that give the scaling model described by equations (15). In this partially penetrated state, the self – field losses are given by

$$G_p(t) = \frac{\mu_0 I_c^2}{S}\frac{n(n+1)}{(2n+1)(3n+2)}\left(\frac{n+1}{n}\right)^{\frac{2n+1}{n+1}}\left(\frac{a}{4b}\right)^{\frac{n}{n+1}}\left(\frac{I(t)}{I_c}\right)^{\frac{3n+2}{n+1}}\left[\frac{E_c}{\mu_0\, dI/dt}\right]^{\frac{1}{n+1}}, \quad I_c = J_{c0}S \tag{70}$$

according to [23]. Note that limiting case $G_p \to G_{csm}$ takes places at $n \to \infty$.

Curve 4 depicts the AC-calculations, which were made in the framework of the 1D-approximation in accordance with the non-isothermal model described by equations (1) – (8).

The results presented clearly show the temperature influence on the simulation features of the self-field losses in the superconductors during non-intensive cooling conditions, which connect with the conclusions formulated above. First, an isothermal approximation may be used during partially penetrated ($I < I_f$) and in the fully penetrated sub-critical states ($I_f < I < I_c$). Here,

$$I_f = \frac{n}{n+1}\frac{\mu_0 J_{c0}}{E_c^{1/n}} a^{\frac{n+1}{n}}\left(\frac{\mu_0}{4b}\frac{dI}{dt}\right)^{\frac{1-n}{n}}\frac{dI}{dt}$$

according to (15). However, the isothermal calculations of losses during fully penetrated sub-critical mode depends on the cross section of superconductor (the higher a, the higher stable value of temperature) and the current charging rate, as calculations show. Second, stable self-field losses become more intensive before current instability onset. The non-isothermal model must be used in these cases to describe correctly the evolution of the thermo-electrodynamics states. Third, Figure 26 demonstrates also the role of the smooth character of the $E(J)$ dependence of superconductor on the self-field losses during fully penetrated mode even in the sub-critical states ($I_f < I < I_c$). In these regimes, the Bean model does not allow one to calculate correctly the stable losses in superconductor even at very high n-values (Figure 27).

As calculations show, the numerical error following from the CSM increases with decreasing n-value and current charging rate. The later conclusion is a result of the estimator $G_p \sim dI/dt^{-1/(n+1)}$ that follows from (70). It clearly demonstrates the dependence of losses on the current charging rate for the given applied current. As mentioned above, the CSM cannot give this dependence in principle.

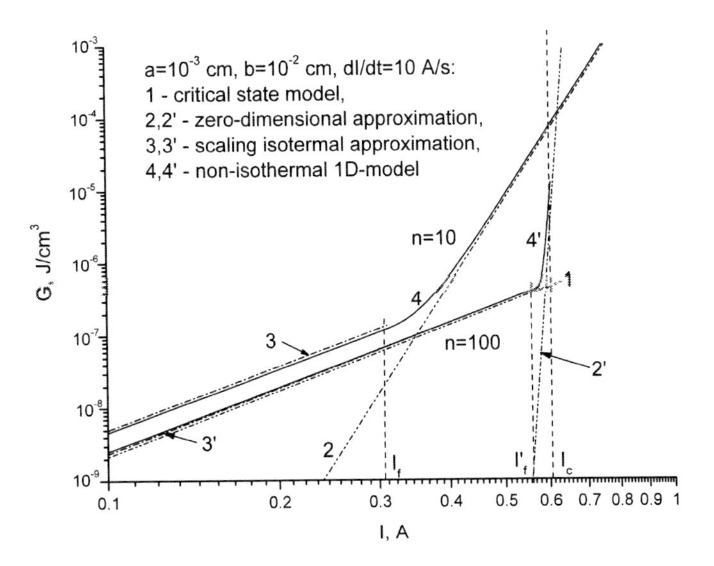

Figure 27. Effect of n-value on the self-field losses.

The numerical error between the calculations using the step-like and real $E(J)$ dependences depends also on the critical current density and cross section of superconductor, according to the (68) and (70).

As a result, to decrease such errors during AC-calculations, the real value of the critical current of a superconductor is not used in the approximations based on the CSM. Usually one utilizes the current after which the fully penetrated mode starts ($I_c = I_f$). In this case, the CSM allows one to get reasonable approximation of the self-field losses during partially penetrated modes (curve 5 in Figure 26(a)). However, such non-physical approximation of the effective I_c-value depends on the n-value, current charging rate and cross section of a superconductor. In the meantime, analytical formulae proposed above allow one to describe the self-field losses both during partially and fully penetrated continuous current charging with constant rate taking into account the non-linear nature of the voltage-current characteristic of superconductor. Figure 28 shows the results of the simulations of self-field losses during current charging with break when the transport current changes as follows

$$I(t) = \begin{cases} \dfrac{dI}{dt}t, & t < t_m, \quad dI/dt = const \\ I_m, & t \geq t_m \end{cases}$$

To calculate the self-field losses taking correctly into consideration the continuous losses in the partially and fully penetrated states the following formulae were used.

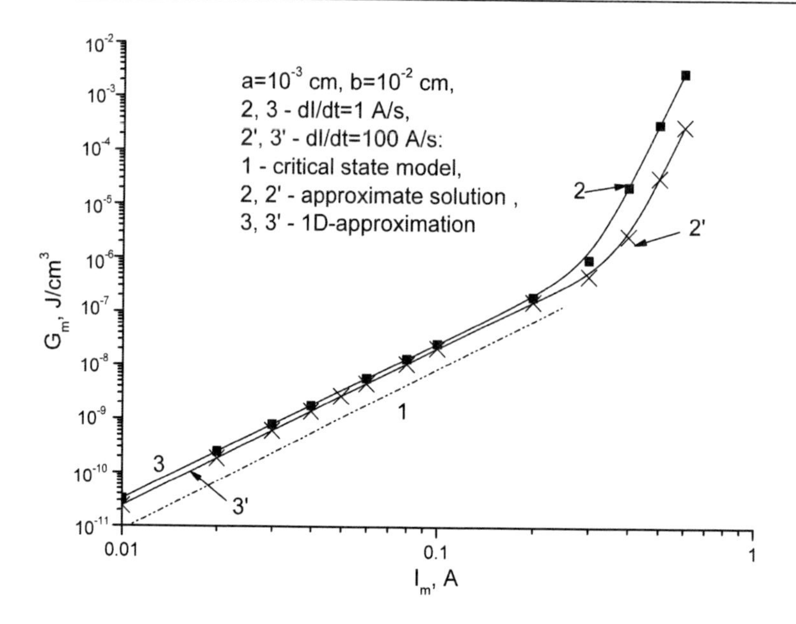

Figure 28. Numerical and analytical calculations of self-field losses using formulae (68), (71), (72).

$$G(I_m)=\frac{\mu_0 I_c^2}{S}\frac{n(n+1)}{(2n+1)(3n+2)}\left(\frac{n+1}{n}\right)^{\frac{2n+1}{n+1}}\left(\frac{a}{4b}\right)^{\frac{n}{n+1}}\left(\frac{I_m}{I_c}\right)^{\frac{3n+2}{n+1}}\left[\frac{E_c}{\mu_0 \, dI/dt}\right]^{\frac{1}{n+1}}, \qquad (71)$$

$$I_m \le I_f$$

$$G(I_m)=\frac{\mu_0 I_c^2}{S}\frac{n(n+1)}{(2n+1)(3n+2)}\left(\frac{n+1}{n}\right)^{\frac{2n+1}{n+1}}\left(\frac{a}{4b}\right)^{\frac{n}{n+1}}\left(\frac{I_f}{I_c}\right)^{\frac{3n+2}{n+1}}\left[\frac{E_c}{\mu_0 \, dI/dt}\right]^{\frac{1}{n+1}}$$

$$+\frac{I_c^2}{S(n+2)}\left[\left(\frac{I_m}{I_c}\right)^{n+2}-\left(\frac{I_f}{I_c}\right)^{n+2}\right]\frac{E_c}{dI/dt}, \qquad I_m > I_f, \qquad (72)$$

relatively.

It is easy to understand that these formulae allow one also to find the self-field losses as a function of time during continuous current charging at constant rate. In this case, it is necessary to execute the replacement $I_m \rightarrow I(t)$.

Figure 28 obviously proves the non-linear effect of the critical current density of superconductor, its cross section, charging rate on the self-field losses. They also clearly demonstrate essential difference between the self-field losses in the partially and fully penetrated modes, which have various dependences on the induced current and charging rate during stable sub-critical states. In the first case, the self-field loss variation occurs as

$$G_p \sim \left(\frac{I}{I_c}\right)^{\frac{3n+2}{n+1}} \left(\frac{dI}{dt}\right)^{-1/(n+1)} \tag{73}$$

In the fully penetrated states, the stable self-field losses changes as

$$G_f \sim \left(\frac{I}{I_c}\right)^{n+2} \left(\frac{dI}{dt}\right)^{-1}. \tag{74}$$

As a whole, these estimates demonstrate the physical reasons, firstly, why the self-field losses during partially penetrated mode may be smaller or higher than ones during fully penetrated mode and, secondly, why the critical state model does not allow to calculate the self-field losses during fully penetrated current mode since $G \to \infty$ at $n \to \infty$.

Thus, from the physical point of view, the critical state model does not allow to calculate correctly the self-field losses in high-T_c superconductors even in the isothermal partially penetrated states for the given values of J_c and E_c. The numerical error depends on the n-value, current charging rate, the critical current density, cross section of superconductor. At the same time, above-proposed formulae allow one to calculate the self-losses both during partially and fully penetrated states with good accuracy, if the conditions (66) or (67) take place. It is important to note that, as a role, the self-field losses are small during partially penetrated states. They practically will not change the temperature of superconducting filament even when it does not have stabilized matrix. Otherwise, the stable temperature rise effects on the losses during fully penetrated over-critical states, which must be calculated in the framework of the non-isothermal models. In others words, the isothermal self-field loss calculations will be broken, when the cross section of a superconductor, the critical current density, the temperature margin that is equal to T_{cB}-T_0 and the heat transfer coefficient do not satisfy the conditions (66) or (67). Non-isothermal self-field losses may be also observed at low value of the current charging rate or low frequency of the cycle regimes.

10. Self-Field Losses During Serrated (Triangular-Wave) Current Charging Regime

The formulae presented above allow one to investigate the self-field losses in the more general cases, in particular, during serrated regimes. A typical operating current pattern is shown in Figure 29(a). In this mode, the current as a function of time is calculated as

$$I(t) = \begin{cases} \dfrac{dI}{dt} t, & t \le t_1 \\ \pm I_m \mp \dfrac{dI}{dt} t, & t_{2k-1} < t < t_{2k-1}, \quad k = 1, 2, 3, ... \end{cases}.$$

Figure 29(b) show typical evolution of the electric field on the surface of the superconducting slab during partially penetrated serrated mode at $I_m = 0.2$ A and dI/dt = 10

A/s. The calculations were made by means of the 1D isothermal model ($T = T_0$) described by equations (2) – (4), (7), (8) using the following parameters:

$$n = 10, \quad E_c = 10^{-6}\ \text{V/cm}, \quad J_{c0} = 1.52 \times 10^4\ \text{A/cm}^2,$$
$$T_{cB} = 26.1\ \text{K}, \quad T_0 = 4.2\ \text{K}, a = 10^{-3}\text{cm}, \quad b = 10^{-2}\ \text{cm}. \quad (75)$$

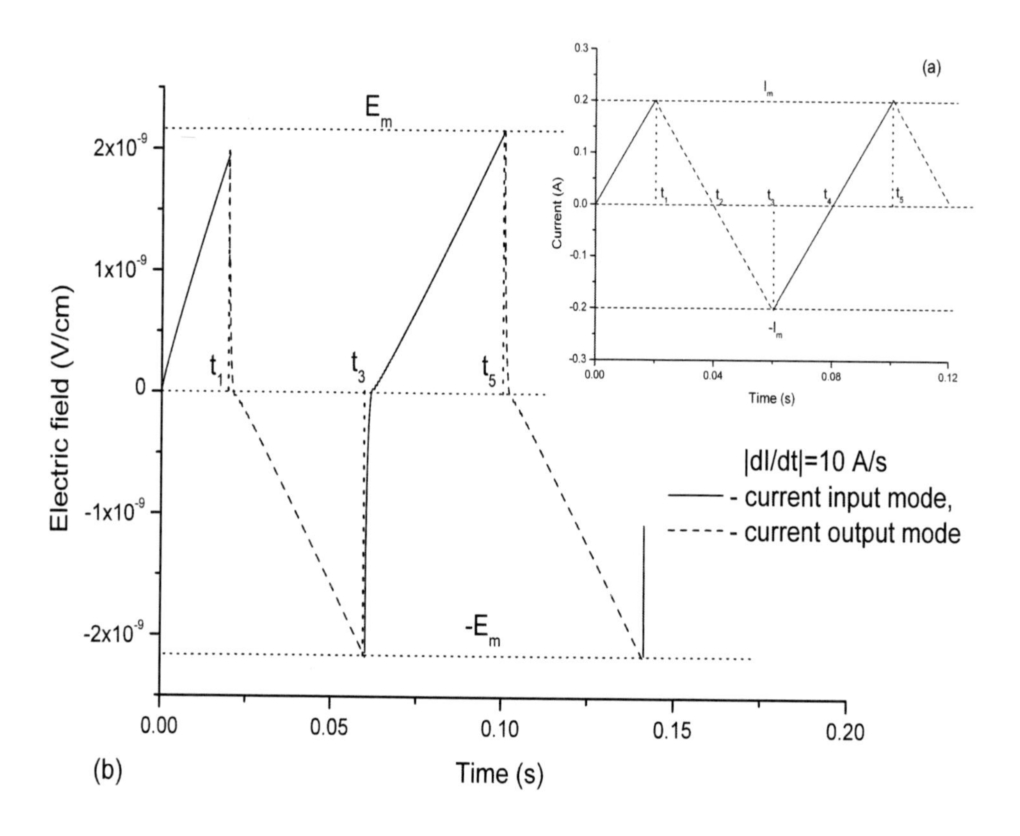

Figure 29. Time windows of serrated incomplete regime ($I_m < I_f = 0.305$ A).

As follows from Figure 29(b), there exist the initial and repeated modes during which the electric field variation occurs. Namely, the initial stage takes places when the applied current increases from initial zero value to I_m ($0 < t < t_1 = t_m$, $t_m = I_m/|dI/dt|$). Note that in this stage the maximum value of the electric field less than the subsequent maximum value $|E_m|$ that takes places at $t_{2k-1} < t < t_{2k+1}$, $k = 1, 2, 3,...$ when it changes in the constant range from E_m to $-E_m$. It is important to underline that these subsequent modes have the following peculiarity. First, the electric field variation has two stages. Initially, it changes very quickly. So, the time window when the electric field decreases from $|E_m|$ to zero is very small. But its variation has practically constant rate in the periods $t_{2k-1} < t < t_{2k+1}$, $k = 1, 2, 3, ...$. Assuming that the electric field on the surface of superconductor does not change, the effective value of dI^*/dt in formula (70) may be defined as

$$\frac{dI^*}{dt} = \frac{1}{\sqrt{2}}\frac{dI}{dt}.$$

Second, the value of the applied current I is the constantly increase with increasing time in formula (70). Therefore, to use it and calculate the self-field losses during time window $t_{2k-1} < t < t_{2k+1}$, $k = 1, 2, 3, \ldots$, the following equivalent value

$$I^*(t) = \frac{1}{2}\frac{dI}{dt}(t - t_{2k-1}), \quad t_{2k-1} = (2k-1)t_m < t < (2k+1)t_m, \quad k = 1,2,3,...$$

may be utilized in both output and input modes. As a result, the following formula

$$G_{1,p}(t) = \frac{\mu_0 I_c^2}{S}\frac{n(n+1)}{(2n+1)(3n+2)}\left(\frac{n+1}{n}\right)^{\frac{2n+1}{n+1}}\left(\frac{a}{4b}\right)^{\frac{n}{n+1}}\left(\frac{1}{I_c}\frac{dI}{dt}t\right)^{\frac{3n+2}{n+1}}\left[\frac{E_c}{\mu_0 \, dI/dt}\right]^{\frac{1}{n+1}} \quad (76)$$

may be used to calculate the self-field losses during initial stage of the partially penetrated regime. Accordingly, the self-field losses may be described analytically by formula

$$G_{k+1,p}(t) = 2\frac{\mu_0 I_c^2}{S}\frac{n(n+1)}{(2n+1)(3n+2)} \times\left(\frac{n+1}{n}\right)^{\frac{2n+1}{n+1}}\left(\frac{a}{4b}\right)^{\frac{n}{n+1}}\left[\frac{dI}{dt}\frac{t-t_{2k-1}}{2I_c}\right]^{\frac{3n+2}{n+1}}\left[\frac{\sqrt{2}E_c}{\mu_0 \, dI/dt}\right]^{\frac{1}{n+1}} \quad (77)$$

during partially penetrated modes in the periods $t_{2k-1} < t < t_{2k+1}$, $k = 1, 2, 3, \ldots$. Then the total value of the self-field losses is equal to $G_p(t) = G_{1,p}(t) + G_{2,p}(t) + G_{3,p}(t) + \ldots$.

Numerical and analytical results of the loss simulations are depicted in Figure 30 for different current charging rates. For comparison, the losses, which give the CSM at $I = I_m$, are also shown. It is seen that numerical and analytical calculations coincide with a good degree of accuracy and the error decreases with increasing dI/dt. At the same time, the CSM model leads to the essential error when the value of I_c is defined at $E_c = 10^{-6}$ V/cm during fully penetrated state.

Using the approximation given by relations (76) and (77) and formula (69), it is easy to find analytically the self-field losses during fully penetrated states when the parameters of superconductor satisfy the conditions (66) or (67). The existence of practically isothermal modes shows Figure 31 for the superconductor under consideration ($h = 10^{-1}$ W/cm^{2}K). To illustrate this possibility, Figure 32 demonstrates the loss evolution in the superconducting slab with parameters (73).

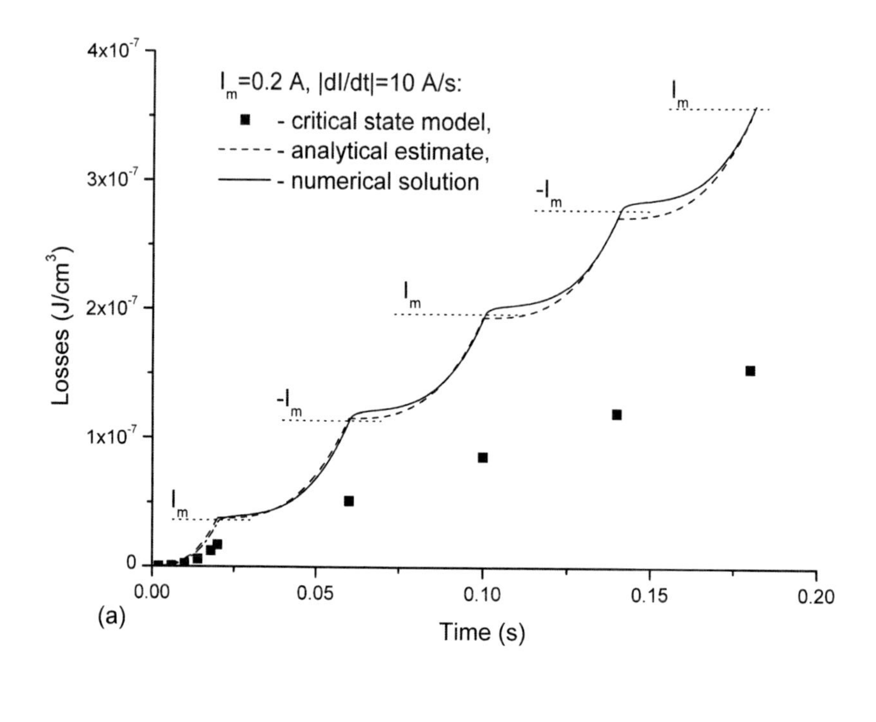

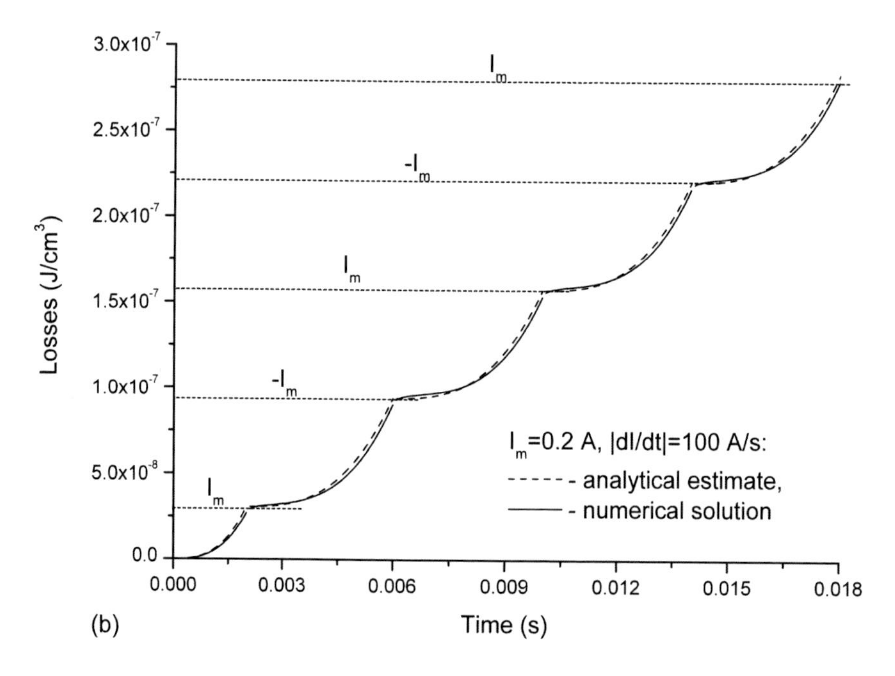

Figure 30. Numerical and analytical calculations of self-field losses during partially penetrated serrated regime (I_f = 0.305 A).

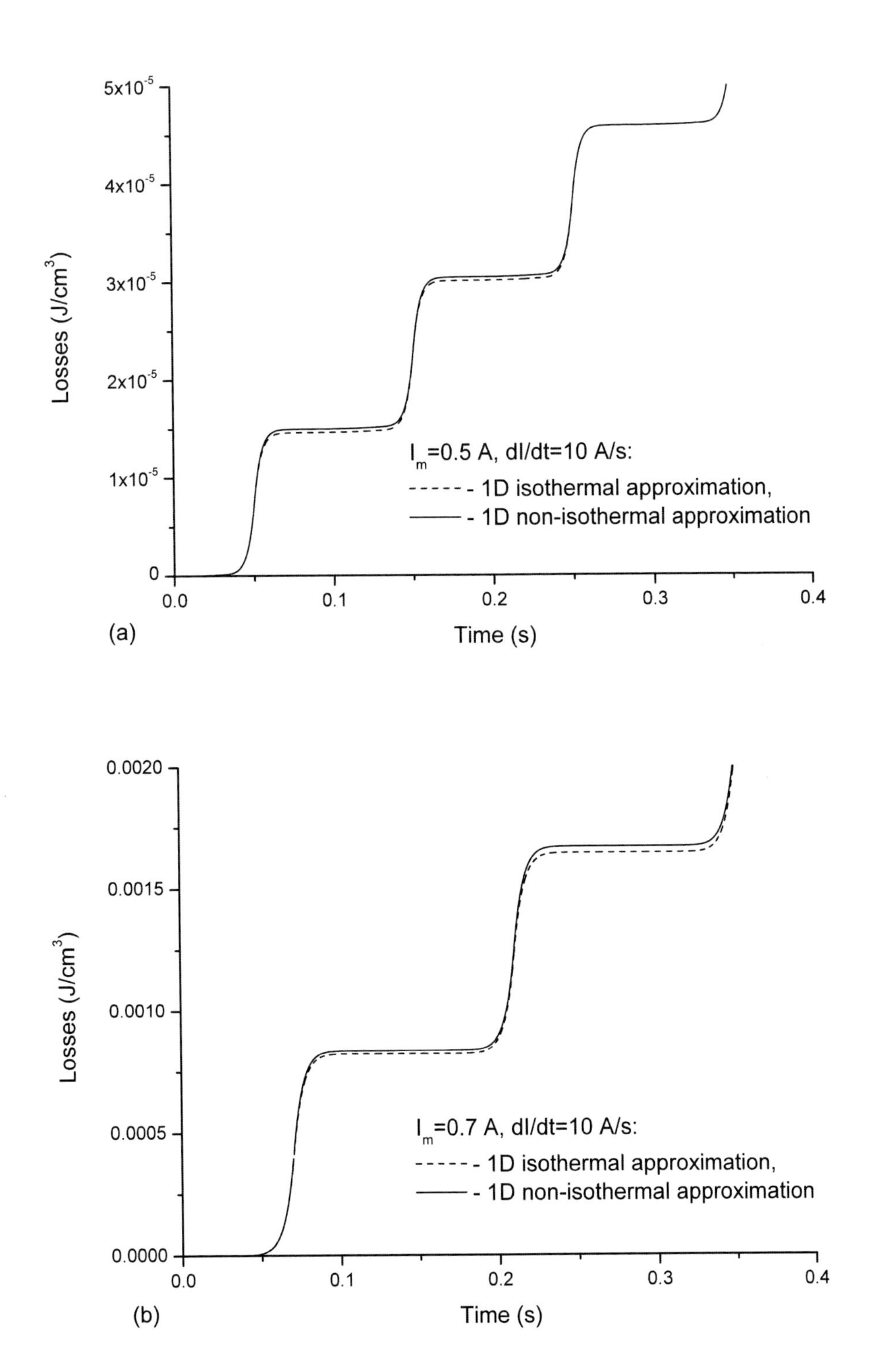

Figure 31. Practically isothermal modes during fully penetrated losses (I_c = 0.61 A, I_q = 0. 87 A).

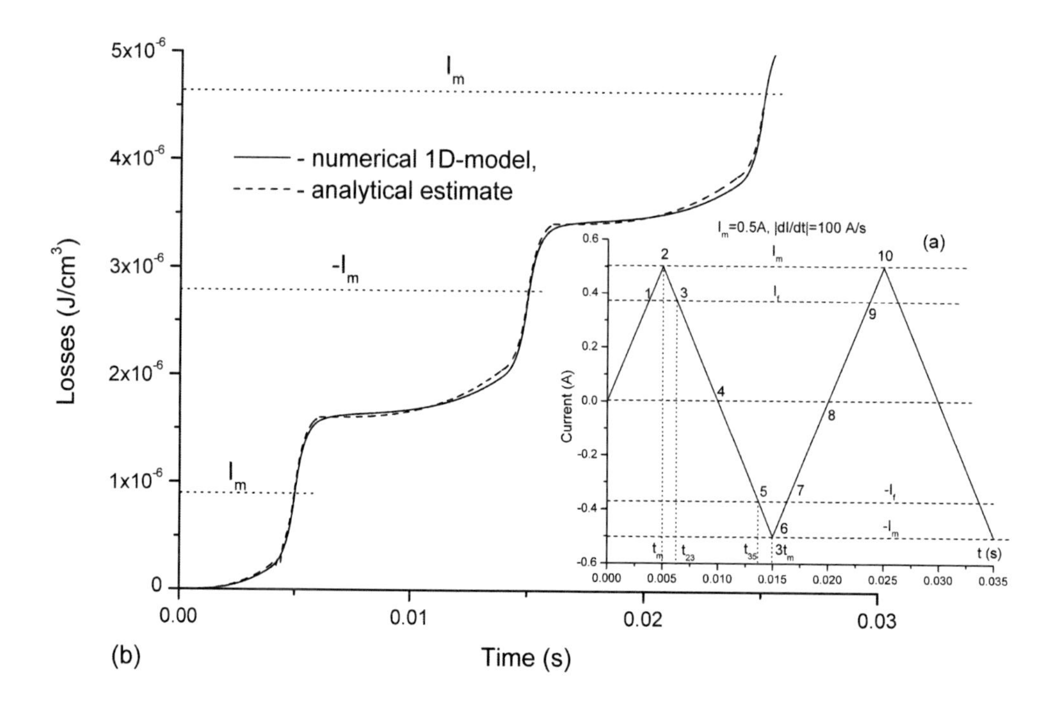

Figure 32. Numerical and analytical calculations of self-field losses during fully penetrated serrated cycle.

The following formulae

$$mode\left(0-1\right):G_{0,1}\left(t\right)=\frac{\mu_0 I_c^2}{S}\frac{n\left(n+1\right)}{\left(2n+1\right)\left(3n+2\right)}$$

$$\times\left(\frac{n+1}{n}\right)^{\frac{2n+1}{n+1}}\left(\frac{a}{4b}\right)^{\frac{n}{n+1}}\left(\frac{1}{I_c}\frac{dI}{dt}t\right)^{\frac{3n+2}{n+1}}\left[\frac{E_c}{\mu_0 \, dI/dt}\right]^{\frac{1}{n+1}},$$

$$mode\,(1-2):G_{1,2}\left(t\right)=\frac{\mu_0 I_c^2}{S}\frac{n((n+1))}{\left(2n+1\right)\left(3n+2\right)}$$

$$\times\left(\frac{n+1}{n}\right)^{\frac{2n+1}{n+1}}\left(\frac{a}{4b}\right)^{\frac{n}{n+1}}\left(\frac{I_f}{I_c}\right)^{\frac{3n+2}{n+1}}\left[\frac{E_c}{\mu_0 \, dI/dt}\right]^{\frac{1}{n+1}}$$

$$+\frac{E_c I_c^2}{(n+2)S \, dI/dt}\left[\left(\frac{1}{I_c}\frac{dI}{dt}t\right)^{n+2}-\left(\frac{I_f}{I_c}\right)^{n+2}\right],$$

$$mode\,(2-3): G_{2,3}(t) = G_{1,2}(t_2)$$

$$+\frac{E_c I_c^2}{(n+2)S\, dI/dt}\left\{\left(\frac{I_m}{I_c}\right)^{n+2} - \left[\frac{1}{I_c}\left(I_m - \frac{dI}{dt}(t-t_m)\right)\right]^{n+2}\right\},$$

$$mode\,(3-4-5): G_{3,5}(t) = G_{2,3}(t_{23}) + 2\frac{\mu_0 I_c^2}{S}\frac{n(n+1)}{(2n+1)(3n+2)}$$

$$\times\left(\frac{n+1}{n}\right)^{\frac{2n+1}{n+1}}\left(\frac{a}{4b}\right)^{\frac{n}{n+1}}\left(\frac{1}{2I_c}\frac{dI}{dt}(t-t_{23})\right)^{\frac{3n+2}{n+1}}\left[\frac{\sqrt{2}E_c}{\mu_0\, dI/dt}\right]^{\frac{1}{n+1}},$$

$$t_{23} = t_m + \frac{I_m - I_f}{dI/dt}$$

$$mode\,(5-6): G_{5,6}(t) = G_{3,5}(t_{35}) +$$

$$+\frac{E_c I_c^2}{(n+2)S\, dI/dt}\left\{\left[\frac{1}{I_c}\left(I_f - \frac{dI}{dt}(t-t_{35})\right)\right]^{n+2} - \left(\frac{I_f}{I_c}\right)^{n+2}\right\},$$

$$t_{35} = t_m + \frac{I_m + I_f}{dI/dt},$$

were used during analytical calculations in the initial stage of the $[(I_m) \div (-I_m) \div (I_m) \div ....]$-cycles. Numerical calculations were based on the 1D-approximation in accordance with the non-isothermal model described by equations (1) – (8). It is seen that the analytical approximation proposed is reasonable during both partially and fully penetrated modes. As a result, they allow one to find the total value of the self-field losses in the $[(I_m) \div (-I_m)]$-cycle. It is equal to

$$G_{\max}\left(I_m, \frac{dI}{dt}\right)$$

$$= 2\frac{\mu_0 I_c^2}{S}\begin{cases} \dfrac{n(n+1)}{(2n+1)(3n+2)}\left(\dfrac{n+1}{n}\right)^{\frac{2n+1}{n+1}}\left(\dfrac{a}{4b}\right)^{\frac{n}{n+1}}\left(\dfrac{I_m}{I_c}\right)^{\frac{3n+2}{n+1}}\left[\dfrac{\sqrt{2}E_c}{\mu_0\, dI/dt}\right]^{\frac{1}{n+1}}, & I_m \le I_f \\ \dfrac{n(n+1)}{(2n+1)(3n+2)}\left(\dfrac{n+1}{n}\right)^{\frac{2n+1}{n+1}}\left(\dfrac{a}{4b}\right)^{\frac{n}{n+1}}\left(\dfrac{I_f}{I_c}\right)^{\frac{3n+2}{n+1}}\left[\dfrac{\sqrt{2}E_c}{\mu_0\, dI/dt}\right]^{\frac{1}{n+1}} + & \\ +\dfrac{E_c}{\mu_0(n+2)S\, dI/dt}\left[\left(\dfrac{I_m}{I_c}\right)^{n+2} - \left(\dfrac{I_f}{I_c}\right)^{n+2}\right], & I_m > I_f \end{cases}$$

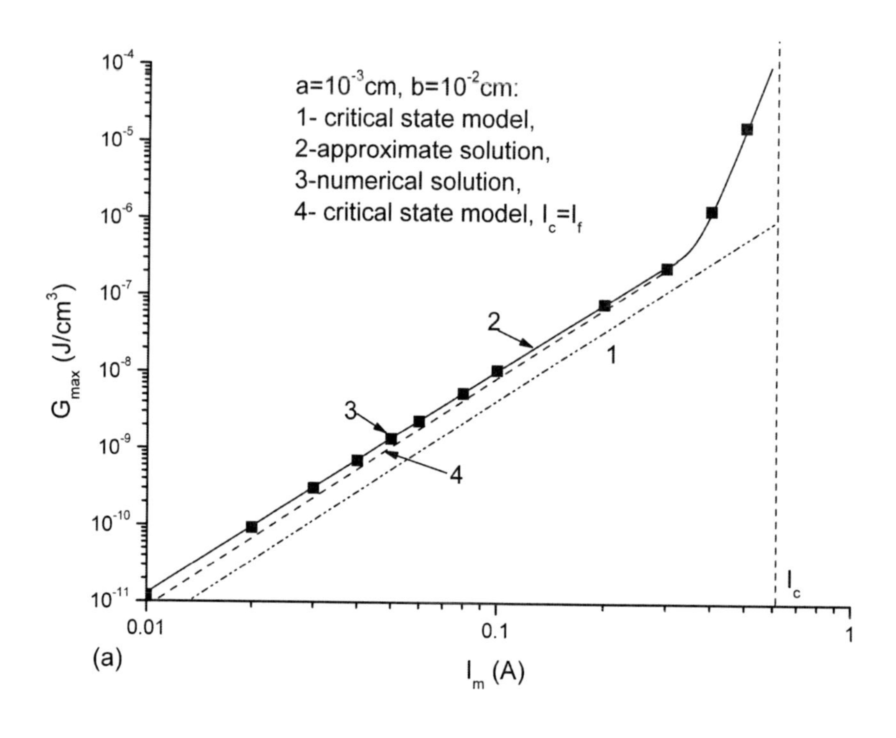

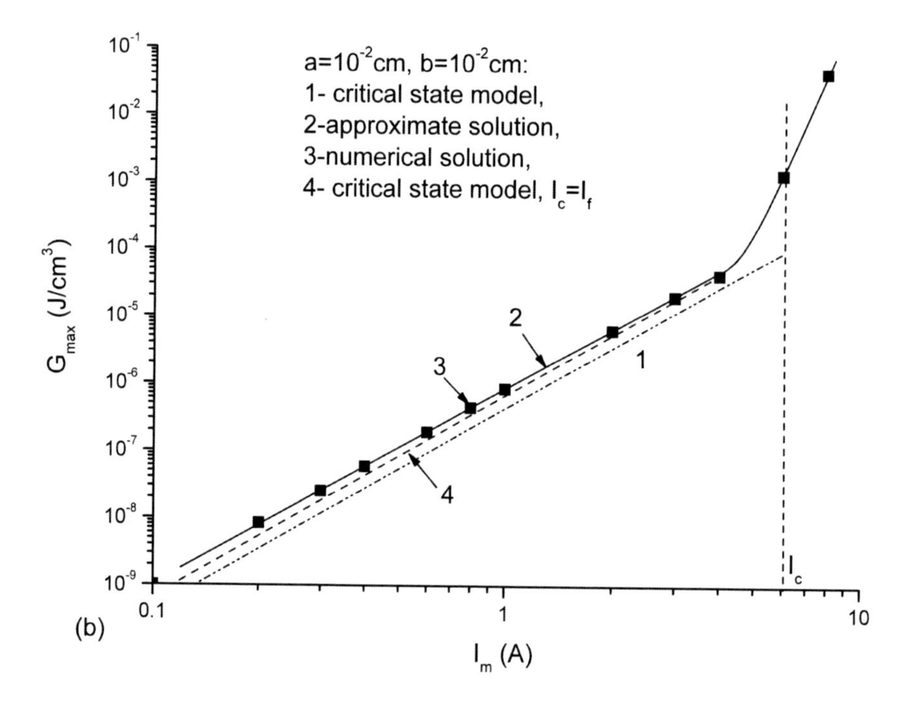

Figure 33. Losses during $(I_m) \div (-I_m)$ cycle.

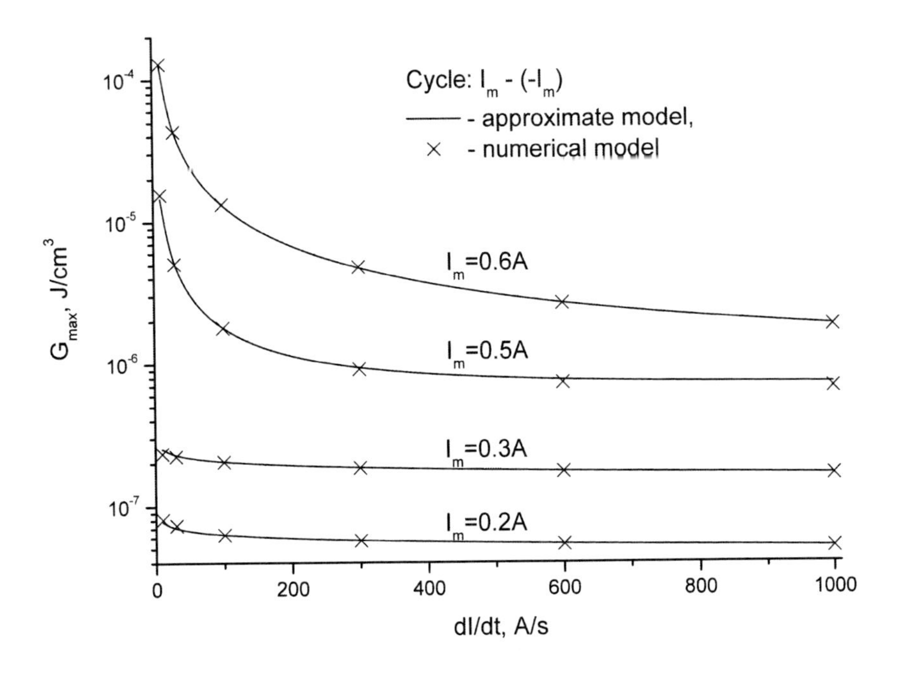

Figure 34. Effect of current charging rate on the losses.

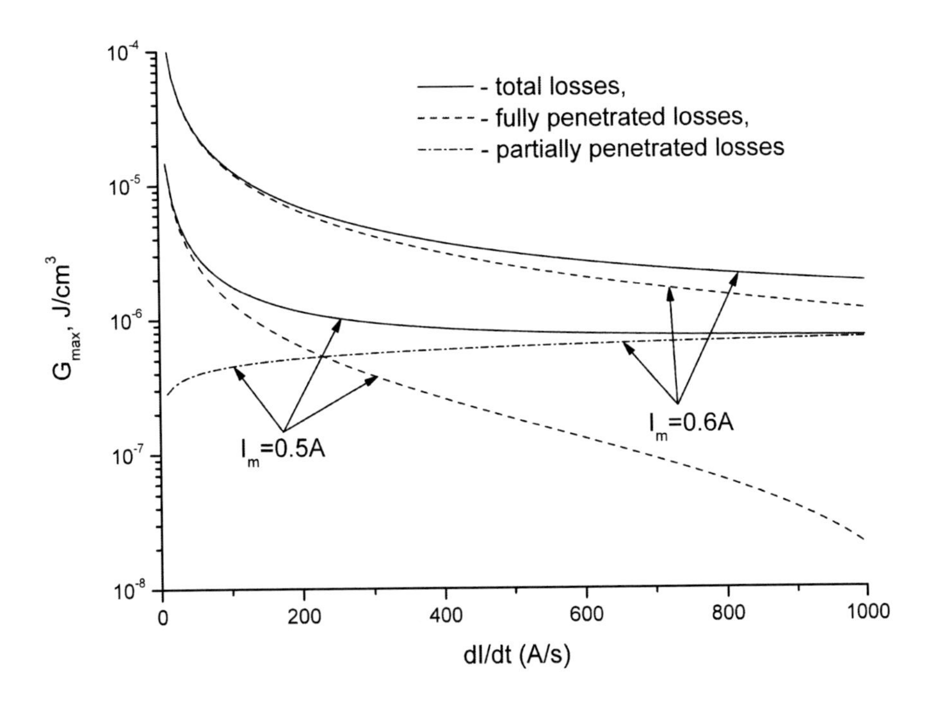

Figure 35. Influence of the partially and fully penetrated modes on the total losses.

Correspondingly, the maximum losses during full cycle $[(I_m) \div (-I_m) \div (I_m)]$ are equal to $2G_{max}$. The results of the G_{max}-simulation are depicted in Figure 33. It presents the dependence of self-field losses on the charged current during cycle from I_m to $-I_m$ in the wide range of I_m-variation at $dI/dt=10$ A/s for different cross sections of the superconducting slab with parameters (75). The total losses as a function of current charging rate are presented in Figure 34. Using above-written analytical formulae, it is easy to compare the losses that take place during partially and fully penetrated states. Figure 35 shows the dependence of the self-field losses on the operating mode during fully penetrated cycles. It is seen that the higher dI/dt, the more important the role of the losses during partially penetrated mode according to the estimates (73) and (74).

Thus, it is developed new analytical tools allowing one to analyze the self-field losses in a superconductor with power law voltage-current characteristic during partially and fully penetrated cycle modes. They are very useful to calculate the self-field losses at different current charging conditions without adjustment utilizing of the effective value of the critical current and, therefore, can explain the basic features of the self-field loss phenomenon in high temperature superconductors, in particular, frequency dependence of the losses. It is shown that there exists the boundary value of the current charging rate or frequency after which the losses during partially penetrated states may exceed the loss-constituent in the fully penetrated mode during cycles. This quantity depends on the superconductor's properties and decreases with decreasing I_m.

11. Stable Overloaded Current Regimes

As above-presented results show, to design the high-T_c superconducting magnets with high operational properties, it is important also to investigate their thermal and electromagnetic characteristics in the over-critical regimes when applied currents are larger than the superconductor's critical current. In particular, it is known that working regimes of high-T_c superconductors may be stable during over-current pulses or sinusoidal over-currents. Serious theoretical and experimental studies about such behaviors have been carried out [59]-[63]. The obtained results are as follows. Superconducting devices may experience over-currents exceeding the critical current. In case of low-T_c superconductors, the over-currents exceeding the critical current means immediate quench. However, in case of high-T_c superconductors, it is not true. Actually, high-T_c superconducting tapes and coils can carry over-currents without thermal runaway when the charged currents exceed the critical value in many times. In these cases, two possible thermo-electrodynamics states are possible. Namely, there exist stable and unstable modes as they exist in the DC modes. It may be completely superconducting after the conclusion of the over-critical pulse and continue quenching even after the current returns to the operating level. The final states are related to the peak current and its pulse length or the frequency of the charged current. As a result, the current stability range tends to increase as the amplitude of the transport current increased. However, basic physical reasons rooted in the existence of such overloaded regimes are not discussed. To understand them, let us model the operational conditions of a superconducting slab under consideration during overloaded AC regimes.

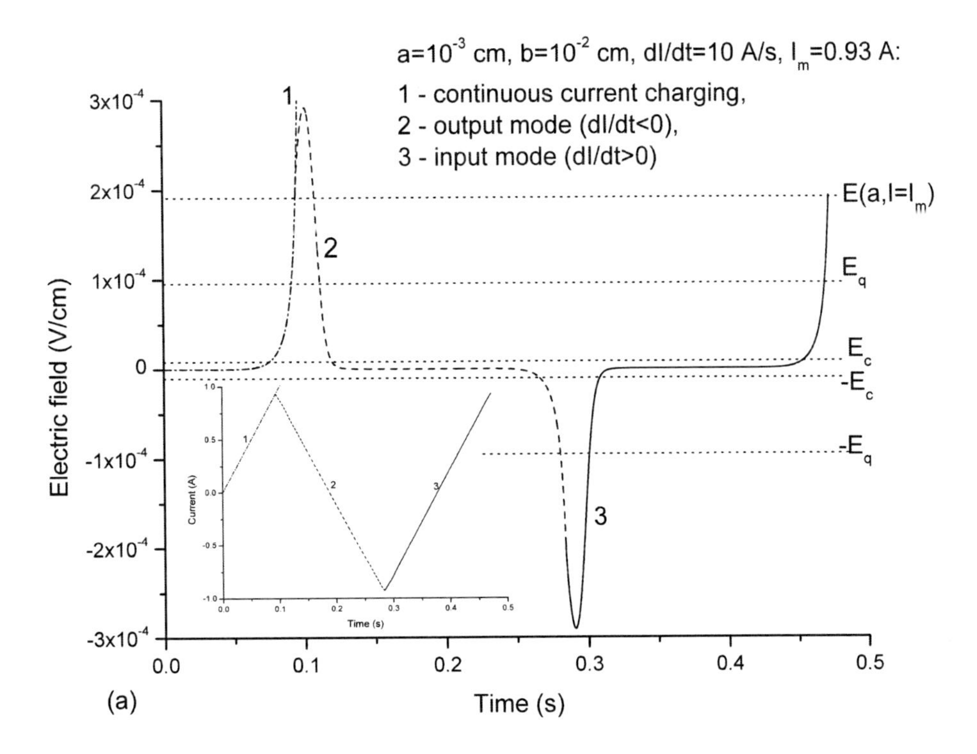

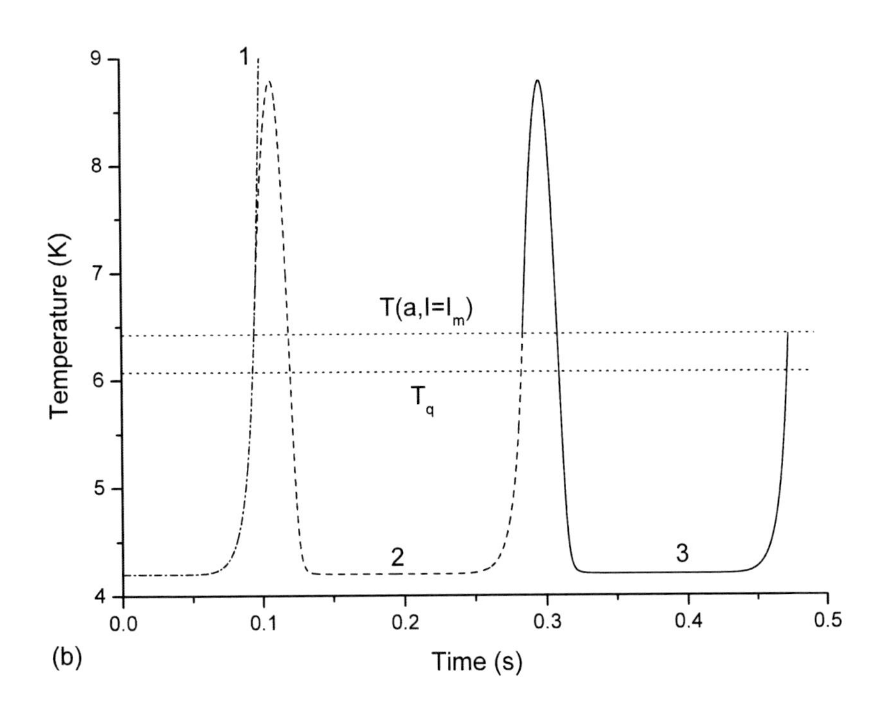

Figure 36. Overloaded electric field and temperature evolution.

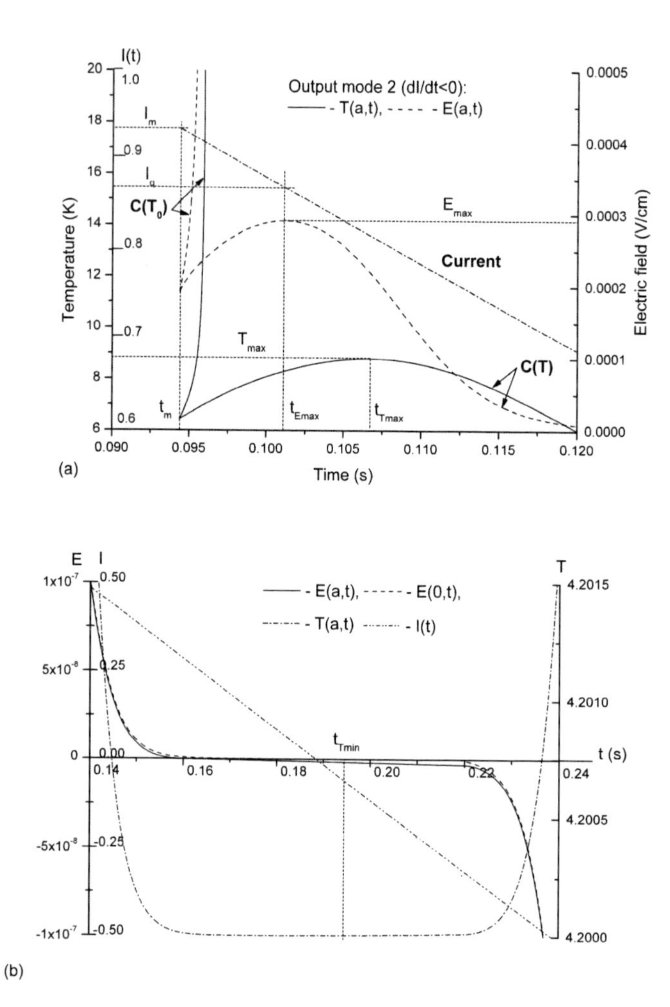

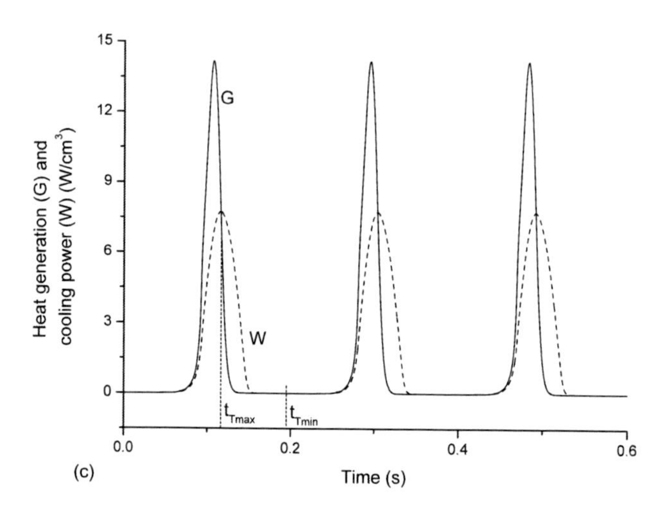

Figure 37. Thermo-electrodynamics peculiarities based on the existence of the overloaded regimes.

Figure 36 shows typical waveform evolution of the electric field and temperature on the surface of the conduction-cooled superconductor (h = 10^{-3} W/cm^{2}K) during serrated regime when the maximum value of the applied current exceeds not only the critical current of superconductor but also its quench current. In the case under consideration, they are equal to I_c = 0.61 A and I_q = 0.87 A, respectively. The corresponding values of the E_c and E_q also are shown. The results presented were obtained numerically under parameters (75) using 1D non isothermal model (1) – (8). For comparison, curve 1 depicts the computed time-varying electric field and temperature in the continuous current charging. As follows from Figure 36, there exist very high stable electric field and temperature. Accordingly, the peak current is over 1.5 times greater than the critical current and the stable overheating is practically equal to 5 K in the case under consideration when the superconductor has no the stabilizing matrix.

Figure 37 allow one to explain the existence of the overloaded regimes. The physical reasons are as follows.

First, according to Figure 37(a) depicted the initial stage of the cycles, there exist three characteristic times: t_m, t_{Emax}, t_{Tmax}. They define the characteristic time windows of the overloaded regimes. The electric field and the temperature still increase as the applied current starts to decrease after peak value I_m. This stage occurs during $t_m < t < t_{Emax}$ when the charged current does not stable because $I > I_q$. However, the formation of this state is stable as the applied current decreases and the electrodynamics state development depends on the temperature dependence of the specific heat capacity as discussed above. The latter feature plays exclusive role in the stable rise of the electric field at $t > t_m$ because the specific heat capacity essentially increases in this state. For comparison, the simulation results of the electric field evolution is shown in Figure 37(a), which were obtained when the specific heat capacity of superconductor also was assumed constant, namely, $C = C(T_0)$ at $t > t_m$. This numerically simulated state has unstable character despite the fact that the applied current decreases.

Second, the induced electric field starts to decrease at $t > t_{Emax}$ while the temperature of the superconductor continues to rise. This stage occurs because decreasing applied current passes into the stable range ($I < I_q$). However, the heat generation still exceeds the heat removal in this case (Figure 37(c)).

Third, the joint reduction of the field and current leads to such thermal dissipation, which becomes less then the heat flux to the coolant. This stage starts to exist at $t > t_{Tmax}$ (Figure 37(a)). As a result, the stable decrease of the current, electric field and temperature take place at $t > t_{Tmax}$. Clearly, this stage also exists at the subsequent mode of the output of applied current in the range $E_{max} > E > 0$ and when the current and electric field becomes negative (Figure 37(b)). At the same time, the losses inevitably start to exceed the heat removal at the further development of the given stage of a cycle, namely, at $t > t_{Tmin}$, as it is visible from Figures 37(b) and 37(c). Therefore, the temperature of superconductor starts to increase. In the end of the output mode 2, the temperature rise gets sharply accruing character. The next input mode of the applied current repeats above-discussed periods, and the cycle goes on. The corresponding cyclic values of the electric field and temperature induced by serrated charged current are shown in Figure 38.

These mechanisms take place at arbitrary current charging rate. The simulation results of temperature waveforms formed in the superconducting slab under consideration are shown in Figure 39 for different values of the dI/dt. The presented results were carried out under zero-dimensional transient model described by equations (3), (4) and (12), to decrease the value of calculations. The average value of the temperature defined as $T_{av} = \frac{1}{t}\int_0^t T dt$ is represented in Figures 39(a) and 39(b). Two characteristic values of the peak current were found during calculations. They identified the stability boundary of the serrated current regimes at the given current charging rate. Accordingly, minimum value of I_m corresponds to the upper current stability boundary of the serrated current stable charged into the superconductor despite its high stable overheating and hence high stably losses. Maximum peak current in Figure 39 defines the peak current when the serrated regimes are unstable.

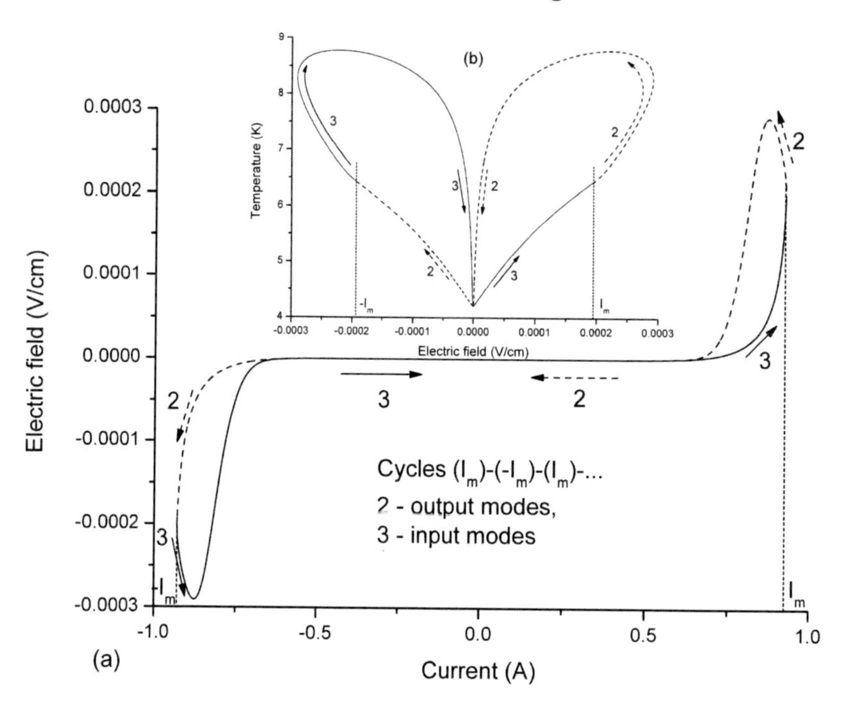

Figure 38. Typical stable variation of the electric field and temperature during serrated overloaded regimes.

As follows from Figure 39, the allowable temperature range of the fluctuations is not monotonous function of the d*I*/d*t*. There exists the maximum value of the temperature peak to which superconductor may be stably heated. The reason is as follows. The temperature of superconductor is directly proportional to the heat dissipation, i.e. *EJ*, which has maximum according to the above-discussed mechanisms of the waveform formation. However, the maximum value of the electric field not only depends on the decay rate of the applied current but also will be non-monotonous. Namely, the induced electric field increases with increasing current charging rate. Therefore, the higher the current charging rate, the higher electric field at small value of d*I*/d*t*. At the same time, it will be decreasing function of d*I*/d*t* at their high values because the higher d*I*/d*t*, the smaller the time during which the maximum value of the electric field is reached. The corresponding time-variation of the electric field and average temperature induced in the superconductor under consideration during serrated regimes with different d*I*/d*t* are shown in Figure 40. Figure 41 shows the d*I*/d*t*–dependences of the limiting average temperature around which the temperature fluctuations occurs and maximum peak current below which the serrated regimes are stable. The corresponding quench values of the temperature and current also are depicted in Figure 41. It is seen that allowable overheating is small at small value of dI/dt. Thereby, the value of I_{max} is close to I_q. Increase in a current charging rate leads to the corresponding rise of the stable range of the temperature and overloaded currents. However, the further increase of dI/dt is accompanied by reduction of the stable range of the temperature fluctuation according to the corresponding change of the self-field losses, as discussed above. At the same time, the increase of I_{max} occurs monotonously even when limiting temperature of the fluctuations decreases and tends to the its limit that equals T_q at high current charging rate.

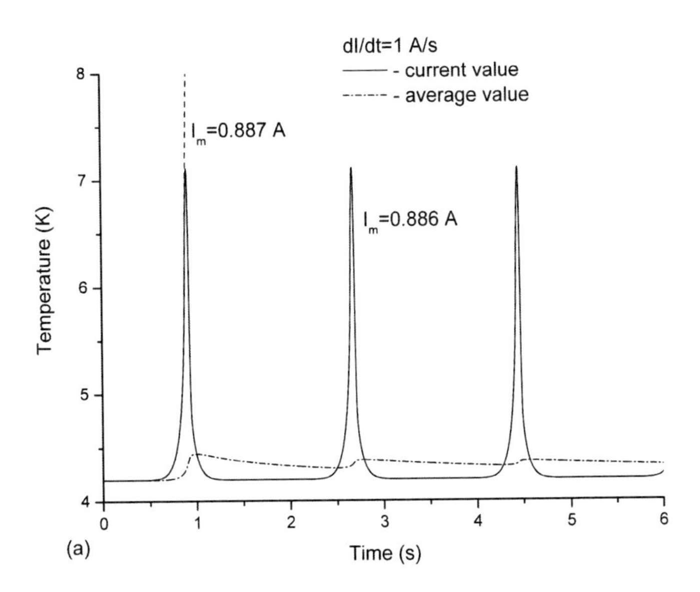

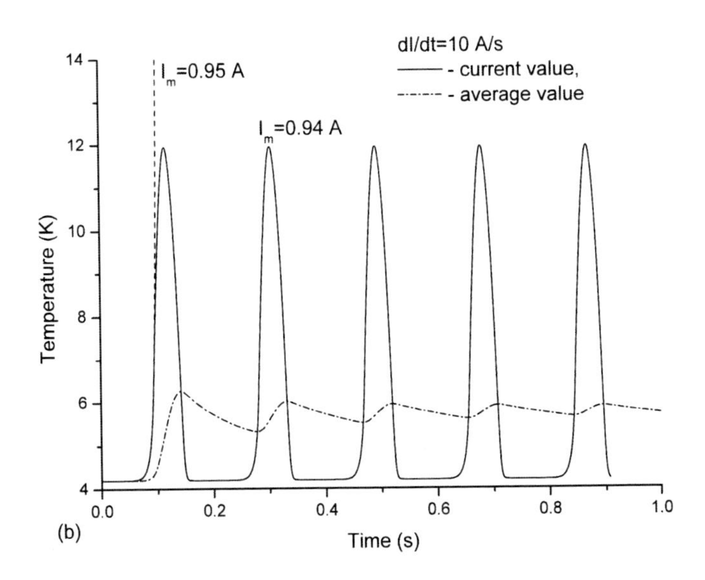

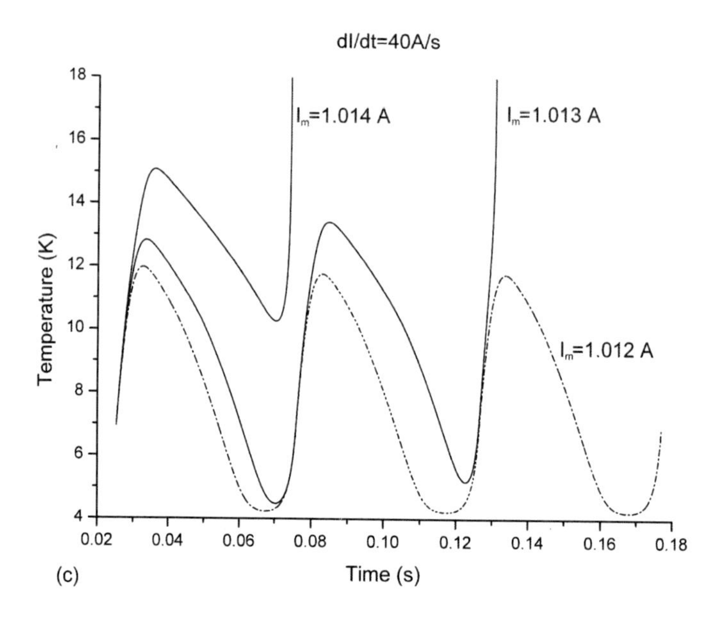

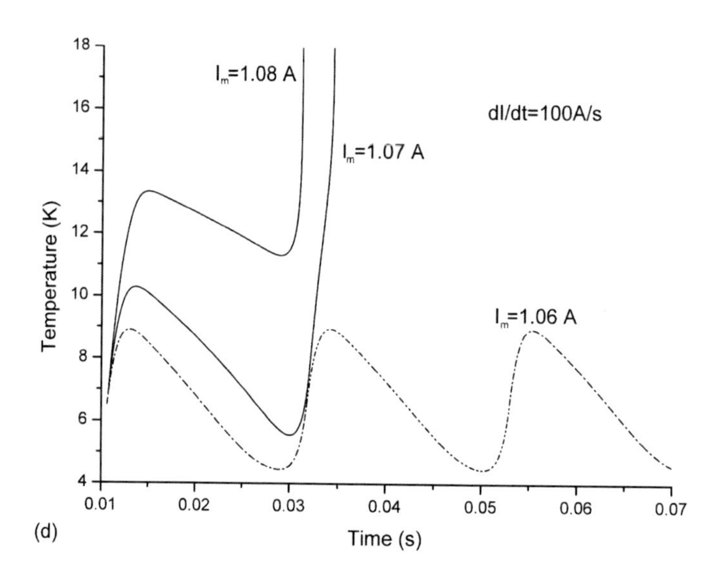

Figure 39. Stable and unstable formation of overloaded regimes.

The discussed features are kept at input of sine-wave currents into superconductors. However, in these regimes the value of dI/dt is not constant. Figure 42 shows the fluctuation of the temperature induced in superconducting slab under consideration by sinusoidal current $I = I_m \sin(2\pi f t)$ at different frequencies, which close to stability boundary. The simulation was made in accordance with zero-dimensional transient model. The corresponding dependences of the fluctuation temperature and maximum value of allowable stable peak current on the current charging rate are presented in Figure 43.

Thereby, the performed analysis allowed to explain why the overloaded currents exist and depend on the peak of the applied current and the applied time. These results will be helpful to study quench processes or detecting quench signal when HTS devices are operated under AC conditions.

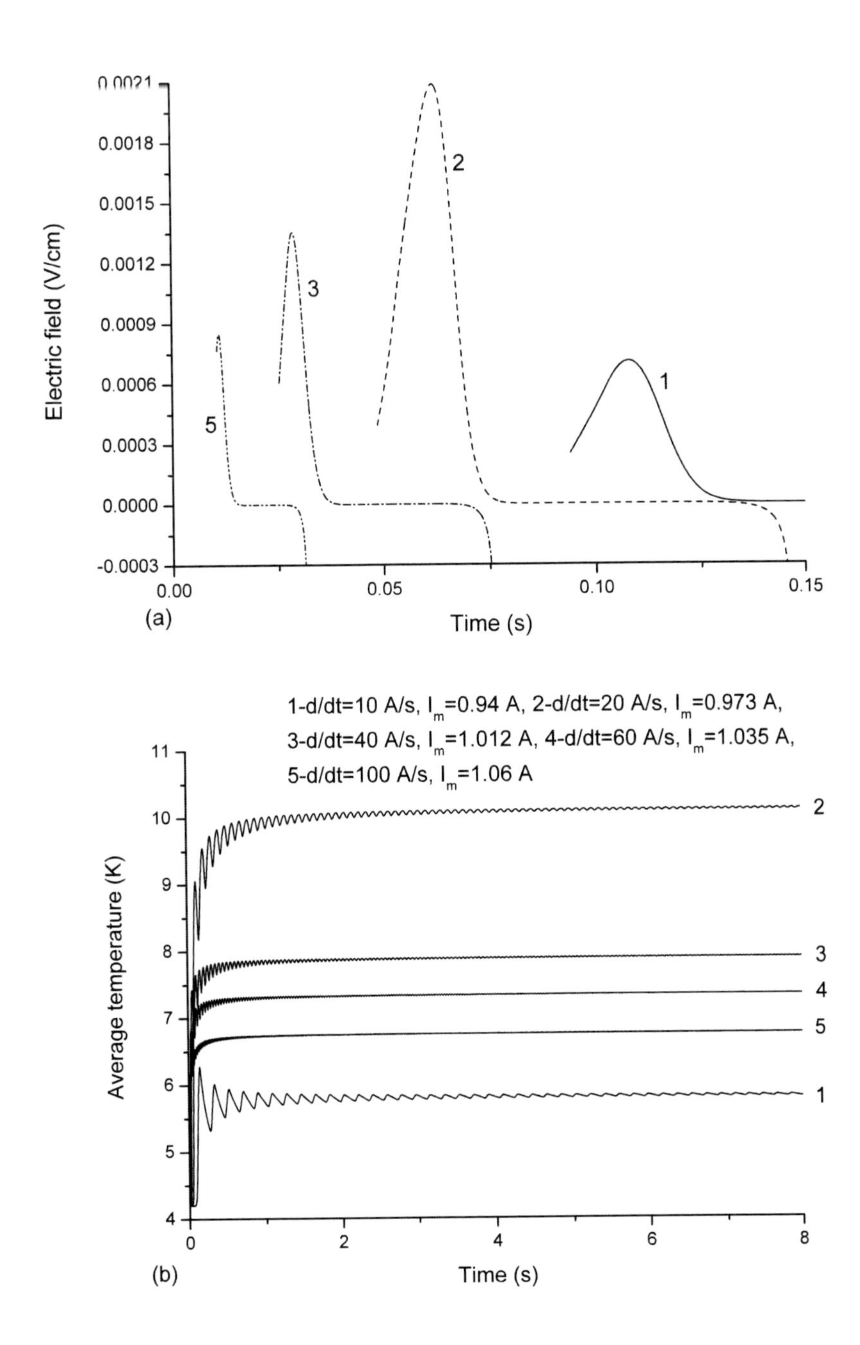

Figure 40. Time-varying electric field on the surface of superconducting slab and its average temperature.

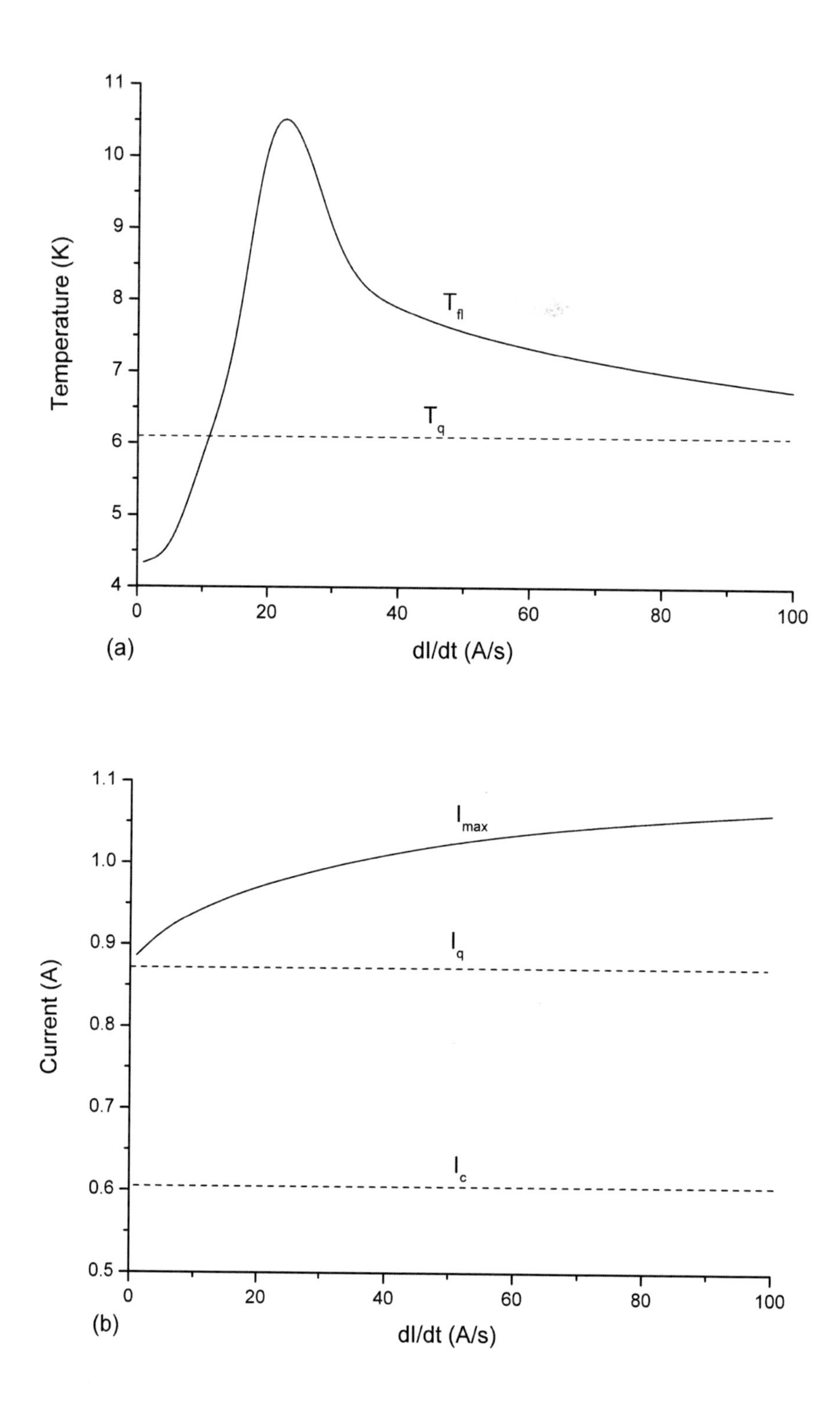

Figure 41. Effect of the current charging rate on the stable values of the fluctuation temperature and limiting overloaded currents.

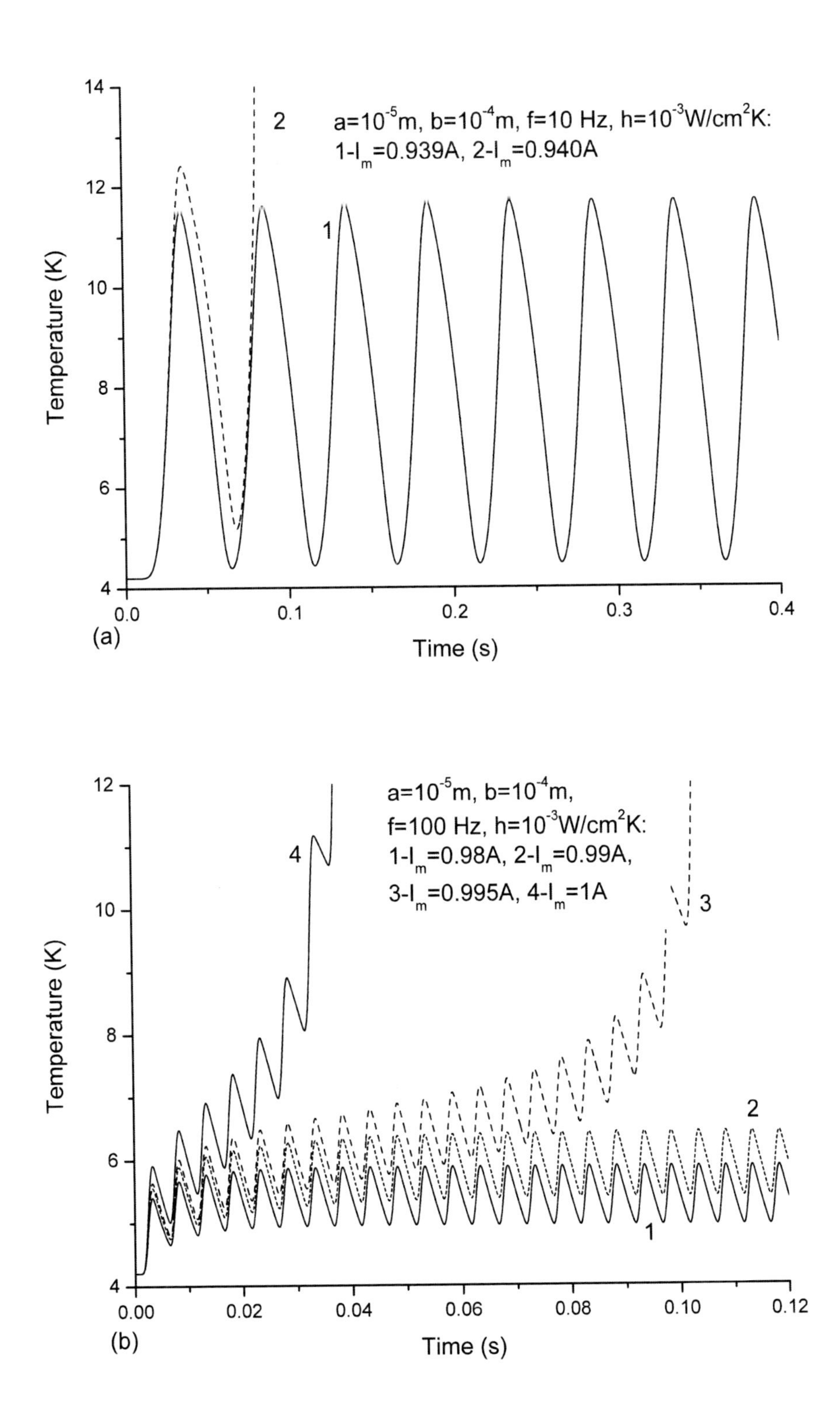

Figure 42. Stable and unstable sine-wave overloaded currents.

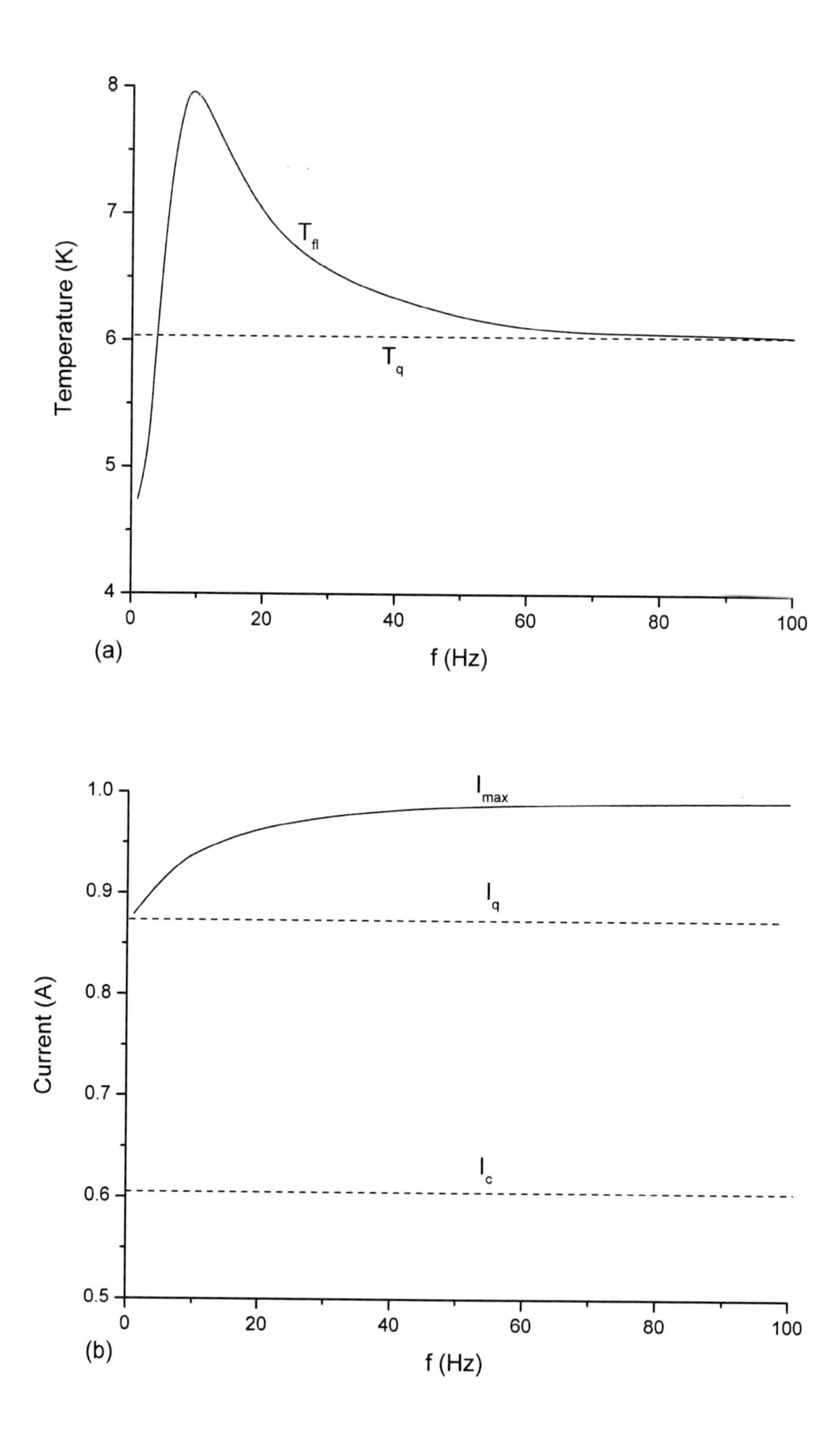

Figure 43. Boundary of stable overloaded currents as a function of frequency.

CONCLUSION

Discussed peculiarities show that the definition of the current stability condition of high-T_c superconductors must take into account the development of the interconnected thermal and electrical states when the temperature of superconductor may stably vary from the temperature of coolant to its critical temperature. This approach allows one to determine the current stability conditions taking into consideration the possible changes in temperature of superconductor up to its critical temperature including the possible emergence of multi-stable resistive states. The results obtained show as follows.

The temperature of a high-T_c superconductor may essentially differ from the coolant temperature before current instability onset. It depends on the current charging rate, cross section of superconductor, cooling conditions. Therefore, the voltage-current characteristic of a high-T_c superconductor during continuous current charging has only a positive slope in the low electric field region. As a result, it does not allow one to find the boundary between stable and unstable operating states. This peculiarity has to be considered during experiments at which the critical current of high-T_c superconductors is defined.

The calculated current of the instability determining the maximum allowable value of the charged current is reduced, if the thermal heterogeneity of the electrodynamics states is taken into consideration in the theoretical analysis of the stability conditions. As a result, the limiting stable values of the electric field and current depend nonlinearly on the thickness of the superconductor, its critical properties as well as on the cooling conditions. Therefore, the current of instability will not increase proportionally to the increase in the thickness of superconductor or its critical current density. This leads to the thermal degradation of the current-carrying capacity of superconductor. The effect of the thermal degradation must be considered in investigations of current-carrying capacity of superconductors having a high critical current density.

The allowable stable values of electric field and current can be below or above those determined by a priori chosen critical parameters of the superconductor. The criteria allowing one to estimate the boundary of these stability areas (sub-critical and over-critical) are written taking into account the size effect.

The violation features of the steady current distribution in high-T_c superconductors ($Bi_2Sr_2CaCu_2O_8$, $Bi_2Sr_2Ca_2Cu_3O_8$ and $YBa_2Cu_3O_7$) cooled by liquid refrigerant (helium, hydrogen and nitrogen) are studied. It is shown that the mechanism of the current instability depends on the type of coolant. Firstly, the destruction of stable current states may occur after trivial transition of the cooling conditions on the surface of superconductor from the nucleate to film boiling regimes. This instability mechanism is more probable for superconductors cooled by liquid helium. Secondly, the current instability may occur due to the disruption of the stable formation of the voltage-current characteristic of a superconductor even in the nucleate boiling regime of cooling. Such stability violation of the charged current is most probable when liquid nitrogen cools the superconductor. The necessary criteria allowing one to determine the influence of the properties of superconductor and coolant on the current instability mechanism are written.

The thermal and electric modes of high-T_c superconductors without the stabilizing matrix may have the nontrivial regimes in the over-critical field range when the n-value decreases with temperature. They are conditioned by the modification of the voltage-current and temperature-current characteristics when stable and unstable branches are not single. In this case, the static thermo-electric modes may lead to the jump of the electric field and temperature without the transition of a superconductor into the normal state. Their development depends on the coolant temperature: the lower this quantity, the more probably the existence of an additional stable mode. This phenomenon is due to the static change of the voltage-current and temperature-current characteristics of a superconductor that may occur according to the corresponding temperature variation of the quantities $J_c(T)$ and $n(T)$. These additional stable modes may be observed in the over-critical electric field range after the high stable temperature rise of a superconductor.

Thus, the correct investigation of the thermal and electric regimes of high-T_c superconductors must be done under the discussed nontrivial formation mechanism of the thermo-electric modes considering that stable temperature variation of superconductors may occur in the wide range.

REFERENCES

[1] Wilson, M. N.; *Superconducting magnets.* Clarendon Press, Oxford, 1983.

[2] Gurevich, A. V.; Mints, R. G.; Rakhmanov, A. L. *The physics of composite superconductors.* Beggel House, NY, 1997.

[3] Tinkham, M. *Inroduction to superconductivity.* McGraw-Hill, New York, 1975.

[4] Nishijima, G.; Awaji, S.; Murase, S.; Shimamoto, S.; Watanabe, K. *IEEE Trans. Appl. Supercond.* 2002, 12, 1155-1158.

[5] Nishijima, G.; Awaji, S.; Watanabe K. *IEEE Trans. on Appl. Supercon.* 2003, 13, 1576-1579.

[6] Romanovskii, V. R.; Watanabe, K.; Awaji, S.; et al. *Supercond. Sci. Technol.* 2004, 17, 1242-1246.

[7] Romanovskii, V. R.; Watanabe, K.; Awaji, S.; et al. *Physica C–Superconductivity and its applications,* 2004, 416, 126-136.

[8] Romanovskii, V. R.; Awaji, S.; Nishijima, G.; Watanabe, K. *J of Applied Physics,* 2006, 100, 063905.

[9] Zeldov, E.; Amer, N. M.; Koren, G.; et al. *Appl. Phys. Lett.* 1990, 56, 680-682.

[10] Polak, M.; Hlasnik, I.; Krempasky, L. *Cryogenics*, 1973, 13, 702-711.

[11] Kalsi, S. S.; Aized, D.; Connor, B.; et al. *IEEE Trans. on Appl. Supercon.* 1997, 7, 971-975.

[12] Kumakura, H.; Kitaguchi, H.; Togano, K.; et al. *Cryogenics,* 1998, 38, 163-167.

[13] Kumakura, H,; Kitaguchi, H.; Togano, K.; et al. *Cryogenics,* 1998, 38, 639-643.

[14] Wetzko, M.; Zahn, M.; Reiss, H. *Cryogenics*, 1995, 35, 375-386.

[15] Kiss, T.; Vysotsky, V. S.; Yuge, H.; et al. *Physica C – Superconductivity and its applications*, 1998, 310, 372-376.

[16] Lehtonen, J.; Mikkonen, R.; Paasi, J. *Physica C – Superconductivity and its applications*, 1998, 310, 340-344.

[17] Lehtonen, J.; Mikkonen, R.; Paasi, J. *Supercond. Sci. Techno.* 2000, 13, 251-258.

[18] Rakhmanov, A. L.; Vysotsky, V. S.; Ilyin, Y. A.; et al. *Cryogenics*, 2000, 40, 19-27.

[19] Anagawa, K.; Yamada, Y.; Shibanchi, T.; et al. *Appl. Phys. Lett.* 2003, 83, 2381-2383.

[20] Krasnov, V. M.; Sandberg, M.; Zogaj, I. *Phys. Review Lett.* 2005, 94, 077003.

[21] Majoros, M.; Glowacki, B. A.; Campbell, A. M. *Physica C–Superconductivity and its applications*, 2002, 372-376, 919-922.

[22] Rettelbach, T.; Schmitz, G. J. *Supercond. Sci. Technol.* 2003, 16, 645-655.

[23] Romanovskii, V. R. *Cryogenics*, 2002, 42, 29-37.

[24] Romanovskii, V. R. *Physica C–Superconductivity and its applications*, 2003, 384, 458-468.

[25] Wesche, R. *Physica C–Superconductivity and its applications*, 1995, 246, 186-194.

[26] van der Laan, D. C.; van Eck, B.; ten Haken, H. J. N.; Schwartz, J.; ten Kate H. H. J. *IEEE Trans. App. Superc.* 2001, 11, 3345-3348.

[27] van der Laan, D. C.; van Eck, H. J. N.; Schwartz, J.; ten Haken, B.; ten Kate, H. H. J. *Physica C – Superconductivity and its applications*, 2002, 372-376, 1024-1027.

[28] Gioacchino, D. D.; Tripodi, P.; Gambardella, U. *Physica C-Superconductivity and its applications*, 2002, 372-376, 945-948.

[29] Kiss, T.; Inoue, M.; Kuga, T.; et al. *Physica C-Superconductivity and its applications*, 2003, 392-396, 1053-1062.

[30] Inoue, M.; Kiss, T.; Kuga, T.; et al. *Physica C-Superconductivity and its applications*, 2003, 392-396, 1078-1082.

[31] Bottura, L. *Note-CRYO/02/027*, CryoSoft library, CERN, 2002.

[32] Seto, T.; Murase, S.; Shimamoto S.; et al. *Teion Kogaku,* 2001, 36,. 60-67 (in Japanese).

[33] Romanovskii, V. R.; Watanabe, K.; Awaji, S.; Nishijima, G. *Supercond Sci Technol,* 2006, 19, 703-710.

[34] Newson, M. S.; Ryan, D. T.; Wilson, M. N.; Jones, H. *IEEE Trans. on Appl. Supercon.* 2002, 12, 725-728.

[35] Watanabe, K.; Awaji, S.; Motokawa, M. *Physica B,* 2003, 329-333, 1487-1488.

[36] Bellis, R.H.; Iwasa, Y. *Cryogenics*, 1994, 34, 129-144.

[37] Junod, A.; Wang, K. O.; Tsukamoto, T.; et al. *Physica C-Superconductivity and its applications*, 1994, 229,. 209-230.

[38] Herrmann, P. F.; Albrecht, C.; Bock, J.; et al. *IEEE Trans Appl Supercon*, 1993, 3, 876-880.

[39] Keilin, V. E.; Romanovskii, V. R. *IEEE Trans on Mag*, 1992, 12, 771-774.

[40] Keilin, V. E.; Romanovskii, V. R. *Cryogenics*, 1993, 33, 986-994.

[41] Lykov, A. V. *Theory of thermal conductivity* (in Russian). Vysshayashkola, Moscow, 1967.

[42] Trit, N. M. *Thermal conductivity: theory, properties and applications.* Kluwer Academic Publishers, NY, 2005.

[43] Brentari, E. G.; Smith, R. *Adv. Cryo. Engn.* 1965, 10, 325-341.

[44] Awaji, S.; Watanabe, K.; Kobayashi, N.; et al. *IEEE Trans. Appl. Supercond.* 1996, 32, 2776-2779.

[45] Awaji, S.; Watanabe, K. *Jpn. J Appl. Phys.* 2001, 40, L1022-L1025.

[46] Uher, C. *J of Superconductivity and Novel Magnetism*, 1990, 3, 337-389.

[47] Romanovskii, V. R.; Watanabe, K. *Supercond. Sci. and Technol.* 2005, 18, 407-416.

[48] Prigozhin, Leonid *IEEE Trans. on Appl. Supercon.* 1997, 4, 3866-3873.

[49] Däumling M. *Supercond. Sci. Techno.* 1998, 11, 590-593.

[50] Karmakar, D.; Bhagwat, K. V. *Physica C-Superconductivity and its applications*, 2003, 398, 20-30.

[51] Carr, W. J. Jr. *Physica C-Superconductivity and its applications*, 2004, 415, 109-117.

[52] Ruiz-Alonso, D.; Coombs, T. A.; Campbell, A. M. *Supercond. Sci. Techno.* 2005, 18, S209-S214.

[53] Hong, Z.; Campbell, A. M.; T A Coombs, T. A. *Supercond. Sci. Techno.* 2006, 19, 1246-1252.

[54] Norris, W. T. *J. Phys. D: Appl. Phys.* 1970, 3, 489-507.

[55] Brandt, E. H. *Physical Review B*, 1996, 54, 4246-4264.

[56] Nibbio, N.; Stravrev, S.; Dutoit, B. *IEEE Trans. on Appl. Supercon.* 2001, 11, 2631-2633.

[57] Stavrev, S.; Grilli, F.; Dutoit, D.; et al, *IEEE Trans. on Magnetic.* 2001, 38, 849-852.

[58] Enomoto, N.; Amemiya, N. *Physica C-Superconductivity and its applications*, 2004, 412-414, 1050-1055.

[59] Burkhardt, E. E.; Schwartz, J. *IEEE Trans. Appl. Supercond.* 1997, 7, 199-202.

[60] Duckworth, R.; Pfotenhauer, J.; Lue, J. et al., *Advances in Cryogenic Engineering.* (In: Proc. of the Int. Cryogenic Materials Conf.-ICMC), 2003, 48, 313-319.

[61] Vysotsky, V. S.; Sytnikov, V. E.; et al, *IEEE Trans on Appl Supercon.* 2005, 15, 1655-1658.

[62] Vysotsky, V. S.; et al, *J. Phys.: Conf. Ser.*, 2006, 43, 877-880.

[63] Vysotsky, V. S. et al. , *J. Phys.: Conf. Ser.*, 2008, 97, 012015.

In: Superconductivity
Editor: Vladimir Rem Romanovskií

ISBN 978-1-61324-843-0

Chapter 3

STUDY OF SUPERCONDUCTING STATE PARAMETERS OF METALLIC COMPLEXES USING PSEUDOPOTENTIAL THEORY

Aditya M. Vora[1,2*]
[1] Humanities and Social Science Department, S. T. B. S. College of Diploma Engineering, Opp. Spinning Mill, Varachha Road, Surat 395 006, Gujarat, India
[2] Parmeshwari 165, Vijaynagar Area, Hospital Road
Bhuj–Kutch, Gujarat, India

ABSTRACT

In this chapter, the theoretical investigation of five superconducting state parameters (SSP) viz; electron-phonon coupling strength λ, Coulomb pseudopotential μ^*, transition temperature T_C, isotope effect exponent α and effective interaction strength N_OV of metallic complexes using model potential formalism with the pseudo-alloy-atom (PAA) model is discussed. Various local field correction functions are used for the first time in the present investigation to study the screening influence on the aforesaid properties. It is observed that the electron-phonon coupling strength λ and the transition temperature T_C are quite sensitive to the selection of the local field correction functions, whereas the Coulomb pseudopotential μ^*, isotope effect exponent α and effective interaction strength N_OV show weak dependences on the local field correction functions. The present results of the SSPs are found in qualitative agreement with the available experimental or theoretical data wherever exist. The T_C equation has been proposed, which provide successfully the T_C values of metallic complexes under

[*] E-mail address: voraam@yahoo.com

consideration. Also, the present study confirms the superconducting phase in metallic superconductors. A strong dependency of the SSPs on the valence 'Z' is found.

Keywords: Pseudopotential; pseudo-alloy-atom (PAA) model; superconducting state parameters; metallic complexes.

PACS Nos.: 61.43.Dq; 71.15.Dx; 74.20.-z; 74.70.Ad

1. INTRODUCTION

During last several years, the superconductivity remains a dynamic area of research in condensed matter physics with continual discoveries of novel materials and with an increasing demand for novel devices for sophisticated technological applications. A large number of metals and amorphous alloys are superconductors, with critical temperature T_C ranging from 1-18K. The pseudopotential theory has been used successfully in explaining the superconducting state parameters (SSP) for metallic complexes by many workers [1-83]. Many of them have used well known model pseudopotential in the calculation of the superconducting state parameters (SSP) for the metallic complexes [5-66]. Recently, we have studied the superconducting state parameters (SSP) of some metallic superconductors using single parametric model potential formalism [28-66]. The study of the superconducting state parameters (SSP) of the metallic superconductors such as metals, alloys and metallic glasses may be of great help in deciding their applications; the study of the dependence of the transition temperature T_C on the composition of metallic elements is helpful in finding new superconductors with high T_C. The application of pseudopotential to binary metallic superconductors involves the assumption of pseudoions with average properties, which are assumed to replace three types of ions in the binary systems, and a gas of free electrons is assumed to permeate through them. The electron-pseudoion is accounted for by the pseudopotential and the electron-electron interaction is involved through a dielectric screening function. For successful prediction of the superconducting properties of the alloying systems, the proper selection of the pseudopotential and screening function is very essential [28-66].

Therefore, in the present chapter, we have used well known McMillan's theory [67] of the superconductivity for predicting the superconducting state parameters (SSP) viz. the electron-phonon coupling strength λ, Coulomb pseudopotential μ^*, transition temperature T_C, isotope effect exponent α and effective interaction strength $N_O V$ of some metallic complexes using model potential formalism. To see the impact of various exchange and correlation functions on the aforesaid properties, we have employed here five different types of local field correction functions proposed by Hartree (H) [84], Taylor (T) [85], Ichimaru-Utsumi (IU) [86], Farid *et al.* (F) [87] and Sarkar *et al.* (S) [88]. We have incorporated for the first time more advanced local field correction functions due to IU, F and S with Ashcroft's empty core (EMC) model potential [89] in the present computation.

In most of the theoretical studies of superconductivity of bulk metallic complexes, the Vegard's law was used to calculate electron-ion interaction from the potential of the pure components. Also in bulk metallic glasses, the translational symmetry is broken, and therefore, the momentum (or quasi-momentum) should not be used to describe the state of the system. The virtual crystal approximation enables us to keep the concept of the momentum only in an approximate way. But, it is well established that pseudo-alloy-atom (PAA) is more meaningful approach to explain such kind of interactions in binary systems [28-66]. In the PAA approach a hypothetical monoatomic crystal is supposed to be composed of pseudo-alloy-atoms, which occupy the lattice sites and from a perfect lattice in the same way as pure metals. In this model the hypothetical crystal made up of PAA is supposed to have the same properties as the actual disordered alloy material and the pseudopotential theory is then applied to studying various properties of alloy systems. The complete miscibility in the glassy alloy systems is considered as a rare case. Therefore, in such binary systems the atomic matrix elements in the pure states are affected by the characteristics of alloys such as lattice distortion effects and charging effects. In the PAA model, such effects are involved implicitly. In addition to this it also takes into account the self-consistent treatment implicitly [28-66].

The well known Ashcroft's empty core (EMC) model potential [89] used in the present computations of the SSP of metals is of the form,

$$V(q) = \frac{-8\pi Z}{\Omega_O q^2} \cos(q r_C), \tag{1}$$

here, Z, Ω_O and r_C are the valence, atomic volume and parameter of the model potential of metals, respectively. The Ashcroft's empty core (EMC) model potential is a simple one-parameter model potential, which has been successfully found for various metallic complexes [5-66]. When used with a suitable form of dialectic screening functions, this potential has also been found to yield good results in computing the SSP of metallic elements [5-66]. Here, the model potential parameter r_C is fitted in such a way that the calculated values of the transition temperature T_C agrees well with the theoretical or experimental value of the T_C as close as possible.

2. Computational Methodology

The most exhaustive study of the relationship between microscopic theory and observed superconducting transition temperature T_C was made by McMillan [67]. His work was based on the Eliashberg gap equation [90], which are extensions of the original BCS theory [4]. The formulation of the McMillan [67] has been reanalyzed particularly for the case of the strong coupling superconductors, by Allen and Dynes [75]. For a theoretical estimation of the material properties λ and μ^*, the essential ingredients are the electron-ion pseudopotential, dielectric screening and the phonon frequencies. While the phonon spectra are available experimentally from the inelastic scattering of the neutrons for number of the metallic glasses, otherwise it can be calculated from the Debye temperature relation given by Butler [91]. Since the theoretical prediction of the SSP depends on the appropriate representation of the electron-ion interaction potential, it is interesting to study the pseudopotential dependence of the electron-phonon coupling strength parameter λ and hence of T_C. According the McMillan [67], the electron-phonon coupling strength is given by

$$\lambda = 2\int_0^{\infty} d\omega\left[\alpha^2(\omega)F(\omega)/\omega\right]. \tag{2}$$

where, $\alpha^2(\omega)F(\omega)$ is the spectral function, which when appropriately evaluated in the plane wave approximation for the scattering on the Fermi surface yields [5-66],

$$\lambda = \frac{m_b\,\Omega_0}{4\pi^2\,k_F\,M\langle\omega^2\rangle}\int_0^{2k_F} q^3\,|\,W(q)|^2\,dq\,. \tag{3}$$

Here m_b is the band mass, M the ionic mass, Ω_O the atomic volume, k_F the Fermi wave vector, $W(q)$ the screened pseudopotential and $\langle\omega^2\rangle$ the averaged square phonon frequency, of the ternary metallic glasses, respectively. The $\langle\omega^2\rangle$ is calculated using the relation given by Butler [85], $\langle\omega^2\rangle^{1/2} = 0.69\,\theta_D$, where θ_D is the Debye temperature of the ternary systems.

Using $X = q/2k_F$ and $\Omega_O = 3\pi^2 Z/(k_F)^3$, we get Eq. (3) in the following form,

$$\lambda = \frac{12\,m_b\,Z}{M\,\langle\omega^2\rangle}\int_0^1 X^3\,|\,W(X)|^2\,dX\,, \tag{4}$$

where Z and $W(X)$ are the valence and the screened EMC pseudopotential [89], respectively.

The BCS theory gives a relation $T_C \approx \theta_D \exp(-1/N(0)V)$ for the superconducting transition temperature T_C in terms of the Debye temperature θ_D. The electron-electron interaction V consists of the attractive electron-phonon-induced interaction minus the repulsive Coulomb interaction. The notation is used $\lambda = N(0)V_{e-ph}$. And the Coulomb repulsion $N(0)V_C$ is called μ, so that $N(0)V = \lambda - \mu^*$, where μ^* is a “renormalized” Coulomb repulsion, reduced in value from μ to $\mu/[1+\mu\ln(\omega_P/\omega_D)]$. This suppression of the Coulomb repulsion is a result of the fact that the electron-phonon attraction is retarded in time by an amount $\Delta t \approx 1/\omega_D$ whereas the repulsive screened Coulomb interaction is retarded by a much smaller time, $\Delta t \approx 1/\omega_P$ where ω_P is the electronic plasma frequency. Therefore, μ^* is bounded above by $1/\ln(\omega_P/\omega_D)$ which for conventional metals should be $\leq$

0.2. Values of λ are known to range from ≤ 0.10 to ≥ 2.0. Also, The parameter μ^* is assigned a value in the range 0.10-0.15, consistent with tunneling and with theoretical guesses. Calculations of μ or μ^* are computationally demanding and are not yet under theoretical control. Calculations of λ are slightly less demanding, are under somewhat better theoretical control, and have been attempted for many years. Prior to 1990, calculations of λ generally required knowing the phonon frequencies and eigenvectors as input information, and approximating the form of the electron-ion potential. McMillan [67] and Hopfield [68] pointed out that one could define a simpler quantity, $\eta = N(0)\langle I^2 \rangle$ with $\langle \omega^2 \rangle = \frac{2}{\lambda}\int_0^\infty d\Omega\Omega\alpha^2 F(\Omega)$. The advantage of this is that η and $\langle I^2 \rangle$ are purely "electronic" quantities, requiring no input information about phonon frequencies or eigenvectors. Gaspari and Gyorffy [92] then invented a simplified algorithm for calculating η, and many authors have used this. These calculations generally require a "rigid ion approximation" or some similar guess for the perturbing potential felt by electrons when an atom has moved. Given η, one can guess a value for $\langle \omega^2 \rangle$ (for example, from θ_D). In the weak coupling limit of the electron-phonon interaction, the fundamental equations of the BCS theory should be derived from the Eliashberg equations. This conversion is possible upon some approximation of the phonon frequency $|\omega| \geq \omega_D$ with ω_D denote the Debye frequency [79]. Morel and Anderson [93] are given the relation of the transition temperature $\lambda - \mu^* = \lambda - \mu/[1 + \mu \ln(E_F/\omega_l)]$, which is nearly equal to the factor 6 for monovalent, bivalent and tetravalent metals. Where $E_F = k_F^2$ is the Fermi energy and ω the phonon frequency of the metallic substances. The effect of phonon frequency is very less in comparison with the Fermi energy. Hence, the overall effect of the Coulomb pseudopotential is reduced by the large logarithmic term. Therefore, Rajput and Gupta [94] have introduced the new term $10\,\theta_D$ in place of the phonon frequency ω_l from the Butler's relation [79] for the sake of simplicity and ignoring the lattice vibrational effect, which is generated consistent results of the Coulomb pseudopotential. The parameter μ^* represents the effective interelectronic Coulomb repulsion at the Fermi surface [79]. Hence, in the present case, we have adopted relation of the Coulomb pseudopotential given by Rajput and Gupta [94]. Therefore, the Coulomb pseudopotential μ^* is given by [5-66]

$$\mu^* = \frac{\dfrac{m_b}{\pi k_F}\displaystyle\int_0^1 \frac{dX}{\varepsilon(X)}}{1 + \dfrac{m_b}{\pi k_F}\ln\left(\dfrac{E_F}{10\,\theta_D}\right)\displaystyle\int_0^1 \frac{dX}{\varepsilon(X)}}. \tag{5}$$

As it is evident from the equation (5) that, which was originally derived by Bogoliubov *et al.* [95], the Coulomb repulsion parameter μ^* is essentially weakened owing to a large logarithmic term in the denominator. Here, $\varepsilon(X)$ the modified Hartree dielectric function, which is written as [84],

$$\varepsilon(X) = 1+(\varepsilon_H(X) - 1)(1 - f(X)). \tag{6}$$

$\varepsilon_H(X)$ is the static Hartree dielectric function [84] and $f(X)$ the local field correction function. In the present investigation, the local field correction functions due to H [84], T [85], IU [86], F [87] and S [88] are incorporated to see the impact of exchange and correlation effects. The details of all the local field correction functions are as follows,

The Hartree screening function [84] is purely static, and it does not include the exchange and correlation effects. The expression of it is,

$$f(q) = 0. \tag{7}$$

Taylor (T) [85] has introduced an analytical expression for the local field correction function, which satisfies the compressibility sum rule exactly. This is the most commonly used local field correction function and covers the overall features of the various local field correction functions proposed before 1972. According to Taylor (T) [85],

$$f(q) = \frac{q^2}{4k_F^2}\left[1+\frac{0.1534}{\pi k_F^2}\right]. \tag{8}$$

The Ichimaru-Utsumi (IU) local field correction function [86] is a fitting formula for the dielectric screening function of the degenerate electron liquids at metallic and lower densities, which accurately reproduces the Monte-Carlo results as well as satisfies the self consistency condition in the compressibility sum rule and short range correlations. The fitting formula is

$$\begin{aligned} f(q) &= A_{IU}Q^4 + B_{IU}Q^2 + C_{IU} \\ &+\left[A_{IU}Q^4+\left(B_{IU}+\frac{8\ A_{IU}}{3}\right)Q^2 - C_{IU}\right]\left\{\frac{4-Q^2}{4Q}\ln\left|\frac{2+Q}{2-Q}\right|\right\}. \end{aligned} \tag{9}$$

On the basis of Ichimaru-Utsumi (IU) local field correction function [86], Farid *et al.* (F) [87] have given a local field correction function of the form

$$\begin{aligned} f(q) &= A_F Q^4 + B_F Q^2 + C_F \\ &+\left[A_F Q^4 + D_F Q^2 - C_F\right]\left\{\frac{4-Q^2}{4Q}\ln\left|\frac{2+Q}{2-Q}\right|\right\}. \end{aligned} \tag{10}$$

Based on Eqs. (9-10), Sarkar *et al.* (S) [88] have proposed a simple form of local field correction function, which is of the form

$$f(q) = A_S\{1-(1+B_SQ^4)\exp(-C_SQ^2)\}. \tag{11}$$

Where $Q = q/k_F$. The parameters A_{IU}, B_{IU}, C_{IU}, A_F, B_F, C_F, D_F, A_S, B_S and C_S are the atomic volume dependent parameters of IU, F and S-local field correction functions. The mathematical expressions of these parameters are narrated in the respective papers of the local field correction functions [86-88]. After evaluating λ and μ^*, the transition temperature T_C and isotope effect exponent α are investigated from the McMillan's formula [5-67]

$$T_C = \frac{\theta_D}{1.45}\exp\left[\frac{-1.04(1+\lambda)}{\lambda-\mu^*(1+0.62\lambda)}\right], \tag{12}$$

$$\alpha = \frac{1}{2}\left[1-\left(\mu^*\ln\frac{\theta_D}{1.45T_C}\right)^2\frac{1+0.62\lambda}{1.04(1+\lambda)}\right]. \tag{13}$$

The expression for the effective interaction strength N_OV is studied using [5-66]

$$N_OV = \frac{\lambda-\mu^*}{1+\frac{10}{11}\lambda}. \tag{14}$$

3. Results and Discussion

In the present section, the superconducting state parameters (SSP) of metallic complexes are discussed briefly, which indicates superconducting phase in the metallic complexes.

3.1. SSP of $In_{1-x}Tl_x$ binary alloys

The input parameters and constants used in the present calculations are taken from our earlier paper [62]. Table 1 shows the presently calculated values of the SSP of $In_{1-x}Tl_x$ $(0 \le x \le 1)$ binary alloys with other such available theoretical [30, 69, 70] as well as experimental findings [30].

The calculated values of the electron-phonon coupling strength λ for $In_{1-x}Tl_x$ $(0 \le x \le 1)$ binary alloys, using five different types of the local field correction functions with EMC model potential, are shown in Table 1 with other theoretical data [5, 8, 9]. It is noticed from the present study that, the percentile influence of the various local field correction functions with respect to the static H-screening function on the electron-phonon coupling strength λ is 24.32%-46.39%. Also, the H-screening yields lowest values of λ, whereas the values obtained from the F-function are the highest. It is also observed from the Table 1 that, λ obtained from H-function goes on decreasing from the values of 1.1835 $\rightarrow$ 0.7449 as the concentration 'x' of 'Tl' is increased from 0.0$\rightarrow$0.50, while for concentration 'x' of 'Tl' increases, λ goes on increasing. Similar trends are also observed for most of local field correction functions. The increase or decrease in λ with concentration 'x' of 'Tl' shows a gradual transition from weak coupling behaviour to intermediate coupling behaviour of electrons and phonons, which may be attributed to an increase of the hybridization of sp-d electrons of 'Tl' with increasing concentration (x). This may also be attributed to the increase role of ionic vibrations in the Tl-rich region. Here we can consider only range of concentration 'x' of 'Tl'. The present results are found in qualitative agreement with the available theoretical data [30, 69, 70]. The $In_{1-x}Tl_x$ $(0 \le x \le 1)$ binary alloys are type-II superconductors.

The computed values of the Coulomb pseudopotential μ^*, which accounts for the Coulomb interaction between the conduction electrons, obtained from the various forms of the local field correction functions are tabulated in Table 1. It is observed from the Table 1 that for all binary alloys, μ^* lies between 0.11 and 0.14, which is in accordance with McMillan [67], who suggested $\mu^* \approx 0.13$ for simple and non-simple metals. The weak influence of the screening functions shows on the computed values of μ^*. The percentile influence of the various local field correction functions with respect to the static H-screening function on μ^* for the $In_{1-x}Tl_x$ binary alloys is observed in the range of 5.64%-11.77%. Again the H-screening function yields lowest values of μ^*, while the values obtained from the F-function are the highest. The present results are found in qualitative agreement with the available theoretical data [30, 69, 70].

Table 1 contains calculated values of the transition temperature T_C for $In_{1-x}Tl_x$ $(0 \le x \le 1)$ binary alloys computed from the various forms of the local field correction functions along with the experimental or theoretical findings [30, 69, 70]. From the Table 1 it can be noted that, the static H-screening function yields lowest T_C whereas the F-function yields highest values of T_C. The present results obtained from the H-local field correction functions are found in good agreement with available experimental or theoretical data [30, 69, 70]. It is also observed that the static H-screening function yields lowest T_C whereas the F-function yields highest values of T_C. The calculated results of the transition temperature T_C for In, $In_{0.90}Tl_{0.10}$, $In_{0.75}Tl_{0.25}$, $In_{0.73}Tl_{0.27}$, $In_{0.67}Tl_{0.33}$, $In_{0.57}Tl_{0.43}$, $In_{0.50}Tl_{0.50}$, $In_{0.27}Tl_{0.73}$, $In_{0.17}Tl_{0.83}$, $In_{0.07}Tl_{0.93}$ and Tl deviate in the range of 11.07%-98.48%, 4.17%-91.41%, 9.83%-71.10%, 11.01%-69.45%, 11.42%-70.13%, 6.33%-106.23%, 7.61%-109.36%, 25.30%-43.57%, 11.86%-66.63%, 7.04%-97.72% and 30.26%-135.80% from the experimental findings [30], respectively.

The values of the isotope effect exponent α for $In_{1-x}Tl_x$ $(0 \le x \le 1)$ binary alloys are tabulated in Table 1. The computed values of the α show a weak dependence on the dielectric screening, its value is being lowest for the H-screening function and highest for the F-function. Since the experimental value of α has not been reported in the literature so far, the present data of α may be used for the study of ionic vibrations in the superconductivity of alloying substances. Since H-local field correction function yields the best results for λ and T_C, it may be observed that α values obtained from this screening provide the best account for the role of the ionic vibrations in superconducting behaviour of this system.

The values of the effective interaction strength $N_O V$ are listed in Table 1 for different local field correction functions. It is observed that the magnitude of $N_O V$ shows that the $In_{1-x}Tl_x$ $(0 \le x \le 1)$ binary alloys under investigation lie in the range of weak coupling superconductors. The values of the $N_O V$ also show a feeble dependence on dielectric screening, its value being lowest for the H-screening function and highest for the F-screening function.

Table 1. Superconducting state parameters $In_{1-x}Tl_x$ binary alloys

Alloys	SSP	Present results					Expt. [30]	Others [30, 69, 70]
		H	T	IU	F	S		
In	λ	0.8224	1.1187	1.1675	1.1835	1.0224	–	0.26, 0.69, 0.67, 0.76, 0.88, 0.89, 1.16, 0.80, 1.09, 1.14, 1.15, 1.02, 0.72, 0.95, 0.98, 1.02, 1.03, 0.91
	μ^*	0.1248	0.1364	0.1380	0.1390	0.1319	–	0.09, 0.10, 0.11, 0.12, 0.11, 0.12, 0.13, 0.13, 0.12, 0.11, 0.12, 0.12, 0.13, 0.13, 0.12
	T_C (K)	3.7809	6.2820	6.6530	6.7562	5.5455	3.404	4.77x10^{-3}, 2.8, 3.4, 4.0, 3.85, 6.38, 6.76, 8.67, 5.8, 3.02, 5.12, 5.35, 5.71, 5.82, 4.81
	α	0.4445	0.4560	0.4572	0.4572	0.4541	0.45	0.45, 0.47, 0.45,0.46,0.44,0.46
	$N_O V$	0.3991	0.4870	0.4994	0.5032	0.4615	–	0.12, 0.4, 0.49,0.50,0.47,0.37, 0.44, 0.45, 0.46, 0.47, 0.43
$In_{0.90}Tl_{0.10}$	λ	0.7943	1.0818	1.1295	1.1449	0.9884	–	1.09,0.80,0.92,1.68,1.31, 1.32, 1.25,1.50, 1.46, 0.71, 0.85
	μ^*	0.1241	0.1356	0.1372	0.1382	0.1311	–	0.14, 0.13, 0.13, 0.15, 0.14, 0.14, 0.15, 0.14, 0.11, 0.12
	T_C (K)	3.4168	5.8166	6.1785	6.2781	5.1058	3.28	5.84, 3.33, 4.44, 9.68, 7.48, 7.56, 7.05, 8.69, 8.46, 2.75
	α	0.4422	0.4546	0.4559	0.4559	0.4525	–	0.44, 0.45, 0.45, 0.46, 0.47
	$N_O V$	0.3892	0.4771	0.4896	0.4933	0.4515	–	0.39, 0.43, 0.36

Table 1. (Continued)

Alloys	SSP	Present results					Expt. [30]	Others [30, 69, 70]
		H	T	IU	F	S		
$In_{0.80}Tl_{0.20}$	λ	0.7732	1.0543	1.1012	1.1162	0.9631	–	1.06,0.79,0.89,1.63,1.28, 1.29, 1.22,1.46,1.43,0.69,0.68, 0.85
	μ^*	0.1233	0.1348	0.1364	0.1374	0.1304	–	0.14, 0.13, 0.13, 0.15, 0.14, 0.14, 0.14, 0.14, 0.11, 0.11
	T_C (K)	3.1399	5.4523	5.8061	5.9026	4.7642	–	5.47,3.2 4.11,9.20, 6.83, 6.91, 6.66, 8.25, 8.03, 2.51, 2.72
	α	0.4405	0.4536	0.4550	0.4550	0.4514	–	0.44, 0.45, 0.44, 0.45
	N_0V	0.3816	0.4695	0.4821	0.4858	0.4440	–	0.39, 0.42, 0.35, 0.36
$In_{0.75}Tl_{0.25}$	λ	0.7650	1.0437	1.0903	1.1053	0.9533	–	0.93
	μ^*	0.1230	0.1344	0.1360	0.1370	0.1300	–	–
	T_C (K)	3.0296	5.3038	5.6539	5.7490	4.6257	3.36	4.79, 2.47
	α	0.4399	0.4532	0.4547	0.4547	0.4510	–	–
	N_0V	0.3787	0.4666	0.4793	0.4830	0.4410	–	–
$In_{0.73}Tl_{0.27}$	λ	0.7621	1.0400	1.0866	1.1015	0.9499	–	1.05, 0.79, 0.88, 1.61, 1.26, 1.27, 1.20, 1.44, 1.40, 0.68
	μ^*	0.1228	0.1343	0.1359	0.1369	0.1299	–	0.14, 0.13, 0.13, 0.15, 0.14, 0.14, 0.14, 0.14, 0.11
	T_C (K)	2.9902	5.2502	5.5990	5.6935	4.5760	3.36	5.27,3.14,3.93,8.92,7.07, 7.14, 6.44, 7.99, 7.78, 3.36, 2.38
	α	0.4397	0.4531	0.4546	0.4546	0.4508	–	0.44, 0.45, 0.44, 0.46
	N_0V	0.3776	0.4656	0.4783	0.4820	0.4400	–	0.39, 0.42, 0.35
$In_{0.70}Tl_{0.30}$	λ	0.7582	1.0351	1.0816	1.0964	0.9453	–	–
	μ^*	0.1226	0.1341	0.1356	0.1367	0.1297	–	–
	T_C (K)	2.9361	5.1758	5.5226	5.6163	4.5070	–	–
	α	0.4395	0.4530	0.4545	0.4545	0.4507	–	–
	N_0V	0.3763	0.4642	0.4770	0.4806	0.4387	–	–
$In_{0.67}Tl_{0.33}$	λ	0.7548	1.0309	1.0773	1.0920	0.9414	–	1.04,0.79,0.87,1.59,1.25, 1.26, 1.19, 1.43, 1.39, 0.67, 0.90
	μ^*	0.1224	0.1338	0.1354	0.1364	0.1295	–	0.14, 0.13, 0.13, 0.15, 0.14, 0.14, 0.14, 0.14, 0.11
	T_C (K)	2.8877	5.1084	5.4533	5.5463	4.4448	3.26	5.12,3.13,3.80,8.71,6.66, 6.74, 6.28,7.81,7.60,2.30, 4.51, 2.21
	α	0.4393	0.4529	0.4544	0.4544	0.4506	–	0.44, 0.44, 0.44, 0.46
	N_0V	0.3751	0.4631	0.4758	0.4795	0.4375	–	0.39, 0.41, 0.35

Alloys	SSP	Present results					Expt. [30]	Others [30, 69, 70]
		H	T	IU	F	S		
$In_{0.65}Tl_{0.35}$	λ	0.7529	1.0284	1.0748	1.0895	0.9391	–	0.847
	μ^*	0.1223	0.1337	0.1353	0.1363	0.1293	–	–
	T_C (K)	2.8583	5.0673	5.4110	5.5035	4.4069	2.60	3.56, 1.68
	α	0.4392	0.4529	0.4544	0.4544	0.4505	–	–
	N_OV	0.3744	0.4624	0.4752	0.4789	0.4368	–	–
$In_{0.60}Tl_{0.40}$	λ	0.7489	1.0236	1.0700	1.0846	0.9345	–	1.03,0.80,0.86,1.58,1.24, 1.25, 1.18, 1.42, 1.38, 0.67, 0.67
	μ^*	0.1219	0.1333	0.1349	0.1359	0.1290	–	0.13, 0.13, 0.13, 0.14, 0.14, 0.14, 0.14, 0.11, 0.11
	T_C (K)	2.7956	4.9772	5.3182	5.4096	4.3244	–	4.98, 3.12, 3.68, 8.51, 6.51, 6.58, 6.13, 7.64, 7.43, 2.22, 2.46, 3.41, 1.57
	α	0.4391	0.4529	0.4544	0.4544	0.4505	–	0.44, 0.45, 0.44, 0.45
	N_OV	0.3730	0.4611	0.4740	0.4777	0.4355	–	0.39, 0.41, 0.34, 0.35
$In_{0.57}Tl_{0.43}$	λ	0.7471	1.0216	1.0680	1.0826	0.9326	–	1.02, 0.80, 0.86, 1.57, 1.23, 1.24, 1.18, 1.41, 1.38, 0.66
	μ^*	0.1217	0.1331	0.1347	0.1357	0.1288	–	0.13, 0.13, 0.13, 0.14, 0.14, 0.14, 0.14, 0.14, 0.11
	T_C (K)	2.7647	4.9318	5.2713	5.3621	4.2831	2.60	4.94,3.14,3.65,8.43,6.45,6.52, 6.08,7.59,7.37,2.20,3.56, 1.68
	α	0.4392	0.4529	0.4545	0.4545	0.4506	–	0.44, 0.45, 0.44
	N_OV	0.3724	0.4607	0.4735	0.4772	0.4350	–	0.39, 0.41, 0.34
$In_{0.50}Tl_{0.50}$	λ	0.7449	1.0195	1.0661	1.0807	0.9305	–	1.02,0.81,0.86,1.56,1.23, 1.24, 1.18,1.41,1.37,0.66,0.67, 0.84
	μ^*	0.1213	0.1326	0.1342	0.1352	0.1283	–	0.13, 0.13, 0.13, 0.14, 0.14, 0.14, 0.14, 0.14, 0.11, 0.11
	T_C (K)	2.7118	4.8498	5.1862	5.2758	4.2096	2.52	4.86,3.18,3.58,8.29,6.35, 6.41, 5.99,7.45,7.26,2.16, 3.41, 1.57
	α	0.4395	0.4532	0.4548	0.4548	0.4509	–	0.44, 0.45, 0.44
	N_OV	0.3718	0.4603	0.4732	0.4769	0.4346	–	0.39, 0.41, 0.34
$In_{0.40}Tl_{0.60}$	λ	0.7464	1.0228	1.0700	1.0847	0.9332	–	1.02,0.83,0.86,1.56,1.23, 1.24, 1.18, 1.41, 1.38, 0.66, 0.68
	μ^*	0.1206	0.1320	0.1335	0.1346	0.1276	–	0.13, 0.13, 0.13, 0.14, 0.14, 0.14, 0.14, 0.14, 0.11, 0.11
	T_C (K)	2.6810	4.7894	5.1225	5.2106	4.1586	–	4.80,3.29,3.53,8.14,6.26, 6.32, 5.92, 7.34, 7.15, 2.14, 2.44
	α	0.4405	0.4541	0.4556	0.4556	0.4517	–	0.44, 0.45, 0.44, 0.45
	N_OV	0.3728	0.4616	0.4747	0.4784	0.4358	–	0.40, 0.41, 0.35, 0.36

Table 1. (Continued)

Alloys	SSP	Present results					Expt. [30]	Others [30, 69, 70]
		H	T	IU	F	S		
$In_{0.30}Tl_{0.70}$	λ	0.7535	1.0339	1.0820	1.0969	0.9430	–	–
	μ^*	0.1199	0.1313	0.1329	0.1339	0.1270	–	–
	T_C (K)	2.7024	4.7943	5.1252	5.2122	4.1699	–	–
	α	0.4423	0.4553	0.4568	0.4568	0.4531	–	–
	N_OV	0.3760	0.4653	0.4785	0.4822	0.4394	–	–
$In_{0.27}Tl_{0.73}$	λ	0.7568	1.0388	1.0873	1.1023	0.9474	–	1.092
	μ^*	0.1198	0.1311	0.1327	0.1337	0.1268	–	–
	T_C (K)	2.7192	4.8085	5.1390	5.2258	4.1856	3.64	5.59, 2.76
	α	0.4429	0.4558	0.4573	0.4573	0.4536	–	–
	N_OV	0.3774	0.4668	0.4801	0.4838	0.4409	–	–
$In_{0.20}Tl_{0.80}$	λ	0.7668	1.0536	1.1030	1.1183	0.9606	–	1.05,0.89,0.88,1.60,1.27, 1.28, 1.22, 1.45, 1.42, 0.69, 0.71
	μ^*	0.1193	0.1307	0.1322	0.1332	0.1263	–	0.13, 0.12, 0.12, 0.14, 0.13, 0.13, 0.14, 0.11, 0.11
	T_C (K)	2.7778	4.8653	5.1950	5.2812	4.2448	–	4.87,3.68,3.62,8.08,6.30, 6.36, 5.99, 7.34, 7.17, 2.26, 2.65
	α	0.4447	0.4570	0.4585	0.4584	0.4549	–	0.45, 0.45, 0.46
	N_OV	0.3815	0.4714	0.4847	0.4885	0.4454	–	0.42, 0.42, 0.35, 0.37
$In_{0.17}Tl_{0.83}$	λ	0.7721	1.0613	1.1113	1.1267	0.9676	–	1.06,0.90,0.88,1.61,1.28, 1.29, 1.23, 1.46, 1.43, 0.70, 0.89
	μ^*	0.1191	0.1305	0.1320	0.1331	0.1261	–	0.13, 0.12, 0.12, 0.14, 0.13, 0.13, 0.14, 0.11
	T_C (K)	2.8116	4.8999	5.2294	5.3155	4.2801	3.19	4.92,3.76,3.65,8.09,6.24, 6.39, 6.02,7.37,7.18,2.37, 3.47, 1.79
	α	0.4456	0.4576	0.4590	0.4590	0.4556	–	0.45, 0.45, 0.47
	N_OV	0.3837	0.4738	0.4871	0.4909	0.4477	–	0.43, 0.42, 0.36
$In_{0.10}Tl_{0.90}$	λ	0.7872	1.0831	1.1344	1.1502	0.9872	–	–
	μ^*	0.1187	0.1300	0.1316	0.1326	0.1257	–	–
	T_C (K)	2.9120	5.0057	5.3349	5.4206	4.3868	–	–
	α	0.4478	0.4592	0.4605	0.4605	0.4572	–	–
	N_OV	0.3897	0.4802	0.4937	0.4975	0.4540	–	–

Alloys	SSP	Present results					Expt. [30]	Others [30, 69, 70]
		H	T	IU	F	S		
$In_{0.07}Tl_{0.93}$	λ	0.7950	1.0942	1.1462	1.1622	0.9972	–	1.09, 0.94, 0.91, 1.66, 1.31, 1.32, 1.26, 1.50, 1.49, 0.71
	μ^*	0.1185	0.1298	0.1314	0.1324	0.1255	–	0.13, 0.12, 0.12, 0.14, 0.13, 0.13, 0.14, 0.13,0.11
	T_C (K)	2.9649	5.0621	5.3913	5.4768	4.4434	2.77	5.09, 4.08, 3.82, 8.19, 6.48, 6.54, 6.19, 7.45, 7.41, 2.46
	α	0.4488	0.4599	0.4612	0.4611	0.4580	–	0.46, 0.46, 0.45, 0.46
	$N_O V$	0.3927	0.4835	0.4970	0.5007	0.4572	–	0.44, 0.43, 0.37
Tl	λ	0.8162	1.1246	1.1783	1.1948	1.0246	–	0.71, 0.72, 1.07, 1.07, 0.88, 1.22, 1.27, 1.29, 0.96
	μ^*	0.1181	0.1294	0.1310	0.1320	0.1251	–	0.10, 0.11, 0.11, 0.12, 0.11
	T_C (K)	3.1132	5.2213	5.5504	5.6357	4.6027	2.390	2.34, 2.36, 4.39, 4.8, 3.87, 6.09, 6.42, 6.52, 4.45
	α	0.4514	0.4617	0.4629	0.4628	0.4599	–	0.46, 0.47
	$N_O V$	0.4008	0.4921	0.5057	0.5095	0.4657	0.263	0.22, 0.48, 0.43, 0.52, 0.53, 0.54, 0.45

In Table 1, the comparison of presently obtained data of SSP with experimentally available values [30] and other various theoretical data [30, 69, 70] is reported for pure metals like In and Tl. The experimental values of transition temperature T_C for pure metals In and Tl are 3.404K and 2.39K, respectively. Our present results of T_C show good agreement with them. The experimental data of α for pure In is 0.45.The excellent agreement with this number is achieved in the present investigations for the pure metal. The presently reported values of $N_O V$ for pure Tl give the satisfactory results in comparison with the observed experimental data 0.263.It is also seen that for proper reproduction of experimental values of SSP the choice of dielectric function is also an important parameter.

3.2. SSP of $Al_{1-x}Li_x$ binary alloys

Al-Li based alloys can be distinguished from other aluminium alloys due to their higher strength at a lower density and they are, therefore, promising structural materials for aviation and space engineering. Recently, their static and elastic properties have been the subject of numerous papers. The studies were usually performed at cryogenic temperatures, mainly covering the range down to 77.3 K. Only some of them were done at 4.2 K and therefore not all the studies were concerned with the possible superconductivity of the Al-Li alloys and the specific plastic effects which show up during the superconducting transition [71, 72] Very recently, Ou *et al.* [71] have been reported superconducting transition parameters using fitting of the density of states as well as Coulomb pseudopotential.

Table 2. Superconducting state parameters of $Al_{1-x}Li_x$ binary alloys

Glass	SSP	Present results					Expt. [71, 72]
		H	T	IU	F	S	
Li	λ	0.5161	0.8227	0.8859	0.8919	0.7066	0.382
	μ^*	0.1664	0.1868	0.1896	0.1901	0.1796	0.10
	T_c (K)	1.1610	7.1197	8.6427	8.7714	4.5256	1.16
	α	0.1718	0.3292	0.3442	0.3449	0.2950	–
	N_0V	0.2380	0.3638	0.3857	0.3876	0.3209	–
$Al_{0.01}Li_{0.99}$	λ	0.5084	0.8064	0.8678	0.8733	0.6943	0.377
	μ^*	0.1658	0.1859	0.1888	0.1892	0.1789	0.10
	T_c (K)	1.0710	6.7526	8.2289	8.3485	4.2734	1.07
	α	0.1636	0.3249	0.3404	0.3411	0.2902	–
	N_0V	0.2343	0.3580	0.3796	0.3814	0.3160	–
$Al_{0.03}Li_{0.97}$	λ	0.5048	0.7956	0.8554	0.8606	0.6867	0.375
	μ^*	0.1648	0.1846	0.1874	0.1878	0.1777	0.10
	T_c (K)	1.0506	6.5829	8.0270	8.1399	4.1753	1.05
	α	0.1644	0.3242	0.3396	0.3403	0.2899	–
	N_0V	0.2330	0.3545	0.3758	0.3775	0.3134	–
$Al_{0.038}Li_{0.962}$	λ	0.5066	0.7945	0.8536	0.8586	0.6873	–
	μ^*	0.1640	0.1836	0.1863	0.1867	0.1767	–
	T_c (K)	1.1005	6.6380	8.0730	8.1811	4.2516	1.10
	α	0.1732	0.3268	0.3418	0.3424	0.2939	–
	N_0V	0.2346	0.3547	0.3757	0.3773	0.3143	–
$Al_{0.05}Li_{0.95}$	λ	0.5075	0.7936	0.8523	0.8572	0.6871	0.378
	μ^*	0.1634	0.1828	0.1855	0.1859	0.1760	0.10
	T_c (K)	1.1309	6.6855	8.1176	8.2249	4.2991	1.13
	α	0.1784	0.3286	0.3434	0.3440	0.2964	–
	N_0V	0.2354	0.3548	0.3757	0.3773	0.3146	–
$Al_{0.10}Li_{0.90}$	λ	0.5119	0.7886	0.8449	0.8494	0.6866	0.385
	μ^*	0.1608	0.1792	0.1818	0.1823	0.1727	0.10
	T_c (K)	1.2900	6.8745	8.2813	8.3786	4.5289	1.29
	α	0.2029	0.3365	0.3500	0.3505	0.3078	–
	N_0V	0.2397	0.3549	0.3750	0.3765	0.3164	–
$Al_{0.104}Li_{0.896}$	λ	0.5082	0.7811	0.8366	0.8409	0.6809	–
	μ^*	0.1605	0.1789	0.1815	0.1819	0.1724	–
	T_c (K)	1.2407	6.6936	8.0772	8.1702	4.4025	1.24
	α	0.1991	0.3344	0.3481	0.3486	0.3054	–
	N_0V	0.2378	0.3521	0.3721	0.3735	0.3141	–

The input parameters and constants used in the present investigation are taken from our earlier paper [55]. The superconducting state parameters of $Al_{1-x}Li_x$ binary alloys are displayed in Table 2 and Figures 1-5 along with other such experimental findings [71, 72].

The present results of superconductivity are found to be in qualitative agreement with the available experimental data [71, 72] in the literature.

It is seen from the Table 2 and Figures 1-5 that, among all five screening functions, the H-screening function gives the minimum value of the superconducting state parameters (SSP) while the F-screening function gives the maximum value. The present findings due to T, IU and S-local field correction functions are lying between these two screening functions. These local field correction functions are able to generate consistent results regarding the SSP of $Al_{1-x}Li_x$ alloys. The numerical values of the aforesaid properties are found to be quite sensitive to the selection of the local field correction function and showing a significant variation with the change in the function. Also λ goes on increasing from the values of $0.5066 \rightarrow 0.8733$ as the concentration 'x' of 'Li' is increased from $0.01 \rightarrow 0.104$.

The increase in λ shows a gradual transition from weak coupling behaviour to intermediate coupling behaviour of electrons and phonons, which may be attributed to an increase of the hybridization of sp-d electrons. A small initial drop is shown in the graph of λ near 3 at. % 'Li' then increase by almost linearly at 10 at. % Li, same nature was also observed by Ou *et al.* [71]. Generally, 'Li' exhibits non-superconducting nature in normal laboratory condition. But, in the present case it exhibits superconducting nature. The computed results of the electron-phonon coupling strength λ for $Al_{1-x}Li_x$ alloys deviate in the range of 32.96%-133.48% from the experimental findings [71, 72]. With respect to the static H-dielectric function, the influence of various local field correction functions on λ is 33.98%-72.82%. Such influence on μ^* is observed in the range of 7.40%-14.24%. These changes in λ and μ^* make drastic variation on T_C, α and $N_O V$. While μ^* accounts for the Coulomb interaction between the conduction electrons lies between 0.16 and 0.19, which is in accordance with McMillan [67]. Ou *et al.* [71] have fitted the values of μ^* in the range of 0.09 to 0.11 for obtaining better λ from the experimental data of T_C. But, we have avoided such type of fitting in the computation. The higher values of μ^* in comparison of those of Ou *et al.* [71] may be due to the screening effects.

The calculated results of the transition temperature T_C for $Al_{1-x}Li_x$ alloys deviate in the range of 0.05%-680.23% from the experimental findings [71, 72]. A small initial drop is observed in the graph of T_C near 3 at. % Li then increase by almost linearly at 10 at. % Li, same nature was also observed by Ou *et al.* [71]. They have noted that, it may be due to the exact value of λ certainly relies on the selection of μ^*. More important to this study, however, is to elucidate the alloying or Li-content effect on T_C. It is also seen from the graphical nature of T_C, the composition dependence can be described by linear regression of the data obtained for H-dielectric function for different values of the concentration (x), which yields

$$T_C(K) = 0.0171x + 1.0684. \tag{15}$$

The graph of the fitted T_C equation is displayed in Figure 4, which indicates that T_C drops and increases almost linearly with increasing 'Li' content with a slope $dT_C/dC = 0.0171$. Wide extrapolation predicts a T_C = 1.0684K for the hypothetical case of '$Al_{0.01}Li_{0.99}$'.

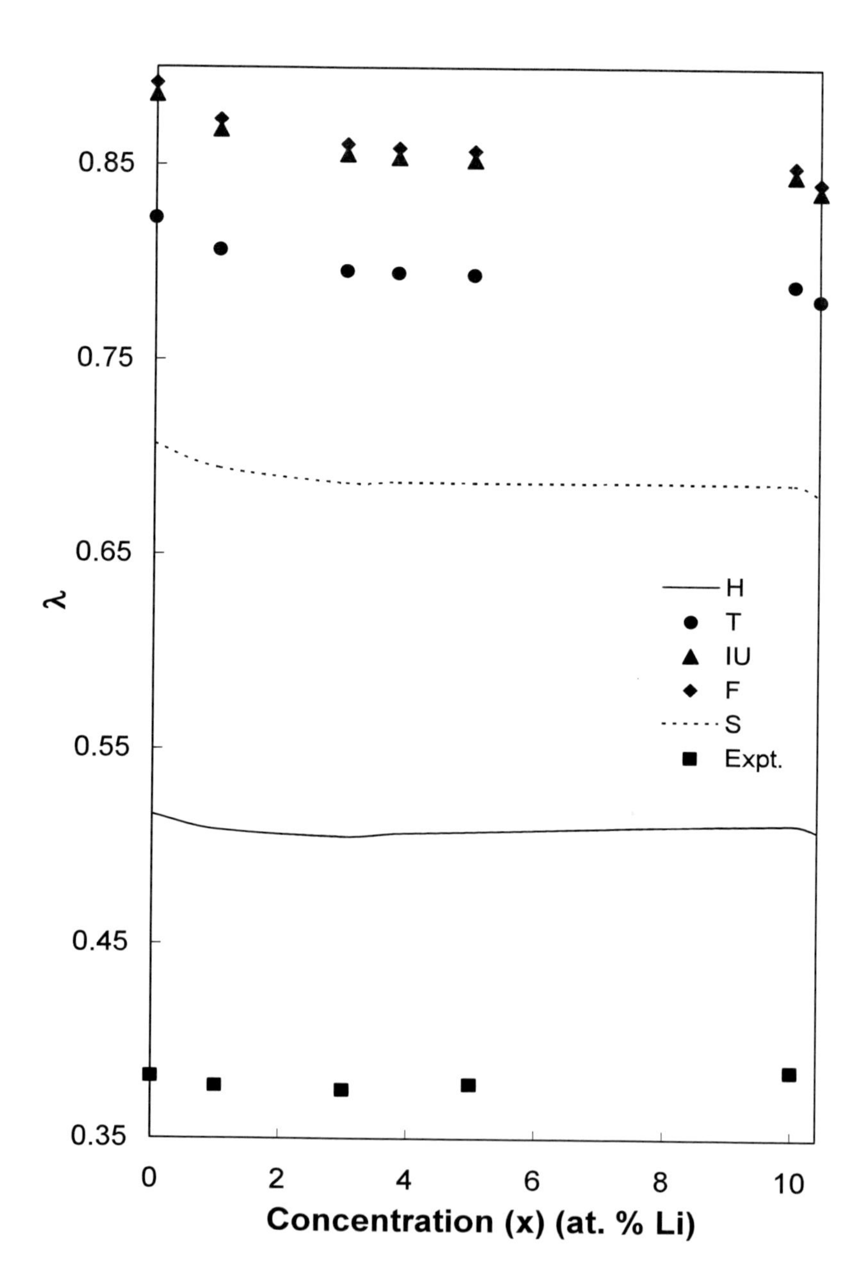

Figure 1. Variation of electron-phonon coupling strength (λ) with Li-concentration (x) (in at %).

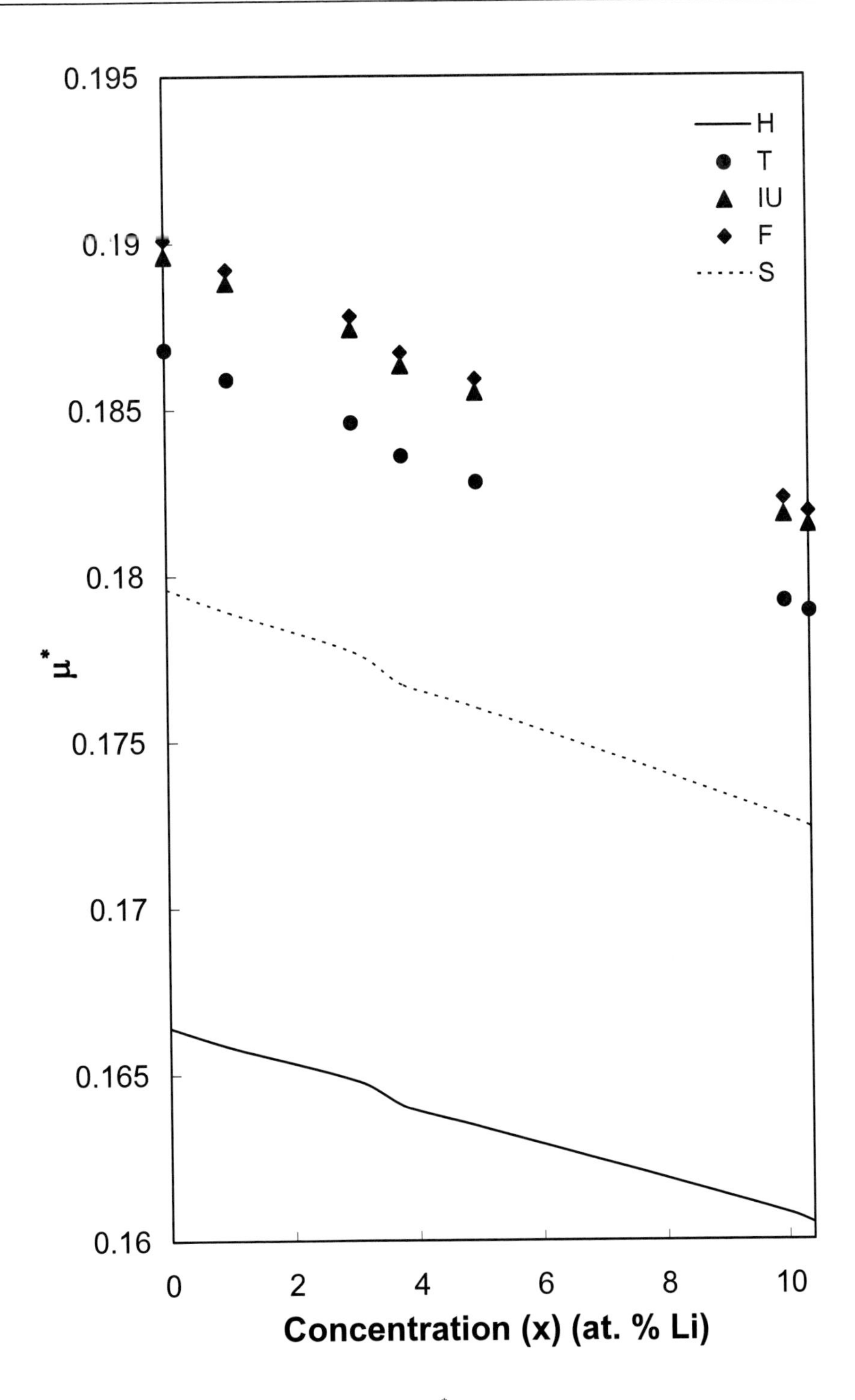

Figure 2. Variation of Coulomb pseudopotential (μ^*) with Li-concentration (x) (in at %).

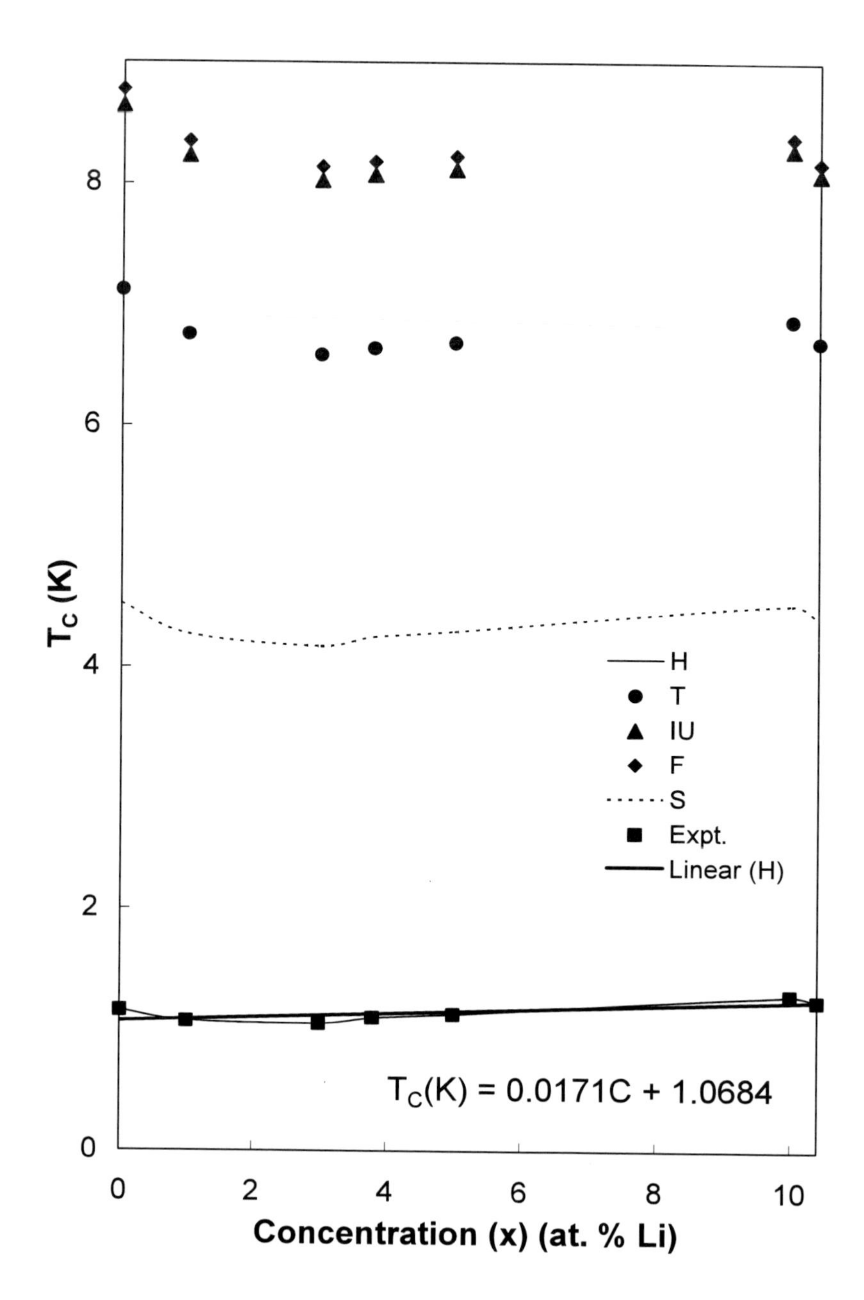

Figure 3. Variation of transition temperature (T_C) with Li-concentration (x) (in at %).

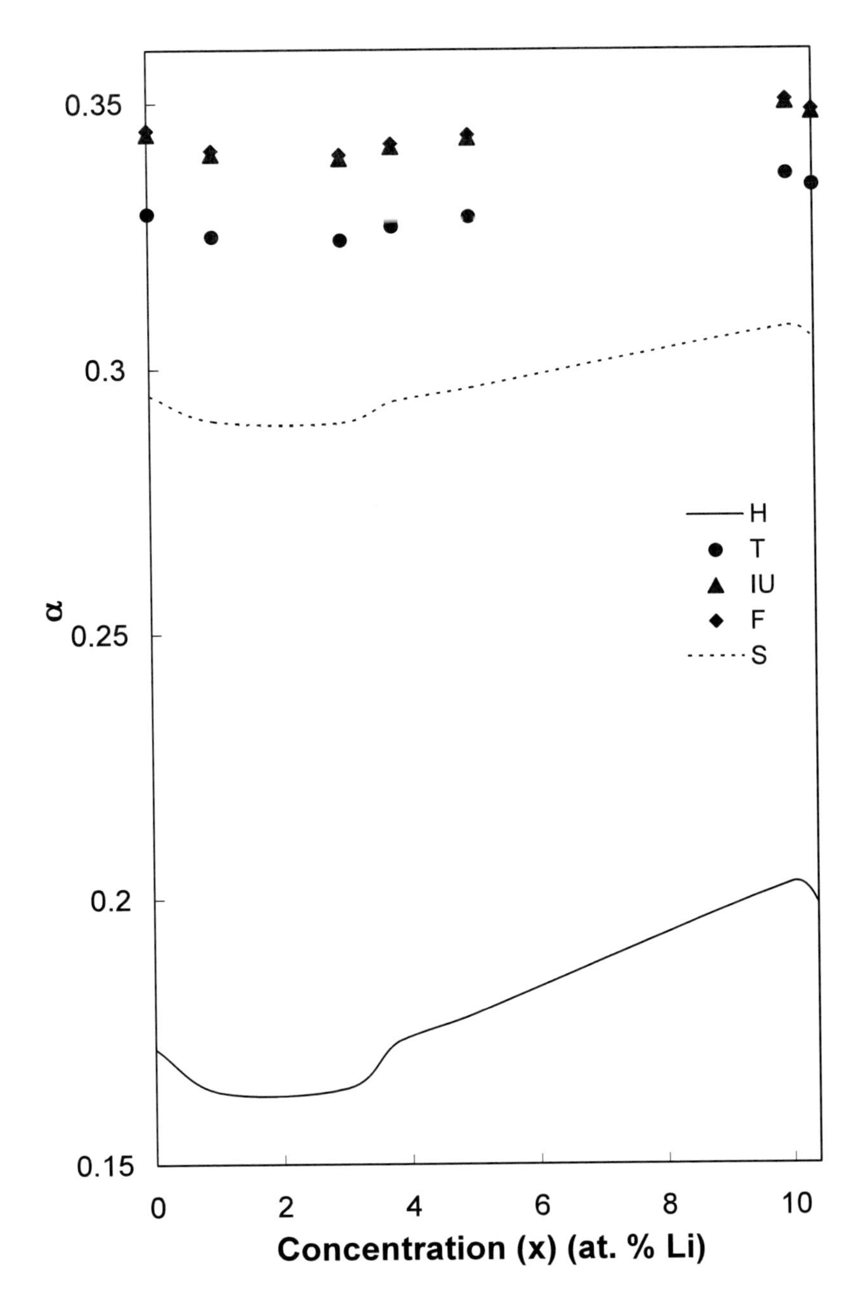

Figure 4. Variation of isotope effect exponent (α) with Li-concentration (x) (in at %).

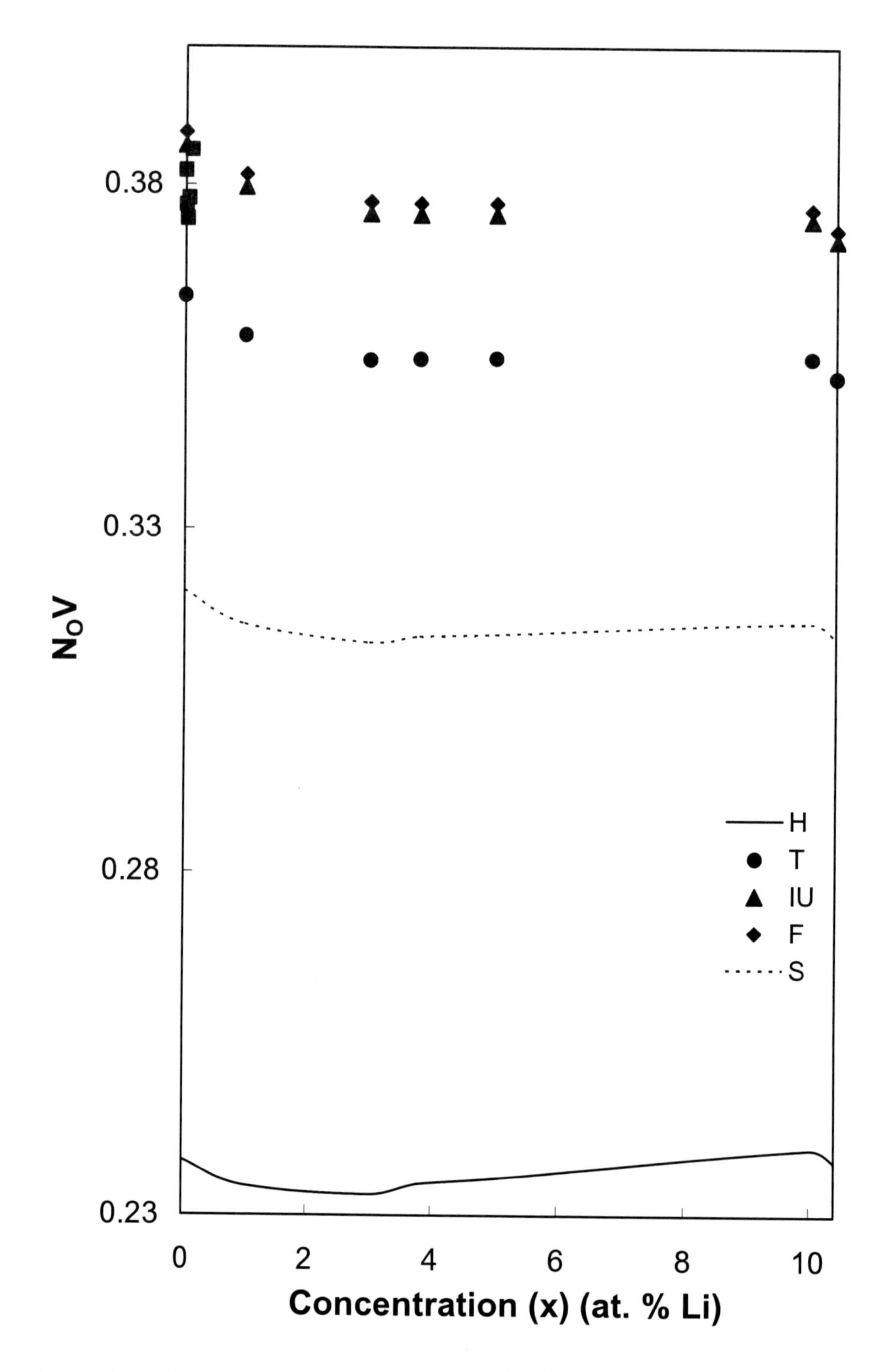

Figure 5. Variation of effective interaction strength ($N_O V$) with Li-concentration (x) (in at %).

The computed values of the α shows a weak dependence on the dielectric screening. Since the experimental value of α has not been reported in the literature so far, the present data of α may be used for the study of ionic vibrations in the superconductivity of alloying substances. Since H-dielectric function yields the best results for λ and T_C, it may be observed that α values obtained from this screening provide the best account for the role of the ionic vibrations in superconducting behaviour of this system. It is also observed that the magnitude of $N_O V$ lie in the range of weak coupling superconductors and also show a feeble dependence on dielectric screening. The effective interaction strength $N_O V$ represents combined effect of the electronic density of states at the Fermi surface, which is related to the Coulomb pseudopotential μ^* and pairing potential arising from the electron-phonon interaction related to the electron-phonon coupling strength λ, respectively. Hence, the graphical nature of the effective interaction strength $N_O V$ is same as those of the electron-phonon coupling strength λ. In the absence of experimental data for the SSP, the presently computed values of these parameters may be considered to form reliable data for aforesaid alloys, as they lie within the theoretical limits of the Eliashberg-McMillan formulation [67].

3.3. SSP of Ag-Zn-Al alloys

The input parameters and other constants used in the present computation are taken from our earlier paper [52]. Table 3 show the presently calculated values of the SSP viz. electron-phonon coupling strength λ, Coulomb pseudopotential μ^*, transition temperature T_C, isotope effect exponent α and effective interaction strength $N_O V$ at various concentrations for binary alloys with available experimental findings [73, 74].

The calculated values of the electron-phonon coupling strength λ for Ag_xZn_{1-x} and Ag_xAl_{1-x} binary alloys, using five different types of the local field correction functions with EMC model potential, are shown in Table 3 with the experimental data [73, 74]. It is noticed from the present study that, the percentile influence of the various local field correction functions with respect to the static H-screening function on the electron-phonon coupling strength λ is 23.70%-41.17%, 23.79%-41.32%, 23.89%-41.50%, 23.95%-41.61%, 23.99%-41.69%, 24.15%-42.00%, 24.24%-42.16%, 24.41%-42.48%, 24.91%-43.52%, 25.49%-44.79%, 28.79%-53.47%, 28.23%-52.03%, 27.69%-50.66%, 27.52%-50.25%, 27.35%-49.81%, 27.17%-49.36%, 26.51%-47.71% and 26.03%-46.55% for $Ag_{0.13}Zn_{0.87}$, $Ag_{0.14}Zn_{0.86}$, $Ag_{0.15}Zn_{0.85}$, $Ag_{0.155}Zn_{0.845}$, $Ag_{0.16}Zn_{0.84}$, $Ag_{0.175}Zn_{0.825}$, $Ag_{0.185}Zn_{0.815}$, $Ag_{0.20}Zn_{0.80}$, $Ag_{0.25}Zn_{0.75}$, $Ag_{0.305}Zn_{0.695}$, $Ag_{0.76}Al_{0.24}$, $Ag_{0.73}Al_{0.27}$, $Ag_{0.70}Al_{0.30}$, $Ag_{0.69}Al_{0.31}$, $Ag_{0.68}Al_{0.32}$, $Ag_{0.67}Al_{0.33}$, $Ag_{0.63}Al_{0.37}$ and $Ag_{0.60}Al_{0.40}$ binary alloys, respectively. Also, the H-screening yields lowest values of λ, whereas the values obtained from the F-function are the highest. It is also observed from the Table 3 that, λ goes on decreasing from the values of 0.2756→0.4260 as the concentration 'x' of 'Ag' is increased from 0.13→0.305, while for concentration 'x' of 'Al' increases, λ goes on increasing from 0.2666→0.4489. The increase or decrease in λ with concentration 'x' of 'Zn' and 'Al' shows a gradual transition from weak coupling behaviour to intermediate coupling behaviour of electrons and phonons, which may be attributed to an increase of the hybridization of sp-d electrons of 'Zn' and 'Al' with increasing or decreasing concentration (x). This may also be attributed to the increase role of ionic vibrations in the 'Zn' or 'Al' metals-rich region. The present results are found in qualitative agreement with available experimental data [73, 74]. The calculated results of the electron-phonon coupling strength λ for $Ag_{0.13}Zn_{0.87}$, $Ag_{0.14}Zn_{0.86}$, $Ag_{0.15}Zn_{0.85}$, $Ag_{0.155}Zn_{0.845}$, $Ag_{0.16}Zn_{0.84}$, $Ag_{0.175}Zn_{0.825}$, $Ag_{0.185}Zn_{0.815}$, $Ag_{0.20}Zn_{0.80}$, $Ag_{0.25}Zn_{0.75}$, $Ag_{0.305}Zn_{0.695}$, $Ag_{0.76}Al_{0.24}$, $Ag_{0.73}Al_{0.27}$, $Ag_{0.70}Al_{0.30}$, $Ag_{0.69}Al_{0.31}$, $Ag_{0.68}Al_{0.32}$, $Ag_{0.67}Al_{0.33}$, $Ag_{0.63}Al_{0.37}$ and $Ag_{0.60}Al_{0.40}$ deviate in the range of 3.35%-45.90%, 3.26%-45.94%, 3.04%-45.80%, 2.83%-45.62%, 2.46%-45.17%, 0.35%-41.50%, 2.06%-39.24%, 4.32%-36.32%, 9.46%-29.94%, 12.79%-26.27%, 6.47%-43.55%, 9.14%-38.14%, 8.01%-38.60%, 7.06%-39.65%, 6.34%-40.30%, 5.92%-40.52%, 1.02%-49.21% and 4.53%-53.73% binary alloys from available experimental findings [73, 74], respectively. The presently computed results of λ from H-screenings are found in qualitative agreement with the available experimental data [73, 74].

The computed values of the Coulomb pseudopotential μ^*, which accounts for the Coulomb interaction between the conduction electrons, obtained from the various forms of the local field correction functions are tabulated in Table 3. It is observed from the Table 3 that for Ag_xZn_{1-x} and Ag_xAl_{1-x} binary alloys, the μ^* lies between 0.13 and 0.16, which is in accordance with McMillan [67], who suggested $\mu^* \approx 0.13$ for simple and non-simple metals. The weak screening influence shows on the computed values of μ^*. The percentile influence of the various local field correction functions with respect to the static H-screening function on μ^* for the binary alloys is observed in the range of 5.51%-10.48%, 5.52%-10.49%, 5.53%-10.51%, 5.54%-10.54%, 5.54%-10.52%, 5.56%-10.55%, 5.57%-10.56%, 5.59%-10.58%, 5.66%-10.66%, 5.73%-10.75%, 6.13%-11.28%, 6.06%-11.19%, 5.99%-11.10%, 5.97%-11.07%, 5.95%-11.04%, 5.92%-11.01%, 5.84%-10.91% and 5.78%-10.83% for $Ag_{0.13}Zn_{0.87}$, $Ag_{0.14}Zn_{0.86}$, $Ag_{0.15}Zn_{0.85}$, $Ag_{0.155}Zn_{0.845}$, $Ag_{0.16}Zn_{0.84}$, $Ag_{0.175}Zn_{0.825}$, $Ag_{0.185}Zn_{0.815}$, $Ag_{0.20}Zn_{0.80}$, $Ag_{0.25}Zn_{0.75}$, $Ag_{0.305}Zn_{0.695}$, $Ag_{0.76}Al_{0.24}$, $Ag_{0.73}Al_{0.27}$, $Ag_{0.70}Al_{0.30}$, $Ag_{0.69}Al_{0.31}$, $Ag_{0.68}Al_{0.32}$, $Ag_{0.67}Al_{0.33}$, $Ag_{0.63}Al_{0.37}$ and $Ag_{0.60}Al_{0.40}$ binary alloys, respectively. Again the H-screening function yields lowest values of μ^*, while the values obtained from the F-function are the highest. The theoretical or experimental data of μ^* is not available for the further comparisons.

Table 3 contains calculated values of the transition temperature T_C for Ag_xZn_{1-x} and Ag_xAl_{1-x} binary alloys computed from the various forms of the local field correction functions along with the experimental findings [73, 74]. From the Table 3 it can be noted that, the static H-screening function yields lowest T_C whereas the F-function yields highest values of the T_C. The present results obtained from the H-local field correction functions are found in good agreement with available experimental data [73, 74]. The calculated results of the transition temperature T_C for binary alloys viz. $Ag_{0.13}Zn_{0.87}$, $Ag_{0.14}Zn_{0.86}$, $Ag_{0.15}Zn_{0.85}$, $Ag_{0.155}Zn_{0.845}$, $Ag_{0.16}Zn_{0.84}$, $Ag_{0.175}Zn_{0.825}$, $Ag_{0.185}Zn_{0.815}$, $Ag_{0.20}Zn_{0.80}$, $Ag_{0.25}Zn_{0.75}$, $Ag_{0.305}Zn_{0.695}$, $Ag_{0.76}Al_{0.24}$, $Ag_{0.73}Al_{0.27}$, $Ag_{0.70}Al_{0.30}$, $Ag_{0.69}Al_{0.31}$, $Ag_{0.68}Al_{0.32}$, $Ag_{0.67}Al_{0.33}$, $Ag_{0.63}Al_{0.37}$ and $Ag_{0.60}Al_{0.40}$ deviate in the range of 38.30%-277.18%, 41.09%-292.66%, 32.49%-274.95%, 29.61%-269.37%, 24.29%-257.90%, 6.36%-176.32%, 21.12%-137.87%,

37.55%-96.27%, 18.92%-98.43%, 7.77%-99.22%, 42.67%-242.83%, 26.48%-112.82%, 35.31%-118.33%, 43.71%-137.76%, 37.97%-151.85%, 36.89%-146.51%, 12.16%-277.21% and 85.32%-462.80% binary alloys from experimental findings [73, 74], respectively.

Table 3. Superconducting state parameters of the Ag-Zn-Al binary alloys

Alloys	SSP	Present results					Expt.
		H	T	IU	F	S	[73, 74]
$Ag_{0.13}Zn_{0.87}$	λ	0.3018	0.4081	0.4252	0.4260	0.3733	0.292
	μ^*	0.1379	0.1503	0.1520	0.1523	0.1455	–
	T_C (K)	0.0120	0.2764	0.3747	0.3772	0.1383	0.10
	α	-0.3008	0.0709	0.1017	0.1011	0.0065	–
	N_0V	0.1286	0.1881	0.1970	0.1973	0.1701	–
$Ag_{0.14}Zn_{0.86}$	λ	0.2995	0.4054	0.4224	0.4232	0.3707	0.290
	μ^*	0.1380	0.1504	0.1522	0.1524	0.1456	–
	T_C (K)	0.0103	0.2571	0.3511	0.3534	0.1270	0.09
	α	-0.3259	0.0610	0.0930	0.0923	-0.0057	–
	N_0V	0.1269	0.1863	0.1953	0.1955	0.1684	–
$Ag_{0.15}Zn_{0.85}$	λ	0.2978	0.4035	0.4206	0.4214	0.3689	0.289
	μ^*	0.1381	0.1506	0.1523	0.1526	0.1457	–
	T_C (K)	0.0092	0.2439	0.3355	0.3375	0.1192	0.09
	α	-0.3458	0.0536	0.0867	0.0860	-0.0149	–
	N_0V	0.1257	0.1850	0.1941	0.1944	0.1672	–
$Ag_{0.155}Zn_{0.845}$	λ	0.2972	0.4029	0.4201	0.4209	0.3684	0.289
	μ^*	0.1381	0.1506	0.1523	0.1526	0.1458	–
	T_C (K)	0.0088	0.2396	0.3306	0.3324	0.1166	0.09
	α	-0.3535	0.0510	0.0847	0.0839	-0.0182	–
	N_0V	0.1252	0.1846	0.1938	0.1940	0.1668	–
$Ag_{0.16}Zn_{0.84}$	λ	0.2961	0.4016	0.4188	0.4195	0.3671	0.289
	μ^*	0.1382	0.1507	0.1524	0.1527	0.1458	–
	T_C (K)	0.0082	0.2313	0.3203	0.3221	0.1119	0.09
	α	-0.3664	0.0460	0.0804	0.0795	-0.0243	–
	N_0V	0.1244	0.1838	0.1929	0.1932	0.1659	–

Table 3. (Continued)

Alloys	SSP	Present results					Expt.
		H	T	IU	F	S	[73, 74]
$Ag_{0.175}Zn_{0.825}$	λ	0.2940	0.3993	0.4167	0.4174	0.3650	0.295
	μ^*	0.1383	0.1509	0.1526	0.1529	0.1460	–
	T_C (K)	0.0071	0.2162	0.3025	0.3040	0.1030	0.11
	α	-0.3944	0.0363	0.0723	0.0714	-0.0368	–
	N_0V	0.1228	0.1823	0.1915	0.1918	0.1644	–
$Ag_{0.185}Zn_{0.815}$	λ	0.2919	0.3968	0.4142	0.4149	0.3626	0.298
	μ^*	0.1384	0.1510	0.1527	0.1530	0.1461	–
	T_C (K)	0.0061	0.2015	0.2841	0.2854	0.0947	0.12
	α	-0.4217	0.0261	0.0635	0.0625	-0.0495	–
	N_0V	0.1213	0.1807	0.1900	0.1902	0.1628	–
$Ag_{0.20}Zn_{0.80}$	λ	0.2899	0.3948	0.4124	0.4131	0.3607	0.303
	μ^*	0.1386	0.1512	0.1529	0.1532	0.1463	–
	T_C (K)	0.0052	0.1890	0.2693	0.2704	0.0874	0.14
	α	-0.4503	0.0167	0.0558	0.0547	-0.0618	–
	N_0V	0.1198	0.1793	0.1887	0.1889	0.1614	–
$Ag_{0.25}Zn_{0.75}$	λ	0.2825	0.3869	0.4048	0.4054	0.3529	0.312
	μ^*	0.1391	0.1518	0.1536	0.1539	0.1469	–
	T_C (K)	0.0028	0.1460	0.2159	0.2165	0.0633	0.18
	α	-0.5694	-0.0226	0.0232	0.0219	-0.1130	–
	N_0V	0.1141	0.1739	0.1836	0.1838	0.1559	–
$Ag_{0.305}Zn_{0.695}$	λ	0.2756	0.3800	0.3984	0.3990	0.3458	0.316
	μ^*	0.1397	0.1526	0.1544	0.1547	0.1477	–
	T_C (K)	0.0015	0.1134	0.1748	0.1752	0.0456	0.19
	α	-0.7061	-0.0630	-0.0096	-0.0111	-0.1682	–
	N_0V	0.1087	0.1690	0.1792	0.1793	0.1508	–
$Ag_{0.76}Al_{0.24}$	λ	0.2666	0.3850	0.4079	0.4091	0.3433	0.285
	μ^*	0.1433	0.1572	0.1591	0.1594	0.1521	–
	T_C (K)	0.0003	0.0973	0.1684	0.1714	0.0287	0.05
	α	-1.0876	-0.1089	-0.0341	-0.0334	-0.2760	–
	N_0V	0.0992	0.1687	0.1815	0.1820	0.1458	–

Alloys	SSP	Present results					Expt.
		H	T	IU	F	S	[73, 74]
$Ag_{0.73}Al_{0.27}$	λ	0.2735	0.3920	0.4147	0.4158	0.3507	0.301
	μ^*	0.1427	0.1565	0.1584	0.1587	0.1514	–
	T_C (K)	0.0008	0.1265	0.2099	0.2128	0.0417	0.10
	α	-0.8884	-0.0647	0.0012	0.0013	-0.2072	–
	N_0V	0.1047	0.1736	0.1862	0.1866	0.1512	–
$Ag_{0.70}Al_{0.30}$	λ	0.2806	0.3992	0.4217	0.4227	0.3583	0.305
	μ^*	0.1422	0.1558	0.1576	0.1579	0.1507	–
	T_C (K)	0.0016	0.1624	0.2592	0.2620	0.0591	0.12
	α	-0.7225	-0.0245	0.0335	0.0332	-0.1464	–
	N_0V	0.1103	0.1786	0.1909	0.1913	0.1566	–
$Ag_{0.69}Al_{0.31}$	λ	0.2835	0.4024	0.4249	0.4259	0.3615	0.305
	μ^*	0.1420	0.1555	0.1574	0.1577	0.1505	–
	T_C (K)	0.0021	0.1795	0.2825	0.2853	0.0676	0.12
	α	-0.6652	-0.0090	0.0462	0.0458	-0.1241	–
	N_0V	0.1125	0.1808	0.1930	0.1934	0.1588	–
$Ag_{0.68}Al_{0.32}$	λ	0.2856	0.4045	0.4270	0.4279	0.3638	0.305
	μ^*	0.1418	0.1553	0.1572	0.1575	0.1503	–
	T_C (K)	0.0026	0.1924	0.2995	0.3022	0.0744	0.12
	α	-0.6232	0.0017	0.0548	0.0543	-0.1081	–
	N_0V	0.1142	0.1822	0.1944	0.1947	0.1605	–
$Ag_{0.67}Al_{0.33}$	λ	0.2879	0.4067	0.4291	0.4300	0.3661	0.306
	μ^*	0.1417	0.1551	0.1569	0.1573	0.1500	–
	T_C (K)	0.0031	0.2064	0.3179	0.3205	0.0820	0.13
	α	-0.5824	0.0124	0.0634	0.0628	-0.0923	–
	N_0V	0.1159	0.1837	0.1958	0.1961	0.1621	–
$Ag_{0.63}Al_{0.37}$	λ	0.2980	0.4173	0.4393	0.4402	0.3770	0.295
	μ^*	0.1410	0.1542	0.1561	0.1564	0.1492	–
	T_C (K)	0.0069	0.2800	0.4127	0.4149	0.1234	0.11
	α	-0.4272	0.0567	0.0997	0.0990	-0.0289	–
	N_0V	0.1235	0.1907	0.2024	0.2027	0.1696	–

Table 3. (Continued)

Alloys	SSP	Present results					Expt.
		H	T	IU	F	S	[73, 74]
$Ag_{0.60}Al_{0.40}$	λ	0.3063	0.4263	0.4481	0.4489	0.3860	0.292
	μ^*	0.1405	0.1536	0.1555	0.1558	0.1487	–
	T_C (K)	0.0121	0.3532	0.5045	0.5065	0.1668	0.09
	α	- 0.3254	0.0887	0.1263	0.1254	0.0152	–
	N_0V	0.1297	0.1965	0.2079	0.2082	0.1757	–

The values of the isotope effect exponent α for binary alloys are tabulated in Table 3. The computed values of the α show a weak dependence on the dielectric screening, its value is being lowest for the H-screening function and highest for the F-function. Since the experimental value of α has not been reported in the literature so far, the present data of α may be used for the study of ionic vibrations in the superconductivity of alloying substances. Since H-local field correction function yields the best results for λ and T_C, it may be observed that α values obtained from this screening provide the best account for the role of the ionic vibrations in superconducting behaviour of this system. The negative value of the α is observed for most of the binary alloys, which indicates that the electron-phonon coupling in these metallic complexes do not fully explain all the features regarding their superconducting behaviour. The theoretical or experimental data of α is not available for the further comparisons.

The values of the effective interaction strength N_OV are listed in Table 3 for different local field correction functions. It is observed that the magnitude of N_OV shows that the Ag_xZn_{1-x} and Ag_xAl_{1-x} binary alloys under investigation lie in the range of weak coupling superconductors. The values of the N_OV also show a feeble dependence on dielectric screening, its value being lowest for the H-screening function and highest for the F-screening function. The theoretical or experimental data of N_OV is not available for the further comparisons.

3.4. SSP of Pb-Tl-Bi alloys

The input parameters and other constants used in the present computation are taken from our earlier paper [50]. Table 4 shows the presently calculated values of the SSP viz. electron-phonon coupling strength λ, Coulomb pseudopotential μ^*, transition temperature T_C, isotope effect exponent α and effective interaction strength N_OV at various concentrations for Pb-Tl-Bi alloys with available experimental findings [75].

The calculated values of the electron-phonon coupling strength λ for Pb-Tl-Bi alloys, using five different types of the local field correction functions with EMC model potential, are shown in Table 4 with the experimental data [75]. It is noticed from the present study that, the percentile influence of the various local field correction functions with respect to the static H-screening function on the electron-phonon coupling strength λ is 26.52%-49.52%, 26.42%-51.81%, 26.08%-52.21%, 25.67%-51.91%, 25.72%-60.06%, 25.20%-51.46%, 25.14%-51.69%, 25.11%-52.15%, 25.11%-52.67% and 25.28%-51.24% for $Tl_{0.90}Bi_{0.10}$, $Pb_{0.40}Tl_{0.60}$, $Pb_{0.60}Tl_{0.40}$, $Pb_{0.80}Tl_{0.20}$, $Pb_{0.60}Tl_{0.20}Bi_{0.20}$, $Pb_{0.90}Bi_{0.10}$, $Pb_{0.80}Bi_{0.20}$, $Pb_{0.70}Bi_{0.30}$, $Pb_{0.65}Bi_{0.35}$ and $Pb_{0.45}Bi_{0.55}$ alloys, respectively. Also, the H-screening yields lowest values of λ, whereas the values obtained from the F-function are the highest. It is also observed from the Table 4 that, λ goes on increasing from the values of 0.9788 $\rightarrow$ 1.8834 as the concentration 'x' of 'Tl' is decreased from 0.60 $\rightarrow$ 0.20, while for concentration 'x' of 'Bi' increases except $\alpha Pb_{0.45}Bi_{0.55}$ alloys, λ goes on increasing. The increase or decrease in λ with concentration 'x' of 'Tl' and 'Bi' shows a gradual transition from weak coupling behaviour to intermediate coupling behaviour of electrons and phonons, which may be attributed to an increase of the hybridization of sp-d electrons of 'Tl' and 'Bi' with increasing or decreasing concentration (x). This may also be attributed to the increase role of ionic vibrations in the Tl or Bi-rich region. The present results are found in qualitative agreement with the available experimental data [75]. The calculated results of the electron-phonon coupling strength λ for $Tl_{0.90}Bi_{0.10}$, $Pb_{0.40}Tl_{0.60}$, $Pb_{0.60}Tl_{0.40}$, $Pb_{0.80}Tl_{0.20}$, $Pb_{0.60}Tl_{0.20}Bi_{0.20}$, $Pb_{0.90}Bi_{0.10}$, $Pb_{0.80}Bi_{0.20}$, $Pb_{0.70}Bi_{0.30}$, $Pb_{0.65}Bi_{0.35}$ and $Pb_{0.45}Bi_{0.55}$ deviate in the range of 8.68%-36.54%, 7.60%-29.21%, 3.40%-24.83%, 1.84%-23.10%, 23.27%-52.06%, 0.81%-20.78%, 1.36%-29.07%, 1.58%-31.30%, 2.70%-32.31% and 33.28%-55.88% alloys from available experimental findings [75], respectively.

Table 4. Superconducting state parameters of the Pb-Tl-Bi alloys

Alloys	SSP	Present results					Expt.
		H	T	IU	F	S	[75]
$Tl_{0.90}Bi_{0.10}$	λ	0.7123	1.0067	1.0621	1.0650	0.9012	0.78
	μ^*	0.1195	0.1288	0.1300	0.1303	0.1252	–
	T_C (K)	2.3011	4.5437	4.9372	4.9540	3.7799	2.30
	α	0.4372	0.4558	0.4580	0.4580	0.4515	–
	N_0V	0.3598	0.4584	0.4742	0.4749	0.4265	–
$Pb_{0.40}Tl_{0.60}$	λ	0.9788	1.3999	1.4795	1.4859	1.2374	1.15
	μ^*	0.1185	0.1276	0.1288	0.1290	0.1241	–
	T_C (K)	4.6004	7.2982	7.7222	7.7522	6.3648	4.60
	α	0.4627	0.4719	0.4730	0.4730	0.4695	–
	N_0V	0.4553	0.5599	0.5760	0.5772	0.5239	–
$Pb_{0.60}Tl_{0.40}$	λ	1.1317	1.6220	1.7141	1.7226	1.4269	1.38
	μ^*	0.1179	0.1269	0.1281	0.1283	0.1234	–
	T_C (K)	5.9021	8.6916	9.1119	9.1465	7.7143	5.90
	α	0.4697	0.4764	0.4773	0.4773	0.4746	–
	N_0V	0.4997	0.6042	0.6199	0.6213	0.5674	–

Table 4. Superconducting state parameters of the Pb-Tl-Bi alloys

Alloys	SSP	Present results					Expt.
		H	T	IU	F	S	[75]
$b_{0.80}Tl_{0.20}$	λ	1.2398	1.7739	1.8738	1.8834	1.5581	1.53
	μ^*	0.1174	0.1263	0.1275	0.1277	0.1228	–
	T_C (K)	6.8014	9.6015	10.0139	10.0497	8.6140	6.80
	α	0.4733	0.4788	0.4794	0.4795	0.4772	–
	N_0V	0.5277	0.6306	0.6460	0.6473	0.5940	–
$Pb_{0.60}Tl_{0.20}Bi_{0.20}$	λ	0.8677	1.2976	1.3726	1.3888	1.0909	1.81
	μ^*	0.1276	0.1382	0.1396	0.1398	0.1340	–
	T_C (K)	7.2606	13.4096	14.3079	14.4995	10.6459	7.26
	α	0.4460	0.4625	0.4641	0.4645	0.4564	–
	N_0V	0.4137	0.5319	0.5485	0.5520	0.4804	–
$Pb_{0.90}Bi_{0.10}$	λ	1.3151	1.8770	1.9815	1.9919	1.6465	1.66
	μ^*	0.1172	0.1260	0.1272	0.1274	0.1225	–
	T_C (K)	7.6521	10.5294	10.9469	10.9849	9.5032	7.65
	α	0.4753	0.4800	0.4806	0.4806	0.4787	–
	N_0V	0.5456	0.6470	0.6619	0.6633	0.6104	–
$Pb_{0.80}Bi_{0.20}$	λ	1.3335	1.9055	2.0117	2.0228	1.6688	1.88
	μ^*	0.1174	0.1262	0.1274	0.1276	0.1228	–
	T_C (K)	7.9501	10.8947	11.3199	11.3602	9.8361	7.95
	α	0.4756	0.4802	0.4808	0.4808	0.4789	–
	N_0V	0.5497	0.6512	0.6661	0.6676	0.6142	–
$Pb_{0.70}Bi_{0.30}$	λ	1.3809	1.9782	2.0888	2.1011	1.7277	2.01
	μ^*	0.1176	0.1265	0.1277	0.1279	0.1230	–
	T_C (K)	8.4518	11.4578	11.8872	11.9309	10.3650	8.45
	α	0.4765	0.4808	0.4814	0.4814	0.4795	–
	N_0V	0.5601	0.6617	0.6765	0.6781	0.6243	–
$Pb_{0.65}Bi_{0.35}$	λ	1.4418	2.0711	2.1875	2.2012	1.8039	2.13
	μ^*	0.1177	0.1266	0.1278	0.1280	0.1231	–
	T_C (K)	8.9511	11.9763	12.4032	12.4496	10.8663	8.95
	α	0.4776	0.4816	0.4821	0.4821	0.4804	–
	N_0V	0.5730	0.6745	0.6892	0.6908	0.6367	–

Alloys	SSP	Present results					Expt.
		H	T	IU	F	S	[75]
$\alpha\ Pb_{0.45}Bi_{0.55}$	λ	1.1426	1.6289	1.7195	1.7281	1.4315	2.59
	μ^*	0.1188	0.1278	0.1291	0.1293	0.1243	–
	T_C (K)	7.0042	10.2267	10.7090	10.7508	9.0714	7.0
	α	0.4696	0.4761	0.4769	0.4770	0.4743	–
	N_0V	0.5022	0.6051	0.6205	0.6219	0.5680	–

The computed values of the Coulomb pseudopotential μ^*, which accounts for the Coulomb interaction between the conduction electrons, obtained from the various forms of the local field correction functions are tabulated in Table 4. It is observed from the Table 4 that for all binary alloys, the μ^* lies between 0.11 and 0.14, which is in accordance with McMillan [67], who suggested $\mu^* \approx 0.13$ for simple and non-simple metals. The weak screening influence shows on the computed values of μ^*. The percentile influence of the various local field correction functions with respect to the static H-screening function on μ^* for the Pb-Tl-Bi alloys is observed in the range of 4.77%-9.04%, 4.73%-8.86%, 4.66%-8.82%, 4.60%-8.77%, 5.02%-9.56%, 4.52%-8.70%, 4.60%-8.69%, 4.59%-8.76%, 4.59%-8.75% and 4.63%-8.84% for $Tl_{0.90}Bi_{0.10}$, $Pb_{0.40}Tl_{0.60}$, $Pb_{0.60}Tl_{0.40}$, $Pb_{0.80}Tl_{0.20}$, $Pb_{0.60}Tl_{0.20}Bi_{0.20}$, $Pb_{0.90}Bi_{0.10}$, $Pb_{0.80}Bi_{0.20}$, $Pb_{0.70}Bi_{0.30}$, $Pb_{0.65}Bi_{0.35}$ and $Pb_{0.45}Bi_{0.55}$ alloys, respectively. Again the H-screening function yields lowest values of the μ^*, while the values obtained from the F-function are the highest. The theoretical or experimental data of μ^* is not available for the further comparisons.

Table 4 contains calculated values of the transition temperature T_C for Pb-Tl-Bi alloys computed from the various forms of the local field correction functions along with experimental findings [75]. From the Table 8 it can be noted that, the static H-screening function yields lowest T_C whereas the F-function yields highest values of the T_C. The present results obtained from the H-local field correction functions are found in good agreement with available experimental data [75]. The calculated results of the transition temperature T_C for Pb-Tl-Bi alloys viz. $Tl_{0.90}Bi_{0.10}$, $Pb_{0.40}Tl_{0.60}$, $Pb_{0.60}Tl_{0.40}$, $Pb_{0.80}Tl_{0.20}$, $Pb_{0.60}Tl_{0.20}Bi_{0.20}$, $Pb_{0.90}Bi_{0.10}$, $Pb_{0.80}Bi_{0.20}$, $Pb_{0.70}Bi_{0.30}$, $Pb_{0.65}Bi_{0.35}$ and $Pb_{0.45}Bi_{0.55}$ deviate in the range of 0.05%-115.39%, 0.01%-68.53%, 0.04%-55.03%, 0.02%-47.79%, 0.01%-99.72%, 0.03%-43.59%, 0.00%-42.90%, 0.02%-41.19%, 0.01%-41.10% and 0.06%-53.58% from experimental findings [75], respectively.

The values of the isotope effect exponent α for Pb-Tl-Bi alloys are tabulated in Table 4. The computed values of the α show a weak dependence on the dielectric screening, its value is being lowest for the H-screening function and highest for the F-function. Since the experimental value of α has not been reported in the literature so far, the present data of α may be used for the study of ionic vibrations in the superconductivity of alloying substances. Since H-local field correction function yields the best results for λ and T_C, it may be observed that α values obtained from this screening provide the best account for the role of the ionic vibrations in superconducting behaviour of this system. The theoretical or experimental data of α is not available for the further comparisons.

The values of the effective interaction strength $N_O V$ are listed in Table 4 for different local field correction functions. It is observed that the magnitude of $N_O V$ shows that the Pb-Tl-Bi alloys under investigation lie in the range of weak coupling superconductors. The values of $N_O V$ also show a feeble dependence on dielectric screening, its value being lowest for the H-screening function and highest for the F-screening function. The theoretical or experimental data of $N_O V$ is not available for the further comparisons.

3.5. SSP of $Cu_x Zr_{100-x}$ metallic glasses

The input parameters and other constants used in the present computation are taken from our earlier paper [50]. The presently calculated results of the SSP are tabulated in Table 5 with the experimental [76] and other such theoretical findings [14, 32]. Also, the graphical analyses of the SSP of $Cu_x Zr_{100-x}$ systems are also plotted in Figures 6-10.

The calculated values of the electron-phonon coupling strength λ for ten $Cu_x Zr_{100-x}$ binary metallic glasses, using five different types of the local field correction functions with EMC model potential, are shown in Table 5 with other theoretical data [14, 32]. Also, the graphical variation of λ with the concentration (x) of 'Cu' (in at %) is shown for different types of the local field correction function in Figure 6. The graphical natures of the λ have the same form as suggested by Bakonyi [76]. It is noticed from the Table 5 and Figure 6 that, λ vales are quite sensitive to the local field correction functions. The graph obtained for the H-screening is in good agreement with the results of Bakonyi [76]. It is noticed from the present study that, the percentile influence of the various local field correction functions with respect to the static H-screening function on the electron-phonon coupling strength λ is 19.57%-58.99%. Also, the H-screening yields lowest values of λ, whereas the values obtained from the F-function are the highest. It is also observed from the Table 5 that, λ goes on increasing from the values of 0.2789→0.6055 as the concentration 'x' of 'Zr' is increased from 0.40-0.75. The increase in λ with concentration x of 'Zr' shows a gradual transition from weak coupling behaviour to intermediate coupling behaviour of electrons and phonons, which may be attributed to an increase of the hybridization of sp-d electrons of Zr with increasing concentration (x), as was also observed by Minnigerode and Samwer [78].

This may also be attributed to the increase role of ionic vibrations in the Zr-rich region [13, 14]. The most important feature noted here is that in the series of Cu_xZr_{100-x} metallic glasses, as the concentration (x) of 'Cu' (in at %) increases the present results of λ decreases.

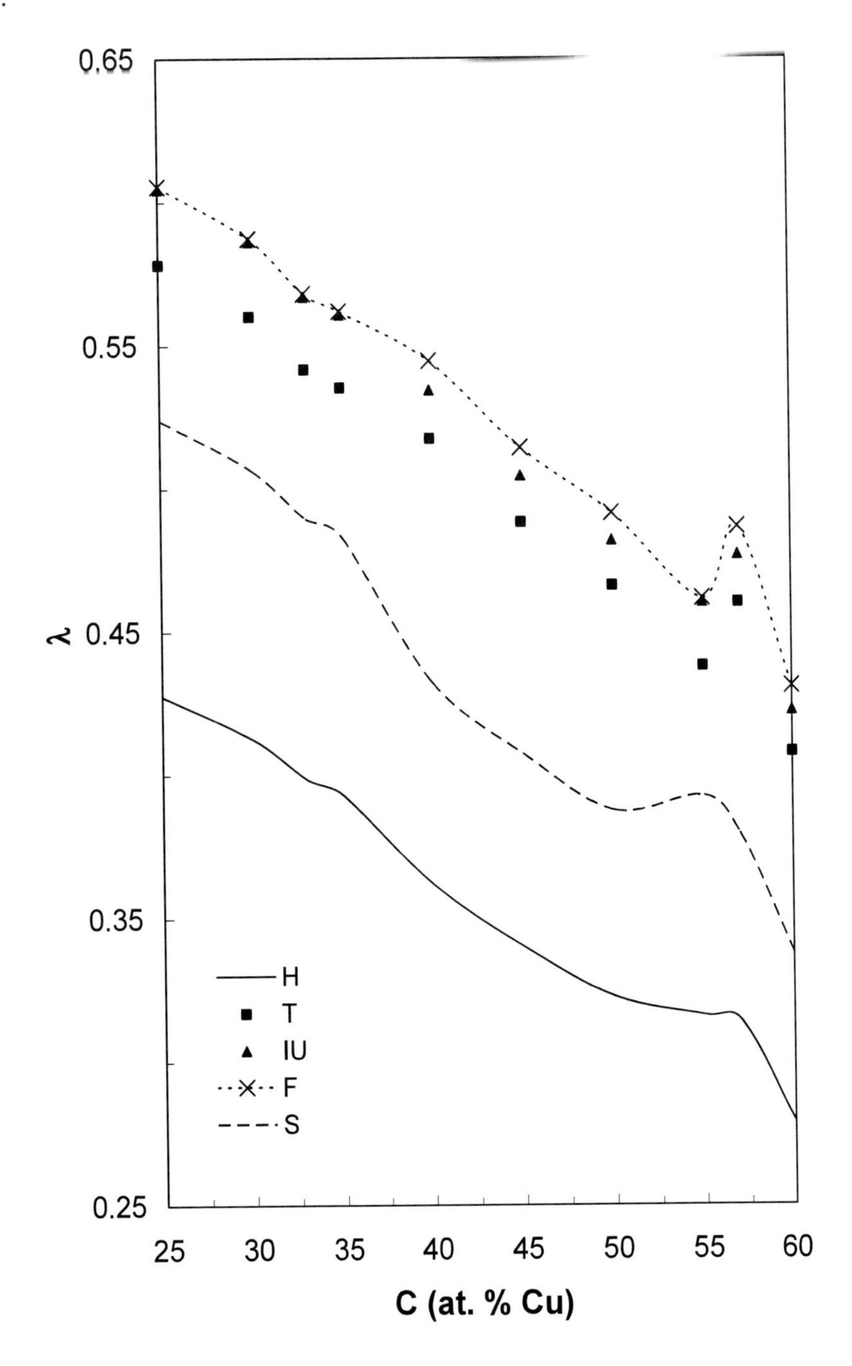

Figure 6. Variation of electron-phonon coupling strength (λ) with Cu-concentration (x) (at %).

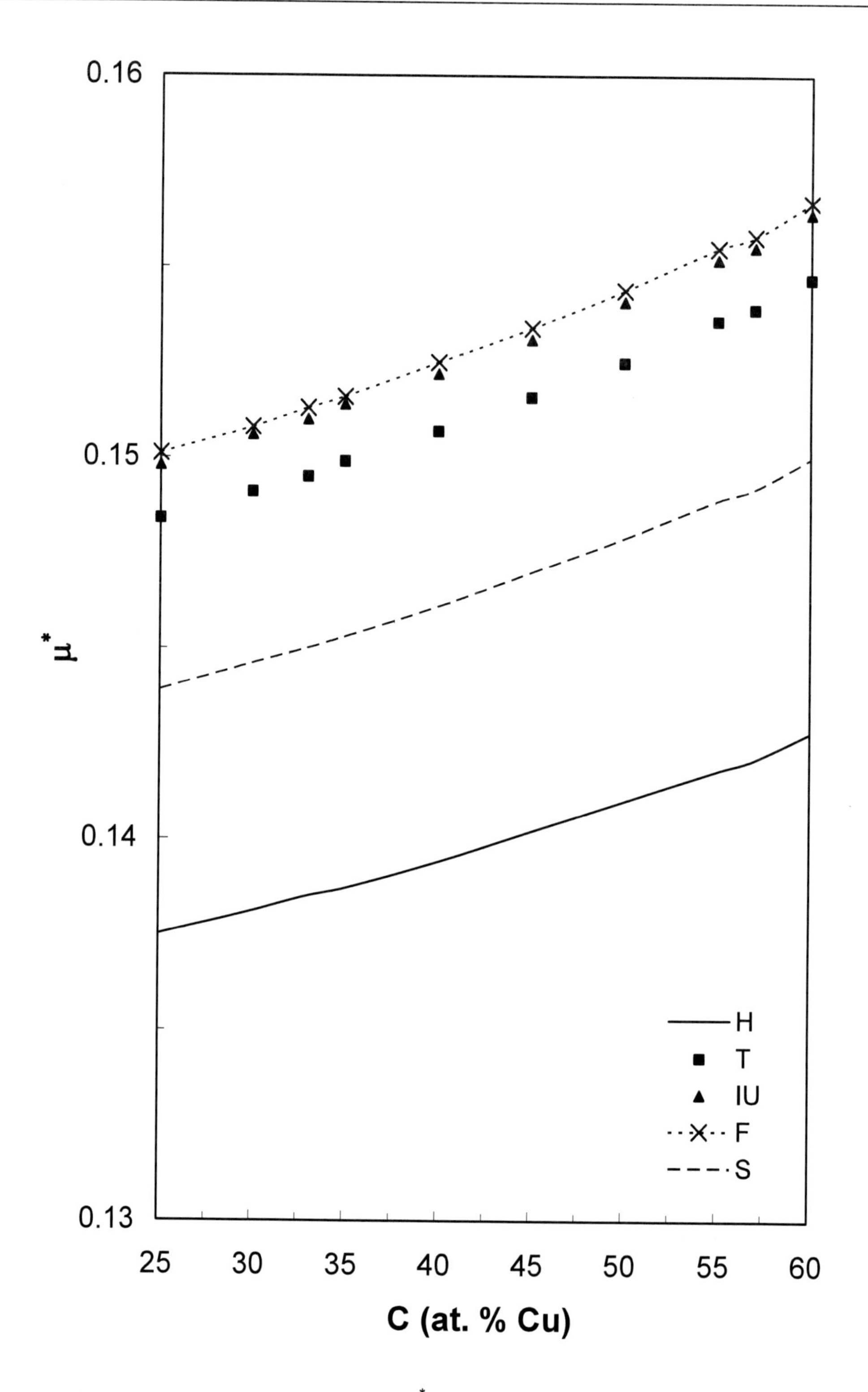

Figure 7. Variation of Coulomb pseudopotential (μ^*) with Cu-concentration (x) (at %).

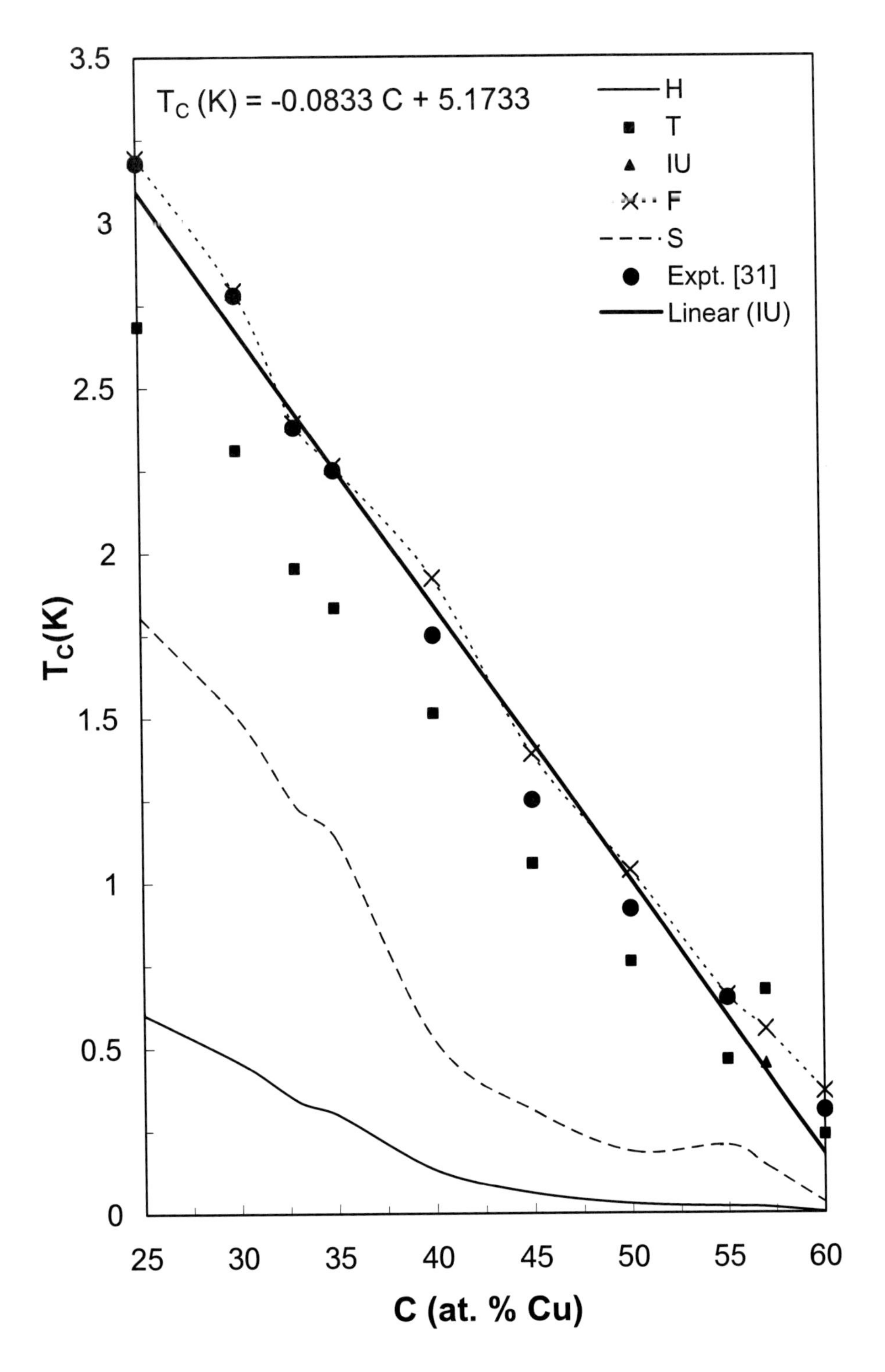

Figure 8. Variation of transition temperature (T_C) with Cu-concentration (x) (at %).

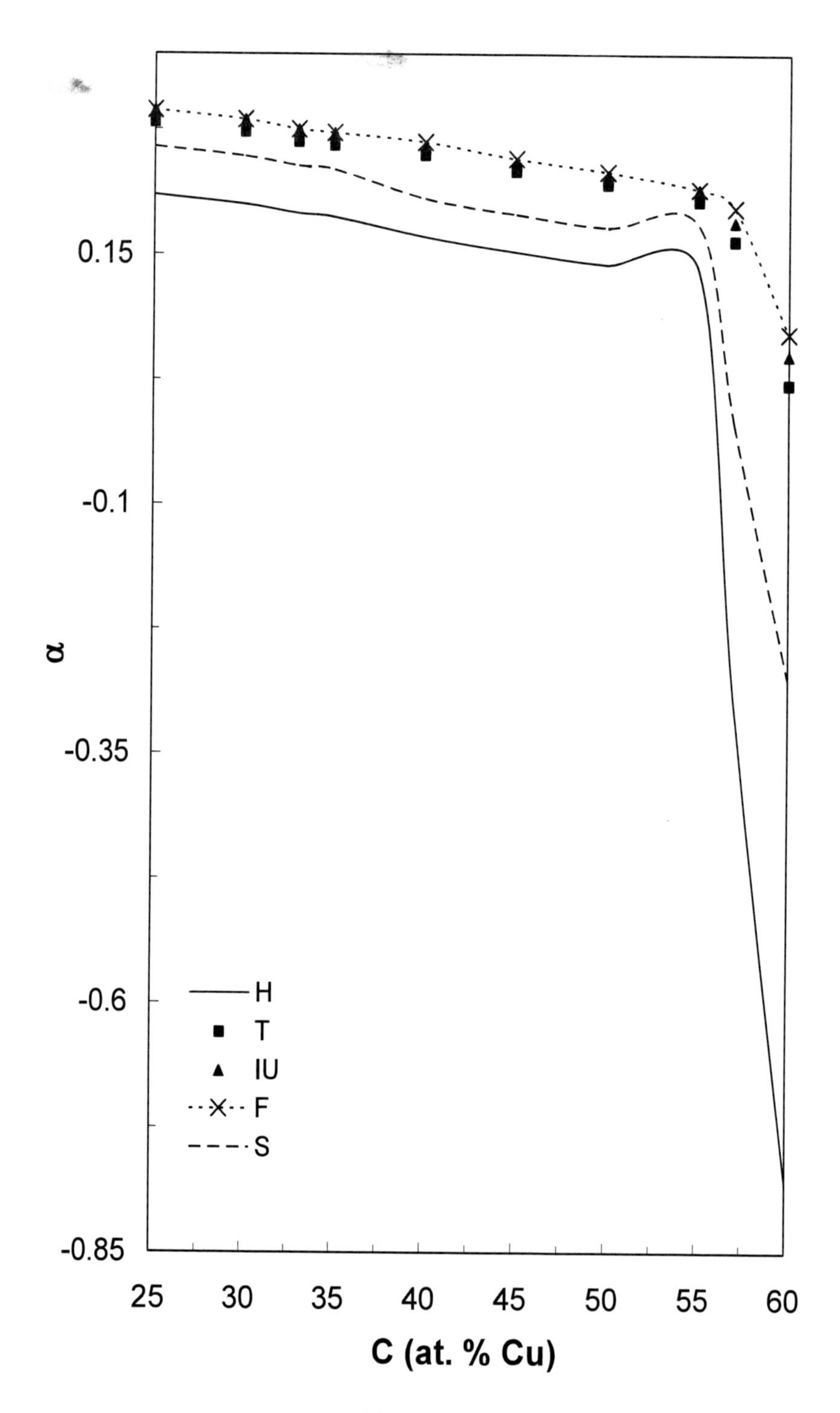

Figure 9. Variation of isotope effect exponent (α) with Cu-concentration (x) (at %).

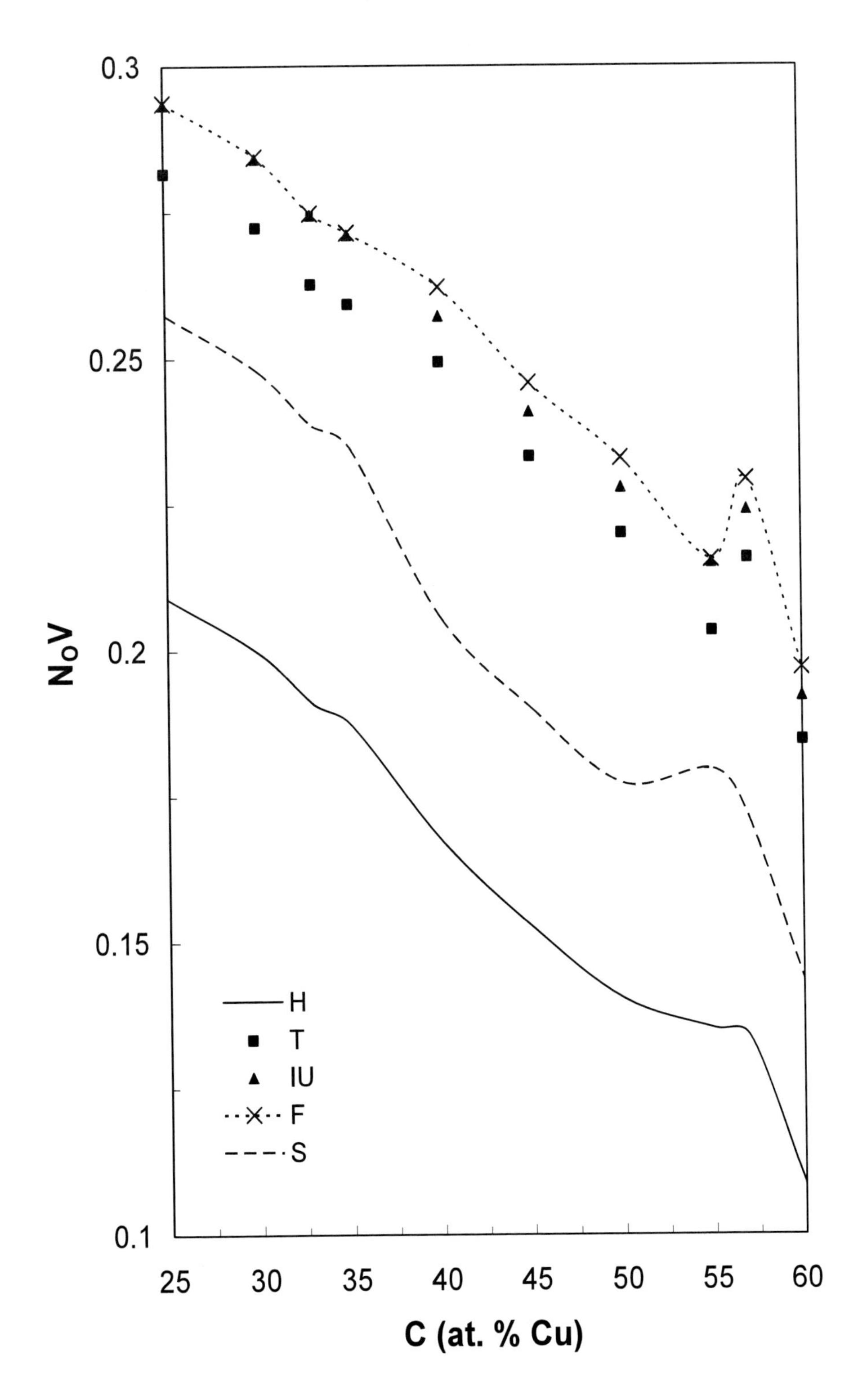

Figure 10. Variation of effective interaction strength ($N_O V$) with Cu-concentration (x) (at %).

The computed values of the Coulomb pseudopotential μ^*, which accounts for the Coulomb interaction between the conduction electrons, obtained from the various forms of the local field correction functions are tabulated in Table 5 with other theoretical data [14, 32]. It is observed from the Table 5 that for all metallic glasses, the μ^* lies between 0.13 and 0.16, which is in accordance with McMillan [67], who suggested $\mu^* \approx 0.13$ for transition metals. The weak screening influence shows on the computed values of the μ^*. The graphs of μ^* versus concentration (x) for different local field correction functions are plotted in Figure 7, which shows the weak dependence of μ^* on the local field correction functions. The percentile influence of the various local field correction functions with respect to the static H- screening function on μ^* for the metallic glasses is observed in the range of 4.65%-10.19%. Again the H-screening function yields lowest values of μ^*, while the values obtained from the F-function are the highest. The present results are found in good agreement with the available theoretical data [14, 32]. Here also, as the concentration (x) of 'Cu' (in at %) increases the present results of μ^* decreases.

Table 5 contain calculated values of the transition temperature T_C for Cu_xZr_{100-x} binary metallic glasses computed from the various forms of the local field correction functions along with the experimental [76] and theoretical findings [14, 32]. From the Table 5 it can be noted that, the static H-screening function yields lowest T_C whereas the F-function yields highest values of T_C. The present results obtained from the IU and F-local field correction functions are found in good agreement with available experimental [76] and theoretical data [14, 32]. The experimental data of T_C for $Cu_{57}Zr_{43}$ metallic glass is not available in the literature. The variation of the computed values of the transition temperature T_C for Cu_xZr_{100-x} metallic glasses with the atomic concentration (x) of 'Cu' (at %), using five different types of the local field correction functions with EMC potential are shown in Figure 8. The graph also includes the experimental values due to Altounian and Strom-Olsen [76]. The calculated results of the transition temperature T_C for $Cu_{60}Zr_{40}$, $Cu_{55}Zr_{45}$, $Cu_{50}Zr_{50}$, $Cu_{45}Zr_{55}$, $Cu_{40}Zr_{60}$, $Cu_{35}Zr_{65}$, $Cu_{33}Zr_{67}$, $Cu_{30}Zr_{70}$ and $Cu_{25}Zr_{75}$ amorphous alloys deviate in the range of 0%-100%, 0%-97%, 0%-97%, 0%-95%, 0%-93%, 0%-87%, 0%-86%, 0%-84% and 0%-81% from experimental findings [76], respectively. It is seen that T_C is quite sensitive to the local field correction functions, and the results of T_C by using IU-screening are in best agreement with the experimental data for the Cu_xZr_{100-x} metallic glasses under investigation, as the relevant curves for IU-screening almost overlaps the experimental curves. It is also observed that the static H-screening function yields lowest T_C whereas the F-function yields highest values of T_C. It is also seen from the graphical nature, T_C decreases almost linearly with increasing Cu-concentration (x).

The composition dependence can be described by linear regression of the data obtained for IU-screening for different values of the concentration 'x', which yields

$$T_C(K) = -0.0833x + 5.1733. \quad (16)$$

The graph of the fitted T_C equation is displayed in Figure 8, which indicates that T_C drops almost linearly with increasing 'Cu' content with a slope $dT_C/dx = -0.0833$. Wide extrapolation predicts a T_C = 5.1733 K for the hypothetical case of 'amorphous pure Zr'. The linear T_C Eq. (16) obtained in the present study closely resembles the linear T_C equations

$$T_C(K) = -0.0893x + 5.35, \quad (17)$$

$$T_C(K) = -0.08369x + 5.1903 \quad (18)$$

suggested by Bakonyi [77] and Sharma *et al.* [14] on the basis of experimental and theoretical data for the Cu_xZr_{100-x} system.

The values of the isotope effect exponent α for Cu_xZr_{100-x} metallic glasses are tabulated in Table 5. Figure 9 depicts the variation of α with Cu-concentration (x) increases (or as the 'Zr' concentration decreases). The computed values of α show a weak dependence on the dielectric screening, its value is being lowest for the H-screening function and highest for the F-function. The negative value of α is observed in the case of metallic glasses, which indicates that the electron-phonon coupling in these metallic complexes do not fully explain all the features regarding their superconducting behaviour. It may be due to the magnetic interactions of the atoms in these metallic complexes. The comparisons of present results with other such theoretical data [14, 32] are highly encouraging. Since the experimental value of α has not been reported in the literature so far, the present data of α may be used for the study of ionic vibrations in the superconductivity of amorphous substances. Since IU-local field correction function yields the best results for λ and T_C, it may be observed that α values obtained from this screening provide the best account for the role of the ionic vibrations in superconducting behaviour of this system. The most important feature noted here is that as the concentration (x) of 'Cu' (in at %) increases the present results of α decreases sharply.

Table 5. Superconducting state parameters of the Cu_xZr_{100-x} binary metallic glasses

Glass	SSP	Present results					Exp. [76]	Others [14, 32]
		H	T	IU	F	S		
$Cu_{60}Zr_{40}$	λ	0.2789	0.4077	0.4222	0.4307	0.3377	–	0.44,0.40,0.39,0.39,0.39
	μ^*	0.1428	0.1547	0.1564	0.1567	0.1500	–	0.15,0.14,0.14, 0.14,0.14
	T_C (K)	0.0015	0.2364	0.3095	0.3668	0.0303	0.31	0.52,0.31,0.31,0.30,0.27
	α	-0.780	0.020	0.049	0.071	-0.275	–	0.16,0.12,0.14,0.14,0.14
	N_0V	0.1086	0.1846	0.1921	0.1970	0.1436	–	0.21,0.19,0.19,0.19,0.18
$Cu_{57}Zr_{43}$	λ	0.3138	0.4599	0.4767	0.4864	0.3795	–	0.6911
	μ^*	0.1421	0.1539	0.1555	0.1558	0.1492	–	0.2052
	T_C (K)	0.0168	0.6744	0.4510	0.5560	0.1420	–	2.6834
	α	-0.293	0.1636	0.1823	0.1965	-0.017	–	0.1662
	N_0V	0.1336	0.2158	0.2241	0.2293	0.1712	–	0.2984
$Cu_{55}Zr_{45}$	λ	0.3158	0.4378	0.4601	0.4614	0.3925	–	0.47,0.44,0.43,0.43, 0.42
	μ^*	0.1418	0.1536	0.1552	0.1555	0.1489	–	0.15,0.14,0.14,0.14,0.14
	T_C (K)	0.0190	0.4629	0.6505	0.6581	0.2045	0.65	0.98,0.65,0.63,0.61, 0.57
	α	-0.270	0.119	0.154	0.154	0.036	–	0.23,0.20,0.21,0.22,0.22
	N_0V	0.1352	0.2033	0.2150	0.2155	0.1796	–	0.23,0.21,0.21,0.21,0.21
$Cu_{50}Zr_{50}$	λ	0.3223	0.4659	0.4818	0.4913	0.3875	–	0.50,0.46,0.45,0.45,0.44
	μ^*	0.1410	0.1525	0.1541	0.1544	0.1479	–	0.15,0.14,0.14,0.14,0.14
	T_C (K)	0.0275	0.7605	0.9190	1.0371	0.1859	0.92	1.33,0.92,0.88,0.86,0.81
	α	-0.199	0.185	0.200	0.213	0.030	–	0.26,0.24,0.25,0.25, 0.25
	N_0V	0.1402	0.2201	0.2279	0.2329	0.1772	–	0.24,0.22,0.22,0.22,0.22
$Cu_{45}Zr_{55}$	λ	0.3398	0.4882	0.5044	0.5141	0.4073	–	0.52,0.48,0.47,0.47, 0.46
	μ^*	0.1402	0.1516	0.1531	0.1534	0.1470	–	0.14,0.14,0.14,0.13, 0.13
	T_C (K)	0.0601	1.0577	1.2492	1.3902	0.3084	1.25	1.74,1.25,1.18,1.15, 1.09
	α	-0.073	0.226	0.238	0.249	0.103	–	0.29,0.27,0.28,0.28, 0.28
	N_0V	0.1525	0.2332	0.2409	0.2458	0.1900	–	0.25,0.24,0.23, 0.23,0.23
$Cu_{40}Zr_{60}$	λ	0.3617	0.5174	0.5343	0.5445	0.4325	–	0.55,0.51,0.50,0.49, 0.49
	μ^*	0.1394	0.1507	0.1522	0.1525	0.1461	–	0.14,0.14,0.14,0.13, 0.13
	T_C (K)	0.1296	1.5137	1.7506	1.9226	0.5192	1.75	2.34,1.75,1.65,1.61, 1.53

	α	0.038	0.267	0.276	0.285	0.170	–	0.32,0.31,0.31,0.31, 0.31
	N_0V	0.1673	0.2494	0.2572	0.2622	0.2056	–	0.27,0.25,0.25,0.25, 0.25
	λ	0.3929	0.5353	0.5607	0.5618	0.4835	–	0.58,0.54,0.52,0.52, 0.51
	μ^*	0.1387	0.1499	0.1514	0.1516	0.1433	–	0.14,0.14,0.13,0.13, 0.13
$Cu_{35}Zr_{65}$	T_C (K)	0.2977	1.8340	2.2521	2.2632	1.1337	2.25	2.93,2.25,2.11,2.07, 1.98
	α	0.143	0.288	0.303	0.303	0.256	–	0.34,0.33,0.34,0.34, 0.34
	N_0V	0.1873	0.2593	0.2711	0.2715	0.2349	–	0.28,0.27,0.26,0.26, 0.26
	λ	0.3983	0.5416	0.5670	0.5681	0.4896	–	0.58,0.54,0.53,0.52, 0.52
	μ^*	0.1385	0.1495	0.1510	0.1513	0.1450	–	0.14,0.14,0.13,0.13, 0.13
$Cu_{33}Zr_{67}$	T_C (K)	0.3374	1.9531	2.3821	2.3928	1.2280	2.38	3.09,2.39,2.23,2.19,2.10
	α	0.159	0.295	0.309	0.309	0.265	–	0.35,0.34,0.35,0.35,0.35
	N_0V	0.1907	0.2627	0.2745	0.2748	0.2385	–	0.29,0.27,0.27,0.26,0.26
	λ	0.4125	0.5601	0.5862	0.5873	0.5066	–	0.47,0.44,0.43,0.43,0.42
	μ^*	0.1381	0.1491	0.1506	0.1508	0.1446	–	0.14,0.14,0.13,0.13,0.13
$Cu_{30}Zr_{70}$	T_C (K)	0.4537	2.3114	2.7814	2.7929	1.4993	2.78	2.53,2.78,2.60,2.56,2.46
	α	0.192	0.312	0.324	0.324	0.285	–	0.36,0.35,0.35,0.35,0.35
	N_0V	0.1996	0.2724	0.2842	0.2845	0.2479	–	0.30,0.28,0.28,0.27,0.27
	λ	0.4275	0.5782	0.6045	0.6055	0.5239	–	0.62,0.58,0.56,0.55,0.55
	μ^*	0.1375	0.1484	0.1498	0.1501	0.1439	–	0.14,0.14,0.13,0.13,0.13
$Cu_{25}Zr_{75}$	T_C (K)	0.6011	2.6848	3.1837	3.1945	1.8046	3.18	3.98,3.18,2.97,2.92,2.81
	α	0.224	0.327	0.338	0.338	0.304	–	0.37,0.36,0.36,0.36,0.36
	N_0V	0.2089	0.2817	0.2934	0.2937	0.2574	–	0.31,0.29,0.28,0.28,0.28

The values of the effective interaction strength N_OV are listed in Table 5 and depicted in Figure 10 for different local field correction functions. It is observed that the magnitude of N_OV shows that the metallic glasses under investigation lie in the range of weak coupling superconductors. The values of N_OV also show a feeble dependence on dielectric screening, its value being lowest for the H-screening function and highest for the F-screening function. The present outcomes are found qualitative agreement with the available theoretical data [14, 32]. Here also, as the concentration (x) of 'Cu' (in at %) increases the present results of N_OV decreases.

3.6. SSP of $Cu_{100-x}Sn_x$ metallic glasses

The input parameters and other constants used in the present computation are taken from our earlier paper [49, 63]. The presently calculated results of the SSP are tabulated in Table 6 with the experimental data [80]. The graphical representations of the SSP of binary systems are also plotted in Figures 11-15.

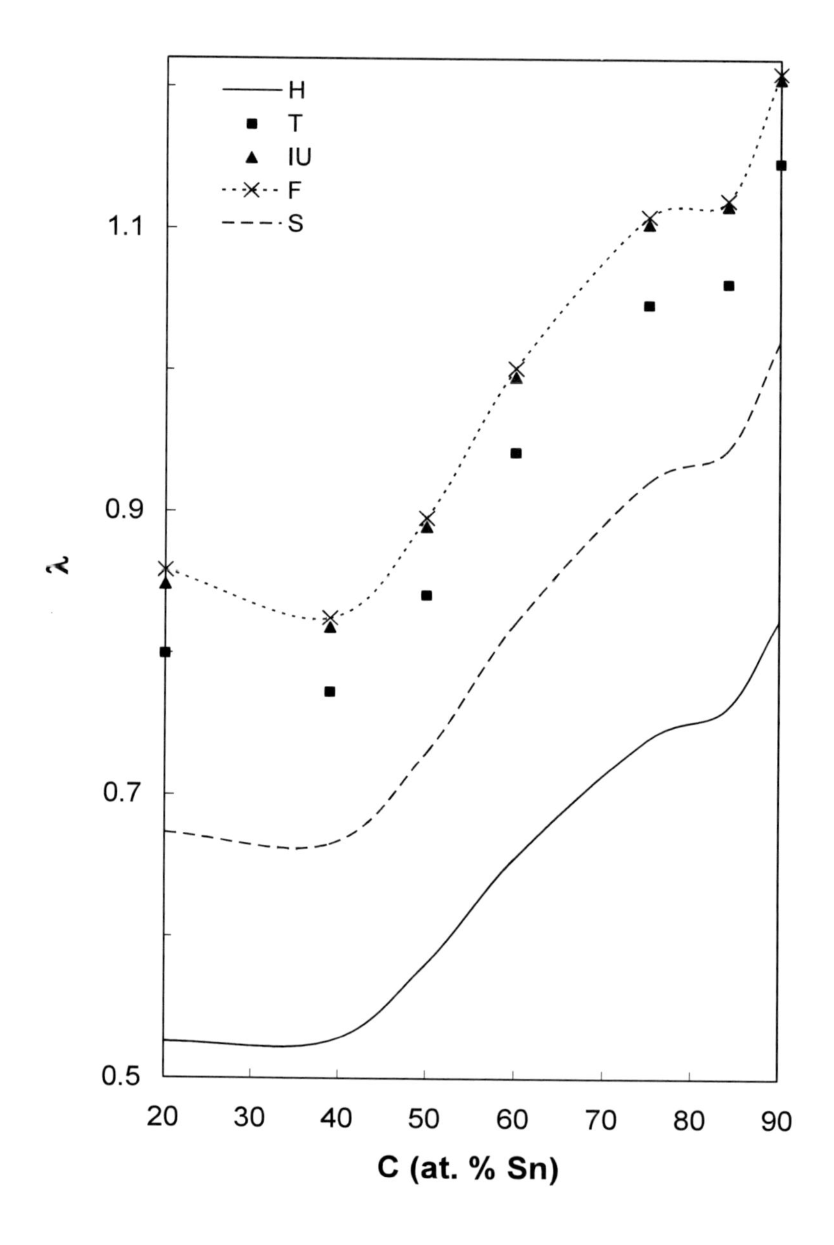

Figure 11. Variation of the electron-phonon coupling strength (λ) with Sn-concentration (x) (in at.%).

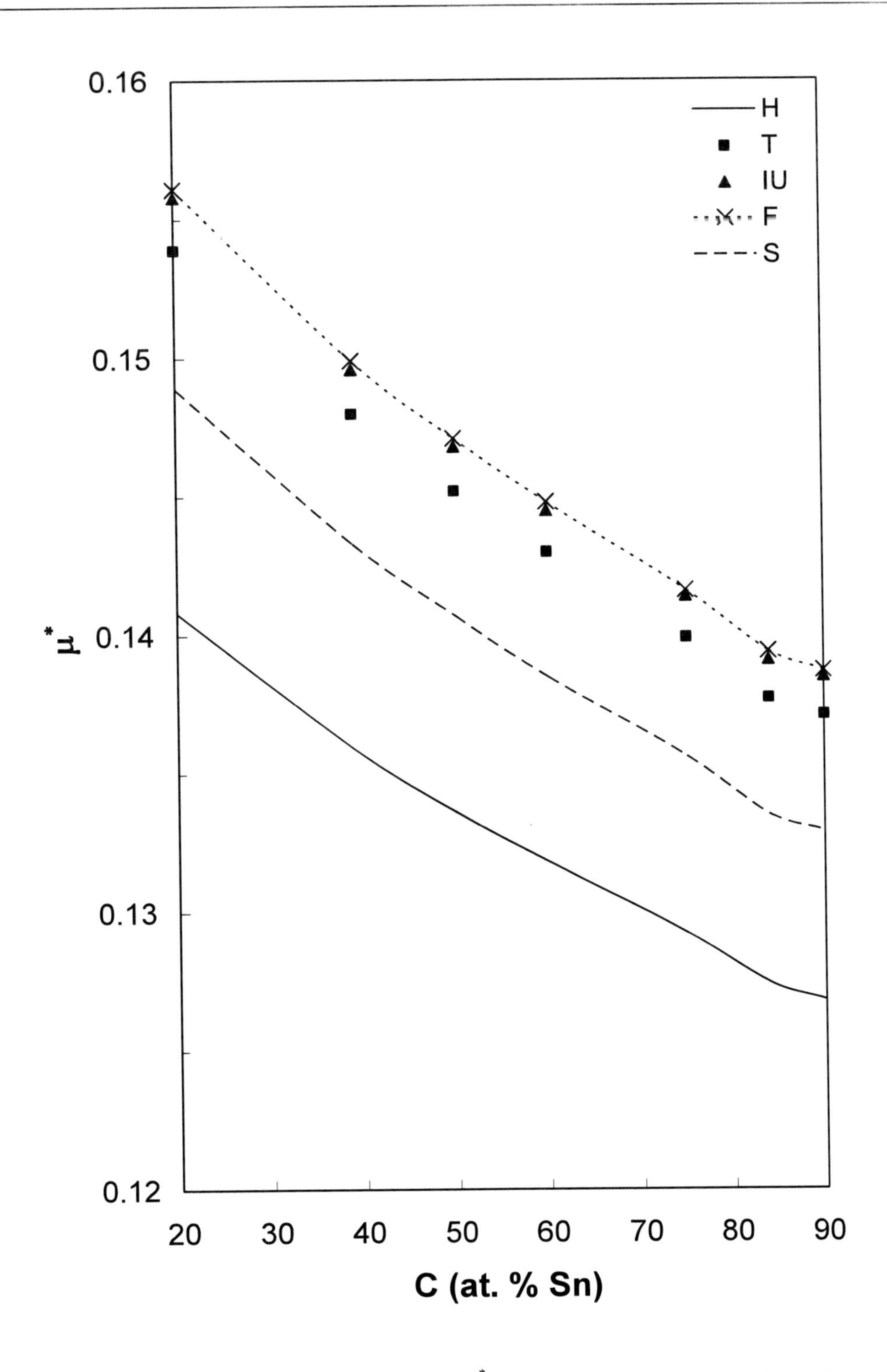

Figure 12. Variations the Coulomb pseudopotential (μ^*) with Sn-concentration (x) (in at.%).

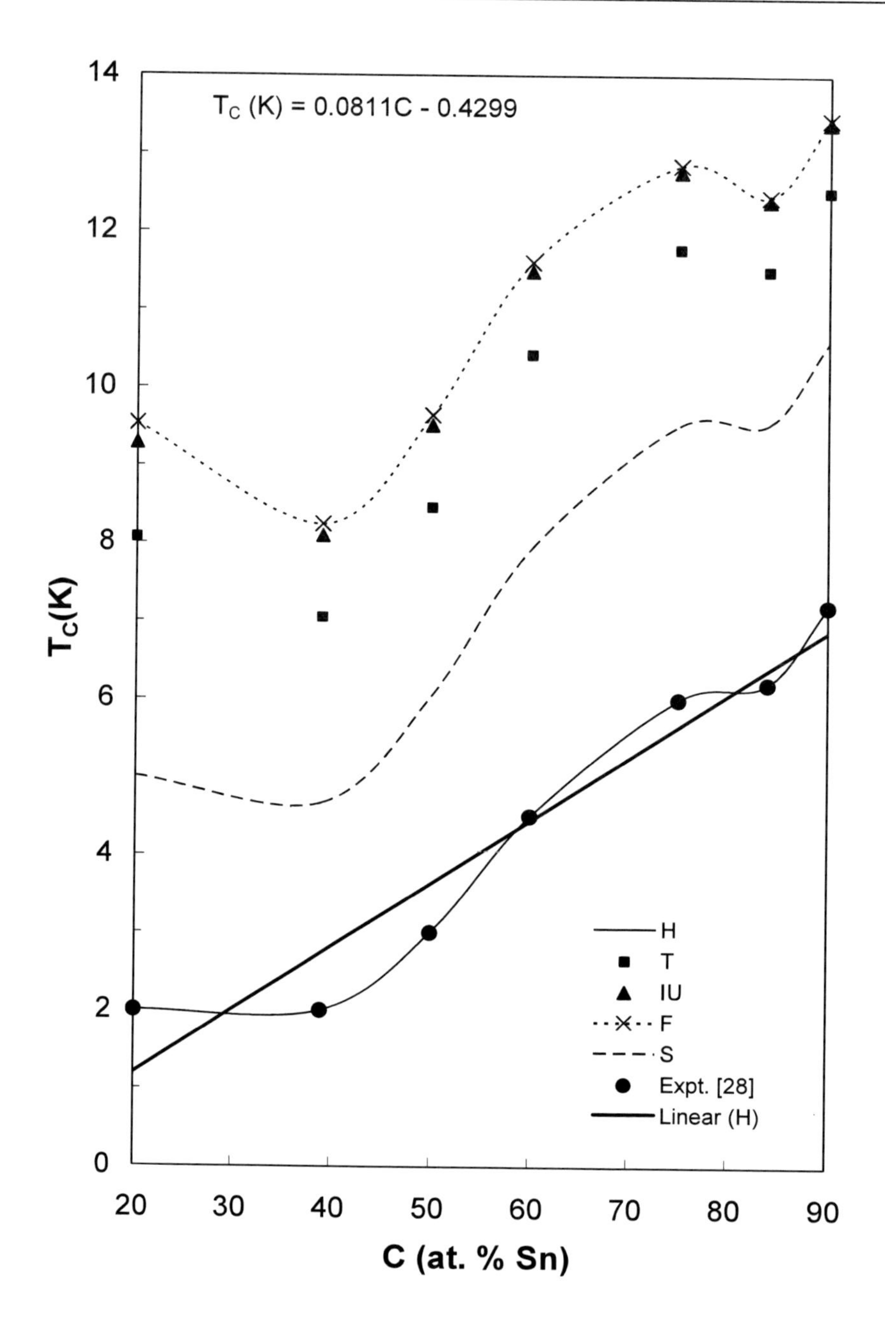

Figure 13. Variation of the transition temperature (T_C) with Sn-concentration (x) (in at.%).

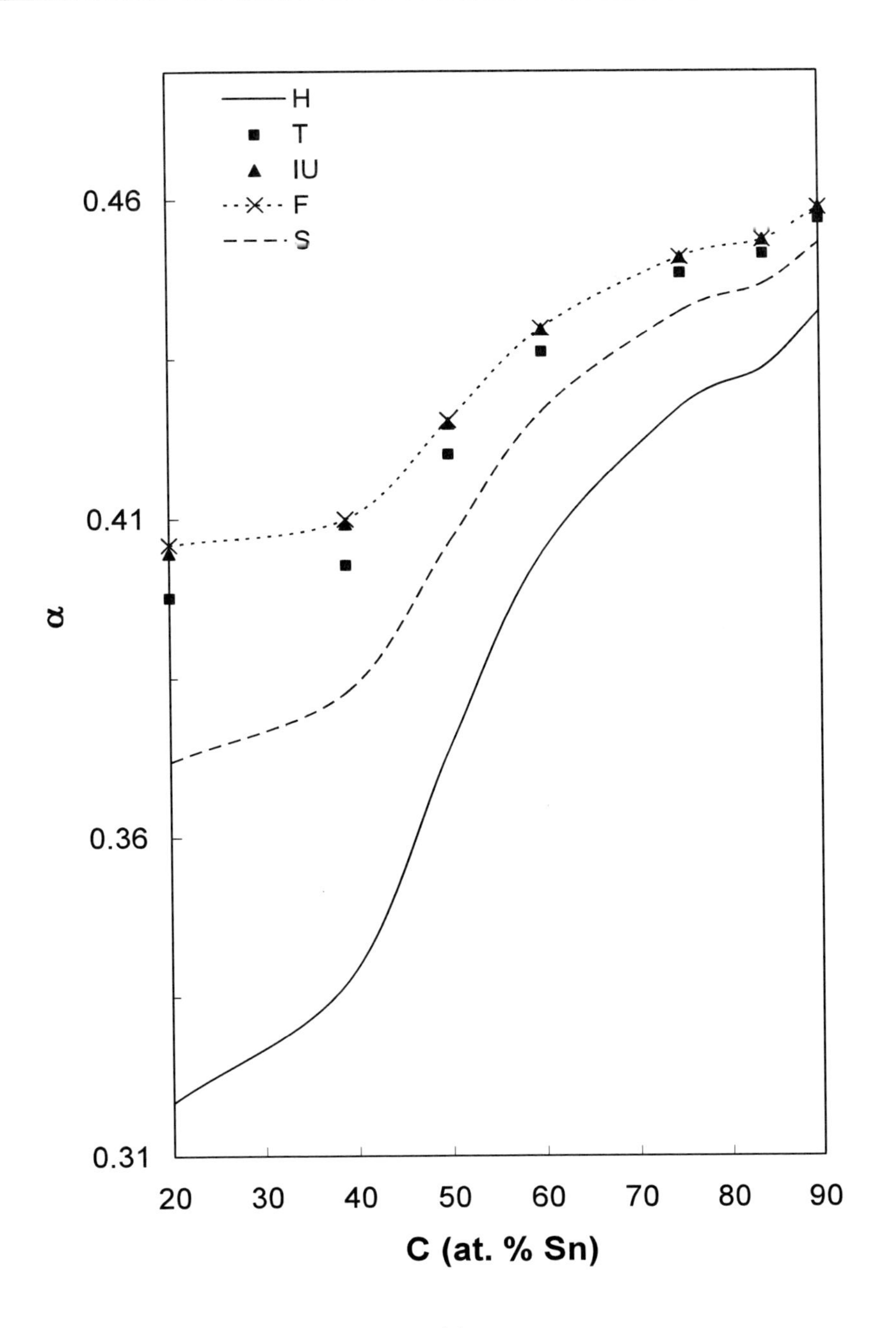

Figure 14. Variation of the isotope effect exponent (α) with Sn-concentration (x) (in at.%).

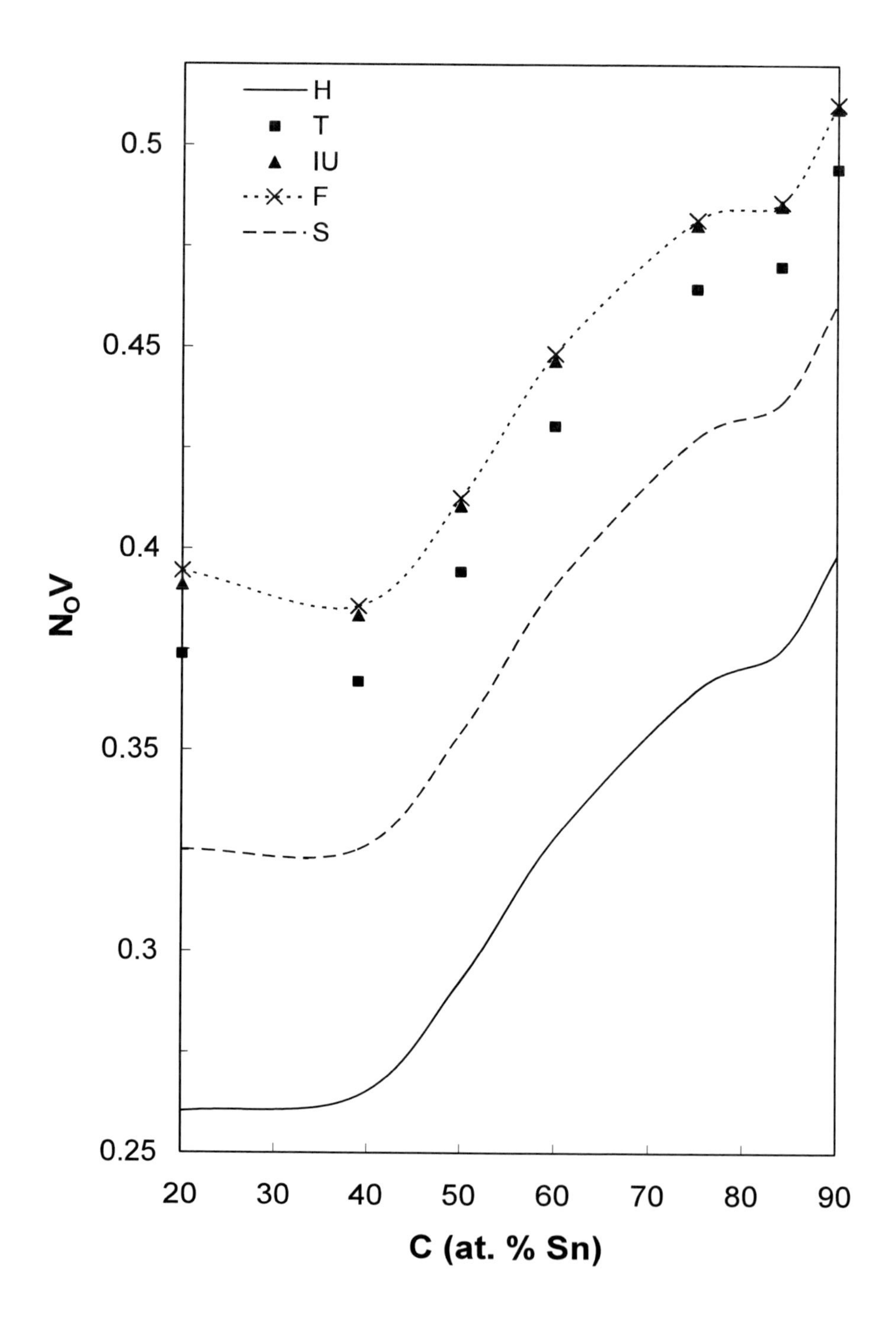

Figure 15. Variation of the effective interaction strength ($N_O V$) with Sn-concentration (x) (in at.%).

The calculated values of the electron-phonon coupling strength λ for seven Cu-based binary metallic glasses, using five different types of the local field correction functions with EMC model potential, are shown in Table 6 with other experimental data [80]. Also, the graphical variation of λ with the concentration (x) of 'Sn' (in at.%) is shown for different types of the local field correction function in Figure 11. It is noticed from the Table 6 and Figure 11 that, λ vales are quite sensitive to the local field correction functions. It is noticed from the present study that, the percentile influence of the various local field correction functions with respect to the static H-screening function on the electron-phonon coupling strength λ is 23.98%-63.33%. Also, the H-screening yields lowest values of λ, whereas the values obtained from the F-function are the highest. It is also observed from the Table 6 that, λ goes on increasing from the values of 0.5255 → 1.2101 as the concentration (x) of 'Sn' is increased from 20 → 90. The increase in λ with concentration (x) of 'Sn' shows a gradual transition from weak coupling behaviour to intermediate coupling behaviour of electrons and phonons, which may be attributed to an increase of the hybridization of sp-d electrons of 'Sn' with increasing concentration (x). This may also be attributed to the increase role of ionic vibrations in the Sn-rich region. The theoretical data of λ is not available for the further comparisons.

The computed values of the Coulomb pseudopotential μ^*, which accounts for the Coulomb interaction between the conduction electrons, obtained from the various forms of the local field correction functions are tabulated in Table 6. It is observed from the Table 6 that for all metallic glasses, μ^* lies between 0.1268 and 0.1561, which is in accordance with McMillan [67], who suggested $\mu^* \approx 0.13$ for transition metals. The weak screening influence shows on the computed values of μ^*. The graphs of μ^* versus concentration (x) for different local field correction functions are plotted in Figure 12, which shows the weak dependence of μ^* on the local field correction functions. The percentile influence of the various local field correction functions with respect to the static H-screening function on μ^* for the metallic glasses is observed in the range of 4.79%-10.87%. Again the H-screening function yields lowest values of μ^*, while the values obtained from the F-function are the highest. Here also, as the concentration (x) of 'Sn' (in at.%) increases the present results of μ^* decreases. The theoretical or experimental data of μ^* is not available for the further comparisons.

Table 6 contains calculated values of the transition temperature T_C for Cu-based binary metallic glasses computed from the various forms of the local field correction functions along with the experimental [80]. From the Table 6 it can be noted that, the static H-screening function yields lowest T_C whereas the F-function yields highest values of T_C. The present results obtained from the H-local field correction function are found in qualitative agreement with available experimental data [80]. The variation of the computed values of the transition temperature T_C for Cu-based metallic glasses with the atomic concentration (x) of 'Sn' (at.%), using five different types of the local field correction functions with EMC potential are shown in Figure 13.

Table 6. Superconducting state parameters of the $Cu_{100-x}Sn_x$ binary metallic glasses

Glass	SSP	Present results					Expt. [80]
		H	T	IU	F	S	
$Cu_{80}Sn_{20}$	λ	0.5255	0.7993	0.8484	0.8583	0.6731	–
	μ^*	0.1408	0.1539	0.1558	0.1561	0.1489	–
	T_C (K)	2.0008	8.0658	9.2817	9.5308	5.0012	0.0, 2.0
	α	0.3184	0.3976	0.4046	0.4059	0.3719	–
	N_0V	0.2603	0.3738	0.3910	0.3944	0.3252	–
$Cu_{61}Sn_{39}$	λ	0.5261	0.7727	0.8182	0.8248	0.6651	–
	μ^*	0.1359	0.1480	0.1496	0.1499	0.1433	–
	T_C (K)	2.0001	7.0515	8.0958	8.2449	4.6626	1.9, 2.0
	α	0.3375	0.4027	0.4092	0.4099	0.3828	–
	N_0V	0.2640	0.3670	0.3834	0.3857	0.3252	–
$Cu_{50}Sn_{50}$	λ	0.5802	0.8411	0.8894	0.8955	0.7295	–
	μ^*	0.1336	0.1452	0.1468	0.1471	0.1407	–
	T_C (K)	3.0006	8.4610	9.5175	9.6462	6.0399	3.0, 3.01
	α	0.3743	0.4201	0.4249	0.4254	0.4063	–
	N_0V	0.2923	0.3943	0.4106	0.4126	0.3540	–
$Cu_{40}Sn_{60}$	λ	0.6554	0.9422	0.9953	1.0015	0.8210	–
	μ^*	0.1318	0.1430	0.1445	0.1448	0.1385	–
	T_C (K)	4.5012	10.4253	11.4941	11.6113	7.9379	4.5, 4.7
	α	0.4049	0.4361	0.4396	0.4398	0.4268	–
	N_0V	0.3281	0.4305	0.4466	0.4484	0.3908	–
$Cu_{25}Sn_{75}$	λ	0.7395	1.0466	1.1035	1.1088	0.9215	–
	μ^*	0.1292	0.1399	0.1414	0.1416	0.1356	–
	T_C (K)	6.0032	11.7797	12.7707	12.8535	9.5265	5.8, 6.0
	α	0.4275	0.4484	0.4508	0.4509	0.4424	–
	N_0V	0.3650	0.4646	0.4803	0.4816	0.4276	–
$Cu_{16}Sn_{84}$	λ	0.7616	1.0615	1.1166	1.1205	0.9442	–
	μ^*	0.1274	0.1377	0.1391	0.1394	0.1335	–
	T_C (K)	6.2018	11.5035	12.4019	12.4558	9.5411	6.2
	α	0.4336	0.4514	0.4536	0.4536	0.4467	–
	N_0V	0.3748	0.4701	0.4851	0.4860	0.4362	–

$Cu_{10}Sn_{90}$	λ	0.8237	1.1467	1.2061	1.2101	1.0212	–
	μ^*	0.1268	0.1371	0.1385	0.1387	0.1329	–
	T_C (K)	7.2008	12.5176	13.3934	13.4434	10.5974	7.2
	α	0.4424	0.4569	0.4586	0.4586	0.4531	–
	N_0V	0.3985	0.4943	0.5093	0.5102	0.4607	–

The graph also includes the experimental values due to Mizutani [80]. It is seen that T_C is quite sensitive to the local field correction functions, and the results of T_C by using H-screening are found in qualitative agreement with the experimental data [80] for the Cu-based metallic glasses under investigation, as the relevant curves for H-screening almost overlaps the experimental curves.

It is noticed from the present study that, the percentile influence of the various local field correction functions with respect to the static H-screening function on the transition temperature T_C is 47.17%-376.35%. The calculated results of the transition temperature T_C for $Cu_{80}Sn_{20}$, $Cu_{61}Sn_{39}$, $Cu_{50}Sn_{50}$, $Cu_{40}Sn_{60}$, $Cu_{25}Sn_{75}$, $Cu_{16}Sn_{84}$ and $Cu_{10}Sn_{90}$ metallic glasses deviate in the range of 0.04%-376.54%, 0.01%-312.25%, 0.02%-221.54%, 0.03%-158.03%, 0.05%-114.23%, 0.03%-100.90% and 0.01%-86.71% from experimental findings [80], respectively. The theoretical data of T_C is not available for the further comparisons.

It is also seen from the graphical nature, T_C increases linearly with increasing Sn-concentration (x). The composition dependence can be described by linear regression of the data obtained for H-screening for different values of the concentration (x), which yields

$$T_C(K) = 0.0811x - 0.4299. \qquad (19)$$

The graph of the fitted T_C equation is displayed in Figure 13, which indicates that T_C drops and increases almost linearly with increasing 'Sn' content with a slope $dT_C/dC = 0.0811$. Wide extrapolation predicts a $T_C =$ 0.4299K for the hypothetical case of 'amorphous pure Sn'.

The values of the isotope effect exponent α for Cu-based metallic glasses are tabulated in Table 6. Figure 14 depicts the variation of α with Sn-concentration (x) increases (or as the 'Cu' concentration decreases). The computed values of α show a weak dependence on the dielectric screening, its value is being lowest for the H-screening function and highest for the F-function. Since the theoretical or experimental value of α has not been reported in the literature so far, the present data of α may be used for the study of ionic vibrations in the superconductivity of amorphous substances. Since H-local field correction function yields the best results for λ and T_C, it may be observed that α values obtained from this screening provide the best account for the role of the ionic vibrations in superconducting behaviour of this system. The most important feature noted here is that as the concentration (x) of 'Sn' (in at.%) increases the present results of α increases.

The values of the effective interaction strength N_0V are listed in Table 6 and depicted in Figure 15 for different local field correction functions. It is observed that the magnitude of

$N_O V$ shows that the metallic glasses under investigation lie in the range of weak coupling superconductors. The values of $N_O V$ also show a feeble dependence on dielectric screening, its value being lowest for the H-screening function and highest for the F-screening function. The theoretical or experimental data of $N_O V$ is not available for the further comparisons. Here also, as the concentration (x) of 'Sn' (in at.%) increases the present results of $N_O V$ increases.

3.7. SSP of $Be_x Zr_{100-x}$ metallic glasses

The input parameters and other constants used in the present computation are taken from our earlier paper [49]. The presently calculated results of the SSP are tabulated in Table 7 with the experimental [81] findings.

The calculated values of the electron-phonon coupling strength λ for $Be_x Zr_{100-x}$ metallic glasses, using five different types of the local field correction functions with EMC model potential, are shown in Table 7 with other experimental data [81]. It is noticed from the Table 1 that, λ vales are quite sensitive to the local field correction functions. It is also observed from the present study that, the percentile influence of the various local field correction functions with respect to the static H-screening function on the electron-phonon coupling strength λ is 21.30%-40.40%, 21.38%-40.53%, 21.40%-39.91% and 21.22%-38.62% for $Be_{45}Zr_{55}$, $Be_{40}Zr_{60}$, $Be_{35}Zr_{65}$ and $Be_{30}Zr_{70}$ metallic glasses, respectively. Also, the H-screening yields lowest values of λ, whereas the values obtained from the F-function are the highest. It is also observed from the Table 7 that, λ goes on increasing from the values of $0.4173 \rightarrow 0.6849$ as the concentration 'x' of 'Zr' is increased from 0.10-0.30. The increase in λ with concentration 'x' of 'Zr' shows a gradual transition from weak coupling behaviour to intermediate coupling behaviour of electrons and phonons, which may be attributed to an increase of the hybridization of sp-d electrons of 'Zr' with increasing concentration (x), as was also observed by Minnigerode and Samwer [78]. This may also be attributed to the increase role of ionic vibrations in the Zr-rich region [14]. The calculated results of the electron-phonon coupling strength λ for $Be_{40}Zr_{60}$, $Be_{35}Zr_{65}$ and $Be_{30}Zr_{70}$ metallic glasses deviate in the range of 40.84%-48.90%, 43.67%-59.74% and 54.34%-67.06% from the experimental findings [23], respectively. The presently computed values of the electron-phonon coupling strength λ are found in the qualitative agreement with the available experimental data [81].

The computed values of the Coulomb pseudopotential μ^*, which accounts for the Coulomb interaction between the conduction electrons, obtained from the various forms of the local field correction functions are tabulated in Table 7 with other experimental data [81]. It is observed from the Table 7 that for all metallic glasses, the μ^* lies between 0.15 and 0.18, which is in accordance with McMillan [67], who suggested $\mu^* \approx 0.13$ for simple and transition metals. Hasegawa and Tanner [81] have also been taken $\mu^* = 0.13$ in their study of $Be_x Zr_{100-x}$ metallic glasses. The weak screening influence shows on the computed values of μ^*. The percentile influence of the various local field correction functions with respect to the static H-screening function on μ^* for the metallic glasses is observed in the range of 5.23%-10.73%, 5.14%-10.55%, 5.14%-10.50% and 5.14%-10.43% for $Be_{45}Zr_{55}$, $Be_{40}Zr_{60}$, $Be_{35}Zr_{65}$ and $Be_{30}Zr_{70}$ metallic glasses, respectively. Again the H-screening function yields lowest values of the μ^*, while the values obtained from the F-function are the highest. The present results are found in good agreement with the available experimental data [81]. Here also, as the concentration (x) of 'Zr' (in at %) increases the present results of μ^* decreases.

Table 7. Superconducting state parameters of the Be_xZr_{100-x} metallic glasses

Glass	SSP	Present results					Expt. [81]
		H	T	IU	F	S	
$Be_{45}Zr_{55}$	λ	0.4173	0.5601	0.5848	0.5859	0.5062	–
	μ^*	0.1454	0.1588	0.1607	0.1610	0.1530	0.13
	T_C (K)	1.0020	4.9874	5.9899	6.0097	3.1961	≤ 1.0
	α	0.1423	0.2698	0.2834	0.2829	0.2413	–
	N_0V	0.1971	0.2659	0.2769	0.2772	0.2419	–
$Be_{40}Zr_{60}$	λ	0.4631	0.6221	0.6496	0.6508	0.5621	1.1
	μ^*	0.1441	0.1572	0.1590	0.1593	0.1515	0.13
	T_C (K)	2.1018	7.7616	9.0308	9.0582	5.3736	2.10
	α	0.2365	0.3233	0.3329	0.3325	0.3036	–
	N_0V	0.2245	0.2969	0.3085	0.3088	0.2717	–
$Be_{35}Zr_{65}$	λ	0.4831	0.6468	0.6748	0.6759	0.5865	1.2
	μ^*	0.1430	0.1559	0.1577	0.1580	0.1504	0.13
	T_C (K)	2.6012	8.5481	9.8081	9.8286	6.1643	2.60
	α	0.2682	0.3410	0.3490	0.3486	0.3252	–
	N_0V	0.2363	0.3091	0.3205	0.3208	0.2845	–
$Be_{30}Zr_{70}$	λ	0.4941	0.6570	0.6839	0.6849	0.5990	1.5
	μ^*	0.1417	0.1544	0.1562	0.1565	0.1490	0.13
	T_C (K)	2.8027	8.4972	9.6286	9.6428	6.3322	2.80
	α	0.2866	0.3504	0.3572	0.3568	0.3375	–
	N_0V	0.2432	0.3146	0.3254	0.3257	0.2913	–

Table 7 contains calculated values of the transition temperature T_C for Be_xZr_{100-x} metallic glasses computed from the various forms of the local field correction functions along with experimental data [81]. From the Table 7 it can be noted that, the static H-screening function yields lowest T_C whereas the F-function yields highest values of T_C. The present results obtained from the H-local field correction functions are found in good agreement with available experimental data [81]. It is seen that T_C is quite sensitive to the local field correction functions, and the results of T_C by using H-screening are in best agreement with the experimental data for the Be_xZr_{100-x} metallic glasses under investigation. The percentile influence of the various local field correction functions with respect to the static H-screening function on T_C for the metallic glasses is observed in the range of 218.79%-499.77%, 155.67%-330.97%, 136.98%-277.85% and 125.93%-244.05% for $Be_{45}Zr_{55}$, $Be_{40}Zr_{60}$, $Be_{35}Zr_{65}$ and $Be_{30}Zr_{70}$ metallic glasses, respectively. The calculated results of the transition temperature T_C for $Be_{45}Zr_{55}$, $Be_{40}Zr_{60}$, $Be_{35}Zr_{65}$ and $Be_{30}Zr_{70}$ metallic glasses deviate in the range of 0.2%-500.97%, 0.09%-331.34%, 0.05%-278.02% and 0.10%-244.39% from experimental findings [81], respectively. Also, the above observations indicate that simple metallic glasses having high valence (more than two) tend to have higher T_C. Perhaps only exception is divalent Be_xZr_{100-x} metallic glasses where high T_C is likely to be due to unusually high Debye temperature. The higher values of T_C may be due to the electron transfer between the transition metal and other metallic element. The increase in T_C has also been attributed to the excitonic mechanism resulting from the granular structure separated by semiconducting or insulating materials [2].

The subtle difference of the shift among the glassy Be_xZr_{100-x} alloys may be reflected in the values of the electron-phonon coupling strength λ. Since both the density and the values of the electron-phonon coupling strength λ are larger for $Be_{30}Zr_{70}$ alloys than those for the $Be_{40}Zr_{60}$ alloy, a larger shift of the centre of gravity of the phonon spectrum toward lower phonon energy for the former than the latter alloy may be expected. If such a trend toward a softer phonon spectrum is related to the degree of the disorder increases as 'Be' content decreases in the Be_xZr_{100-x} glassy system, which was reported by Hasegawa and Tanner [81] from the glass transition temperature T_g. From this, we may thus conclude that the increase of T_C or λ with decreasing 'Be' in the Be_xZr_{100-x} glassy system is due to the increase of the degree of structural disorder as was noted by Hasegawa and Tanner [81].

The values of the isotope effect exponent α for Be_xZr_{100-x} metallic glasses are tabulated in Table 7. The computed values of the α show a weak dependence on the dielectric screening, its value is being lowest for the H-screening function and highest for the F-function. The negative value of the α is observed in the case of metallic glasses, which indicates that the electron-phonon coupling in these metallic complexes do not fully explain all the features regarding their superconducting behaviour. Since the experimental or theoretical values of α has not been reported in the literature so far, the present data of α may be used for the study of ionic vibrations in the superconductivity of amorphous substances. Since H-local field correction function yields the best results for λ and T_C, it may be observed that α values obtained from this screening provide the best account for the role of the ionic vibrations in superconducting behaviour of this system. The most important feature noted here is that as the concentration (x) of 'Zr' (in at %) increases the present results of α increases sharply.

The values of the effective interaction strength $N_O V$ are listed in Table 7 for different local field correction functions. It is observed that the magnitude of $N_O V$ shows that the metallic glasses under investigation lie in the range of weak coupling superconductors. The values of the $N_O V$ also show a feeble dependence on dielectric screening, its value being lowest for the H-screening function and highest for the F-screening function. Here also, as the concentration (x) of 'Zr' (in at %) increases the present results of $N_O V$ increases.

3.8. SSP of $Be_x Al_{100-x}$ metallic glasses

The input parameters and other constants used in the present computation are taken from our earlier paper [41]. The presently calculated results of the SSP are tabulated in Table 8 with the experimental [2] and other such theoretical findings [16, 32].

The calculated values of the electron-phonon coupling strength λ for $Be_x Al_{100-x}$ metallic glasses, using five different types of the local field correction functions with EMC model potential, are shown in Table 8 with other theoretical data [16, 32]. It is noticed from the Table 8 that, λ vales are quite sensitive to the local field correction functions. It is also observed from the present study that, the percentile influence of the various local field correction functions with respect to the static H- screening function on the electron-phonon coupling strength λ is 15.43%-52.06% and 19.85%-46.15% for $Be_{90}Al_{10}$ and $Be_{70}Al_{30}$ metallic glasses, respectively. Also, the H-screening yields lowest values of λ, whereas the values obtained from the F-function are the highest. It is also observed from the Table 8 that, λ goes on increasing from the values of 0.3857→0.3946 as the concentration 'x' of 'Al' is increased from 0.10-0.30. The increase in λ with concentration 'x' of 'Al' shows a gradual transition from weak coupling behaviour to intermediate coupling behaviour of electrons and phonons, which may be attributed to an increase of the hybridization of sp-d electrons of 'Al' with increasing concentration (x), as was also observed by Minnigerode and Samwer [78]. This may also be attributed to the increase role of ionic vibrations in the Al-rich region [14].

The computed values of the Coulomb pseudopotential μ^*, which accounts for the Coulomb interaction between the conduction electrons, obtained from the various forms of the local field correction functions are tabulated in Table 8 with other theoretical data [16, 32]. It is observed from the Table 8 that for all metallic glasses, the μ^* lies between 0.15 and 0.18, which is in accordance with McMillan [67], who suggested $\mu^* \approx 0.13$ for simple and transition metals. The weak screening influence shows on the computed values of μ^*. The percentile influence of the various local field correction functions with respect to the static H-screening function on μ^* for the metallic glasses is observed in the range of 4.75%-10.32% and 4.85%-10.34% for $Be_{90}Al_{10}$ and $Be_{70}Al_{30}$ metallic glasses, respectively. Again the H-screening function yields lowest values of μ^*, while the values obtained from the F-function are the highest.

The present results are found in good agreement with the available theoretical data [16, 32]. Here also, as the concentration (x) of 'Al' (in at %) increases the present results of μ^* decreases.

Table 8 contain calculated values of the transition temperature T_C for Be_xAl_{100-x} metallic glasses computed from the various forms of the local field correction functions along with the experimental [2] and theoretical findings [16, 32]. From the Table 8 it can be noted that, the static H-screening function yields lowest T_C whereas the F-function yields highest values of T_C. The present results obtained from the IU-local field correction functions are found in good agreement with available experimental [2] and theoretical data [16, 32]. It is seen that T_C is quite sensitive to the local field correction functions, and the results of T_C by using IU-screening are in best agreement with the experimental data for the Be_xAl_{100-x} metallic glasses under investigation. The calculated results of the transition temperature T_C for $Be_{90}Al_{10}$ and $Be_{70}Al_{30}$ metallic glasses deviate in the range of 0.08%-93.70% and 0.34%-91.28% from the experimental findings [2], respectively. Also, the above observations indicate that simple metallic glasses having high valence (more than two) tend to have higher T_C. Perhaps only exception is divalent Be-based metallic glasses where high T_C is likely to be due to unusually high Debye temperature. The higher values of T_C may be due to the electron transfer between the transition metal and other metallic element. The increase in T_C has also been attributed to the excitonic mechanism resulting from the granular structure separated by semiconducting or insulating materials [2].

The values of the isotope effect exponent α for Be_xAl_{100-x} metallic glasses are tabulated in Table 8. The computed values of the α show a weak dependence on the dielectric screening, its value is being lowest for the H-screening function and highest for the F-function. The negative value of the α is observed in the case of metallic glasses, which indicates that the electron-phonon coupling in these metallic complexes do not fully explain all the features regarding their superconducting behaviour. The comparisons of present results with other such theoretical data [16, 32] are highly encouraging. Since the experimental value of α has not been reported in the literature so far, the present data of α may be used for the study of ionic vibrations in the superconductivity of amorphous substances. Since IU-local field correction function yields the best results for λ and T_C, it may be observed that α values obtained from this screening provide the best account for the role of the ionic vibrations in superconducting behaviour of this system. The most important feature noted here is that as the concentration (x) of 'Al' (in at %) increases the present results of α increases sharply.

The values of the effective interaction strength N_OV are listed in Table 8 for different local field correction functions. It is observed that the magnitude of N_OV shows that the metallic glasses under investigation lie in the range of weak coupling superconductors. The values of the N_OV also show a feeble dependence on dielectric screening, its value being lowest for the H-screening function and highest for the F-screening function. The present outcomes are found qualitative agreement with the available theoretical data [16, 32]. Here also, as the concentration (x) of 'Al' (in at %) increases the present results of $N_O V$ increases.

Table 8. Superconducting state parameters of the $Be_x Al_{100-x}$ metallic glasses

Glass	SSP	Present results					Exp.	Others
		H	T	IU	F	S	[2]	[16, 32]
$Be_{90}Al_{10}$	λ	0.3857	0.5523	0.5762	0.5865	0.4452	–	0.78,0.64,0.64, 0.64
	μ^*	0.1579	0.1720	0.1739	0.1742	0.1654	–	0.26,0.19,0.19, 0.19
	T_C (K)	0.4535	5.8802	7.2058	7.8827	1.4535	7.2	7.21,7.24,7.56, 7.19
	α	-0.117	0.190	0.209	0.219	0.021	–	- 0.18,0.19,0.19, 0.19
	N_0V	0.1686	0.2532	0.2640	0.2689	0.1992	–	0.29,0.28,0.28, 0.28
$Be_{70}Al_{30}$	λ	0.3946	0.5465	0.5730	0.5767	0.4728	–	0.76,0.62,0.61, 0.61
	μ^*	0.1566	0.1705	0.1724	0.1728	0.1642	–	0.26,0.17,0.17, 0.17
	T_C (K)	0.5317	4.8433	6.1209	6.2992	2.1232	6.1	6.15,6.12,6.15, 6.11
	α	-0.056	0.191	0.214	0.216	0.107	–	-0.14,0.28, 0.28, 0.28
	N_0V	0.1751	0.2512	0.2634	0.2650	0.2159	–	0.29,0.28,0.28, 0.28

3.9. SSP of $Mg_{0.70}Zn_{0.30-x}Ga_x$ metallic glasses

The input parameters and other constants used in the present computation are taken from our earlier paper [47]. The presently calculated results of the SSP are tabulated in Table 9 with the other such experimental findings [82]. Also, the graphical analyses of the SSP of $Mg_{0.70}Zn_{0.30-x}Ga_x$ ternary systems are also plotted in Figures 16-20.

The calculated values of the electron-phonon coupling strength λ for five $Mg_{0.70}Zn_{0.30-x}Ga_x$ ternary amorphous alloys, using five different types of the local field correction functions with EMC model potential, are shown in Table 9 with other experimental data [82]. The graphical nature of λ is also displayed in Figure 16. It is noticed from the present study that, the percentile influence of the various local field correction functions with respect to the static H-screening function on the electron-phonon coupling strength λ is 27.66%-51.07%, 27.38%-50.18%, 27.21%-49.70%, 26.66%-47.76% and 26.62%-47.83% for

$Mg_{0.70}Zn_{0.30}Ga_{0.00}$, $Mg_{0.70}Zn_{0.24}Ga_{0.06}$, $Mg_{0.70}Zn_{0.20}Ga_{0.10}$, $Mg_{0.70}Zn_{0.15}Ga_{0.15}$ and $Mg_{0.70}Zn_{0.10}Ga_{0.20}$, respectively. Also, the H-screening yields lowest values of λ, whereas the values obtained from the F-function are the highest.

Table 9. Superconducting state parameters $Mg_{0.70}Zn_{0.30-x}Ga_x$ amorphous alloys

Amorphous alloys	SSP	Present results					Expt. [82]
		H	T	IU	F	S	
$Mg_{0.70}Zn_{0.30}Ga_{0.00}$	λ	0.3631	0.5176	0.5471	0.5485	0.4635	0.297
	μ^*	0.1432	0.1570	0.1589	0.1592	0.1518	–
	T_C (K)	0.1112	1.3120	1.7254	1.7395	0.7266	0.111, 0.1125
	α	-0.0059	0.2331	0.2566	0.2565	0.1857	–
	$N_O V$	0.1653	0.2452	0.2593	0.2598	0.2193	–
$Mg_{0.70}Zn_{0.24}Ga_{0.06}$	λ	0.3676	0.5216	0.5507	0.5521	0.4682	0.302
	μ^*	0.1427	0.1564	0.1583	0.1586	0.1512	–
	T_C (K)	0.1303	1.3858	1.8041	1.8163	0.7896	0.130, 0.11
	α	0.0194	0.2414	0.2635	0.2633	0.1978	–
	$N_O V$	0.1686	0.2478	0.2615	0.2620	0.2223	–
$Mg_{0.70}Zn_{0.20}Ga_{0.10}$	λ	0.3711	0.5252	0.5543	0.5556	0.4721	0.305
	μ^*	0.1424	0.1560	0.1579	0.1582	0.1509	–
	T_C (K)	0.1461	1.4527	1.8784	1.8899	0.8420	0.146
	α	0.0364	0.2475	0.2687	0.2684	0.2062	–
	$N_O V$	0.1710	0.2499	0.2636	0.2640	0.2247	–
$Mg_{0.70}Zn_{0.15}Ga_{0.15}$	λ	0.3815	0.5345	0.5627	0.5637	0.4832	0.317
	μ^*	0.1412	0.1546	0.1564	0.1567	0.1495	–
	T_C (K)	0.1950	1.5787	1.9963	2.0033	0.9721	0.195
	α	0.0857	0.2651	0.2832	0.2828	0.2313	–
	$N_O V$	0.1784	0.2557	0.2688	0.2691	0.2318	–
$Mg_{0.70}Zn_{0.10}Ga_{0.20}$	λ	0.3980	0.5577	0.5873	0.5884	0.5039	0.333
	μ^*	0.1412	0.1546	0.1564	0.1567	0.1495	–
	T_C (K)	0.2931	1.9873	2.4689	2.4777	1.2668	0.293, 0.306
	α	0.1359	0.2888	0.3046	0.3042	0.2595	–
	$N_O V$	0.1888	0.2678	0.2812	0.2815	0.2433	–

It is also observed from the Table 9 and Figure 16 that, λ goes increasing from the values of 0.3631→0.5884 as the concentration ' x ' of 'Ga' is increased from 0.0 → 0.20. The increase in λ with concentration ' x ' of 'Ga' shows a gradual transition from weak coupling behaviour to intermediate coupling behaviour of electrons and phonons, which may be attributed to an increase of the hybridization of sp-d electrons of 'Ga' with increasing concentration (x). This may also be attributed to the increase role of ionic vibrations in the Ga-rich region. The present results are found in qualitative agreement with available experimental data [82].

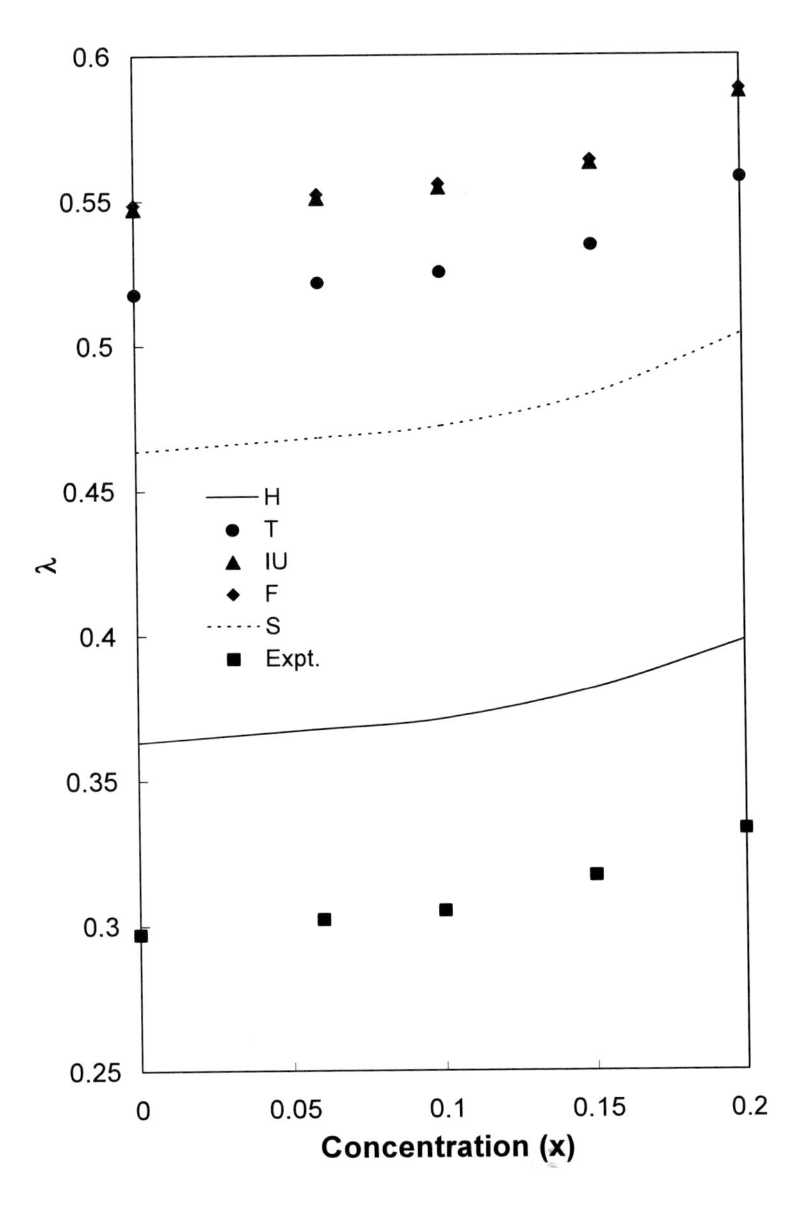

Figure 16. Variation of electron-phonon coupling strength (λ) with Ga-concentration (x) (at %).

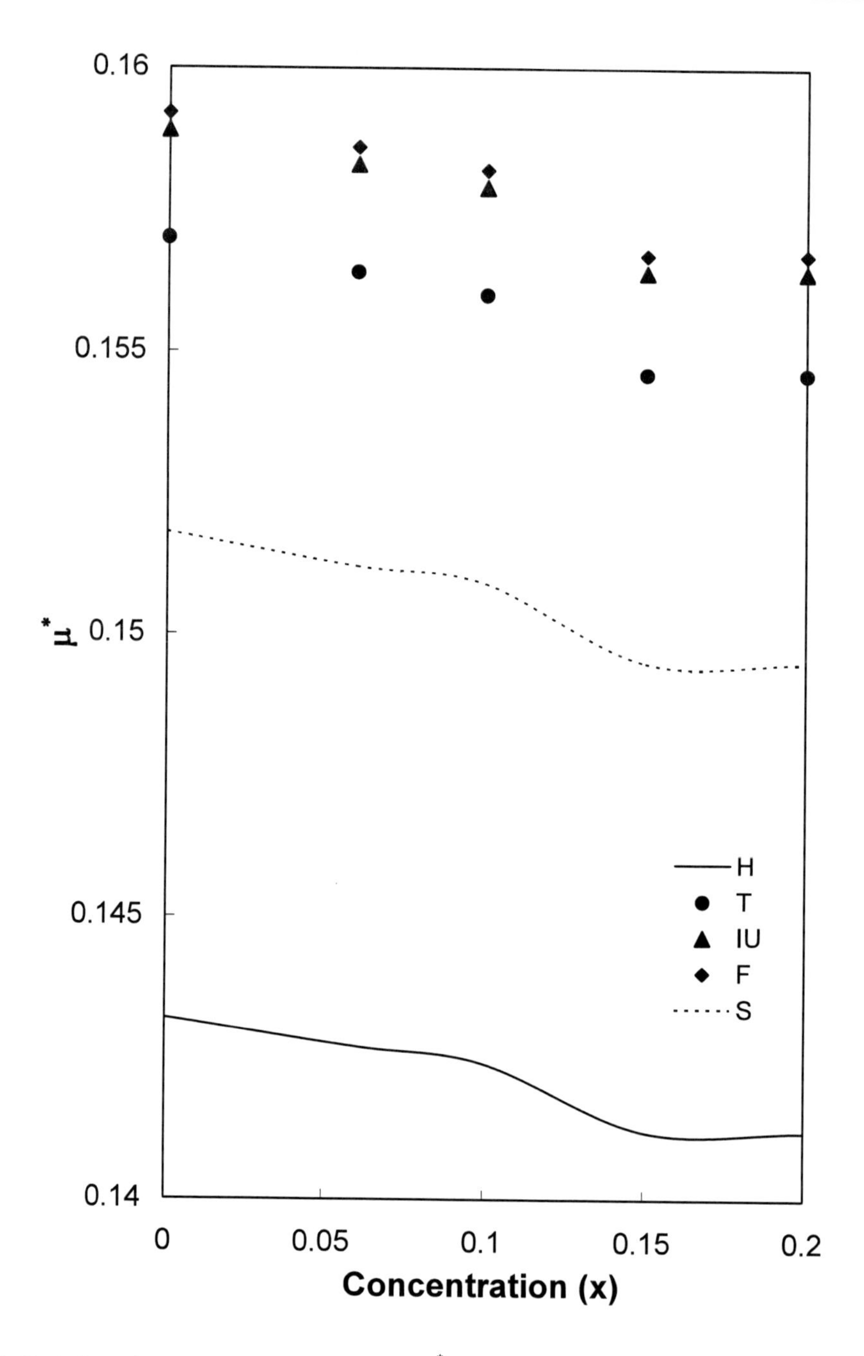

Figure 17. Variation of Coulomb pseudopotential (μ^*) with Ga-concentration (x) (at %).

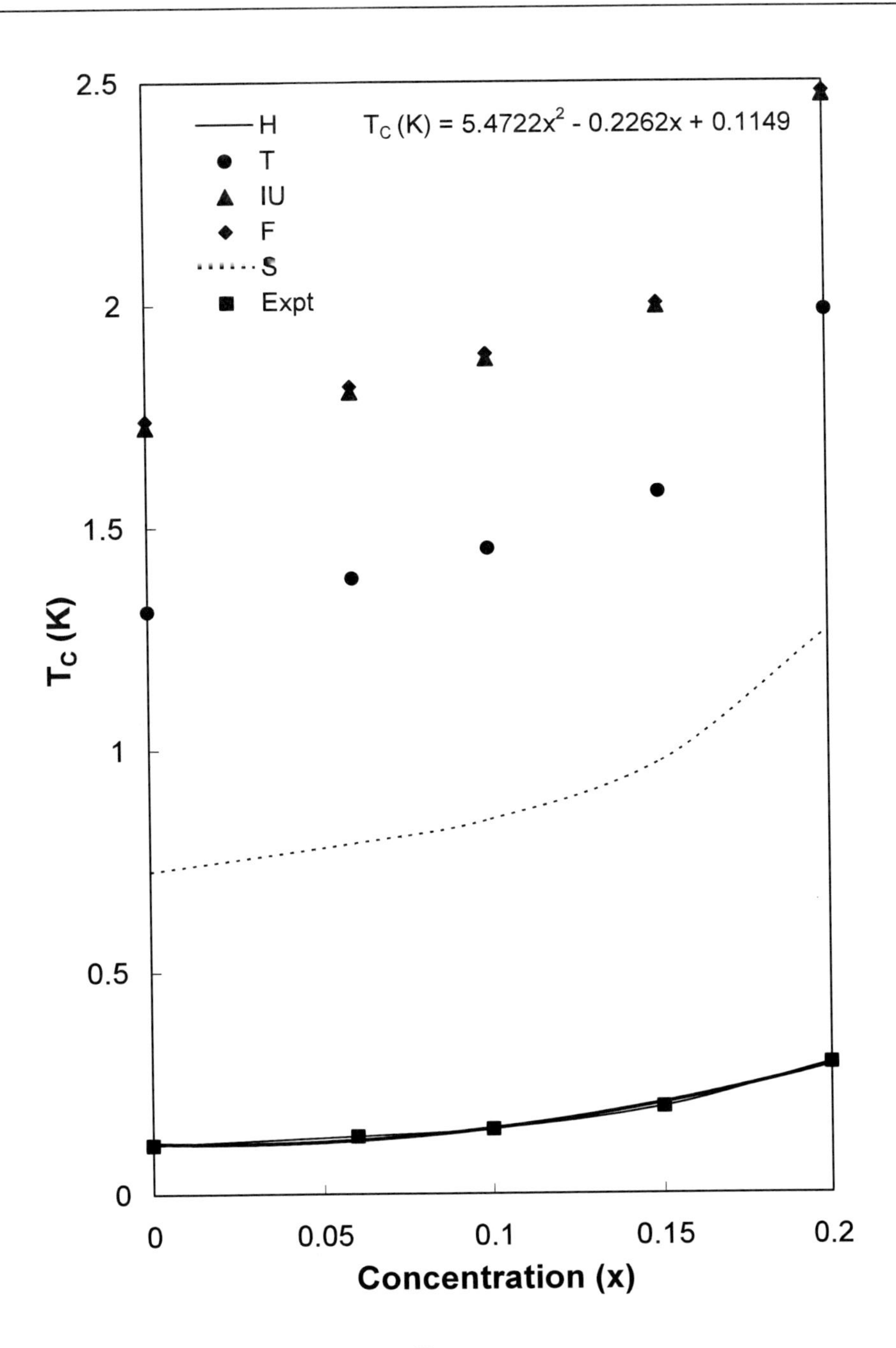

Figure 18. Variation of transition temperature (T_C) with Ga-concentration (x) (at %).

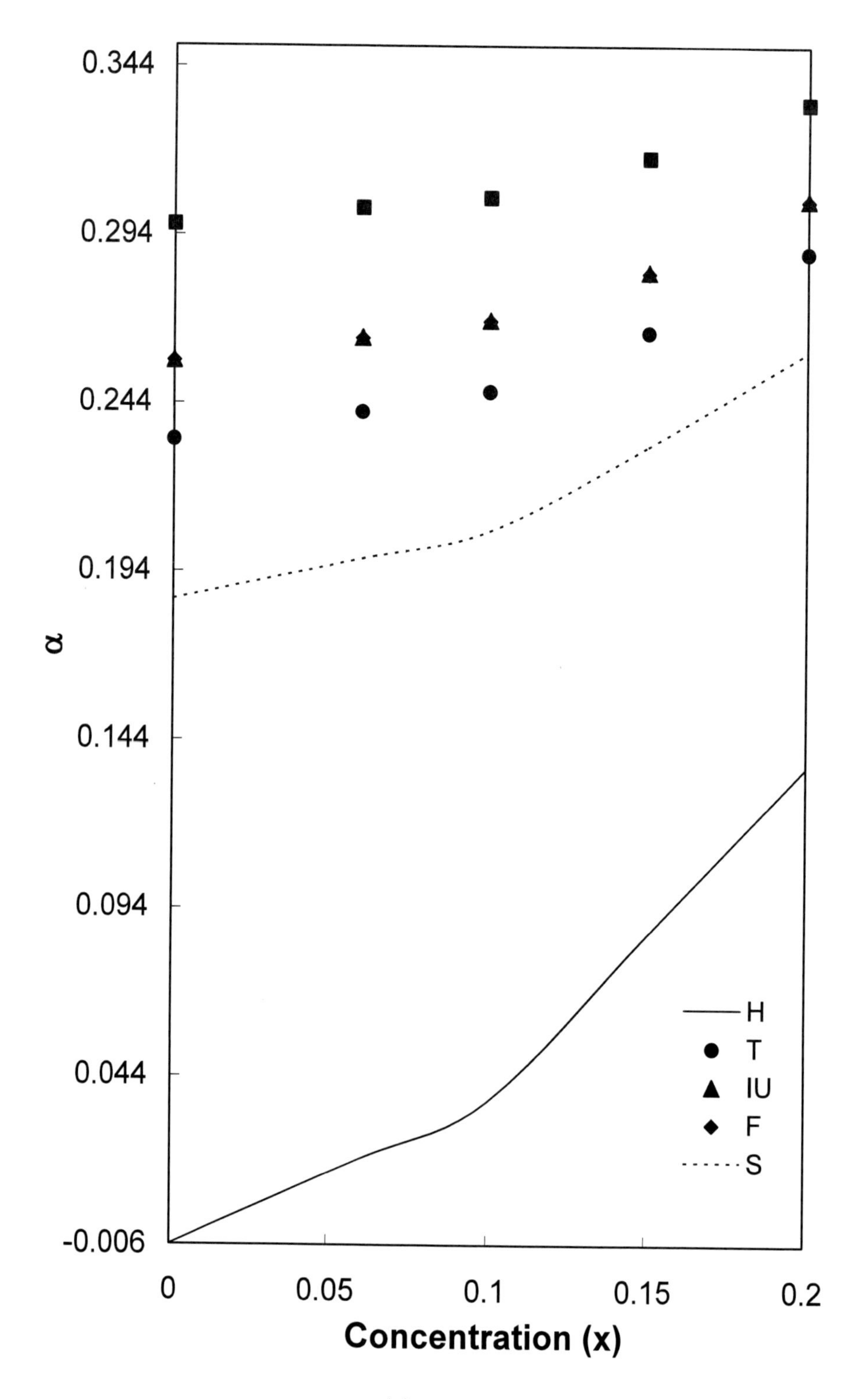

Figure 19. Variation of isotope effect exponent (α) with Ga-concentration (x) (at %).

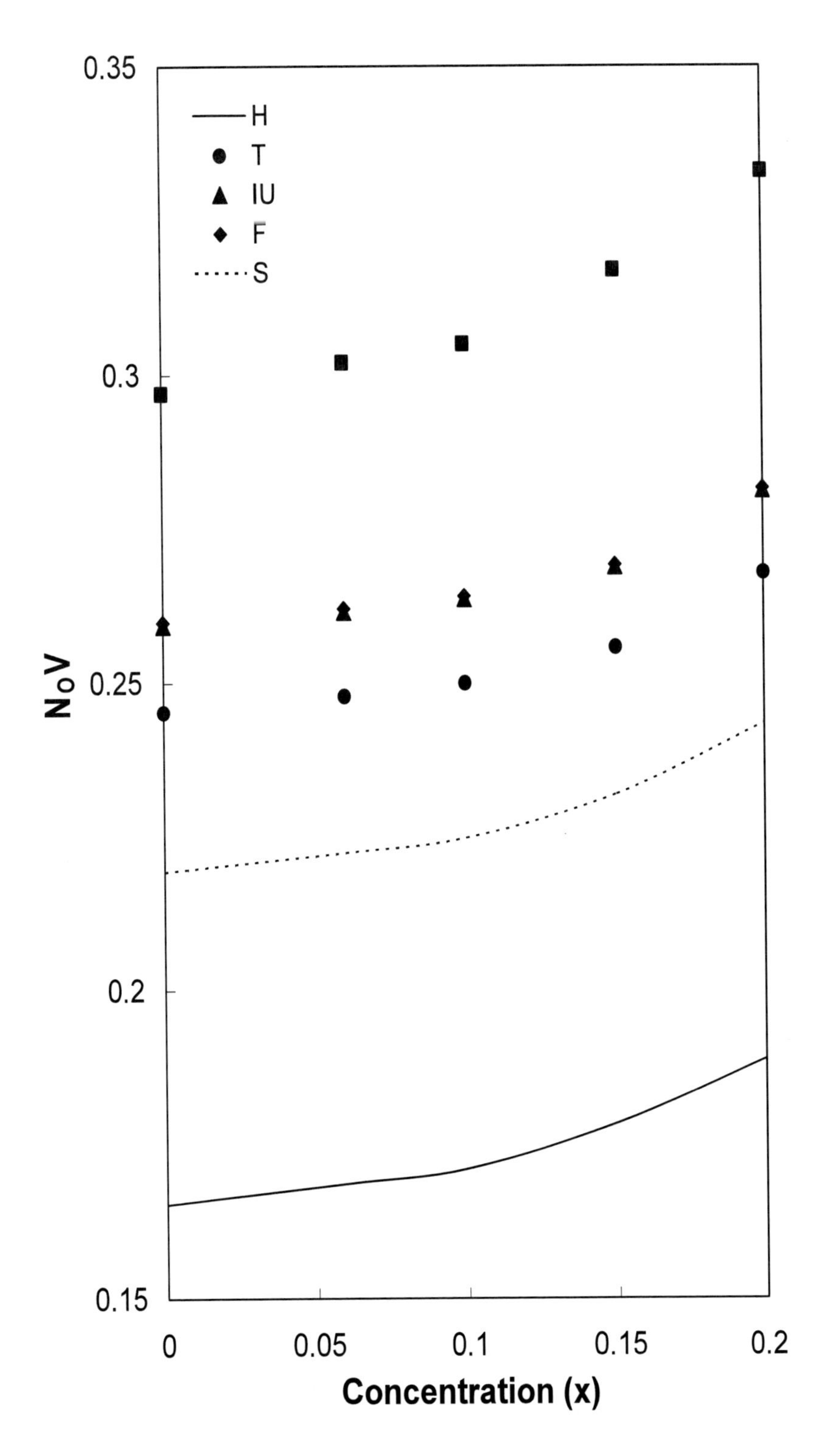

Figure 20. Variation of effective interaction strength (N_OV) with Ga-concentration (x) (at %).

The computed values of the Coulomb pseudopotential μ^*, which accounts for the Coulomb interaction between the conduction electrons, obtained from the various forms of the local field correction functions are tabulated in Table 9. It is observed from that for all five $Mg_{0.70}Zn_{0.30-x}Ga_x$ ternary amorphous alloys, μ^* lies between 0.13 and 0.16, which is in accordance with McMillan [67], who suggested $\mu^* \approx 0.13$ for simple and non-simple metals. The graphs of μ^* versus concentration (x) for different local field correction functions are plotted in Figure 17, which shows the weak dependence of μ^* on the local field correction functions. The percentile influence of the various local field correction functions with respect to the static H-screening function on μ^* for the ternary amorphous alloys is observed in the range of $Mg_{0.70}Zn_{0.30}Ga_{0.00}$, $Mg_{0.70}Zn_{0.24}Ga_{0.06}$, $Mg_{0.70}Zn_{0.20}Ga_{0.10}$, $Mg_{0.70}Zn_{0.15}Ga_{0.15}$ and $Mg_{0.70}Zn_{0.10}Ga_{0.20}$, respectively. Again the H-screening function yields lowest values of μ^*, while the values obtained from the F-function are the highest.

Table 9 contains calculated values of the transition temperature T_C for five $Mg_{0.70}Zn_{0.30-x}Ga_x$ ternary amorphous alloys computed from the various forms of the local field correction functions along with experimental findings [80]. From the Table 9 it can be noted that, the static H-screening function yields lowest T_C whereas the F-function yields highest values of T_C. The present results obtained from the H-local field correction functions are found in good agreement with available experimental data [82]. The theoretical data of T_C for five $Mg_{0.70}Zn_{0.30-x}Ga_x$ ternary amorphous alloys is not available in the literature. It is also observed that the static H-screening function yields lowest T_C whereas the F-function yields highest values of T_C. The calculated results of the transition temperature T_C for $Mg_{0.70}Zn_{0.30}Ga_{0.00}$, $Mg_{0.70}Zn_{0.24}Ga_{0.06}$, $Mg_{0.70}Zn_{0.20}Ga_{0.10}$, $Mg_{0.70}Zn_{0.15}Ga_{0.15}$ and $Mg_{0.70}Zn_{0.10}Ga_{0.20}$ ternary amorphous alloys deviate in the range of 80.56%-100.00%, 75.51%-130.03%, 72.54%-124.47%, 62.40%-106.08% and 94.89%-95.58% from experimental findings [82], respectively.

The variation of the computed values of the transition temperature T_C for $Mg_{0.70}Zn_{0.30-x}Ga_x$ ternary amorphous alloys with the atomic concentration (x) of 'Ga', using five different types of the local field correction functions with EMC potential are shown in Figure 18. The graph also includes the experimental values due to van den Berg *et al.* [80]. It is seen that T_C is quite sensitive to the local field correction functions, and the results of T_C by using H-screening are in best agreement with the experimental data for the $Mg_{0.70}Zn_{0.30-x}Ga_x$ ternary amorphous alloys under investigation, as the relevant curves for H-screening almost overlaps the experimental curves. It is also observed that the static H-screening function yields lowest T_C whereas the F-function yields highest values of T_C. It is also seen from the graphical nature, T_C increases considerably with increasing Ga-concentration (x). The composition dependence can be described by polynomial regression of the data obtained for H-screening for different values of the concentration 'x', which yields

$$T_C(K) = 5.4722\,x^2 - 0.2262\,x + 0.1149\,. \quad (20)$$

The graph of the fitted T_C equation is displayed in Figure 4, which indicates that T_C increases almost considerably with increasing 'Ga' content with a slope $dT_C/dx = -0.2262$. Wide extrapolation predicts a T_C = 0.1149K for the hypothetical case of 'amorphous pure $Mg_{0.70}Zn_{0.30}Ga_{0.00}$ alloy.

The values of the isotope effect exponent α for five $Mg_{0.70}Zn_{0.30-x}Ga_x$ ternary amorphous alloys are tabulated in Table 9. Figure 19 depicts the variation of α with Ga-concentration (x) increases. The computed values of the α show a weak dependence on the dielectric screening, its value is being lowest for the H-screening function and highest for the F-function. The negative value of α is observed in the case of $Mg_{0.70}Zn_{0.30}Ga_{0.00}$ ternary alloy, which indicates that the electron-phonon coupling in these metallic complexes do not fully explain all the features regarding their superconducting behaviour.

Table 10. Superconducting state parameters of $Nb_xTa_yMo_z$ ternary superconductors

Superconductors	SSP	Present results					Others [83]
		H	T	IU	F	S	
$Nb_{0.45}Ta_{0.45}Mo_{0.10}$	λ	0.6894	0.9138	0.9518	0.9553	0.8032	0.823
	μ^*	0.1223	0.1304	0.1314	0.1316	0.1262	0.13
	T_C (K)	6.4950	11.9279	12.8215	12.8953	9.2998	6.491
	α	0.4292	0.4474	0.4495	0.4496	0.4406	–
	N_0V	0.3486	0.4279	0.4398	0.4408	0.3913	–
$Nb_{0.30}Ta_{0.40}Mo_{0.30}$	λ	0.5560	0.7312	0.7605	0.7629	0.6456	0.643
	μ^*	0.1232	0.1313	0.1324	0.1326	0.1271	0.13
	T_C (K)	3.5360	7.9731	8.7725	8.8284	5.7458	3.538
	α	0.3917	0.4227	0.4262	0.4263	0.4115	–
	N_0V	0.2875	0.3603	0.3714	0.3722	0.3267	–
$Nb_{0.40}Ta_{0.30}Mo_{0.30}$	λ	0.5936	0.7796	0.8107	0.8131	0.6885	0.692
	μ^*	0.1230	0.1311	0.1322	0.1324	0.1269	0.13
	T_C (K)	4.5588	9.4901	10.3477	10.4065	7.0505	4.559
	α	0.4052	0.4311	0.4340	0.4341	0.4217	–

Since the experimental value of α has not been reported in the literature so far, the present data of α may be used for the study of ionic vibrations in the superconductivity of alloying substances. Since H-local field correction function yields the best results for λ and T_C, it may be observed that α values obtained from this screening provide the best account for the role of the ionic vibrations in superconducting behaviour of this system.

The values of the effective interaction strength $N_O V$ are listed in Table 9 and depicted in Figure 20 for different local field correction functions. It is observed that the magnitude of $N_O V$ shows that the five $Mg_{0.70}Zn_{0.30-x}Ga_x$ ternary amorphous alloys under investigation lie in the range of weak coupling superconductors. The values of $N_O V$ also show a feeble dependence on dielectric screening, its value being lowest for the H-screening function and highest for the F-screening function.

3.10. SSP of $Nb_xTa_yMo_z$ metallic glasses

The input parameters and other constants used in the present computation are taken from our earlier paper [59]. The presently calculated results of the SSP are tabulated in Table 10 with other such theoretical findings [83]. Also, the graphical analyses of the SSP of $Nb_xTa_yMo_z$ ternary systems are also plotted in Figures 21-25.

The calculated values of the electron-phonon coupling strength λ for five $Nb_xTa_yMo_z$ superconductors, using five different types of the local field correction functions with EMC model potential, are shown in Table 10 with other theoretical data [83]. The graphical nature of λ is also displayed in Figure 21. It is noticed from the present study that, the percentile influence of the various local field correction functions with respect to the static H-screening function on the electron-phonon coupling strength λ is 16.51%-38.56%, 16.11%-37.21%, 15.98%-36.98%, 15.80%-36.47% and 15.18%-33.98% for $Nb_{0.45}Ta_{0.45}Mo_{0.10}$, $Nb_{0.30}Ta_{0.40}Mo_{0.30}$, $Nb_{0.40}Ta_{0.30}Mo_{0.30}$, $Nb_{0.30}Ta_{0.30}Mo_{0.40}$ and $Nb_{0.15}Ta_{0.15}Mo_{0.70}$ superconductors, respectively. It is also observed from the Table 10 that, λ goes decreasing from the values of 0.9553 → 0.4269 as the concentration 'z' of 'Mo' is increased from 0.10 → 0.70. The decrease in λ with concentration 'z' of 'Mo' shows a gradual transition from weak coupling behaviour to intermediate coupling behaviour of electrons and phonons, which may be attributed to an increase of the hybridization of sp-d electrons of 'Mo' with increasing concentration (z). This may also be attributed to the increase role of ionic vibrations in the Mo-rich region. Presently computed λ from S-local field correction function is found an excellent agreement with available theoretical data [83].

The computed values of the Coulomb pseudopotential μ^*, which accounts for the Coulomb interaction between the conduction electrons, obtained from the various forms of the local field correction functions are tabulated in Table 10 with other theoretical data [14].

It is observed from that for all five $Nb_xTa_yMo_z$ superconductors, μ^* lies between 0.12 and 0.14, which is in accordance with McMillan [67], who suggested $\mu^* \approx 0.13$ for transition metals.

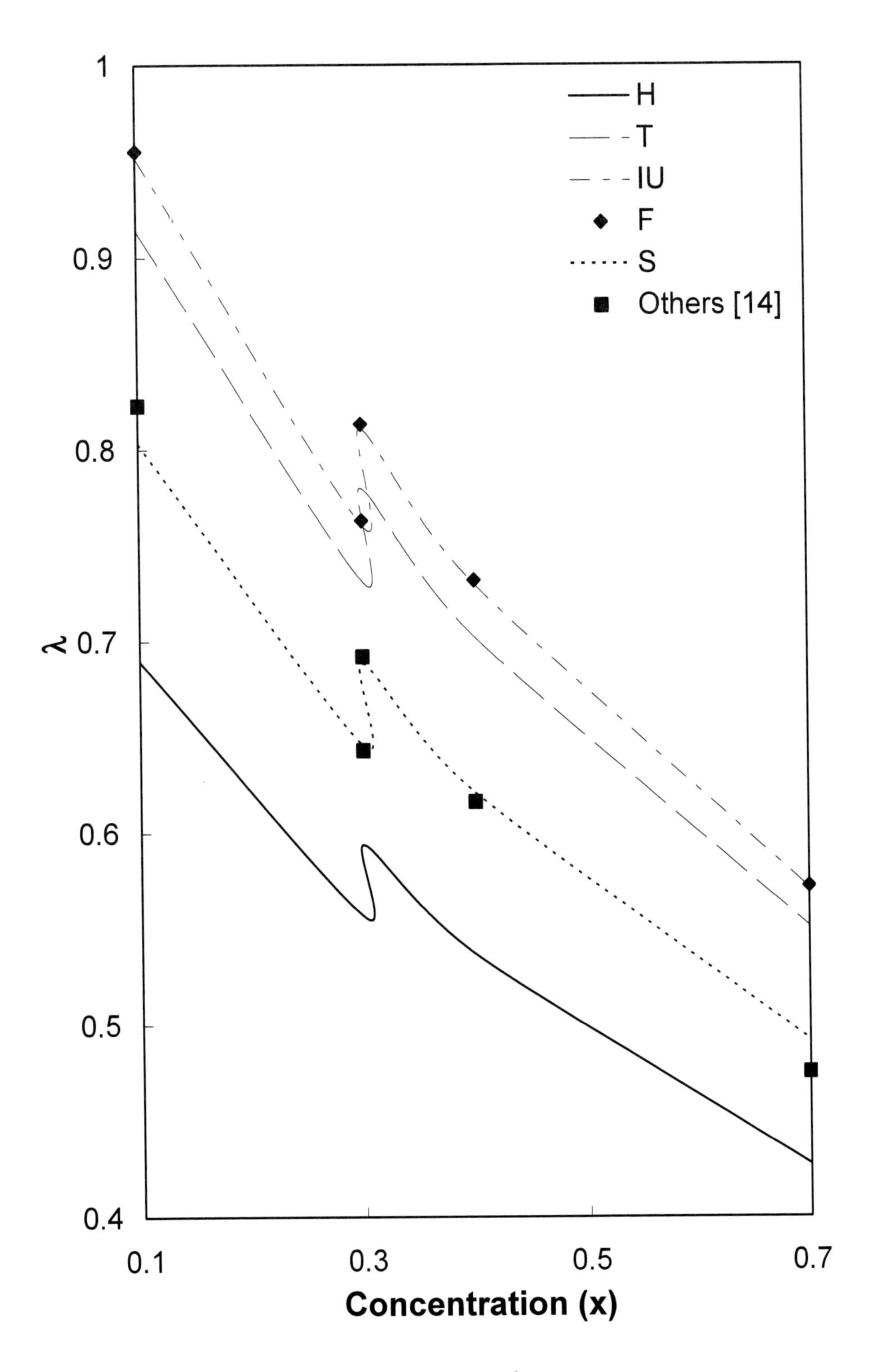

Figure 21. Variation of electron-phonon coupling strength (λ) with Mo-concentration (x) (at %).

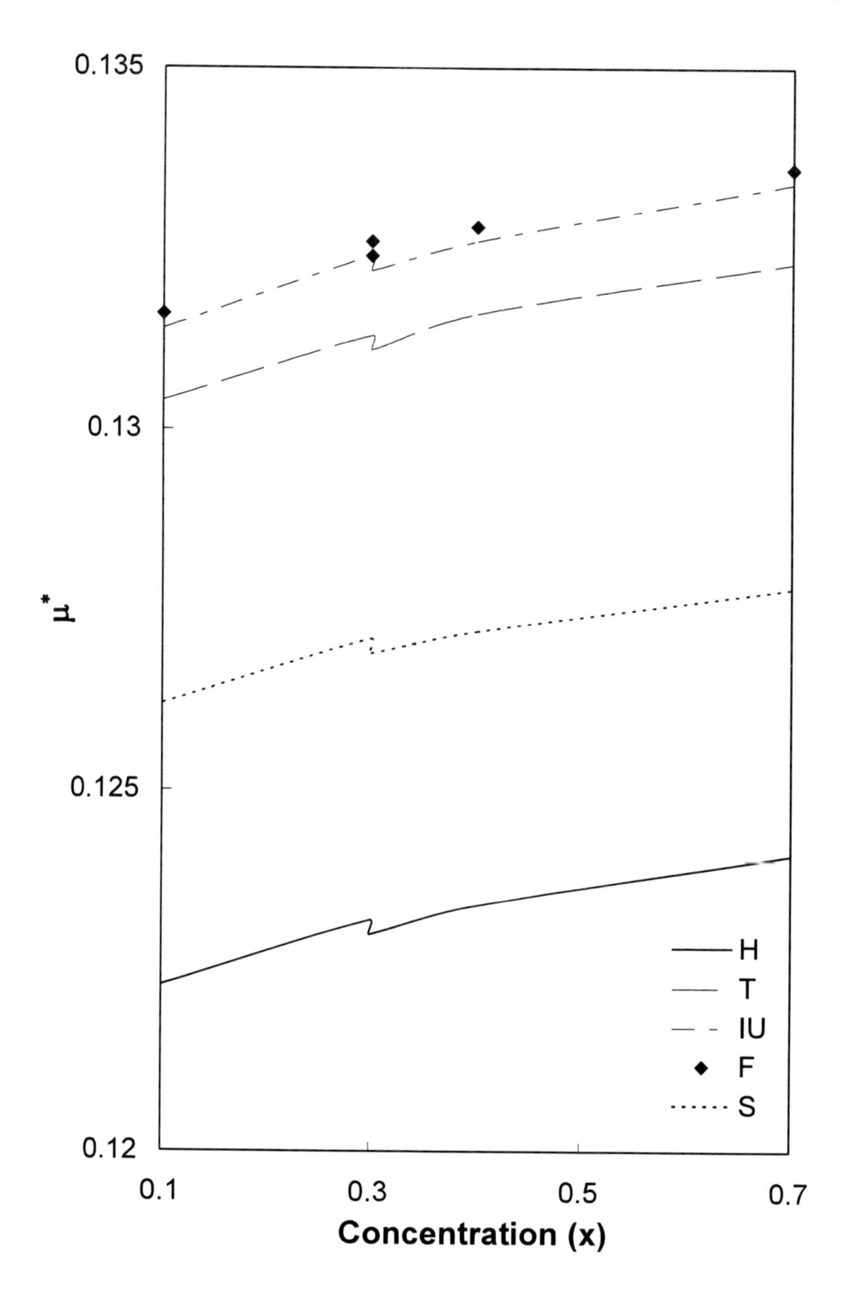

Figure 22. Variation of Coulomb pseudopotential (μ^*) with Mo-concentration (x) (at %).

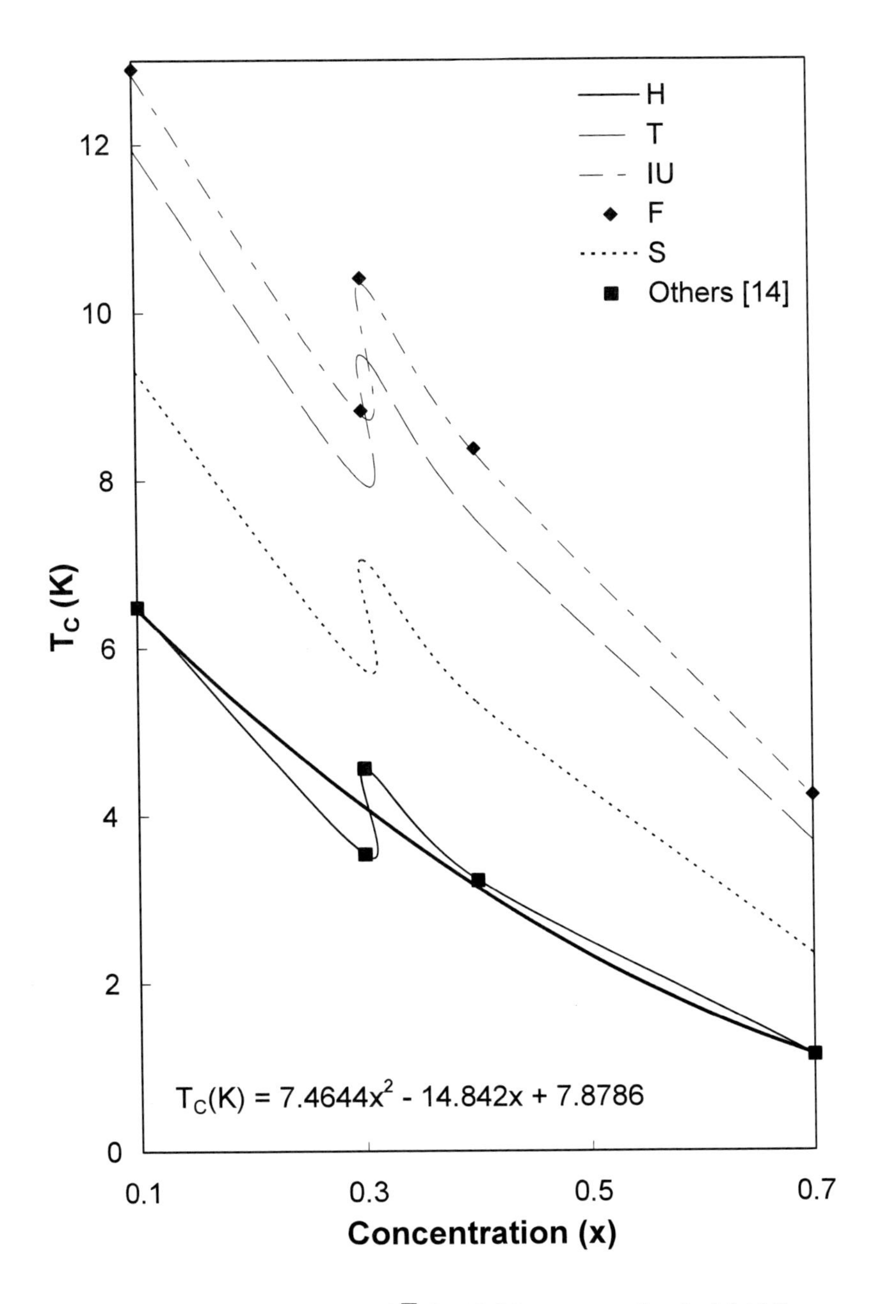

Figure 23. Variation of transition temperature (T_C) with Mo-concentration (x) (at %).

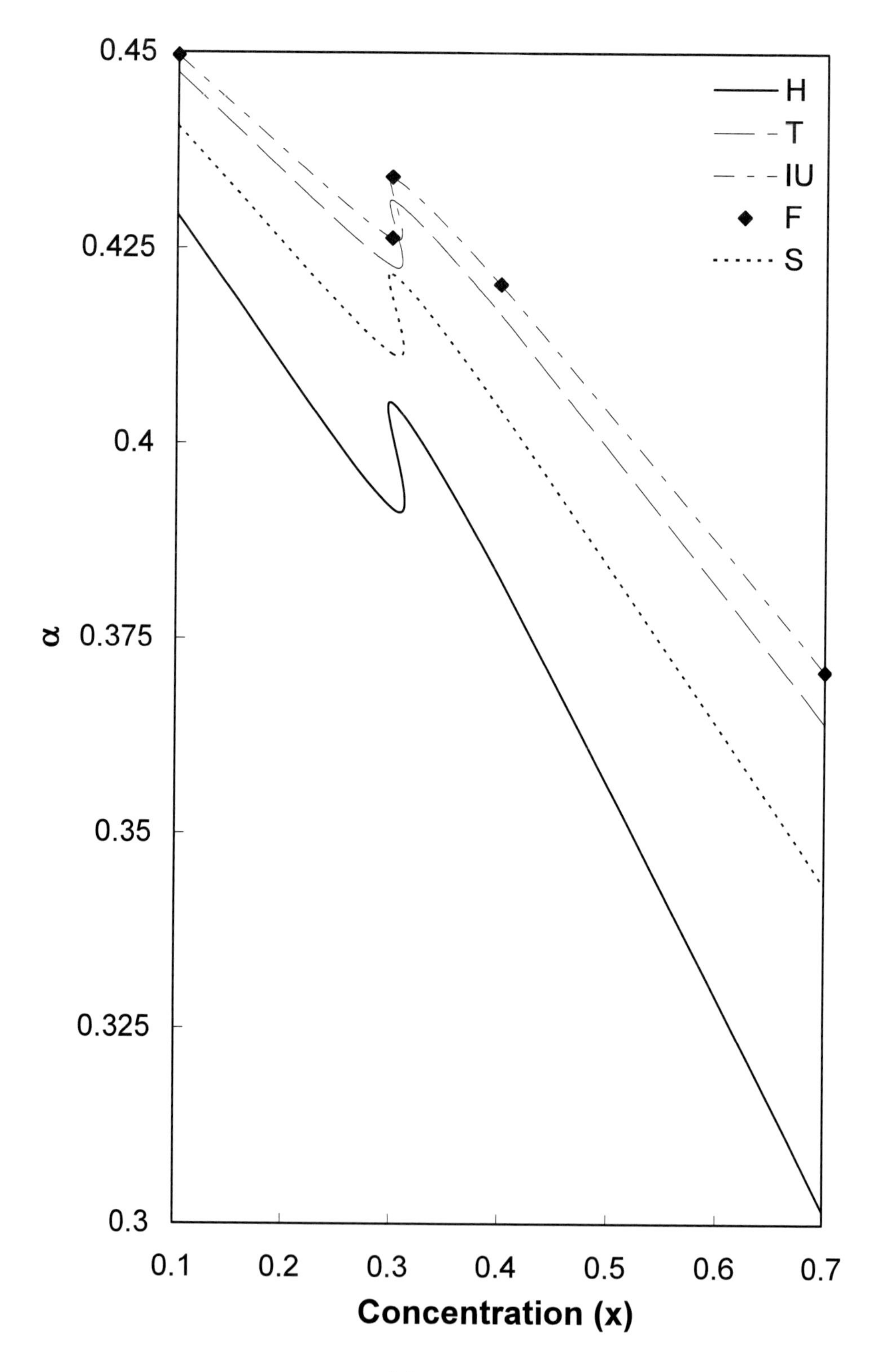

Figure 24. Variation of isotope effect exponent (α) with Mo-concentration (x) (at %).

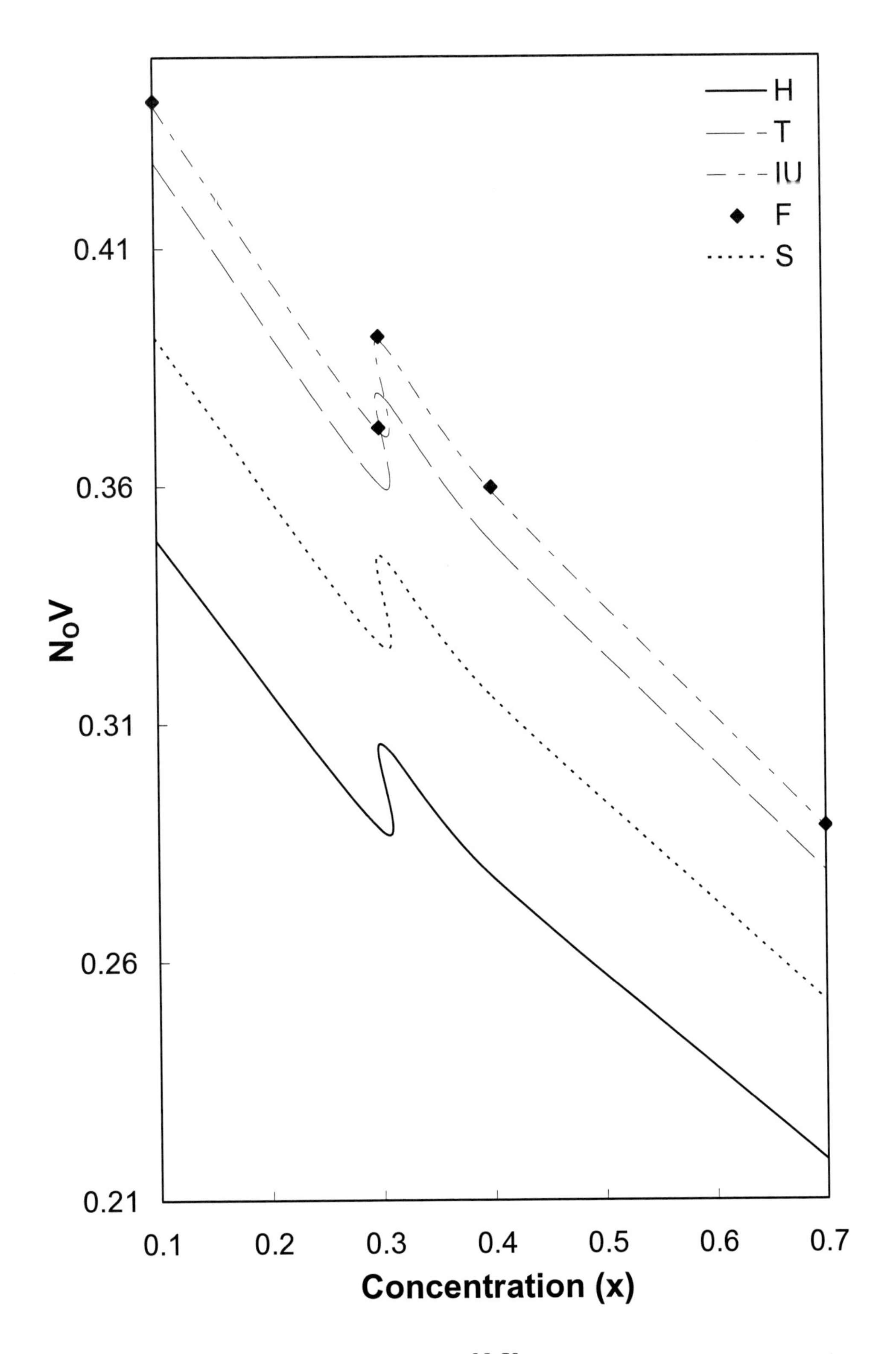

Figure 25. Variation of effective interaction strength (N_OV) with Mo-concentration (x) (at %).

The graphs of μ^* versus concentration (x) for different local field correction functions are plotted in Figure 22, which shows the weak dependence of μ^* on the local field correction functions. The weak screening influence shows on the computed values of μ^*. The percentile influence of the various local field correction functions with respect to the static H-screening function on μ^* for the superconductors is observed in the range of 3.23%-7.63%, 3.15%-7.63%, 3.13%-7.62%, 3.12%-7.65%, 3.00%-7.64% for $Nb_{0.45}Ta_{0.45}Mo_{0.10}$, $Nb_{0.30}Ta_{0.40}Mo_{0.30}$, $Nb_{0.40}Ta_{0.30}Mo_{0.30}$, $Nb_{0.30}Ta_{0.30}Mo_{0.40}$ and $Nb_{0.15}Ta_{0.15}Mo_{0.70}$ superconductors, respectively. The present results are found in good agreement with available theoretical data [83].

Table 10 contains calculated values of the transition temperature T_C for five $Nb_xTa_yMo_z$ superconductors computed from the various forms of the local field correction functions along with theoretical findings [83]. The present results obtained from the H-local field correction functions are found in good agreement with available theoretical data [83]. The experimental data of T_C for five $Nb_xTa_yMo_z$ superconductors is not available in the literature. The calculated results of the transition temperature T_C for $Nb_{0.45}Ta_{0.45}Mo_{0.10}$, $Nb_{0.30}Ta_{0.40}Mo_{0.30}$, $Nb_{0.40}Ta_{0.30}Mo_{0.30}$, $Nb_{0.30}Ta_{0.30}Mo_{0.40}$ and $Nb_{0.15}Ta_{0.15}Mo_{0.70}$ super conductors deviate in the range of 0.06%-98.66%, 0.06%-149.53%, 0%-128.26%, 0.08%-159.75% and 0.04%-272.95% from the theoretical findings [83], respectively.

The variation of the computed values of the transition temperature T_C for $Nb_xTa_yMo_z$ ternary amorphous alloys with the atomic concentration (x) of 'Mo', using five different types of the local field correction functions with EMC potential are shown in Figure 23. The graph also includes the theoretical values due to Chatterjee [83]. It is seen that T_C is quite sensitive to the local field correction functions, and the results of T_C by using H-screening are in best agreement with theoretical data [14] for the $Nb_xTa_yMo_z$ ternary amorphous alloys under investigation, as the relevant curves for H-screening almost overlaps the theoretical curves. It is also observed that the static H-screening function yields lowest T_C whereas the F-function yields highest values of T_C. It is also seen from the graphical nature, T_C increases considerably with increasing Mo-concentration (x). The composition dependence can be described by polynomial regression of the data obtained for H-screening for different values of the concentration 'x', which yields

$$T_C(K) = 7.4644\,x^2 - 14.842x + 7.8786\,. \tag{21}$$

The graph of the fitted T_C equation is displayed in Figure 23, which indicates that T_C increases considerably with increasing 'Mo' content with a slope $dT_C/dx = -14.842$. Wide extrapolation predicts a T_C = 7.8786K for the hypothetical case of 'amorphous pure $Nb_{0.45}Ta_{0.45}Mo_{0.10}$ alloy.

The values of the isotope effect exponent α for five $Nb_xTa_yMo_z$ superconductors are tabulated in Table 10. Figure 24 depicts the variation of α with Mo-concentration (x) increases. The computed values of α show a weak dependence on the dielectric screening function. Since the experimental value of α has not been reported in the literature so far, the present data of α may be used for the study of ionic vibrations in the superconductivity of alloying substances. Since H-local field correction function yields the best results for λ and T_C, it may be observed that α values obtained from this screening provide the best account for the role of the ionic vibrations in superconducting behaviour of this system.

The values of the effective interaction strength N_OV are listed in Table 10 and depicted in Figure 25 for different local field correction functions. It is observed that the magnitude of N_OV shows that the five $Nb_xTa_yMo_z$ superconductors under investigation lie in the range of weak to intermediate superconductors. The values of N_OV also show a feeble dependence on dielectric screening, its value being lowest for the H-screening function and highest for the F-screening function. The theoretical or experimental data of N_OV is not available for the further comparisons.

The arbitrariness is observed in most of the SSP around 0.30 atomic concentration of 'Mo' content. It may be due to the superconducting phase of the substances changed around this point. The $Nb_xTa_yMo_z$ ternary substances are based on the strong superconducting elements of the periodic table. In which, 'Nb' element has high transition temperature T_C than the other two metallic elements i.e. 'Mo' and 'Ta'. Hence, the gradual transition occurs in the $Nb_xTa_yMo_z$ ternary substances from Nb→Mo, i.e. this transition occurs from high to low transition temperature T_C. Which may be affected the nature of the ternary super conductors around 0.30 atomic concentration of 'Mo' content.

As we have mentioned earlier that, Chatterjee [83] has calculated the superconducting properties from average T-matrix approximation (ATA). In which method, the electron-phonon coupling strength λ is computed through APW or KKR method. And then after he has computed transition temperature T_C through λ. But, in the present case we have avoided any type of fitting procedure to compute the superconducting state properties. The present computation is performed in straight forward manner. Also, the present model has considered most of the electron-phonon interaction effects through pseudopotential and its screening. Therefore, the present model is found simple, suitable and more accurate for studying the superconducting properties.

The negative values of the isotope effect exponent α for some of the metallic complexes may be due to the magnetic interactions of the atoms in metallic complexes. The electron-phonon coupling strength λ is dependent of the $D(E_F)$, the total density of states at Fermi energy E_F. For binary alloys superconductors, photoemission measurements showed that the d-band is split into the two components: one crossing the Fermi level which arises from 'A' metallic superconductor, the other due to the 'B' metallic superconductor below E_F.

The relative intensities of these two components vary strongly with concentrations. Such a band splitting is a well known effect in concentrated alloys where nuclear charges or exchange fields of the components differ sufficiently. Then each alloy components has its own d-band., having minimum overlap with the 3d-bands of the other components. The important point is that ' B ' metallic superconductor provides the main contribution to the density of states at E_F [79]. Therefore, as the concentration of the ' B ' metallic superconductor increases, the magnetic interactions of the atoms increase in metallic complexes. It may be the reason of the negative values of the isotope effect exponent α in the present computation. Also, the electron-lattice interactions are not deeply involved in the binary alloys superconductors, which may causes the negative values of the isotope effect exponent α . Since the experimental value of α has not been reported in the literature so far, the present data of the isotope effect exponent α may be used for the study of ionic vibrations in the superconductivity of binary alloys superconductors.

3.11. SSP of $(Ni_{33}Zr_{67})_{1-x}M_x$ $(M = Ti, V, Co, Cu)$ metallic glasses

The input parameters of the $(Ni_{33}Zr_{67})_{1-x}M_x$ $(M = Ti, V, Co, Cu)$ metallic glasses are taken from the literature [11]. The presently calculated results of the superconducting state parameters (SSP) of $(Ni_{33}Zr_{67})_{1-x}M_x$ $(M = Ti, V, Co, Cu)$ ternary metallic glasses are tabulated in Table 11 with other such theoretical [96] and experimental [97] findings. The graphical representations of the superconducting properties state parameters (SSP) are also plotted in Figures 26-30, where these parameters are plotted against the concentration ' x ' of the third element (M) for the four different series of ternary metallic glasses.

It is seen from the present data of the superconducting properties of ternary metallic glasses that, addition of V, Co and Cu as third element (M) to a binary amorphous alloy '$Ni_{33}Zr_{67}$' decreases the parameters λ, α and N_OV where as the Coulomb pseudopotential (μ^*) increases for Ti, Co and Cu-based superconductors while those for V-based superconductors, the Coulomb pseudopotential (μ^*) decreases with concentration (z) of third element (M) and showing that the presence of third element (M) causes the suppression of the superconductivity in particular ternary glasses.

Table 11 that for $(Ni_{33}Zr_{67})_{1-x}M_x$ ($M = Ti, V, Co, Cu$) ternary metallic glasses, μ^* lies between 0.14 and 0.16, which is in accordance with McMillan [67], who suggested $\mu^* \approx 0.13$ for transition metals.

Decrease in λ, α and N_OV suggests weak coupling in these superconductors, the coupling being weakest for Co. This is in conformity with the fact that $(Ni_{33}Zr_{67})_{1-x}Co_x$ ternary metallic glasses do not remain superconducting for higher values of the concentration ' x ' of the third element (M). This may be due to change in influential electron band structure from 4d to 3d as suggested by Varma and Dynes [98].

Table 11. Superconducting state parameters of $(Ni_{33}Zr_{67})_{1-x}M_x$ metallic glasses

Amorphous superconductors	SSP	Present	Others [11]	Expt. [4]
$(Ni_{33}Zr_{67})_1Ti_0$	λ	0.5671	0.553	-
	μ^*	0.1519	0.14	-
	T_C (K)	2.6805	2.6734	2.68
	α	0.3080	0.341	-
	N_OV	0.2753	0.275	-
$(Ni_{33}Zr_{67})_{0.95}Ti_{0.05}$	λ	0.5703	0.558	-
	μ^*	0.1517	0.142	-
	T_C (K)	2.7302	2.7438	2.73
	α	0.3091	0.338	-
	N_OV	0.2757	0.276	-
$(Ni_{33}Zr_{67})_{0.90}Ti_{0.10}$	λ	0.5659	0.558	-
	μ^*	0.1514	0.143	-
	T_C (K)	2.6727	2.6948	2.67
	α	0.3069	0.332	-
	N_OV	0.2737	0.275	-
$(Ni_{33}Zr_{67})_{0.85}Ti_{0.15}$	λ	0.5633	0.558	-
	μ^*	0.1512	0.145	-
	T_C (K)	2.6517	2.6552	2.65
	α	0.3057	0.327	-
	N_OV	0.2725	0.274	-
$(Ni_{33}Zr_{67})_{0.80}Ti_{0.20}$	λ	0.5595	0.557	-
	μ^*	0.1510	0.146	-
	T_C (K)	2.6010	2.6137	2.60
	α	0.3035	0.322	-
	N_OV	0.2707	0.273	-
$(Ni_{33}Zr_{67})_{0.75}Ti_{0.25}$	λ	0.5569	0.557	-
	μ^*	0.1508	0.147	-
	T_C (K)	2.5821	2.5902	2.58
	α	0.3025	0.317	-
	N_OV	0.2696	0.272	-

Table 11. (Continued)

Amorphous superconductors	SSP	Present	Others [11]	Expt. [4]
$(Ni_{33}Zr_{67})_1V_0$	λ	0.5671	0.553	-
	μ^*	0.1519	0.14	-
	T_C (K)	2.6805	2.6734	2.68
	α	0.3080	0.341	-
	N_OV	0.2753	0.275	-
$(Ni_{33}Zr_{67})_{0.95}V_{0.05}$	λ	0.5471	0.545	-
	μ^*	0.1508	0.146	-
	T_C (K)	2.2523	2.2505	2.25
	α	0.2945	0.311	-
	N_OV	0.2647	0.266	-
$(Ni_{33}Zr_{67})_{0.90}V_{0.10}$	λ	0.5358	0.538	-
	μ^*	0.1496	0.151	-
	T_C (K)	2.0804	1.9498	2.08
	α	0.2899	0.285	-
	N_OV	0.2597	0.260	-
$(Ni_{33}Zr_{67})_{0.85}V_{0.15}$	λ	0.5185	0.534	-
	μ^*	0.1485	0.154	-
	T_C (K)	1.7911	1.7836	1.79
	α	0.2784	0.266	-
	N_OV	0.2515	0.256	-
$(Ni_{33}Zr_{67})_1Co_0$	λ	0.5671	0.553	-
	μ^*	0.1519	0.14	-
	T_C (K)	2.6805	2.6734	2.68
	α	0.3080	0.341	-
	N_OV	0.2753	0.275	-
$(Ni_{33}Zr_{67})_{0.95}Co_{0.05}$	λ	0.5556	0.541	-
	μ^*	0.1520	0.141	-
	T_C (K)	2.4024	2.4148	2.40
	α	0.2965	0.330	-
	N_OV	0.2682	0.268	-

Amorphous superconductors	SSP	Present	Others [11]	Expt. [4]
$(Ni_{33}Zr_{67})_{0.90}Co_{0.10}$	λ	0.5404	0.527	-
	μ^*	0.1524	0.142	-
	T_C (K)	2.1004	2.1103	2.10
	α	0.2812	0.316	
	N_OV	0.2602	0.260	-
$(Ni_{33}Zr_{67})_1Cu_0$	λ	0.5671	0.553	-
	μ^*	0.1519	0.14	-
	T_C (K)	2.6805	2.6734	2.68
	α	0.3080	0.341	-
	N_OV	0.2753	0.275	-
$(Ni_{33}Zr_{67})_{0.95}Cu_{0.05}$	λ	0.5596	0.543	-
	μ^*	0.1524	0.141	-
	T_C (K)	2.4326	2.4411	2.43
	α	0.2977	0.332	-
	N_OV	0.2699	0.270	-
$(Ni_{33}Zr_{67})_{0.90}Cu_{0.10}$	λ	0.5430	0.526	-
	μ^*	0.1529	0.141	-
	T_C (K)	2.0707	2.0787	2.07
	α	0.2812	0.317	-
	N_OV	0.2612	0.261	-
$(Ni_{33}Zr_{67})_{0.85}Cu_{0.15}$	λ	0.5339	0.517	-
	μ^*	0.1536	0.142	-
	T_C (K)	1.8701	1.8759	1.87
	α	0.2694	0.307	-
	N_OV	0.2561	0.255	-

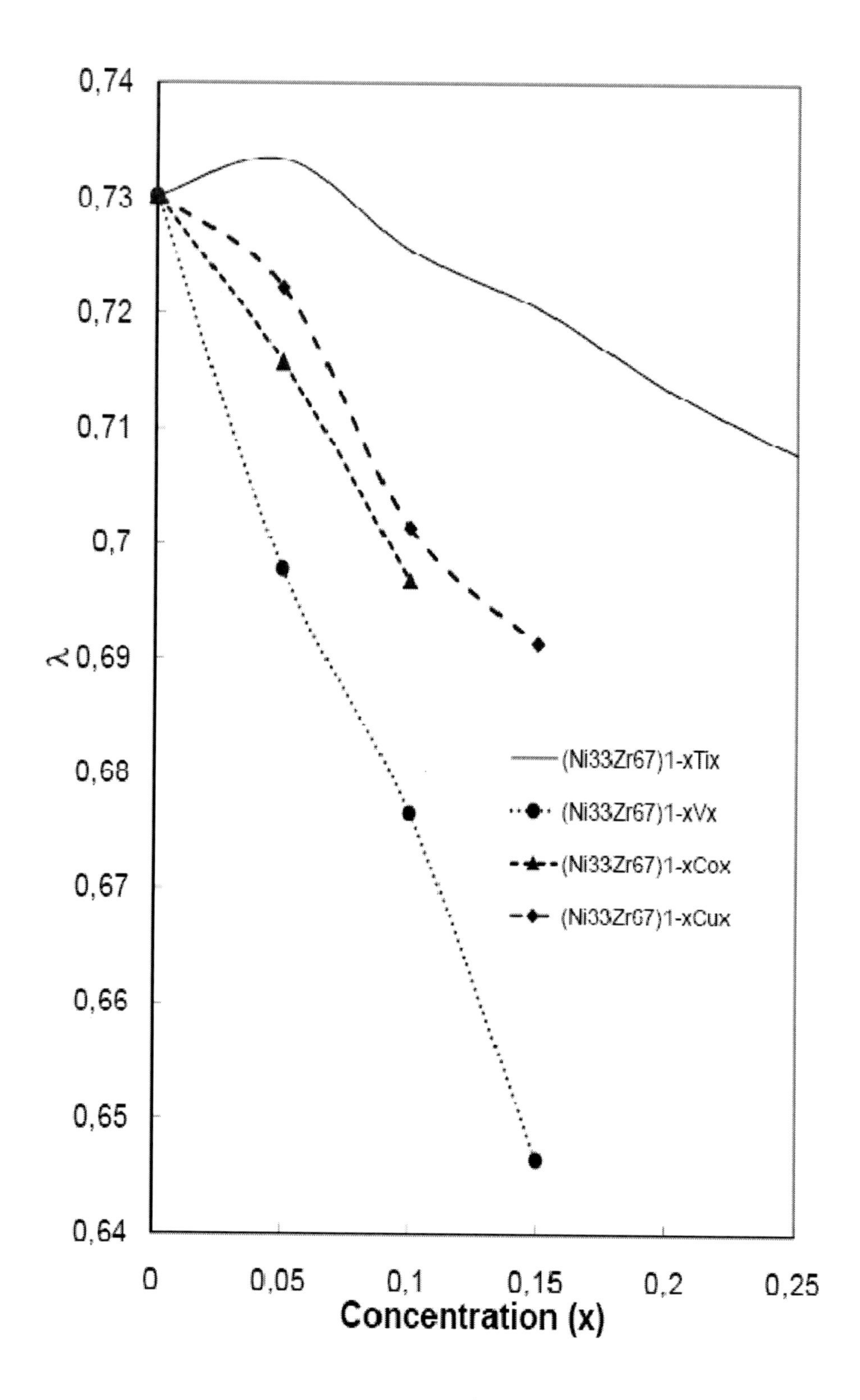

Figure 26. Electron-phonon coupling strength (λ) of $\left(Ni_{33}Zr_{67}\right)_{1-x}M_x$ metallic glasses.

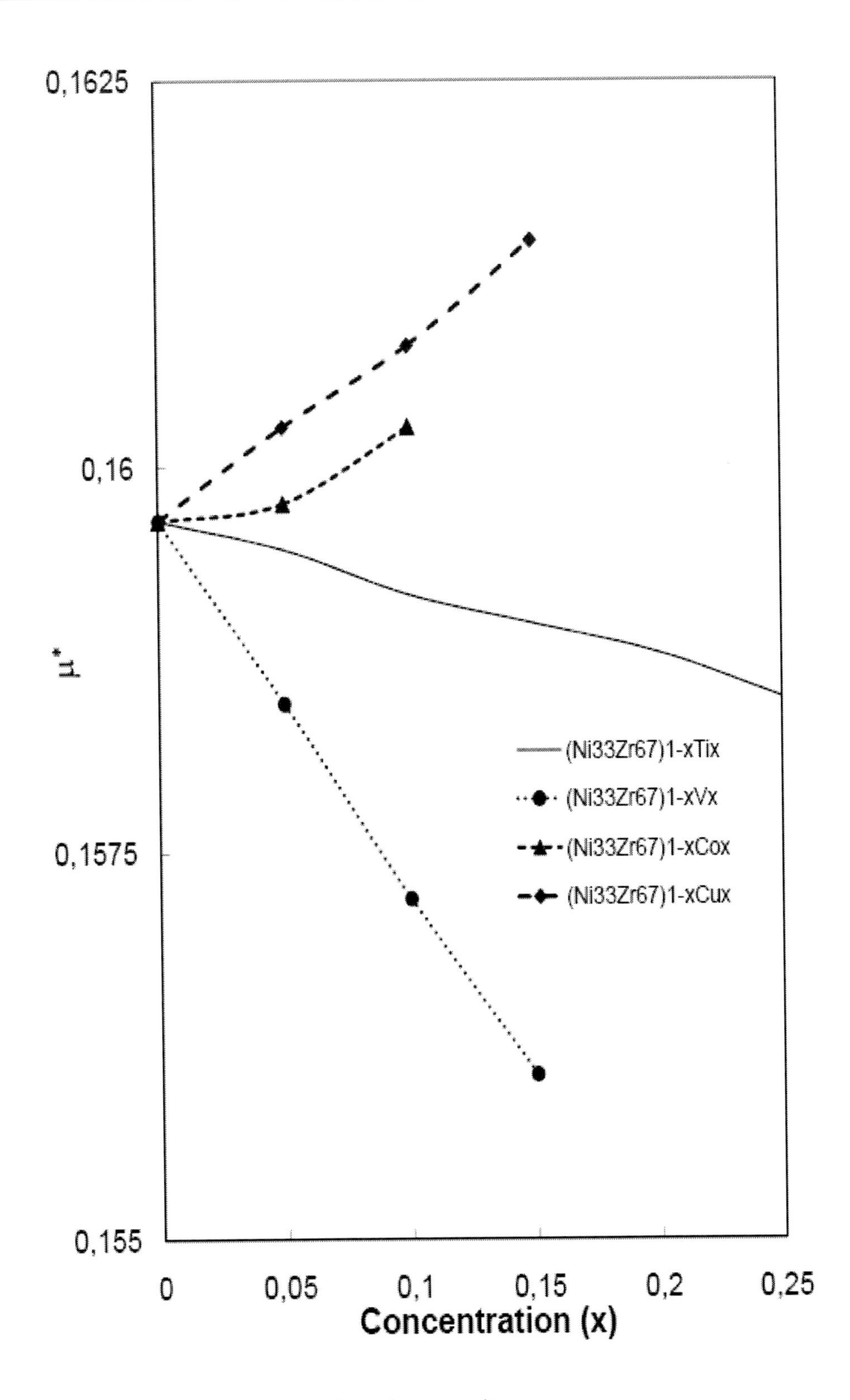

Figure 27. Coulomb pseudopotential (μ^*) of $\left(Ni_{33}Zr_{67}\right)_{1-x}M_x$ metallic glasses.

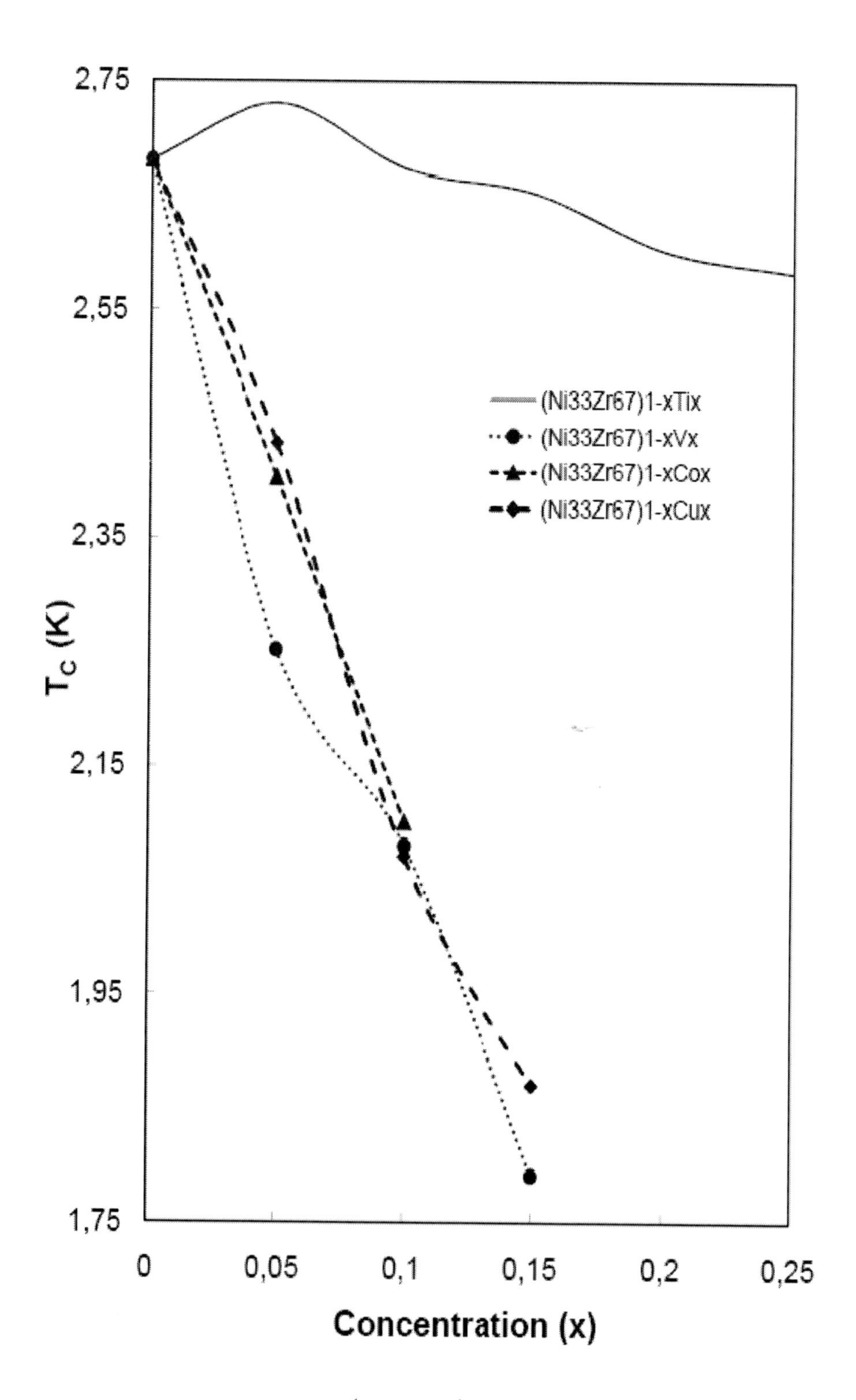

Figure 28. Transition temperature (T_C) of $(Ni_{33}Zr_{67})_{1-x}M_x$ metallic glasses.

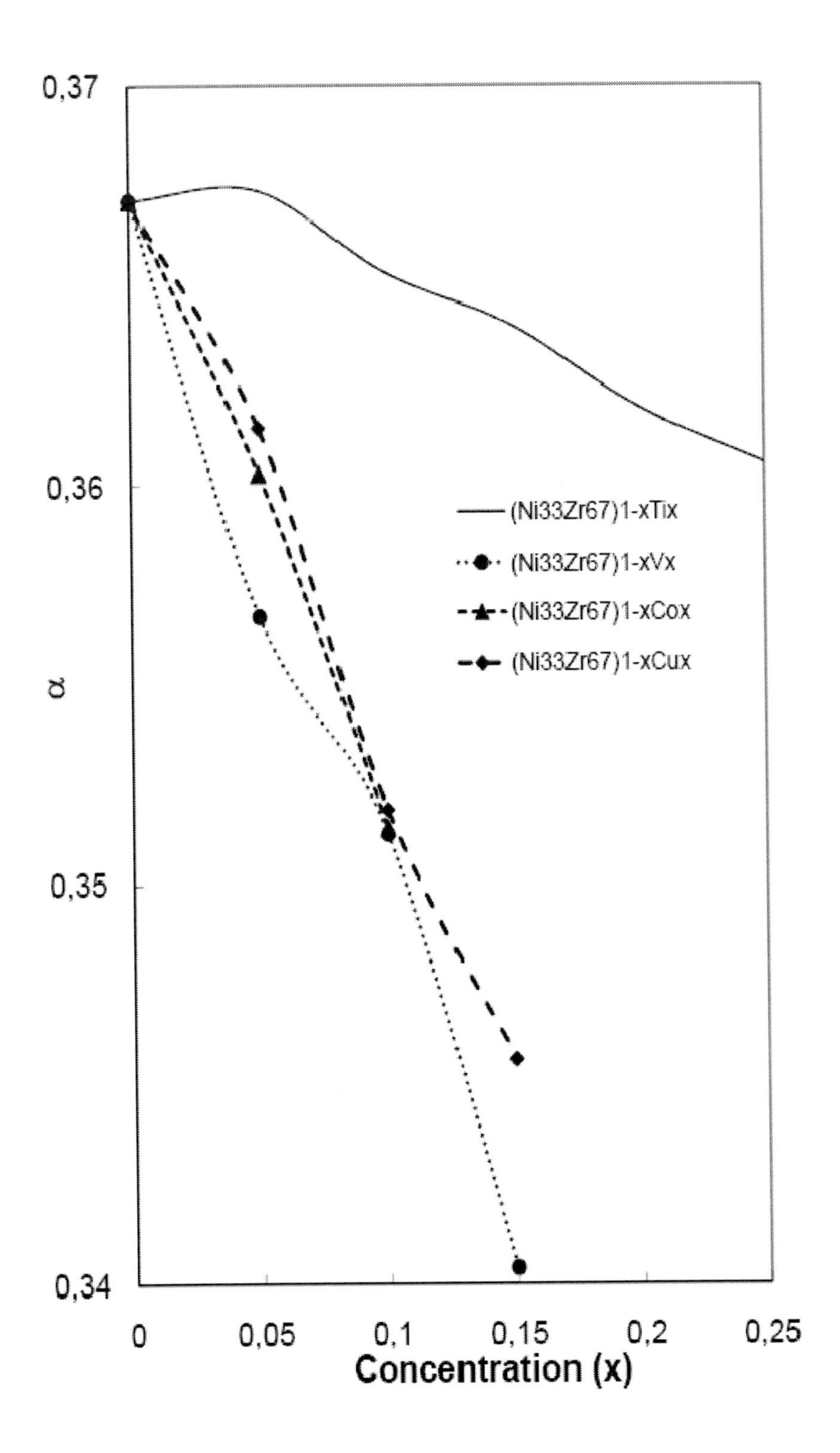

Figure 29. Isotope effect exponent (α) of $(Ni_{33}Zr_{67})_{1-x}M_x$ metallic glasses.

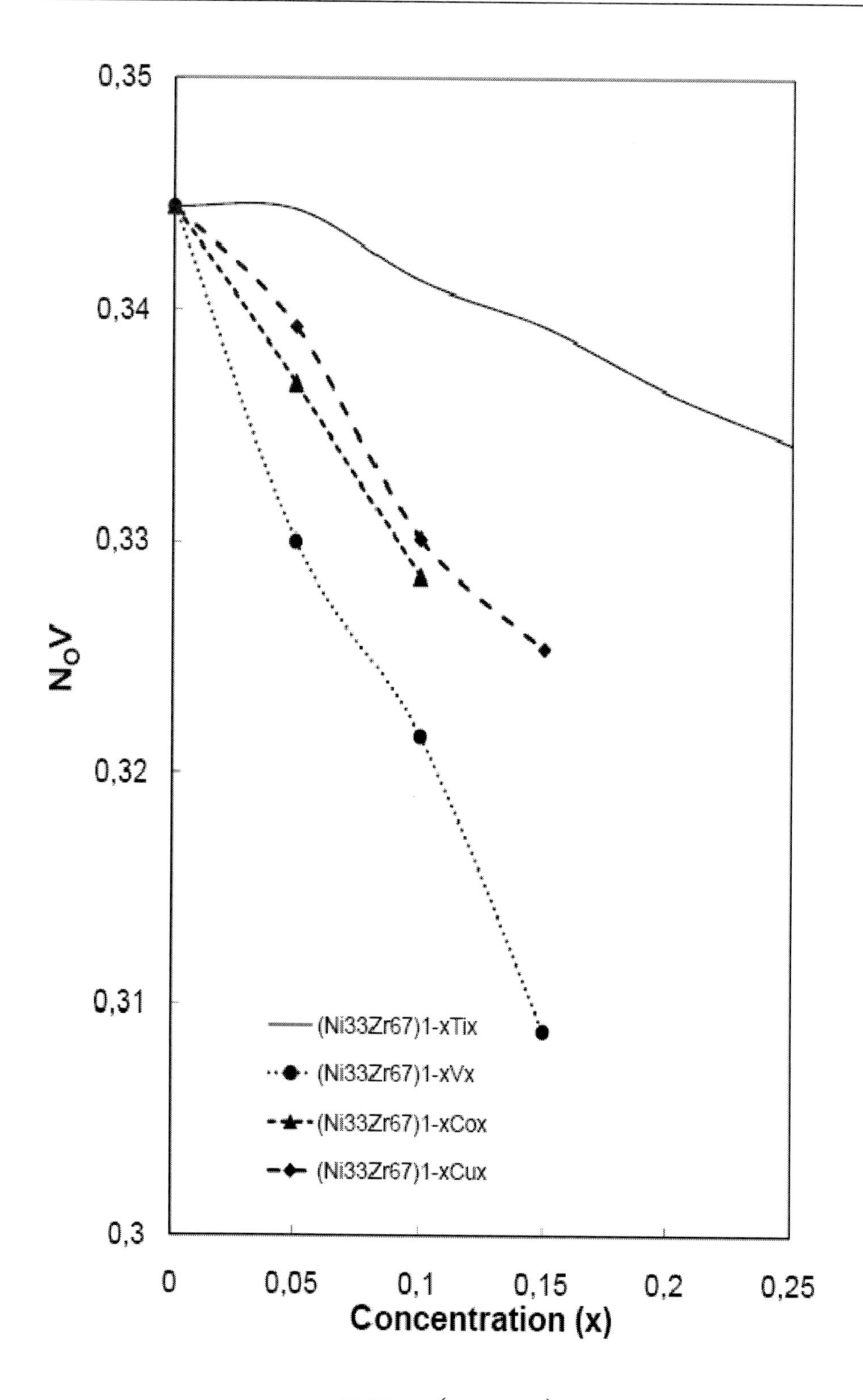

Figure 30. Effective interaction strength (N_OV) of $\left(Ni_{33}Zr_{67}\right)_{1-x}M_x$ metallic glasses.

It is also observed that, the transition temperature T_C decreases as V, Co or Cu is added to binary amorphous alloy '$Ni_{33}Zr_{67}$'. Both specific heat measurements and band structure calculation [99-102] reveal the decrease in density of states at E_F with the addition of the third element (M). Since, T_C is related to the modifications of density of states (DOS) at E_F, $N(E_F)$, decrease in T_C can be related to the modifications of DOS at the Fermi level, $N(E_F)$ [96]. Since, difference is noticed when Ti is added to the binary amorphous alloy '$Ni_{33}Zr_{67}$' (Fig. 28). In this case T_C rises initially and then decreases with the concentration 'x' of the third element (M), showing a peak at about $x = 0.05$. This indicates that on addition of Ti, 3d states grow near E_F and hence contribute substantially to the Fermi level, $N(E_F)$, favouring superconducting behaviour in this case [96].

It is also observed that superconductivity persists only for small values of x (i.e. $x \leq 0.25$) which is because the third element (M) considered here are all 3d-transition metals which have smaller band width and stronger localized character than Zr, thus they causes narrowing of bands in ternary system [96]. These narrow bands have magnetic instabilities which prevent superconductivity as suggested by Allen and Cohen [22]. The present results for the transition temperature T_C show an excellent agreement with the theoretical [96] and experimental [97] findings. Variation of the transition temperature T_C with the concentration 'x' of the third element (M) can be expressed by following quadratic formulae:

$$T_C(K) = -2.214x^2 + 0.033x + 2.698 \quad [((Ni_{33}Zr_{67})_{1-x}Ti_x], \tag{22}$$

$$T_C(K) = 14x^2 - 7.78x + 2.661 \quad [((Ni_{33}Zr_{67})_{1-x}V_x], \tag{23}$$

$$T_C(K) = -4.765x^2 - 5.324x + 2.680 \quad [((Ni_{33}Zr_{67})_{1-x}Co_x], \tag{24}$$

$$T_C(K) = 5x^2 - 6.33x + 2.693 \quad [((Ni_{33}Zr_{67})_{1-x}Cu_x]. \tag{25}$$

These quadratic formulae of T_C Eqs. (22-25) obtained in the present study closely resembles the quadratic T_C equations suggested by Sharma *et al.* [96] on the basis of experimental data for the ternary systems.

$$T_C(K) = -3.1393x^2 + 0.3014x + 2.6961 \quad [((Ni_{33}Zr_{67})_{1-x}Ti_x], \tag{26}$$

$$T_C(K) = 25.6700x^2 - 9.7907x + 2.6740 \quad [((Ni_{33}Zr_{67})_{1-x}V_x], \tag{27}$$

$$T_C(K) = -9.1800x^2 - 4.713x + 2.6734 \quad [((Ni_{33}Zr_{67})_{1-x}Co_x], \tag{28}$$

$$T_C(K) = 1.7190x^2 - 5.7677x + 2.6830 \quad [((Ni_{33}Zr_{67})_{1-x}Cu_x]. \tag{29}$$

These equations may be employed for predicting T_C values of other amorphous superconductors in these systems. The wide extrapolations predicts T_C for the hypothetical case of 'amorphous pure $Ni_{33}Zr_{67}$' alloy. From the overall comparisons of the presently computed results with those of Sharma *et al.* [96], it is noted that, the present results are found to be in good agreement with other experimental data [97]. Therefore, the results obtained from Sarkar *et al.* (S) [88] with EMC model potential [89] is produced consistent results of the superconductivity of ternary metallic glasses.

Sharma *et al.* [96] have fitted the parameter of the model potential r_C is fitted with the RPA form of screening with available experimental data of T_C [97] and this screening function has not included exchange and correlation effects while in the present case we have fitted the parameter of the model potential r_C is fitted with most recent Sarkar *et al.* (S) [88] local field correction functions with available experimental data of T_C [97]. Therefore, in the present case we have incorporated screening effects on the aforesaid properties. This is very much essential for drawing concrete remarks. The present results confirm the applicability of the EMC model potential and Sarkar *et al.* (S) local field correction function [88].

The biggest application right now for superconductivity is in producing the large volume, stable magnetic fields required for MRI and NMR. This represents a multi-billion US$ market for companies such as Oxford Instruments, Siemens etc. The magnets typically use low temperature superconductors (LTS) because high-temperature superconductors (HTS) are not yet cheap enough to cost effectively deliver the high, stable and large volume fields required, notwithstanding the need to cool LTS instruments to liquid helium temperatures. Superconductors are also used in high field scientific magnets because copper has a limit to the field strength it can produce.

The main differences and characteristics of the local field correction functions are played important role in the production of the superconducting state parameters (SSP) of amorphous superconductors. The Hartree (H) dielectric function [84] is purely static and it does not include the exchange and correlation effects. Taylor (T) [85] has introduced an analytical expression for the local field correction function, which satisfies the compressibility sum rule exactly. The Ichimaru-Utsumi (IU) local field correction function [86] is a fitting formula for the dielectric screening function of the degenerate electron liquids at metallic and lower densities, which accurately reproduces the Monte-Carlo results as well as it also satisfies the self consistency condition in the compressibility sum rule and short range correlations. In the present work Hartree (H) local field correction function [84] gives the qualitative agreement with the experimental data with EMC model potential [89] and found suitable in the present case. On the basis of Ichimaru-Utsumi (IU) local field correction function [86], Farid *et al.* (F) [87] and Sarkar *et al.* [88] have given a local field correction function. Hence, F-function represents same characteristic nature.

The effect of local field correction functions plays an important role in the computation of the electron-phonon coupling strength λ and the Coulomb pseudopotential μ^*, which makes drastic variation on the transition temperature T_C, the isotope effect exponent α and the effective interaction strength $N_O V$. The local field correction functions due to IU, F and S are able to generate consistent results regarding the superconducting state parameters (SSP) of the binary amorphous superconductors based on the superconducting (S), conditional superconducting (S') and non-superconducting (NS) elements of the periodic table as those obtained from more commonly employed H and T-functions. Thus, the use of these more promising local field correction functions is established successfully. The computed results of the isotope effect exponent α and the effective interaction strength $N_O V$ are not showing any abnormal values for all amorphous superconductors

According to Matthias rules [96, 97] the metallic superconductors having Z<2 do not exhibit superconducting nature. Hence, the metallic complexes having Z=1 are non-superconductors, but they exhibit superconducting nature in the present case. When we go from Z=1 to Z=8, the electron-phonon coupling strength λ changes with lattice spacing "a". Similar trends are also observed in the values of T_C for most of the metallic superconductors. Hence, a strong dependency of the superconducting state parameters (SSP) of the metallic superconductors on the valence 'Z' is found.

For comparison of superconducting state parameters (SSP), experimental data for most of the metallic superconductors are available in the literature. The presently computed results of the superconducting state parameters (SSP) are found qualitative agreement with the available experimental and theoretical data, which favours applicability of EMC model potential. In contrast with the reported studies, the present study spans the metallic superconductors based on the large number of the superconductors of the periodic table on a common platform of model potential. Hence, the present investigation provides an important set of theoretical data for these metallic superconductors which can be very useful for further comparison either with theory or experiment.

Lastly, we would like to emphasize the importance of involving a precise form for the pseudopotential. It must be confessed that although the effect of pseudopotential in strong coupling superconductor is large, yet it plays a decisive role in weak coupling superconductors i.e. those substances which are at the boundary dividing the superconducting and nonsuperconducting region. In other words, a small variation in the value of electron-ion interaction may lead to an abrupt change in the superconducting properties of the material under consideration. In this connection we may realize the importance of an accurate form for the pseudopotential.

CONCLUSION

The comparison of presently computed results of the superconducting state parameters (SSP) of the metallic superconductors based on the large number of elements of the periodic table with available experimental findings are highly encouraging, which confirms the applicability of the EMC model potential and different forms of the local field correction functions. A strong dependency of the superconducting state parameters (SSP) of metallic superconductors on the valence 'Z' is found. The experimentally observed values of the superconducting state parameters (SSP) are not available for the most of the metallic superconductors therefore it is difficult to drew any special remarks.

However, the comparison with other such theoretical and experimental data supports the present approach. Such study on superconducting state parameters (SSP) of other metallic complexes is in progress.

References

[1] A. V. Narlikar and S. N. Ekbote, Superconductivity and Superconducting Materials, South Asian Publishers, New Delhi - Madras (1983).

[2] S. V. Vonsovsky, Yu. A. Izyumov and E. Z. Kurmaev, Superconductivity of Transition Metals, their Alloys and Compounds, Springer-Verlag, Berlin-Heidelberg-New York (1982).

[3] P. B. Allen, Handbook of Superconductivity, Eds. C. P. Poole, Jr., Academic Press, New York (1999), p. 478.

[4] J. Bardeen, L. N. Cooper and J. R. Scrieffer, Phys. Rev. 108 (1957) 1175.

[5] R. Sharma and K. S. Sharma, Ind. J. Pure & Appl. Phys. 21 (1983) 725.

[6] R. Sharma and K. S. Sharma, Czech. J. Phys. B34 (1984) 325.

[7] R. Sharma, K. S. Sharma and L. Dass, Ind. J. Phys. A60 (1986) 373.

[8] R. Sharma, K. S. Sharma and L. Dass, Czech. J. Phys. B36 (1986) 719.

[9] R. Sharma, K. S. Sharma and L. Dass, Phys. Stat. Sol. (b) 133 (1986) 701.

[10] K. N. Khanna and P. K. Sharma, Phys. Stat. Sol. (b) 91 (1979) 251.

[11] K. N. Khanna and P. K. Sharma, Acta Phys. Pol. A57 (1980) 335.

[12] S. Sharma, K. S. Sharma and H. Khan, Czech J. Phys. 55 (2005) 1005.

[13] S. Sharma, K. S. Sharma and H. Khan, Supercond. Sci. Technol. 17 (2004) 474.

[14] S. Sharma, H. Khan and K. S. Sharma, Phys. Stat. Sol. (b) 241 (2004) 2562.

[15] S. Sharma and H. Khan, in Solid State Physics, Vol. 46, Eds. Sharma *et al.*, Allied Pub., New Delhi (2003), p. 635.

[16] S. Sharma, H. Khan and K. S. Sharma, Ind. J. Pure & Appl. Phys. 41 (2003) 301.

[17] M. Gupta, P. C. Agarwal, K. S. Sharma and L. Dass, Phys. Stat. Sol. (b) 211 (1999) 731.

[18] K. S. Sharma, R. Sharma, M. Gupta, P. C. Agarwal and L. Dass, in The Physics of Disordered Materials, Eds. Saksena *et al.*, NISCOM, New Delhi (1997), p. 95.

[19] P. C. Agarwal, M. Gupta, K. S. Sharma and L. Dass, in The Physics of Disordered Materials, Eds. Saksena *et al.*, NISCOM, New Delhi (1997), p. 102.

[20] M. Gupta, K. S. Sharma and L. Dass, Pramana -J. Phys. 48 (1997) 923.

[21] R. Sharma and K. S. Sharma, Supercond. Sci. Technol. 10 (1997) 557.

[22] P. B. Allen and M. L. Cohen, Phys. Rev. 187 (1969) 525.

[23] S. C. Jain and C. M. Kachhava, Phys. Stat. Sol. (b) 101 (1980) 619.

[24] S. C. Jain and C. M. Kachhava, Ind. J. Pure & Appl. Phys. 18 (1980) 489.

[25] S. C. Jain and C. M. Kachhava, Ind. J. Phys. A55 (1981) 89.

[26] K. S. Sharma and C. M. Kachhava, Solid Stat. Commun. 30 (1979) 719.

[27] K. S. Sharma, N. Bhargava, R. Jain, V. Goyal, R. Sharma and S. Sharma, Ind. J. Pure & Appl. Phys. 48 (2010) 59.

[28] P. N. Gajjar, A. M. Vora and A. R. Jani, Indian J. Phys. 78 (2004) 775.

[29] A. M. Vora, M. H. Patel, P. N. Gajjar and A. R. Jani, Pramana-J. Phys. 58 (2002) 849.

[30] P. N. Gajjar, A. M. Vora, M. H. Patel and A. R. Jani, Int. J. Mod. Phys. B17 (2003) 6001.

[31] A. M. Vora, M. H. Patel, S. R. Mishra, P. N. Gajjar and A. R. Jani, in Solid State Physics, Vol. 44, Eds. Chaplot *et al.*, Narosa, New Delhi (2001), p. 345.

[32] P. N. Gajjar, A. M. Vora and A. R. Jani, Mod. Phys. Lett. B18 (2004) 573.

[33] Aditya M. Vora, Physica C450 (2006) 135.

[34] Aditya M. Vora, Physica C458 (2007) 21; 43.

[35] Aditya M. Vora, J. Supercond. Novel Magn. 20 (2007) 355; 373; 387.

[36] Aditya M. Vora, Phys. Scr. 76 (2007) 204.

[37] Aditya M. Vora, Comp. Mater. Sci. 40 (2007) 492.

[38] Aditya M. Vora, J. Optoelec. Adv. Mater. 9 (2007) 2498.

[39] Aditya M. Vora, Chinese Phys. Lett. 24 (2007) 2624; 25 (2008) 2162.

[40] Aditya M. Vora, Front. Phys. China 2 (2007) 430.

[41] Aditya M. Vora, J. Tech. Phys. 48 (2007) 3.

[42] Aditya M. Vora, Central Euro. J. Phys. (2008) 223; 238; 253.

[43] Aditya M. Vora, J. Cont. Phys. (Armenian Acad. Sci.) 43 (2008) 231; 293.

[44] Aditya M. Vora, Int. J. Theor. Phys., Group Theory and Non. Opt. 12 (2008) 283; 13 (2008) 1.

[45] Aditya M. Vora, J. Ovonic Res. 4 (2008) 13.

[46] Aditya M. Vora, J. Phys. Chem. Sol. 69 (2008) 1841.

[47] Aditya M. Vora, Physica C468 (2008) 937, 2292.

[48] Aditya M. Vora, Sci. Tech. Adv. Mater. 9 (2008) 025017.

[49] Aditya M. Vora, Turkish J. Phys. 32 (2008) 199; 219.

[50] Aditya M. Vora, Tech. Phys. Lett. 34 (2008) 740.

[51] Aditya M. Vora, Moroccans J. Conden. Matter 10 (2008) 15.

[52] Aditya M. Vora, J. Shanghai Univ. (English Edition) 12 (2008) 311.

[53] Aditya M. Vora, Romanian J. Phys. 53 (2008) 885.

[54] Aditya M. Vora, J. Non Cryst. Sol. 354 (2008) 5022.

[55] Aditya M. Vora, Turkish J. Phys. 33 (2009) 57.

[56] Aditya M. Vora, Modern Phys. Lett. 22 (2009) 2881; 23 (2009) 217; 1443.

[57] Aditya M. Vora, Commun. Theor. Phys. 51 (2009) 533.

[58] Aditya M. Vora, EJTP 6 (2009) 357.

[59] Aditya M. Vora, Armenian J. Phys. (Armenian Acad. Sci.) 2 (2009) 213.

[60] Aditya M. Vora, Int. J. Theor. Phys., Group Theory and Non. Opt. 13 (2009)-in press.

[61] Aditya M. Vora, Chinese Phys. Lett. 27 (2010) 026102 (1).

[62] Aditya M. Vora, High Temperature 47 (2009) 635.

[63] Aditya M. Vora, Physica C469 (2009) 241.

[64] Aditya M. Vora, Physica C470 (2010) 475; African Phys. Rev. 4 (2010) 95.

[65] Aditya M. Vora, Bull. Mater. Sci. (2011)-in press; FIZIKA A (2011) - in press; J. Supercond. Novel Magn. (2011) - in press; J. Supercond. Novel Magn. (2011) - in press.

[66] Aditya M. Vora, In: Advances in Materials Science Research, Ed. Maryann C. Wythers, Vol. 3, Nova Science Publishers, Inc., New York (2011)-in press; J. Non-Cryst. Sol. 357 (2011) 2039.

[67] W. L. McMillan, Phys. Rev. 167 (1968) 313.

[68] J. J. Hopfield, Phys. Rev. 186 (1969) 443.

[69] V. Singh, H. Khan and K. S. Sharma, Indian J. Pure & Appl. Phys. 32 (1994) 915.

[70] R. C. Dynes, Phys. Rev. B2 (1970) 644.

[71] M. N. Ou, T. J. Yang, B. J. Chen, Y. Y. Chen and J. C. Ho, Solid State Commun. 142 (2007) 421.

[72] V. V. Pustovalov, N. V. Isaev, V. S. Fomenko, S. E. Shumilin, N. I. Kolobnev and I. N. Fridlyander, Cryogenics 32 (1992) 707.

[73] U. Mizutani and T. B. Massalski, Proc. R. Soc. Lond. A343 (1975) 375.

[74] U. Mizutani and T. B. Massalski, J. Phys. F: Met. Phys. 5 (1975) 2262.

[75] P. B. Allen and R. C. Dynes, Phys. Rev. B12 (1975) 905.

[76] Z. Altounian and J. O. Strom-Olsen, Phys. Rev. B27 (1983) 4149.

[77] I. Bakonyi, J. Non-Cryst. Sol. 180 (1995) 131.

[78] G. von Minnigerode and K. Samwer, Physica B108 (1981) 1217.

[79] R. Hasegawa, Glassy Metals: Magnetic, Chemical and Structural Properties, CRC Press, Florida, 1980.

[80] U. Mizutani, Prog. Mater. Sci. 28 (1983) 97.

[81] R. Hasegawa and L. E. Tanner, Phys. Rev, B16 (1977) 3925.

[82] R. van den Berg, H. von Löhneysen, A. Schröder, U. Mizutani and T. Matsuda, J. Phys. F: Met. Phys. 16 (1986) 69.

[83] P. Chatterjee, Can. J. Phys. 58 (1980) 1383.

[84] W. A. Harrison, Elementary Electronic Structure, (World Scientific, Singapore, 1999).

[85] R. Taylor, J. Phys. F: Met. Phys. 8 (1978) 1699.

[86] S. Ichimaru and K. Utsumi, Phys. Rev. B24 (1981) 7386.

[87] B. Farid, V. Heine, G. Engel and I. J. Robertson, Phys. Rev. B48 (1993) 11602.

[88] A. Sarkar, D. Sen, H. Haldar and D. Roy, Mod. Phys. Lett. B12 (1998) 639.

[89] N. W. Ashcroft, Phys. Lett. 23, 48 (1966).

[90] G. M. Eliashberg, Zh. Eksp. Teor. Fiz. 38 (1960) 966; Zh. Eksp. Teor. Fiz. 39 (1960) 1437; Sov. Phys. JEPT 11 (1960) 696; Sov. Phys. JEPT 12 (1961) 1000.

[91] W. H. Butler, Phys. Rev. B15 (1977) 5267.

[92] G. D. Gaspari and B. L. Gyorffy, Phys. Rev. Lett. 28 (1972) 801.

[93] P. Moral and P. W. Anderson, Phys. Rev. 125 (1962) 1263.

[94] J. S. Rajput and A. K. Gupta, Phys. Rev. 181 (1969) 743.

[95] N. P. Kovalenko, Yu. P. Krasny and U. Krey, Physics of Amorphous Metals (Wiley-VCH, Berlin, 2001).

[96] S. Sharma, H. Khan and K. S. Sharma, Czech. J. Phys. 55 (2005) 1005.

[97] Y. Yamada, Y. Itoh and U. Mizutani, Mater. Sci. Engg. 99 (1988) 289.

[98] C. M. Varma and R. C. Dynes, in: Superconductivity in d- and f-Band Metals, Eds. D. H. Douglass, Plenum Press: New York (1976) p. 507.

[99] U. Mizutani, C. Mishima and T. Goto, J. Phys. Cond. Matter 1 (1989) 1831.

[100] R. Zehringer, P. Oelhafen, H.-J. Guntherodt, Y. Yamada and U. Mizutani, Mater. Sci. Engg. 99 (1988) 317.

[101] U. Mizutani, U. Mizutani and C. Mishima, Solid State Commun. 62 (1987) 641.

[102] Y. Yamada, Y. Itoh and U. Mizutani, Mater. Sci. Engg. 99 (1988) 289.

[103] B. T. Matthias, Progress in Low Temperature Physics, Eds. C. J. Gorter (North Holland, Amsterdam, 1957), Vol. 2.

[104] B. T. Matthias, Physica 69 (1973) 54.

In: Superconductivity
Editor: Vladimir Rem Romanovskii

ISBN 978-1-61324-843-0

Chapter 4

NORMAL-STATE BAND SPECTRUM IN CHAIN-FREE HIGH-TEMPERATURE SUPERCONDUCTORS: MECHANISMS OF MODIFICATION UNDER CHANGING SAMPLE COMPOSITION AND INFLUENCE OF THE NORMAL-STATE PARAMETERS ON THE CRITICAL TEMPERATURE

V. E. Gasumyants and O. A. Martynova
St. Petersburg State Polytechnical University
Polytechnicheskaya, 29, St. Petersburg, Russia

ABSTRACT

We present the results of the systematic comparative study of the thermopower, S, normal-state band spectrum, and superconducting properties for chain-free HTSC's of Bi-, Tl-, and Hg-systems. We have analyzed a large set of the experimental results on the $S(T)$ dependences in samples of all systems in case of increasing both number of copper-oxygen layers, n, and level of non-isovalent doping for different lattice positions. Based on systematization of all these data, we have revealed the common tendencies in $S(T)$ modification under doping both in the underdoped and overdoped regimes.

All the experimental data on thermopower were quantitatively analyzed within a narrow-band model allowing us to determine the main parameters of the band spectrum and charge-carrier system (the effective bandwidth, W_D, degree of the band filling with electrons, degree of state localization, and band asymmetry degree). We have identified the above parameters for all the studied samples and then analyzed the change in their values under variation of sample composition within different HTSC-families.

We have observed that for the samples of near-optimally doped compositions the values of the band parameters for all the systems are close to each other, that points to the fact that the band spectrum structures in Bi-, Tl-, and Hg-based HTSC's are rather alike in general. Besides, a correlation between the variations of the bandwidth and the T_c value with n was observed for all the studied systems.

We have also scaled the parameters of the band spectrum and charge-carrier system, as well as the T_c behavior under doping which level was estimated for samples with different types of impurities based on the thermopower value at T=300K. For all the three systems these values change in an analogous way, but differently in the underdoped and overdoped regimes indicating various mechanisms of the band spectrum modification in these two doping regimes. A qualitative distinction in influence of different impurities is discussed and explained with respect to their type and positions occupied in a lattice.

Analysis of the results has shown that the T_c value in all cases is directly related to the value of the density-of-states at the Fermi level, $D(E_F)$. We have discussed two different mechanisms of the band spectrum modification affecting the critical temperature in an opposite way. Fist, the disordering in the lattice induced by a rise of the defect number leads to a band broadening according to the Anderson localization mechanism that results in decreasing $D(E_F)$. Second, increasing number of copper-oxygen layers forming the band leads to a general rise in the density-of-states and, as a result, $D(E_F)$ increases. The relative contribution of these two mechanisms in band spectrum modification and T_c variation under increasing n and/or doping level is discussed based on all the results obtained.

Finally, we have observed the scaling behavior of T_c in the underdoped regime for each of studied HTSC-families. The variation of the correlation curve $T_c(W_D)$ with increasing number of CuO_2 layers, as well as with a rise of the maximal T_c in the system is briefly discussed.

1. INTRODUCTION

Since the discovery of high-temperature superconductivity in 1986, more than 50 various oxide systems based on yttrium, bismuth, thallium, mercury, neodymium and some metals have been found to be superconducting. Like in case of other materials, the doping method is widely applied for studying the properties of high-temperature superconductors. The bismuth-based $Bi_2Sr_2Ca_{n-1}Cu_nO_{2n+4+\delta}$, thallium-based $Tl_2Ba_2Ca_{n-1}Cu_nO_{2n+4+\delta}$ and mercury-based $HgBa_2Ca_{n-1}Cu_nO_{2n+2+\delta}$ HTSC-systems are very interesting objects to study the changes in different properties under various modifications of material structure and a relationship between parameters of HTSC-materials in the normal and superconducting states. First, different methods of oxygen content change are rather easily realized technologically for these systems. In particular, the annealing regimes variation allows obtaining the compositions with oxygen excess that is especially important. This gives the possibility to trace properties transformation in the above systems in a wide range of doping, namely, from the underdoped regime characterized by the appearance of superconductivity and consecutive improvement of superconducting properties to a superconductor with the maximum temperature of the superconducting transition, T_c (optimally doped compositions), and further to the overdoped regime where superconductivity is suppressed and then totally disappears. Secondly, presence of phases with various number of copper-oxygen layers n in each of these HTSC-families allows to carry out the comparative analysis of changes in their properties both in the normal and in the superconducting state at varying number of these layers. Note that this number is known to be the main structural element responsible for the high-temperature superconductivity effect in all HTSC-materials. Both these directions are rather important both for studying the genesis of the band responsible for the properties of these compounds and for clarifying the influence of the normal-state parameters on the T_c value in high-temperature superconductors.

A huge set of experimental data on the influence of different types of deviations from the optimally doped compositions of Bi , Tl-, and Hg-based HTSC's on their properties has been published. However, the mechanisms of their modification accompanying the transition from the underdoped regime to the overdoped one, as well as the conditions needed for optimal superconducting properties in each of these HTSC-systems remain still unclear. In particular, the mechanism of doping effect on the electron transport phenomena for various impurities within a specific phase is not understood yet, and numerous available experimental data for each of the above systems are not systematized. It should be noted that the quantitative analysis of the experimental data is quite difficult owing to the complicated crystal and electronic structure of Bi-, Tl-, and Hg-based HTSC's and a high probability of the defect formation in these system, especially in case of phases with $n>1$. For this reason the exact theoretical calculations of the band spectrum structure for the considered HTSC-families represent an extremely difficult problem, so most of authors are restricted only to a qualitative analysis of the doping influence on the peculiarities of the electron transport. At the same time it is well-known that the reliable information about band spectrum parameters can be obtained by quantitative analysis of the temperature and concentration dependences of transport coefficients. However, for realizing such an approach one should have some preliminary information on basic features of the energy spectrum because they strongly affect the method to be used for the electron transport description. In the absence of such data this approach can be realized with help of phenomenological models based on some realistic assumptions and giving a possibility to explain the observed transport coefficients' behaviour. If such a model contains a small number of parameters with a clear physical meaning, being applied to the analysis of the experimental results, it allows to determine quantitative characteristics of the band spectrum.

All the transport coefficients in HTSC-material demonstrate unusual behaviour when compared to the classical objects of the solid state physics (first of all, metals and semiconductors). This was observed since the very beginning of investigations of these compounds and a description of transport coefficients change with temperature and/or sample composition can be found in the reviews published long ago (see, for example, [1]-[3]). To describe the temperature dependences of transport coefficients in the normal state, it has been proposed to use the approaches based on various assumptions including variants of single- and two-band models of the band spectrum (see, for example, [4]-[14]), as well as rather unconventional models that suggest the existence of specific features in the properties of the charge-carrier system, for example, the presence of bosons at temperatures above the superconducting transition temperature [15]-[17]. These models contain parameters which are different in a physical meaning and, in a number of cases, lead to different conclusions regarding structural features of the band spectrum and the character of its modification under doping. Some of the above models contain a lot of fitting parameters, just a few of which have a clear physical meaning. Besides, they often aim to describe a single transport coefficient or can be applied only to a specific HTSC-system.

Earlier we have proposed and developed the approach to the electron transport analysis based on a narrow-band model [18], [19]. The advantages of this model are: (i) it contains only three or four parameters and each of them has a specific physical interpretation being directly related to the parameters of the band spectrum or charge-carrier system; (ii) it allows to describe qualitatively the temperature dependences of the four transport coefficients (the resistivity, ρ, thermopower, S, Hall, R_H, and Nernst, Q, coefficients) and to determine the model parameters values from the quantitative analysis of $S(T)$ and $Q(T)$ dependences; (iii) it can be successfully applied to different HTSC-families.

In connection with the above matter, we present here the results of systematic comparative analysis of the thermopower behavior for the chain-free HTSC's of Bi-, Tl-, and Hg-based systems with different changes in sample compositions and various number of CuO_2 layers. All the experimental results are analyzed in the framework of the narrow-band model in order to determine the parameters of the band spectrum and charge-carrier system for each sample with a given composition. A comparative quantitative analysis of all the results obtained allows us to reveal both common peculiarities and specific features of the band spectrum structure in the studied materials as well as the mechanisms of its transformation under different deviations from the optimally doped compositions in the underdoped and overdoped regimes and with increasing number of CuO_2 layers. The results on modification of the normal-state parameters are discussed in comparison with the changes in the critical temperature in order to establish a correlation between them and to elucidate the possible mechanisms of the influence of parameters of the charge-carrier system on the superconducting properties of the studied HTSC-families.

2. The Narrow-Band Model

Our approach to the analysis of the transport coefficients in HTSC-materials was first proposed in [20] for explanation of the main peculiarities in the temperature dependences of resistivity and thermopower characteristic of near optimally doped $YBa_2Cu_3O_y$ superconductors – linearity of $\rho(T)$ and a weak temperature dependence of S. Then the narrow-band model was shown to be applicable for qualitative description of the $\rho(T)$, $S(T)$, and $R_H(T)$ dependences for samples of this system with an oxygen content varied in a wide range, as well as for determination of some band spectrum parameters based on the quantitative analysis of thermopower [21]. Afterwards, we have repeatedly and successfully used this model for analyzing the behavior of these transport coefficients in $YBa_2Cu_3O_y$ with different types of non-isovalent doping [18], [19], [22]-[33], as well as in other HTSC-families including Bi- [19], [34]-[36], Tl- [37], [38], and Hg-based [19], [39], [40] systems, though with an additional assumption. Moreover, based on the same assumptions on the principal peculiarities of the band spectrum structure we have developed a method of quantitative analysis of the normal-state Nernst coefficient in HTSC-materials [41]. Applying this approach to the experimental results for doped $YBa_2Cu_3O_y$ we were able not only to explain the observed $Q(T)$ dependences, but also to analyze them together with the $S(T)$ results in order to obtain information on additional peculiarities of the band spectrum structure and the charge-carrier mobility values [42]. The validity of our model was confirmed by other authors who applied it to the analysis of the experimental data on the thermopower [43]-[56] and Hall coefficient [57]-[59] in HTSC's of different systems. All the referred results make it possible to consider the narrow-band model as an efficient tool for comparative analysis of the changes in the normal-state parameters under different variations of sample composition in the studied HTSC-systems.

As mentioned above, the detailed description of the narrow-band model and its application to the quantitative analysis of the experimental data on the transport properties of HTSC's can be found elsewhere [18], [19]. Here we will describe briefly only the key points of this model and introduce its parameters which will be determined and discussed hereafter in case of studied systems.

The narrow-band model is based on the assumption that the band structure of high-T_c superconductors contains a sharp peak of the density-of-states (DOS) function, located near the Fermi level, E_F. This may be either a single narrow band or a sharp DOS peak on the background of a wide band. In the latter case if the magnitude of this peak is considerably superior to the background DOS value and the Fermi level is located inside this peak, it is its narrowness that mainly determines the peculiarities of the transport phenomena. Note that a similar assumption about the narrowness of the energy interval responsible for the conduction process in HTSC-materials has been repeatedly used in its different aspects by other authors to explain qualitatively the transport properties (see, for example, [4], [10]-[12], [60], [61]). Besides, different experimental results and some theoretical calculations point out the possibility of existence of such a narrow DOS peak in the band spectrum of HTSC-materials. A detailed list of such papers confirming indirectly the validity of our approach to the transport coefficients' analysis can be found elsewhere [18], [19].

The presence of the above features in the band spectrum should be taken into account when calculating the transport coefficients using the standard kinetic integrals.

$$\rho = \frac{1}{\langle\sigma\rangle}\cdot\frac{1+\exp\left(-2\mu^*\right)+2\exp\left(-\mu^*\right)\cosh W_\sigma^*}{2\exp\left(-\mu^*\right)\sinh W_\sigma^*}, \tag{1}$$

$$S = -\frac{k_B}{e}\left\{\frac{W_\sigma^*}{\sinh W_\sigma^*}\left[\exp\left(-\mu^*\right)+\cosh W_\sigma^* - \frac{1}{W_\sigma^*}\left(\cosh\mu^* + \cosh W_\sigma^*\right)\times\right.\right.$$
$$\left.\left.\times\ln\frac{\exp\left(\mu^*\right)+\exp\left(W_\sigma^*\right)}{\exp\left(\mu^*\right)+\exp\left(-W_\sigma^*\right)}\right]-\mu^*\right\}, \tag{2}$$

$$R_H = \frac{\langle\sigma_H\rangle}{\langle\sigma\rangle^2}\frac{\exp\left(\mu^*\right)-1}{\exp\left(\mu^*\right)+1}\frac{\cosh\left(\mu^*/2+W_\sigma^*/2\right)\cosh\left(\mu^*/2-W_\sigma^*/2\right)}{\cosh^2\left(W_\sigma^*/2\right)}, \tag{3}$$

where k_B is the Boltzmann constant; e is the elementary charge; $W_D^* \equiv W_D/2k_BT$, $W_\sigma^* \equiv W_\sigma/2k_BT$, W_D is the total effective bandwidth, W_σ is the effective width of an energy interval of the electrons responsible for the conduction process; $\langle\sigma\rangle$ and $\langle\sigma_H\rangle$ are the band-average values of the $\sigma(E)$ and $\sigma_H(E)$ functions, correspondingly; $\mu^* = \mu/k_BT$, μ is the chemical potential calculated as

$$\mu^* = \ln\frac{\sinh\left(FW_D^*\right)}{\sinh\left[\left(1-F\right)W_D^*\right]} - bW_D, \quad (4)$$

According to our previous results [18], [19], [21] if the band responsible for the conduction process is narrow, one can use the simplest approximations for the DOS, $D(E)$, differential conductivity, $\sigma(E)$, and Hall conductivity, $\sigma_H(E)$, functions by rectangles of different widths (see Figure 1). Within those approximations it is easy to derive the following analytical expressions describing the temperature dependences of transport coefficients:
F is the degree of the band filling with electrons that is equal to the ratio of the total number of electrons, n, to the total number of states, N, in the band; b is the band asymmetry degree which is a normalized difference between the centers of rectangles approximating the $D(E)$ and $\sigma(E)$ functions . For simplicity, we assume in Eq. (3) that $W_\sigma = W_{\sigma H}$. In the general case these parameters can be different and the $W_{\sigma H}$ value will additionally influence the calculated $R_H(T)$ dependence. Besides, we do not present here the formula for the $Q(T)$ calculations. As shown in [41], when analyzing the Nernst coefficient it is necessary to take into account the character of the dispersion law that modifies the Eq. (3) used for the $R_H(T)$ description. Detailed theoretical analysis of the Nernst coefficient in the case of a narrow conduction band is presented in [41]. The results of its application to the experimental data analysis for doped $YBa_2Cu_3O_y$ can be found in [42].

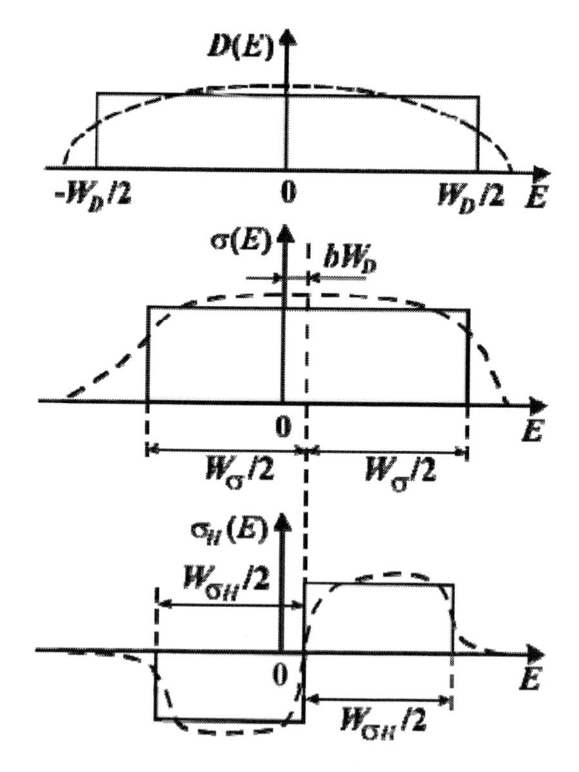

Figure 1. Model approximations of the density-of-states $D(E)$, differential conductivity $\sigma(E)$, and Hall conductivity $\sigma_H(E)$ functions in the framework of the narrow-band model.

As seen from Eqs. (1)-(4), the temperature dependences of three transport coefficients can be calculated by using four model parameters (F, W_D, W_σ, b). As shown in [18], [19], these formulas allowed to obtain the calculated $\rho(T)$, $S(T)$, and $R_H(T)$ curves well describing all their specific features characteristic of HTSC-materials of different systems. It is necessary to emphasize that in the framework of the above approach the temperature dependence of thermopower can be calculated in absolute units, while the temperature dependences of resistivity and Hall coefficient can only be approximated within constant and generally unknown factors. Besides, due to the nature of the Seebeck effect the value and temperature dependence of thermopower are characteristic of the material, but not of a specific sample with its defect structure influencing the $\rho(T)$ and $R_H(T)$ results. These are the reasons why the thermopower is the most informative coefficient when the narrow-band model is used. By varying the model parameters we can achieve a good quantitative agreement of the experimental and calculated results [18], [19] for samples of different compositions. This provides a possibility to determine the values of all the model parameters characterizing the band structure and properties of the charge-carrier system of HTSC-materials in the normal state. Note that when discussing the effect of doping influence it is preferably to use the parameter $C \equiv W_\sigma/W_D$ instead of W_σ because its change is related to the variation of the degree of the charge carrier localization, namely the less the C value the more the fraction of localized carriers.

According to our previous results, the conduction band in the $YBa_2Cu_3O_y$ system is close to be symmetric ($b = 0$) [18], [19], [23]-[25], [27], [29]. In such a case each of the three other model parameters influences different peculiarities of the $S(T)$ dependence so that their values can be determined unambiguously from the experimental data for samples with a given composition. However, for explaining the thermopower results for other HTSC-systems it is necessary to take into account a slight band asymmetry, i.e. to use the four-parameter version of the model [19], [34]-[40], [62]. Introduction of the fourth parameter b broadens the range within which the other parameters can be varied, thus resulting in an ambiguity of their determination. This is why we will present below the calculation results not only by points indicating the parameters for the best-fitted $S(T)$ curves, but also by error bars presenting a range of possible parameter values giving a satisfactory agreement of the experimental and calculated $S(T)$ dependences for each specific sample. This can be considered as an error in determination of the model parameters values when using the asymmetric narrow-band model. Let us note that these errors are quite small so that we can unambiguously establish the tendencies in variation of the band spectrum and charge-carrier system parameters under different changes in the sample composition as shown below.

3. EXPERIMENTAL DATA FOR THE ANALYSIS

For the analysis of the band spectrum structure and transformation in chain-free Bi-, Tl-, and Hg-based HTSC's we have used the available experimental data on the thermopower behavior and the T_c values for the samples of those systems with different compositions. Specifically, we have analyzed the $S(T)$ dependences for the following systems: $Bi_2Sr_2Ca_{n-1}Cu_nO_y$ (n = 1,2,3) [34], [63]-[65], $Bi_2Sr_{2-x}La_xCuO_y$ [66]-[68], $(Bi,Pb)_2(Sr,La)_2CuO_y$ [69], $Bi_2Sr_2Ca_{1-x}R_xCu_2O_y$ (R = Y,Pr,Gd) [34], [70]-[72], $Bi_2Sr_{2-x}K_xCaCu_2O_y$ [73],

$Bi_{2-x}Pb_xSr_2Ca_2Cu_3O_y$ [74], $Tl_2Ba_2Ca_{n-1}Cu_nO_y$ (n = 1,2,3) [75]-[80], $Tl_2Ba_2Ca_{1-x}R_xCu_2O_y$ (R = Y,Pr) [81], [82], $Tl_2Ba_2Ca_{1-x}Y_xCu_{2-y}Co_yO_z$ [83], $Tl_{1.7}Ba_2Ca_{2.3}Cu_3O_y$ [78], $Tl_{1.7}Ba_2Ca_{3.3}Cu_4O_y$ [84], $HgBa_2Ca_{n-1}Cu_nO_y$ (n = 1,2,3) [85], $HgBa_2CuO_y$ [86], $HgBa_2Cu_{0.97}Zn_{0.03}O_y$ [87], [88], $Hg_{0.5}Fe_{0.5}Sr_2Ca_{1-x}Y_xCu_2O_y$ [89]. In most of cases the authors of the above papers used ceramics samples prepared by the standard for each system technology, excluding single-crystal samples of $Bi_2Sr_2Ca_{1-x}R_xCu_2O_y$ (R = Pr,Gd) [72], $(Bi,Pb)_2(Sr,La)_2CuO_y$ [69], and $Bi_2Sr_2Ca_2Cu_3O_y$ [63] systems. Note that various regimes of the final sample annealing resulted in a different oxygen subsystem condition in the investigated samples.

As seen from the above list of HTSC-systems, for our analysis we have used the experimental results obtained for the phases of Bi-, Tl-, and Hg-based HTSC's with different numbers of CuO_2 layers. Besides, within each phase we used a large data set obtained for the samples with different oxygen content and/or different type and level of cation doping. This provided a possibility to perform the systematic investigation of transport and superconducting properties in the studied systems. Then, based on the analysis of the $S(T)$ dependences within the narrow-band model we obtained the reliable information on the band spectrum structure and charge-carrier system parameters as well as on their modification both in different doping regimes and with increasing number of CuO_2 layers.

Table 1. Compositions of the studied samples for different doping regimes

	Bi-based HTSC	Tl-based HTSC	Hg-based HTSC
overdoped regime	$Bi_2Sr_2CuO_y$, $Bi_2Sr_{2-x}La_xCuO_y$, $Bi_2Sr_2Ca_{1-x}Y_xCu_2O_y$, $Bi_2Sr_2Ca_{1-x}R_xCu_2O_y$ (R = Y, Pr, Gd), $Bi_2Sr_{2-x}K_xCaCu_2O_y$, $(Bi,Pb)_2(Sr,La)_2CuO_y$	$Tl_2Ba_2CuO_y$	$HgBa_2CuO_y$, $HgBa_2CaCu_2O_y$, $Hg_{0.5}Fe_{0.5}Sr_2Ca_{1-x}Y_xCu_2O_y$
near-optimally doped regime	$Bi_2Sr_2CuO_y$, $Bi_2Sr_{2-x}La_xCuO_y$, $Bi_2Sr_2Ca_{1-x}Y_xCu_2O_y$, $Bi_2Sr_2Ca_{1-x}R_xCu_2O_y$ (R = Y, Pr, Gd), $Bi_2Sr_{2-x}K_xCaCu_2O_y$, $Bi_2Sr_2Ca_2Cu_3O_y$, $(Bi,Pb)_2(Sr,La)_2CuO_y$	$Tl_2Ba_2Ca_{n-1}Cu_nO_y$ (n = 1,2,3,4), $Tl_2Ba_2Ca_{1-x}R_xCu_2O_y$ (R = Y, Pr), $Tl_2Ba_2Ca_{1-x}Y_xCu_{2-y}Co_yO_z$, $Tl_{1.7}Ba_2Ca_{2.3}Cu_3O_y$, $Tl_{1.7}Ba_2Ca_{3.3}Cu_4O_y$	$HgBa_2Ca_{n-1}Cu_nO_y$ (n = 1,2,3), $HgBa_2Cu_{0.97}Zn_{0.03}O_y$, $Hg_{0.5}Fe_{0.5}Sr_2Ca_{1-x}Y_xCu_2O_y$
underdoped regime	$Bi_2Sr_{2-x}La_xCuO_y$, $Bi_2Sr_2Ca_{1-x}R_xCu_2O_y$ (R = Y, Pr, Gd), $Bi_2Sr_2Ca_2Cu_3O_y$	$Tl_2Ba_2Ca_{1-x}R_xCu_2O_y$ (R = Y, Pr), $Tl_2Ba_2Ca_{1-x}Y_xCu_{2-y}Co_yO_z$, $Tl_{1.7}Ba_2Ca_{3.3}Cu_4O_y$,	$HgBa_2Ca_{n-1}Cu_nO_y$ (n = 1,2,3), $HgBa_2Cu_{0.97}Zn_{0.03}O_y$, $Hg_{0.5}Fe_{0.5}Sr_2Ca_{1-x}Y_xCu_2O_y$

For a comparative analysis of the changes in the normal-state properties and the T_c value under doping all samples were divided into three groups: optimally doped, underdoped, and overdoped compositions. As an optimally doped sample we considered the one with the maximal critical temperature within each phase. To obtain more representative data the samples with insignificant deviations from the optimal composition and the T_c values close to the maximal one for a given phase (i.e. so-called near-optimally doped compositions) were also included into this group. As an additional criterion for this selection we used the absolute thermopower value at T = 300K, S_{300}, which is known to be directly related to the doping level (see, for example, [76]). As a result, we considered the samples with the S_{300} value being in the range from – 2 to 10 μkV/K as near-optimally doped ones. The samples with the S_{300} value being above or below these values were considered as the underdoped or overdoped ones, correspondingly.

Such a classification is also useful since we investigated the systems with various type of doping that made it impossible to use the content of a specific impurity for a comparative analysis of changes in the normal-state and superconducting properties. The S_{300} value for the systems with various substitutions is a universal parameter identifying a degree of the deviation of the sample composition from the optimally doped one and represents the only reasonable criterion of the doping level in this case.

As a result, the samples of all investigated systems can be divided into groups according to the doping regime. The specific compositions for each of these regimes in case of bismuth-, thallium-, and mercury-based systems are given in Table 1.

4. THERMOPOWER VARIATION UNDER DOPING

All the temperature dependences of thermopower in Bi-, Tl-, and Hg-based HTSC's are characterized by similar peculiarities. Figures 2-7 present the typical $S(T)$ dependences for different doped HTSC-systems of the studied families – bismuth-based $Bi_2Sr_2CuO_y$, $Bi_2Sr_2Ca_2Cu_3O_y$, and $Bi_2Sr_{2-x}La_xCuO$; thallium-based $Tl_2Ba_2Ca_{1-x}Y_xCu_{2-y}Co_yO_z$; mercury-based $HgBa_2Cu_{1-x}Zn_xO_y$ and $Hg_{0.5}Fe_{0.5}Sr_2Ca_{1-x}Y_xCu_2O_y$.

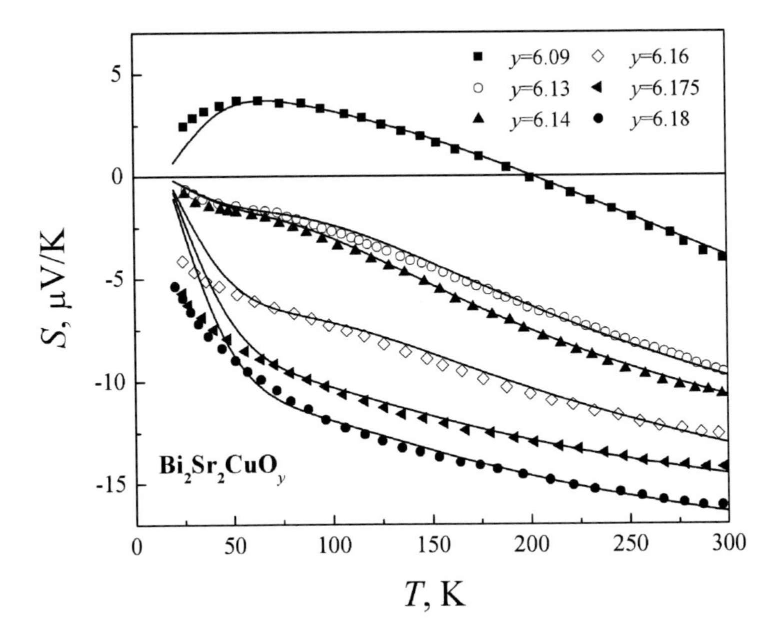

Figure 2. Temperature dependences of thermopower in $Bi_2Sr_2CuO_y$ samples with various oxygen content. Symbols are the experimental data, solid lines represent the results of calculations within the narrow-band model.

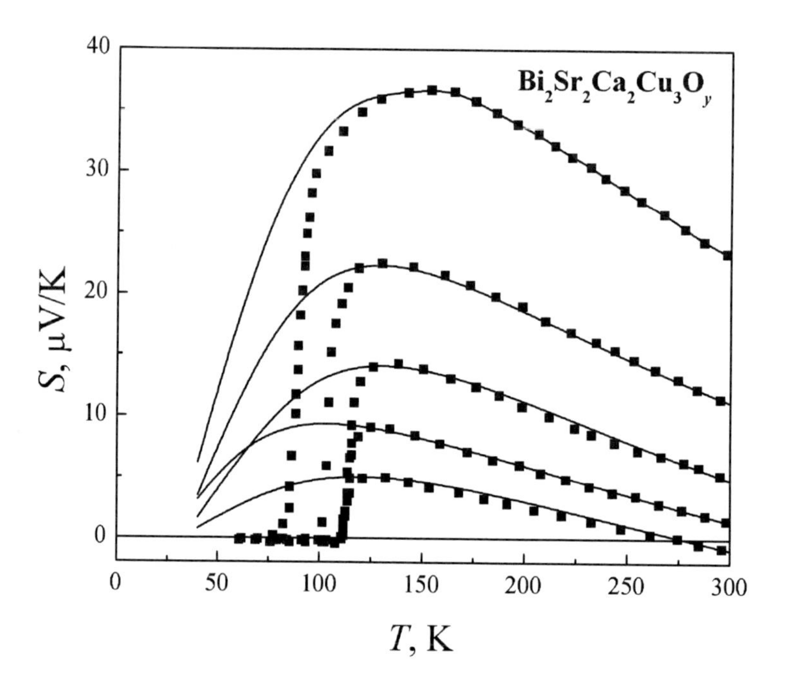

Figure 3. Temperature dependences of thermopower in $Bi_2Sr_2Ca_2Cu_3O_y$ samples with various oxygen content. The oxygen content decreases with increasing thermopower value. Symbols are the experimental data, solid lines represent the results of calculations within the narrow-band model.

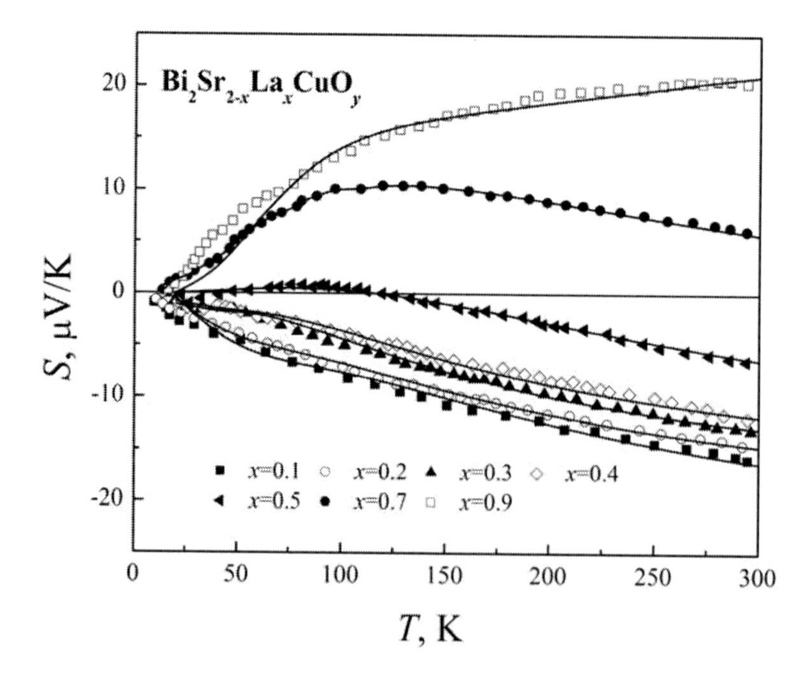

Figure 4. Temperature dependences of thermopower in $Bi_2Sr_{2-x}La_xCuO_y$ system. Symbols are the experimental data, solid lines represent the results of calculations within the narrow-band model.

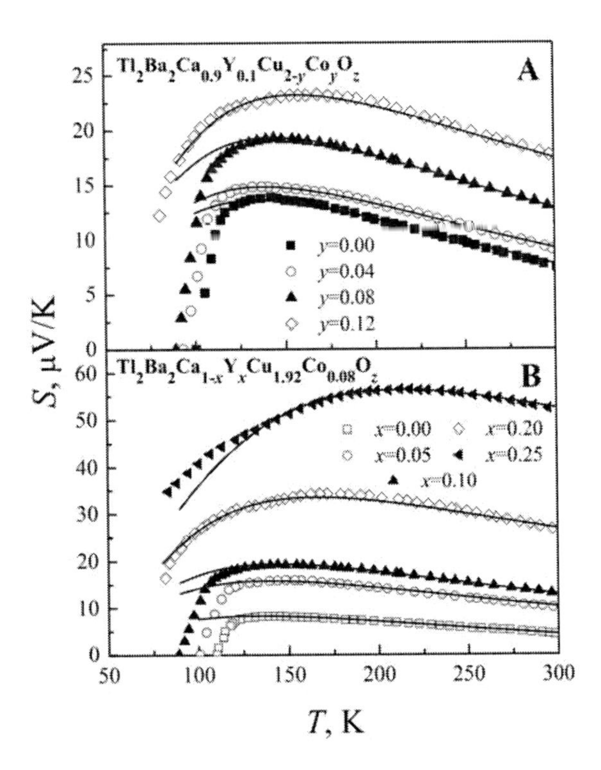

Figure 5. Temperature dependences of thermopower in $Tl_2Ba_2Ca_{1-x}Y_xCu_{2-y}Co_yO_z$ system. Symbols are the experimental data, solid lines represent the results of calculations within the narrow-band model.

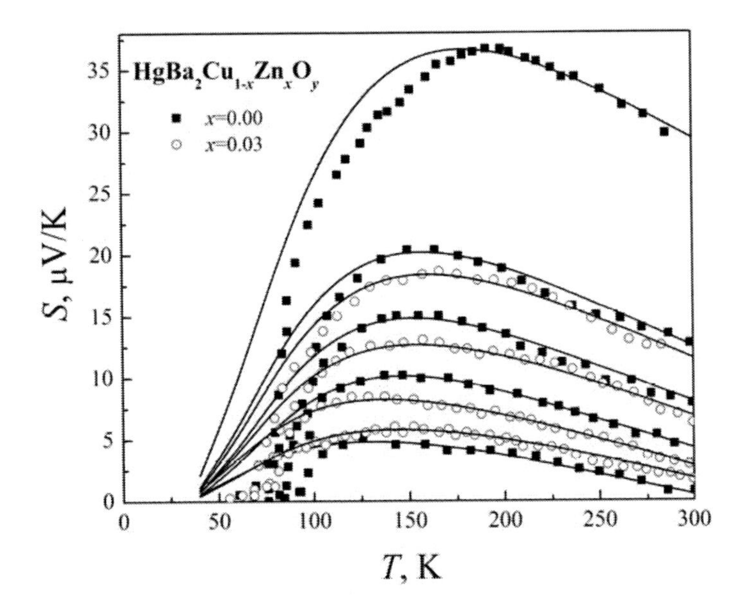

Figure 6. Temperature dependences of thermopower in $HgBa_2Cu_{1-x}Zn_xO_y$ system with various oxygen content. At a fixed zinc concentration the oxygen content decreases with increasing thermopower value. Symbols are the experimental data, solid lines represent the results of calculations within the narrow-band model.

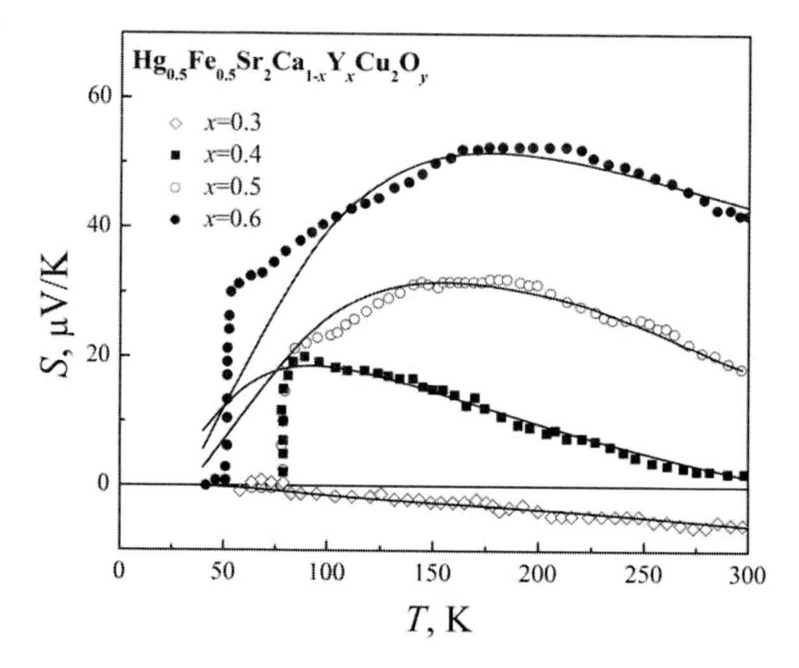

Figure 7. Temperature dependences of thermopower in $Hg_{0.5}Fe_{0.5}Sr_2Ca_{1-x}Y_xCu_2O_y$ system. Symbols are the experimental data, solid lines represent the results of calculations within the narrow-band model.

The main distinctive feature of $S(T)$ dependences for slightly-doped samples of all the systems (or near-optimally doped compositions with the S_{300} values being in the range from –2 to 10 μV/K) is an extended region of almost linear decrease of the thermopower with increasing temperature. At the same time, another important feature characteristic of Y-based HTSC's, namely the presence of a broad maximum on a $S(T)$ curve at temperatures well above the superconducting transition, exists for most of the compositions studied.

Increasing doping level in all the systems leads to a significant modification of $S(T)$ dependences. As one can see from Figures 4, 5, and 7, non-isovalent cation substitutions (La → Sr in the Bi-2201 system, Y → Ca and/or Co → Cu in the Tl-2212 system and Y → Ca in the Hg-1212 system) result in an increase in the thermopower values. Besides, linearity of $S(T)$ dependences gradually vanishes with increasing impurity content, and their maximum broadens and shifts to higher temperatures. Such a transformation of $S(T)$ curves is a general characteristic of a transition from optimally doped compositions to the underdoped regime and it coincides qualitatively with that observed in Y-based HTSC's [18], [19].

The described changes in a shape of $S(T)$ dependences are also characteristic of a case of oxygen doping taking into account that oxygen is a unique anion in HTSC-materials. For this reason, the increasing oxygen content should lead to a rise in the mobile holes number and, as a result, the transition to the underdoped regime occurs with decreasing oxygen content. As seen from Figures 3 and 6 by the example of $Bi_2Sr_2Ca_2Cu_3O_y$ and $HgBa_2Cu_{1-x}Zn_xO_y$ systems, the maximum of $S(T)$ curves becomes smoother with increasing oxygen deficit and shifts to higher temperatures. The only difference from the cation doping is that the $S(T)$ linearity remains evident for the underdoped regimes, i.e. for samples with a low oxygen content.

The transformation of thermopower at the transition from optimally doped to overdoped compositions is characterized by other specific features. For both cation and oxygen doping the linearity of $S(T)$ curves is essentially distorted, and their maximum characteristic of optimally doped and underdoped compositions gradually vanishes. The S_{300} value at the transition to the overdoped regime consistently decreases and reaches – (10 ÷ 15) μV/K. These features are clearly seen in Figures 2 and 4 where the experimental data on transformation of the $S(T)$ dependences in $Bi_2Sr_2CuO_y$ system under increasing oxygen content and in $Bi_2Sr_{2-x}La_xCuO_y$ system under decreasing lanthanum content are presented.

The described features of the temperature dependences of thermopower and their modification under doping are characteristic of all the discussed families of chain-free HTSC's.

It is necessary to pay special attention to the thermopower behavior in the $Bi_2Sr_{2-x}K_xCaCu_2O_y$ system (see Figure 8). These samples can be carried rather conventionally to the case of transition from optimally doped compositions to slightly overdoped regime (the S_{300} values for $Bi_2Sr_{2-x}K_xCaCu_2O_y$ samples with different x are in a range of – (1.5 ÷ 4) μV/K). However, an increase in potassium content leads to an anomalously strong rise in the $S(T)$ slope whereas S_{300} changes insignificantly. The reasons for such behavior of thermopower call for special discussion.

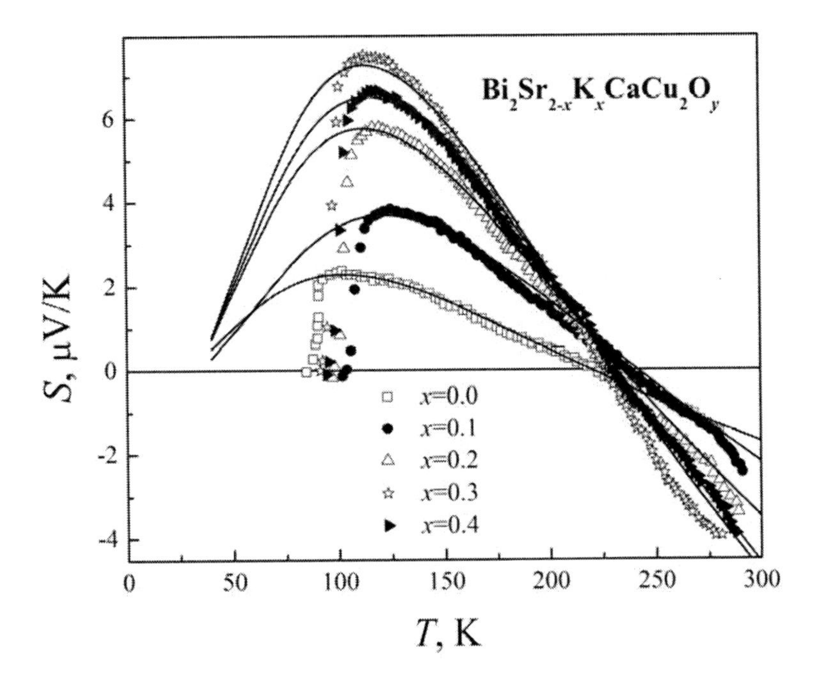

Figure 8. Temperature dependences of thermopower in $Bi_2Sr_{2-x}K_xCaCu_2O_y$ system. Symbols are the experimental data, solid lines represent the results of calculations within the narrow-band model.

Let us consider the influence of various types of non-isovalent cation substitutions on the absolute thermopower values in the studied systems in more details. The concentration dependences of S_{300} for doped Bi- and Tl-based systems are presented in Figures 9 and 10, correspondingly. It is seen that the change in S_{300} with increasing doping level has a different character for the studied HTSC-families and depends on the doping type, as well as on the number of copper-oxygen layers in a given system.

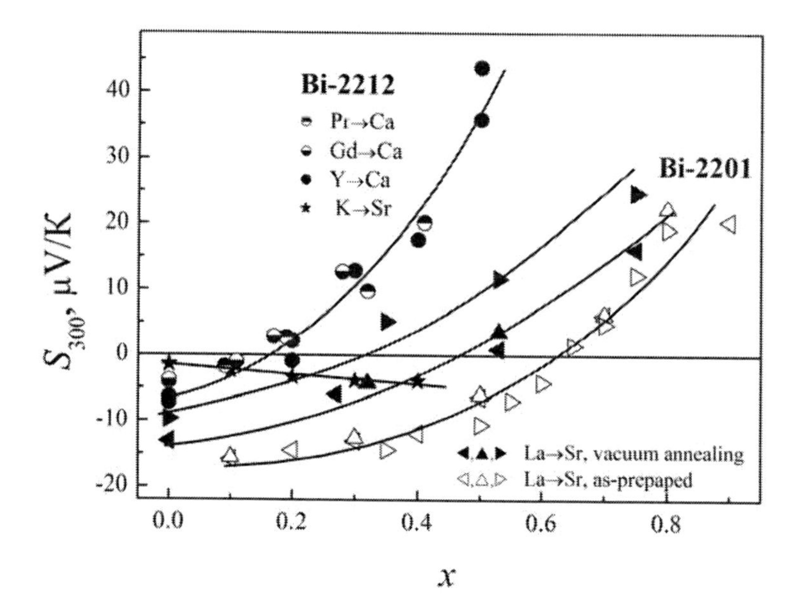

Figure 9. Room-temperature thermopower in $Bi_2Sr_{2-x}La_xCuO_y$ (Bi-2201), $Bi_2Sr_2Ca_{1-x}(Pr,Gd,Y)_xCu_2O_y$ (Bi-2212), and $Bi_2Sr_{2-x}K_xCaCu_2O_y$ (Bi-2212) systems.

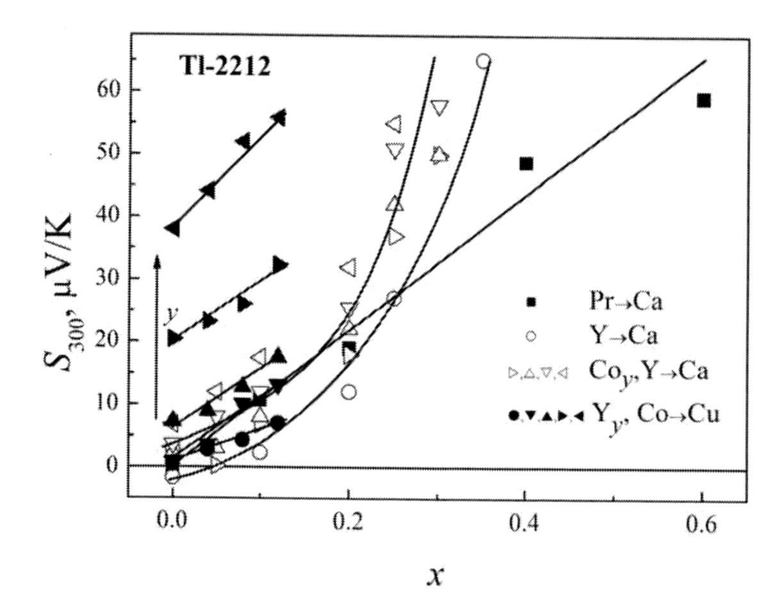

Figure 10. Room-temperature thermopower in $Tl_2Ba_2Ca_{1-x}(Pr,Y)_xCu_2O_y$, $Tl_2Ba_2Ca_{1-x}Y_xCu_{2-y}Co_yO_z$, and $Tl_2Ba_2Ca_{1-y}Y_yCu_{2-x}Co_xO_z$ systems.

First of all one should note that the $S_{300}(x)$ dependence for the $Bi_2Sr_{2-x}K_xCaCu_2O_y$ system is different from the ones for other considered substitutions that is obviously related to a specific influence of the $K^{1+} \rightarrow Sr^{2+}$ substitution on the normal-state properties of the Bi-2212 system (first associated with a relation of valences of potassium and strontium ions). Unlike all other systems, the valence of the substituting element in this case is lower than the valence of the substituted one. This apparently leads to a decrease in the oxygen content in $Bi_2Sr_{2-x}K_xCaCu_2O_y$ with increasing x that compensates the charge influence of the impurity. As a result, S_{300} changes very insignificantly with increasing doping level from $S_{300} = -1.5$ μV/K for $x = 0$ to $S_{300} = -4$ μV/K for $x = 0.4$.

For all other systems a considerable and successive increase in the thermopower value under doping is observed, and the only distinction in the shape of $S_{300}(x)$ dependences for specific substitutions consists in a range of S_{300} values and degree of their rise. Such a change in the thermopower value caused by a type of the studied substitutions (in all cases the valence of a substituting element is higher compared to the valence of a substituted one, e.g. $La^{3+} \rightarrow Sr^{2+}$; Y^{3+}, Pr^{3+}, $Gd^{3+} \rightarrow Ca^{2+}$, $Co^{3+} \rightarrow Cu^{2+}$) points to the fact that the number of mobile holes decreases with impurity content. Let us consider the tendencies in the S_{300} variation for the studied systems in more details.

The $La^{3+} \rightarrow Sr^{2+}$ substitution in the Bi-2201 phase for sample series with various oxygen content leads to almost identical rise of S_{300} under doping (see Figure 9). Thus, the overall level of these values depends on the level of oxygen content in different series that is varied using additional annealing. As seen from Figure 9, a stronger rise of S_{300} is observed in the case of Y^{3+}, Pr^{3+}, $Gd^{3+} \rightarrow Ca^{2+}$ substitutions in the Bi-2212 phase. The S_{300} value changes qualitatively similar to the case of $Y^{3+} \rightarrow Ca^{2+}$ substitution in the Tl-based system (Tl-2212 phase), both in case of a single doping and in series with a different fixed content of an additional cobalt impurity in the copper sites (see Figure 10). Likewise, an increase in the S_{300} value under cobalt doping is observed in the samples of Tl-2212 system with a fixed amount of an additional yttrium impurity in calcium sites, and the degree of this rise becomes stronger as the yttrium content increases.

Quantitative comparison of the doping influence on the S_{300} value in different phases of the bismuth HTSC-systems shows that despite the same relation between valences of substituting and substituted elements, the rate of S_{300} rise under lanthanum doping in the Bi-2201 phase is less than in cases of yttrium, praseodymium, and gadolinium doping in the Bi-2212 phase (see Figure 9). On the other hand, a comparison of the degree of increase in the absolute thermopower value under doping by yttrium and praseodymium in cases of similar phases of bismuth- and thallium-based HTSC's with $n = 1$ (namely, Bi-2212 and Tl-2212 compounds) testifies to an essentially stronger influence of those impurities on the S_{300} value in the Tl-2212 system (compare Figure 9 and Figure 10). Thus, the S_{300} values in the Bi-2212 phase with 30% of calcium substituted by yttrium or praseodymium are about 13 μV/K and 10 μV/K correspondingly, while they reach values of ~ 41 μV/K and ~ 33 μV/K in the Tl-2212 phase at the same impurity concentrations. Besides, the Pr → Ca and Y → Ca substitutions in the Bi-2212 system lead to an identical increase in the S_{300} value but there is an essential distinction between their effects in the Tl-2212 system. In the latter, the degree of S_{300} rise under Y → Ca substitution increases with the doping level (by analogy with the Bi-2212 system), whereas S_{300} increases linearly in case of the Pr → Ca substitution and, as a whole, essentially weaker as compared with the Bi-2212 system.

All the above tendencies in S_{300} variation under doping in the studied systems are undoubtedly related to a specific effect of each impurity on the band spectrum structure in the normal state which will be discussed below.

In conclusion of this section we will discuss qualitatively the change in superconducting properties of the studied HTSC-systems.

The most important feature of the chain-free Bi-, Tl-, and Hg-based HTSC's is a rise of the maximal T_c with increasing number of copper-oxygen layers for the case of $n \leq 3$ [19], [63], [69], [70], [75], [77], [78], [82], [90], as seen from Figure 11. At the same time, the further increase in n leads to a decrease in critical temperature. For any number of the CuO_2 layers the critical temperature in the Hg-based system demonstrates the highest values and it is higher in the Tl-based system than in the Bi-based one. The maximally reached T_c value in case of single-layered systems is about 35K for the Bi-2201 phase while it is considerably higher for the Tl-2201 and Hg-1201 phases reaching 90K and 98K correspondingly. Such a strong difference in the maximal T_c values for the phases of different systems with a larger number of copper-oxygen layers is not observed. Let us also note that such a high value of the critical temperature in the single-layered $Tl_2Ba_2CuO_y$ and $HgBa_2CuO_y$ systems almost coincides with the maximal T_c value observed in near-stoichiometric $YBa_2Cu_3O_7$ compounds containing two CuO_2 layers (see Figure 11).

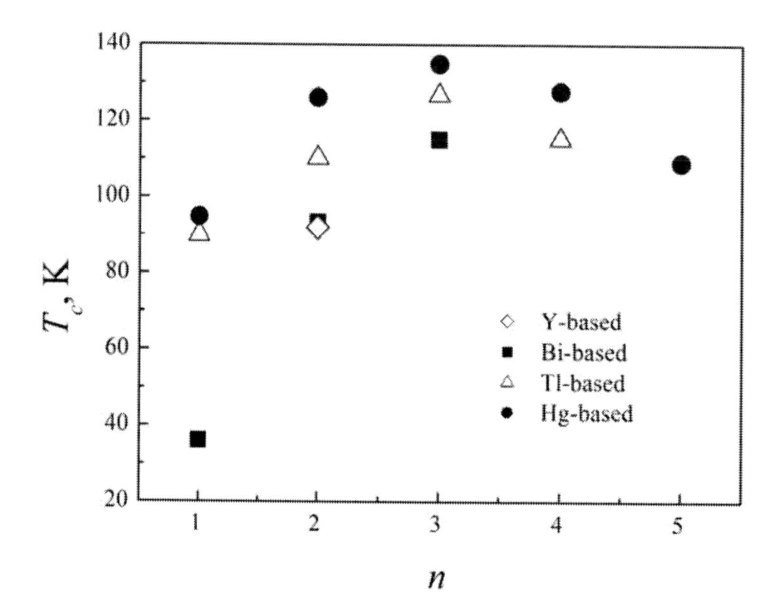

Figure 11. Maximal critical temperature for phases with various number of CuO_2 layers in different HTSC-families.

The doping influence on the T_c value is illustrated in Figure 12 demonstrating the relationship between the critical temperature for the Bi-, Tl-, and Hg-based HTSC's and the S_{300} value chosen as a universal criterion of the doping level in samples with various types of substitutions. One can see that in all cases the $T_c(S_{300})$ dependence is non-monotonic. The near-optimally doped samples (with $S_{300} = -2 \div 10$ μV/K) demonstrate the highest T_c values for each phase of all studied systems. A maximum on the $T_c(S_{300})$ dependence is situated approximately at $S_{300} = 0$ in all cases. When going to the underdoped or overdoped regimes (that corresponds to an increase or a decrease in the S_{300} value, correspondingly), there is a deterioration of superconducting properties, though in varying degree for different systems. Thus, it is possible to track the T_c value change under increasing doping level for the Bi-2201 and Tl-2201 systems in the overdoped regime down to the total suppression of the superconducting properties at $S_{300} = -20$ μV/K and $S_{300} = -11$ μV/K, correspondingly. It is necessary to stress that the critical temperature decreases to a greater extent in the overdoped regime than in the underdoped one. This is apparent for the systems with one or two CuO_2 layers but not obvious when $n = 3$ that can be related to a narrow range of overdoped compositions for which the data on the superconducting properties and thermopower are available.

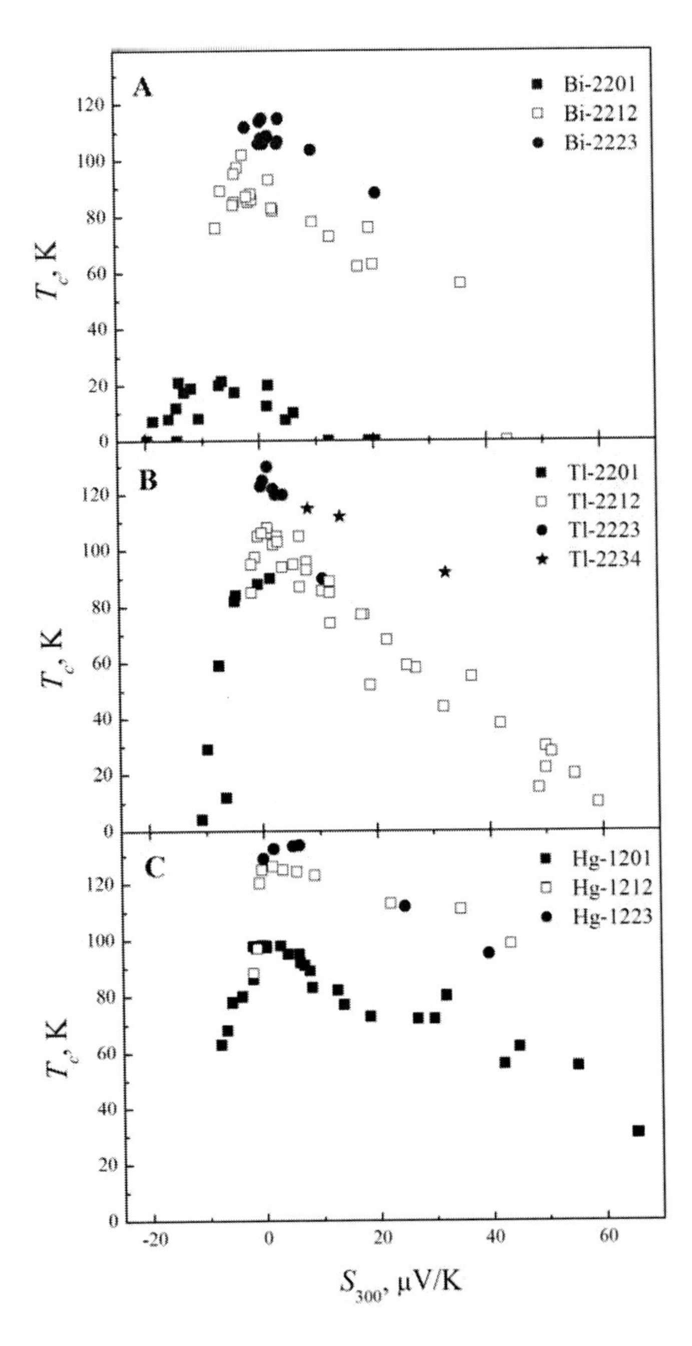

Figure 12. Dependence of critical temperature on room-temperature thermopower for various phases of Bi- (A), Tl- (B), and Hg-based (C) HTSC's.

5. Parameters of the Band Spectrum Structure and Charge-Carrier Systems and their Transformation Under Variations of Sample Composition

All the experimental data on the temperature dependences of thermopower for various phases of doped Bi-, Tl-, and Hg-based HTSC-materials described and discussed above have been analyzed in the framework of the narrow-band model [18], [19]. The common peculiarities of the $S(T)$ dependences established in Section 4 when analyzing all the experimental data discussed here as well as the results we obtained earlier for some compositions of the studied systems [19], [34]-[40] indicate that for the quantitative descriptions of the thermopower behavior the asymmetric version of the narrow-band model should be used. For this reason, before describing and analyzing the modification of parameters of the band spectrum and charge-carrier system in the studied systems under both doping influence and increasing number of copper-oxygen layers, it is necessary to discuss a question on the uniqueness of their determination within the narrow-band model using Eqs. (2), (4).

Obviously, introducing a fourth parameter b (characterizing the degree of the band asymmetry) results in ambiguous determination of other parameters. To analyze the extent of such an ambiguity for each experimental $S(T)$ dependence we have first made multiple calculations of two types fixing values of either b or $C \equiv W_\sigma/W_D$. This allowed us to reveal a "physically meaningful" range of possible variation of those parameters. Based upon these results, we have then performed a series of calculations for each $S(T)$ curve in order to determine all possible sets of the model parameters allowing to describe satisfactorily this experimental dependence. Thus, for each of the given sample the whole possible range of variations of all the four model parameters (F, W_D, C, and b) was determined. These results will be presented below in the form of error bars in figures showing the concentration dependences of the model parameters, whereas the points will indicate the calculation results giving the best-fitted curves. Note that despite these errors, the tendencies in variation of the band spectrum and charge-carrier system parameters can be well revealed that allows us to discuss the mechanisms of their modification under doping.

The results on the thermopower data analysis obtained by applying Eqs. (2), (4) are shown in Figures 2-8, where we present the experimental $S(T)$ dependences together with the calculation results corresponding to the optimal values of the model parameters. In most cases a good agreement of calculated and experimental curves is observed over the whole measured temperature range of $T = T_c \div 300$ K. In case of high thermopower values there is a certain discrepancy between those dependences at low temperatures ($T < 100 \div 130$ K). This is a consequence of the main narrow-band assumption on comparability of the Fermi smearing and the total effective bandwidth W_D by an order of magnitude used by us when deriving Eqs. (1)-(4). In case of high thermopower values (that corresponds to a wider band as it will be shown below) and low temperatures this relation is violated, therefore the model becomes too rough and calculations can give an appreciable error.

Before discussing the results of the experimental data analysis in details, let us establish two common characteristics of the band spectrum of all studied HTSC-families without taking into account the specific features of its transformation for each system. The first of them is the similarity of parameters of the band spectrum and charge-carrier system for near-optimally doped compositions. The second one is an analogous modification of the band spectrum for phases with various number of CuO_2 layers and for different substitutions.

All the results obtained show that for the samples of near-optimally doped compositions the values of all model parameters for different systems are close enough to each other. The degree of the band filling with electrons, F, is slightly less than a half, its values lie in a range of $0.43 \div 0.5$, the total effective bandwidth, W_D, is in the range of $50 \div 210$ meV, and values of the parameter characterizing the degree of state localization, C, lie in a range of $0.2 \div 0.45$. Besides, the presence of a slight asymmetry of the band is characteristic for all the investigated systems, the values of the b parameter are about $-(0.02 \div 0.05)$. Thus, the main parameters of the band spectrum structure and charge-carrier system for near-optimally doped samples of chain-free HTSC of bismuth, thallium, and mercury systems are close to those observed in the case of Y-based superconductors [18], [19], [23]-[25], [27], [29], excluding the existence of a slight band asymmetry.

Varying the doping level (both by cation substitutions and by oxygen content change) leads to a variation of model parameters. This variation in case of transition from the optimally doped compositions to the underdoped regime includes a slight change of the band asymmetry degree, a strong broadening of the band accompanied by a rise in the degree of state localization and an increase in the band filling with electrons. As for transition to the overdoped compositions, the comparative analysis of the available results allows one to reveal the presence of some common tendencies in model parameters variation despite substantially smaller amount of the experimental data on the $S(T)$ dependences in this case. When going to the overdoped regime, the band filling decreases and the total effective bandwidth drops sharply that is accompanied by an essential increase in fraction of localized states and the degree of the band asymmetry.

Thus, all main features of the band spectrum structure and charge-carrier system including their variation under doping are analogous in all the three investigated HTSC-families for phases with different numbers of copper-oxygen layers. Having briefly described these peculiarities, we will now turn to more detailed consideration and analysis of the results obtained for each of studied systems.

5.1. Bismuth-based HTSC's

We will first consider the results obtained for Bi-based HTSC's. Figures 13-15 demonstrate the dependences of the band filling degree and the total effective bandwidth on the S_{300} value in cases of oxygen content variation (see Figure 13 for the Bi-2201, Bi-2212, and Bi-2223 phases) and different types of non-isovalent cation doping (see Figure 14 for the La → Sr substitution for the Bi-2201 phase; see Figure 15 for the Y, Pr, Gd → Ca and K → Sr substitutions for the Bi-2212 phase). As one can see, the obtained data are related not only to cases of the near-optimally doped and underdoped regimes, but also to the slightly overdoped regime for the Bi-2212 phase and a wide range of this regime for the Bi-2201 one.

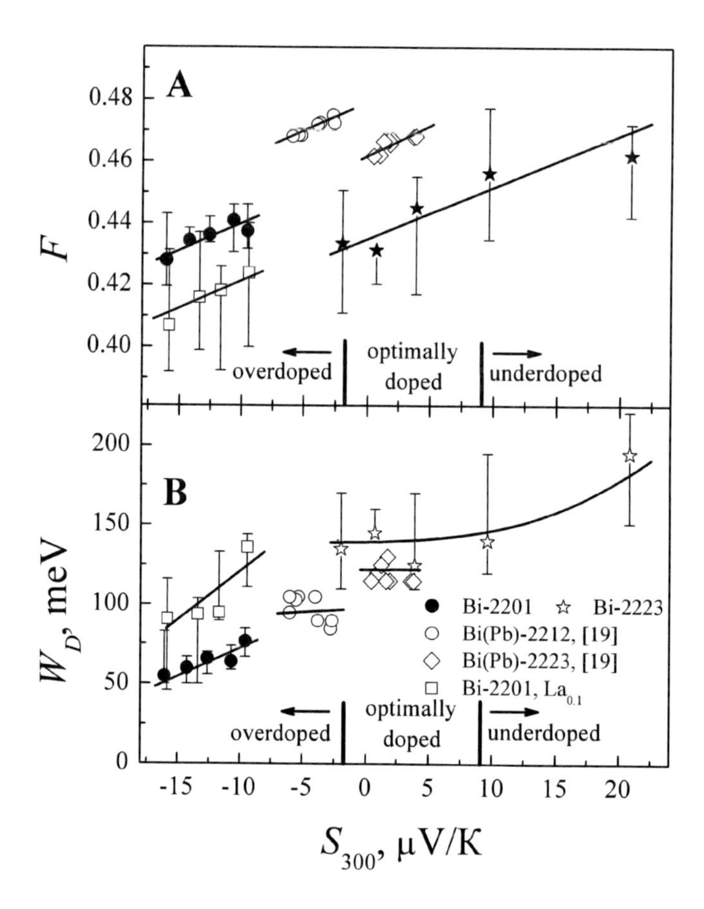

Figure 13. Doping dependences of band filling degree (A) and total effective bandwidth (B) in Bi-2201, Bi-2212, and Bi-2223 systems with various oxygen content.

Let us start from a discussion of the dynamics of the F value. First of all we would like to note that transition from optimally doped compositions to underdoped regime (i.e. as the S_{300} value changes from 0 μV/K to higher positive values) can be characterized by an increase in F in all cases though the degrees of this increase are different. This fact reflects the donor activity of both oxygen deficit and the substitutions studied as expected from crystal-chemical considerations. Indeed, a decrease in the content of oxygen being a single anion in the lattice of all HTSC-materials or an increase in the content of non-isovalent impurity having a higher valence than a substituted cation ($La^{3+} \rightarrow Sr^{2+}$ or Pr^{3+}, Gd^{3+}, $Y^{3+} \rightarrow Ca^{2+}$ substitutions) lead to the same type of disturbance of the charge balance in the lattice that should result in increasing number of the mobile electrons inducing an increase in the F value. Note that the comparison of the $F(S_{300})$ dependences for series of $Bi_2Sr_{2-x}La_xCuO_y$ samples with various oxygen content (obtained by different synthesis procedures or by annealing under various conditions) and different impurity amounts in the underdoped regime shows that an overall decrease in the oxygen content does not lead to a change in the shape of the $F(S_{300})$ dependence causing only a rise of a level of the band filling values (see Figure 14A) within each series as a whole, that additionally testifies to donor activity of oxygen vacancies. Besides, a character of the F value changing with oxygen content in the Bi-2223 phase does not depend on their cation composition. As seen from Figure 13A, the $F(S_{300})$ dependences for $Bi_2Sr_2Ca_2Cu_3O_y$ and $(Bi,Pb)_2Sr_2Ca_2Cu_3O_y$ systems in the optimally doped and underdoped regimes are similar and just shifted relative to each other.

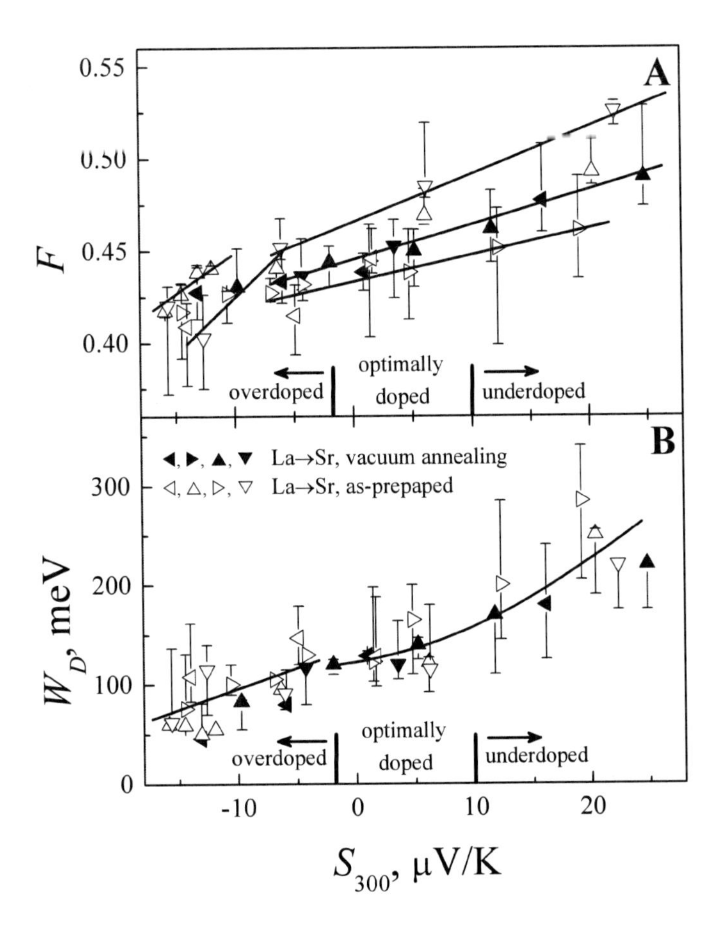

Figure 14. Doping dependences of band filling degree (A) and total effective bandwidth (B) in Bi-2201 system with the La → Sr substitution.

As to the distinctions observed for doping of various types, it is necessary to note the following. Whereas the $F(S_{300})$ dependence is close to be linear with increasing oxygen deficit (the Bi-2223 phase) and partial substitution for strontium by lanthanum (the Bi-2201 phase), it is non-linear in case of substitutions for calcium by yttrium, praseodymium or gadolinium in the Bi-2212 phase and the rate of the F value increase rises with increasing S_{300}. This tendency is most distinctly visible in case of the Y→ Ca substitution due to a wide range of the presented underdoped compositions, however it is also observed in case of the Pr, Gd → Ca substitutions, though less clearly (see Figure 15).

This feature can be explained with help of the results and conclusions on the mechanism of the calcium influence on the band spectrum in the $YBa_2Cu_3O_y$ system obtained earlier. As shown in [19], [22], [26], [28], [30], [32], [33], the doping of that system with calcium leads to the formation of an additional peak in the DOS function causing thereby the band asymmetry and consecutive increase in its degree. According to the results obtained for the $Bi_2Sr_2Ca_{1-x}R_xCu_2O_y$ (R = Y, Pr, Gd) system, a decrease in the calcium content leads to a consecutive decrease in the band asymmetry degree (though this effect is rather weak) while the b value for the $Bi_2Sr_{2-x}La_xCuO_y$ samples in the underdoped regime remains almost unchanged with increasing lanthanum content.

By an analogy with Y-based HTSC-system it is possible to suppose that calcium introduces additional states into the conduction band in case of Bi-based HTSC's as well. In such a case a decrease in its content in the $Bi_2Sr_2Ca_{1-x}R_xCu_2O_y$ system should lead to reduction of the total number of states in the band. As a result, two factors should influence simultaneously the $F(x)$ or $F(S_{300})$ dependences in case of the Y, Pr, Gd → Ca substitutions, namely a rise in the number of electrons (due to increasing amount of impurity with a higher valence) and a decrease in the number of states in the band (due to decreasing calcium content). It is obvious, that in that case the above dependences should be non-linear and their slope should increase indeed with the doping level that corresponds well to the results of our calculations. In case of the La → Sr substitution in the Bi-2201 phase the latter factor affecting the F value disappears, therefore the $F(S_{300})$ curves should be linear that is also in agreement with the results obtained for this system.

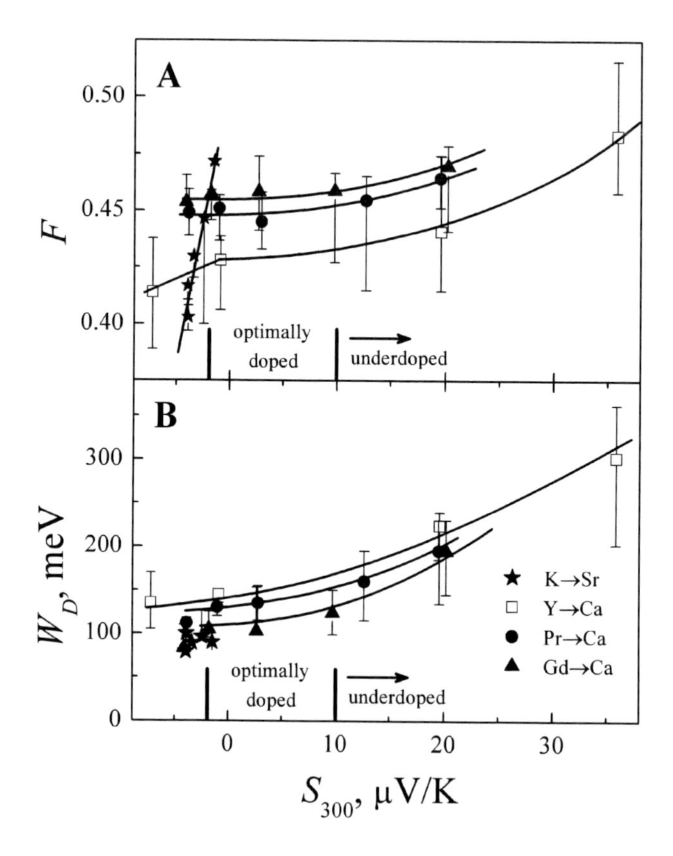

Figure 15. Doping dependences of band filling degree (A) and total effective bandwidth (B) in Bi-2212 system with the K → Sr and (Y,Pr,Gd) → Ca substitutions.

As to the overdoped regime, in case of most studied systems there is a tendency to a decrease in the band filling with electrons when going from the optimally doped compositions. This tendency appears the more distinctly the higher the maximal doping level achieved in this regime is. This fact corresponds to an increase in the concentration of mobile holes, i.e. to surplus doping by mobile carriers in comparison with the optimally doped compositions.

Special attention should be paid to the results obtained for the $Bi_2Sr_{2-x}K_xCaCu_2O_y$ system (see Figure 15A) showing an anomalously strong decrease in the F value in a slightly overdoped regime. The degree of the band filling with electrons decreases under increasing potassium content from 0.47 down to 0.4 in a very small doping range close to the near-optimally doped compositions (the S_{300} value changes from -1.5 to -4 μV/K). We should note that simultaneously the band asymmetry degree increases sharply from $b = -0.03$ to $b = -0.09$. It is possible to speculate that the observed strong changes in these parameters are related to a specific effect of the potassium impurity on the band spectrum structure whose nature remains unclear.

The results on the variation of the total effective bandwidth are also presented in Figures 13-15. As one can see, for samples of near-optimally doped compositions the W_D values fall within the range of 95 ÷ 200 meV, 90 ÷ 150 meV, and 110 ÷ 175 meV for the Bi-2201, Bi-2212, and Bi-2223 phases, correspondingly. Let us first consider the band modification in the underdoped regime. Figure 13 shows that in case of increasing oxygen deficit (samples of the Bi-2223 phase annealed in an oxygen-deficit atmosphere) the W_D value changes insignificantly with a weak tendency to an increase. At the same time, the C parameter changes rather slightly demonstrating a weak tendency to a decrease only when going to a sample with the maximal doping level corresponding to $S_{300} = 20.8$ μV/K. On the contrary, non-isovalent cation substitutions both for strontium by lanthanum (the Bi-2201 phase, see Figure 14B) and for calcium by yttrium, praseodymium or gadolinium (the Bi-2212 phase, see Figure 15B) lead to a strong band broadening and a marked increase in the degree of state localization at the band edges (the C parameter decreases). Note that despite distinctions in the overall level of the bandwidth values for different substitutions in the Bi-2212 phase or at various oxygen content in the Bi-2201 phase, the values of W_D and C change under increasing content of impurities practically in the same way in different studied systems. Besides, in the underdoped regime the degree of the band asymmetry changes rather weakly in all phases of Bi-based HTSC's.

Thus, the observed character of the band spectrum transformation in the Bi-based HTSC-systems in the underdoped regime is qualitatively identical to that revealed earlier for Y-based HTSC's [18], [19], [23]-[25], [27], [29]. It suggests the mechanism of the modification of the conduction band in those two HTSC-families to be rather the same. Deviations from optimally doped compositions through increasing either oxygen deficit or content of a non-isovalent impurity result in a rising degree of the system disordering. This is caused both by the direct effect of the introduction of impurity atoms into the lattice and by simultaneous changes in the oxygen subsystem due to a charge effect of non-isovalent cation substitutions (variation of the oxygen content and redistribution of its atoms in a lattice caused by introducing an additional positive charge). In its turn, according to the Anderson localization mechanism this leads to a band broadening accompanied by an increase in the fraction of localized states at the conduction band edges in full accordance with the results of our calculations. Some distinctions in the level of W_D values for series of $Bi_2Sr_{2-x}La_xCuO_y$ samples with various oxygen content, as well as for series of $Bi_2Sr_2Ca_{1-x}R_xCu_2O_y$ samples with different type of R → Ca substitutions are related to a difference in oxygen subsystem conditions in these series caused by used regimes of sample synthesis or additional annealing.

As for the observed weak change of the W_D values in samples of the Bi-2223 phase with increasing oxygen deficit (see Figure 13B), this fact can be explained by the features of oxygen composition in Bi-based HTSC's. It is known that unlike the $YBa_2Cu_3O_y$ system, where the oxygen content in the samples with maximal T_c is close to its stoichiometric value $y \approx 7$, the presence of excess oxygen is characteristic of optimally doped samples of various phases of the bismuth-based systems. In particular, according to the results presented in [91], in case of the Bi-2223 system the maximal value of $T_c \approx 108$ K is observed at $y \approx 10.19 \div 10.28$. If so, the oxygen subsystem in optimally doped samples of the Bi-based phases is already disordered and a minor decrease in the oxygen content does not lead to an essential rise of the degree of this disordering. As a result, the effective bandwidth and the degree of state localization in $Bi_2Sr_2Ca_2Cu_3O_y$ samples change insignificantly with decreasing oxygen content and start to increase considerably only when going to the sample with a maximal level of the oxygen deficit.

Let us consider now the results obtained for the overdoped regime. As seen from Figures 13-15, the experimental data for this case are presented mostly for the Bi-2201 phase (in case of an excess oxygen content and a low level of the La → Sr doping), whereas the overdoped regime in the Bi-2212 phase is represented only by three samples with the Y, Pr, Gd → Ca substitutions and four samples with the K → Sr substitution.

Transition from the optimally doped compositions to the overdoped regime (evidenced by an increase in negative S_{300} values) is accompanied by a strong band broadening for all studied substitutions except for K → Sr in the Bi-2212 system though its degree depends on a doping type and a considered phase. Simultaneously, a decrease in the C parameter indicating a rise in the degree of state localization, as well as a consecutive increase in the band asymmetry degree are observed in all cases.

Like the case of the band filling, a special attention should be paid to the results on the $Bi_2Sr_{2-x}K_xCaCu_2O_y$ system. In the presented narrow range of the overdoped regime the W_D and C values remain almost unchanged. The possible range of their variations for all $Bi_2Sr_{2-x}K_xCaCu_2O_y$ samples is $W_D = 80 \div 124$ meV (see Figure 15) and $C = 0.4 \div 0.6$. Let us remind that those samples are characterized also by a strong increase in the band asymmetry degree and a sharp fall of the band filling with doping. All these data testifies to a special effect of the K → Sr substitution on the Bi-2212 properties in comparison with other types of doping. The most probable reason for this is that potassium is the only considered impurity with a valence lower than a valence of the substituted atom. However, additional investigations are needed to clarify the physical mechanisms of potassium effect on the Bi-2212 properties.

Thus, the modification of the band spectrum parameters in the overdoped regime is characterized by the distinctive features as compared to that observed in the underdoped regime. This allows one to assume that mechanisms of the band spectrum transformation in those two regimes are rather different. In additional, let us note that the above difference in the band spectrum transformation correlates with the experimentally observed distinction in the T_c value change in the underdoped and overdoped regimes mentioned in Section 4 (see Figure 12A).

5.2. Thallium-based HTSC's

The results obtained for Tl-based HTSC's with various types and levels of doping are shown in Figures 16, 17 and Table 2. Unlike Bi-based superconductors, most of results on Tl-based system concern samples belonging to the case of the optimally doped and underdoped regimes while the data for the overdoped regime are presented only for the Tl-2201 phase in case of oxygen doping.

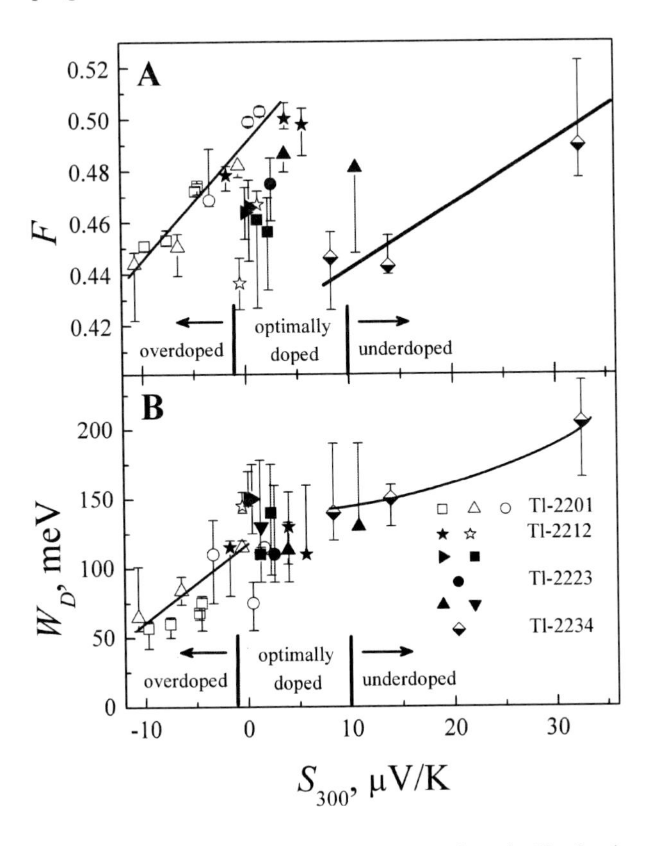

Figure 16. Doping dependences of band filling degree (A) and total effective bandwidth (B) in Tl-2201, Tl-2212, Tl-2223, and Tl-2223 systems with various oxygen content.

Let us first discuss the band filling dynamics. As seen from Figure 16A, in case of oxygen doping (change in the oxygen content in samples by their annealing in different atmospheres), like the case of Bi-based HTSC's considered above, the value of the F parameter for various phases of thallium system changes almost linearly with increasing oxygen deficit for the Tl-2201 phase in the overdoped regime, the Tl-2212 phase in the near-optimally doped regime, and the Tl-2234 phase in the underdoped regime. It is necessary to emphasize that the band filling in the overdoped regime changes much stronger than in the underdoped regime at quantitatively identical variations in the doping level.

Contrary to the oxygen doping, in case of cation substitutions in the Tl-2212 phase the F value increases non-linearly with increasing S_{300} value (see Figure 17A). We would like to remind that a similar F dependence on the doping level has been observed for the Bi-2212 system in case of the Y, Pr, Gd → Ca substitutions (see Figure 15). This allows us to suppose that in the Tl-2212 system calcium ions also participate in formation of the conduction band. As a result, partial substitutions for calcium lead to decreasing number of band states that is combined with the charge influence of non-isovalent doping by trivalent ions $(Y, Pr)^{3+} \rightarrow Ca^{2+}$ resulting in increasing number of electrons in the band. This can explain a non-linear character of the $F(S_{300})$ dependence.

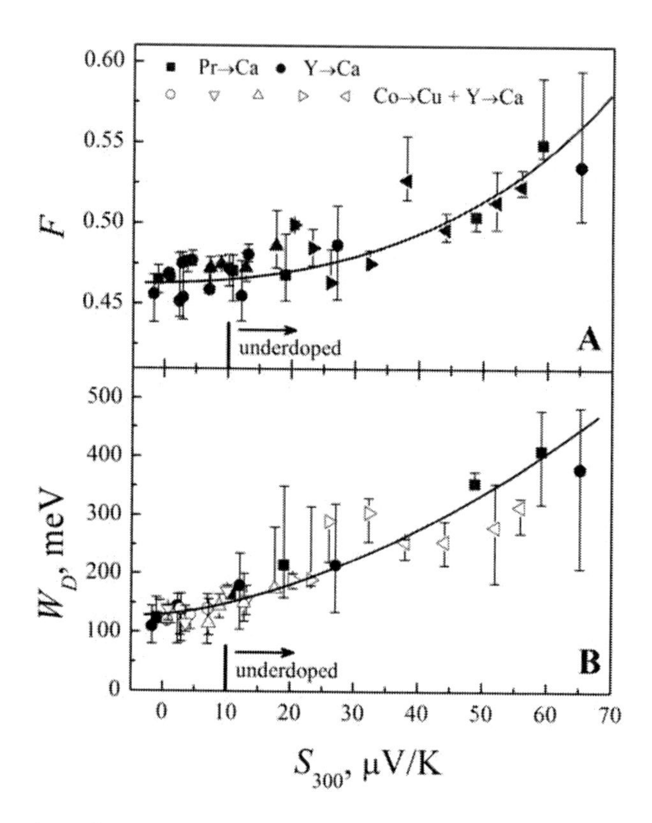

Figure 17. Doping dependences of band filling degree (A) and total effective bandwidth (B) in Tl-2212 system with the (Y,Pr) → Ca and simultaneous Co → Cu and Y → Ca substitutions.

Let us discuss in more details the F behavior in the $Tl_2Ba_2Ca_{1-x}Y_xCu_{2-y}Co_yO_z$ system because the most extensive and systematic experimental data on the temperature dependences of the thermopower are available for this system [82] (for 20 samples with different levels of Y → Ca and/or Co → Cu doping sintered in a uniform fabrication cycle).

As seen from the presented data, at the fixed cobalt content the Y → Ca substitution leads to consecutive and non-linear rise in the F value and the rate of this rise increases with yttrium content. As a result, a character of the $F(x)$ dependence in the $Tl_2Ba_2Ca_{1-x}Y_xCu_{2-y}Co_yO_z$ system coincides with the case of the single Y, Pr → Ca substitutions and can be accordingly explained by the same way. At the same time, at the fixed yttrium content the $F(y)$ dependence is nonmonotonic that demands separate consideration.

It is obvious that the Co → Cu substitution, like the Y → Ca one, changes the charge balance in the lattice because of an excess charge introduced into it by cobalt (the valence of cobalt is higher than that of copper it substitutes for) and, hence, causes an increase in number of free electrons, which favors an increase in F. Despite this fact, the calculations do not reveal a successive increase in the F value. However, as cobalt substitutes directly for copper participating in formation of the band responsible for the conduction process, it is possible to guess that in this case the effect of destruction of the conduction band with decreasing copper content gives the additional essential contribution to the observed $F(y)$ dependence. It is a combined effect of the two above factors that results in a nonmonotonic increase in F in the case of the Co → Cu substitution.

Table 2. Critical temperature, thermopower at $T = 300$ K, band filling with electrons, and total effective bandwidth in $Tl_2Ba_2Ca_{1-x}Y_xCu_{2-y}Co_yO_z$ samples

x	T_c, K	S_{300}, μV/K	F	W_D, meV
$Tl_2Ba_2Ca_{1-x}Y_xCu_2O_z$				
0.0	108	1.0	0.469 ÷ 0.471	114 ÷ 121
0.05	106	0.9	0.467 ÷ 0.47	119 ÷ 145
0.1	105	7.5	0.461 ÷ 0.479	125 ÷ 165
0.2	77	19	0.497 ÷ 0.501	175 ÷ 199
0.25	55	36	0.51 ÷ 0.55	224 ÷ 260
$Tl_2Ba_2Ca_{1-x}Y_xCu_{1.96}Co_{0.04}O_z$				
0.0	102	2.7	0.475 ÷ 0.482	94 ÷ 136
0.05	103	3.7	0.469 ÷ 0.477	99 ÷ 144
0.1	96	9.1	0.47 ÷ 0.475	125 ÷ 174
0.2	68	23	0.474 ÷ 0.497	179 ÷ 313
0.25	38	43	0.49 ÷ 0.506	215 ÷ 289
$Tl_2Ba_2Ca_{1-x}Y_xCu_{1.92}Co_{0.08}O_z$				
0.0	94	4.1	0.473 ÷ 0.483	115 ÷ 130
0.05	93	9.8	0.46 ÷ 0.481	140 ÷ 181
0.1	89	12	0.463 ÷ 0.482	156 ÷ 201
0.2	59	26	0.483 ÷ 0.51	219 ÷ 290
0.25	28	51	0.496 ÷ 0.532	186 ÷ 353
$Tl_2Ba_2Ca_{1-x}Y_xCu_{1.88}Co_{0.12}O_z$				
0.0	87	7.1	0.457 ÷ 0.469	110 ÷ 195
0.05	85	11.5	0.476 ÷ 0.487	128 ÷ 203
0.1	77	17.5	0.472 ÷ 0.507	173 ÷ 278
0.2	44	32	0.475 ÷ 0.483	256 ÷ 330
0.25	20	53	0.52 ÷ 0.533	272 ÷ 329

As for the overdoped regime realized in samples of the Tl-2201 phase with the excess oxygen content (see Figure 16), in this case, as already mentioned, the F value falls sharply with increasing doping level that is similar to the case of Bi-based HTSC's.

The values of the total effective bandwidth for near-optimally doped samples of Tl-based HTSC's are 55 ÷ 130 meV for Tl-2201, 80 ÷ 165 meV for Tl-2212, 90 ÷ 180 meV for Tl-2223, and 120 ÷ 190 meV for Tl-2234 phases, correspondingly. Note that for this family we can discuss the results on the W_D values in case of $n = 4$, i.e. then the critical temperature becomes lower as compared to the phase with $n = 3$.

For discussing the W_D variation in the underdoped regime we can use the results obtained for the three $Tl_{1.7}Ba_2Ca_{3.3}Cu_4O_z$ samples with different oxygen content (see Figure 16B), single Y, Pr → Ca doping in the Tl-2212 phase (see Figure 17B), and also double-substituted $Tl_2Ba_2Ca_{1-x}Y_xCu_{2-y}Co_yO_z$ samples (see Figure 17B and Table 2). Some qualitative conclusions can also be drawn for the Tl-2223 phase, while, according to the data used, in this case we achieve only a very slightly underdoped regime (see Figure 16B). As one can see, like the case of Bi-based HTSC's, the W_D value increases in oxygen deficit samples essentially slower that is also related to the presence of surplus oxygen in optimally doped samples of thallium-based system. The band broadening is accompanied by a consecutive increase in the fraction of localized states at its edges while the degree of the band asymmetry in the underdoped regime changes weakly enough. Thus, all tendencies in variation of the band spectrum parameters in the Tl-based HTSC-systems in the underdoped regime are identical to the case of Bi-based ones. This allows one to draw a conclusion that the Anderson localization mechanism is realized in this system as well. As discussed above, an increase in the oxygen deficit or a level of non-isovalent cation doping leads to an enhancing disordering in the lattice. As a consequence, the conduction band broadens and the degree of state localization at its edges rises.

Let us discuss in details the band spectrum transformation in $Tl_2Ba_2Ca_{1-x}Y_xCu_{2-y}Co_yO_z$ samples within the above mechanism. The analysis of the data presented in Table 2 shows that in case of the Co → Cu substitution (at the fixed yttrium content) an increase in the W_D values is distinctly observed even in the range of very small y values while for the Y → Ca substitution (at the fixed cobalt content) it becomes appreciable only at $x > 0.1$. A stronger cobalt effect on the conduction band at a low impurity content in comparison with the Y → Ca substitution can be explained by taking into account the fact that cobalt directly replaces plane copper and therefore significantly destroys the conduction band formed by overlapping copper and oxygen orbitals. Thus, the W_D value increases under the cobalt influence owing to two mechanisms: (i) the band distortion due to substitutions for plane copper positions and (ii) the band broadening caused by a rise in the disordering degree. It is obvious that the combined effect of these two mechanisms should lead to a noticeable increase in the W_D value even at small cobalt concentrations ($y \leq 0.1$) contrary to the case of the Y → Ca substitution where the only reason for the band broadening is realization of the Anderson localization mechanism.

Concluding this Section, let us analyze the band spectrum modification in Tl-based HTSC's in the overdoped regime. Unlike the Bi-based system, according to the experimental data only seven samples of the Tl-2201 phase with surplus oxygen content can be considered as belonging to this doping regime (see Figure 16). Nevertheless, it is obvious that at transition to the overdoped regime the W_D value decreases sharply (approximately twofold as the S_{300} value changes from -3 μV/K to -10.5 μV/K). According to our calculations, at the same time there observed a strong rise in the degree of state localization (the C parameter changes from $0.35 \div 0.55$ to $0.1 \div 0.25$) and the degree of the band asymmetry (the b parameter changes from -0.025 to -0.046). Thus, a character of the band spectrum modification in Tl-based HTSC-systems in the overdoped regime is the same as in case of bismuth-based superconductors considered above. Besides, it differs essentially from the case of the underdoped regime that correlates with a different character of the T_c value variation in those two regimes (see Figure 12B). This confirms the above assumption that mechanisms of the band spectrum transformation at transitions from optimally doped compositions to the underdoped and overdoped regimes are appreciably different.

5.3. Mercury-based HTSC's

The calculation results for Hg-based HTSC's are shown in Figure 18. It is necessary to note that unlike bismuth- and thallium-based superconducting systems discussed above, the thermopower data are presented in this case mainly for samples where a change of the doping level is reached by variations of the oxygen content in the Hg-1201, Hg-1212, and Hg-1223 phases. The only exception is the $Hg_{0.5}Fe_{0.5}Sr_2Ca_{1-x}Y_xCu_2O_y$ system with the non-isovalent $Y^{3+} \rightarrow Ca^{2+}$ cation substitution (that is similar to cases of the Bi-2212 and Tl-2212 systems) in samples with preliminary substitution for mercury by iron.

Figure 18A presents the results on the variation of the band filling. These data show that for all systems increasing doping level in the underdoped regime leads to a rise in the F values, like the case of other chain-free HTSC-families considered above. This can be obviously explained by a charging state of removed oxygen or a relation between valences of the calcium and yttrium ions (in case of the $Hg_{0.5}Fe_{0.5}BaCa_{1-x}Y_xCu_2O_y$ system). It is necessary to note that in case of both the $Y^{3+} \rightarrow Ca^{2+}$ substitution in the Hg-1212 phase and oxygen deficit increase in $HgBa_2Cu_{0.97}Zn_{0.03}O_y$ samples of the Hg-1201 phase the overall level of the F values is less as compared to other studied systems. This fact is obviously related to the effect of iron or zinc ions preliminarily introduced into mercury or copper positions in corresponding systems that leads to a disturbance of the charge balance in the lattice in the first case and to a partial distortion of the conduction band in the second one. Both these effects cause overall decrease in the F values in the above systems.

Results for the overdoped regime in case of Hg-based HTSC's are limited by six samples of the Hg-1201 and Hg-1212 phases. Nevertheless, the tendency to a sharp fall of the F values at transition to this doping regime observed for Bi- and Tl-based HTSC's can be revealed in this case as well (see Figure 18A).

The W_D values for near-optimally doped samples of phases with different number of copper-oxygen layers are $70 \div 140$ meV (Hg-1201), $80 \div 180$ meV (Hg-1212), and $90 \div 210$ meV (Hg-1223). The only exception is the $HgBa_2Cu_{0.97}Zn_{0.03}O_y$ samples in which the conduction band is a little wider than in case of other studied systems that can be explained by its partial distortion by zinc impurity as pointed above when discussing the band filling.

As seen from Figure 18B, the underdoped regime can be characterized by an appreciable band broadening. According to our calculations, the fraction of the localized states also increases (the C value decreases consistently with increasing S_{300}, though in a smaller degree as compared to other studied families). An exception is again the $HgBa_2Cu_{0.97}Zn_{0.03}O_y$ system where both W_D and C values remain almost unchanged. However, since the data for this system are small in number and predominantly presented by the near-optimally doped samples, they keep within a common tendency of the W_D and C parameters variation characteristic of the underdoped regime. Thus, a character of the modification of the band spectrum structure in this regime is analogous to cases of Bi- and Tl-based HTSC-systems and can be explained, in our opinion, by realization of the Anderson localization mechanism.

Although the data for the overdoped regime are limited in number, one can see from Figure 18B that there is an obvious tendency toward band narrowing in this doping range (the W_D value for the Hg-1201 phase decreases sharply from 145 meV at $S_{300} = 0$ μV/K to 80 meV at $S_{300} = -8$ μV/K). Thus, a character of the band spectrum modification in Hg-based HTSC's in this regime coincides with the ones established above for bismuth- and thallium-based superconductors as well.

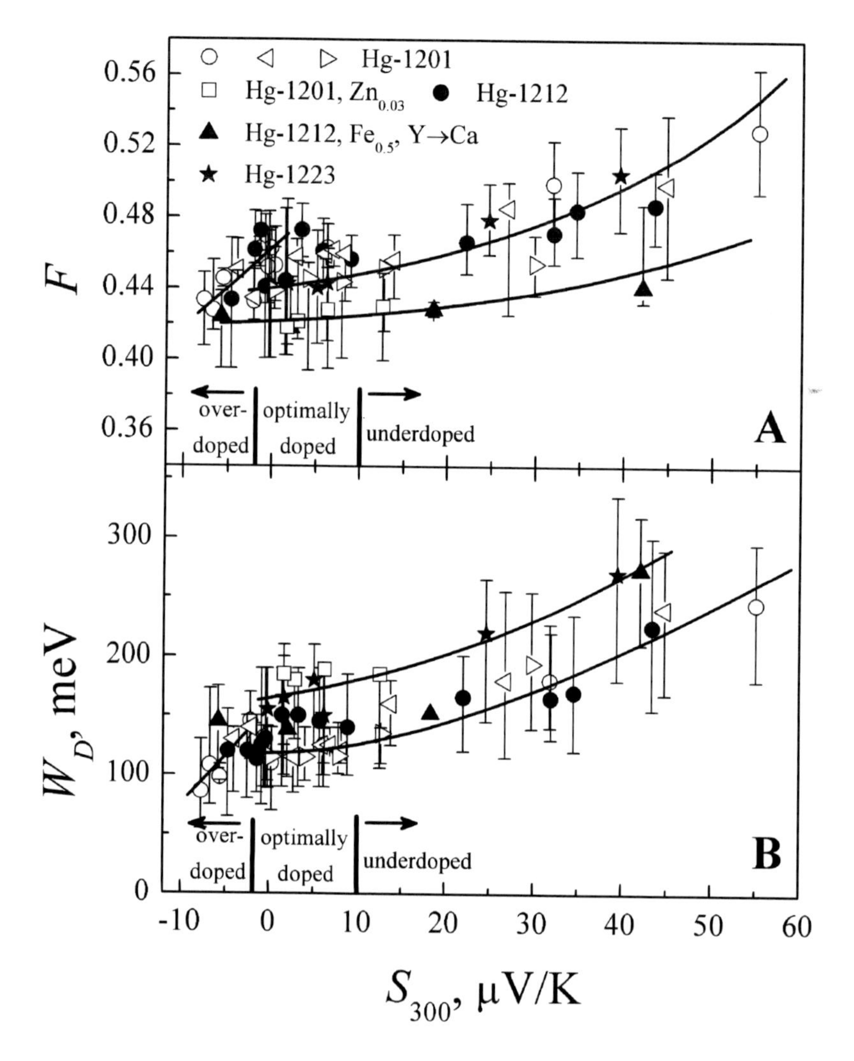

Figure 18. Doping dependences of band filling degree (A) and total effective bandwidth (B) in Hg-1201, Hg-1212, Hg-1223, and $HgBa_2Cu_{0.97}Zn_{0.03}O_y$ systems with various oxygen content and in $Hg_{0.5}Fe_{0.5}Sr_2Ca_{1-x}Y_xCu_2O_y$ system with different yttrium content.

6. Common Peculiarities of the Band Spectrum Transformation and Influence of its Parameters on the Critical Temperature Value

Summarizing all the results obtained we will establish in this Section the overall peculiarities of the band spectrum structure in the chain-free Bi-, Tl-, and Hg-based HTSC-systems and its transformation under both doping influence and increasing number of CuO_2 layers. Besides, we will compare the variations of the band parameters in the normal state and experimentally observed changes in the critical temperature.

It should be preliminary noted that we have earlier observed [19] the existence of the universal correlation between the T_c and W_D values in Y-based HTSC-systems doped by oxygen deficit or non-isovalent impurities in different lattice positions (in case they do not affect directly the plane copper atoms forming along with the oxygen ones the band responsible for the conduction process). The existence of this correlation was interpreted as follows. Since the increasing level of doping causes a band broadening, it initiates a general fall of the DOS function including its value at the Fermi level, $D(E_F)$. Though the mechanism of high-temperature superconductivity remains unclear, the T_c value increases with rising $D(E_F)$ in all the proposed models of this phenomenon, like the case of the classical Bardeen-Cooper-Shrieffer (BCS) theory. This allowed us to suppose that it is the doping-induced broadening of the conduction band leading to a decrease in the $D(E_F)$ value that is the main reason for suppression of superconductivity in the $YBa_2Cu_3O_y$ system under doping influence [19].

Keeping this finding in mind, let us consider the results obtained for Bi-, Tl-, and Hg-based HTSC-systems.

First of all we would like to stress that the values of the total effective bandwidth and the degree of the band filling with electrons characteristic for all studied systems are close to the data obtained earlier for Y-based HTSC's [18], [19], [23]-[25], [27], [29]. This testifies to a generality of the basic features of the band spectrum structure in HTSC-materials of different systems and allows one to assume that the existence of a narrow density-of-states peak in the band spectrum in an immediate neighborhood of the Fermi level is the feature having a great importance for realizing the high-temperature superconductivity effect. It is very likely that this peak arises owing to the van Hove singularity which presence in the HTSC's band spectrum was discussed by many authors (see, for example, review [92]). If so, an additional sharp peak in the density-of-states appears on the background of a "standard" wide band and it is this peak that predominantly determines electron transport phenomena in the studied HTSC-systems.

Let us first analyze a character and mechanisms of the influence of increasing number of CuO_2 layers in near-optimally doped samples on the parameters of the normal and superconducting states. As indicated above, the degree of the band filling is close to the $F \approx 0.5$ value and depends slightly on both the number of CuO_2 layers and a specific HTSC-family. Contrary to this, the total effective bandwidth in all the three investigated HTSC-families changes noticeably when going to the phases with a larger n. The ranges of W_D values, characteristic of the samples of near-optimally doped compositions are presented in Table 3. One can see that there is a tendency to a band broadening with increasing number of copper-oxygen layers. The only exception is the Bi-2201 phase where the W_D values are a little higher than for the Bi-2212 phase. In this respect we would like to remind that this system is also distinguished by superconducting properties. As mentioned in Section 4, the T_c value in the Bi-2212 phase is extremely low as compared to the Tl- and Hg-based HTSC's with a single CuO_2 layer while so considerable distinctions in T_c for different HTSC-families at $n = 2,3$ are not observed (see Figure 11). One can suppose that both these factors are related to the presence of additional specific features in the band spectrum structure of the Bi-2212 system.

Table 3. Values of the total effective bandwidth in near-optimally doped samples of studied HTSC-families with different number of copper-oxygen layers

Number of CuO_2 layers, n	W_D, meV		
	$Bi_2Sr_2Ca_{n-1}Cu_nO_y$	$Tl_2Ba_2Ca_{n-1}Cu_nO_y$	$HgBa_2Ca_{n-1}Cu_nO_y$
1	95 ÷ 200	55 ÷ 130	70 ÷ 140
2	90 ÷ 150	80 ÷ 165	80 ÷ 190
3	110 ÷ 175	90 ÷ 180	90 ÷ 210
4	–	120 ÷ 190	–

In our opinion, the modification of the band spectrum with increasing number of CuO_2 layers occurs owing to two alternative mechanisms (see Figure 19).

On the one hand, an increase in the number of copper-oxygen layers responsible for the formation of a narrow conduction band results in an increase in the total number of states in the band, i.e., in an increase in the DOS peak on the whole and, in particular, in its broadening. As a result, the $D(E_F)$ value increases (see Figure 19A). Note that the observed constancy of the band filling for optimally doped samples for the phases with different n implies that the ratio of the number of electrons to the total number of states in the band remains unchanged, i.e., an increase in the number of states in the band is accompanied by an increase in the number of free carriers. On the other hand, the overall complication of the crystal structure with increasing number of copper-oxygen layers renders the system with a larger n potentially more defective due to phase inhomogeneity and various structural defects. An additional evidence for an increase in the concentration of defects in more complicated phases is provided by our calculations of the state localization degree, which slightly increases when going from phases with $n = 1$ to those with a larger n (the C parameter decreases). This effect causes the band broadening due to the Anderson localization mechanism that results in a decrease of the $D(E_F)$ values (see Figure 19B).

In case of $n \leq 3$ the first effect appears more pronounced, therefore the $D(E_F)$ value increases that leads, accordingly, to an improvement of superconducting properties. When going to phases with $n > 3$, the second effect becomes dominant. As a result, despite a relative rise of the DOS peak owing to participation of a larger number of CuO_2 layers in its formation, the band broadening caused by a considerable quantity of structural defects leads to a decrease in the $D(E_F)$ values and, accordingly, to a T_c suppression. Thus the presented consideration can fully explain the T_c variation with increasing n for the studied systems (see Figure 11).

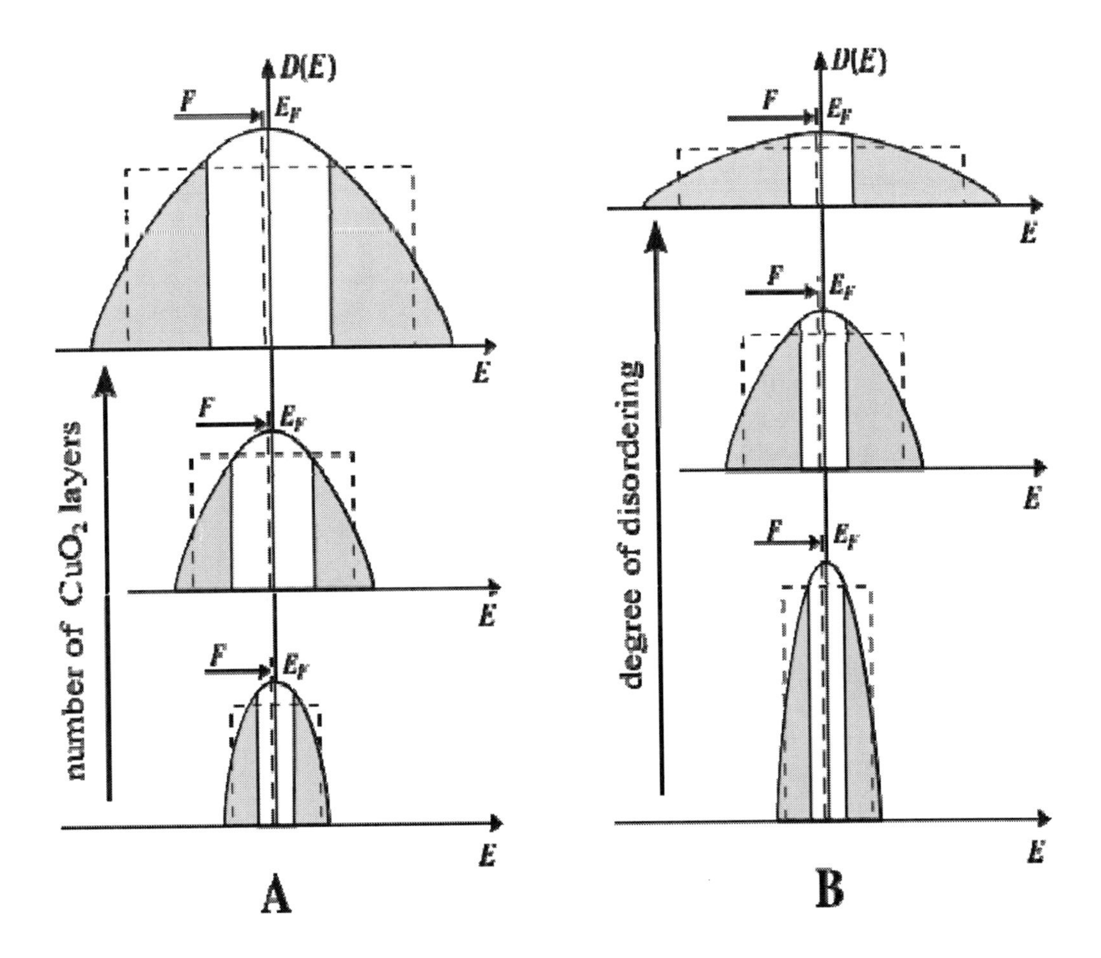

Figure 19. Two mechanisms of the band structure modification in studied HTSC-systems: band broadening and an increase in the DOS peak with increasing number of copper-oxygen layers (A); band broadening and an increase in the DOS peak with increasing disordering (B). The shaded regions correspond to localized states at the conduction-band edges, the rectangles show approximations used for thermopower calculations within the narrow-band model.

The next step consists in analyzing the modification of the band spectrum in studied HTSC's in different regimes of doping. As was pointed out in Section 5, when going from the optimally doped compositions to the underdoped regime, the changes in the band spectrum parameters in all the studied systems include an increase in F and a rather strong rise in W_D accompanied by a decrease in C. Such a modification of the band spectrum is similar to the case of doped samples of the $YBa_2Cu_3O_y$ system [19]. It allows us to conclude that the mechanisms of the band spectrum modification are identical in the studied bismuth-, thallium-, and mercury-based HTSC's and in Y-based superconductors. An increase in oxygen deficit and/or a level of non-isovalent doping lead to an enhancing degree of disordering in lattices and, in accordance with the Anderson mechanism, to the band broadening and an increase in the fraction of localized states on its edges. This corresponds well to a diagram in Figure 19B, the only difference from the effect of rising disordering under increasing number of CuO_2 layers is that the degree of the band filling increases in this case that results in a small shift of the Fermi level to higher energies. In its turn, it leads to a fall of the $D(E_F)$ value causing thereby the suppression of superconducting properties in the underdoped regime.

As for the overdoped regime, for all the studied HTSC-systems an increase in the doping level results in a significant band narrowing, an increase of the degree of its asymmetry, and a sharp increase in the fraction of localized states. The last fact points to a strong rise of lattice disordering and formation of a large number of structural defects. It is necessary to mention that according to [93], the most significant factor influencing the T_c value in the overdoped regime in case of thallium based HTSC's is a strong increase in the number of the oxygen defects located between the TlO layers that correlates with our results. We suppose that this circumstance leads to significant changes in the band spectrum structure causing a decrease in the $D(E_F)$ value and, accordingly, a superconductivity suppression. Anyway, the results obtained show that the modification of the band spectrum in the underdoped and overdoped regimes is caused by different mechanisms.

In concluding this Section we will present the data on a relationship between the T_c and W_D values. The $T_c(W_D)$ dependences plotted based on the results obtained for all the investigated systems in the underdoped regime are shown in Figure 20 (with error bars for the W_D value for each specific sample) together with the universal curve observed earlier for doped samples of the $YBa_2Cu_3O_y$ system [19]. The presented results allow us to draw two important conclusions. First, irrespectively of the doping type in all studied systems and for the phases with different number of CuO_2 layers within each of them the $T_c(W_D)$ dependence, like the case of Y-based HTSC's, has a nearly universal character. Moreover, increasing number of CuO_2 layers in each of studied HTSC-families leads to a uniform shift of this dependence to higher critical temperatures while maintaining its shape. One exception is the Bi-2223 phase that can be related to a limited data available for this system. In fact, the only sample in this case corresponds to an essential deviation from near-optimally doped composition. Thus, we have a well-grounded opportunity to assert that the band spectrum parameters in the normal state in all studied HTSC-systems influence directly the superconducting properties and the superconductivity suppression under doping in the underdoped regime occurs owing to the broadening of the conduction band causing reduction of the $D(E_F)$ value. Secondly, as seen from the data presented above, the T_c value decreases under the band broadening stronger when going from the yttrium system to the bismuth one and further to thallium- and mercury-based HTSC's, i.e. with increase of the maximal T_c value characteristic of the given system. Thereupon we would like to note that in the framework of different proposed mechanisms of high-temperature superconductivity the T_c value is determined by two main parameters, like the case of the classical BCS theory. They are the value of the density-of-states at the Fermi level and the value of the matrix element of an interaction leading to the electron pairing. Therefore, the observed difference in $T_c(W_D)$ dependences for various chain-free HTSC-systems allows to guess that in the systems demonstrating higher critical temperatures the increasing doping level induces, along with a decrease in the $D(E_F)$ value, a stronger fall of the value of a matrix element of pairing.

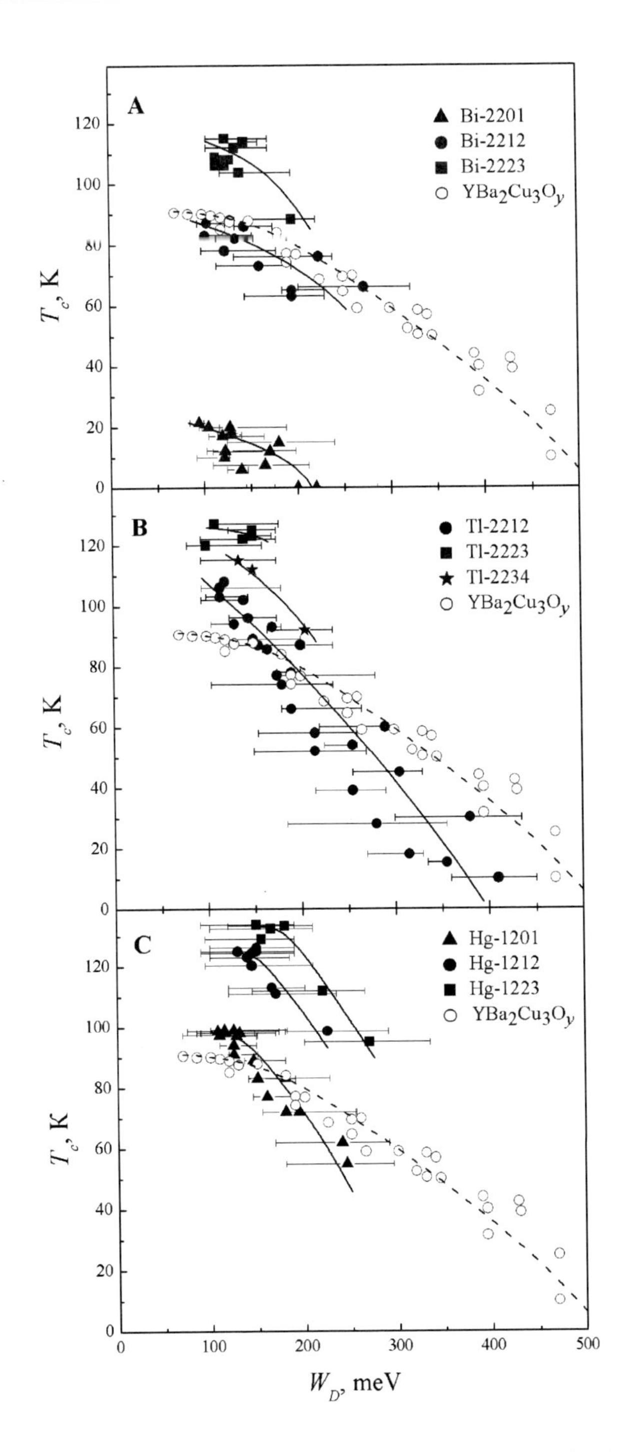

Figure 20. Correlation dependences of critical temperature on total effective bandwidth for different phases of Bi-, Tl-, and Hg-based HTSC's in the underdoped regime in comparison with data for doped $YBa_2Cu_3O_y$.

CONCLUSION

Thus, we have presented the results of the systematic study of the thermopower behavior in chain-free HTSC's of Bi-, Tl-, and Hg-based systems. All the experimental data on the $S(T)$ dependences were quantitatively analyzed within the narrow-band model that allowed us to determine the main parameters of the band spectrum and charge-carrier system and then to discuss their variation under both doping influence and increasing number of CuO_2 layers. As a result, the common peculiarities characteristic of the band spectrum structure and its transformation under variation of the sample composition in the studied systems have been revealed and analyzed together with the variation of the critical temperature value. The main results obtained can be summarized as follows.

1. An increase in the doping level in all the systems leads to a suppression of superconductivity both in the underdoped and overdoped regimes, although the T_c value changes to a different degree decreasing considerably stronger for the case of the overdoped samples. Increasing number of CuO_2 layers for all the studied HTSC-families leads to a rise of the T_c values for $n \leq 3$, further increase in n results in a suppression of superconductivity.

2. The thermopower values at T =300K for the samples of near-optimally doped compositions of all studied HTSC-systems are about 0 μV/K. The main distinctive features of the $S(T)$ dependences in this case are the presence of an extended region of almost linear decrease in thermopower with increasing temperature and a broad maximum on the $S(T)$ curves at temperatures well above T_c. In the underdoped regime the thermopower value increases under doping, the linearity of its temperature dependence, in most cases, gradually vanishes, and its maximum broadens and shifts to higher temperatures that coincides qualitatively with transformation of the $S(T)$ dependences under doping in Y-based HTSC's. In the overdoped regime an increasing level of doping leads to a rise of negative thermopower values, a distortion of the linearity of the $S(T)$ curves and disappearance of their maximum.

3. The temperature dependences of the thermopower for all the studied systems of chain-free HTSC's can be well described and quantitatively analyzed within the asymmetric narrow-band model. Despite using four model parameters, their values in all cases can be determined with rather insignificant errors allowing one to reveal the tendencies of changes in the band spectrum and charge-carrier system parameters both under varying number of copper-oxygen layers and increasing doping level in the underdoped and overdoped regimes.

4. The degree of the band filling with electrons for the samples of near-optimally doped compositions is about $0.43 \div 0.5$, the total effective bandwidth is in the range of $50 \div 210$ meV, the values of the C parameter characterizing the degree of a state localization lie in the range of $0.2 \div 0.45$, and the degree of the band asymmetry is small enough being in the range of $-(0.02 \div 0.05)$. Thus, the values of all parameters of the band spectrum and charge-carrier system for all the studied HTSC's differ insignificantly and are close to the values characteristic of Y-based superconductors. This points to a generality of the basic features of the band spectrum in different HTSC-systems and makes it possible to believe that a narrow peak of the density-of-states in an immediate vicinity of the Fermi level is the feature of the band spectrum structure being of great importance for realization of the high-temperature superconductivity effect.

5. Increasing number of copper-oxygen layers leads to a broadening of the conduction band while other parameters of the band spectrum remain almost unchanged. In case of the underdoped regime increasing level of both oxygen and cation doping results in a slight increase of the band asymmetry degree, a rise of the band filling with electrons, as well as an essential band broadening accompanied by increasing fraction of localized states at the band edges. At transition from the optimally doped compositions to the overdoped regime the degree of band filling decreases, the total effective bandwidth falls sharply, the fraction of localized states and the band asymmetry degree rise significantly. These results clearly indicate that the modification of the band spectrum in the underdoped and overdoped regimes is induced by physically different mechanisms.

6. The comparative analysis of the results obtained for the studied HTSC-systems with different changes in their compositions shows that in all cases variations of the critical temperature are directly related to the changes in the density-of-states at the Fermi level caused by the band spectrum transformation. This distinctly indicates the normal-state band spectrum and charge-carrier system parameters to impact directly the superconducting properties of HTSC-materials.

7. Modification of the band spectrum under various types of changes in samples composition is related to two different mechanisms influencing the superconducting properties in the opposite ways. First, the lattice disordering arising due to an increase in number of crystal structure defects leads to a band broadening and states localization at the band edges owing to the Anderson localization mechanism that causes a decrease in the $D(E_F)$ value. Secondly, an increasing number of copper-oxygen layers forming the conduction band results in an overall increase in the density-of-states in the band that induces a rise of the $D(E_F)$ value. The first mechanism is realized when going from optimally doped compositions to the underdoped regime and it leads to suppression of superconductivity under doping. When the number of copper-oxygen layers increases, the effects of both mechanisms are combined. At $n \leq 3$ the second mechanism is dominant in comparison with the first one, therefore the values of $D(E_F)$ and T_c increase with n. At $n > 3$ the second mechanism becomes more essential that leads to a decrease in the values of $D(E_F)$ and T_c.

8. In the underdoped regime the correlation dependence between the critical temperature and the total effective bandwidth has a universal character for each of the studied systems irrespectively of a doping type. This indicated that the main reason for the superconductivity suppression in this doping regime is the band broadening leading to a fall of the $D(E_F)$ value. The observed difference in the $T_c(W_D)$ dependences for different studied systems allows one to assume that non-isovalent doping gives rise to a stronger decrease in the matrix element of pairing in HTSC-systems with higher maximal values of the critical temperature.

ACKNOWLEDGMENTS

This work was supported by grant from the Ministry of Education and Science of the Russian Federation (Federal Target Program "Scientific and scientific-pedagogical personnel of innovative Russia" for 2009 – 2013 years, Contract No. P 1237). The authors are deeply indebted to Dr. E.V.Vladimirskaya for her help during the preparation of the manuscript.

References

[1] Ong N. P. In *Physical Properties of High Temperature Superconductors II*; Ginsberg D. M.; Ed.;World Scientific: Singapore, 1990; pp. 459-507.

[2] Kaiser A. B.; Uher C. In *Studies of High Temperature Superconductors*; Narlikar A.V.; Ed.; Nova Science Publishers: New York, 1991; Vol. 7, pp. 353-392.

[3] Iye Y. In *Physical Properties of High Temperature Superconductors III*; Ginsberg D.M.; Ed.; World Scientific: Singapore, 1992; pp. 285-361.

[4] Bar-Ad S.; Fisher B.; Ashkenazi J.; Genossar J. *Physica C.* 1988, 156, 741-749.

[5] Das A. N.; Ghosh B.; Choudhury P. *Physica C.* 1989, 158, 311-325.

[6] Eagles D. M.; Savvides N. *Physica C.* 1989, 158, 258-264.

[7] Genossar J.; Fisher B.; Ashkenazi J. *Physica C.* 1989, 162-164, 1015-1016.

[8] Trugman S. A. *Phys. Rev. Lett.* 1990, 65, 500-503.

[9] Kresin V. Z.; Wolf S. A. *Phys. Rev. B.* 1990, 41, 4278-4285.

[10] Moshchalkov V. V. *Physica B.* 1990, 163, 59-62.

[11] Cohn J. L.; Skelton E. F.; Wolf S. A.; Liu J. Z. *Phys. Rev. B.* 1992, 45, 13140-13143.

[12] Forro L.; Lukatela J.; Keszei B. *Solid State Commun.* 1990, 73, 501-505.

[13] Xin Y.; Wong K. W.; Fan C. X.; Sheng Z. Z.; Chan F. T. *Phys. Rev. B.* 1993, 48, 557-561.

[14] Newns D. M.; Tsuei C. C.; Huebener R. P.; van Bentum P. J. M.; Pattnaik P. C.; Chi C. C. *Phys. Rev. Lett.* 1994, 73, 1695-1698.

[15] Nagaosa N.; Lee P. A. *Phys. Rev. Lett.* 1990, 64, 2450-2453.

[16] Ikegawa S.; Wada T.; Yamashita T.; Ichinose A.; Matsuura K.; Kubo K.; Yamauchi H.; Tanaka S. *Phys. Rev. B.* 1991, 43, 11508-11511.

[17] Alexandrov A. S.; Bratkovsky A. M.; Mott N. F. *Phys. Rev. Lett.* 1994, 72, 1734-1737.

[18] Gasumyants V. E.; Kaidanov V. I.; Vladimirskaya E. V. *Physica C.* 1995, 248, 255-275.

[19] Gasumyants V. E. In *Advances in Condensed Matter and Materials Research*; Gerard F.; Ed.; Nova Science Publishers: New York, 2001; Vol. 1, pp. 135-200.

[20] Kaz`min S. A.; Kaidanov V. I.; Leising G. *Sov. Phys. Solid State.* 1988, 30, 1703-1705.

[21] Gasumyants V. E.; Kaz'min S. A.; Kaidanov V. I.; Smirnov V. I.; Baikov Yu. A.; Stepanov Yu. P. *Superconductivity*. 1991, 4, 1184-1203.

[22] Vladimirskaya E. V.; Gasumyants V. E.; Patrina I. B. *Phys. Solid State.* 1995, 37, 1084-1087.

[23] Vladimirskaya E.; Gasumyants V.; Patrina I. *Superlattices and Microstructures*. 1997, 21, Suppl. A, p.71-77.

[24] Gasumyants V. E.; Vladimirskaya E. V.; Patrina I. B. *Phys. Solid State*. 1997, 39, 1352-1357.

[25] Gasumyants V.E.; Vladimirskaya E.V.; Patrina I.B. Phys. Solid State. 1998, 40, 14-18.

[26] Gasumyants V. E.; Vladimirskaya E. V.; Elizarova M. V.; Ageev N. V. *Phys. Solid State*. 1998, 40, 1943-1949.

[27] Gasumyants V. E.; Vladimirskaya E. V.; Elizarova M. V. *Phys. Solid State*. 1999, 41, 350-354.

[28] Elizarova M. V.; Gasumaynst V. E. *Phys. Solid State*. 1999, 41, 1248-1255.

[29] Gasumyants V. E.; Elizarova M. V., Suryanarayanan R. *Phys. Rev. B*. 2000, 61, 12404-12411.

[30] Gasumyants V. E.; Elizarova M. V.; Vladimirskaya E. V.; Patrina I. B. *Physica C*. 2000, 341-348, 585-588.

[31] Gasumyants V. E.; Elizarova M. V.; Patrina I. B. *Supercond. Sci. Technol*. 2000, 13, 16000-1606.

[32] Elizarova M. V.; Matrynova O. A.; Potapov D. V.; Gasumyants V. E.; Mezentseva L. P. *Phys. Solid State*. 2005, 47, 434-438.

[33] Matrynova O. A.; Gasumyants V. E. *Phys. Solid State*. 48, 1223-1229.

[34] Ageev N. V.; Gasumaynts V. E.; Kaydanov V. I. *Phys. Solid State*. 1995, 37, 1171-1177.

[35] Gasumyants V. E.; Ageev N. V.; Vladimirskaya E. V.; Smirnov V. I.; Kazanskiy A. V.; Kaydanov V. I. *Phys. Rev. B*. 1996, 53, 905-910.

[36] Gasumyants V. E.; Ye M.; Vladimirskaya E. V.; Stolypina L. Yu.; Deltour R. *Physica C*. 1997, 282-289, 1267-1268.

[37] Matrynova O. A.; Gasumyants V. E. *Phys. Solid State*. 2007, 49, 1611-1616.

[38] Martynova O. A.; Gasumyants V. E. *Physica C*. 2008, 468, 394-400.

[39] Elizarova M. V.; Lukin A. O.; Gasumyants V. E. *Physica C*. 2000, 341-348, 1825-1828.

[40] Elizarova M. V.; Lukin A. O.; Gasumyants V. E. *Phys. Solid State*. 2000, 42, 2188-2196.

[41] Ageev N. V.; Gasumyants V. E. *Phys. Solid State*. 2001, 43, 1834-1844.

[42] Gasumyants V. E.; Ageev N. V.; Elizarova M. V. *Phys. Solid State*. 2005, 47, 202-213.

[43] Chen X. H.; Li T. F.; Yu M.; Ruan K. Q.; Wang C. Y.; Cao L. Z. *Physica C*. 1997, 290, 317-322.

[44] Sita D. R.; Singh R. *Physica C*. 1998, 296, 21-28.

[45] Sita D. R.; Singh R. *Mod. Phys. Lett*. 1998, 22, 475-488.

[46] Samuel E. I.; Bai V. S.; Sivakumar K. M.; Ganesan V. *Phys. Rev. B*. 1999, 59, 7178-7183.

[47] Singh R.; Sita D. R. *Physica C*. 1999, 312, 289-298.

[48] Ghorbani S. R.; Andersson M.; Rapp O. *Phys. Rev. B*. 2002, 66, 104519 (7 p.).

[49] Neeleshwar S.; Muralidhar M.; Murakami M.; Venugopal Reddy P. *Physica C*. 2003, 391, 131-139.

[50] Lal R.; Awana V. P. S.; Peurla M.; Rawat R.; Ganesan V.; Kishan H.; Narlikar A. V.; Laiho R. *J. Phys.: Condens. Matter*. 2006, 18, 2563-2571.

[51] Mori M.; Kameyama T.; Enomoto H.; Ozaki H.; Takano Y.; Sekizawa K. J. *Alloys & Compounds*. 2006, 408-412, 1222-1225.

[52] Gahtori B.; Lal R.; Agarwal S. K.; Ahsan M. A. H.; Rao A.; Lin Y. F.; Sivakumar K. M.; Kuo Y.-K. *J. Phys.: Condens. Matter*. 2007, 19, 256212 (12 p.).

[53] Ghorbani S. R.; Rostamabadi E. *Physica C*. 2008, 468, 60-65.

[54] Rao A.; Das A.; Chakraborty T.; Gahtori B.; Agarwal S. K.; Kumar Sarkar C.; Sivakumar K. M.; Wu K. K.; Kuo Y. K. *J. Phys.: Condens. Matter*. 2008, 20, 485212 (6 p.).

[55] Sun C.; Yang H.; Cheng L.; Wang J.; Xu X.; Ke S.; Cao L. *Phys. Rev. B*. 2008, 78, 104518 (7 p.).

[56] Gahtori B.; Agarwal S. K.; Chakraborty T.; Rao A.; Kuo Y.-K. *Physica C*. 2009, 469, 27-29.

[57] Ghorbani S. R.; Andersson M.; Rapp O. *Physica C*. 2003, 390, 160-166.

[58] Ghorbani S. R.; Andersson M.; Rapp O. *Physica C*. 2005, 424, 159-168.

[59] Ghorbani S. R.; Abrinaey F. *Int. J. Modern Phys. B*. 2009, 23, 5779-5788.

[60] Krylov K. R.; Ponomarev A. I.; Tsidilkovski I. M.; Tsidilkovski V. I.; Basuev G. V.; Kozhevnikov V. L.; Cheshnitski S. M. *Phys. Lett. A*. 1988, 131, 203-207.

[61] Fisher B.; Genossar J.; Patlagan L.; Reisner G. M. *Phys. Rev. B*ѣ 1993, 48, 16056-16060.

[62] Elizarova M. V.; Gasumyants V. E. *Phys. Rev. B*. 2000, 62, 5989-5996.

[63] Fujii T.; Terasaki I.; Watanabe T.; Matsuda A. *Phys. Rev. B*. 2002, 66, 024507 (5 p.).

[64] Konstantinovic Z.; Le Bras G.; Forget A.; Colson D.; Jean F.; Collin G.; Ocio M.; Ayache C. *Phys. Rev. B*. 2002, 66, 020503 (4 p.).

[65] Namgung C.; Irvine J. T. S.; Binks J. H.; Lachowski E. E.; West A. R. *Supercond. Sci. Technol*. 1989, 2, 181-184.

[66] Choi M.; Kim J. *Phys. Rev. B*. 2000, 61, 11321-11323.

[67] Okada Y.; Ikuta H.; Kondo Y.; Mizutani U. *Physica C*. 2005, 426-431, 386-389.

[68] Dumont Y.; Ayache C.; Collin G. *Phys. Rev. B*. 2000, 62, 622-625.

[69] Kondo T.; Takeuchi T.; Mizutani U.; Yokoya T.; Tsuda S.; Shin S. *Phys. Rev. B*. 2005, 72, 024533 (9 p.).

[70] Gaojie X.; Qirong P.; Zejun D.; Li Y.; Yuheng Z. *Phys. Rev. B*. 2000, 62, 9172-9178.

[71] Munakata F.; Matsuura K.; Kubo K.; Kawano T.; Yamauchi H. *Phys. Rev. B*. 1992, 45, 10604-10608.

[72] Zhao X.; Sun X.; Wu W.; Li X. G. *J. Phys.: Condens. Matter*. 2001, 13, 4303-4311.

[73] Sekhar M. C.; Suryanarayana S. *Physica C*. 2004, 415, 209-219.

[74] Awana V. P. S.; Moorthy V. N.; Narlikar A. V. *Phys. Rev. B*. 1994, 49, 6385-6387.

[75] Tanatar M.; Yefanov V.; Dyakin V.; Akimov A. I.; Chernyakova A. P. *Physica C*. 1991, 185-189, 1247-1248.

[76] Obertelli S.; Cooper J.; Tallon J. *Phys. Rev. B*. 1992, 46, 14928-14931.

[77] Martin C.; Hejtmanek J.; Simon Ch.; Maignan A.; Raveau B. *Physica C*. 1995, 235-239.

[78] Kaneko T.; Hamada K.; Adachi S.; Yamauchi H. *Physica C*. 1992, 197, 385-388.

[79] Shu-Yuan L.; Li L.; Dian-Lin Z.; Duan H.-M.; Hermann A. *Europhys. Lett*. 1990, 12, 641-646.

[80] Alcacer L.; Almeida M.; Braun U.; Goncalves A.; Green S.; Lopes E.; Luo H.; Politis C. *Modern Phys. Lett. B*. 1988, 2, 923-928.

[81] Bhatia S. N.; Chowdhury P.; Gupta S.; Padalia B. D. *Phys. Rev. B*. 2002, 66, 214523 (8 p.).

[82] Keshri S.; Mandal J.; Mandal P.; Poddar A.; Das A.; Ghosh B. *Phys. Rev. B*. 1992, 47, 9048-9054.

[83] Poddar A.; Bandyopadhyay B.; Chattopadhyay B. *Physica C*. 2003, 390, 120-126.

[84] Kaneko T.; Hamada K.; Adachi S.; Yamauchi H.; Tanaka S. *J. Appl. Phys*. 1992, 71, 2347-2350.

[85] Chen F.; Xiong Q.; Xue Y. Y.; Huang Z. J.; He Z. H.; Lin Q. M.; Clayhold J. A.; Chu C. W. *Texas Center for Supercond*. 1996, preprint, No. 96:006.

[86] Yamamoto A.; Hu W.-Z.; Yajima S. *Phys. Rev. B*. 2000, 63, 024504 (6 p.).

[87] Yamamoto A.; Minami K.; Hu W.-Z.; Izumi M.; Tajima S. *Phys. Rev. B*. 2002, 65, 104505 (7p.).

[88] Yamamoto A.; Minami K.; Hu W.-Z.; Izumi M.; Tajima S. *Physica C*. 2001, 357-360, 34-38.

[89] Kandyel E. *Physica C*. 2005, 422, 102-111.

[90] Chu C. W. *Texas Center for Supercond*. 1996, preprint No. 96:011.

[91] Karppinen M.; Lee S.; Lee J. M.; Poulsen1 J.; Nomura1 T.; Tajima S.; Chen J. M.; Liu R. S.; Yamauchi H. *Phys. Rev. B*. 2003, 68, 054502 (5 p.).

[92] Markiewicz R. S. *J. Phys. Chem. Solids*. 1997, 58, 1179-1310.

[93] Hermann A. M.; Duan H. M.; Kiehl W.; Paranthaman M. *Physica C*. 1993, 209, 199-202.

In: Superconductivity
Editor: Vladimir Rem Romanovskiĭ

ISBN 978-1-61324-843-0

Chapter 5

INVESTIGATIONS OF THE OVERDOPED STATE IN POLYCRYSTALLINE $R_{1-X}CA_XBA_2CU_3O_{7-\Delta}$ SAMPLES (R=Y, EU, GD, ER)

E. Nazarova*[1]*, K. Nenkov*[2,3]*, A. Zaleski*[4]*,
***K. Buchkov*[1] *and A. Zahariev*[1]**

[1]Georgi Nadjakov Institute of Solid State Physics
Bulgarian Academy of Sciences
72 Tzarigradsko Chaussee Blvd.,
Sofia, Bulgaria
[2]IFW, Leibniz Institute for Solid State and Materials Research
P.O. BOX 270016,
Dresden, Germany
[3]International Laboratory of High Magnetic Fields and Low Temperatures
95 Gajowicka str.,
Wroclaw, Poland
[4]Institute of Low Temperature and Structure Research
PAS, 2 Okolna str.,
Wroclaw, Poland

ABSTRACT

Chemical substitution is a simple technological decision for the generation of a large number of nano-sized defects. Replacement of some atoms in the structure with other ones of different valence, ion radius and magnetic moment can result in deformation of lattice, charge distribution, appearance of oxygen and other atoms vacancies or disorder. Partial replacement of rare earth ion in $RBa_2Cu_3O_{7-\delta}$ by Ca^{2+} ion with similar ionic radius but lower valence value, provides additional holes and makes the overdoped region of the T(p) phase diagram accessible for study. Investigation of intra- and inter-granular effects in overdoped polycrystalline $R_{1-x}Ca_xBa_2Cu_3O_{7-\delta}$ (R=Y, Eu, Gd, Er and x=0, 0.025, 0.05,

0.10, 0.20, 0.30) samples was carried out by using different experimental technique. X-ray powder diffraction analysis and SEM were used for the examination of phase formation and microstructure. AC magnetic susceptibility measurements (of fundamental and third harmonics) as a function of temperature, DC magnetic field, frequency and AC magnetic field amplitude were exploited for the investigation of a large number of properties: differentiation between inter- and intra- granular effects, estimation of intergranular J_{cinter}, activation energy for TAFF, irreversibility line or non-linear dissipation processes. DC magnetization measurements were performed at fixed temperatures as a function of magnetic field and intragranular J_{cintra} was obtained. Transport measurements (resistivity vs. temperature and I-V characteristics) at different magnetic fields were used for the establishment of vortex-glass-vortex-liquid phase transition and scaling parameters.

It was established that low level overdoping leads to the improvement of intragranular critical current, flux pinning and irreversibility field at 77 K making it higher than in non substituted, fully oxygenated YBCO samples. Temperature dependence of intergranular critical current showed that it is governed by the S-I-S type joints between the grains. For highly overdoped samples the suppression of intragranular critical current and flux pinning has been observed. The intergranular critical current is characterized by S-N-S type. Indirect evidence suggests that this is a result of carriers' phase separation supporting the idea that the quality of superconducting condensate is strongly influenced by overdoping. The field dependence of activation energy for TAFF shows that 2D pancake vortices are characteristic of underdoped samples, while 3D vortex system exists in overdoped ones. Hole concentration is an essential parameter that controls many properties in HTS. We also investigated how the vortex dynamics was influenced by the doping effect. By using the third harmonics signal of AC magnetic susceptibility the irreversibility line was determined. The existence of vortex glass-vortex liquid phase transition was confirmed also by transport measurements. The scaling behavior of E-J data in Ca substituted samples is similar to the other polycrystalline YBCO samples. Previously established morphology dependence of dynamic exponent (**z**) was confirmed. However, **z** values are smaller than the usually reported for non-substituted YBCO. Static exponent (ν) shows a tendency for field dependence. These observations have been explained with the peculiarities of Ca substituted samples.

1. INTRODUCTION

Revealing the mechanism of nonconventional superconductivity is a goal stimulating the progress of condensed matter physics since the discovery of high temperature super conductors (HTSC) in 1986 [1]. Investigations have been focused predominantly on a different family of cuprate perovskites. However, recently discovered iron-based superconductors [2] and their modifications, with critical temperature (T_c) reaching almost 55 K, also attract researchers' attention. In cuprates and iron pnictides chemical substitution plays a key role in inducing the superconducting phase. Therefore it is often used as a method for studying essential problems in superconducting materials. Replacement of some atoms in the structure with the other of different valence, ion radius and magnetic moment can result in deformation of the lattice, charge redistribution, appearance of vacancies or disorder. All these reflect the macroscopic properties of superconductors like T_c, pinning, critical current

(J_c), irreversibility field (H_{irr}). In iron-based materials it is believed that magnetism could be suppressed either by doping or applying pressure in order bulk-phase superconductivity to appear [3].

In spite of their widely varying structure (different number of CuO_2 planes) HTS have a universal phase behavior. The scaled transition temperature, T_c/T_{cmax} is a parabolic function of the plane hole concentration (p) with onset, maximum and termination for p = 0.05, 0.16 and 0.27, respectively [4]. The optimal doping at p_{opt} = 0.16 is connected with the maximal critical temperature, T_{cmax}, while at critical doping (p = 0.19) the superconductivity is most robust due to the maximization of the condensation energy and superfluid density at T = 0 [5]. Samples with hole concentration lower than the optimal (p_{opt}) are underdoped and in the extreme limit $p < 0.05$ –antifferomagnetic insulators. By variation of oxygen content in the range of $6.4 \leq z \leq 6.92$ underdoped 123-type samples can be obtained. Samples with $p>p_{opt}$ are overdoped. They become metals behind the upper limit of $p>0.28$. Non-oxygen deficient $YBa_2Cu_3O_7$ is slightly overdoped [6], while for heavily overdoped 1-2-3 superconductors additional hole supply is needed. The overdoped side of the phase diagram ($T/T_{cmax}(p)$) is less studied and not accessible without introduction of additional carriers. Partial replacement of rare earth ion in $RBa_2Cu_3O_{7-\delta}$ by Ca^{2+} ion with similar ionic radius but lower valence value, provides extra holes and makes the overdoped region of the T(p) phase diagram attainable for study. This is the situation for highly oxygenated samples. In the opposite case of insufficient carriers Ca doping is crucial for the appearance of superconductivity. In oxygen deficient samples ($z \leq 6.4$) Ca substitution increases the number of carriers and superconductivity with T_c=34 K was observed in $Y_{0.7}Ca_{0.3}Ba_2Cu_3O_{6.02}$ polycrystalline sample [7]. Ca doping induces also superconductivity in $PrBa_2Cu_3O_{7-\delta}$ which is a semiconductor [8].

It is surprising, however, that an increase of the charge concentration in planes leads to a decrease of T_c on the overdoped side. Different explanations of this fact exist. It is proposed that the reduction of T_c is a result of increasing the 3D electron dynamics when superconductivity can be strongly suppressed by electron-electron scattering and scattering on structural disorder (M-O interlayer between the CuO_2 sheets) [9]. In the presence of nonmagnetic substituents (like Zn on the Cu site and Ca on the R site) the impurities also are strongly scattering centers. Zn, for example, induces local magnetic moments upon its neighboring copper atoms [10] and the spin vacancy creates a perturbation of the local antiferromagnetic correlations [11]. Another reliable explanation of the T_c suppression is the appearance of phase separation in the superconducting and normal – metal ground state.

Two types of doped holes appear in the overdoped region. In one of them holes condensed into low energy superconducting state. The other extra holes are expelled from the superconducting islands to the surrounding area forming non-superconducting metallic sea. The bulk superconductivity is established due to the Josephson coupling (or proximity effect) between the islands [12], [13]. In this sense the overdoped HTSC systems are fundamentally different from conventional BCS superconductors where all the normal-state charge carriers participate in the superfluid [12]-[14]. This picture is supported by observation of percolative superconductivity, for example in overdoped single crystals (LSCO, Tl-2201, Bi2212) [13].

We carry out a detailed study of overdoped state in polycrystalline $R_{1-x}Ca_xBa_2Cu_3O_{7-\delta}$ superconducting samples, where R = Y, Eu, Gd, Er and x = 0; 0.025; 0.05; 0.10; 0.20 and 0.30. The influence of overdoping on the inter- and intra-granular effects, critical current, irreversibility line, flux pinning, activation energy for thermally assisted flux flow (TAFF) and AC losses was investigated.

2. Samples Preparation and Experimental Details

Samples preparation: $R_{1-x}Ca_xBa_2Cu_3O_z$ samples were prepared from the high purity $BaCO_3$, R_2O_3, CuO and $CaCO_3$ powders. The obtained mixture was ground and three times sintered by the standard solid state reaction. The first sintering was at 900°C in the flowing oxygen for 21 hours. After grinding the powder was sintered at 930°C for the second time at the same atmosphere followed by slow cooling and additional annealing at 450°C for 2 hours. Tablets were pressed, sintered for the third time at 950°C for 23 hours and subsequently annealed at 450°C for 23 hours. In order to prepare highly overdoped samples final annealing at 450°C sometimes has been carried out for 48 or 100 hours.

In order to prepare underdoped sample $Y_{0.8}Ca_{0.2}Ba_2Cu_3O_{7-\delta}$ for comparison it has been treated at nitrogen atmosphere.

Thermally quenched samples were produced during the third procedure. After 21 hours of sintering at 950°C in oxygen, the temperature was reduced to 500°C (for YCaBCO-500); 400°C (for YCaBCO-400) and 320°C (for YCaBCO-320), respectively. Annealing was performed for 48 hours followed by rapid quenching of the tablet from the corresponding temperature to room temperature.

Samples investigations: Standard X-ray powder diffraction analysis was used for examination of specimens' structure. The samples microstructure was investigated by SEM Philips 515.

The AC magnetic susceptibility was investigated with commercial 7000 Lake Shore Cryogenics, Inc. susceptometer, MagLab-Oxford 7000 susceptometer and PPMS – Quantum Design at different AC field amplitudes, frequencies and temperatures.

The DC magnetization was investigated using a Quantum Design SQUID Magnetometer. The full hysteresis cycle was recorded after the specimen was zero field cooled to the desired temperature.

The transport measurements were performed on Quantum Design PPMS. In order to prevent the sample from Joule heating effect the DC current was applied for a very short time - 0.002 sec. Thick current leads have been used, soldered to the sample's surface. The boundary values for I, V and power have been specified and the measurement was impossible in case that some of these values were exceeded. The voltages were detected with an error of several nanovolts. The applied magnetic field was perpendicular to the current direction. I-V measurements used for investigation of scaling behavior have been performed on sample with $x = 0.025$ with dimensions: $S = 0.116x0.211$ cm^2 and $L = 0.420$ cm, and sample with $x = 0.20$ with $S = 0.184 \times 0.283$ cm^2 and $L = 0.340$ cm, where S is the sample's cross-section and L is the distance between the voltage leads. The electric field E, current density J and resistivity ρ have been found according to the relations: $E = V/L$; $J = I/S$ and $\rho = E/J$.

The values of the critical current, J_c, were obtained by the E-J characteristics at given temperature according to the offset criterion. The tangent was drawn to E-J curve at interception point with the electric field criterion 10 μV/cm. The critical current is determined as the current where this tangent extrapolates to zero electric field.

Two types of measurements have been performed: resistivity vs. temperature and I-V curves at constant temperatures. Both measurements were done at different magnetic fields ranging from 0.1 T to 6.9 T.

Temperature dependence of critical current density at zero magnetic field was also investigated for some samples.

3. SAMPLES CHARACTERIZATION

The X-ray diffraction analysis at room temperature was performed for all investigated samples $R_{1-x}Ca_xBa_2Cu_3O_z$ (R = Y with x = 0; 0.025; 0.05; 0.10; 0.2; 0.25 and 0.3 and R = Eu, Gd, Er with x = 0; 0.2; 0.25 and 0.3). Independently of R element all samples with x=0 - 0.2 show a single phase which can be indexed on the basis of 1-2-3 orthorhombic type structure. By detailed analysis in the range $2\theta \approx 30°- 40°$, where the main peaks of $BaCuO_2$ phase occur it is shown that detectable amount of this impurity phase exists only in samples with $x \geq 0.25$ [15]-[16]. The appearance of $BaCuO_2$ phase indicates that Ca starts to substitute not only for R element but also for Ba [17], [18]. In this case some cation deficiency on R and Cu positions will take place. In our samples $BaCuO_2$ was not found in the X-ray detection limit in the specimens with x = 0 - 0.2, indicating that Ca substitutes predominantly for R element in them. In spite of that we do not expel the limited Ca substitution on the Ba site for x = 0.20. It is shown by neutron refinement that the occupancy of Ca on the Y site is close to 100% up to x = 0.15 but it is reduced for higher Ca content, for example x_{eff}= 0.18 in case of x = 0.20 [19]. In polycrystalline samples with increasing the Ca content substitution on Y and Ba position is moved from 3 : 1 towards 1 : 1 [18].

On increasing x the double peaks at $2\theta \approx 54.5°$ show tendency towards a triplet peak structure [16], which has been identified with the appearance of ortho-II phase [20].

The unit cell lattice parameters **a**, **b** and **c** are determined and orthorhombicity **(b-a)/(b+a)** was calculated. It has been found that independently of the type of R species **a** lattice parameter remains almost constant. The **b** parameter and orthorhombicity decrease on increasing Ca concentration (Figure 1) in consistency with the appearance of ortho-II phase. It means that the structure loses oxygen from the chains and total oxygen decreases monotonically with increasing x. This is supported also by the increasing value of **c** lattice parameter [16] and has been established by direct oxygen measurements for different $R_{1-x}Ca_xBa_2Cu_3O_z$ compounds [15], [21]-[22].

According to XRD analysis fully deoxygenated Ca-doped compound $Y_{0.8}Ca_{0.2}Ba_2Cu_3O_z$ (denoted as Y0.8-N) has tetragonal crystal structure. It has been indicated by the following peaks: (103), (110), (200), (006), (213) and (116). Lattice parameters of tetragonal structure are a = 3.849 Å and c = 11.818 Å [23]. Using the correlation between the **c** lattice parameter and the oxygen content in Ca substituted samples [24], [25], we estimate the oxygen content in Y08-N sample to be 6.0 ÷ 6.2. In fact, the sample consists of two types of tetragonal unit cells: some of them Ca substituted, others non substituted but both without chain oxygen. Thus, the number of carriers per CuO_2 plane is determined only from Ca concentration (p = x/2 = 0.1), i.e. the sample is highly underdoped. Its critical temperature, determined as the onset of $\chi'(T)$ dependence, is between 40 and 50 K and also confirms the underdoping.

Microstructure of ceramic samples depends on many technology conditions like stoichiometry, rates of heating and cooling, temperatures of sintering and annealing, duration of these procedures, environment, grinding, pressing of tablets and others. In spite of the fact that so many factors are important it is found that Ca substitution leads to finer grain structure crystallization in polycrystalline $RBa_2Cu_3O_{7-\delta}$ samples [26], [27] and ruthenocuprates as well [28]. Our SEM investigations on polycrystalline samples with R = Y and Gd also show that increasing the Ca content leads to the formation of smaller grains. In Figure 2(a, b, c) SEM images (with equal magnification x5000) are presented for $Y_{1-x}Ca_xBa_2Cu_3O_z$ samples with x = 0.025; 0.10 and 0.20, respectively.

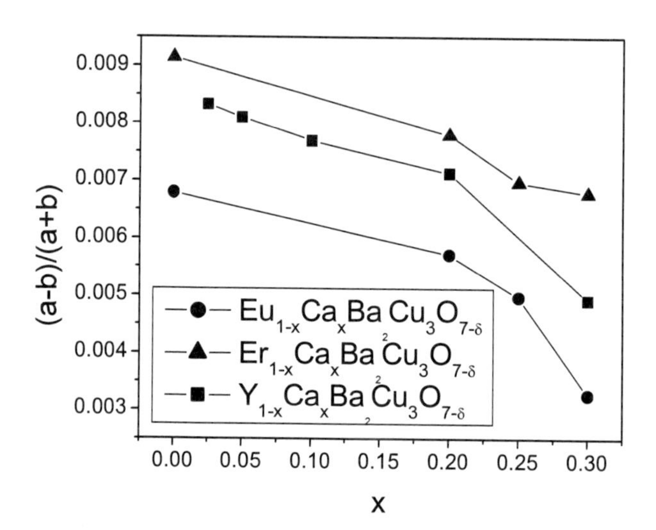

Figure 1. Orthorhombicity (a-b)/(a+b) vs. Ca concentration (x) in different superconducting systems.

Figure 2. (a, b, c) SEM images of $Y_{1-x}Ca_xBa_2Cu_3O_z$ samples with x=0.025; 0.10 and 0.20, respectively. In all micrographs the magnification is 5000 and marker is 10 μm.

It is seen from Figure 2 that on increasing the Ca content the grain size decreases and for x = 0.20 grains are the smallest with irregular shape.

4. INTRAGRANULAR PROPERTIES

Intragranular critical current

In small magnetic fields (~ 100 Oe) the low intergranular critical current (J_{cinter}) in polycrystalline HTSC samples is rapidly depressed to the values $\leq$ 10 A/cm^2. Thus $J_c(B)$ is successfully described by a model of an array with strongly superconducting grains joined by Josephson junctions. M(H) measurements are very useful for the determination of intragranular critical current (J_{cintra}) as the applied magnetic field breaks down the intergrain connections. By applying Bean critical state model [29] J_{cintra} is calculated according to the relationship $J_c = 15\Delta M/d$, where $\Delta M = M^+ - M^-$ is the difference of the magnetic moments between ascending and descending field branches of the hysteresis (in emu/cm^3) and **d** is the average grain size (in cm) within the ceramic samples.

In Figure 3 M(H) dependences for $Y_{1-x}Ca_xBa_2Cu_3O_{(7-\delta)}$ (x = 0.025 and 0.05) , $Y_{1-y}Pr_yBa_2Cu_3O_{(7-\delta)}$ (y = 0.05) and non substituted YBCO are presented at T = 4.2 K.

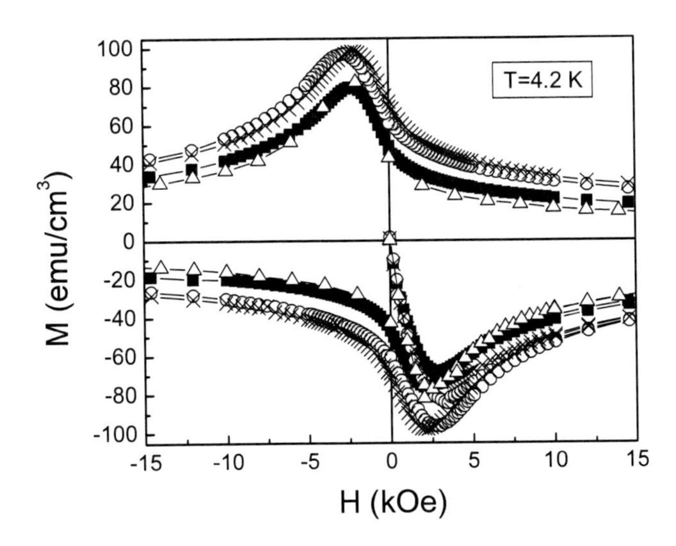

Figure 3. M-H curves for ■ – non substituted YBCO, Δ - Y(Ca)BCO – with x = 0.05, ○ - Y(Ca)BCO – with x = 0.025 and x - Y(Pr)BCO – with y = 0.05.

The following important observations should be mentioned: (A) The two branches of the hysteresis loops for ascending and descending field are almost symmetric about the axis M = 0 for all samples. (B) Small amount of Pr or Ca substitution improved the pinning ability of the 1-2-3 system and enlarged the hysteresis curve when compared with the non-substituted YBCO. (C) The small amount of Ca substitution (x = 0.025) in the YBCO sample produced a hysteresis loop with almost the same ΔM as Pr (y = 0.05) substitution did [30]. A question arises from this fact. If the pinning generated by Pr substitution is connected with the magnetic moment of the Pr atom then what is the pinning mechanism obtained by substitution with non-magnetic Ca which gives the similar result?

Actually two factors determined the intragranular critical current: the pinning force of the vortices arising from the inhomogeneity introduced by substitutions (J_c) and depairing critical current density $J_{co} = 4H_c/3\sqrt{6}\mu_o\lambda$, where λ is the London penetration depth and H_c is the thermodynamic critical field [31]. It has been shown that the increased amount of Pr substitution increases London penetration depth and suppresses H_c [32].

In case of Ca substitution it is established that λ(5K) increases when the level of overdoping is enhanced [33]. Thus the depairing critical current will decrease when Pr or Ca concentration increases resulting in the suppression of ΔM and J_{cintra}, respectively. We observe this effect when M(H) curves for $R_{1-x}Ca_xBa_2Cu_3O_{(7-\delta)}$ (R = Y, Eu, Gd, Er) samples are recorded with $x \geq 0.2$ [16, 34]. In Figure 4 and Figure 5 M(H) curves for samples with R=Er (x = 0; 0.3) and R = Y (x = 0; 0.2) are presented, respectively. Large amount of Ca substitution shrinks the hysteresis curves when compared with that for non substituted samples.

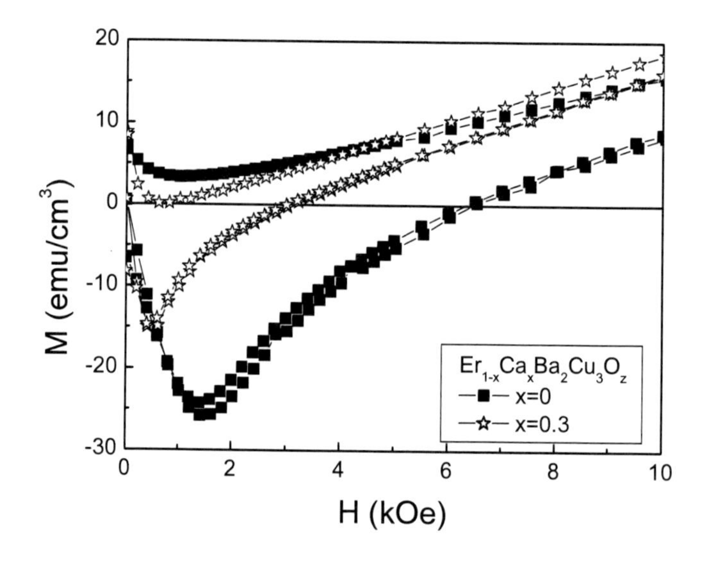

Figure 4. M-H curves for $Er_{1-x}Ca_xBa_2Cu_3O_{7-\delta}$ samples with x = 0 and x = 0.30 at T = 20 K.

In the Figure 5 hysteresis loops for two different $Y_{0.8}Ca_{0.2}Ba_2Cu_3O_{(7-\delta)}$ samples are presented: (a)-48 hours oxygenated (Y08-48) and (b) previous sample oxygenated for another 52 hours at 450°C (Y08-100). Sample (b) is more overdoped with $T_{cmidpoint}$ approximately 10 K lower than sample (a). Nevertheless it shows smaller ΔM resulting in weaker J_{cintra} as two samples should have almost identical grain size. The penetration peak (maximal initial magnetization due to full penetration) for sample (b) coincides with the central peak (maximal magnetization during the magnetic field cycle) suggesting percolative superconductivity. In HTSC, due to small coherent length (ξ), the energy cost for the creation of phase boundaries is small, what facilitates phase separation [12], [35]-[36]. In the case of slight overdoping, the non-superconducting regions are small and their dimensions are probably comparable with the coherent length. As a result, a broad M(H) curve is obtained due to the effective pinning. When the doping considerably exceeds the optimal concentration the non-superconducting clusters coalesce. The increasing of their dimensions reduced the pinning and the critical current is significantly suppressed.

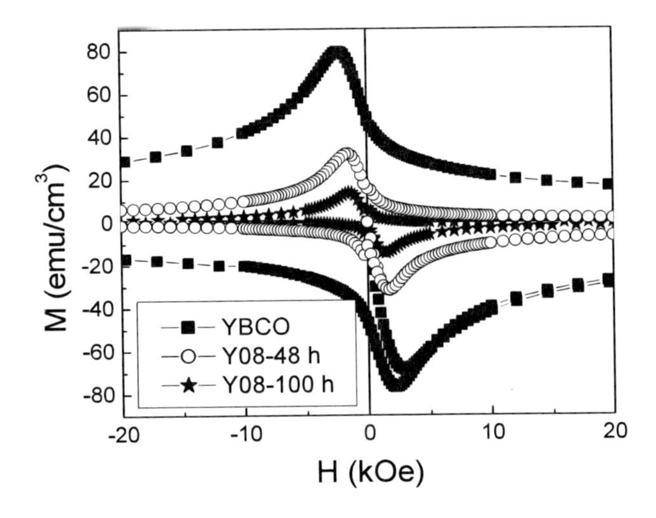

Figure 5. M-H curves for $Y_{1-x}Ca_xBa_2Cu_3O_{7-\delta}$ samples with x = 0 and x = 0.20 at temperature 4.2 K. Samples $Y_{0.8}Ca_{0.2}Ba_2Cu_3O_{7-\delta}$ are different time (48 h and 100 h) oxygenated.

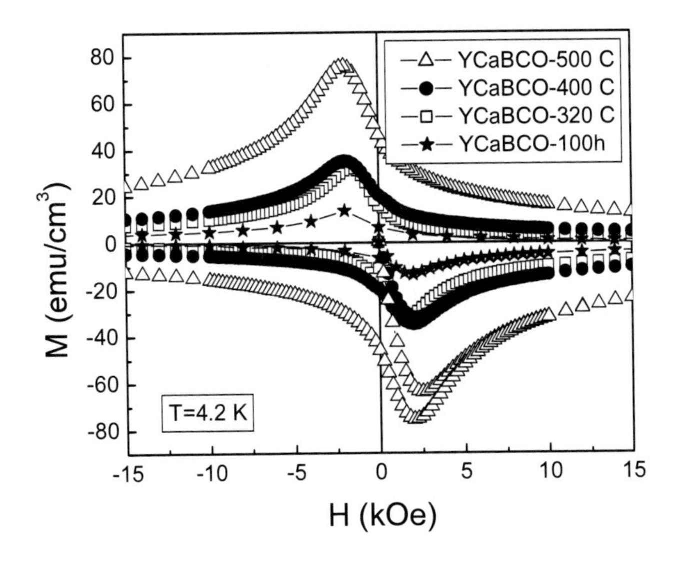

Figure 6. M-H curves for $Y_{0.8}Ca_{0.2}Ba_2Cu_3O_{7-\delta}$ samples quenched at indicated temperatures Sample (b) $Y_{0.8}Ca_{0.2}Ba_2Cu_3O_{7-\delta}$ oxygenated 100 h is also presented.

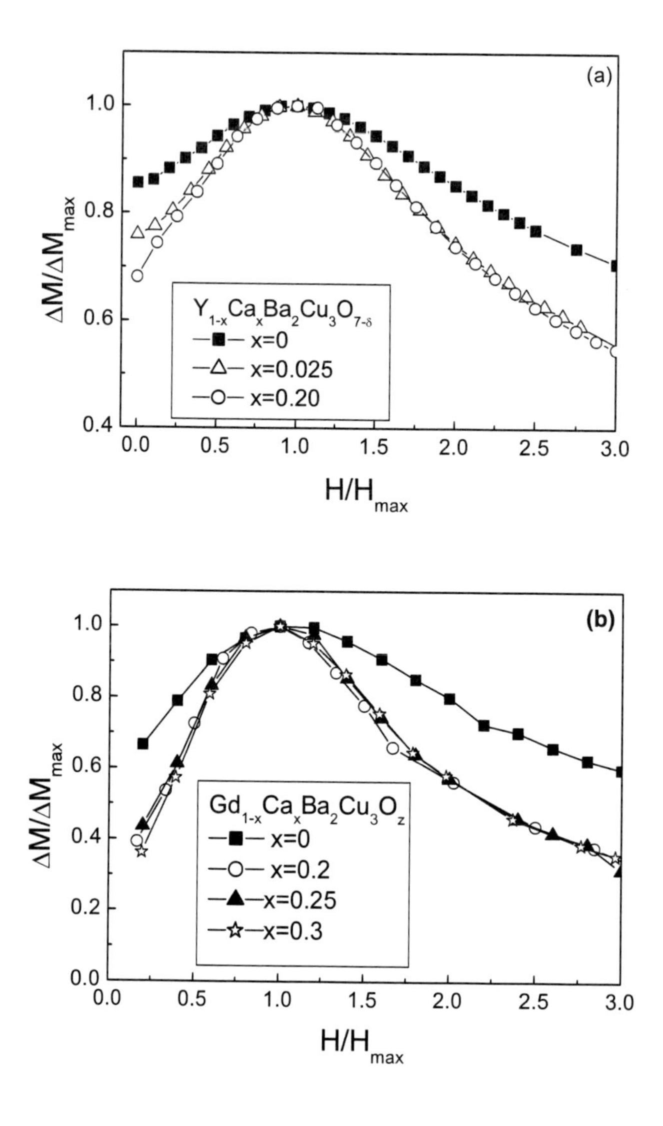

Figure 7. Plots of $\Delta M/\Delta M_{max}$ versus H/H_{max} for $Y_{1-x}Ca_xBa_2Cu_3O_{7-\delta}$ samples with x = 0, 0.025. and 0.20 (a) and for $Gd_{1-x}Ca_xBa_2Cu_3O_{7-\delta}$ with x = 0.20, 0.25 and 0.30 at T = 4.2 K.

As a verification of this hypothesis we investigated the M(H) curves of samples with significant Ca amount (x = 0.2) but quenched at different temperatures (320°C, 400°C, 500°C) [37]. The higher the quenching temperature, the more defects (oxygen vacancies, disorder) are formed in the sample. In spite of the fact that oxygen loss decreases carrier concentration, ΔM for quenched samples increases (Figure 6) when compared with the strongly overdoped sample (Y08-100). Thus quenching decreases the overdoping, resulting in T_c enhancement. It is known that the critical current is a parameter highly dependent upon the material processing. However, quenching ensures not only different materials characteristics, but strongly influences the level of overdoping and the quality of the superconducting condensate, respectively.

In order to separate different flux pinning mechanisms in substituted and non-substituted samples, the normalized loop with $\Delta M/\Delta M_{max}$ was plotted against the reduced field b = H/H_{max}, with H_{max} - the field of maximum loop width ΔM_{max} for two superconducting systems $Y_{1-x}Ca_xBa_2Cu_3O_{7-\delta}$ and $Gd_{1-x}Ca_xBa_2Cu_3O_{7-\delta}$ (Figure 7a, b). From this figures we see that all substituted samples are scaled on a single curve different from the curve for the non-substituted sample. This means that the pinning in Ca-substituted specimens has a common nature. Most likely those are not oxygen vacancies as their number increases with x and should enhance ΔM.

Irreversibility line

According to the Bean critical state model the hysteretic non-linear relationship between the magnetization and the external magnetic field exists in type II superconductors as a result of flux pinning. This brings about higher harmonics generation of AC magnetic susceptibility. However, the AC magnetic response is frequency dependent, which is not accounted for in the initial model. In its extended version, dynamic losses related to the thermally activated vortex motion are included [38], [39].

The vortex motion generates a resistive state described by the Ohmic dependence $E = \rho J$, where $\rho(B,T)$ is J independent and in the E-J characteristics zones appear with linear behavior. This is the so-called linear response of type II superconductors to an AC magnetic field. It is connected to the thermally assisted flux flow (TAFF) regime, characterized by thermally activated vortex hopping and flux flow (FF) regime – dominated by viscous motion of the vortex liquid. In fact, FF regime could also be non-linear at high vortex velocities due to the non-equilibrium distribution of quasi-particles in the vortex core and to the decrease of viscosity [40]-[43]. The E-J characteristics have a non-linear behaviour, when the vortex motion is governed by thermally activated flux creep (FC) and sample's resistivity is J-dependent.

In general, the AC susceptibility technique is a useful tool for the investigation of different flux-dynamic regimes and especially the third harmonics signal which is a very sensitive method for their identification [44]-[46]. The AC susceptibility signal could be investigated as a function of temperature, DC magnetic field intensity, AC magnetic field amplitude and frequency. By a variation of these parameters different dynamic regimes could be studied. For their proper identification the experimental results are compared to susceptibility curves simulated by numerical calculations of the non-linear diffusion equation for the magnetic field [47]. The observations of high harmonics signals (χ_n) imply the non-linear response of the flux system. Non-linearity is not always a precondition to irreversibility (for example, non-linear flux flow). The irreversibility in the magnetization behaviour can be a result of bulk pinning, surface or edge barrier effects and contributes to the third harmonics (χ_3) signal appearance. At certain conditions (superposition of DC and AC magnetic fields when $H_{ac} \ll H_{dc}$) the flux creep is non-linear and the irreversible dynamical regime determines the χ_3 signal [44], [46]. In this case the third harmonics signal is closely related to the flux depinning (irreversibility) line (IL) and is often used for its determination [48]-[50].

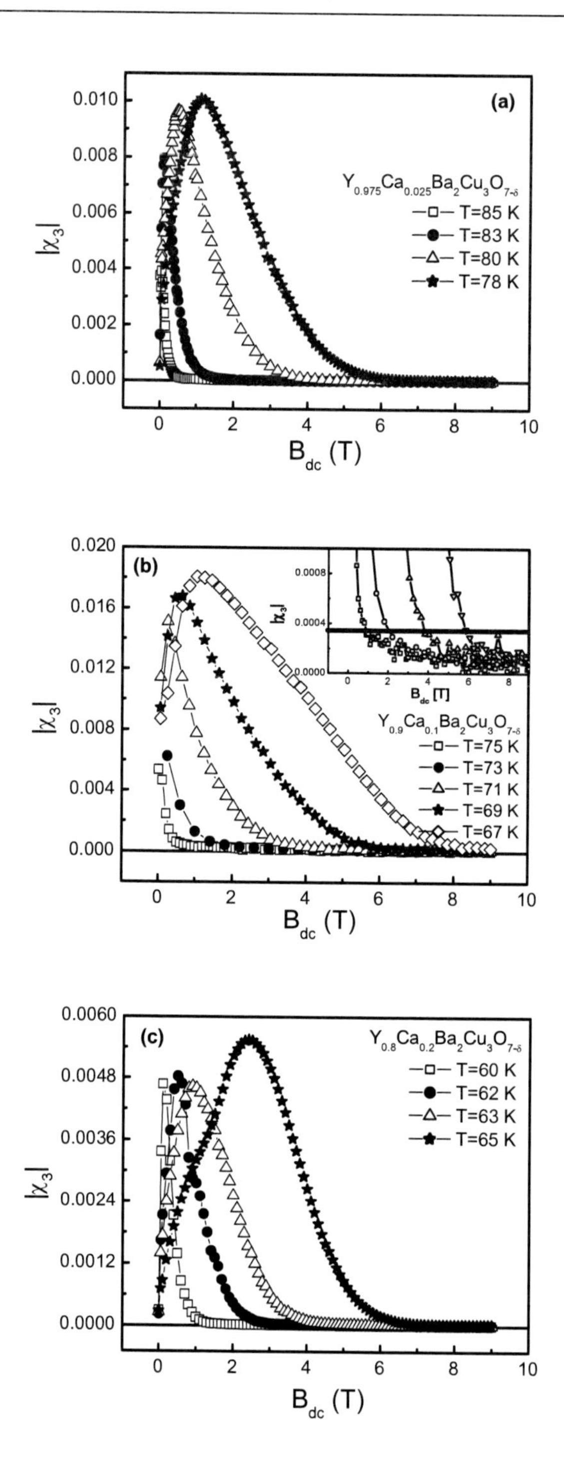

Figure 8. Module of third harmonics AC magnetic susceptibility as a function of DC field at different temperatures (H_{ac} = 0.1 Oe and f = 1000 Hz) for the sample with x = 0.025 (a), for the sample with x = 0.10 (b) and for the sample with x = 0.20 (c). The inset at (b) shows the noise level two orders of magnitude smaller than the signal at peak position.

For the determination of IL we used the module of the third harmonics signal, ($|\chi_3|$), as a function of a DC magnetic field (H_{dc}) at fixed temperature and small AC magnetic field amplitude ($H_{ac} < H_{dc}$) [51]. In Figure 8(a,b.c) $|\chi_3|$ vs. H_{dc} dependences are presented for all investigated samples (with x = 0.025, 0.10 and 0.20). The criterion used to determine the H_{irr} value (above which the vortex liquid occur) is the deviation of $|\chi_3|$ from the noise signal at fixed temperature. The signal which is two orders of magnitude smaller than that in the peak position is referred to noise signal (see inset of Figure 4b). In this manner the value where the $|\chi_3|$ signal starts to increase is determined within one percent error. This responds to no more than 0.1 T error when the H_{irr} is determined.

The obtained irreversibility lines are presented in Figure 9(a). On increasing the overdoping they are shifted to lower temperatures. Thus for a given magnetic field the position of IL for the highly overdoped sample (x = 0.20) is by about 20 K lower than that for the sample with x = 0.025. The irreversibility field for $Y_{0.975}Ca_{0.025}Ba_2Cu_3O_{7-\delta}$ at 77 K is estimated to be about 7 T. This value is higher than the reported for YBCO polycrystalline sample, which is about 5 T [52]. Usually, the IL is fitted by the dependence:

$$B_{irr}(T) = B_{irr}(0)(1-T/T_c)^n \quad (1)$$

predicted by the collective pinning theory [53], where **n** is a model dependent parameter. The assumption for a 3D-2D flux-line transition due to the decoupling of conducting layers leads to n = 3/2 [54], while flux-line melting model using a Lindemann type melting criterion predicts n = 2 [55]. The double logarithmic plots of the irreversibility field versus $[1-(T/T_c)]$ are presented for all samples in Figure 9(b). The slope of the lines, which yields the values of **n**, grows on increasing the overdoping. The value of n for $Y_{0.975}Ca_{0.025}Ba_2Cu_3O_{7-\delta}$ is close to 2 (2.23 ± 0.08) which might indicate the existence of glass-liquid phase transition. The **n** value increases when the overdoping grows and the values for the other two samples (with x=0.10 and x = 0.20) are large (3.96 ± 0.05 and 6.36 ± 0.150, respectively) within such interpretation. In fact, using only the **n** value, it is impossible to identify what kind of irreversibility is determinant in different samples. Additional research have been performed to this effect.

Magnetization measurements could be helpful to discriminate bulk pinning from surface barrier effects [56]. The lightly overdoped sample (x = 0.025) exhibits a broad, almost symmetric with respect to M = 0 curve (Figure 3) The highly overdoped sample (x = 0.20, oxygenated for 100 hours) exhibits a significant reduction of the hysteresis loop width as compared to the non-substituted one (Figure 5). The other important fact is the almost zero magnetization of the descending branch of M(H) curve for the $Y_{0.8}Ca_{0.2}Ba_2Cu_3O_{7-\delta}$ sample. This is a telling demonstration of the presence of Bean-Livingston surface barrier [57], [58]. As a result of the competition between the attractive force acting between the vortex and its mirrored antivortex, and repulsive force between the vortex at the surface and the shielding currents, a surface barrier appears. It prevents the vortex penetration in the sample up to field H_p, (identical to the thermodynamic critical field H_c for the smooth surface) and does not prevent their exit. Surface imperfections allow partial penetration thus reducing the penetration field down to H_{c1} [59]. In our case the highly overdoped sample $Y_{0.8}Ca_{0.2}Ba_2Cu_3O_{7-\delta}$ (oxygenated for 100 hours) shows M(H) curve determined by surface barrier effects.

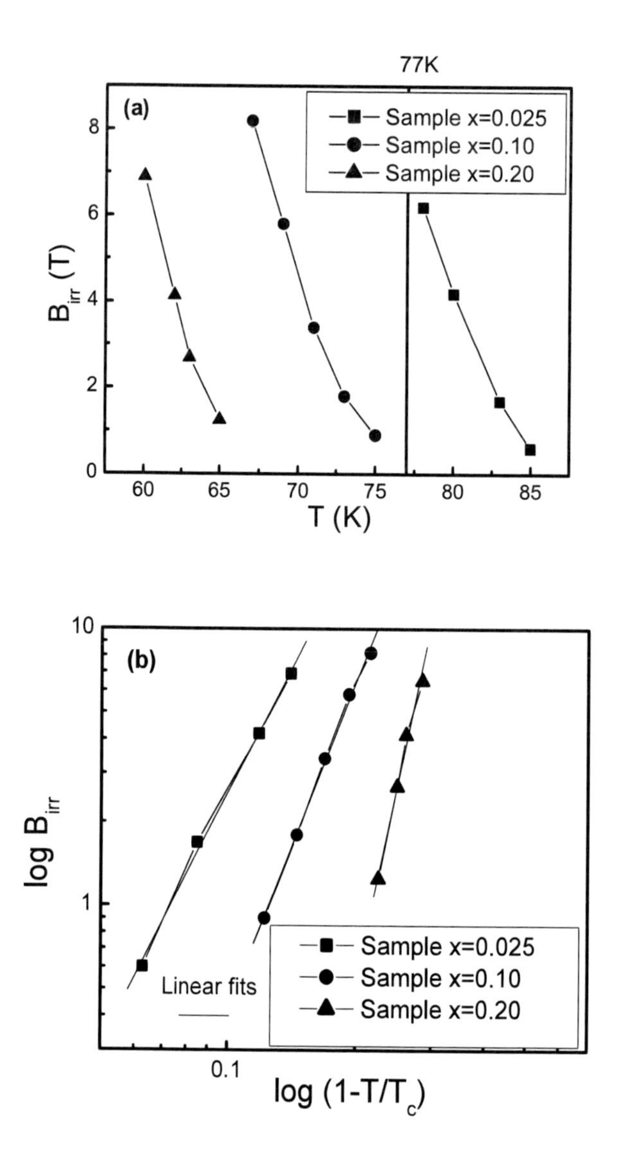

Figure 9. Plot of the irreversibility fields against temperature (irreversibility line) for $Y_{1-x}Ca_xBa_2Cu_3O_{7-\delta}$ samples with x = 0.025, 0.10 and 0.20. The 77 K boundary is explicitly indicated (a). The double logarithmic plots of irreversibility field vs. $[1-(T/T_c)]$ for all samples (b).

In order to determine $H_p(T)$ for the overdoped sample $Y_{0.8}Ca_{0.2}Ba_2Cu_3O_{7-\delta}$ (oxygenated for 48 hours) the offsets of M(H) curves at different temperatures (4 K, 10 K, 20 K, 30 K, 40 K, 50 K and 60 K) were recorded. It has been established that penetration field monotonically decreases when the temperature increases (600 Oe, 535 Oe, 398 Oe, 270 Oe, 188 Oe, 155 Oe and 70 Oe, respectively). According to [59]

$$H_p(T) = H_c(0)[(1 - t^2)(1 + t^2)^{1/2}]\exp(-2T/T_0), \tag{2}$$

where $H_c(0)$ is thermodynamic critical field at T = 0, t = T/T_c and T_0 is a characteristic temperature. By plotting $\ln[H_p/(1 - t^2)(1 + t^2)^{1/2}]$ versus T, both $H_c(0)$ and T_0 were determined to be 810 Oe and 10.44 K, respectively. The irreversibility field, in the case of vortex penetration through a surface barrier (at $T > T_0$) [60], is determined by the expression:

$$H_{irr} \approx H_{c2}(T_0/2T)\exp(-2T/T_0), \tag{3}$$

where H_{c2} is the upper critical field. The experimental data for H_{irr} were collected at $T>T_0$. In Figure 10 these data are presented and fitted using the above-mentioned dependence. This is another consideration supporting the assumption that the experimental results for the overdoped $Y_{0.8}Ca_{0.2}Ba_2Cu_3O_{7-\delta}$ sample are influenced by the surface barrier effects. As the specimen surfaces have similar quality, the appearance of the surface barrier effect is associated with the overdoping of the sample and decreasing of grain size. The last resulted in growing of grains boundary surfaces.

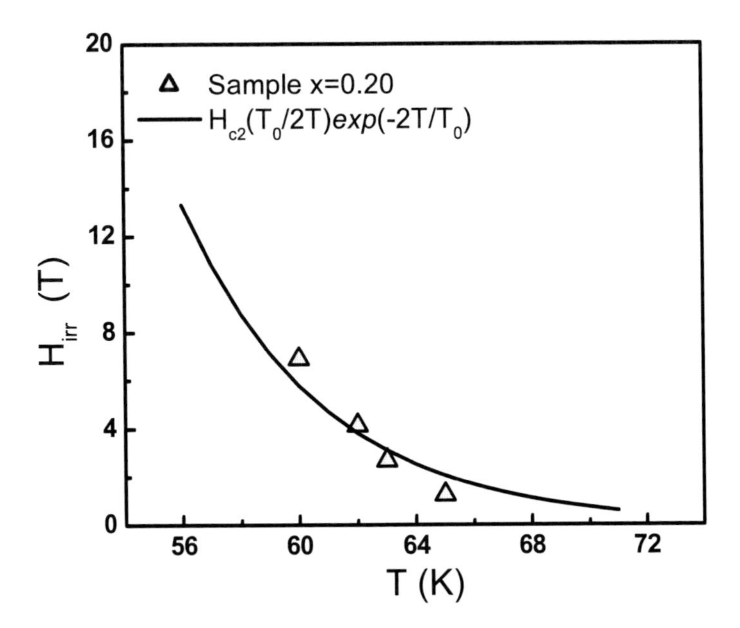

Figure 10. Plot of the irreversibility fields vs. temperature for the sample $Y_{0.8}Ca_{0.2}Ba_2Cu_3O_{7-\delta}$ (48 h oxygenated) fitted by exponential law $H_{irr} \approx H_{c2}(T_0/2T)\exp(-2T/T_0)$, where T_0 = 10.44 K and the upper critical field H_{c2} is the fitting parameter.

In fact low Ca substitution (2.5 at. %) in polycrystalline YBCO sample enhances the irreversibility field up to 7 T at 77 K and makes the hysteresis loop (at 4.2 K) broader than that for non substituted YBCO. Further increasing of Ca content shifts the IL to lower temperatures, deteriorates bulk pinning, and observed irreversibility is dominated by the surface barrier effects.

Scaling behavior of current-voltage characteristics

Originally the evidence for presence of vortex-glass phase in HTSC comes from transport experiments [61], [62]. It has been shown that at given temperature $T = T_g$ a power law describes the I-V curve. At $T < T_g$ a negative curvature on the log I vs. log V plot appeared indicating the presence of vortex-glass state. On increasing the temperature at $T > T_g$ the typical for vortex creep phenomena positive curvature is observed.

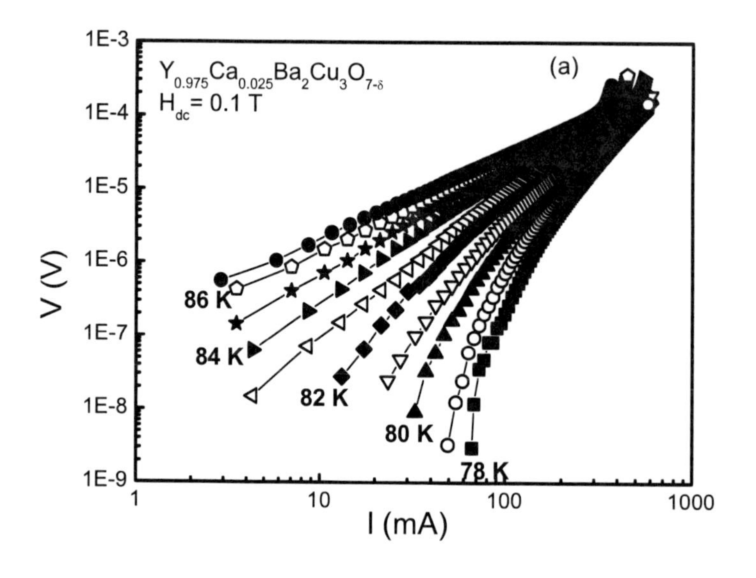

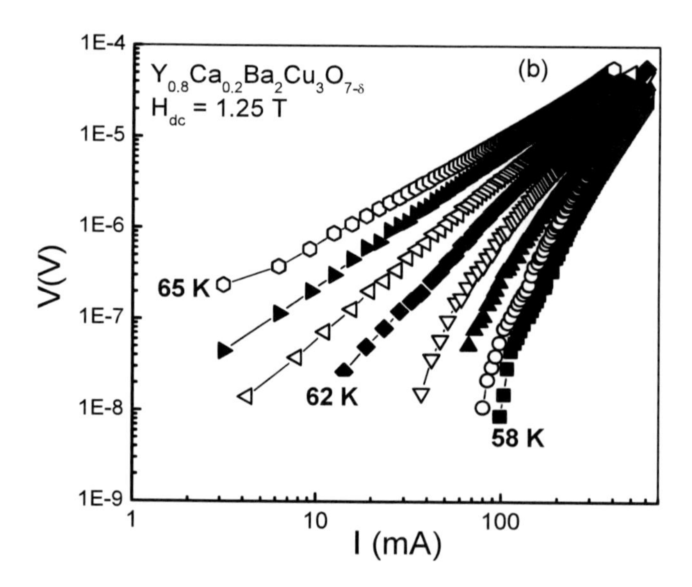

Figure 11. The I-V curves at given constant temperatures for a sample with x = 0.025 at B = 0.1 T (a) and for a sample with x = 0.20 at B = 1.25 T (b).

In order to verify the existence of vortex glass state in the investigated here samples the I-V characteristics have been measured at fixed magnetic field and different temperatures. For these measurements it is very important to keep constant temperature and to protect sample from the Joule heating. In Figure 11 (a) and Figure 11 (b) logI-logV curves are presented for the $Y_{0.975}Ca_{0.025}Ba_2Cu_3O_z$ and $Y_{0.8}Ca_{0.2}Ba_2Cu_3O_{7-\delta}$ samples at B = 0.1 T and 1.25 T, respectively, and different temperatures. It is seen from these figures that the negative curvatures, characteristic for vortex-glass state, exist in both samples. For the $Y_{0.8}Ca_{0.2}Ba_2Cu_3O_{7-\delta}$ specimen we found that the best linear fit with the least square deviation is reached for the I-V dependence at $T_g = 64$ K and for $Y_{0.975}Ca_{0.025}Ba_2Cu_3O_z$ at $T_g = 85$ K, respectively. At temperatures higher than T_g the vortex liquid phase appeared. The I-V curves in Figure 11(b) were obtained at magnetic field 1.25 T determined from the first point of irreversibility line (1.25 T; 65 K) for the sample with x = 0.20. The obtained result for T_g coincides very well with the previously determined IL and suggests that the flux creep contribution cannot be excluded from the $|\chi_3(H_{dc})|$ signal.

The best evidence of a phase transition near T_g is the scaling behavior existence [62]. The I-V isotherms should collapse onto two master curves above and below T_g, representing the liquid and glass states of the flux lines. The vortex correlation length, ξ_g and the relaxation time τ_g in the vortex glass state are the quantities diverging at the transition temperature T_g according to the relations:

$$\xi_g \propto \left|T - T_g\right|^{-\nu} \quad \text{and} \quad \tau_g \propto \xi_g^z, \tag{4}$$

where v and z are the static and dynamic exponents, respectively. The vortex-glass-vortex-liquid phase transition is analyzed by using the scaling relation [63]:

$$E(J) \approx J\xi_g^{D-2-z}\varepsilon_\pm\left(J\xi_g^{D-1}\Phi_0 / k_B T\right), \tag{5}$$

where E is the electric field, J is the current density, Φ_0 is the flux quantum, k_B is Boltzmann constant, D is the dimensionality of the system under consideration and $\varepsilon_\pm$ are the scaling functions above and below the glass transition temperature T_g. After determining the proper value of $T_g(H)$ and critical exponents from the experiment the scaling should be provided. Following the described analysis experimental results for many different samples have been scaled.

Recently non-power law I-V dependences have been observed in similar $Y_{1-x}Ca_xBa_2Cu_3O_{7-\delta}$ samples [26]. Experiments have been carried out at very small magnetic fields (up to 20 Oe) and only intergranular flux pinning is probed. We used higher magnetic fields in the range 0.1 T – 6.9 T, when the intragranular pinning is active. As it has been already discussed small amount of Ca substitution increased the pinning ability of the YBCO samples (Figure 3). It was shown also that Ca substitution leads to finer grain structure crystallization in polycrystalline $RBa_2Cu_3O_{7-\delta}$ samples (Figure 2). But the interrelations among the average distance between the vortex lines ($\alpha = (\Phi_0/B)^{1/2}$), the vortex correlation length, ξ_g, in glass state, and grain size in the sample are important for the vortex-glass-vortex-liquid phase transition. The above discussed characteristics of Ca substituted samples make them different in a sense from ordinary YBCO polycrystalline specimens. Establish ment of a scaling behavior in them will confirm the method's generality and will show that overdoped Ca substituted samples in spite of their peculiarities present no exception. It is important to mention yet that the examination of the vortex dynamics in HTSC is essential not only from fundamental but from practical point of view as well. The vortex movement generates dissipation in type II superconductors, which should be controlled when the practical application of these materials is considered.

The transport measurements have been performed for a wide temperature range at an interval of 1K and fixed magnetic fields: for sample with x = 0.025 (0.1 T; 0.6 T; 1.69 T; 4.14 T and 6.9 T) and for sample with x = 0.20 (0.1 T; 1.25 T; 2.69 T; 4.14 T and 6.9 T) [64]. According to the model at phase transition temperature a power-law behavior is expected between voltage and current $E(J, T = T_g) \sim J^{(z+1)/(D-1)}$ or $\rho(J, T = T_g) \sim J^{(z-1)/(D-1)}$ [62]. The latter dependence was used for determination of the critical exponent z.

Samples' dimensionality is a question under discussion [61], [65]. A 3D phase transition is accepted in a finite temperature range around the transition temperature when analytical treatments of layered system are made. Thus, at T_g, layers are coupled at all length scales, but start to decouple at higher temperatures [65].

Following the model predictions, the region of constant resistivity (ρ_{lin}) at small current density is investigated above $T_g(H)$. The resistivity should vanish at $T_g(H)$ according to the relation $\rho_{J\to 0} \sim (|T-T_g|/T_g)^{\nu(z-1)}$. The experimental ρ vs. T dependences at different magnetic fields, have been plotted as ρ_{lin} against $(T-T_g)/T_g$ on a log-log scale. The plot is linear in the critical regime with a slope ν(z-1). Thus the ν value has been determined when the z is known. In Figure 12(a, b) the ρ vs.($|T-T_g|/T_g$) dependences measured at small current (20 mA) and indicated magnetic fields are presented for the both samples. The ν values are obtained with an error of 1-5% .

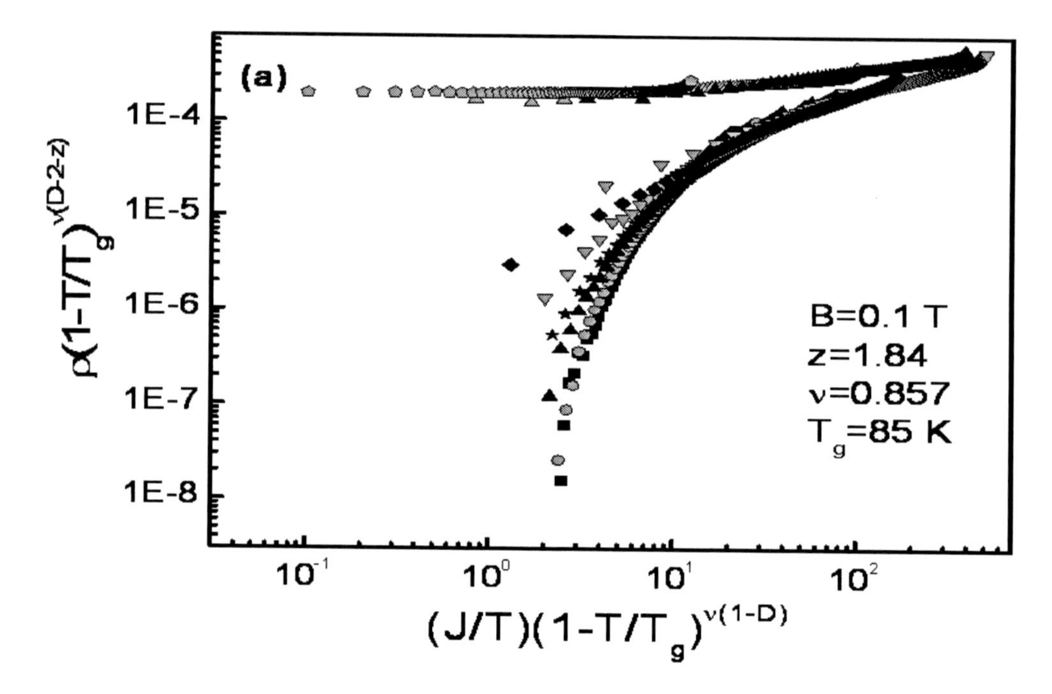

Figure 12. (Continued).

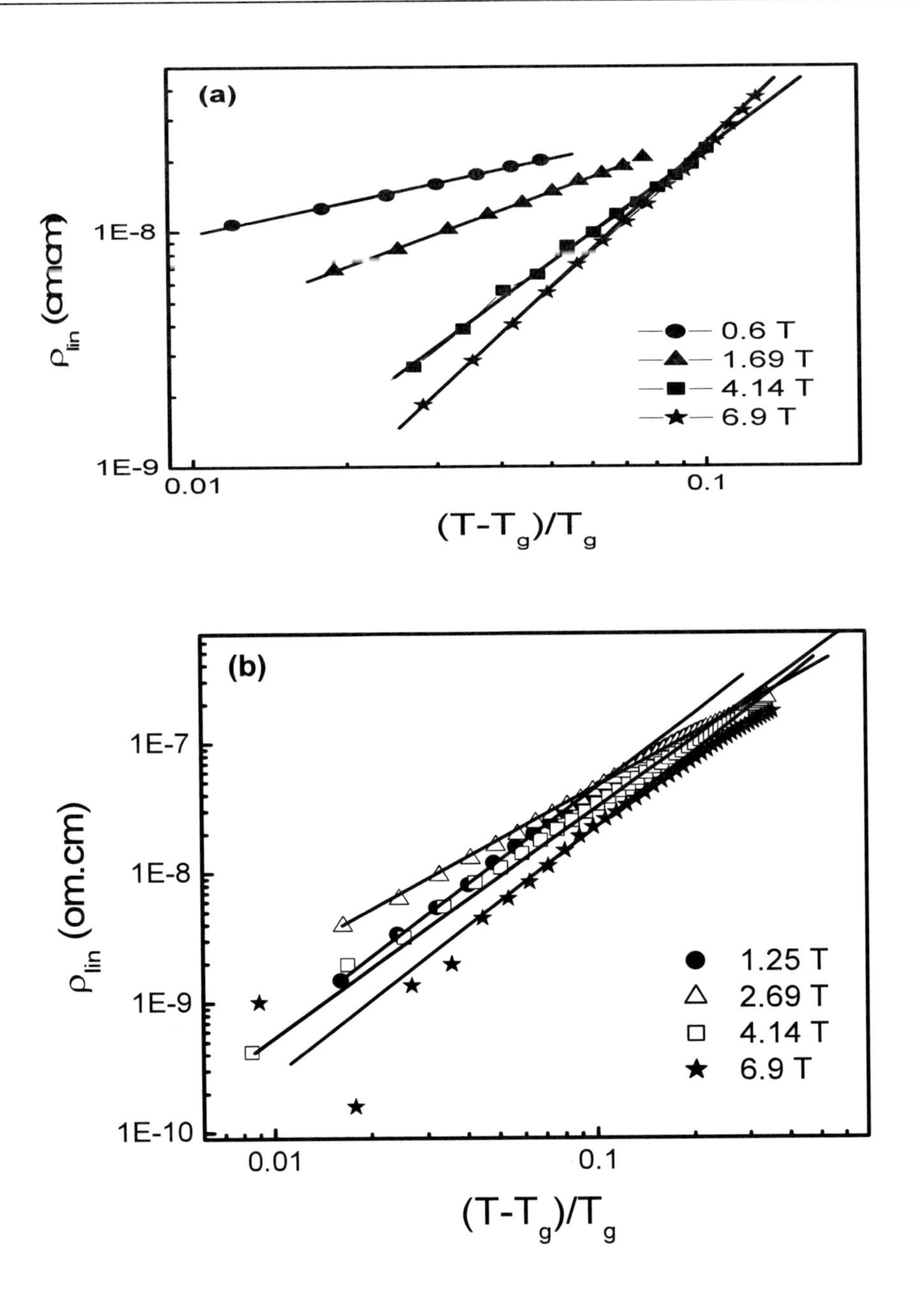

Figure 12. Low current resistivity (ρ_{lin}) as a function of $(T-T_g)/T_g$ for: (a) sample $Y_{0.975}Ca_{0.025}Ba_2Cu_3O_{7-\delta}$ and (b) sample $Y_{0.8}Ca_{0.2}Ba_2Cu_3O_{7-\delta}$, at different magnetic fields.

The width of temperature interval (ΔT) where the critical regime develops for a given magnetic field is determined from the presentation in Figure 12. The critical regime persists at narrow temperature interval for sample $Y_{0.975}Ca_{0.025}Ba_2Cu_3O_{7-\delta}$ but this interval is extended for both samples when the magnetc field increases for both samples.

By comparing the values of dynamic exponent z for two investigated samples it was established that the sample with smaller grains ($Y_{0.8}Ca_{0.2}Ba_2Cu_3O_{7-\delta}$) has larger z ($\sim 3$), while the sample with the larger grains ($Y_{0.975}Ca_{0.025}Ba_2Cu_3O_{7-\delta}$) shows smaller z ($\leq 2$). This confirms previously reported morphology dependence of dynamic exponent z [66]. Comparison of z values for Ca substituted and non substituted YBCO samples [66] showed that substituted samples have smaller z values for the similar grain size and magnetic fields range. The z suppression is more pronounced (about 1) for the $Y_{0.975}Ca_{0.025}Ba_2Cu_3O_{7-\delta}$ sample, while for $Y_{0.8}Ca_{0.2}Ba_2Cu_3O_{7-\delta}$ it is only several tenths. The reduction of the z exponent indicates that the vortex relaxation time is enhanced and confirms the improved pinning in Ca substituted samples in comparison with non-substituted.

The scaling behavior for both samples is presented in Figure 13 (a, b, c, d, e) and Figure 14 (a, b, c, d, e).

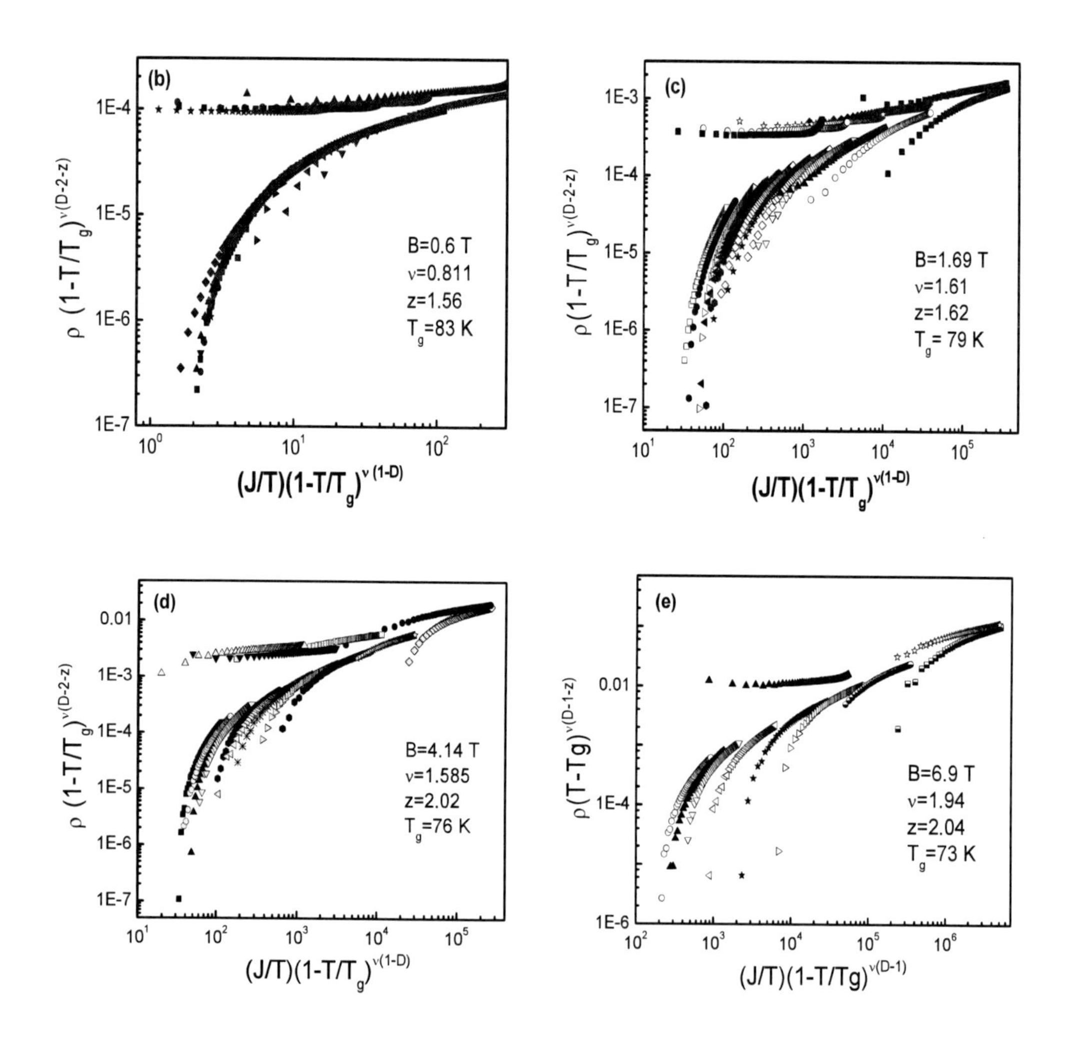

Figure 13. (a, b, c, d, e) Scaling collapse of E-J data for sample $Y_{0.975}Ca_{0.025}Ba_2Cu_3O_{7-\delta}$ at indicated magnetic fields for given static (ν) and dynamic (z) exponents at corresponding temperature intervals (a) 78-83 K (b) 70-87 K (c) 66-84 K (d) 64-74 K and (e) 61-75 K. Samples' dimensionality is D=3.

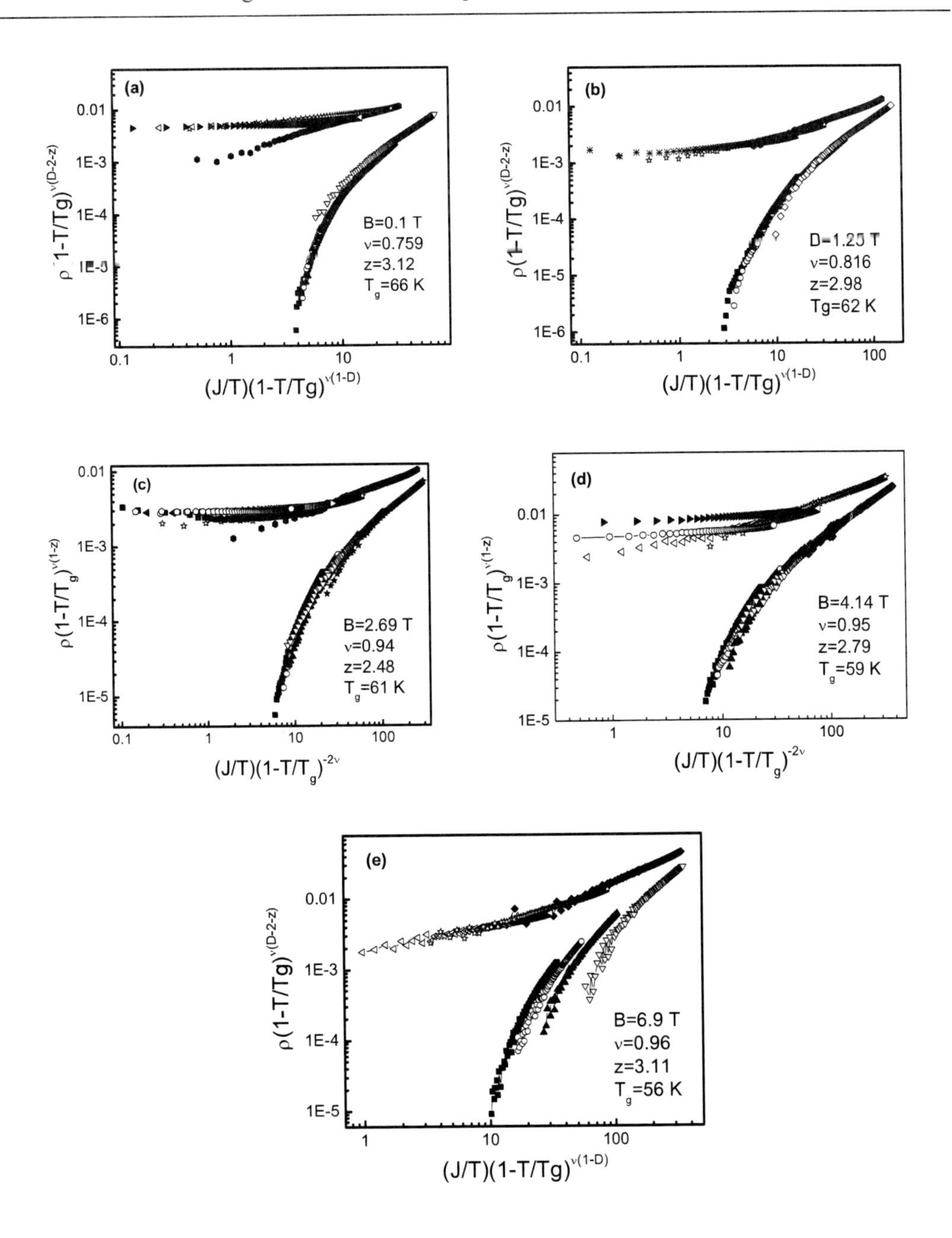

Figure 14. (a, b, c, d, e) Scaling collapse of E-J data for sample $Y_{0.8}Ca_{0.2}Ba_2Cu_3O_{7-\delta}$ at indicated magnetic fields for given static (ν) and dynamic (z) exponents at corresponding temperature intervals (a) 63-70 K (b) 58-65 K (c) 56-68 K (d) 54-63 K and (e) 52-58 K. Samples' dimensionality is D = 3.

Static exponent ν is almost field independent for $Y_{0.8}Ca_{0.2}Ba_2Cu_3O_{7-\delta}$ in consistency with the model predictions. More frequently it is close to but smaller than 1. For comparison ν is 1.13-1.15 at 0.1 T for the previously discussed polycrystalline YBCO samples [66]. For a sample with x = 0.025 ν is in the range 0.86 -1.94 and increases more than 2 times when the field grows from 0.1T to 6.9 T. The static exponent is related to the vortex correlation length. Their possible field dependence is connected with its suppression to a value smaller than the distance between the vortices [61, 67].

The vortex correlation length (ξ_g) has been estimated by using the relation [61]:

$$J_c = k_B T / \Phi_0 \xi_g^{(D-1)}, \tag{6}$$

where J_c is the critical current at $T < T_g$ and D = 3. For a sample with x = 0.025 at T = 81 K (T_g = 85 K) the critical current has been determined to be $4.5.10^4$ A/m^2 and the correlation length is about 3.5 μm. For the other sample with x = 0.20 the critical current at T = 62 K (also by 4 K lower than T_g = 66 K) is 6.10^4 A/m^2 and the corresponding ξ_g value is found to be ~2.7 μm. At this field (0.1 T) inter-vortex spacing is found to be ~14.4 μm, which is about 4 - 5 times higher than obtained ξ_g values confirming the presence of vortex-glass phase below T_g.

We determine also ξ_g at temperatures 2K higher than T_g according to the relation (6) and now J is the current density at which the resistivity starts to deviate from its constant, low-current value. For the sample with x = 0.20 at the highest field (6.9 T) $\xi_g > \alpha$. However, for the other sample $\xi_g \approx \alpha$ at the highest field (6.9 T) and $\xi_g < \alpha$ at all other fields. This difference could be the reason for a non-identical behaviour of scaling parameter ν in both samples.

The critical exponents for Ca substituted samples at given fields have been determined and scaling collapse of the E-J data established. For Ca substituted samples the z values are smaller than usually reported for non substituted YBCO. The reasonable explanation of this fact is the higher pinning, which has been established earlier independently [34], [51]. The obtained field dependence for static exponent ν (more pronounced for $Y_{0.975}Ca_{0.025}Ba_2Cu_3O_{7-\delta}$) is connected with the special interrelation between the vortex correlation length and intervortex spacing ($\xi_g \leq \alpha$) for all magnetic fields above T_g.

5. Intergranular Properties

Chemical doping with Ca is important also with respect to the intergranular current. It is known that grain boundaries (GBs) in YBCO superconductors are depleted of carriers [68], [69], what leads to a reduced intergrain critical current density. Schmehl et al. [70] reported that by substituting Ca^{2+} at Y^{3+} site, the weak link effect of high angle GBs in YBCO multilayer films was significantly reduced and the critical current density was enhanced more than seven times. An improvement of GBs transport in bulk melt-processed [71] and sintered [15] YBCO samples by a Ca substitution has also been established.

Hole concentration is an essential parameter that controls many properties in HTSC. How the vortex dynamics is influenced by the doping effect? Samples belonging to underdoped region have lower charge concentration (p) in CuO_2 plane, and almost empty charge reservoirs. In the overdoped side the charge concentration in planes increases, 3D carrier dynamics is enhanced and the anisotropy decreases. Generally, doping strongly affects coupling between the CuO_2 layers influencing the vortex dynamics in grains. In this study we investigate how the doping influences the intergranular flux dynamics, comparing the properties of the samples from the opposite sides of the $T_c(p)$ phase diagram [4].

The peak in the imaginary part of the first harmonic AC susceptibility, $\chi''_1(T)$, occurs when the measuring frequency is of the order of the inverse relaxation time of the vortex system [72]. The peak position depends logarithmically on AC field frequency. This dependence is used for the determination of activation energy, E_a, needed for triggering the flux creep, when thermally activated fluctuations overcome the pinning [73]. The expression for E_a is given by:

$$E_a = k_B T_p \ln(f/f_0), \qquad (7)$$

where T_p is the temperature corresponding to the χ''_1 peak position and f_0 is a characteristic frequency in the range $10^9 - 10^{12}$ Hz.

It has been established that in all investigated samples intergranular $\chi''_1(T)$ dependences are shifted to higher temperatures when the frequency is increased at constant AC field amplitude. The peaks of $\chi''_1(T)$ dependences are shifted to lower temperatures when the magnetic field amplitude increases at constant low frequency. These experimental results indicate that the thermally activated flux creep influences intergranular flux dynamics. This is in agreement with the previous reports [74]. Therefore, based on the above equation, we have calculated the flux creep activation energies for all examined samples.

The underdoped sample $Y_{0.8}Ca_{0.2}Ba_2Cu_3O_{7-\delta}$ treated in nitrogen athmosphere (denoted as Y08-N) has a very low T_c and the maxima in $\chi''_1(T)$ dependences are not well expressed down to 2K even for amplitude $H_{ac} = 0.1$ Oe. For E_a determination we used the curves displacement with frequency on the level possibly very close to the maximum. We chose this procedure after a careful investigation of its applicability on the other samples where the maximum of $\chi''_1(T)$ is well manifested. Once the $1/T_p$ vs. ln(f) dependence has been built taking the values of T_p at the maximum, then the same dependence is obtained but for T_p, we took the value at which the line y = m [max $\chi''_1(T)$], m = 0.7 ÷ 0.9 intercepts the $\chi''_1(T)$ plots for different frequencies. Thus we obtain a series of parallel lines (for different m) with the same slopes and independent of "m", the calculated E_a value remains constant and equal to that determined from the maximum of $\chi''_1(T)$. This result shows that such an approach is justified and we use it also to determine E_a for the Y08-N sample. The defined value of E_a at H_{ac}=0.1 Oe is very low (E_a = 0.004 eV). Most likely it is related to intragranular flux pinning and the corresponding intergranular value should be even smaller [23].

On Figure 15 the $1/T_p$ vs. ln(f) dependences are presented for all investigated samples. It is seen from the figure that underdoped samples (Y08-N and $GdBa_2Cu_3O_{7-\delta}$ denoted as Gd123) have higher slope of the $1/T_p$ vs. ln(f) dependence and hence lower activation energy than the overdoped samples $YBa_2Cu_3O_{7-\delta}$, $Y_{0.8}Ca_{0.2}Ba_2Cu_3O_{7-\delta}$ oxygenated 48 and 100 h (denoted respectively Y123, Y08-48 and Y08-100). Increasing the level of overdoping raises the activation energy. The exact values of E_a for H_{ac}=0.1 Oe are given in Table 1.

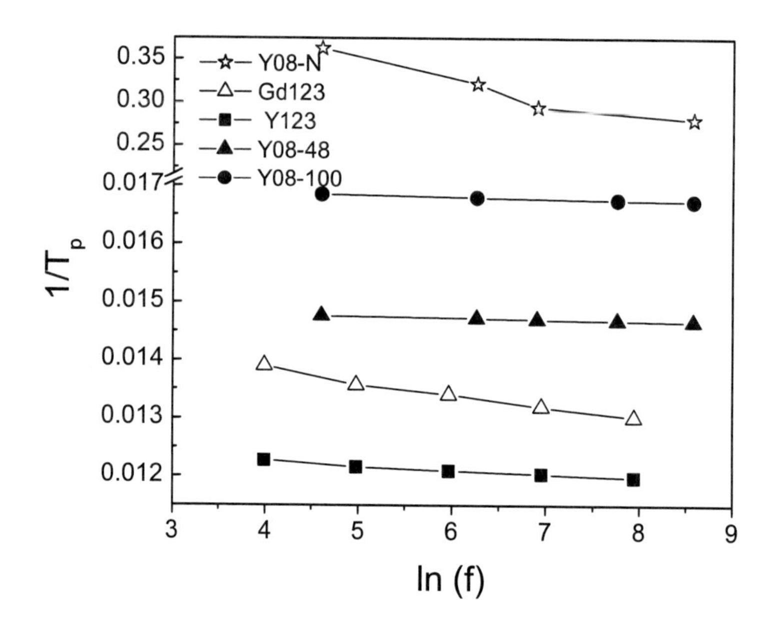

Figure 15. $1/T_p$ versus ln (f) dependences for indicated samples. The magnetic field amplitude is 0.1 Oe and the frequency is varied in the interval 145 -5300 Hz.

Table 1. Sample composition, symbols, intra- and intergranular critical temperatures and activation energy for TAFF

Sample	Symbol	$T_{c,intra}$ (K)	$T_{c,inter}$ (K)	E_a (eV)
$Y_{0.8}Ca_{0.2}Ba_2Cu_3O_z$	Y08-N	~50.0	-	0.004
$Gd_1Ba_2Cu_3O_z$	Gd123	91.7	85.8	0.397
$Y_1Ba_2Cu_3O_z$	Y123	92.0	90.1	1.118
$Y_{0.8}Ca_{0.2}Ba_2Cu_3O_z$	Y08-48	82.7	69.4	2.646
$Y_{0.8}Ca_{0.2}Ba_2Cu_3O_z$	Y08-100	81.5	66.6	2.837

In order to make clear whether different vortex dimensionality exists in underdoped (Y08-N) and overdoped (Y08-100) samples we performed experiments for the determination of activation energy at four different AC field amplitudes:7.95 A/m; 79.57 A/m; 795.78 A/m and 1591.59 A/m. In Figure 16 (a, b) the $E_a(H_{ac})$ dependences are presented for both samples, respectively. For underdoped Y08-N sample the $E_a(H_{ac})$ dependence is found to be logarithmic, which is the typical behavior in 2D vortex system [75]. This correlates with the small carrier concentration, large **c** lattice parameter (11.818 Å) and high anisotropy of this sample. In spite of the fact that this result probably refers to the intragranular pinning it is reasonable to expect similar behavior for intergranular pinning too. For an overdoped Y08-100 sample the experimentally found points are well approximated by the $E_a \sim H_{ac}^{-2/3}$ dependence shown in the inset of Figure 16b. According to [76] this dependence is characteristic of a 3D vortex system.

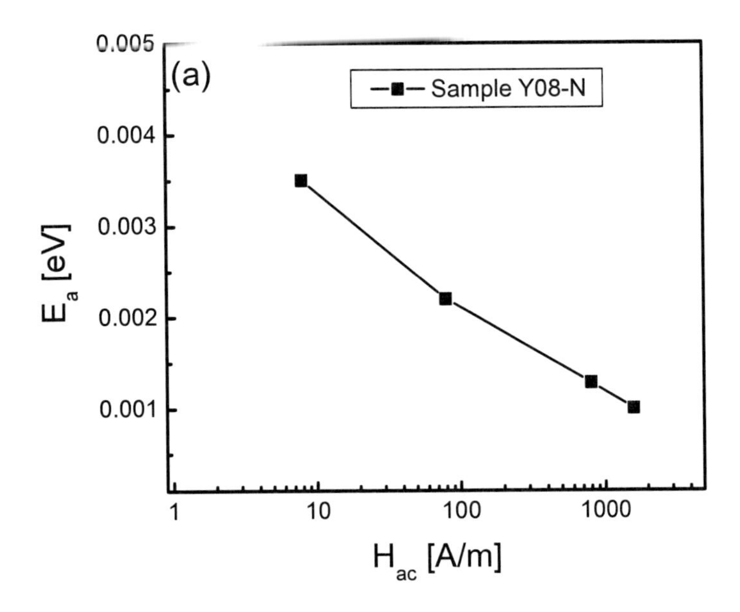

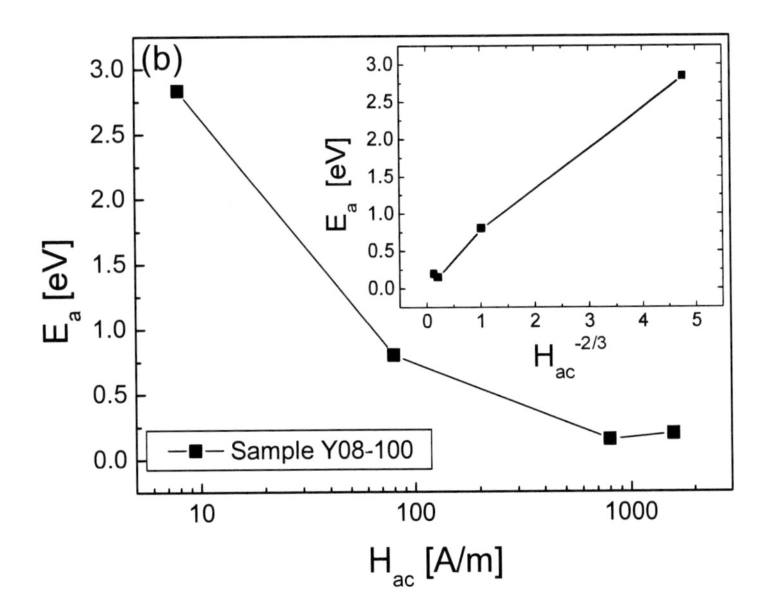

Figure 16. Activation energy (E_a) vs. magnetic field amplitudes (H_{ac}) for an underdoped sample Y08-N (a) and for an overdoped sample Y08-100 (b). Inset E_a is presented as a function of $H_{ac}^{-2/3}$.

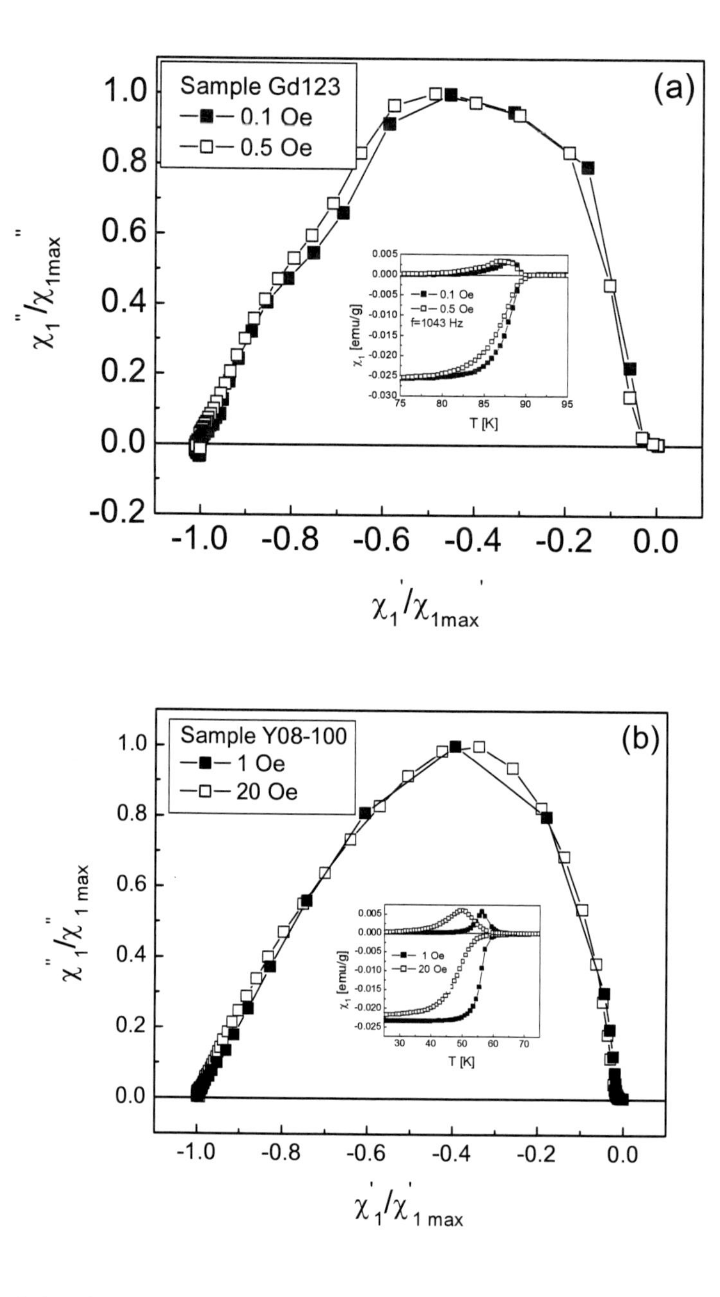

Figure 17. Coles-Coles plots for underdoped Gd123 sample (a) and for overdoped Y08-100 sample (b). Initial AC susceptibility data are presented in the insets.

In order to compare intergranular behavior in underdoped and overdoped samples, which have different T_{cinter} and different geometry, we used the Coles-Coles (χ'' versus χ') plot of the magnetic susceptibility. The advantage of this presentation is the elimination of the geometry dependent factors [77], [78] as well as the possibility to compare different samples without knowing the exact J(T) dependence [77]. In Figure 17 (a, b) Coles-Coles plots for underdoped Gd123 and overdoped Y08-100 samples are presented, respectively. The values of abscissas are normalized by using the maximum value of $|\chi'_1|$ for the corresponding sample, where the Meissner state is reached. χ''_1 is also normalized to the corresponding maximum value. It has been previously shown [77] that this presentation gives the same results when the fundamental susceptibility is measured as a function of temperature at constant magnetic field or the measurement is performed at varying magnetic fields and constant temperature. We also established this experimentally. In Figure 17 experimental results of type $\chi_1(T)_{Hac=const}$ are used and initial AC susceptibility data for investigated samples are presented at insets. For underdoped Gd123 sample (Figure 17a) two dependences are presented for $H_{ac} = 0.1$ Oe and $H_{ac} = 0.5$ Oe, f = 1048 Hz. It is seen from the Coles-Coles plots that the maximum is shifted to the (-1) side of χ'_1 axis. According to Shantsev et al. [79] the steeper slope at $\chi' \rightarrow -1$ and maximum shift to the negative -1 side is an indication of flux creep existence in the sample. For an overdoped sample Y08-100 the curves presented (Figure 17b) are measured for stronger magnetic fields $H_{ac} = 1$ Oe and $H_{ac} = 20$ Oe. In both fields the maximum occurs at $\chi'_1/\chi'_{1\,max} = 0.34$ which is very close to the value 0.38 predicted on the basis of Bean critical state model [78]. Therefore, J_c is almost independent of magnetic field in the overdoped sample up to the highest field used in the investigations, while for an underdoped sample the flux creep is present even in lower magnetic fields.

Maximum of $\chi''_1(T)$ is reached when the applied magnetic field penetrates to the center of the sample. The increase of the magnetic field amplitude shifts the maximum position towards lower temperatures [74]. This can be used for the determination of J(T) dependence when the sample's shape is known. The samples have been approximated with long cylinders (l=10R, where R is the radius and l is the length of the cylinder) and the critical current density at T_p has been calculated according to a relation $J(T) = H_{ac}/R$. The $\chi''_1(T)_{Hac=const}$ dependences at four different H_{ac} amplitudes indicated above are used for $J_c(T)$ determination in underdoped Gd123 and overdoped Y08-48, Y08-100 samples. The corresponding $J_c(T)$ dependences are presented in Figure 18(a, b). The results are well approximated (with least square deviation) with linear fit for underdoped specimen and with quadratic fit for overdoped samples.

Similar behavior was already observed in direct $J_c(T)$ measurements in a series of Ca substituted $GdBa_2Cu_3O_z$ samples – Figure 19 [16]. The $J_c(T)$ dependence for an underdoped sample is understandable. Within the individual grain CuO_2 planes are separated from insulating layers. In the frame of polycrystalline sample large angle grain boundaries depleted of carriers or unfavourably oriented grains behave as Josephson junctions. Thus the observed linear temperature dependence of J_c may be ascribed to the superconductor-insulator-superconductor (S-I-S) type joints existence in grains and between them. More unusual is quadratic $J_c(T)$ dependence for overdoped sample, which is an indication of superconductor-normal metal-superconductor (S-N-S) type joints presence. Similar behavior is well known for superconducting samples with various metallic additions (Ag, Pt) [80].

Using this analogy we can suppose that normal carriers and/or small regions of normal state are present at grain boundaries of overdoped samples. It is known that Ca segregates at the grain boundary regions. We can speculate that Ca substituted regions with dimensions comparable to the unit cell with different oxygen content and smaller T_c become normal at the conditions of experiment. It is also possible that phase separation occurs in overdoped samples and normal carriers appear. These might be expected explanations of the observed S-N-S type behavior of $J_c(T)$.

In fact it is established that a correlation exists between the intergranular critical current and flux pinning activation energy. In underdoped samples the intergranular current shows S-I-S behaviour and the activation energy is small, while in overdoped samples the intergranular current is changed to S-N-S type and activation energy increases. It was found out that Ca substitution not only increases carrier concentration and improves intergranular current but is also significant for intra- and intergranular pinning.

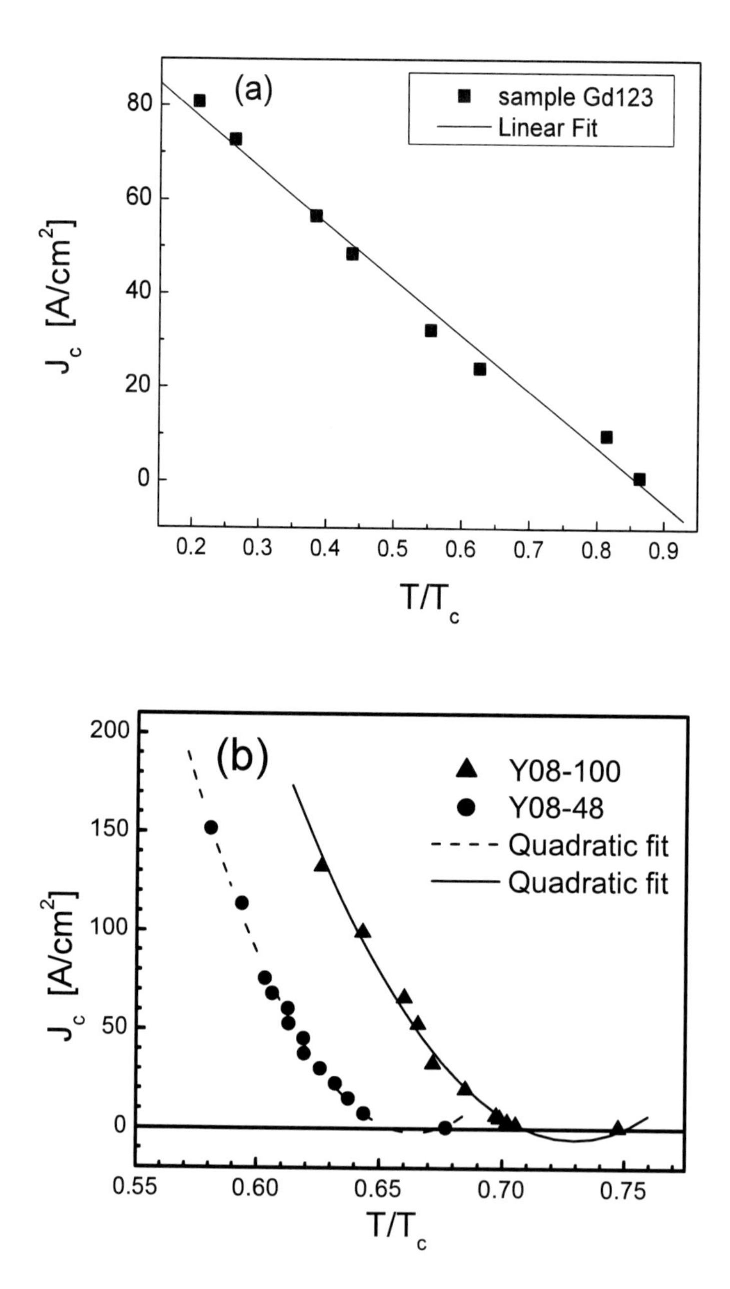

Figure 18. Temperature dependences of intergranular critical current density for underdoped Gd123 sample (a) and overdoped Y08-48 and Y08-100 samples (b).

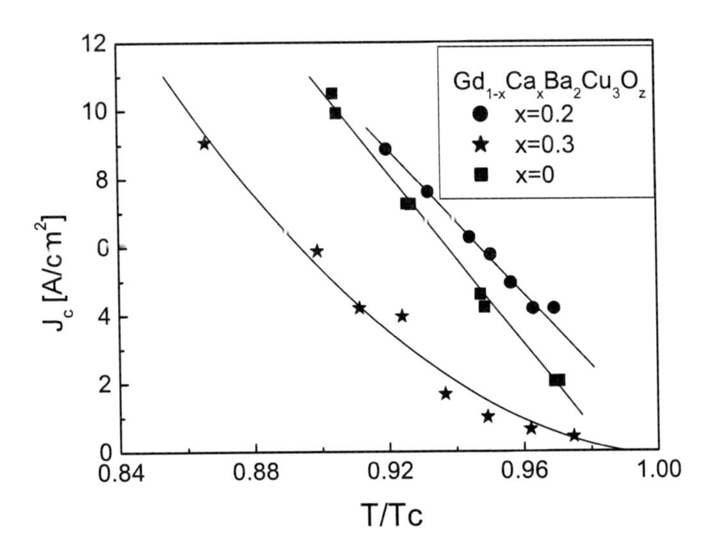

Figure 19. J_{cinter} vs. T/T_{conset} for $Gd_{1-x}Ca_xBa_2Cu_3O_{7-\delta}$ samples. The lines present the best linear fit to the data for samples with x = 0 and x = 0.2 and the best quadratic fit for the sample with x = 0.3.

6. Application

Since the discovery of high $-T_c$ superconductors many efforts have been made to develop the superconducting tapes for technical applications. The so-called first generation tapes were produced by the oxide powder in tube (OPIT) technology using the BSCCO superconducting system mainly. It has been later realized that the YBCO compound is more promising due to its critical parameters (J_c, H_{irr} and H_{c2}), lower anisotropy and simple phase synthesis. The development of second generation coated conductors method is essentially connected with 1-2-3 superconducting system. The coated conductor technology meets serious difficulties to achieve the in-plane grain alignment necessary for high intergranular critical current. Different methods, such as ion beam assisted deposition (IBAD) of buffer layers [81], [82] and biaxially textured substrates [83], [84] were examined for orientation of superconducting grains. In OPIT technology partial grain alignment is naturally obtained during the rolling process. In spite of that alignment and grain connections are found to be crucial for YBCO tapes obtained by OPIT technology [85].

Our investigations show that large amount of Ca substitution ($x \geq 0.10$) in YBCO system decreases the melting temperature down to 990 C. This is important for the hot rolling process of the tapes and final sintering process at 920 C and helps for better texturing. Obtained Ag-sheathed $Y_{1-x}Ca_xBa_2Cu_3O_{7-\delta}$ tape with x = 0.3 shows partial texturing [86], [87] and short samples have ~600-700 A/cm^2 critical current at 77 K and zero magnetic field.

We have investigated the χ (T) dependencies for tapes with non substituted YBCO core and substituted with Ca (x = 0.3). The AC magnetic field amplitude of 0.1 Oe and different frequencies in the range of 145 Hz-10 000 Hz have been used. In Figure 20 the results for a tape with $Y_{0.7}Ca_{0.3}Ba_2Cu_3O_{7-\delta}$ core are presented. The double step transition in χ' (T) dependence is connected with superconducting transition at higher temperature (~80 K) and with the silver signal at lower temperature (~20-40 K). The high temperature maximum of χ'' (T) (~ 80 K), shifts to higher temperatures on increasing the frequency. The shift in the low frequency range (145 – 1000 Hz) is associated with the superconducting core while that at higher frequency range (5313-10000 Hz) is associated with silver sheath. It has been established that an $Y_{0.7}Ca_{0.3}Ba_2Cu_3O_{7-\delta}$ based tape has narrow superconducting transition and is less influenced by the frequency of AC magnetic field than the YBCO tape.

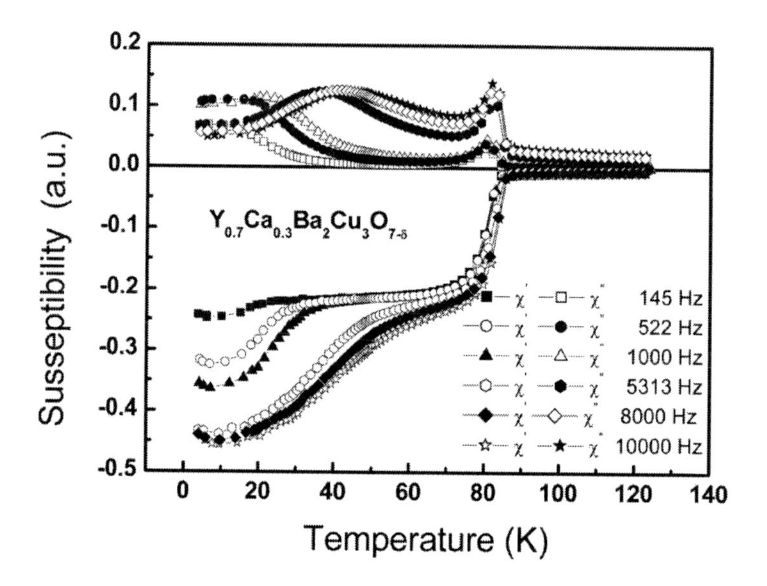

Figure 20. AC magnetic susceptibility vs. temperature at H_{ac}=79.57 A/m and different frequencies for an $Y_{0.7}Ca_{0.3}Ba_2Cu_3O_{7-\delta}$ based tape.

We hope that further work will contribute to the improvement of the parameters of $Y_{1-x}Ca_xBa_2Cu_3O_{7-\delta}$ based tapes. In spite of the result of these efforts the obtained dependences concerning the overdoped state are readily applicable for the coated conductors.

CONCLUSION

In summary, we investigated systematically the influence of Ca substitution on different properties of $R_{1-x}Ca_xBa_2Cu_3O_{7-\delta}$ polycrystalline samples – intra- and intergranular critical current, flux pinning, irreversibility line, activation energy for TAFF, scaling behavior. The main results and conclusions could be summarized as follows:

Small amount of Ca substitution (2.5 at %) in 1-2-3 superconducting system improves the intragranular critical current and flux pinning. Large amount of Ca (≥ 20 at %) deteriorates the J_{cintra} and influences the quality of the superconducting condensate causing phase separation. Pinning in Ca substituted $R_{1-x}Ca_xBa_2Cu_3O_{7-\delta}$ samples has a common nature independently of the Ca concentration.

H_{irr}(77K) has been increased up to 7 T for $Y_{0.975}Ca_{0.025}Ba_2Cu_3O_{7-\delta}$, which is higher than the previously reported for the non-substituted YBCO. For this sample the behavior of irreversibility line is determined by bulk pinning, while for $Y_{0.8}Ca_{0.2}Ba_2Cu_3O_{7-\delta}$ sample the irreversibility is due to the surface barrier effects.

The scaling collapse of the E-J data in Ca substituted samples was established, similar to the other polycrystalline YBCO samples. Previously observed morphology dependence of dynamic exponent z was confirmed. For Ca substituted samples the z values are smaller than usually reported for non substituted YBCO. The static critical exponent is field dependent (especially ν for $Y_{0.975}Ca_{0.025}Ba_2Cu_3O_{7-\delta}$), which is connected with the special interrelation between the vortex correlation length and intervortex spacing for all magnetic fields above T_g. The reasonable explanation of this fact is the higher pinning, resulting in enhanced relaxation time and narrow critical region.

It has been established that a correlation exists between the intergranular critical current and flux pinning activation energy. In underdoped samples the activation energy is small and intergranular current shows S-I-S behaviour, while in overdoped samples the activation energy increases and intergranular current is changed to S-N-S type.

It has been found that Ag-sheathed tape with the overdoped $Y_{0.7}Ca_{0.3}Ba_2Cu_3O_{7-\delta}$ superconducting core, has narrow superconducting transition and is less influenced by the AC magnetic field frequency than similar tape with non substituted YBCO core.

ACKNOWLEDGMENT

The authors are grateful to G. Fuchs, K.-H. Muller, E. Burzo, V. Kovachev, T. Mydlarz, H. Ignatov, J. Georgiev, I. Balasz, A. Stoyanova-Ivanova, S. Terzieva and K. Zalamova who helped in the realization of separate ideas, concerning this work.

This work is partially supported through the EURATOM Project FU07-CT-2007-00059.

REFERENCES

[1] Bednorz, J. G.; Muller, K. A. *Z. Phys. B.* 1986, 64, 189-193.

[2] Kamihara, Y.; Watanabe, T.; Hirano, M.; Hosono, H. *J. Am. Chem. Soc.* 2008, 130, 3296-3297.

[3] Paglione, J.; Greene, R. L. *Nature Physics.* 2010, 6, 645-658.

[4] Tallon, J. L.; Bernhard, C.; Shaked, H.; Hitterman, R. L.; Jorgensen, J. D. *Phys. Rev. B* 1995, 51, 12911-12914.

[5] Tallon, J. L.; Loram, J. W.; Williams, G. V.; Cooper, J. R.; Fisher, I. R.; Johnson, J. D.; Staines, M. P.; Bernhard, C. *Phys. St. Solidi B* 1999, 215, 531-543.

[6] Tallon, J. L.; Flower, N. E. *Physica C.* 1993, 204, 237-246.

[7] Lergos-Gledel, C.; Marucco, J. F.; Vincent, E.; Favrot, E. D.; Poumellec, B.; B. Touzelin, B.; Gupta, M.; Alloul, H. *Physica C.* 1991, 175, 279-284.

[8] Norton, D. P.; Lowndes D. H.; Sales, B. C.; Budai, J. D.; Chakoumakos, B. C.; Kerchner, H. R. *Phys. Rev. Lett.* 1991, 66, 1537-1540.

[9] Bernhard, C.; Tallon, J. L.; Bucci, C.; De Renzi, R.; Guidi, G.; Williams, G.V.; Niedermayer, Ch. *Phys. Rev. Lett.* 1996, 77, 2304-2307.

[10] Mahajan, A.V.; Alloul, H.; Collin, G.; Marucco, J. F. *Phys. Rev. Lett.* 1994, 72, 3100-3103.

[11] Bobroff, J.; MacFarlane, W. A.; Alloul, H.; Mendels, P.; Blanchard, N.; Collin, G.; Marucco, J. F. *Phys. Rev. Lett.* 1999, 83, 4381-4384.

[12] Uemura, Y. J. *Physica C* 1997, 194, 282-287.

[13] Wen, H. H.; Li, S. L.; Zhao, Z. W.; Liu, Z. Y.; Yang, H. P.; Zheng, D. N.; Zhao, Z. X. *Europhys. Lett.* 2002, 57, 260-266.

[14] Uemura, Y. J. http://arxiv.org/abs/cond-mat/0406301

[15] Nazarova, E. K.; Zaleski, A. J.; Zahariev, A. L.; Stoyanova-Ivanova, A. K.; Zalamova, K. N. *Physica C,* 2004, 403, 283-289.

[16] Nazarova, E. K.; Nenkov, K.; Fuchs, G.; Muller, G. K.; *Physica C* 2006, 436, 25-31.

[17] Buckley, R. G.; Tallon, J. L.; Pooke D. M.; Presland, M. R. *Physica C* 1990, 165, 391-396.

[18] Buckley, R. G.; Pooke, D. M.; Tallon, J. L.; Presland, M. R.; Flower, N. E.; Staines, M. P.; Johnson, H. L.; Meylan, M.; Williams, G. V.; Bowden, M. *Physica C* 1991, 174, 383-393.

[19] Niedermayer, Ch.; Bernhard, C.; Blasius, T.; Golnik, A.; Moodenbaugh, A.; Budnick, J. L. *Phys. Rev. Lett.* 1998, 80, 3843-3846.

[20] Cava, R. J.; Hewat, A. W.; Hewat, E. A; Batlogg, B.; Marezio, M.; Rabe, K. M.; Krajewski, J. J.; Peck Jr. W. F.; Rupp Jr. L.W. *Physica C* 1990, 165, 419-433.

[21] Sedky, A.; Gupta, A.; Awana, V. P.; Narlikar, A. V. *Phys. Rev. B.* 1998, 58, 12495-12502.

[22] Bichile, G. K.; Raibagkar, R. L.; Jadhav, K. M.; Kuberkar D. G.; Kulkarni, R.G. *Sol. St. Commun.* 1993, 88, 629-632.

[23] Nazarova, E. K.; Zaleski, A. J.; Nenkov, K. A.; Zahariev, A. L. *Physica C* 2008, 468, 955-960.

[24] Starowicz, P.; Sokolowski, J.; Balanda, M.; Szytula, A. *Physica C* 2001, 363, 80-90.

[25] Hejtmanek, J.; Jirak Z.; Knizek, K.; Dlouha, M.; Vratislav, S. *Phys. Rev. B.* 1996, 54, 16226-16233.

[26] Xu, K.; Qiu, J. ; Shi, L. *Supercond Sci Technol.* 2006, 19, 178-183.

[27] Laval, J.; Orlova, T.S. *Supercond Sci Technol,* 2002, 15, 1244-1251.

[28] Balchev, N.; Nenkov, K.; Mihova, G.; Pirov, J.; Kunev, B. *Physica C*, 2010, 470, 178-182.

[29] Bean, C. P. *Rev. Mod. Phys.* 1964, 36, 31-39.

[30] Zahariev, A.; Nazarova, E.; Nenkov, K.; Mydlarz, T.; Kovachev, V. *7th International Conference of the Balkan Physical Union*, ed. By A. Angelopoulos and T. Fildisis, 2009 AIP 978-0-7354-0740-4/09, pp. 367-372.

[31] Harada, T.; Yoshida, K. *Physica C* 2003, 387, 411-418.

[32] Zhuo, Yi.; Choi, J. H.; Kim, M. S.; Park, J. N.; Bae, M. K.; Lee, S. I. *Phys. Rev. B.* 1997, 56, 8381-8385.

[33] Lai, L. S.; Juang, J. Y.; Wu, K. H.; Uen, T. M.; Lin Y. J.; Gou, Y. S. *Physica C* 2004, 415, 133-138.

[34] Nazarova, E.; Zaleski, A.; Zahariev, A.; Buchkov, K.; Kovachev, V. *J. Optoelectr. Adv. Mater.* 2009, 11, 1545-1548.

[35] Uemura, Y. J.; Keren, A. L.; Le, P.; Luke, G. M.; Wu, W. D.; Kubo, Y.; Manako, T.; Simakawa, Y.; Subramanian, M.; Cobb, J. L.; Market, J. T. *Nature* 1993, 364, 605-607.

[36] Mourachkine, A. *Room-Temperature Superconductivity*, Cambridge International Science Publishing, Cambridge, 2004, p.175.

[37] Nazarova, E.; Zaleski, A.; Zahariev, A.; Buchkov, K. In *Nanoscience & Nanotechnology*; Balabanova, E.; Dragieva, I.; Ed.; Prof. Marin Drinov Academic Publishing House, Sofia, 2008, vol 8. pp 158-161.

[38] Fabbricatore, P.; Farinon, S.; Gemme, G.; Musenich, R.; Parodi R.; Zhang, B. *Phys. Rev. B.* 1994, 50, 3189-3199.

[39] Zhang, Y. J.; Qin, M. J.; Ong C. K. *Physica C* 2001, 351, 395-401.

[40] Farrell, D. E.; Dinewitz, I.; Chandrasekhar, B.S. *Phys. Rev. Lett.* 1966, 16, 91-94.

[41] Lefloch, F.; Hoffmann, C.; Demolliens, O. *Physica C* 1999, 319, 258-266.

[42] Pace, S.; Filatrella, G.; Grimadi, G.; Nigro A.; M. G. Adesso, M. D. *AIP Conference Proceedings* 2006, 850, 873- 874.

[43] Huebener, R. P.; *J. Phys.: Condens. Matter.* 2009, 21, 254208-254209.

[44] Senatore, C.; Clayton, N.; Lezza, P.; Polichetti, M.; Pace, S.; Flukiger,R. *IEEE Trans. Appl. Supercond.* 2005, 15, 3329-3332.

[45] Senatore, C.; Policheti, M.; Clayton, N.; Flukiger, R.; Pace, S. *Physica C* 2004, 401, 182-186.

[46] Polichetti, M.; Adesso, M. G.; Pace, S.; *European Phys. Jour. B: Condens. Matter.* 2003, 36, 27-36.

[47] Di Gioacchino, D.; Celani, F.; Tripodi, P.; Testa, A. M.; Pace, S. *Phys. Rev. B* 1999, 59, 11539-11550.

[48] Di Gioacchino, D.; Marcelli, A.; Zhang, S.; Fratini, M.; Poccia, N.; Ricci, A.; Bianconi, A. http://arxiv.org/abs/0905.1633.

[49] Deak, J.; McElfresh, M.; Clem, J. R.; Hao, Z.; Konczykowski, M.; Muenchausen, R. ; Foltyn, S.; Dye, R. *Phys. Rev. B* 1994, 49, 6270-6279.

[50] Waki, K.; Higuchi, T.; Yoo, S. I.; Murakami, M. *Appl. Supercond.* 1997, 5, 133-138.

[51] Nazarova, E.; Zaleski, A.; Buchkov, K. *Physica C* 2010, 470, 421-427.

[52] Nakane, T.; Karppinen, M.; Yamauchi, H., *Physica C* 2001, 357-360, 226-229.

[53] Larkin, A. I.; Ovchinnikov, Y. N. *J. Low Temp. Phys.* 1979, 34, 409-428.

[54] Yeshurun, Y.; Malozemoff, A.P. *Phys. Rev. Lett.* 1988, 60, 2202-2205.

[55] Houghton, A.; Pelcovits, R. A.; Sudbo, A. *Phys. Rev. B* 1989, 40, 6763-6770.

[56] Bean, C. P.; Livingston, J. D. *Phys. Rev. Lett.* 1964, 12, 14-16,

[57] Campell, A. M.; Evetts, J.E. *Critical Currents in Superconductors,* Taylor & Francis, London, 1972, p. 142.

[58] Konczykowski, M.; Burlachkov, L. I.; Yeshurun, Y.; Holtzberg, F. *Phys. Rev. B* 1991, 43, 13707-13710.

[59] Reissner, M.; *Physica C* 1997, 290, 173-187.

[60] Burlachkov, L.; Geshkenbein, V. B.; Koshelev, A. E.; Larkin, A. I.; Vinokur, V. M.; *Phys. Rev. B* 1994, 50, 16770-16773.

[61] Koch, R. H.; Foglietti, V.; Gallagher, W. J.; Koren, G.; Gupta A.; Fisher, M. P. *Phys Rev Lett.* 1989, 63, 1511-1514.

[62] Huse, D.; Fisher M. P.; Fisher, D. S. *Nature* 1992, 358, 553-559.

[63] Fisher, M. Phys Rev Lett 1989, 62, 1415-1418.

[64] Nazarova, E.; Nenkov, K.; Buchkov K.; Zahariev,A.http://arxiv.org/abs/1012.5267.

[65] Pierson, S.; *Phys. Rev. Lett.* 1995, 75, 4674-4677.

[66] Joshi, R.; Hallok, R.; Taylor, J. *Phys. Rev. B* 1997, 55, 9107-9119.

[67] Roberts, J.; Brown, B.; Hermann, B.; Tate, J.; *Phys. Rev. B* 1994, 49, 6890-6894.

[68] Hilgenkamp, H.; Schneider, C. W.; Schulz, R. R.; Goetz, B.; Schmehl, A.; Bielefeldt, H.; Mannhart, J.; *Physica C* 1999, 326-327, 7-11.

[69] Hilgenkamp, H.; Mannhart, J. *Rev. Mod. Phys.* 2002, 74, 485-549.

[70] Schmehl, C.A.; Goetz, B.; Schulz, R.R.; Schneider, C.W.; Bielefeldt, H.; Hilgenkamp, H.; Mannhart, J. *Europhys. Lett.* 1999, 47, 110-115.

[71] Shlyk, L.; Krabbes, G.; Fuchs, G.; Nenkov, K. *Physica C* 2002, 383, 175-182.

[72] Geshkenbein, V.B.; Vinokur, V.M.; Fehrenbacher, R. *Phys. Rev. B* 1991, 43, 3748-3751.

[73] Nikolo, M.; Goldfarb, R.B. *Phys. Rev. B* 1989, 39, 6615-6618.

[74] Muller, D.K. *Physica C* 1990, 168, 585-590.

[75] Brunner, O.; Antognazza, L.; Triscone, J.M.; Mieville, L.; Fischer, O.; *Phys. Rev. Lett.* 1991, 67, 1354-1357.

[76] Xiaojun, X.; Lan, F.; Liangbin, W.; Yuheng, Z.; Jun, F.; Xiaowen, C.; Kebin L.; Hisashi, S. Phys. Rev. B 1999, 59, 608-612.

[77] Herzog, T.; Radovan, H. A.; Ziemann, P.; Brandt, E. H. *Phys. Rev. B* 1997, 56, 2871-2881.

[78] Cho, J. H. *Physica C* 2001, 361, 99-106.

[79] Shantsev, D. V.; Galperin, Y.M.; Johansen, T.H. *Phys. Rev. B* 2000, 61, 9699-9706.

[80] Nazarova, E.; Zahariev, A.; Angelow, A.; Nenkov, K. *J. Supercond. Nov. Magn.* 2000, 13, 329-334.

[81] Reade R. P.; Berdahl, P.; Russo, R. E.; Garrison, S. M.; *Appl. Phys. Lett.* 1992, 61, 2231 – 2233.

[82] Wu, X. D.; Foltyn, S. R.; Arendt, P.; Townsend, J.; Adams, C.; Campbell, I.H.; Tiwari, P.; Coulter, Y.; Peterson, D. E. Appl. Phys. Lett. 1994, 65, 1961-1963.

[83] Goyal, A.; Norton, D.P.; Budai, J. D.; Paranthaman, M.; Specht, E.D.; Kroeger, D. M.; Christen, D.K.; He, Q.; Saffian, B.; List, F. A.; Lee, D. F.; Martin, P. M.; Klabunde, C. E.; Hartfield, E.; Sikka, V.K.; *Appl. Phys. Lett.* 1996, 69, 1795-1797.

[84] Norton, D. P.; Goyal, A.; Budai, J.D.; Paranthaman, M.; Specht, E. D.; Kroeger, D. M.; Christen, D. K.; He, Q.; Saffian, B.; List, F.A.; Lee, D. F.; Martin, P. M.; Klabunde, C. E.; Hartfield, E.; Sikka, V. K.; *Science* 1996, 274, 755-757.

[85] Glowacki, B. A.; *Supercond Sci Technol.* 1998, 11, 989-994.

[86] Ignatov, H.; Nazarova, E.; Zahariev, A.; Lazarova, V.; Georgiev, J.; Stoyanova-Ivanova, A.; Terzieva, S.; Kliavkov, K.; Kovachev, V. *J. Supercond. Nov. Magn.* 2008, 21, 69-73.

[87] Nazarova, E.; Buchkov, K.; Zahariev, A.; Georgiev, J.; Nenkov, K.; Ignatov, H.; Kovachev, V.; Burzo, E.; Balasz, I. *J. Mat. Sci Technol.* 2009, 17, 230-239.

In: Superconductivity
Editor: Vladimir Rem Romanovskií
ISBN 978-1-61324-843-0

Chapter 6

NUMERICAL CALCULATION OF TRAPPED MAGNETIC FIELD FOR BULK SUPERCONDUCTORS

Alev Aydıner [*]
Physics Department
Karadeniz Technical University
Trabzon, Turkey

ABSTRACT

Bulk high-temperature superconductors have significant potential for industrial applications, such as magnetic bearings, flywheel energy storage systems, non-contact transport devices, levitation and trapped-field magnets. The term 'trapped-field magnet' describes a bulk superconductor supporting a trapped magnetic field after a magnetization process. The trapped magnetic field of the superconducting bulk magnet has been reported to be superior to that of a conventional permanent magnet. For further enhancements of the trapped magnetic field of the bulk superconductors, it is important to increase the critical current density (J_c) and the size of the single domain. On the other hand, the trapped magnetic field depends also on several other parameters, such as the shape and the aspect ratio of the sample and the distance between the sample surface and the observation point, which is also important for the future design of electrical machines that utilize high trapped fields, like motors and generators.

For this reason this chapter summarizes the numerical calculation of trapped magnetic field. It is known that the calculation method using the sand-pile model and Biot–Savart law is useful for the determination of the magnetic characteristics of the sample. At the first stage, it was examined and extended earlier calculations of field distribution within the bulk superconductor for constant critical current density. At the second stage it was outlined calculation of field distribution within the bulk superconductor for field dependent critical current density.

[*] E-mail address: alevaydiner@gmail.com

1. Introduction

High-temperature superconductors, especially the bulk superconductors have significant potential for industrial applications, such as magnetic bearings, flywheel energy storage systems and contactless transport devices [1]-[3]. One of the prospective applications of high-temperature bulk superconductors is a superconducting magnet for the magnetically levitated (Maglev) train [4], [5]. Maglev train is a super-high-speed nonadhesive transport system with a combination of superconducting magnets and linear motor technology. The purpose of the Maglev train is super-high-speed running, safety, reliability, low environmental impact and minimum required maintenance. Hence, the trapped magnetic field applications of bulk superconductors have attracted worldwide interests.

The term 'trapped-field magnet' describes a bulk superconductor supporting a trapped magnetic field after a magnetization process. The trapped magnetic field of the superconducting bulk magnet has been reported to be superior (e.g. 17 T at 29 K [6]) to that of conventional permanent magnet [3]. For further enhancements of the trapped magnetic filed of the bulk superconductors, it is important to increase the critical current density (J_c) and the size of the single domain. On the other hand, the trapped magnetic field depends also on several other parameters, such as the shape and the aspect ratio, i.e. the radius and the thickness of the sample and the distance between the sample surface and observation point, which is also important for the future design of electrical machines that utilize high trapped-fields like motors and generators [5], [7]-[9].

For the calculation of the theoretical trapped magnetic field distribution, most previous work has used some models assuming current paths inside the superconductor in the critical state [10]:

First, the simplest model is the one-dimensional model using a disk or a cylindrical sample. Frankel [11] calculated the field profiles for a thin disk assuming a field-independent J_c. Daumling and Larbelestier [12] calculated the field profiles for a thin disk assuming a field-dependent J_c. Conner and Malozemoff [13] extended this work to find the relationship between the sample geometry and the trapped field profile. Frangi et al. [14] used a thin-walled tube model for a cylinder bulk sample [10].

Second, a model for a rectangular shape is more important for a practical use. Tamegai et al. performed numerical calculation of local field profiles on a rectangular sample using the infinite strip model [15] and using the sand-pile model [16]. In the sand-pile model, it is assumed that a current flows along the edge to form a square loop. Grant et al. [17] determined current and field distribution in squares of thin-film sample using an array of equal sized circulating current cells [10], while Xing et al. [18], and Kamijo and Kawano [19] used a computation methodology based on the linearization and matrix inversion which was applied to 2D samples. Keeping the planar approximation to the distribution of current, extensions of the discretization and matrix inversion method were made 3D samples by Wijngaarden et al. [20], [21], Jooss et al. [22] and Amoros et al. [23]. Then, Portabella et al. solved the inversion problem by a finite-element procedure that may be applied to 3D samples without regularity assumption, but with important computational requirements [24].

Higuchi et al. [10] and Nagashima et al. [25] calculated the field distribution for large rectangular bulk and compared with the experimental data. Fukai et al. [26], [27] determined the J_c-B characteristics of disk samples by measuring trapped fields in the increasing and decreasing field process and based on the sand-pile model and Biot-Savart law. It is found that for further enhancements of the trapped magnetic filed of the bulk superconductors, it is important to increase the critical current density (J_c) and the size of the single domain. On the other hand, the trapped magnetic field depends also on several other parameters, such as the shape and the aspect ratio of the sample and the distance between the sample surface and observation point, which is also important for the future design of electrical machines that utilize high trapped-fields like motors and generators [5], [7]-[9]. However, the influence of all these parameters mentioned above on the mapping of the trapped field and the components of the trapped field have not been investigated in a systematic manner. Previous efforts have focused heavily on the radial and axial components of the trapped magnetic fields of the cylindrical sample and z-component of the trapped magnetic field of square sample. However any research on x and y-components of the trapped magnetic field of square sample has been performed. Therefore Aydıner and Yanmaz [28] determined the effects of these parameters on the mapping of the trapped field and the components of the trapped field for the cylindrical and square samples. Than Pina et al. [29] used the calculation of x- and y-component of the trapped field for the design of high temperature superconducting machines.

Fukai et al. [26], [27] measured the trapped magnetic field by varying the thickness of the bulk superconductor, which could well be simulated with the method based on the sand-pile model and Biot-Savart's law under the condition that J_c is constant. However, the simulation based on the constant J_c was not able to reproduce the trapped field of a large bulk superconductor, because the B dependence of J_c was not neglected. Particularly for the RE-Ba-Cu-O with the prominent secondary peak effect on the J_c-B curve, the model with the constant J_c failed to reproduce the empirical results. Then Fukai et al. [30] modified the simulation method by taking account of the field dependence of J_c's which could calculate the trapped field of a large bulk Nd-Ba-Cu-O superconductor with the secondary peak effect.

Therefore, the purpose of this paper is twofold: first to examine and extend earlier calculations of field distribution within the bulk superconductor for constant critical current density, and second to outline calculation of field distribution within the bulk superconductor for field dependent critical current density.

2. Numerical Calculation

2.1. Sand-Pile Model for Trapped Magnetic Field

The sand-pile model [16], [25] and the Biot-Savart law are used for numerical calculation of the trapped magnetic field distribution. According to the sand-pile model large square and cylindrical samples are separated into concentric current loops and it is assumed that the current flows along the edges, as shown schematically in Figures 1 and 2. The current I flowing in each current loop is given by

$$I = J_c \, \Delta w \, \Delta t , \tag{1}$$

where J_c is the critical current density and is used as a fitting parameter, and Δw and Δt are the width and the thickness of the current loop, respectively. It is well known that the Biot-Savart law gives the magnetic field at a point only for a small element of the conductor. In this connection, for each conductive loop, the field $\vec{B}$ can be expressed using the Biot–Savart law by the following equation:

$$\vec{B} = \frac{\mu_0 I}{4\pi} \int \frac{d\vec{s} \times \hat{r}}{r^2}. \quad (2)$$

In equation (2), μ_0 is the permeability of free space, $d\vec{s}$ is a current element vector, and r and $\hat{r}$ are the distance and the unit vector directed from the current element to the observation point.

For square symmetry (Figure 1(b)), according to equation (2), at observation point P, the magnetic field vector $\vec{B}_z$ generated by the straight line current I is represented by [28]

$$\vec{B}_z = \frac{\mu_0 I}{4\pi}\left(x\hat{k}\right)\left[\frac{L-y}{\left(x^2+z^2\right)\left[x^2+(L-y)^2+z^2\right]^{1/2}} + \frac{y}{\left(x^2+z^2\right)\left(x^2+y^2+z^2\right)^{1/2}}\right], \quad (3)$$

where L is the length of current line. And also, x- and y-components of the trapped magnetic field can be calculated as upwards.

For cylindrical symmetry (Fig. 2(b)), at observation point P, the magnitude of the axial magnetic field (B_z) generated by the current I flowing are given by [31]

$$B_r = \frac{\mu_0 I}{2\pi} \frac{z}{r\left[(a+r)^2+z^2\right]^{1/2}}\left[-K(k) + \frac{a^2+r^2+z^2}{(a-r)^2+z^2}E(k)\right], \quad (4)$$

$$B_z = \frac{\mu_0 I}{2\pi} \frac{1}{\left[(a+r)^2+z^2\right]^{1/2}}\left[K(k) + \frac{a^2-r^2-z^2}{(a-r)^2+z^2}E(k)\right], \quad (5)$$

where K and E are complete elliptic integrals of the first and second kind, with $k^2 = 4ar\left[(a+r)^2+z^2\right]^{-1}$.

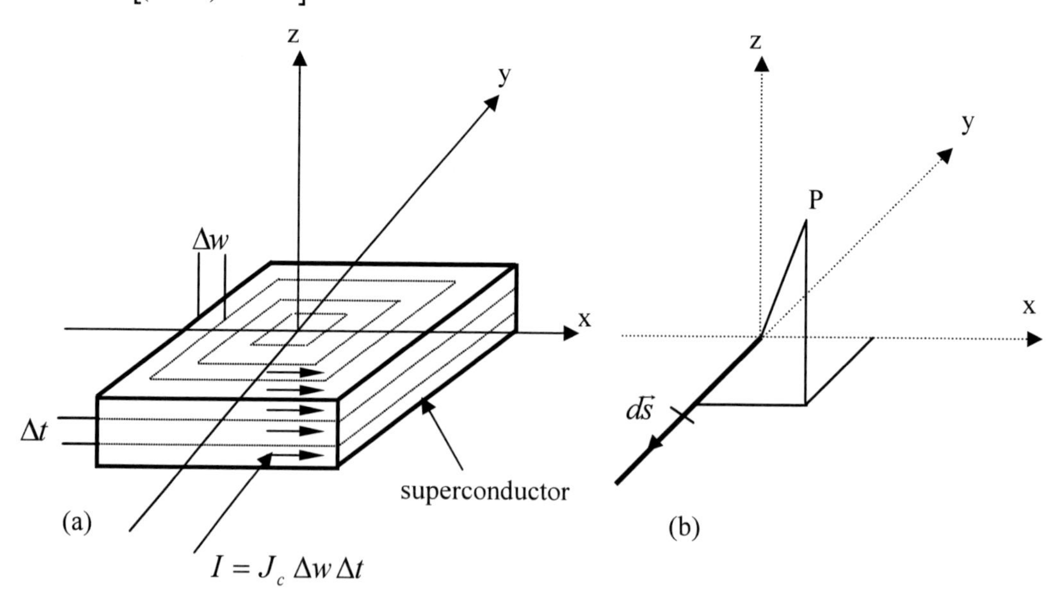

Figure 1. (a) Schematic illustration of the current loops, (b) coordinate system for the square superconductor sample used for numerical calculation [28].

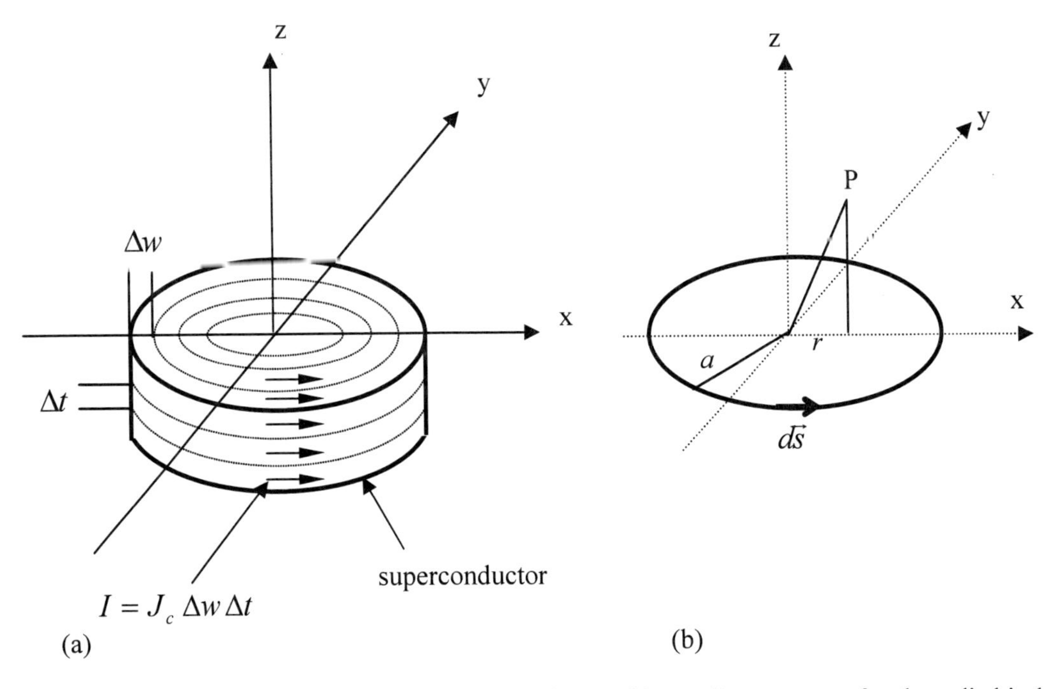

Figure 2. (a) Schematic illustration of the current loops, (b) coordinate system for the cylindrical superconductor sample used for numerical calculation [28].

2.2. Field Dependent Critical Current Density (Combined J_c Model)

Modified the simulation method by taking account of the field dependence of J_c's which could calculate the trapped field of a bulk superconductor with secondary peak effect as follows: On the basis of an analysis of experimental data in terms of this model, the pinning potential in RE-123 single crystal was identified as logarithmic. The commonly observed scaling of the $J_c(B)$ curves with temperature and electric field were shown to be associated with power-law field dependences of the characteristics parameters $J_0(B) \propto B^m$ and $U_0(B) \propto B^{-n}$ in the expression fort the pinning energy

$$U = (T, B, E, J) = U_0(T, B, E)\ln(J / J_0)$$

with m and n positive [32]. This model resulted in [33, 34]

$$J_c(B) \propto B^m \exp\left(cB^n\right). \tag{6}$$

The equation (6) exhibits a maximum at $B = B_p$. From the condition for a extreme at $B = B_p$, $\left.\frac{dJ_c}{dB}\right|_{B=B_p} = 0$, the constant c can be calculated, $c = -\left(\frac{m}{n}\right)B_p^{-n}$, which helps to transform equation (6) into [32], [35]

$$J_c(B) = J_p b_p^m \exp\left[\left(1 - b_p^n\right)m / n\right] \tag{7}$$

with $J_p = J_c\left(B_p\right)$ and $b_p = \frac{B}{B_p}$.

Jirsa et al. [35] showed that in Dy-123 single crystal the central peak exhibits a different temperature scaling to the second peak. This indicates that the underlying pinning mechanisms are different. It was also shown that the field dependence of the central peak after the separation of the second peak is an exponentially decaying function [32],

$$J_c(B) \propto \exp(-\omega B). \tag{8}$$

A similar behavior has been also observed in other RE-123 [32]. Assuming that in the studied melt-textured samples the central peaks have similar characters, Jirsa et al. [32] combined the two contributions (6) and (8), and obtained

$$J_c(B) = J_{p1} \exp(-\omega b_p) + J_{p2} b_p^m \exp[(1 - b_p^n) m / n], \tag{9}$$

where J_{p1} and J_{p2} denote the respective heights of the central and second peaks, other parameters are fitting parameters representing the kind and the effectiveness of pinning centers. The first term of the above equation corresponds to the component which has normal field dependence in that J_c decreases with increasing B, while the second term for the J_c component which first increases with B and decreases again after reaching its maximum J_{p2}.

3. Result and Discussion

3.1. Fitting of Trapped Magnetic Field Curves for Constant Critical Current Density

In order to conformity of sand-pile model and Biot-Savart law to trapped magnetic field calculation, Aydiner and Yanmaz [28] adopted experiment and calculation. They picked out experiment from Nariki et all. [36]. The axial trapped magnetic field distribution of the Gd-Ba-Cu-O cylindrical sample with 32 mm diameter and 17 mm thickness containing 10 wt% Ag_2O at 77 K [36] is shown in Figure 3(a). The maximum trapped field at 1.2 mm above the surface is reported to be 2.05 T, as shown in Figure 3(a). Using the sand-pile model and Biot-Savart law, the axial trapped magnetic field distribution was calculated for a cylindrical sample with 32 mm diameter and 17 mm thickness, and with fitting parameter J_c = 31150 A cm^{-2}; see Figure 3(b). Whereas the fitting parameter J_c was chosen in this study J_c-B curves were used in [36]. In the mentioned reference, the mean value of J_c was in the range of 30 000–35 000 A cm^{-2} at around 2 T. hence, it was concluded that the experimentally fitted value of J_c = 31 150 A cm^{-2} reasonably reflects the overall J_c value of he sample. Comparing the numerical result with the measured field distribution, it can be seen that the measured trapped field and calculated trapped field are in good agreement. Therefore, it is thought that this calculation method is useful for the determination of the magnetic characteristics of the sample [28].

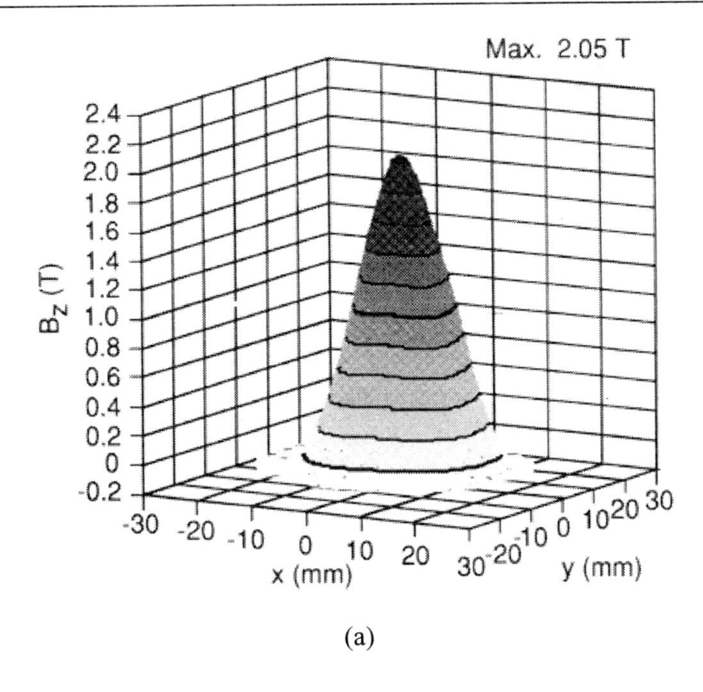

(a)

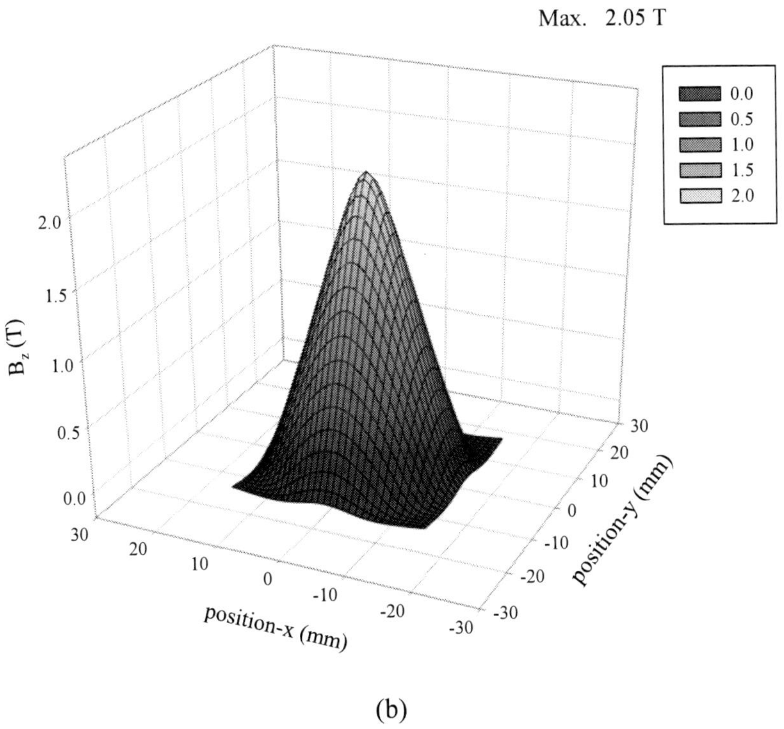

(b)

Figure 3. (a) The axial trapped magnetic field distribution of the Gd-Ba-Cu-O cylindrical sample with 32 mm diameter and 17 mm thickness containing 10 wt% Ag_2O at 77 K. This bulk was fabricated employing fine Gd-211 powder with a particle size of 0.2 μm as a starting material [36]. (b) The calculated axial trapped magnetic field distribution using the sand-pile model and the Biot-Savart law for a cylindrical sample with 32 mm diameter and 17 mm thickness (fitting parameter J_c = 31 150 A cm^{-2}) [28].

3.2. Shape Dependence of Axial Trapped Magnetic Field

The effect of the thickness change of the axial trapped magnetic field at 1 mm above the top surface of the square and cylindrical superconductor samples are shown in Figure 4. The trapped field was calculated by varying the thickness of the superconductor samples with dimensions of 20 × 20 mm^2 square and a cylindrical with 20 mm in diameter under the condition that $\Delta t = \Delta w = 0.1$ mm and $J_c = 20\ 000$ A/cm^2. Figure 4 illustrates that the trapped magnetic field values of the samples of both shapes are in similar characteristics. It can be seen on these curves that the trapped field rapidly increased with the increasing the thickness of the bulk superconductor, and then remained stable when the square and cylindrical sample thickness reached at a certain value. It is noted that this value is equal to the dimension of the square sample and the diameter of the cylindrical sample.

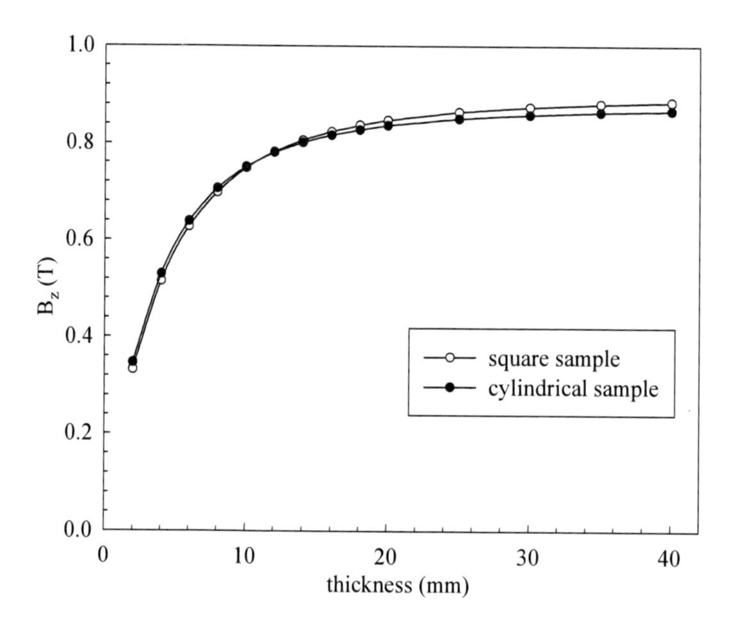

Figure 4. The thickness dependence of the axial trapped magnetic field for the square and cylindrical superconductor samples [28].

3.3. Diameter and Thickness Dependence of Axial Trapped Magnetic Field for Square Sample

In order to demonstrate the effect of the size on axial trapped magnetic field for bulk superconductors, samples having various dimensions were selected. The axial trapped magnetic field at 1 mm above the top surface of the square superconductors is shown in Figure 5. The trapped magnetic field was calculated by varying the thickness of the square superconductor samples with dimensions 10 × 10, 20 × 20 and 30 × 30 mm^2 under the condition that $\Delta t = \Delta w = 0.1$ mm and $J_c = 20\ 000$ A/cm^2. It can be seen on this figure that the trapped magnetic field rapidly increases with increasing the thickness of the bulk superconductor and remains stable when the sample thickness reaches at a certain value. It is noted that this value is equal to the dimensions of the square sample. The effect of the sample size is also shown in Figure 5. When the dimension of the sample is less, the trapped magnetic field is small and it increases with increasing dimension of the bulk sample [28]. This is understandable if one considers the effects of the dimension and thickness on the B_z in Eq. (3).

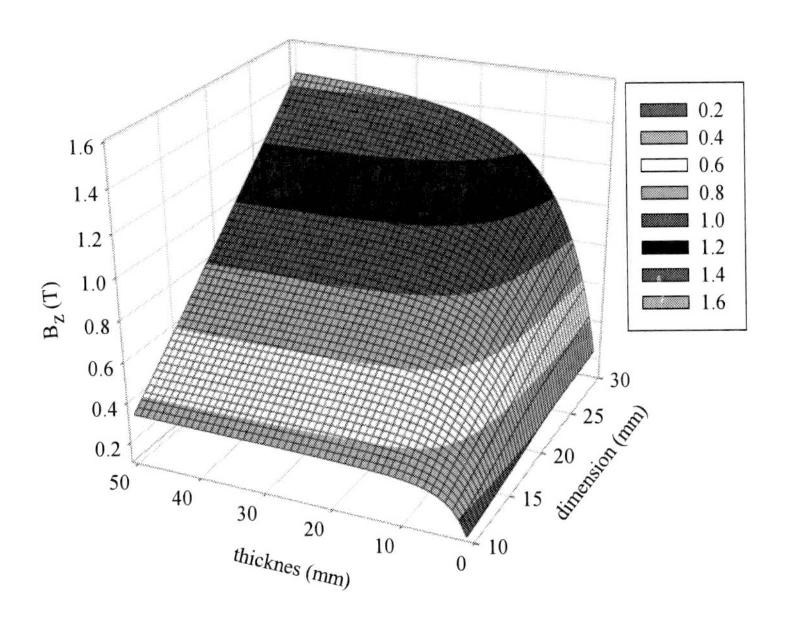

Figure 5. The dimension and thickness dependences of the axial trapped magnetic field for the square superconductor sample [37].

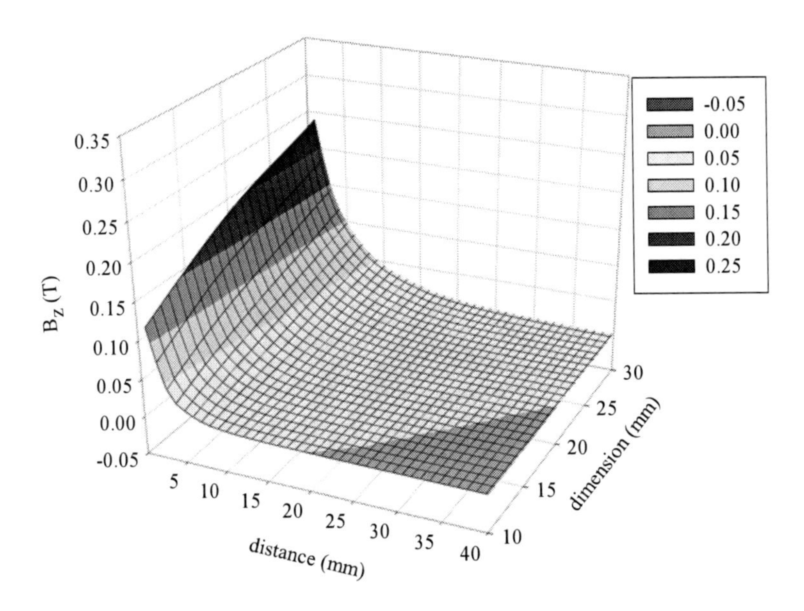

Figure 6. The axial trapped magnetic field distribution as a function of distance between the sample and observation point for a square superconductor sample [37].

In Figure 6, the axial trapped magnetic field is plotted as a function of distance between the top surface of the square samples and the observation point which is along the axis passing through the centre of the sample. The calculation was performed for square samples with dimensions 10 × 10, 20 × 20 and 30 × 30 mm^2 and 1 mm thickness under the condition that $\Delta t = \Delta w = 0.1$ mm and $J_c = 20\ 000$ A/cm^2. It can be seen that the trapped magnetic field decreases with increasing distance and then reduces to zero when the distance reached the sample dimension [28]. This figure also exhibits that the trapped magnetic field increases with increasing dimensions of the square sample, as also shown in Figure 5.

3.4. Diameter and Thickness Dependence of Axial Trapped Magnetic Field for Cylindrical Sample

The effect of the diameter and thickness change on the axial trapped magnetic field at 1 mm above the top surface of the cylindrical superconductors is shown in Figure 7. The trapped magnetic field was calculated by varying the thickness of the cylindrical superconductor samples with 10, 20, 30, 40 and 50 mm diameter under the condition that $\Delta t = \Delta w = 0.1$ mm and $J_c = 20\ 000$ A/cm^2. It can be seen on this figure that the trapped magnetic field rapidly increased on the increasing the thickness of the bulk superconductor and the remained stable when the sample thickness reached at a certain value. It is noted that this value is equal to the diameter of the cylindrical sample. The effect of the sample size is also shown in Figure 7.

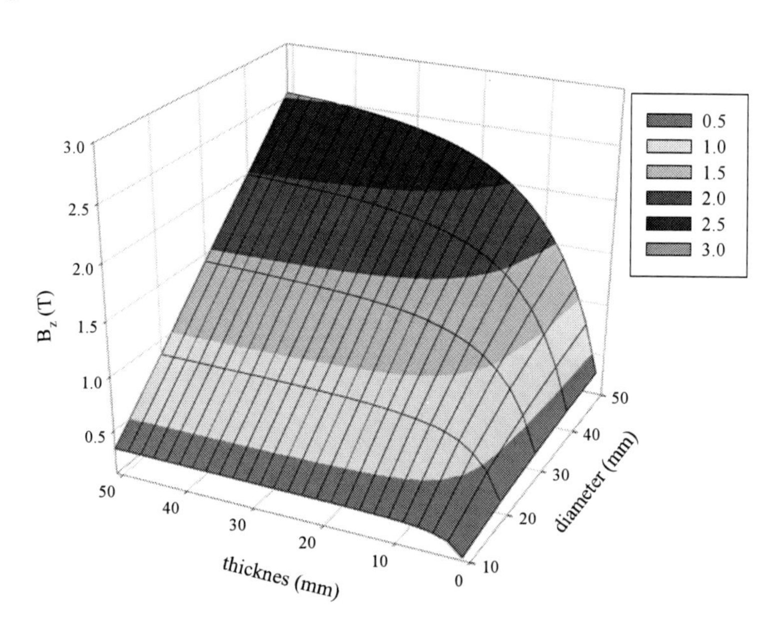

Figure 7. The diameter and thickness dependences of the axial trapped magnetic field for the cylindrical superconductor samples [38].

When the diameter of the sample is less, the trapped magnetic field is small and it increases with increasing diameter of the bulk sample [28]. This is understandable if one considers the effects of the diameter and thickness on the B_z in Eq. (5). The results in this figure show that the thickness or the aspect ratio (R / D) strongly affects the trapped magnetic field values, where R is the radius and D is the thickness of the cylindrical sample. In this manner, the trapped magnetic field is given by the following equation [1]:

$$B_{trap} = AJ_c\ell, \tag{10}$$

where A is a geometrical constant that depends on the shape and aspect ratio of the bulk superconductor sample, J_c is the critical current density and ℓ is the length of the sample.

In Figure 8, the axial trapped magnetic field is plotted as a function of distance between the top surface of the cylindrical samples and the observation point which is on the centre of the axis. The calculation was performed for cylindrical samples with 10, 20, 30, 40 and 50 mm diameter and 1 mm thickness under the condition that $\Delta t = \Delta w = 0.1$ mm and $J_c = 20\,000$ A/cm^2. It can be seen that the trapped magnetic field decreased with the increasing distance and then reduced to zero when the distance reached the sample dimension [28]. This figure also exhibits that the trapped magnetic field increased with increasing diameter of the cylindrical sample, as also shown in Figure 7.

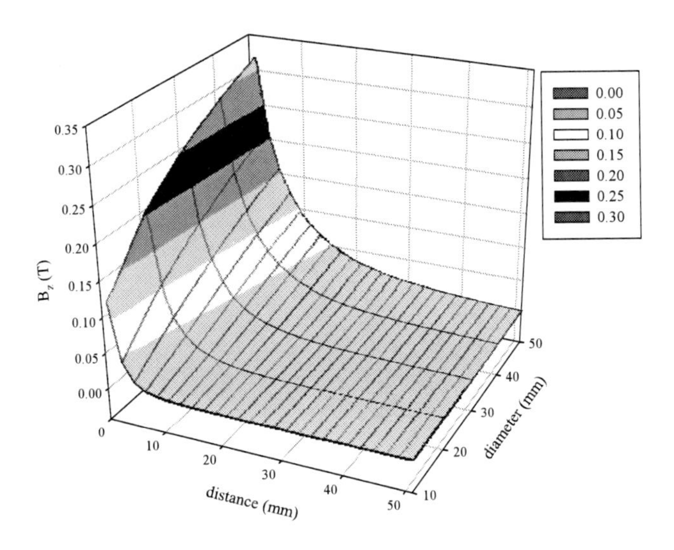

Figure 8. The axial trapped magnetic field at each gap from the surface of the cylindrical superconductor samples to the observation point on the center axis [38].

3.5. Trapped Magnetic Field Distributions in 3D

The x-, y- and z-components of the trapped magnetic field distribution in 3D are shown in Figure 9. The calculation was carried out for 1 mm above the top surface of the square sample with dimensions 30 × 30 × 1 mm^3 under the condition that $\Delta t = \Delta w = 0.1$ mm and $J_c = 20\,000$ A/cm^2. Figures 9(a) and (b) show that the magnitudes of the x- and y-components of the magnetic field are equal, but the directions of the trapped field are transverse.

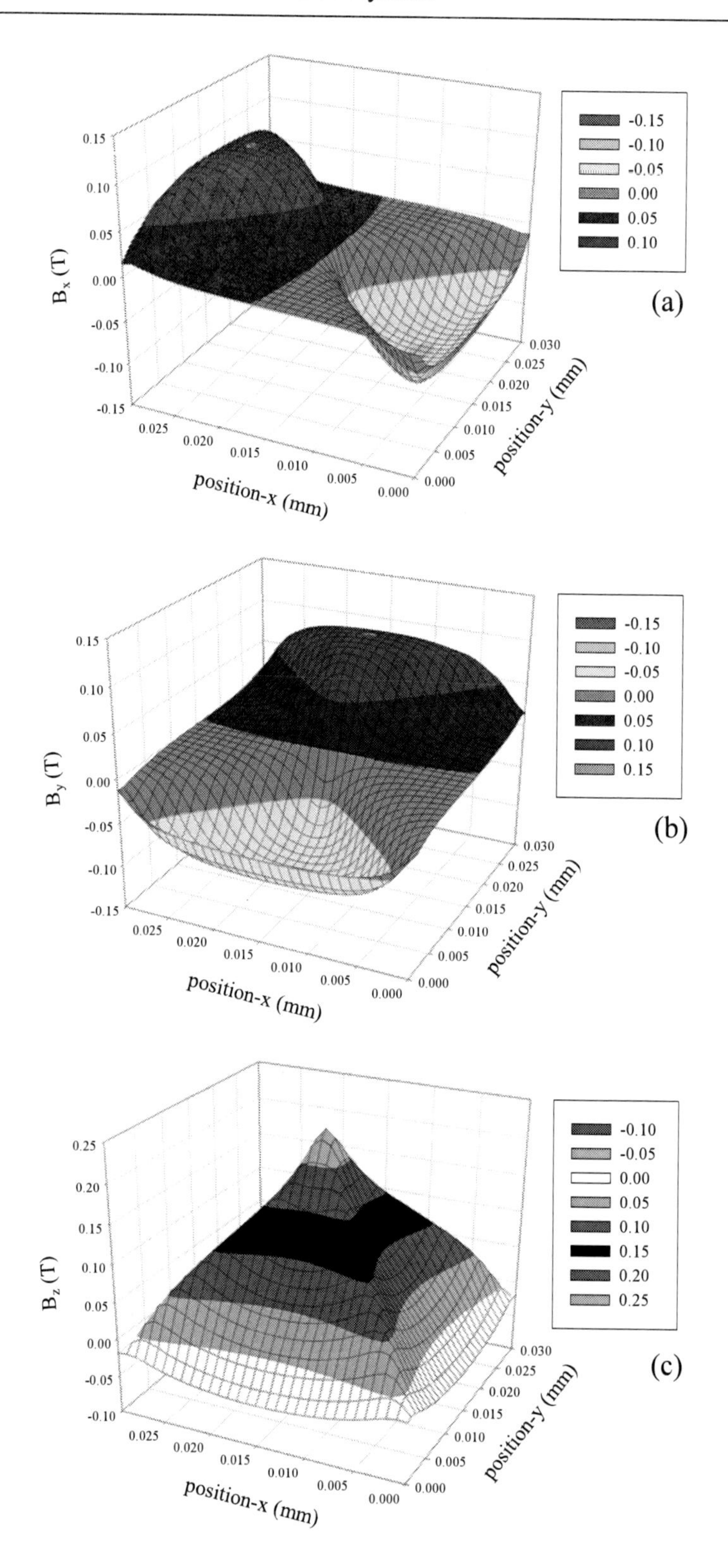

Figure 9. (a) *x*-component, (b) *y*-component and (c) *z*-component of the trapped magnetic field distribution in 3D for a square sample with dimensions 30 × 30 × 1 mm^3 [28].

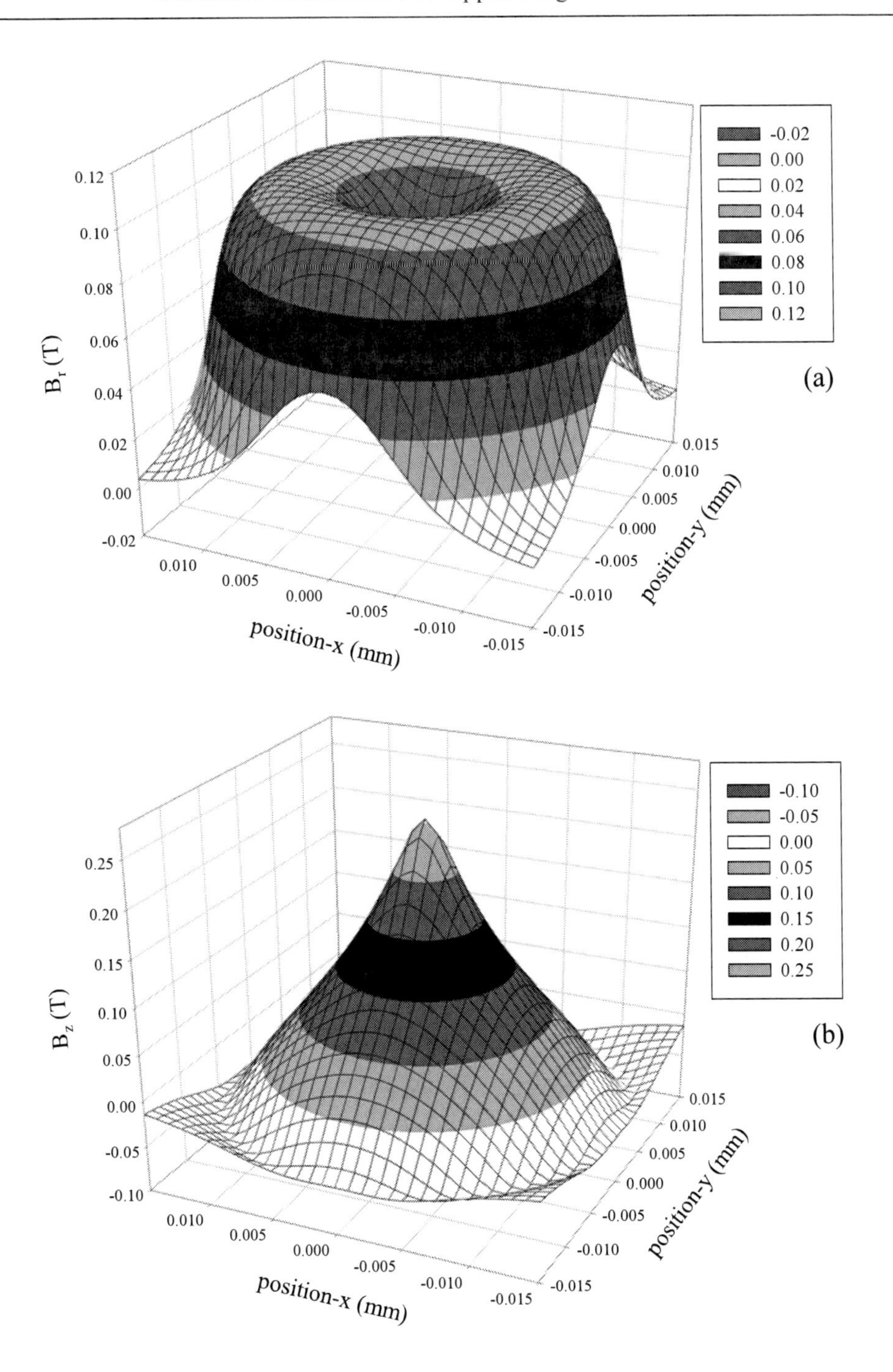

Figure 10. (a) The radial and (b) the axial trapped magnetic field distribution in 3D for the cylindrical sample with 30 mm diameter [28].

In addition, the axial trapped field distribution has a pyramidal shape with a single peak, as shown in Figure 9(c). It is also shown that topological features of the trapped magnetic field maps are quite different for the all components.

Figures 10 (a) and (b) show the radial and axial trapped magnetic field distribution in 3D for the cylindrical sample with 30 mm diameter and 1 mm thickness, at 1 mm above the top surface of the sample under the condition that $\Delta t = \Delta w = 0.1$ mm and $J_c = 20\,000$ A/cm^2. At 1 mm above the surface centre of the sample, which is the observation point, the axial trapped magnetic field is maximum, while the radial trapped magnetic field is minimum.

3.6. Fitting of $J_c - B$ Curves

Recently, the model of general thermally activated creep has been widely discussed [32]-[35], [39]; this model elucidated the relationship between the static and dynamic approaches to the fishtail problem. On the basis of an analysis of experimental data in terms of this model, the pinning potential in a Tm123 single crystal and, later, also other RE123 single crystal was identified as logarithmic [32]. The details of this model are described in "Section 2.2". For the fitting of $J_c - B$ curves with the secondary peak effect, it assumed that the $J_c(B)$ has two components as in equation (9). Fukai et al. [30] calculated the $J_c - B$ curve for Nd-Ba-Cu-O superconductor block, firstly $m = 1$. Then they determined the other parameters, ω , n, J_{p1}, J_{p2} and B_{peak}, in that two different components are assumed to contribute to the total J_c values. As a result, they calculated the best fitted parameters to be $\omega = 4$, $n = 1$, $J_{p1} = 3.2\times10^4$ A/cm^2, $J_{p2} = 2.1\times10^4$ A/cm^2 and $B_{peak} = 1.5$ T, shown in Figure 11. One can see that the fitting curve can reproduce the empirical results.

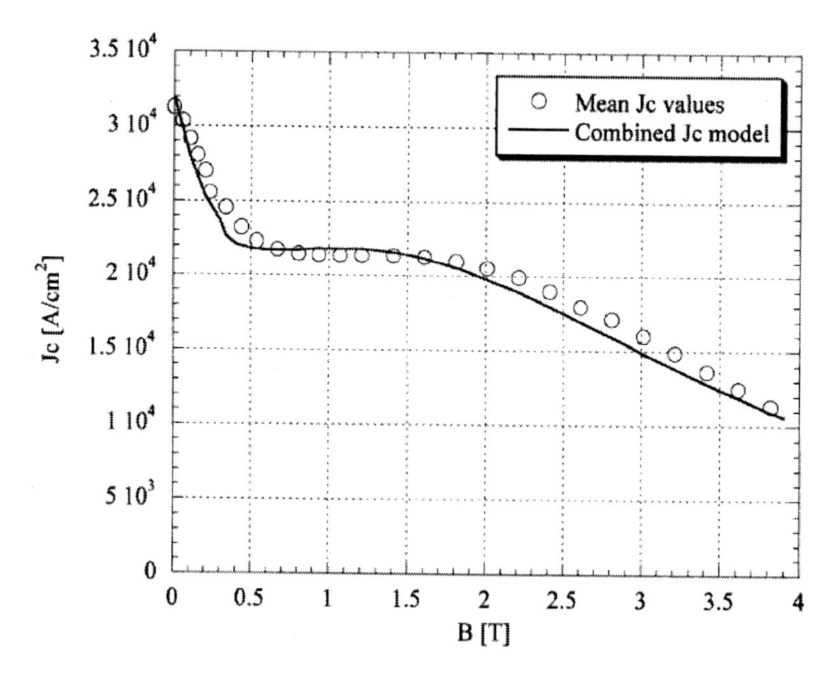

Figure 11. The open circles represent the mean J_c values averaged over several data obtained from different locations of a massive Nd-Ba-Cu-O superconductor block. The solid line represents the $J_c - B$ curve calculated using the parameters: $m = 1$, $\omega = 4$, $n = 1$, $J_{p1} = 3.2\times10^4$ A/cm^2, $J_{p2} = 2.1\times10^4$ A/cm^2 and $B_{peak} = 1.5$ T [30].

It is likely that both the shape of the universal curve at high fields and the corresponding values of m and n are specific to each sample and are determined by a distribution and type of the microscopic pinning sites in the sample. By an artificial change of the distribution and quality of the pinning sites as done, e.g., by a variation of oxygen content or by an irradiation of the samples, the shape of the universal curve can be significantly modified [35]. In Ref. [33], the authors found, for $TmBa_2Cu_3O_7$, and in Ref. [30] for $NdBa_2Cu_3O_7$, m = n = 1. Also, in Ref. [35], the authors found, for $DyBa_2Cu_3O_7$, m = 1.01 and n = 0.5.

An example of a typical fitting procedure by means of equation (9) is the fit of the data measured on MTG-YBCO sample (in Figure 12) and TSMG-YBCO sample (in Figure 13). One can see that the fitting curve can reproduce the empirical results for both samples. A special feature of equation (9) in the fits of experimental data of different RE123 crystals was that the parameter m could be set equal to one [32], [33], [35], [40]. In the present work it is also used this simplification. Thus, keeping $m = 1$, the other parameters, ω , n, J_{p1}, J_{p2} and B_{peak} , was established from the experimental curves. In Figure 12, melt-processed YBCO sample were fabricated at 1050°C of growth temperature. The details of fabrication are described in elsewhere [41]. It can be seen the best fitted parameters to be $\omega = 4.6$, $n = 1$, $J_{p1} = 7090$ A/cm^2, $J_{p2} = 1825$ A/cm^2 and $B_{peak} = 0.9$ T. Also for TSMG-YBCO sample (Figure 13) the best fitted parameters is $\omega = 3.5$, $n = 1.15$, $J_{p1} = 52250$ A/cm^2, $J_{p2} = 13485$ A/cm^2 and $B_{peak} = 1.0$ T.

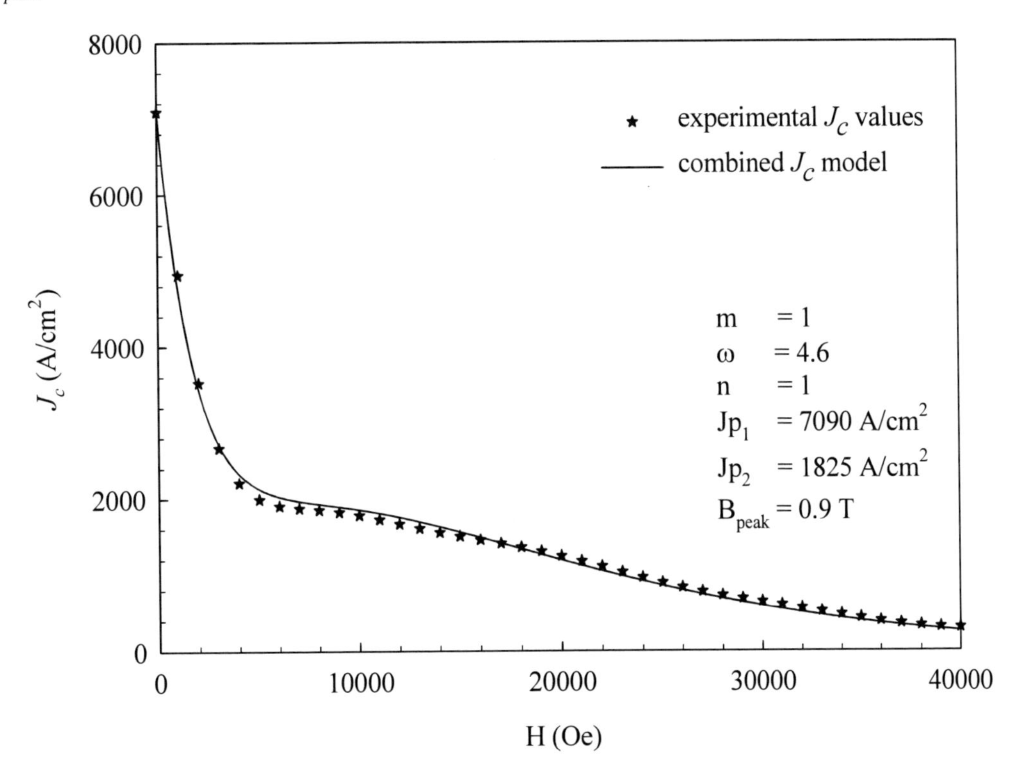

Figure 12. Star sign represents the magnetic field dependence of critical current density of the MTG-YBCO superconductor measured 77 K. The solid line represents the calculated $J_c - B$ curve.

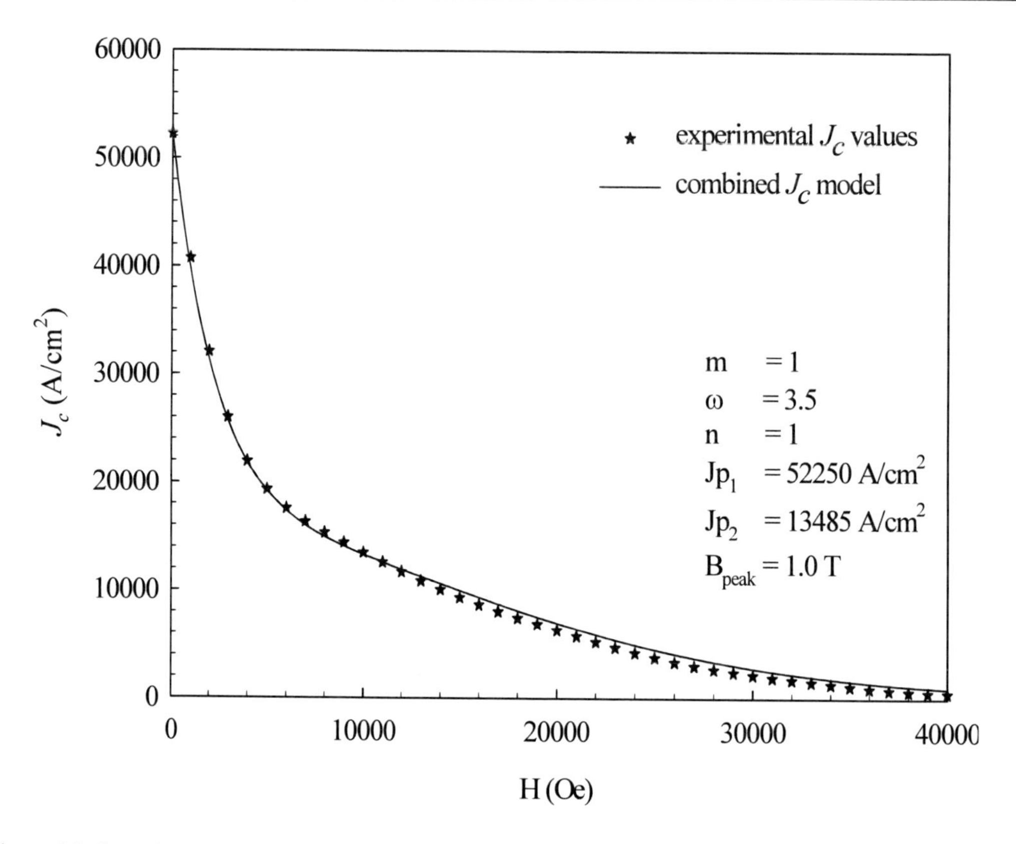

Figure 13. Star sign represents the magnetic field dependence of critical current density of the TSMG-YBCO superconductor measured 77 K. The solid line represents the calculated $J_c - B$ curve.

CONCLUSION

1. The calculation method, using the sand-pile model and the Biot–Savart law, is useful for the determination of the magnetic characteristics of the sample.
2. The trapped magnetic field was almost identical for square and cylindrical samples with the same dimensions.
3. When the sample thickness exceeded the sample dimension, no further change occurred for the magnitude of the trapped magnetic field.
4. The trapped magnetic field value decreased with increasing distance between the sample surface and the observation point. When the distance reached the sample thickness, the trapped magnetic field became very close to zero.
5. It should take account of the B dependence of J_c for the simulation of the trapped magnetic field of large grain sample with the secondary peak effect.

ACKNOWLEDGMENT

This study was supported by Turkish Scientific and Research Council (TUBITAK) research grant (TBAG- 107T751) and Karadeniz Technical University research grant (BAP-2008.111.001.8). I would like to thank Prof. Dr. Ekrem YANMAZ for valuable comments and useful discussions in the Physics Department of Karadeniz Technical University

REFERENCES

[1] Murakami, M. *Physica C*, 2001, 357-360, 751-754.

[2] Hull, J. R. *Supercond. Sci. Technol.*, 2000, 13, R1-R15.

[3] Yamachi, N.; Sakai, N.; Sawa, K.; Murakami, M. *Supercond. Sci. Technol.*, 2005, 18, S67-S71.

[4] Fujimoto, H.; Kamijo, H.; Higuchi, T.; Nakamura, Y.; Nagashima, K.; Murakami, M. *IEEE Trans. Appl. Supercond.*, 1999, 9, 301-304.

[5] Fujimoto, H. Supercond. Sci. Technol., 2000, 13, 827-829.

[6] Tomita, M.; Murakami, M. *Nature*, 2003, 421, 517-520.

[7] Nariki, S. ; Sakai, N. ; Murakami, M. *Supercond. Sci. Technol.*, 2002, 15, 648-652.

[8] Zeisberger, M.; Habisreuther, T.; Litzkendorf, D.; Müller, R.; Surzhenko, O.; Gawalek, W. *Physica C*, 2002, 372-376, 1890-1893.

[9] Fukai, H.; Tomita, M.; Matsui, M.; Murakami, M.; Nagatomo, T. *Physica C*, 2002, 378-381, 738-741.

[10] Higuchi, T.; Sakai, N; Murakami, M.; Hashimoto, M. *IEEE Trans. Appl. Supercond.*, 1995, 5 1818-1821.

[11] Frankel, D. J. J. *Appl. Phys.*, 1979, 50, 5402-5407.

[12] Daumling, M.; Larbalestier, D. C. *Phys. Rev. B*, 1989, 40, 9350-9353.

[13] Conner, L. P.; Malozemoff, A. P. *Phys. Rev. B*, 1991, 43, 402-407.

[14] Frangi, F.; Jansak, L.; Majoros, M.; Zannella, S. *Physica C*, 1994, 224, 20-30.

[15] Tamegai, T.; Krusin-Elbaum, L.; Civale, L.; Santhanam, P.; Brady, M. J.; Masselink, W. T.; Holtzberg, F. *Phys. Rev. B*, 1992, 45, 8201-8204.

[16] Tamegai, T.; Iye, Y.; Oguro, I.; Kishio, K. *Physica C*, 1993, 213, 33-42.

[17] Grant, P. D.; Denhoff, M. W.; Xing, W.; Brown, P.; Gorokov, S.; Irwin, J. C.; Heinrich, B.; Zhou, H.; File, A. A.; Cragg, A. R. *Physica C*, 1994, 229, 289-300.

[18] Xing, W.; Heinrich, B.; Hu, Z.; Fife, A.; Cragg, A. *J. Appl. Phys.*, 1994, 76, 4244-4255.

[19] Kamijo, H.; Kawano, K. *IEEE Trans. Appl. Supercond.*, 1997, 7, 1228-1231.

[20] Wijngaarden, R. J.; Spoelder, H. J. W.; Surdeanu, R.; Griessen, R. *Phys. Rev. B*, 1996, 54, 6742-6749.

[21] Wijngaarden, R. J.; Heeck, K.; Spoelder, H. J. W.; Surdeanu, R.; Griessen, R. *Physica C*, 1998, 295, 177-185.

[22] Jooss, C.; Forkl, A.; Warthmann, R.; Kronmuller, H. *Physica C*, 1998, 299, 215-230.

[23] Amoros, J.; Carrera, M.; Granados, X.; Fontcuberta, J.; Obradors, X. Applied Superconductivity 1997, *Inst. Phys. Conf. Series* No. 158, 1997, 1639-1642.

[24] Carrera, M.; Amoros, J.; Carrillo, A. E.; Obradors, X.; Fontcuberta, J. *Physica C*, 2003, 385, 539-543.

[25] Nagashima, K.; Higuchi, T.; Sok, J.; Yoo, S.I.; Fujimoto, H.; Murakami, M. *Cryogenics*, 1997, 37, 577-581.

[26] Fukai, H.; Tomita, M., Murakami, M.; Nagatomo, T. *Supercond. Sci. Technol.*, 2000, 13, 798-801.

[27] Fukai, H.; Tomita, M.; Murakami, M.; Nagatomo, T. *Physica C*, 2001, 357-360, 774-776.

[28] Aydıner, A.; Yanmaz, E. *Supercond. Sci. Technol.*, 2005, 18, 1010-1015.

[29] Pina, J. M.; Neves, M. V.; Rodrigues, A. L. *Powering*, 2007, 12-14, 185-190.

[30] Fukai, H.; Tomita, M.; Murakami, M.; Nagatomo, T. *Supercond. Sci. Technol.*, 2002, 15, 1054-1057.

[31] Smythe, W. R. *Static and Dynamic Electricity*, McGraw-Hill: New York, 1968, 291.

[32] Jirsa, M.; Muralidhar, M.; Murakami, M.; Noto, K.; Nishizaki, T.; Koyabashi, N. *Supercond. Sci. Technol.*, 2001, 14, 50-57.

[33] Perkins, G. K.; Cohen, L. F.; Zhukov, A. A.; Caplin, A. D. *Phys. Rev. B*, 1995, 51, 8513- 8520.

[34] Perkins, G. K.; Caplin, A. D. *Phys. Rev. B*, 1996, 54, 12551-12556.

[35] Jirsa, M.; Pust, L.; Dlouhy, D.; Koblischka, M. R. *Phys. Rev. B*, 1997, 55, 3276-3283.

[36] Nariki, S.; Sakai, N.; Murakami, M. *Supercond. Sci. Technol.*, 2002, 15, 648-652.

[37] Aydıner, A.; Yanmaz, E. *Physica C*, 2010, 470, 1193-1197.

[38] Aydıner, A.; Yanmaz, E. J. *Supercond. Nov. Magn.*, 2010, 23, 457-463.

[39] Jirsa, M.; Koblischka, M. R.; Muralidhar, M.; Higuchi, T.; Murakami, M. *Physica C*, 2000, 338, 235-245.

[40] Jirsa, M.; Pust, L. *Physica C*, 1997, 291, 17-24.

[41] Aydıner, A.; Çakır, B.; Seki, H.; Başoğlu, M.; Wongsatanawarid, A.; Murakami M.; Yanmaz, E. J. *Supercond. Nov. Magn.* (in pres; 2011) DOI: 10.1007/s10948-010-0841-6.

In: Superconductivity
Editor: Vladimir Rem Romanovskiĭ
ISBN 978-1-61324-843-0

Chapter 7

ION MODIFIED HIGH-TC - JOSEPHSON JUNCTIONS AND SQUIDS

S. S. Tinchev*
Institute of Electronics
Bulgarian Academy of Sciences
Sofia, Bulgaria

ABSTRACT

After the discovery of high-T_c superconductivity there has been considerable effort in developing Josephson junctions and SQUIDs from these materials. During this time only very limited numbers of fabrication techniques were developed, because of the difficult material properties of high-T_c superconductors. Therefore the most successful type junctions remain grain boundary junctions. Twenty years ago I proposed and successfully fabricated ion modified high-T_c Josephson junctions and SQUIDs. In this review all published results are summarized from our first demonstration in 1990 until the recent successfully fabrication in Berkeley of more then 15000 junctions with a critical current spread of only about 16 %.

1. INTRODUCTION

Almost 25 years after discovery of high-T_c superconductivity it is still a problem to fabricate Josephson junctions and SQUIDs from these materials. During this period of time different fabrication techniques were developed. The most successful are grain boundary junctions (bicrystal and step-edge) as well as ramp junctions. However, they all have drawbacks – large variation of junction's parameters and limited large scale integration capability.

* E-mail address: stinchev@ie.bas.bg

1990 I proposed and demonstrated [1] an alternative technology – ion modified Josephson junctions. In this technology accelerated particle irradiation is used to reduce T_c in a narrow region of a superconducting bridge and thus a SNS (superconducting-normal-superconducting) operation was possible. In our first experiments [1] electron irradiation in an ordinary scanning electron microscope was used. Better results were obtained, however, by ion irradiation. In this technique a microbridge is patterned from an YBCO film by standard photolithography and covered by PMMA mask*. Next a narrow channel crossing the microbridge was opened by electron beam lithography – Figure 1. During the ion modification actually this narrow unprotected region of the bridge is exposed to the ion beam and is modified.

Important point was the choice to use oxygen ions for this irradiation, which are inherent to YBCO and do not introduce additional effects. Thus operation at liquid nitrogen temperature was possible. However, later other kinds of ions (Ar^+, Ne^+ and H^+) were also successfully used [2]-[4].

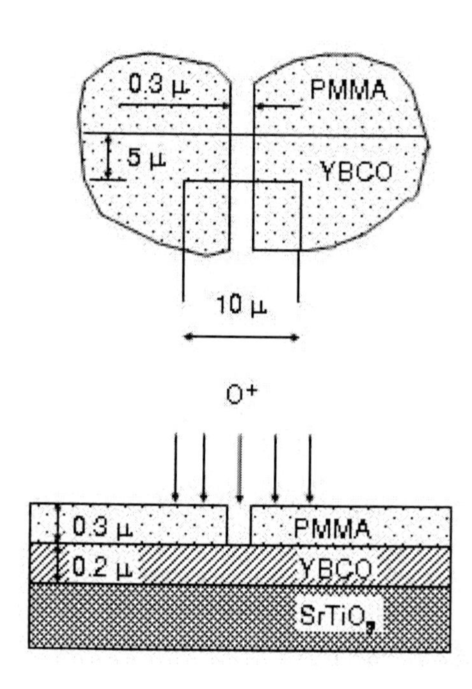

Figure 1. Shemetic diagram of the weak link showing the masking techniques.

Figure 2 and Figure 3 are reprinted from our first published paper [1]. Figure 2 shows the superconducting transition of a 200 nm YBCO film before and after 100 keV oxygen ion irradiation with an irradiation dose of $2x10^{13}$ Ions/cm^2. One can see that the shape of the superconducting transition does not change significantly; only the critical temperature decreases. This measurement was carried out contactlessly, monitoring the resonant frequency of the SQUID tank circuit during the change of the temperature.

*In this paper only the masked ion beam modification will be considered, although focused ion beam was also successfully used to fabricate Josephson junctions and SQUIDs from high- T_c materials.

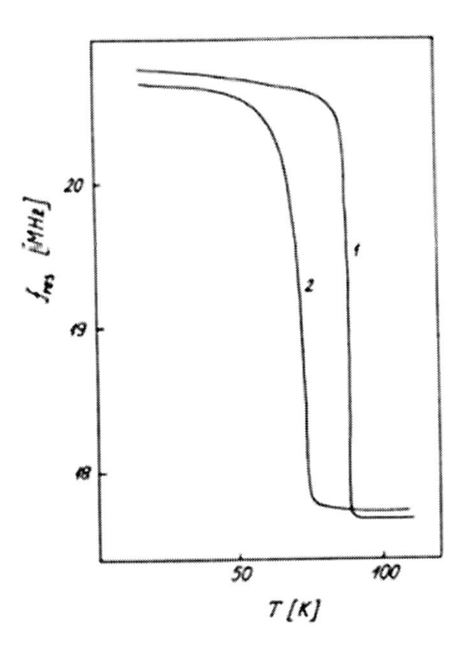

Figure 2. Superconducting transition of the YBCO film before (curve 1) and after (curve 2) 100 keV oxygen ion modification with dose of $2x10^{13}$ O^{+}/cm^{2}. Figure reprinted from [1]. Copyright 1990 by the IOP Publishing.

Figure 3 is an optical micrograph of the bridge region. The line across the bridge is the non-protected area.

Figure 3. Micrograph of the bridge region of a SOUID after ion irradiation. The line across the bridge is the non-protected area. Figure reprinted from [1]. Copyright 1990 by the IOP Publishing.

The general idea of our technology is to fabricate superconductor-normal-superconductor (SNS) junctions, where the normal interlayer is a superconductor above its transition temperature. The suppressed T_c and thus the operation temperature of the devices produced by this technique can be easily controlled by adjusting the radiation dose as Figure 4 shows.

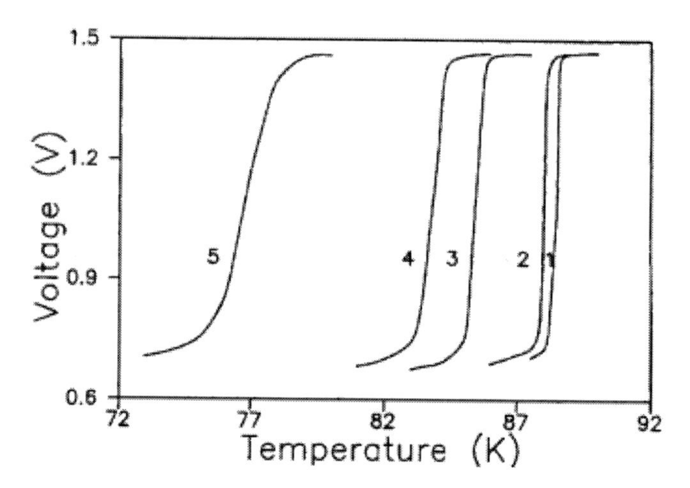

Figure 4. Superconducting transition for different irradiation doses at 100 keV O^+ ions [5]: curve (1) before irradiation, curve (2) $1x10^{12}$, (3) $5x10^{12}$, (4) $1x10^{13}$, (5) $3x10^{13}$. Copyright 1996 by Elsevier.

2. Properties of YBCO Weak Links Prepared by Local Oxygen Ion Modification

2.1. Current-voltage characteristics

Figure 5 shows the I-V characteristic of a weak link with a critical current of approximately 10 μA as typically used in our RF SQUIDs [5]. From the figure it can be seen that the resistive branch of the current-voltage characteristic shows almost a RSJ-like (resistively shunted junction model) behavior with thermal noise rounding and normal-state resistance $R_n \sim 1\ \Omega$. For these junctions the I_cR_n product is typically around 10 μV. For weak links with larger critical current or the same junction at lower temperatures the I-V curves were more flux-flow-like see Figure 6. This change is caused by a transition from SNS to SS'S type junction.

In our first experiment the slit was rather wide, typically about 300 nm. Later other groups reduced significantly the slit dimensions. In a paper [4] of the group of Professor Seidel in Jena, Germany a double layer resist mask was used which allows fabrication of narrowed slits down to 120 nm. In their experiments oxygen ions were also used with an ion energy of 30 keV or 100 keV and varying the dose between 10^{13} ions/cm^2 and 10^{14} ions/cm^2. The influence of film thickness in the range 30–110 nm and slit width between 120- 250 nm on the superconducting properties of these junctions was investigated.

Significant smaller slit of 50 nm in the implantation mask was fabricated in the Juelich – Aachen groups in Germany [6] using three layers resist. RSJ-like behavior properties of junctions in YBCO thin films, fabricated by oxygen irradiation at 200 keV through a 50 nm wide slit was observed [6] within the temperature window shown in Figure 7. Further smaller slit down to 20 nm was recently reported by Bergeal et al. [7] and nice RSJ-characteristics were demonstrated.

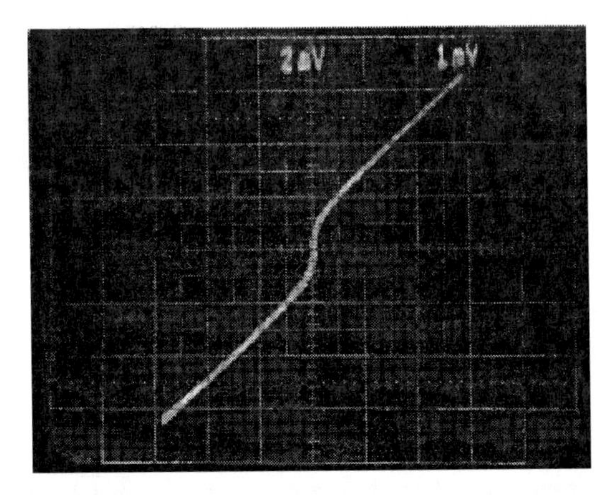

Figure 5. Typical I-V characteristic of low-I_c weak link at 77 K [5]. The vertical scale is 10 μA/div, the horizontal scale is 10 μV/div. Copyright 1996 by Elsevier.

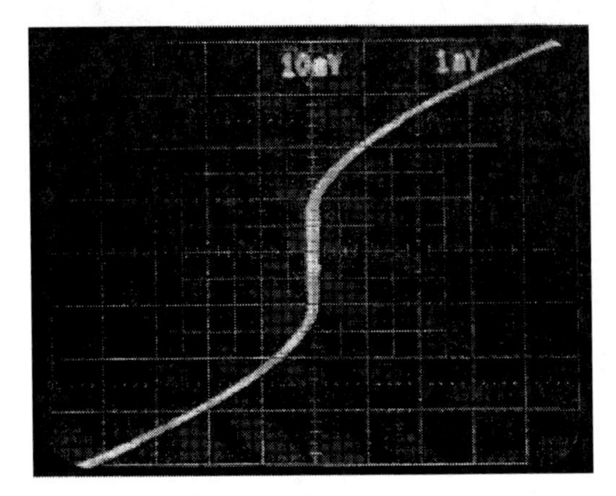

Figure 6. I-V curve for microbridge showing flux flow [5]. The vertical axis is 50 μA/div, the horizontal axis is 10 μV/div. Copyright 1996 by Elsevier.

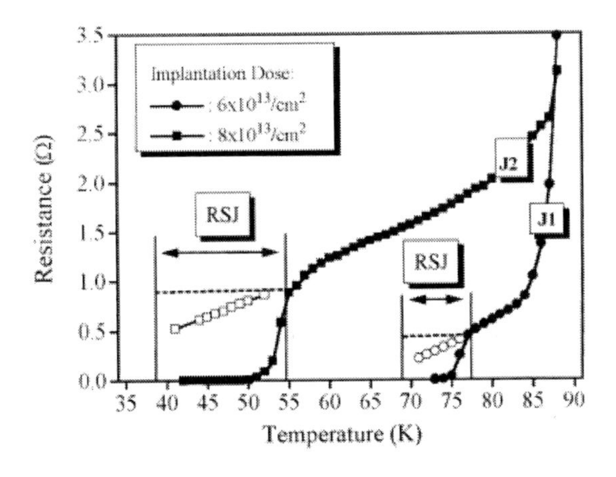

Figure 7. Resistance vs. temperature for two 3 μm wide junctions irradiated with different fluences. Marked is the temperature window for RSJ current voltage-characteristics. Figure reprinted with permission from [6]. Copyright 1999 by 1999 Elsevier Science B.V.

2.2. Microwave behavior

Under microwave irradiation our weak links with small critical currents ($I_c \sim 10$ μ.A) did not show Shapiro steps. Only the critical current of the bridge was suppressed as shown in Figure 8(a). In weak links having a larger critical current, rounded steps were observed (Figure 8(b)).

Later other groups using smaller slits observed clear Shapiro steps on the I-V characteristics of the irradiated junctions [7]. At 77 K they are similarly rounded by the thermal fluctuation. At 4.2 K, however, the steps are well defined, appearing at equal distance and with oscillating amplitudes according to the standard theory of the Josephson effect.

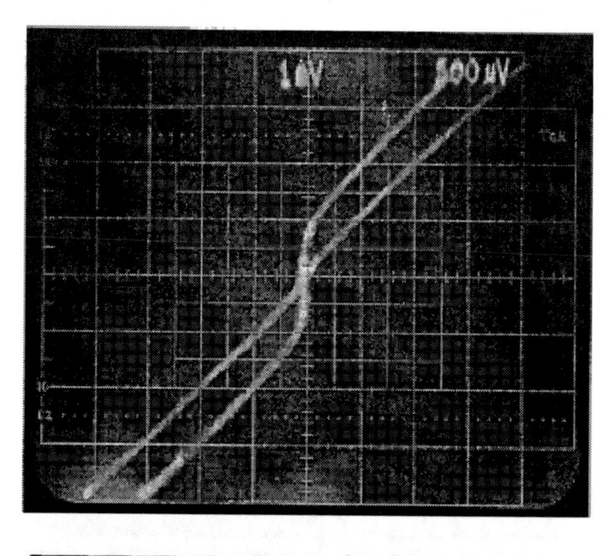

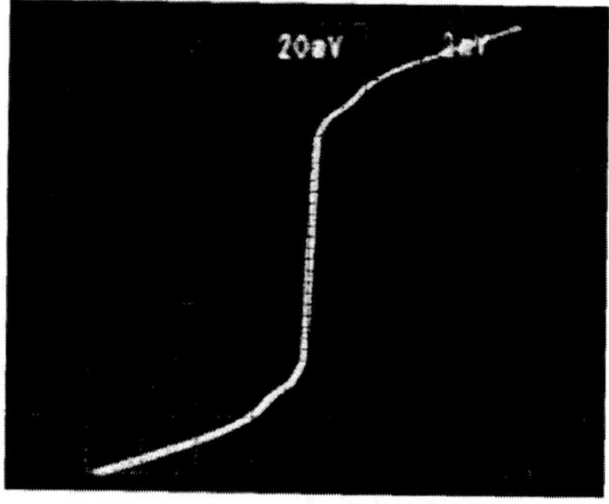

Figure 8. (a) I-V characteristics of oxygen ion modified weak links [5] (a) I-V characteristics of low-I_c microbridge at 77 K without and under microwave irradiation (10 GHz, 10.5 dBm). The vertical grid is 5 μ/div, the horizontal grid is 5 μV/div. (b) I-V characteristics of the mid-I_c weak links showing rounded steps under microwave irradiation at 9. 41 GHz. Copyright 1996 by Elsevier.

2.3. Magnetic-field results (proof of the junction quality by Fraunhofer diffraction patterns)

The dependence of the critical current on applied magnetic field yields information on the current distribution through the junction area. In our first samples two types of characteristics were observed. In one group, see Figure 9, the critical current decreases nearly monotonically up to a field of about 1 mT. This is far from the ideal Fraunhofer diffraction pattern, indicating a non-uniform distribution of the current across the junction. The deviation from the ideal characteristic probably be due to a low quality of the used film.

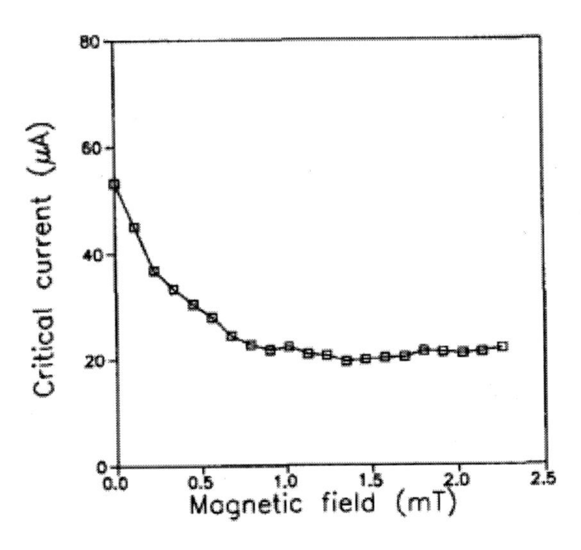

Figure 9. Critical current as a function of the applied magnetic field [5] for a weak link indicating a non-uniform distribution of the current across the junction. Copyright 1996 by Elsevier.

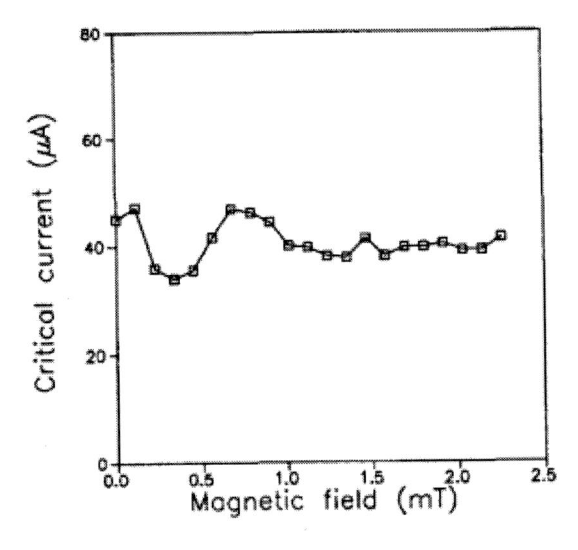

Figure 10. $I_c(B)$ curve of a weak link [5] showing more uniform distribution of the current across the junction. Copyright 1996 by Elsevier.

The second type of $I_c(B)$ characteristics, Figure 10, exhibits deep first minima suggesting a more uniform distribution of the current across the junction. There is a good agreement between the experimental and the expected position of the minima, where the theoretical value is calculated by $B = 1.84\ \Phi_0/w^2$, here w is the width of the junction and Φ_0 is the flux quantum. The observed high secondary peaks could be explained by at least two points in the bridge with enhanced current density acting as an intrinsic DC SQUID. In both cases the critical current is not completely suppressed to zero in Earth field and also up to relatively high fields (~ 0.1 T).

Nearly ideal Fraunhofer pattern were reported later by other groups [6], see Figure 11, indicating a homogeneous critical current density across the barrier and therefore a homogeneous defect distribution throughout the implanted region of the microbridge. This is probably result of using high quality YBCO films and junctions with narrower slits (~50 nm in this case).

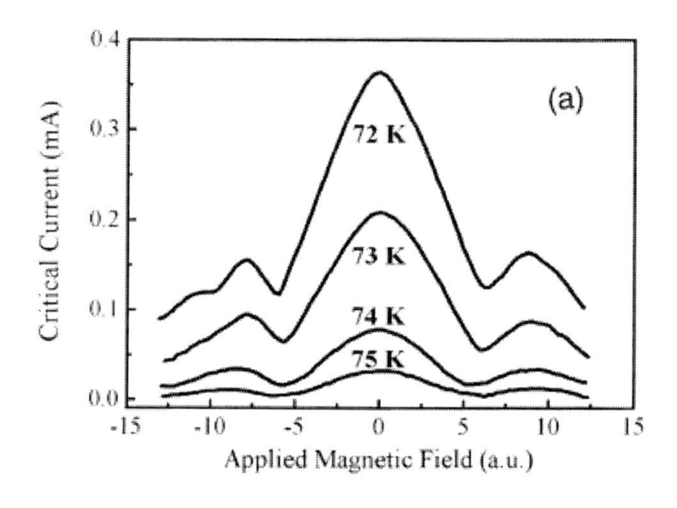

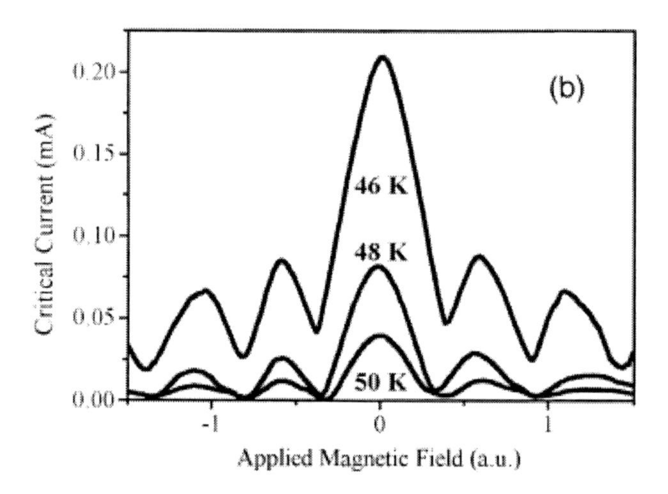

Figure 11. Critical current as a function of applied magnetic field for two junctions at different temperatures. Figure reprinted with permission from [6]. Copyright 1999 by Elsevier Science B.V.

2.4. Current-phase relation

The relation between the superconducting current and phase-difference across the junction is of considerable scientific and technical importance in weak links, because it is well known than just in weak links the current-phase relation can be nonsinusoidal. Despite its importance, the current-phase relation of high-T_c junctions was unknown until 1994 and our measurement of current-phase of high-T_c weak link fabricated by oxygen-ion irradiation [8] was the first measured current-phase relation of any high-T_c junction.

We have used an indirect technique to obtain current phase-relation from measurements of the AC impedance of a weak link incorporated in a superconducting ring. In this method actually a nonhysteretic mode of RF SQUID operation is used, where the resonant frequency of the tank circuit is magnetic field dependent – Figure 12. From this dependence the current phase-relation was extracted. No significant deviations from sinusoidal Josephson relation were found – Figure 13, which indicates that oxygen-ion irradiated junctions are probably S-N-S type.

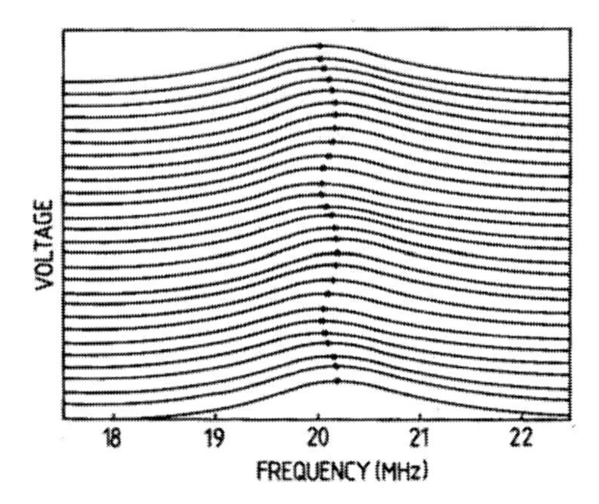

Figure 12. Tank circuit resonsnt frequency shift of the of a RF SQUID in nonhysteretic mode for different magnetic field [8]. Copyright 1994 by Elsevier.

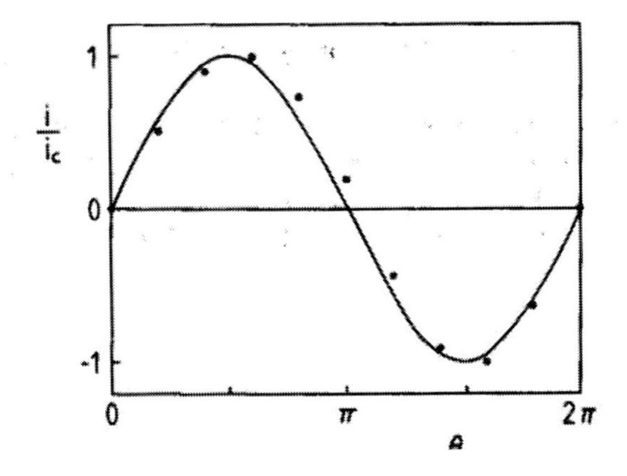

Figure 13. The ideal sinusoidal Josephson current-phase relation (solid line) and the measured one (dots) [8]. Copyright 1994 by Elsevier.

2.5. Temperature dependence of the critical current

A drawback of these junctions is the small temperature range of operation, because of the strong temperature dependence of the critical current – Figure 14. This behavior is evident also from Figure 7, where the temperature window for RSJ current voltage-characteristics is marked.

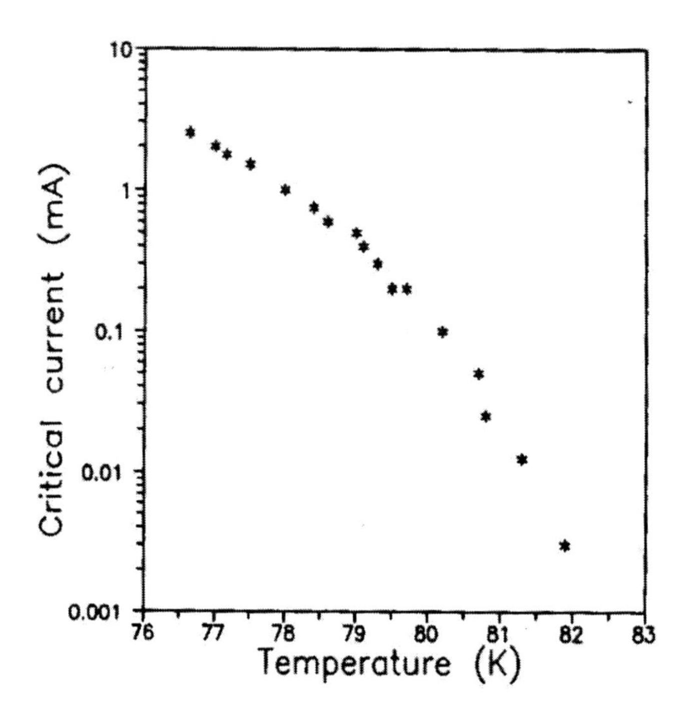

Figure 14. Critical current as a function of the temperature near 77K [5] for a junction irradiated with $2x10^{13}$ O^+/cm^2 at 100 keV. Copyright 1996 by Elsevier.

3. MODELING OF THE ION MODIFIED JOSEPHSON JUNCTIONS

A very important point is that the reduction of the critical temperature caused by the ion irradiation and thus the operation temperature of these weak links can be predicted theoretically and well controlled [9], [10]. Figure 15 shows of the calculated T_c profile. In this calculation the Monte Carlo code TRIM was used together with the dependence of T_c on the deposited energy introduced by Summers [11].

Similar calculation were carried out later by others groups using the TRIM (SRIM) or other computer programs [12], [13]. Figure 16 shows two dimensional plots for 50 keV proton beam irradiation damage process in Au masked YBCO on $LaAlO_3$ substrate using the Monte Carlo simulation code CRYSTAL.

Moreover one can calculate [14] also the annealing effects on the T_c and thus the operation temperature of these devices after a thermal annealing – Figure 17.

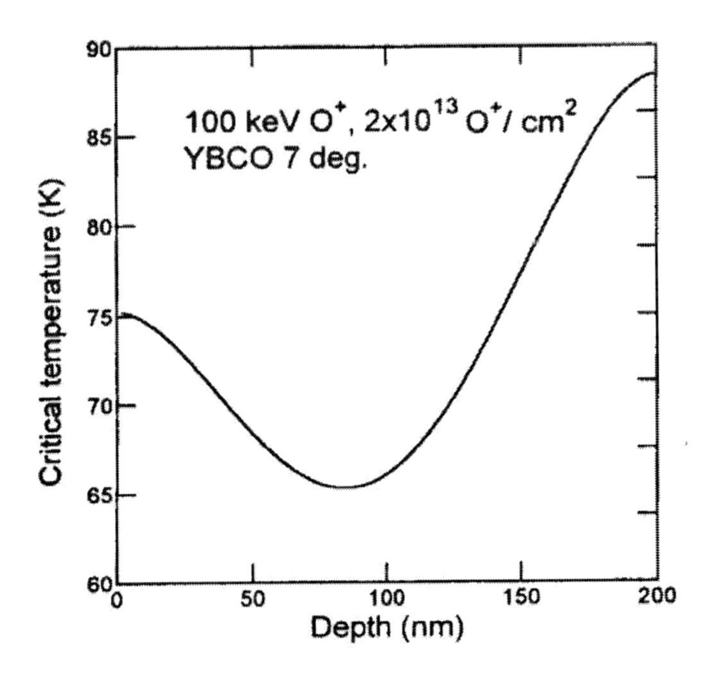

Figure 15. Calculated T_c as a function of the YBCO film depth for a single modification with 100 KeV oxygen ions. Figure reprinted with permission from [9]. Copyright 2006 by the American Institute of Physics.

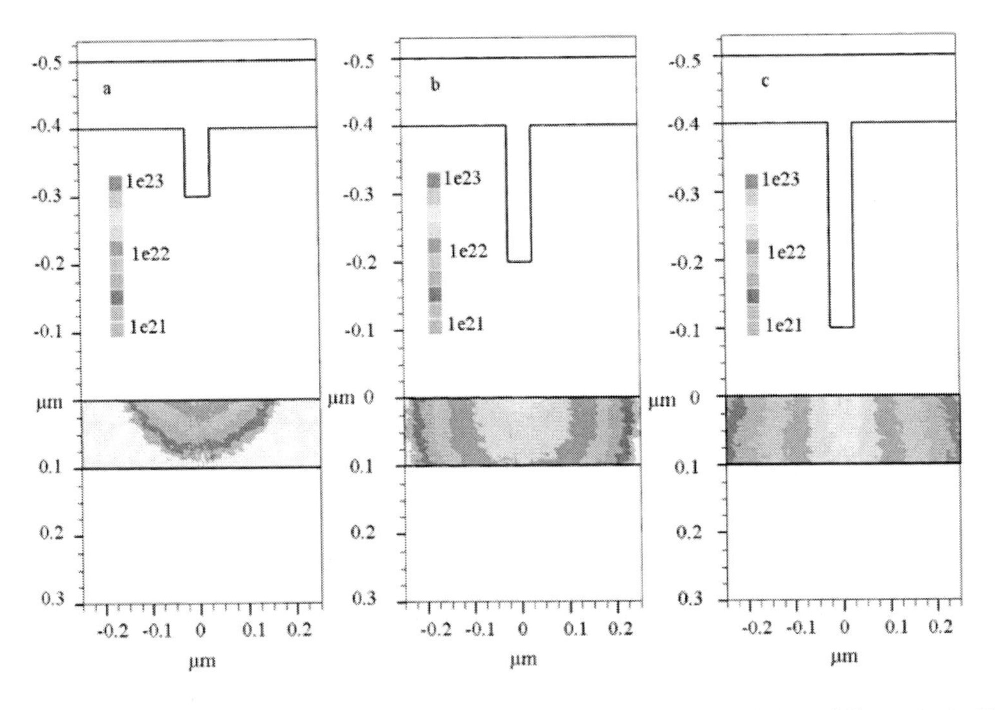

Figure 16. Simulated 30 keV proton beam irradiation for 50 nm slit mask and three different Au buffer layers. Figure reprinted with permission from [12]. Copyright 1999 by Elsevier Science B.V.

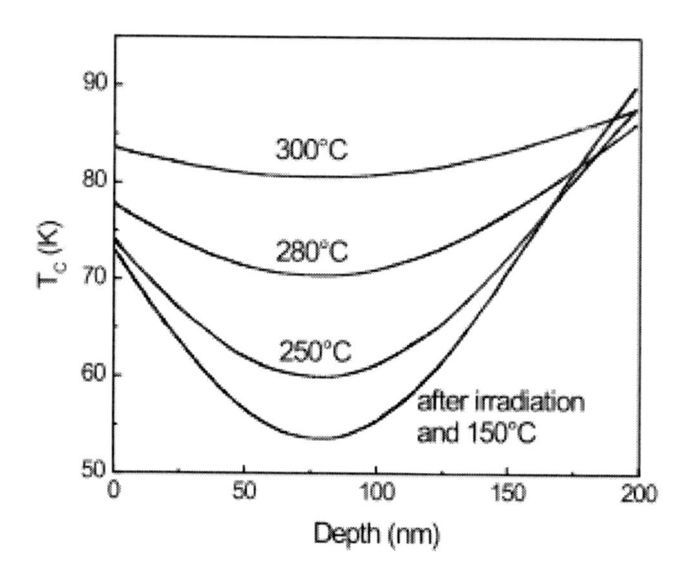

Figure 17. Calculated critical temperature depth profiles [14] for single energy implantation with 100 keV, $3x10^{13}$ $O^+cm/^2$ after a 20 min thermal annealing at different temperatures. Copyright 2002 by Elsevier.

4. MgB_2 Ion Modified Josephson Junctions

Using ion modification Josephson junctions and SQUIDs were successfully fabricated not only from YBCO but also from MgB_2 thin films [15]-[17]. In the [15] 100-nm-thick MgB_2 films with T_c of 38 K were used to fabricate superconductor-normal-superconductor type (SNS) Josephson junctions. The chips were exposed to a 100 keV H_2 monitoring the barrier resistance in situ following by RTA annealing at 300 °C. The junctions showed nonhysteretic current-voltage characteristics between 36 K and 4.2 K. Magnetic field dependence of the I_c showed deviation from the ideal Fraunhofer diffraction at higher magnetic fields indicating some inhomogeneities in the junctions. Shapiro steps were observed at the expected voltages. In the next paper [16] of the same group much better modulation of the critical current by applied magnetic fields and the Shapiro steps under microwave irradiation were observed.

In the University of California at San Diego Cybart et all [17] fabricated also planar thin-film MgB_2 Josephson junctions and 20-junction series arrays using 200-keV Ne^+ ion implantation with a dose of $1x10^{13}$ ions/cm^2 and electron-beam lithography. Resistively shunted junction I-V characteristics were observed in the temperature range of 34–38 K – Figure 18. Flat Shapiro steps in arrays – Figure 19 suggest good junction uniformity with a small spread in junction parameters. A conclusion was made that the higher operating temperature and close spacing of these junctions make junctions fabricated by this technology promising candidates for quantum voltage standards and other devices.

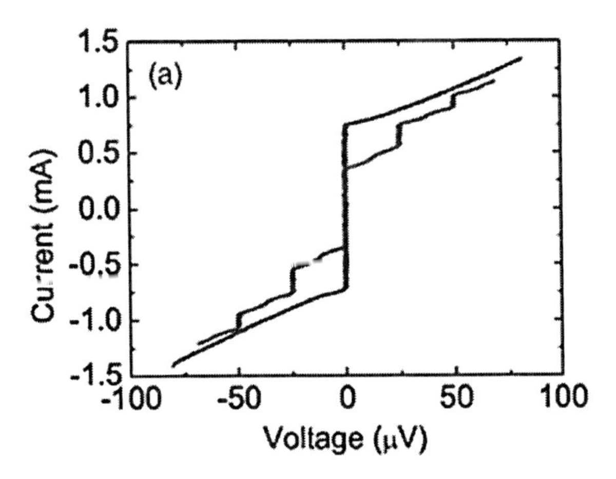

Figure 18. I-V characteristics for a single junction measured at 37.2 K, with and without 12 GHz microwave radiation. Figure reprinted with permission from [17]. Copyright 2006 by the American Institute of Physics.

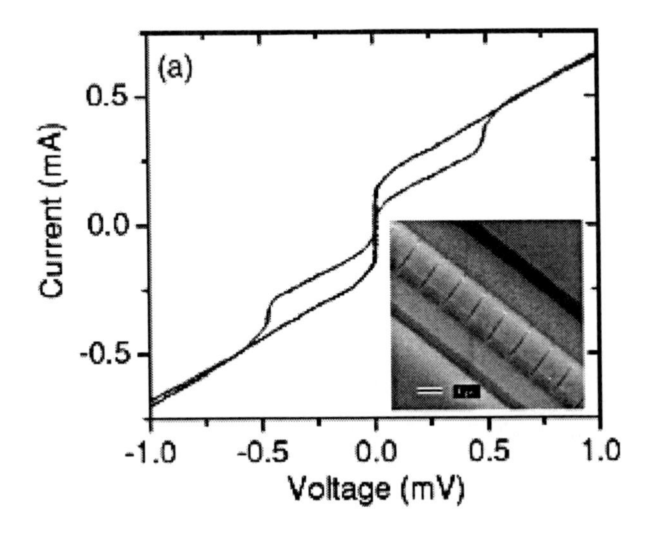

Figure 19. I-V characteristics for a 20-junction array measured at 37.5 K, with and without 12 GHz microwave radiation. Figure reprinted with permission from [17]. Copyright 2006 by the American Institute of Physics.

5. HIGH-T$_C$ SQUIDS BASED ON ION MODIFIED JOSEPHSON JUNCTIONS

Both RF and DC SQUIDs were successfully fabricated by using ion modified high-T$_c$ Josephson junctions. Most of the RF SQUIDs [1, 18] that we made were washer type SQUIDs operating at frequency of about 25 MHz. A typical layout is shown in the Figure 20(a). The washer was 1400 μm x 1400 μm with a SQUID loop of 100 μm x 100 μm.

The bridge was typical 10 μm long and 5 μm wide. These SQUIDs operate at liquid nitrogen temperature in hysteretic mode although non-hysteretic mode was used to measure the current-phase relation of the weak links as already mention above. In some magnetometer design a big single input coil was directly connected to the washer – Figure 20(b) and Figure 20(c). First order planar gradiometer – Figure 20(d) was also designed and tested [19]. The noise these SQUIDs was white about (2 - 4).10^{-4} $\Phi_0/Hz^{-1/2}$for frequencies above 0.2 Hz. Below this frequency 1/f noise was usually observed.

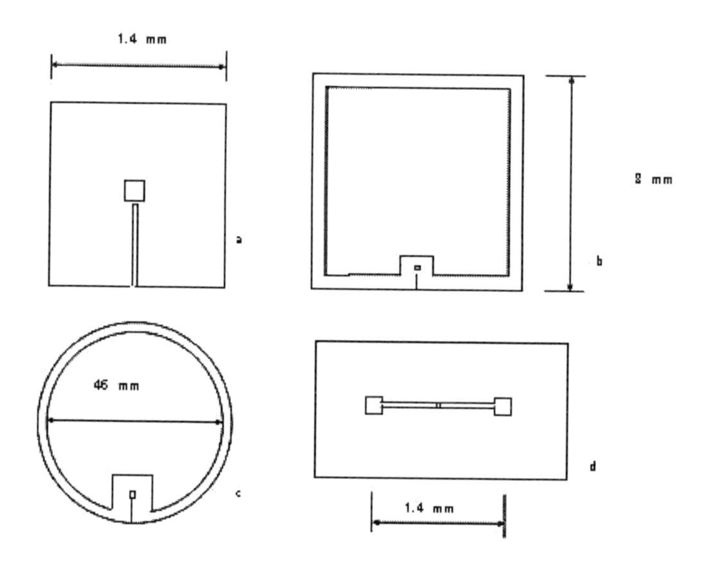

Figure 20. Typical layouts of RF SQUIDs.

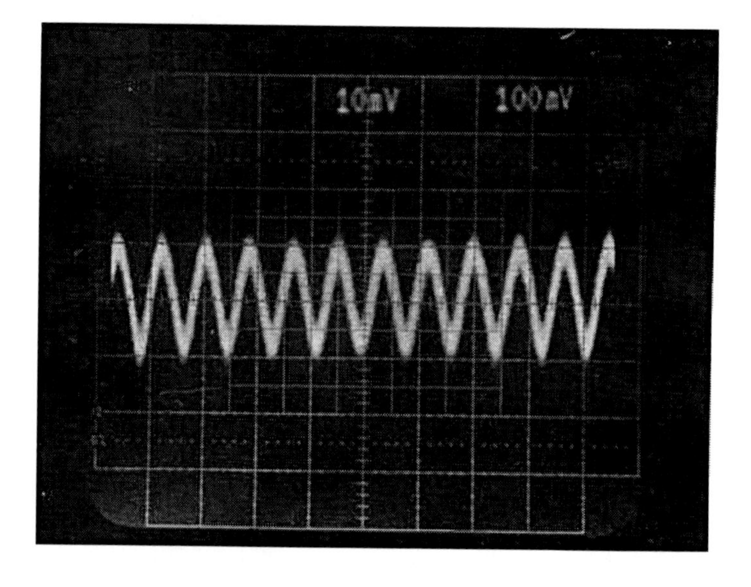

Figure 21. Typical pattern of RF SQUID fabricated by oxygen ion modification measured at temperature 77 K.

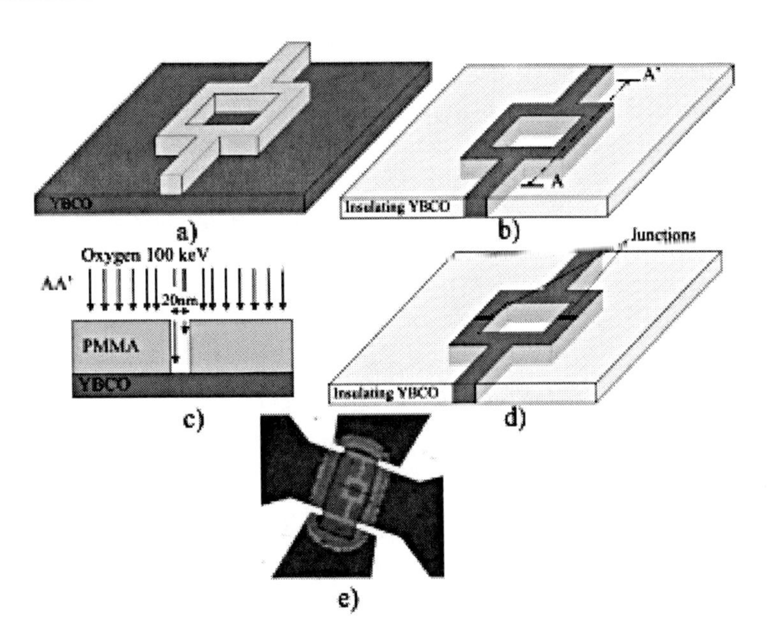

Figure 22. Fabrication steps of YBa_2Cu3O_7 DC SQUIDs by 100 keV oxygen implantation. Figure reprinted with permission from [20]. Copyright 2006 by the American Institute of Physics.

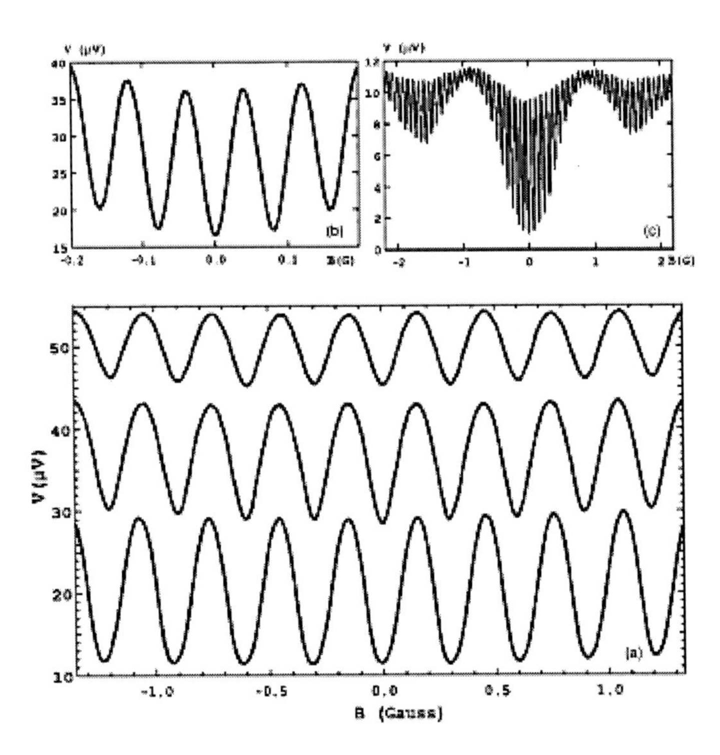

Figure 23. Voltage vs applied magnetic field characteristics for YBa_2Cu3O_7 DC SQUID made by by 100 keV oxygen implantation. Figure reprinted with permission from [20]. Copyright 2006 by the American Institute of Physics.

Recently a DC SQUIDs were fabricated [20] from 150 nm thick YBa_2Cu3O_7 by 100 keV oxygen implantation. A washer geometry – Figure 22 have been used with loops 100 μm^2 or 36 μm^2 and inductance 32 pH or 17 pH respectively. The SQUIDs were patterned by using a gold mask without material removing. The sample was irradiated with a high fluence $5x10^{15}$ O^+/cm^2 of 110 keV oxygen ion such that the unprotected parts become insulating. Two very narrow slits about 20 nm wide were opened by electronic lithography across each arms of the superconducting loop. The SQUIDs were operated at the bias current regime and have nice voltage to flux characteristics (Figure 23) with a transfer function of ~ 60 $\mu V/\Phi_0$.

6. Some Applications of HT_C SQUIDs Used Ion Modified Josephson Junctions

During my work at F.I.T. in Germany some measurement systems based on oxygen-ion implanted high-T_c SQUIDs were developed and tested. Figure 24 shows a scanning SQUID magnetometer [21], where a spatial resolution of about 1 mm for room temperature samples was achieved. This system was used to image spatial variation of weak magnetic fields of different samples. Figure 30 shows the magnetization distribution of a cross section of a drill core taken from a depth of 2306 m from the German Continental Deep Drilling Program (KTB).

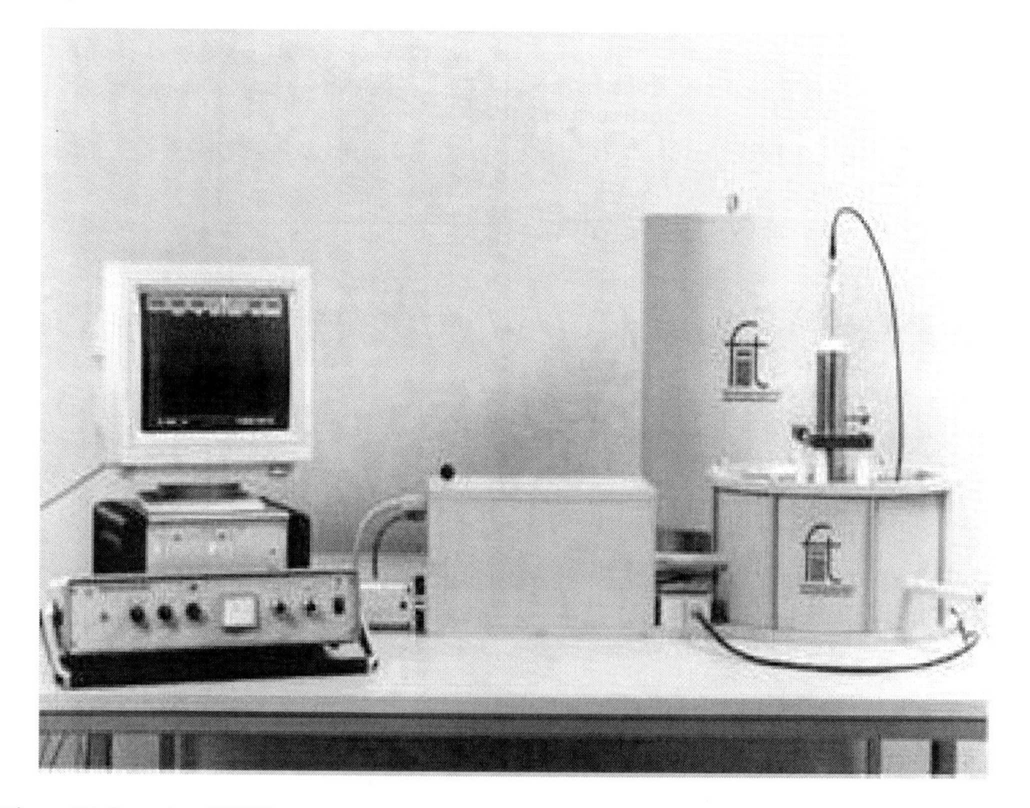

Figure 24. Scanning SQUID magnetometer.

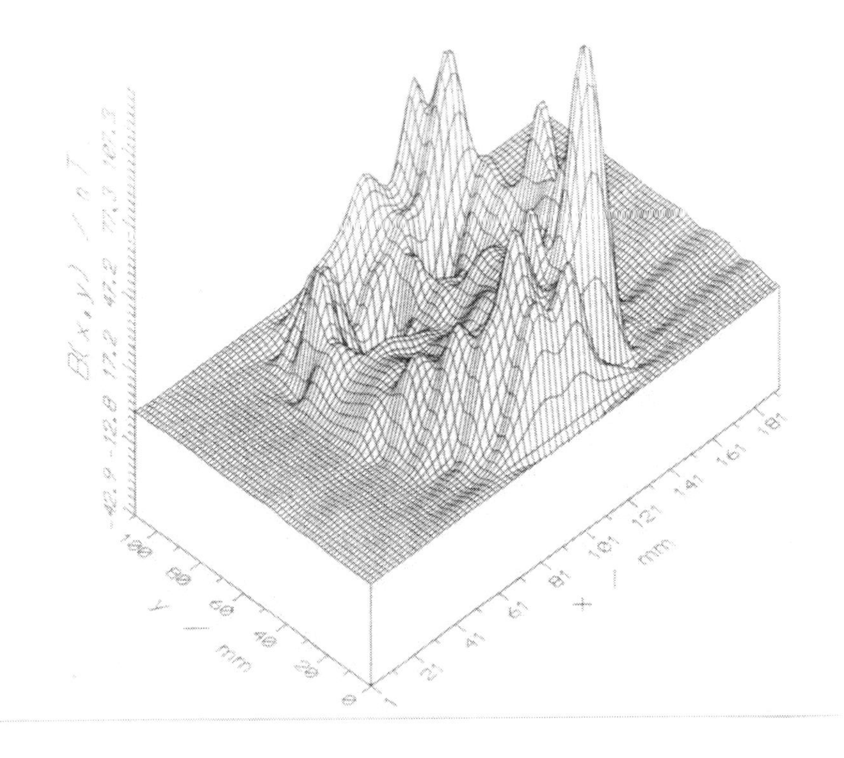

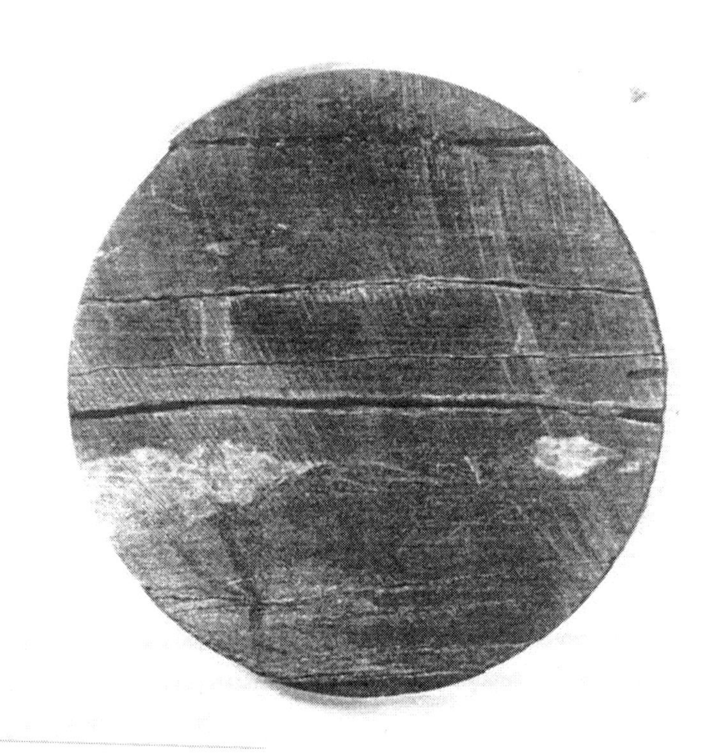

Figure 25. Magnetization distribution (a) of a cross section of a drill core (b) taken from a depth of 2306 m from the German Continental Deep Drilling Program (KTB).

Another very interesting system was the SQUID-spinner magnetometer [22] for measurement of paleomagnetic samples. Spinner magnetometers have been used in geomagnetism for over 30 years. With spinner magnetometer not only the magnetization of the samples, but also magnetization direction can be determinate, which is important in this kind of measurements. These instruments usually use flux-gates or pick-up coils as magnetic field sensors. We developed a spinner magnetometer – Figure 26, where the sensor is a high-T_c RF SQUID. In our system the samples are rotated and the output SQUID signal with rotation frequency is phase-sensitive detected, see Fig 27. After changing the sample position in the sample holder in the following measurements it is possible to calculate the magnitude and the direction of the sample magnetization. A resolution of about 10^{-5} A/m for integration time of 5 s was achieved. To our knowledge this system was the first developed spinner magnetometer based on 77 K SQUID.

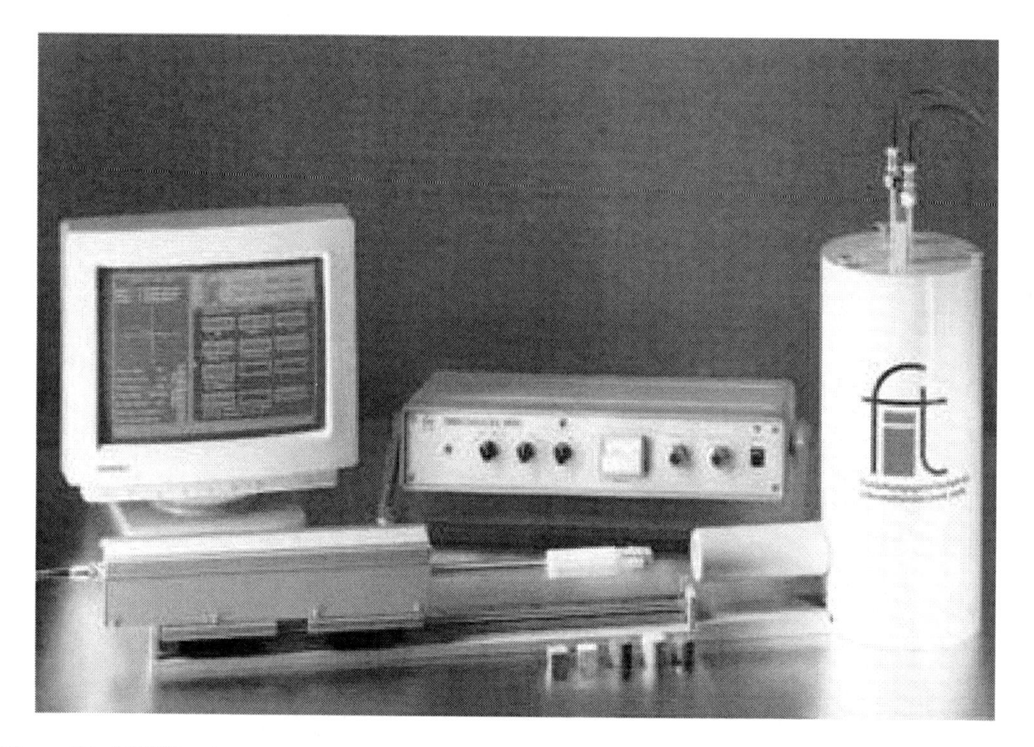

Figure 26. SQUID-spinner magnetometer.

A three axes high-T_c SQUID Magnetometer system [23] is shown in Figure 28. It consist of 3 RF SQUID magnetometers located perpendicularly to each other in order to detect the three orthogonal components of the magnetic field. The 3 RF SQUID chips are washer type with large flux focusing area (1.4 mm x 1.4 mm) and SQUID hole 100 μm x 100 μm (see Figure 20(a)). Flat spiral coils were used to coupe to the SQUID chips. The resonant frequency of the tank circuits were 18, 19 and 20 MHz, respectively. The whole assembly is placed in thin-walled brass cylinder preventing RF interference above 10 KHz. The noise properties of the system were as usual preamplifier-limited in the white noise region. At 10 Hz the magnetic field resolution was ~ 2 pT.Hz $^{-1/2}$.

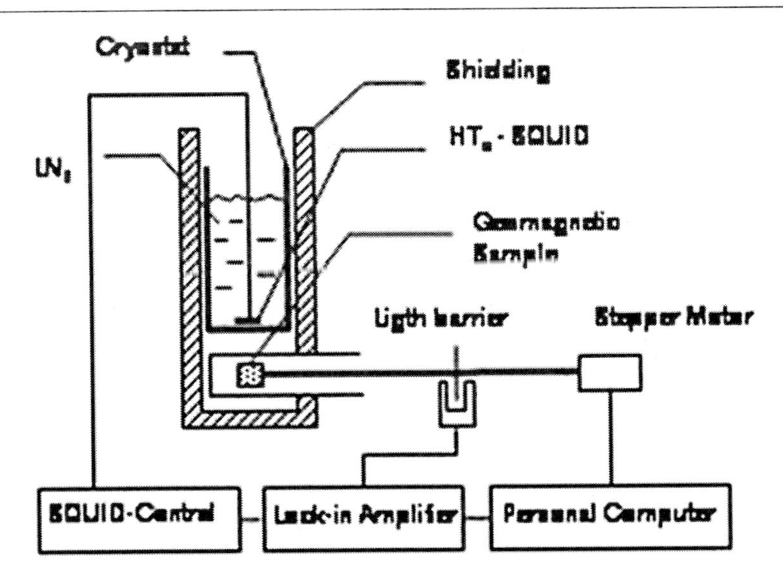

Figure 27. A schematic diagram of the experimental arrangement of the_SQUID-spinner magnetometer.

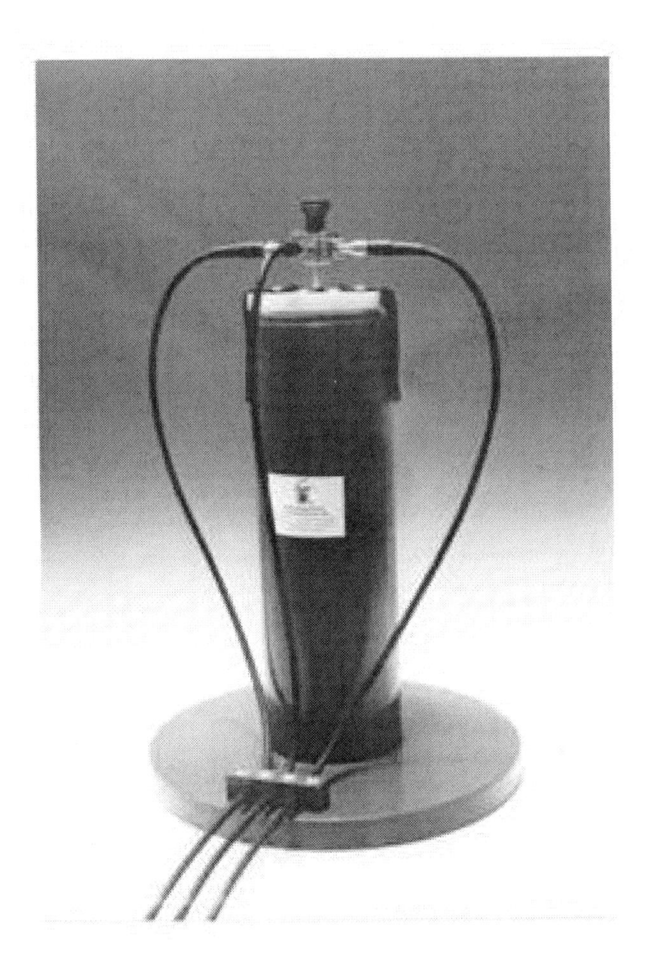

Figure 28. Three axes high-T_c SQUID Magnetometer.

Our SQUIDs with oxygen - ion modified weak links were the first electronic device based on high-T_c superconductivity tested in space [24]. On September 12, 1993, the German satellite ASTROSPAS was launched on the board of the Space Shuttle Discovery, Flight STS-51. The satellite included the SESAM experiment, developed by DFLR in Braunschweig, Germany, to investigate surface effects. One of the 20 SESAM samples was a chip with 16 high-T_c RF SQUIDs made by oxygen ion modification. During the 10 days flight the SQUID chip was exposed to the open space environment. No changes of the SQUID properties were observed after the space mission.

Figure 29. Start of the Space Shuttle "Discovery" on September 12, 1993 with high-T_c -SQUIDs on board.

7. Recent Progress in Ion Modified Josephson Junctions and SQUIDs

In the last years significant contributions in improving properties of ion irradiated high-T_c Josephson junctions are made. In [25] assuming that the source of parameters dispersion is the slit's size variation it was shown that the spread of the junction characteristics can be reduced by increasing energy of the ions – Figure 30. Similarity the same group have shown [26] that long time annealing (10 hours) can reduce the scattering in reduced critical temperature, and therefore in I_c – Figure 31.

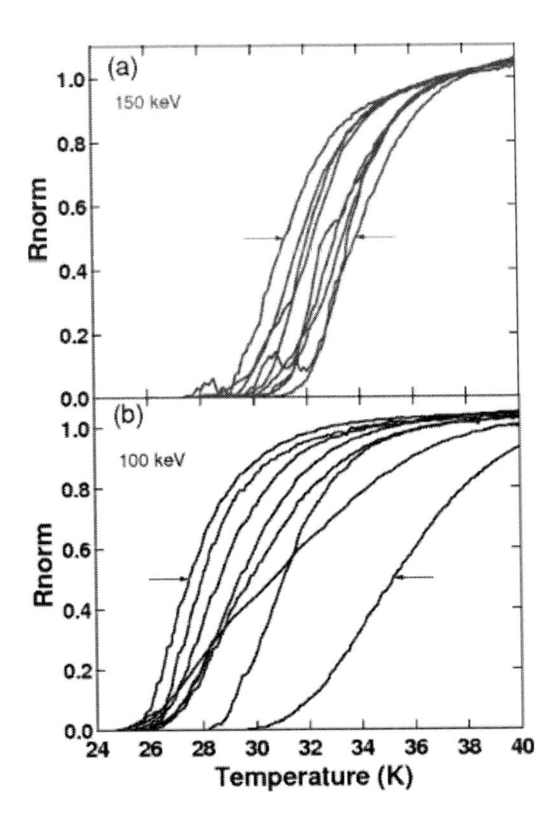

Figure 30. Normalized resistance as a function of temperature for Josephson junctions irradiated with (a) 150 and (b) 100 keV O^+ ions. Figure reprinted with permission from [25]. Copyright 2007 by the American Institute of Physics.

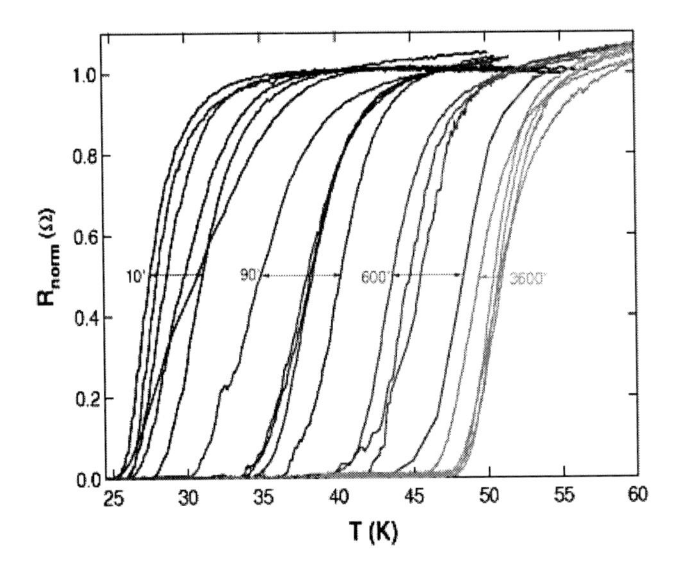

Figure 31. Normalized resistance as a function of temperature [26] for Josephson Junctions irradiated with 100 keV O^+ ions for different annealing times (t = 10, t = 90', t = 600' and 3600').

In the last two years a team in California published some interesting papers in this field. In the first one [27] they fabricated a 280-SQUIDs array with distributed loop areas in order to fabricate SQUIF (Superconducting Quantum Interference Filter). They get such behavior, but most important is that these 280 junctions have a very small standard deviation of the junction critical current of only 12 %.

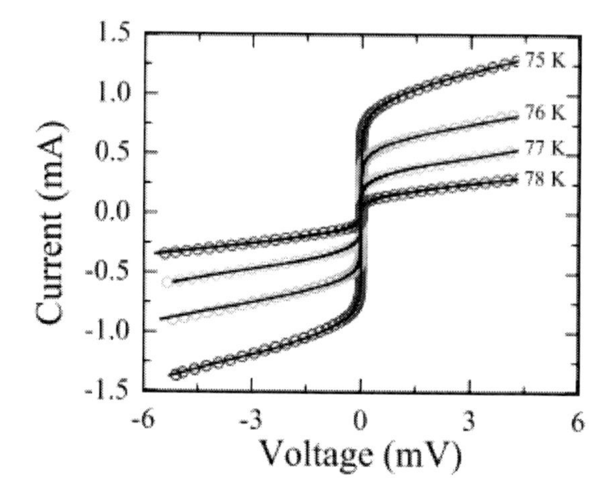

Figure 32. Current-voltage characteristics for the 280-junction array measured at four temperatures around 77 K. Figure reprinted with permission from [27]. Copyright 2006 by the American Institute of Physics.

In a recent paper [28] more than 15000 junctions with 35 nm wide slits were produced with only 16 % standard deviation. This is a very important achievement of this technology.

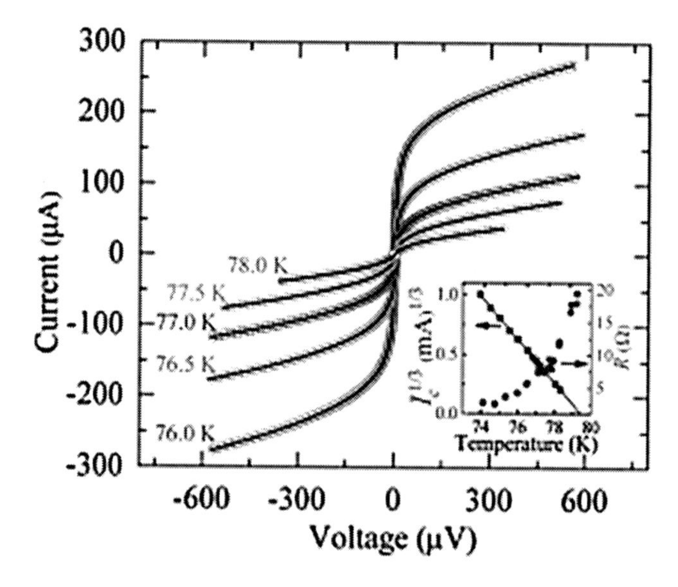

Figure 33. I-V characteristics for five temperatures for the 15820-junctions Josephson array. The inset shows fitted R and Ij vs T. Figure reprinted with permission from [28]. Copyright 2009 by the American Chemical Society.

Conclusion

During the last 20 years after our first demonstration of ion beam modified Josepson junctions and SQUIDs from high-T_c thin films significant improvements of their properties were achieved. The most important is the demonstration that this technology is capable to produce large numbers of junction with very small spread of their parameters. Together with the possibility to fabricate junctions on the arbitrary place of the substrate this make this technology a promising candidates for quantum voltage standards and other devices, where large number of junctions are needed. However, some drawbacks of these junctions, as the small temperature range of operation still remain and need further work.

References

[1] S. S. Tinchev, Investigation of RF SQUIDs made from epitaxial YBCO films, *Supercond. Sci. Technol.*, 3 (1990), 500-503.

[2] S. Katz, S. I. Woods, R. C. Dynes and A. G. Sun, Stability and uniformity of planar high temperature Josephson junctions fabricated using nanolithography and ion damage, IEEE Trans. on Applied Superconductivity, Vol 9, Iss 2, Part 3, (1999), 3005-3007.

[3] N. H. Peng, D. J. Kang, C. Jeynes et al., High quality YBa2Cu3O7-delta Josephson junctions and junction arrays fabricated by masked proton beam irradiation damage, *IEEET Appl. Supercon.*, Vol. 13 (2003), No. 2, 889-892, Part 1.

[4] F. Schmidl, L. Doerrer, S. Wunderlich, F. Machalett, U. Huebner, H. Schmidt, S. Linzen, H. Schneidewind, N. v. Freyhold, and P. Seidel, Superconducting properties of ion beam modified YBCO microbridges, *Journal of Low Temperature Physics*, Vol. 106, Nos. 3/4, 1997, 405-416.

[5] S. S. Tinchev, Properties of YBCO weak links prepared by local oxygen-ion induced modification, *Physica C*, 256 (1996), 191-198.

[6] F. Kahlmann, A. Engelhardt, J. Schubert, W. Zander, Ch. Buchal, J. Hollkott, $YBa_2Cu_3O_{7-}$ Josephson junctions fabricated by oxygen implantation, *Nuclear Instruments and Methods in Physics Research B*, 148 (1999) 803-806.

[7] N. Bergeal, J. Lesueur, M. Sirena, G. Faini, M. Aprili, J. P. Contour and B. Leridon, Using ion irradiation to make high-T_c Josephson junctions, *J. Appl. Phys.*, 102 (2007), 083903.

[8] S. S. Tinchev, Current-phase relation in high-T_c-weak links made by oxigen-ion irradiation, *Physica C*, 222 (1994), 173-176.

[9] S. S. Tinchev, Modeling of $YBa_2Cu_3O_{7-\delta}$ weak links produced by oxygen-ion modification, *J. Appl. Phys.*, 78, No. 9 (1995), 5851-5853.

[10] S.S. Tinchev, Critical temperature depth profiling and improvement of $YBa_2Cu_3O_{7-\delta}$ weak links produced by ion modification, J. Appl. Phys., 81(1997), 324-327.

[11] G. P. Summers, E. A. Burke, D. B. Chrisey, M. Nastasi, J. R. Tesmer, Effect of particle-induced displacements on the critical temperature of $YBa_2Cu_3O_{7-\delta}$, *Appl. Phys. Lett.*, 55 (1989), 1469.

[12] Nianhua Peng, Chris Jeynes, Roger Webb, Ivan Chakarov, Dae Joon Kang, David Moore, Mark Blamire, Monte Carlo simulations of energetic proton beam irradiation damage defect productions in YBCO thin films with Au masks, *Nuclear Instruments and Methods in Physics Research B*, 188 (2002), 189-195.

[13] M. Sirena, N. Bergeal, J. Lesueur, G. Faini, R. Bernard, J. Briatico, D. G. Crete, J. P. Contour, Study and optimization of ion-irradiated high T_c Josephson junctions by Monte Carlo simulations, *Journal of Applied Physics*, 101 (2007), 123925.

[14] U. Barkow, D. Menzel, S. S. Tinchev, Creating homogeneous depth profiles for YBaCuO microbridges with modified ion beam implantation, *Physica C*, 370, No. 4 (2002), 246-252.

[15] D.-J. Kang, N. H. Peng, R. Webb, C. Jeynes J. H. Yun, S. H. Moon, B. OhG. Burnell, E. J. Tarte, D. F. Moore, and M. G. Blamire Realization and properties of Mg2B metal-masked ion damage junctions, *Applied Physics Letters*, Vol. 81, No. 19, 4 November 2002, 3600-3602.

[16] D.-J. Kang, N. H. Peng, C. Jeynes, R. Webb, H. N. Lee, B. Oh, S. H. Moon, G. Burnell, N. A. Stelmashenko, E. J. Tarte, D. F. Moore, M. G. Blamire, Josephson Effects in MgB2 Metal Masked Ion Damage Junctions, *IEEE Transactions on Applied Superconductivity*, Vol. 13, No. 2, June 2003, 1071-1074.

[17] Shane A. Cybart, Ke Chen, Y. Cui, Qi Li, X. X. Xi, R. C. Dynes, Planar MgB_2 Josephson junctions and series arrays via nanolithography and ion damage, *Applied Physics Letters*, 88 (2006), 012509.

[18] S. S. Tinchev, High-T_c-SQUIDs with local oxygen-ion irradiated weak links, *IEEE Trans. Appl. Supercond.*, 3, No. 1 (1993), 28-32.

[19] S. S. Tinchev, YBCO Thin-film Gradiometer, Applied Superconductivity, Ed. H.C. Freyhardt, Volume 2, *DGM Informationsgesellschaft Verlag*, 1993, 1435-1437.

[20] N. Bergeal, J. Lesueur, G. Faini, M. Aprili, J. P. Contour, High T_c superconducting quantum interference devices made by ion irradiation, *Appl. Phys. Lett.*, 89 (2006), 112515.

[21] S. S. Tinchev, J. H. Hinken, M. Stiller, A. Baranyak, D. Hartmann, High T_c RF SQUID magnetometer system for high-resolution magnetic imaging, *IEEE Trans. Appl. Supercond.*, 3, No.1 (1993), 2469-2471.

[22] S. S. Tinchev, High-T_c SQUIDs made by oxygen-ion irradiation - technology, properties and applications (invited), *Proceedings of the Seventh International Symposium on Weak Superconductivity*, June 6-10, 1994, Smolenice Castle, Slovak Republic, 312-323.

[23] S. S. Tinchev, J. H. Hinken, M. Stiller, Three axes high-T_c SQUID Magnetometer, *Proceedings of the ESA/ESTEC workshop on: Space Applications of High Temperature Superconductors*, 27-28 April 1993, ESTEC, Noordwijk, The Netherlands, 147-151.

[24] M. Klinger, J. H. Hinken, S. S. Tinchev, First space test of high-T_c SQUIDs, *IEEE Trans. Appl. Superconductivity*, 5 (1995), No. 2, 2759-2760.

[25] M. Sirena, X. Fabrèges, N. Bergeal, and J. Lesueur, G. Faini, R. Bernard, J. Briatico, Improving the IcRn product and the reproducibility of high T_c Josephson junctions made by ion irradiation, *Applied Physics Letters*, 91 (2007), 262508.

[26] M. Sirena, S. Matzen, N. Bergeal, J. Lesueur, G. Faini, R. Bernard, J. Briatico, D. Crété, Annealing effect on the reproducibility of Josephson Junctions made by ion irradiation, *Journal of Physics: Conference Series*, 97 (2008), 012073.

[27] Shane A. Cybart, S. M. Wu, S. M. Anton, I. Siddiqi, John Clarke, R. C. Dynes, Series array of incommensurate superconducting quantum interference, devices from $YBa_2Cu_3O_{7-\delta}$ ion damage Josephson junctions, *Applied Physics Letters*, 93 (2008), 182502.

[28] Shane A. Cybart, Steven M. Anton, Stephen M. Wu, John Clarke, and Robert C. Dynes, Very Large Scale Integration of Nanopatterned $YBa_2Cu_3O_{7-\delta}$ Josephson Junctions in a Two-Dimensional Array, *Nano Letters*, Vol. 9, No. 10 (2009), 3581-3585.

In: Superconductivity
Editor: Vladimir Rem Romanovskiĭ

ISBN 978-1-61324-843-0

Chapter 8

PROGRESS IN THE STUDY OF FLUX RELAXATION AND VORTEX PENETRATION IN TYPE-II SUPERCONDUCTORS

Rongchao Ma[1] and Yueteng Ma[2]
[1]Department of Physics, University of Alberta
Edmonton, Alberta, Canada
[2]China Construction Eighth Engineering Division.CORP.LTD
Nanning, Guangxi, P. R. China

ABSTRACT

Study on the vortex dynamics of type-II superconductors provides the information about the pinning ability and current-carrying ability of the superconductors. However, the vortex dynamics are currently described with multiple models. In this chapter, we presented a short review on the recently developed general mathematical models of flux relaxation and vortex penetration phenomena in the type-II superconductors. It is shown that the activation energy of vortices in flux relaxation and vortex penetration process can be expanded as series of current density (or internal field). The corresponding time evolution equations of current density (or internal field) can be obtained. These general expressions can be applied to arbitrary vortex systems. It is also shown that, by introducing a time parameter, one can convert a flux relaxation process starting with a current density below the critical value into a process starting with the critical current density. Similarly, by introducing some time parameters, one can convert a vortex penetration process with a non-zero initial internal field into a process with a zero initial internal field. Therefore, one can use a single formula to describe the flux relaxation (or vortex penetration) phenomenon with an arbitrary initial condition.

1. Introduction

A type-II superconductor has two critical magnetic fields, i.e., the lower critical field H_{c1} and upper critical field H_{c2}. When an applied magnetic field H_a increases to a value $H_{c1} < H_a < H_{c2}$, part of H_a can penetrate into the bulk of the superconductor in form of quantized vortices [1], [2]. Each of these vortices carries a single flux quantum $\Phi_0 = 2.068 \times 10^{-7}$ G. A superconductor is in the so-called mixed state after part of H_a penetrates into the bulk of the superconductor. In application, most type-II superconductors work in the mixed state. Thus, it is necessary to study the physical properties of type-II superconductors in this state.

The vortex motion in type-II superconductors is dissipative because of the normal core of the vortices [3]. For a superconductor to carry a zero-resistance current, the vortices must be fixed in the superconductor. This can be achieved by introducing some pinning centers (or defects) into the bulk of the superconductor. The pinning centers can pin down vortices because the superconductor has a lower energy when the vortices sit on the pinning centers. In other words, there is an attractive interaction between the vortices and pinning centers which results in an energy barrier to the vortex motion, i.e., activation energy.

However, it is possible that a vortex can obtain energy higher than the activation energy due to thermal fluctuations [4], mechanical vibration and quantum tunnelling [5]-[7]. Once a vortex obtains energy higher than the activation energy, it will jump to the other adjacent pinning centers. The recent studies [9], [10] have proposed that both flux relaxation and vortex penetration can be regarded as processes of vortices hopping between adjacent pinning centers.

The vortices hopping process can be described by the Arrhenius relation [3], [8] which shows that the vortex hopping rate is related to the activation energy. The activation energy of vortex is a function of current density (or internal field). Various current (or internal field) dependent activation energy were proposed and each has particular physical meaning. In flux relaxation process, there are inverse-power law [11], [12], logarithmic law [13], [14], infinite series law [9], quadratic law [9] and linear law [8]. In vortex penetration process, there are infinite series law, quadratic law and linear law [10]. On the other hand, it can be shown that the activation energy of a vortex is an increasing function of time [9], [10]. By combing the current (or internal field) dependence of activation energy with time dependence of activation energy, one can find out the time evolution equations of current density (or internal field). Thus, the activation energy of vortex plays an important role in the study of vortex dynamics of type-II superconductors.

This chapter includes two parts. In the first part, we discussed the recent progress in the study of flux relaxation phenomenon in type-II superconductors. In the second part, we discussed the recent progress in the study of vortex penetration phenomenon. Since the derivations of these theoretical models do not rely on the detailed microstructures of the superconductors, we expect them to be applicable to both low T_c and high T_c superconductors. This is just a short review based on our own research. Therefore, the interested readers should also consult literatures to learn other newly developed methods and results in this field.

2. Flux Relaxation

For simplicity, let us assume that in flux relaxation process the external field is zero. Thus, the vortices are subjected to the Lorentz force of current density and pinning forces of pinning centers.

In the mixed state, a vortex temporarily sits on one or more pinning centers because of the activation energy U_r. Due to the thermal fluctuation, mechanical vibration and quantum tunnelling, there is a probability that a vortex (or a vortex bundle) [8] can obtain energy higher than U_r. Consequently, the vortex can escape away from the pinning center and jump to the next one. The hopping frequency can be described by the Arrhenius relation [3]

$$\nu = \nu_0 e^{-U_r / kT}, \tag{1}$$

where ν is the hopping frequency of the vortex, ν_0 is the attempting frequency of the vortex, k is the Boltzmann constant and T is temperature.

Since vortex motion is dissipative, the current density in a superconductor is a decreasing function of time in flux relaxation process. In low-T_c superconductors, the time dependence of current density usually shows logarithmic behaviour because of the large elastic modulus and strong pinning [8]. However, in high-T_c superconductors, the time dependence of current density usually shows non-logarithmic behaviour because of the small elastic modulus and random weak pinning [11]-[19]. This can be well explained using the infinite series model [9]. Let us first discuss the current dependence of activation energy in the flux relaxation process.

2.1. Current dependence of activation energy in flux relaxation process

In flux relaxation process, the Lorentz force of current density j can increase vortex's hopping rate. According to the Arrhenius relation, the Lorentz force will reduce vortex's activation energy U_r. This indicates that U_r must be a decreasing function of j.

On the other hand, at the critical current density j_c the vortex lattice is said to be melted and vortices can move in the bulk of the superconductor. This indicates that the activation energy U_r should reduce to zero at j_c.

Let us now summarize the basic properties of the current dependence of activation energy in flux relaxation process as follows:

1) The activation energy U_r is a decreasing function of the current density j, i.e., $dU_r/dj < 0$.

2) At the critical current density j_c, the activation energy U_r is zero, that is, $U_r(j_c) = 0$.

These two constraint conditions must be satisfied by all the mathematical models of current dependence of activation energy in flux relaxation process. Let us first discuss the general expression of the current dependence of activation energy.

2.1.1. Infinite series activation energy

An early study [9] has shown that the activation energy U_r can be expressed as a series of current density j (or internal field B). For the convenience of calculating the critical current density j_c, let us expand U_r as a series of the normalized current density j/j_c, that is,

$$U_r(j) = U_{r0} - \sum_{l=1}^{\infty} a_l \left(\frac{j}{j_c}\right)^l . \tag{2}$$

The higher order terms a_l $(l>2)$ represents the contributions from non-elastic deformation and interaction between vortices. The second order term a_2 represents the contributions from elastic deformation and first order term a_1 represents the contributions from the Lorentz force of current density [9].

The constraint condition $U_r(j_c) = 0$ indicates that the coefficients a_l must satisfy the following equation:

$$U_{r0} = \sum_{l=1}^{\infty} a_l . \tag{3}$$

The inverse of Eq. (3) is important in analysing the experimental data which will be further discussed later in the study of the time dependence of current density.

2.1.2. Quadratic activation energy

In a non-interacting elastic vortex system (with vanishing inelastic deformation), $a_l = 0$ $(l > 2)$. Thus, the current dependent activation energy Eq. (2) reduces to the following quadratic activation energy [9.]

$$U_r(j) = U_{r0} - a_1\left(\frac{j}{j_c}\right) - a_2\left(\frac{j}{j_c}\right)^2 . \tag{4}$$

The constraint condition $U_r(j_c) = 0$ gives $U_{r0}=a_1+a_2$.

This approximation can be applied to a dilute vortex system in which the inelastic deformation and interaction between vortices is small. Therefore, we can ignore the higher order terms $a_l = 0$ $(l > 2)$.

2.1.3. Linear activation energy

In a rigid non-interacting vortex system, the elastic deformation vanishes $(a_2=0)$. Therefore, Eq. (2) reduces to the following linear activation energy [8], [9]

$$U_r(j) = U_{r0} - a_1\left(\frac{j}{j_c}\right) . \tag{5}$$

The constraint condition $U_r(j_c) = 0$ gives $U_{r0}=a_1$.

This approximation can be applied to the dilute vortex systems in low-T_c superconductors with large elastic modulus and some isotropic high-T_c superconductors at lower temperatures, where the higher order terms $a_l = 0$ ($l > 1$) can be ignored.

2.1.4. Inverse-power activation energy

From the collective creep theory [11], [12], one can obtain the inverse-power activation energy, $U_r(j) \sim U_{r0}(j_c / j)^{\mu}$, where μ is a parameter representing the vortex structure. To obtain the so-called interpolation formula of time dependence of current density, the inverse-power activation energy is usually replaced by the following generalized form

$$U_r(j) = \frac{U_{r0}}{\mu}\left[\left(\frac{j_c}{j}\right)^{\mu} - 1\right]. \quad (6)$$

The number 1 on the right side of Eq. (6) is introduced to ensure the constraint condition $U_r(j_c) = 0$.

2.1.5. Logarithmic activation energy

In high-T_c superconductors, sometimes it is more accurate to use the following logarithmic activation energy [13], [14]

$$U_r(j) = U_{r0} \ln\left(\frac{j_c}{j}\right). \quad (7)$$

Eq. (6) and Eq. (7) have singularity at $j = 0$. Therefore, the flux relaxation process at $j = 0$ has to be described by the infinite series activation energy (Eq. (2)), quadratic activation energy (Eq. (4)) or linear activation energy (Eq. (5)).

2.1.6. Other activation energy

Except the above theoretical models, other forms of current dependence of activation energy were also proposed. Such as in a Josephson junction, the current dependence of activation energy is shown to be [3]

$$U_r(j) = U_{r0}\left(1 - \frac{j}{j_c}\right)^{3/2}. \quad (8)$$

This activation energy and the linear activation energy can be generalized to [3].

$$U_r(j) = U_{r0}\left(1 - \frac{j}{j_c}\right)^{\alpha}. \quad (9)$$

An early study [9] has shown that the activation energy of a real vortex system must be nonlinear function of current density. All these theoretical models can be regarded as the special cases of the infinite series model (See Eq. (2)).

On the other hand, one can also determine the activation energy $U_r(j)$ from experimental measurements using Maley's methods [20]. Usually, this method is used with the inverse-power law to determine the vortex structure.

In this section, we discussed the current dependence of activation energy. As mentioned before, the activation energy is also a function of time, which will be discussed in the next section.

2.2. Time dependence of activation energy in flux relaxation process

In flux relaxation process, the vortex activation energy U_r is a decreasing function of current density j, and j is a decreasing function of time t. Therefore, U_r is an increasing function of t. This relationship can be found out using the Arrhenius relation.

Since the reduction in the current density is caused by vortex motion, the rate of change of the current density should be proportional to the vortex hopping rate, that is,

$$\frac{dj}{dt} = -Ce^{-U_r/kT}. \tag{10}$$

With logarithmic accuracy, we have the following equation [9, 21-23]

$$U_r(t) = kT\ln\left(1 + \frac{t_i + t}{\tau}\right), \tag{11}$$

where k is the Boltzmann constant, T is temperature, and $\tau = -kT/[C(dU_r/dj)]$ is a short time scale parameter and

$$t_i = \tau\left(e^{U_i/kT} - 1\right). \tag{12}$$

The physical meaning of t_i is: the time interval during which the activation energy increases from 0 to the initial value U_i. By introducing this time parameter, one can use a single formula to describe the flux relaxation process with arbitrary initial conditions.

Eq. (11) represents the general time dependence of activation energy. One can see that the activation energy in flux relaxation process is an increasing function of time. Combining Eq. (11) with a detailed current dependence of activation energy, one can obtain the time dependence of current density. This is discussed in the next section.

2.3. Time dependence of current density in flux relaxation process

In the study of flux relaxation process, the measureable physical quantity is current density, but not activation energy. Thus, the final purpose in this study is to derive the time dependence of current density, from which one can find out the pinning potential, and therefore, calculating the critical current density.

2.3.1. Time dependence of current density with infinite series activation energy

The infinite series activation energy is described by Eq. (2). Substituting Eq. (2) into Eq. (11), we have

$$\sum_{l=1}^{\infty} a_l \left(\frac{j}{j_c}\right)^l = w_r(t), \tag{13}$$

where

$$w_r(t) = U_{r0} - kT\ln\left(1+\frac{t_i+t}{\tau}\right). \tag{14}$$

Inverting Eq. (13), we have

$$j(t) = j_c \sum_{l=1}^{\infty} b_l \left[w_r(t)\right]^l \tag{15}$$

and the coefficients b_l are

$$b_l = \frac{1}{a_1^l}\frac{1}{l}\sum_{s,t,u\cdots}(-1)^{s+t+u+\cdots}\frac{l(l+1)\cdots(l-1+s+t+u+\cdots)}{s!t!u!\cdots}\left(\frac{a_2}{a_1}\right)^s\left(\frac{a_3}{a_1}\right)^t\left(\frac{a_4}{a_1}\right)^u\cdots, \tag{16}$$

where $s+2t+3u+\cdots=l-1$. On considering the symmetry between Eq. (13) and Eq. (15), we can obtain the inverse coefficients a_l by doing a communication $a_l \leftrightarrow b_l$, that is,

$$a_l = \frac{1}{b_1^l}\frac{1}{l}\sum_{s,t,u\cdots}(-1)^{s+t+u+\cdots}\frac{l(l+1)\cdots(l-1+s+t+u+\cdots)}{s!t!u!\cdots}\left(\frac{b_2}{b_1}\right)^s\left(\frac{b_3}{b_1}\right)^t\left(\frac{b_4}{b_1}\right)^u\cdots. \tag{17}$$

To calculate the activation energy, we can use the following procedure: fitting the experimental data with Eq. (15), we obtain the parameters U_{r0} and b_l. Substituting b_l into Eq. (17), we obtain the coefficients a_l. Substituting a_l and U_{r0} into Eq. (2), we obtain the activation energy $U_r(j)$.

Putting $t = 0$ in Eq. (15), we obtain the initial current density of the flux relaxation process

$$j_i = j_c \sum_{l=1}^{\infty} b_l \left[U_{r0} - kT\ln\left(1+\frac{t_i}{\tau}\right)\right]^l. \tag{18}$$

Putting $t_i = 0$ in Eq. (18) and using the inverse of Eq. (3), $1 = \sum_{l=1}^{\infty} b_l U_{r0}^{\ l}$, we know that $j_i \to j_c$. This means that the physical meaning of t_i can also be interpret as: the time interval during which the current density j reduce from the critical value j_c to the initial value j_i.

2.3.2. Time dependence of current density with quadratic activation energy

Combing Eq. (11) with the quadratic activation energy Eq. (4), we obtain the time dependence of current density in a dilute elastic vortex system

$$j(t) = j_c \frac{a_1}{2a_2}\left[\sqrt{1 + 4\frac{a_2}{a_1^2} w_r(t)} - 1\right], \tag{19}$$

where $w_r(t)$ is defined by Eq. (14).

2.3.3. Time dependence of current density with linear activation energy

Combing Eq. (11) with the linear activation energy Eq. (5), we obtain the time dependence of current density in a dilute rigid vortex system

$$j(t) = \frac{j_c}{a_1} w_r(t), \tag{20}$$

where $w_r(t)$ is defined by Eq. (14).

The linear activation energy Eq. (5) and the corresponding logarithmic time dependence current density Eq. (20) has been well confirmed in lower temperature superconductors and some high-T_c superconductors [30-34]. But in highly layered superconductors, the time dependence of current density usually shows non-logarithmic behaviour, which need to be explained using a model with nonlinear activation energy.

2.3.4. Time dependence of current density with inverse-power activation energy

Combing Eq. (11) with the inverse-power activation energy Eq. (6), we obtain the following interpolation formula

$$j(t) = j_c\left[1 + \frac{\mu kT}{U_{r0}} \ln\left(1 + \frac{t_i + t}{\tau}\right)\right]^{-1/\mu}. \tag{21}$$

The inverse-power law (collective creep theory) has profound physical meanings. It gives good interpretation to the flux relaxation process with random weak pinning centers. This model is appropriate for describing the flux relaxation process in high-T_c superconductors which have short coherence lengths and the small defects like oxygen vacancies can be effective pinning centers.

2.3.5. Time dependence of current density with logarithmic activation energy

Substituting Eq. (11) into the logarithmic activation energy Eq. (7), we obtain the following the time dependence of current density

$$j(t) = j_c\left(1 + \frac{t_i + t}{\tau}\right)^{-kT/U_{r0}}. \tag{22}$$

Eq. (22) has 4 fitting parameters, but Eq. (21) has 5 fitting parameters. Thus, it is more convenient to use Eq. (22) to analysis the experimental data.

2.3.6. Time dependence of current density with other activation energy

Substituting Eq. (11) into Eq. (9), we obtain the following the time dependence of current density

$$j(t) = j_c\left\{1-\left[\frac{kT}{U_{r0}}\ln\left(1+\frac{t_i+t}{\tau}\right)\right]^{1/\alpha}\right\}. \tag{23}$$

Finally it should be mentioned that Xue et. al. [35] have proposed that, in a thin-wall cylinder, the time dependence of current density is

$$j(t) = j_c\left[\ln\left(\frac{t_2+t}{t_1}\right)\right]^{-1/\mu}, \tag{24}$$

where μ, t_1, t_2 are free fitting parameters.

In this section, we discussed the mathematical models of flux relaxation phenomenon. From the time dependence of current density $j(t)$, we see that one can convert a flux relaxation process starting with a current density j_i below the critical value [24-29] into a process starting with the critical current density j_c by introducing the time parameter t_i.

3. Vortex Penetration

In this section, we will study the vortex penetration process [36-45] use some similar concepts and mathematical tools as that used in the study of flux relaxation process. In the study of vortex penetration process, we are more interested in vortex motion than current density. Thus, we will use internal field, not current density.

In vortex penetration process, the repulsive force of internal field B prevents vortex motion and reduces vortex hopping rate. According to the Arrhenius relation, this will increase vortex's activation energy U_p. Thus, U_p is an increasing function of B [10].

On the other hand, under an applied field B_a, the internal field should approach a maximum value B_e when the superconductor reaches an equilibrium state [46]. Also, the activation energy U_p should reach a maximum value U_e at the equilibrium field B_e.

Let us now summarize the basic properties of the field dependence of activation energy in vortex penetration process as follows:

a) The activation energy U_p is an increasing function of the internal field B, i.e., $dU_p/dB > 0$.

b) The equilibrium field B_e should be smaller than the applied field B_a because of the surface screening effect (or Meissner effect) [47], i.e., $B_e < B_a$.

c) Under an applied field B_a, the equilibrium activation energy is the maximum activation energy, i.e., $U_p(B) \le U_e$.

3.1. Field dependence of activation energy in vortex penetration process

3.1.1. Infinite series activation energy

The early studies [10, 48] have shown that U_p can be expressed as a series of internal field B. For the convenience of calculating the equilibrium field B_e, let us expand U_p as a series of the normalized internal field B/B_e, that is,

$$U_p(B) = U_{p0} + \sum_{l=1}^{\infty} c_l \left(\frac{B}{B_e} \right)^l , \tag{25}$$

where $U_{p0} = U_{BL} + U_c$ is the activation energy at vanishing internal field, U_{BL} is the Bean-Livingston surface barrier and U_c is pinning potential.

At the equilibrium field B_e, the activation energy U_p reaches the maximum value U_e, that is,

$$U_e = U_{p0} + \sum_{l=1}^{\infty} c_l . \tag{26}$$

The physical meaning of the coefficients c_l is similar to the coefficients a_l in flux relaxation process (See Eq. (17)).

3.1.2. Quadratic activation energy

In a non-interacting elastic vortex system (with vanishing inelastic deformation), $a_l = 0$ ($l > 2$). Thus, Eq. (25) reduces to the following quadratic activation energy [10], [48]

$$U_p(B) = U_{p0} + c_1 \left(\frac{B}{B_e} \right) + c_2 \left(\frac{B}{B_e} \right)^2 \tag{27}$$

and Eq. (26) reduces to $U_e = U_{p0} + c_1 + c_2$.

3.1.3. Linear activation energy

In a non-interacting rigid vortex system (with vanishing elastic deformation), $a_l = 0$ ($l > 1$). Thus, Eq. (25) reduces to the following linear activation energy [10, 48]

$$U_p(B) = U_{p0} + c_1 \left(\frac{B}{B_e} \right) \tag{28}$$

and Eq. (26) reduces to $U_e = U_{p0} + c_1$.

3.2. Time dependence of activation energy in vortex penetration process

In vortex penetration process, the activation energy U_p is an increasing function of the internal field B, and B is an increasing function of time t. Therefore, U_p is an increasing function of t.

Because the increase of internal field is caused by vortex penetration, the rate of change of the internal field should be proportional to the vortex hopping rate. According to the Arrhenius relation, we have [10], [48]

$$\frac{dB}{dt} = Ce^{-U_p/kT}. \tag{29}$$

With logarithmic accuracy, we have the following equation [48]

$$U_p(t) = kT\ln\left(1+\frac{t_i+t}{\tau}\right) = kT\ln\left(1+\frac{t_0+t_v+t}{\tau}\right), \tag{30}$$

where $\tau = kT/[C(dU_p/dB)]$ and

$$t_i = \tau\left(e^{U_i/kT}-1\right), \tag{31a}$$

$$t_0 = \tau\left(e^{U_{p0}/kT}-1\right), \tag{31b}$$

$$t_v = t_i - t_0 = \tau\left(e^{U_i/kT}-e^{U_{p0}/kT}\right). \tag{31c}$$

The physical meaning of t_v is: the time interval during which the activation energy increase from U_{p0} to the initial value U_i. The parameter t_0 is an equivalence time parameter of the potential U_{p0}. This does not have a counterpart in the flux relaxation process. By introducing these time parameters, one can use a single formula to describe the vortex penetration process with arbitrary initial conditions.

3.3. Time dependence of internal field in vortex penetration process

The time dependence of internal field in a vortex penetration process can be obtained by combing Eq. (30) with a detailed internal field dependent activation energy $U_p(B)$.

3.3.1. Infinite series activation energy

Combing Eq. (30) with the infinite series activation energy Eq. (25), we have the following general time dependence of internal field in a vortex penetration process

$$B(t) = B_e\sum_{l=1}^{\infty} d_l\left[w_p(t)\right]^l, \tag{32}$$

where

$$w_p(t) = kT\ln\left(1+\frac{t_0+t_v+t}{\tau}\right) - U_{p0}. \tag{33}$$

If we are only interested in find out the time dependence of internal field, we can simplify Eq. (32) by absorbing B_e into the coefficients d_l.

Putting $t=0$ in Eq. (32), we obtain the initial internal field in the vortex penetration process

$$B_i = B_e \sum_{l=1}^{\infty} d_l \left[kT \ln\left(1 + \frac{t_0 + t_v}{\tau}\right) - U_{p0} \right]^l . \tag{34}$$

3.3.2. Quadratic activation energy

Combining Eq. (30) with the quadratic activation energy Eq. (27), we obtain the following time dependence of internal field

$$B(t) = B_e \frac{c_1}{2c_2} \left[\sqrt{1 + 4\frac{c_2}{c_1^2} w_p(t)} - 1 \right], \tag{35}$$

where $w_p(t)$ is defined by Eq. (33).

3.3.3. Linear activation energy

Combining Eq. (30) with the linear activation energy Eq. (28), we obtain the following time dependence of internal field

$$B(t) = \frac{B_e}{c_1} w_p(t), \tag{36}$$

where $w_p(t)$ is defined by Eq. (33).

In this section, we have discussed the mathematical models of vortex penetration phenomenon. From the time dependence of internal field $B(t)$, we see that one can convert a vortex penetration process with a non-zero initial internal field B_i into a process with a zero initial internal field by introducing the time parameter t_i.

CONCLUSION

On the basis of different physical consideration, a number of current (or internal field) dependent activation energies of vortices in flux relaxation and vortex penetration process were proposed. From mathematical consideration, these special activation energies can be generalized to the infinite series activation energy. On the other hand, the activation energy of vortex can be also determined from experiments using Maley's methods. Using this method, one can obtain the activation energy of vortex without involving any detailed physical considerations.

Using a detailed current (or internal field) dependence of activation energy and the time dependence of activation energy, one can obtain the corresponding time dependence of the current density (or internal field). In flux relaxation process, by introducing a time parameter, one can convert a flux relaxation process with an initial current density below the critical values into a process with a critical initial current density. In vortex penetration process, by introducing some time parameters, one can convert a vortex penetration process with a non-zero initial internal field into a process with a zero initial internal field. Finally, one can use a single formula to describe the vortex penetration process with an arbitrary initial condition.

Although the vortex dynamics in type-II superconductors has been extensively studied, there are still some open problems:

1) More current dependence of activation energy may still exist because of the complex of vortex structure, and the interaction between vortices and pinning centers. Also, the physical meaning of the activation energy needs further investiogations.

2) The melting transition of the vortex systems needs further study. Especially, the vortex dynamics of highly layered superconductors is complicated and it is hard to use a simple formula to describe the properties of vortex system in these superconductors. Currently, we determine the melting transition using the Lindeman criterion, which is proposed for crystal lattice. However, the experimental measurements have shown that the Lindeman criterion does not apply to the melting transition in the vortex system of highly layered superconductors, for example, Bi and Tl series superconductors.

3) The effects of surface pinning in flux relaxation and vortex penetration process are not well understood. In low-T_c superconductors, the effects of surface pinning are not significant because of the strong bulk pinning. But in high-T_c superconductors, the bulk pinning is random weak pinning and surface pinning dominates at higher temperatures. Some researchers believe that the non-logarithmic flux relaxation behaviour is caused by the bulk-surface pinning transition. Thus, the effects of surface pinning on the flux relaxation and vortex penetration process cannot be ignored. However, the exact physics behind this phenomenon is currently not clear.

REFERENCES

[1] Abrikosov, A. A. *Soviet Phys. JETP* 1957, 5, 1174.
[2] Kleiner, W. H.; Roth, L. M.; Autler, S. H. *Phys. Rev.* 1964, 133, A1226.
[3] Tinkham, Michael. *Introduction to Superconductivity*. McGraw-Hill, New York, 1996.
[4] Landau, I. L. and Ott, H. R. Phys. Rev. B 2001, 63, 184516.
[5] Koren, G.; Mor, Y.; Auerbach, A. and Polturak, E. *Phys. Rev. B* 2007, 76, 134516.
[6] Nicodemi, M.; Jensen, H. J. *Phys. Rev. Lett.* 2001, 86, 4378.
[7] Hoekstra, A. F. Th.; Griessen, R.; Testa, A. M.; Fattahi, J. el; Brinkmann, M.; Westerholt, K.; Kwok, W. K. and Crabtree, G. W. *Phys. Rev. Lett.* 1998, 80, 4293.
[8] Anderson, P. W. *Phys. Rev. Lett.* 1962, 9, 309.
[9] Ma, R. J. *Appl. Phys.* 2010, 108, 053907.
[10] Ma, R. J. *Appl. Phys.* 2011, 109, 013913.
[11] Feigel'man, M. V.; Geshkenbein, V. B. and Vinokur, V. M. *Phys. Rev. B* 1991, 43, 6263.
[12] Feigel'man, M. V.; Geshkenbein, V. B.; Larkin, A. I.; Vinokur, V. M. *Phys. Rev. Lett.* 1989, 63, 2303.
[13] Zeldov, E.; Amer, N. M.; Koren, G.; Gupta, A.; Gambino, R. J. and McElfresh, M. W. *Phys. Rev. Lett.* 1989, 62, 3093.

[14] Zeldov, E.; Amer, N. M.; Koren, G.; Gupta, A.; McElfresh, M. W. and Gambino, R. J. *Appl. Phys. Lett.* 1990, 56, 680.
[15] Fisher, Matthew P. A. *Phys. Rev. Lett.* 1989, 62, 1415.
[16] Burlachkov, L. *Phys. Rev. B* 1993, 47, 8056.
[17] Chikumoto, N.; Konczykowski, M.; Motohira, N.; Malozemoff, A. P. *Phys. Rev. Lett.* 1992, 69, 1260.
[18] Weir, S. T.; Nellis, W. J.; Dalichaouch, Y.; Lee, B. W.; Maple, M. B.; Liu, J. Z.; Shelton, R. N. *Phys. Rev. B* 1991, 43, 3034.
[19] Gurevich, A. and Kupfer, H. *Phys. Rev. B* 1993, 48, 6477.
[20] Maley, M. P. and Willis, J. O. *Phys. Rev. B* 1990, 42, 2639.
[21] Beasley, M. R.; Labusch, R. and Webb, W. W. *Phys. Rev.* 1969, 181, 682.
[22] Geshkenbein, V. B. and Larkin, A. I. *Sov. Phys. JETP* 1989, 68, 639.
[23] Ma, R. arXiv:1101.0442.
[24] Miclea, C. F.; Mota, A. C.; Nicklas, M.; Cardoso, R.; Steglich, F.; Sigrist, M.; Prokofiev, A. and Bauer, E. *Phys. Rev. B* 2010, 81, 014527.
[25] Miu, L.; Miu, D.; Petrisor, T.; Tahan, A. El; Jakob, G. and Adrian, H. *Phys. Rev. B* 2008, 78, 212508.
[26] Harada, K.; Kasai, H.; Kamimura, O.; Matsuda, T.; Tonomura, A.; Okayasu, S. and Kazumata, Y. *Phys. Rev. B* 1996, 53, 9400.
[27] Chikumoto, N.; Konczykowski, M.; Motohira, N. and Malozemoff, A. P. *Phys. Rev. Lett.* 1992, 69, 1260.
[28] Lessure, H. S.; Simizu, S. and Sankar, S. G. *Phys. Rev. B* 1989, 40, 5165.
[29] Sumption, M. D.; Haugan, T. J.; Barnes, P. N.; Campbell, T. A.; Pierce, N. A. and Varanasi, C. *Phys. Rev. B* 2008, 77, 094506.
[30] Kim, Y. B.; Hempstead, C. F. and Strnad, A. R. *Phys. Rev.* 1963, 131, 2486.
[31] Beasley, M. R.; Labusch, R. and Webb, W. W. *Phys. Rev.* 1969, 181, 682.
[32] Miu, L. and Miu, D. *Supercond. Sci. Technol.* 2010, 23, 025033.
[33] Yang, Huan; Ren, Cong; Shan, Lei and Wen, Hai-Hu. *Phys. Rev. B* 2008, 78, 092504.
[34] Reissner, M.; Lorenz, *J. Phys. Rev. B* 1997, 56, 6273.
[35] Xue, Y. Y.; Gao, L.; Ren, Y. T.; Chan, W. C.; Hor, P. H. and Chu, C. W. *Phys. Rev. B* 1991, 44, 12029.
[36] Sela, Eran and Affleck, Ian. *Phys. Rev. B* 2009, 79, 024503.
[37] Hernandez, A. D. and Lopez, A. *Phys. Rev. B* 2008, 77, 144506.
[38] Baelus, B. J.; Kanda, A.; Shimizu, N.; Tadano, K.; Ootuka, Y.; Kadowaki, K. and Peeters, F. M. *Phys. Rev. B* 2006, 73, 024514.
[39] Berdiyorov, G. R.; Cabral, L. R. E. and Peeters, F. M. *J. Math. Phys.* 2005, 46, 095105.
[40] Baelus, B. J.; Kadowaki, K. and Peeters, F. M. *Phys. Rev. B* 2005, 71, 024514.
[41] Erdin, Serkan. *Phys. Rev. B* 2004, 69, 214521.
[42] Elistratov, Andrey A.; Vodolazov, Denis Yu.; Maksimov, Igor L. and Clem, John R. *Phys. Rev. B* 2002, 66, 220506(R).
[43] Wang, Y. M.; Zettl, A.; Ooi, S. and Tamegai, T. *Phys. Rev. B* 2002, 65, 184506.
[44] Mawatari, Yasunori and Clem, John R. *Phys. Rev. Lett.* 2001, 86, 2870.
[45] Pissas, M. and Stamopoulos, D. *Phys. Rev. B* 2001, 64, 134510.
[46] Bean, C. P. *Phys. Rev. Lett.* 1962, 8, 250.
[47] Bean, C. P. and Livingston, J. D. *Phys. Rev. Lett.* 1964, 12, 14.
[48] Ma, R. *J. Appl. Phys.* 2011, 109, 103910.

In: Superconductivity
Editor: Vladimir Rem Romanovskií
ISBN 978-1-61324-843-0

Chapter 9

FAST SUPERCONDUCTIVE JOINING OF Y-BA-CU-O BULKS WITH HIGH QUALITY

***Chai Xiao*[1], *Zou Guisheng*[1], *Guo Wei*[1,2], *Wu Aiping*[1] *and Ren Jialie*[1]**

[1]Department of Mechanical Engineering & Key Laboratory
for Advanced Materials Processing Technology
Ministry of Education of P. R. China
Tsinghua University, Beijing, China
[2]School of Mechanical Engineering & Automation
Beihang University, Beijing, China

ABSTRACT

Y-based Superconductors(Y-Ba-Cu-O) is considered to be one of the promising superconducting materials owing to its good applications in many fields, and these applications need Y-Ba-Cu-O bulks with large dimension and complex shape. However, at present, the mean current density of large bulk decreases significantly compared with that of the small one and the shape of the bulk is very simple, usually in form of cylinder or hexagonal. In order to meet the need of the practical using of Y-Ba-Cu-O, joining technique has been used. Up to now, a lot of researchers had done a great deal of investigation on soldering, however, almost all of them needed a relatively long time to realize superconductive joining. In this paper, fast superconductive joining technology of Y-Ba-Cu-O was investigated. Three kinds of solders, including sintered Yb-Ba-Cu-O, sintered Y-Ba-Cu-O/Ag and melt Y-Ba-Cu-O/Ag solder were synthesized and different soldering thermal cycles were designed respectively. The results show that the soldered bulk using sintered Yb-Ba-Cu-O can partially recover the superconductivity but accumulation of Yb_2BaCuO_5 (Yb211), which results in decrease in superconductivity, was observed in the bonding zone. In comparison, soldering with Y-Ba-Cu-O/Ag solders realized much better superconductive bonding owing to its thinner solder thickness and slower cooling rates during soldering. Moreover, soldering with the melt Y-Ba-Cu-O/Ag solder obtained better performance than soldering with the sintered solder. One plausible reason for the better performance of melt solder is that the dense structure owing to its pretreatment at higher temperature. The microstructure of the bonding zones was investigated by scan electron microscope and the images indicate that neither Ag particles nor Y_2BaCuO_5 (Y211) accumulated during the soldering process. The trapped field image also reveals that the supercurrent is almost not blocked by the bonding zone.

1. INTRODUCTION

High Temperature superconductors (HTS) are considered to be one of promising material on account for its good electro-magnetically properties. There are two types of HTS, one is Bi-based material Bi-Sr-Ca-Cu-O (BSCCO), the other is Y-based material Y-Ba-Cu-O (YBCO). Comparing with BSCCO, YBCO has better superconductivity when external magnetic field was applied. As a result, melt textured YBCO bulk is considered as one of promising materials for its important new technical applications in high power density motors, levitators, flywheels and fault current limiters [1]-[3]. On account for the requirements of these devices, bulks with larger size and complex shape are needed. However, the performances of the large bulks (about100mm in diameter) degrade significantly, as reported by R. Tournier et al. [4]. Consequently, Researchers tried other methods, and joining technique is one promising way to overcome these problems.

To date, there are many types of joining technique to realize enlargement of single domain YBCO bulk. For example, one could classify the joining techniques into two categories, non-filler joining and joining with solder, according to the existence of filler materials. Non-filler joining, mainly solid state diffusion bonding,can join the base material and partly recover the superconductivity, as reported by Salama et al. [5], Cardwell et al. [6] and W.Lo et al. [7]. Nevertheless, this technique seriously relies on a perfect contact between grains before joining which is rarely achievable, consequently, the application of this technique confronted with some obstacles. On the other hand, joining with solder does not require such perfect contact condition as non-filler joining. This encourages scientists and researchers to develop filler materials and correlated processes for joining single domain Y-Ba-Cu-O bulks.

Up to now, researchers and scientists have developed many solder joining methods. On the basis of the solders' status, one could classify theses methods into 4 categories: (a) powder solder, (b) sintered plates as solder, (c) Ag foil as solder, and (d) sliced plates from melt textured bulks as solder. For (a), Re-Ba-Cu-O (Re = Yb, Tm…) powders are used as the solder by either glued or painted on the joining surface, after moistening with organic solvent (ethanol [8] or glycerin [9], [10]), or direct sandwiched by two monoliths [11], [12]. The result of this method is acceptable: the solder and the base material are firmly bonded, and the superconductivity after joining is much better than the initial stage. However, certain defects (mainly voids) are commonly found in the bonding zone owing to the shrinkage during welding and the phase change of the solder during the heating process. These defects degrade the performance of the joint. For (b), conversely, thanks to the preheat effect of the solder before and during the welding process, the shrinkage decreased significantly compared with the powder solder, and there are mainly two types of solders: Re-Ba-Cu-O solders and Y-Ba-Cu-O/Ag solder. H.Zheng et al. [13] used Tm-Ba-Cu-O as the solder to accomplish a self-seeded joining; C. Harnois et al. [14], [15], on the other hand, used Ag added Y-Ba-Cu-O solder to realized the multi-seeded welding and self-seeded welding. Moreover, K.Iida et al. [16]-[19] used both types: Er-Ba-Cu-O and Ag_2O added Y-Ba-Cu-O, both types have lower melting point than Y-Ba-Cu-O, and the performances of the bonding zone, mechanical and superconductive, are extremely good. The type (c), as reported by S.Iliescu et al. [20]-[23], also obtained encouraging result. Besides, T.puig et al. [24] and Th. Hopfinger et al. [25] introduced a new kind of solder (d): melt textured Y-Ba-Cu-O/Ag, into the joining process and obtained a good joint owing to its mass effect in decreasing shrinkage.

To sum up, a lot of researchers have investigated into the joining process with solder. Most of the process required a slow-cooling stage with relatively longtime (about 100h) for about 50 °C at the beginning of the cooling to realize the self-seeded welding except T. A. Prikhna et al. [12]. They cooled the sample with the rate up to 100°C/h but still obtained a good joint. However, relatively large numbers of defects, especially voids are found in the joint due to the shrinkage during the joining process. Aim to solve the large porosity rate in the bonding zone, according to the report of A. Mahmood et al. [25], the preheat process can largely reduced the porosity of the material; consequently, the status of the joining agent should also influence the performance of the joint. As a result, we tried to use milled sintered plate or sliced melt plate instead of the powder to realize fast superconductive joining, and the experiment reveals that fast superconductive joining with high quality is achieved by using sintered Ag doped YBCO solders and melt Ag doped YBCO solders.

2. Preparation of the Joint and Joining Process

$YBa_2Cu_3O_y$ (Y123) and Y_2BaCuO_5 (Y211) powders were prepared from Y_2O_3, $BaCO_3$ and CuO powders via a solid-state reaction. Single domain Y-Ba-Cu-O (Y123 : Y211 = 4 : 1, 0.5wt% of $BaCeO_3$) bulk samples with dimensions of 25mm × 13mm were fabricated by the top-seeded melt-texture growth (TSMG) method.

Optimum performance of the joint in the bonding zone can be expected if the bonding zone reveals a similar microstructure as the base YBCO melt textured bulk. Moreover, aggregation or defect should be avoided in the bonding zone. During the joining process the nucleation is different from that of TSMG process. In the TSMG process, the seed crystal essentially generates a controlled oriented point nucleus in the bulk, which further grows according to the temperature gradients. Whereas during the joining process, the nucleation mechanism is different: it is so called “self-seeded” process. Unlike the point nucleus in TSMG process, during the joining process, the Y123 phase symmetrically nucleates on the whole surfaces of the joining bulks, more importantly, this nucleation happens simultaneously on both interfaces of the solder and base material, and this generate two growth fronts growing towards each other and meet each other at the center of the bonding zone.

Therefore, the process for the realization of “self-seeded” joint between YBCO ceramics is based on the high temperature phase diagram of the YBCO-Ag pseudo phase diagram [26] and the melting temperature of YbBCO. As phase diagram indicates, the peritectic temperature of YBCO/Ag composite is decreased by about 40°C when process is carried out at ambient atmosphere, and the melting temperature of YbBCO is 920 °C, which is much lower than the melting point of YBCO.

The step that we have followed to obtain the superconductive joints was illustrated in Figure 1. The base YBCO was cut by a diamond saw perpendicular to the ab plan and the surface in contact (ac planes) are polished carefully to improve their mechanical contact during the joining process. After finishing the assembling process, the assembly was put into a isothermal furnace. The thermal cycle used to realize fast superconductive joining is schematically represented in Figure 2.

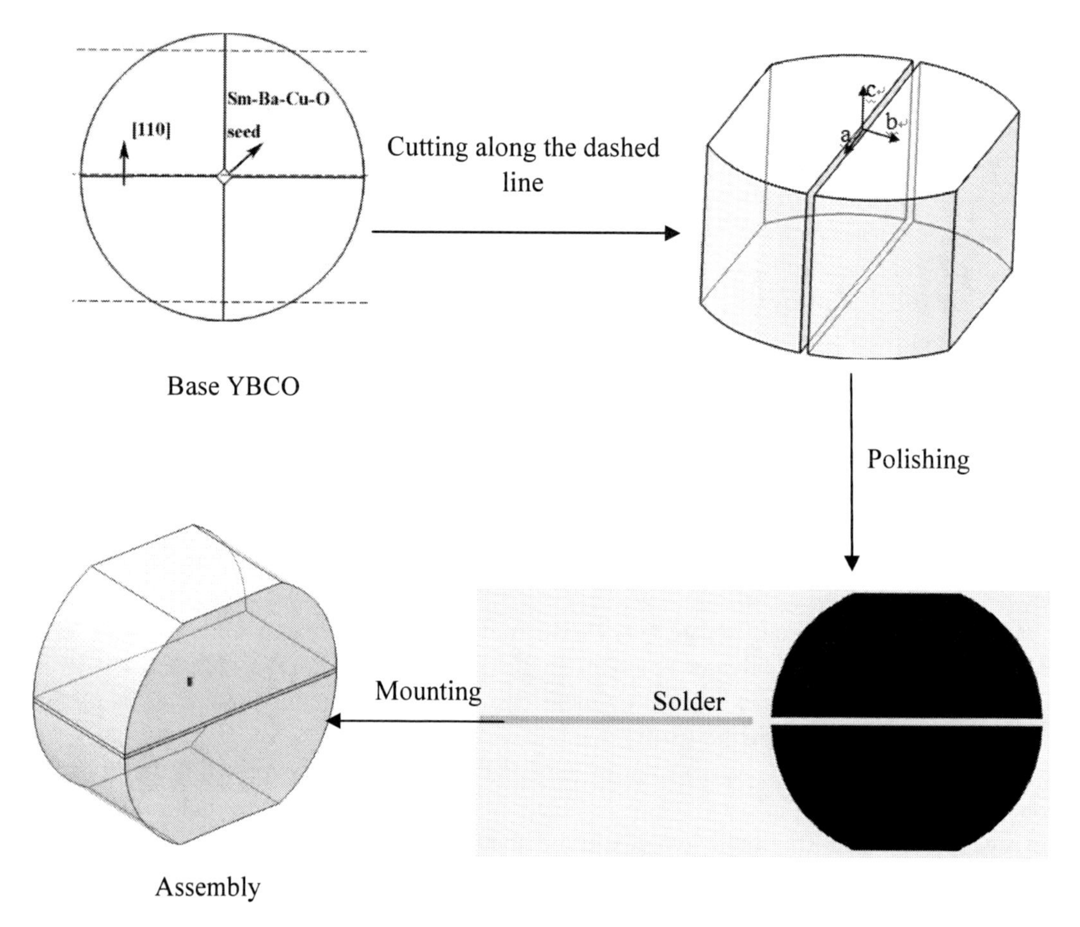

Figure 1. Preparation of the assembly.

The high temperature process has been carried out for all the assemblies using three kinds of solder. In Figure 2, there are three critical temperature points that determine the quality of the joint: T_{max}, T_l and T_{ox}. T_{max} is the maximum temperature during the whole thermal cycle, T_l is the temperature at which the slow cooling process ends, and the gap between T_{max} and T_l is the temperature window that the solder melt and then induce the crystal pattern similar to the base material during the slow cooling process. And the T_{ox} is the temperature of the oxygenation after the joining process.

Therefore, the assembly was heated to T_{max} and dwell for t_l to realize a homogeneous melt of the solder, but lower than the melting point of YBCO base material, so that the at the interface, the YBCO base material can act as the seed during the slow cooling process. Then the assembly was slowly cooled down to T_l, which is lower than the melt temperature of the solder, during this period, the “self-seed” joint could be achieved. But one should noted that

T1 and cooling rate vary according to the solder material, and both of them are critical to the quality of the joint, this will be discussed in following parts. Once the interface and the bonding zone solidified, the assembly is cooled with furnace to the room temperature.

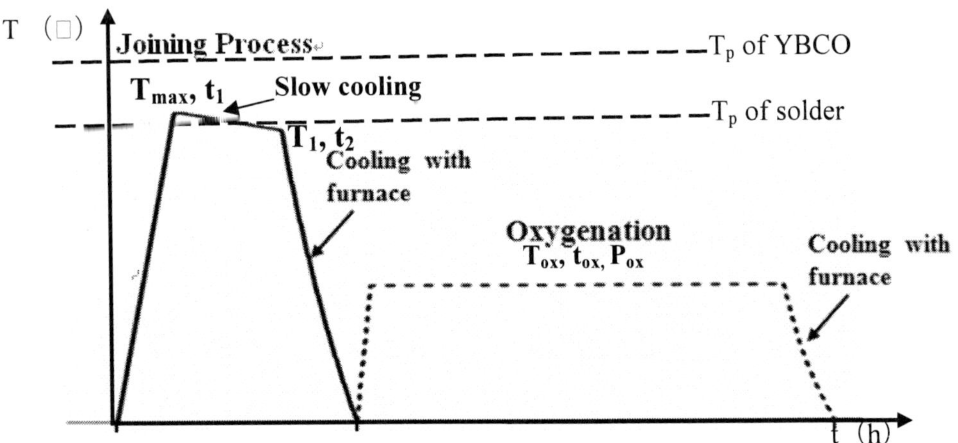

Figure 2. the high temperature process for the assembly.

The high-temperature process reduced the oxygen from the both YBCO and YbBCO and the microstructure of them have changed from orthorhombic, which is superconductive, to tetragonal structure, which is non-superconductive. As a result, an oxygenation process should follow by the joining process to recover the superconductivity for both base material and the joint. There are three parameters that determine the oxygenation quality. In this paper, all the oxygenation process was carried out at oxygen atmosphere with the pressure of 1MPa, and the temperature of 450 °C, the oxygenation time t_{ox} varies according to the size of the sample to be oxygenation, in our joining process, t_{ox} is 48h.

For optimization of the joining process, parameters such as maximum temperature of joining process (T_{max}), the dwell time (t_1), the slow cooling ending temperature T1, the solder composition and solder thickness as well as the cooling rate (r) have been studied.

3. JOINTS PROPERTIES AND ANALYSIS

3.1. Joints using YbBCO as solder

YbBCO is a kind of superconductive material which has same crystal structure and similar crystal parameter with YBCO, but much lower melting point, As a result, YbBCO based solder was developed and used in joining YBCO bulks, in this paper, YbBCO solder was composite of $YbBa_2Cu_3O_7$(Yb123)powder and Yb_2BaCuO_5(Yb211)powder, the mixture powders were pressed into plate with dimension of 70mm in length, 14mm in width as well as 1.2mm in thick and sintered at 885°C, then the sintered plate was machined to 0.5mm in thickness by abrasive paper. The parameters of the thermal cycle are as follows: T_{max} = 980 °C, t_1 = 2h, T_1 = 900 °C, r = 20 - 25 °C /h, and the results show that a considerable high quality joint can be obtained by these parameter, the levitation force recovery rate is good, as indicated in table 1, and the microstructure is also acceptable: An obvious transition layer was observed at the interface, which means the "self-seeded" joining process was at least partially realized during the joining process. The microstructure in the transition layers is similar to that of the base material, as shown in Figure 3. In order to investigate the composition in the transition layers, EDS was taken, the EDS result, as shown in Figure 4 and table 2, indicates that the transition layer was composite of (Y,Yb)123, which act as the matrix , and (Y, Yb)211, which form the pinning center. The average size of the (Y,Yb)211 Particles is around 2μm, which is a little bitter larger than that of Y211 in the base material.

Table 1. levitation recovery rate of the joint using YbBCO solder

Sample No.	Original levitation force (A)	Levitation force after joining (B)	Recovery rate(B/A)
1	70.2N	60N	85.5%
2	68.7N	55.9N	81.4%

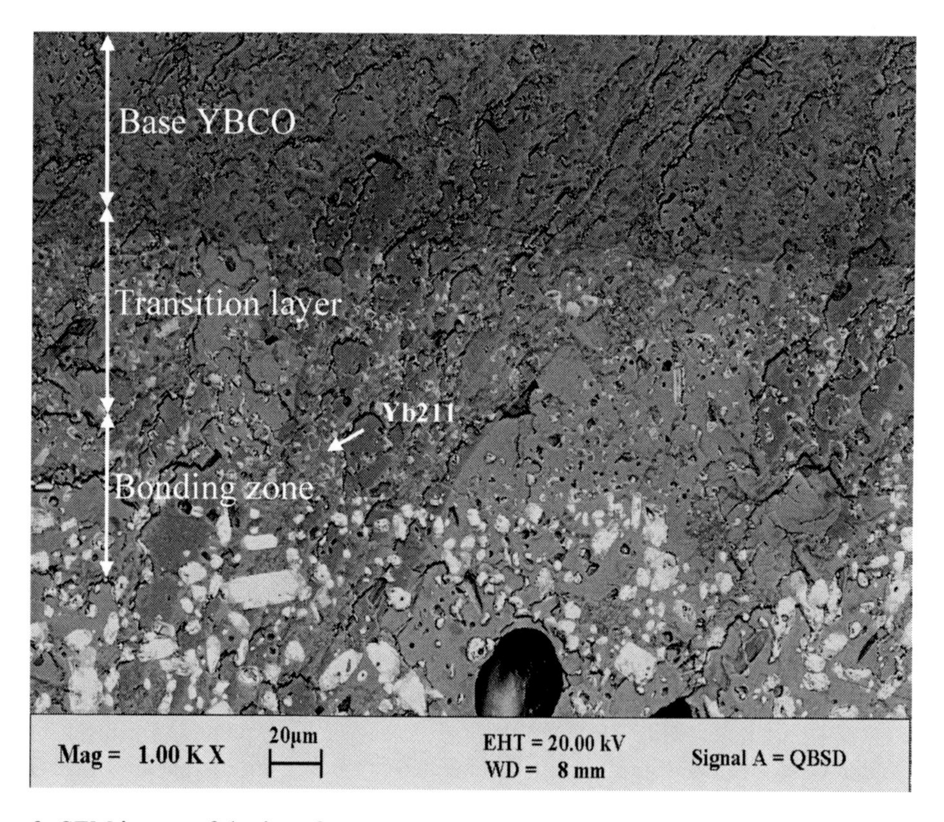

Figure 3. SEM image of the interface.

However, the microstructure in the bonding zone is not similar as that in the base YBCO material; serve accumulation of Yb 211 was found in the bonding zone, as indicted in Figure 3.

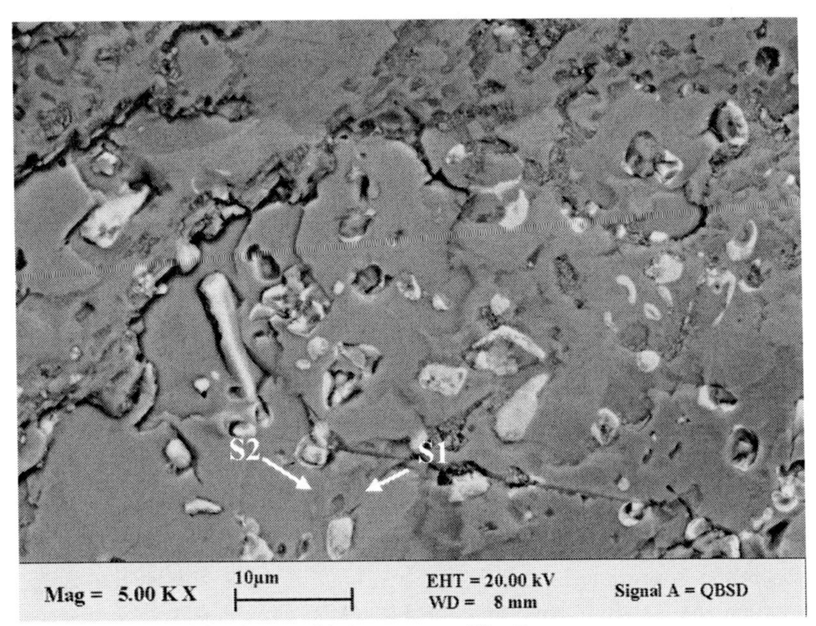

Figure 4. SEM image of transition layer with larger magnification.

Table 2. EDS results of S1 and S2

Sample No.	Element contents (at %)					Estimate phase
	O	Cu	Ba	Yb	Y	
S1	58.62	11.25	10.74	9.51	9.88	(Y,Yb)211
S2	56.37	22.15	14.32	3.1	4.06	(Y,Yb)123

This is very detrimental to the superconductive properties of the joint: the superconductivity of HTS is largely influenced by the pinning effect of the pinning center, Yb 211 particles form the pinning center in the bonding zone, the pinning effect is determined by both the size and the distribution of the pinning center: It is believed that when the size of the RE211(RE = Y,Yb,Tm…) is larger than 0.5μm, the superconductivity is better if the average size of the RE211 is smaller and the distribution of the RE211 is more homogenous. In addition, some defects, such as pores and some cracks were also observed in the bonding zone, hence blocked the supercurrent in the bonding zone and also reduced the superconductivity. The accumulation of the Yb211 resulted from the relatively high cooling rate (r) and the crack should result from the dissimilar crystal parameters with the base material, as indicated by C.Harnois et al. [13]. Meanwhile, regarding on the thickness of the transition layer and the bonding zone, we conclude that in order to obtain similar microstructure as the base YBCO in the bonding zone, the thickness of the solder should be reduced. As a result, new solders were adopted in the joining process.

3.2. Joints using sintered YBCO/Ag as solder

Many researchers have reported the advantage of the YBCO/Ag system, the pseudo phase diagram reported by Wiesner et al. [26] have shown that the eutectic reaction would happened if the Ag content is between 5wt% to 20wt%, and the eutectic temperature is about 980°C, about 40 °C lower than that of YBCO. Moreover, the superconductivity of YBCO/Ag is no worse or even better than YBCO owing to the existence of Ag can fill the residual pores during the growth process. As a result, sintered YBCO/Ag solder plate was used to joining base YBCO material.

YBCO/Ag solder was composed of $YBa_2Cu_3O_7$ (Y123) powders + 25wt%Y_2BaCuO_5 (Y211) powders + 10wt% Ag_2O powders, the mixed powders were pressed into 70mm*14mm*1.2mm plates and sintered at 910°C in air for 10h to achieve homogenous heating after 2h ball milling process, then the solder plates were milled to 0.4 mm in thickness by abrasive paper. The parameters of the thermal cycle are as follows: T_{max} = 995 °C, t_l = 45min, T_l = 940 °C and r = 5.5 °C /h. The joining result is much better than using YbBCO solder, as reported by Guo Wei et al. [27]: from the ab plan, the joint is well bonded, the crystal pattern in the base material and in the bonding zone are almost the same, no distinct boundary between the bonding zone and the base YBCO material was observed, the only difference is that in the bonding zone there are large amount of the white particles. The EDS shows that these white particles are Ag, as indicted in Figure 5. We can also observe that the distribution of Ag particles in the bonding zone is a little bit inhomogeneous, which do little harm to the superconductivity of the joint. As we have discussed before, the size and distribution of the pinning center is critical to the superconductivity of the joint, the pinning center of the bonding zone is greatly influenced by the size and distribution of Y211, which could not be distinguished clearly in Figure 5. Therefore, SEM image with larger magnification was taken, as shown in Figure 6. The average size of Y211 particles is similar to that of in the base YBCO, as indicted in Figure 7, meanwhile, the distribution of Y211 particles is almost homogenous, both of these shows that a well bonding with good superconductivities was achieved.

The SEM image of the ac plan also reveals that good bonding is achieved during the joining process, as shown in Figure 8, we can observe that some microcracks propagate from the base YBCO into the bonding zone and even penetrate through the bonding zone to another half of the base material, which also reveals that the crystals in the bonding zone grew along the same direction as the crystals orient in the base YBCO material. This phenomenon is also reported by K.Iida et al. [17], however, the density of microcracks on the ac plan in their experiment is much larger than that in our experiment. In addition, some accumulation of Ag and Ba-Cu-O phase were discovered at the center of the bonding zone, this should be resulted from the simultaneous crystal growth at both interfaces during the joining process, the Ag and residual Ba-Cu-O phase was pushed toward to the center and meet with each other and finally form the joint. Although the existence of these accumulations is bad for the superconductivity, the supercurrent would not be block because these accumulations are discontinuous.

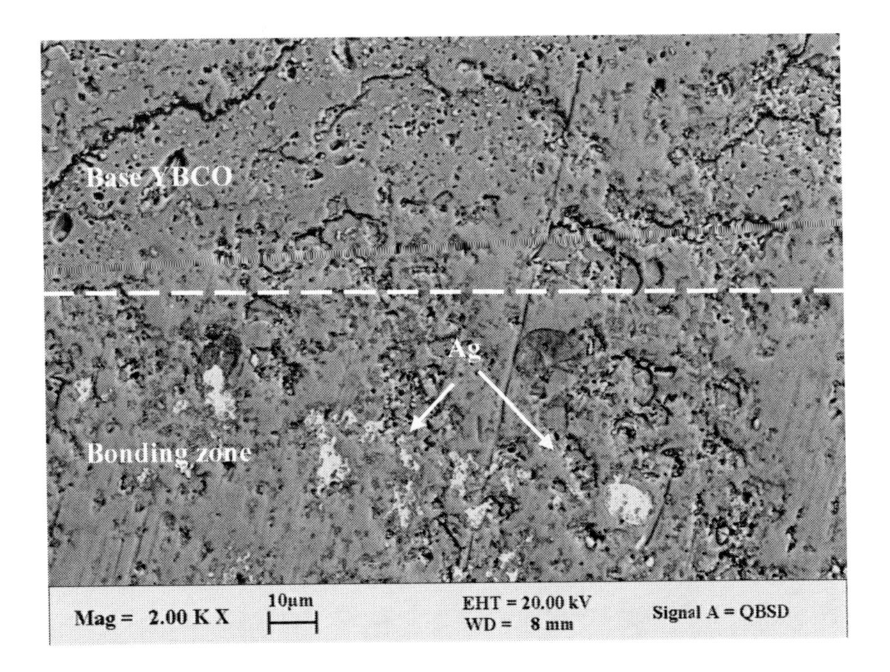

Figure 5. SEM image of the interface between base material and bonding zone on ab plan.

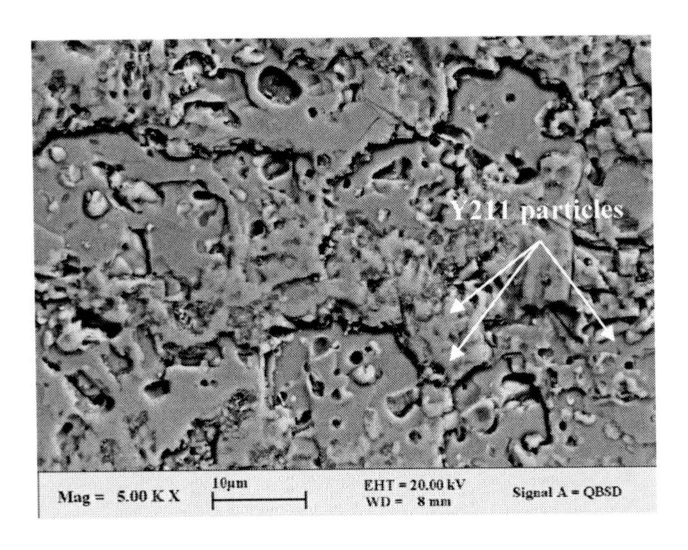

Figure 6. SEM image of the bonding zone.

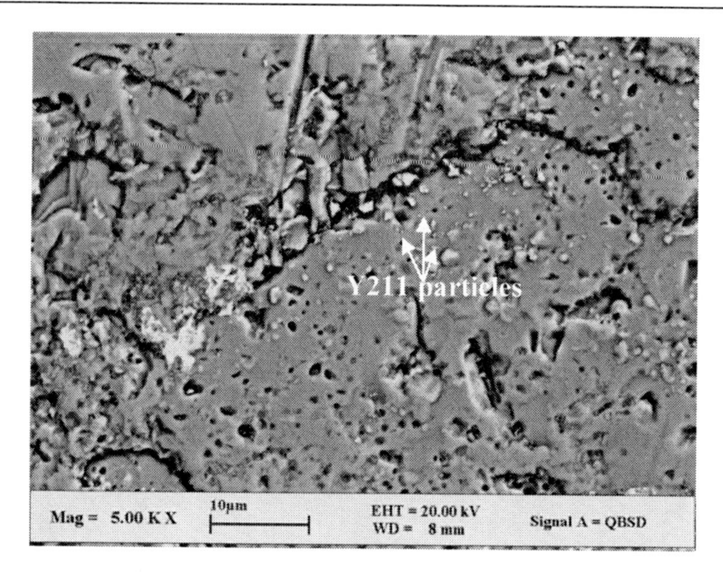

Figure 7. SEM image of the base YBCO.

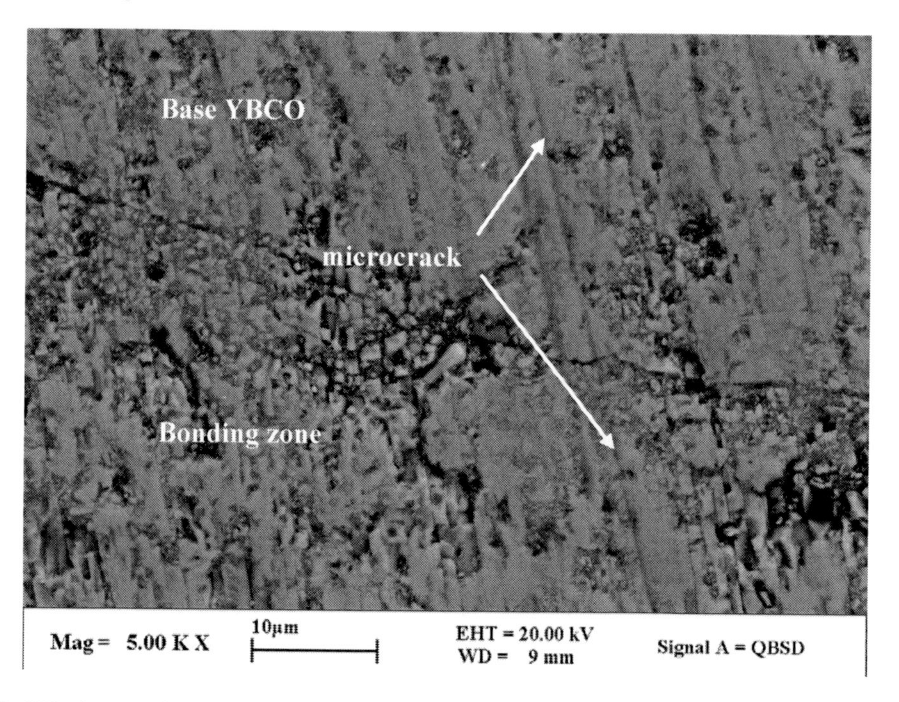

Figure 8. SEM image of the interface between base YBCO and bonding zone on ac plan.

The superconductivity of the joint is also examined by trapped field examination, the trapped-flux contour map is shown in Figure 10, and nearly one peak was achieved, indicating that the supercurrent in the joint recovered very well.

However, a number of voids still exist in the joining, as shown in Figure 11. One plausible reason for this phenomenon is that the decomposition of Ag_2O during the heating process synthesis O2, which results in the voids. As a result, metallic Ag powder was used to replace Ag_2O powder and all the other parameters are the same as that of using Ag_2O powder as additive [28].

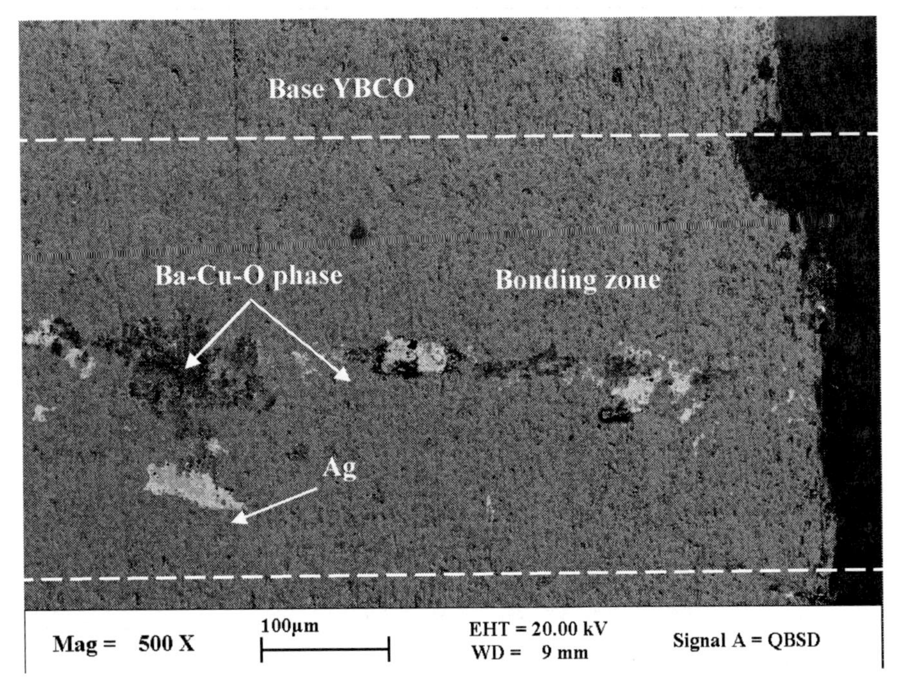

Figure 9. SEM image of bonding zone on ac plan.

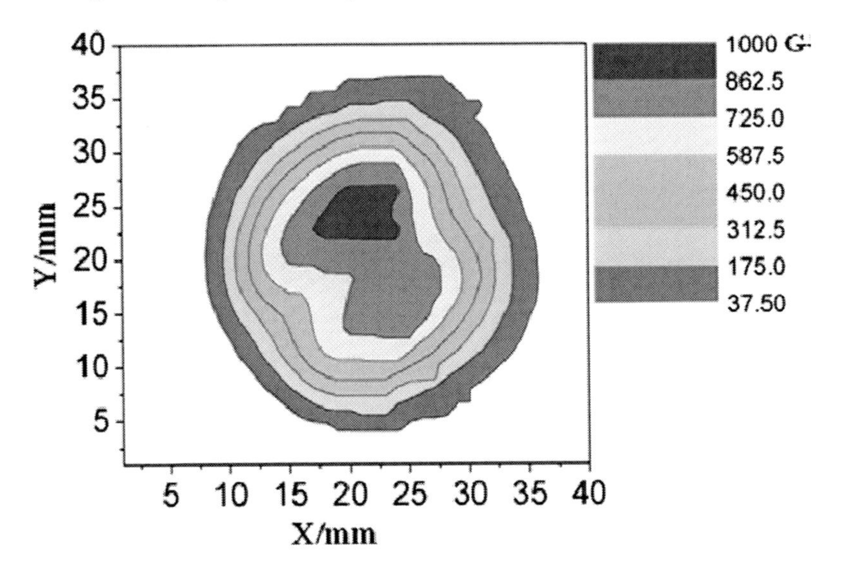

Figure 10. Trapped-flux contour map of the joint.

The improvement is obvious: Based on the image of the bonding zone on ab plan,voids are rarely observed and the Ag particles are finer and more homogenous distribution.As discussed above, this phenomenon is helpful to obtain good superconductivities. In addition, the size and distribution of the Y211 particles are also evaluated, as shown in Figure 12. One can find that the average size of Y211 particles is around 2 μm, which is almost the same as that in the base YBCO, and the distribution of Y211 particles is homogenous. As a result, the superconductivity of this joint is better than that using solder with Ag_2O powder as additive; this is also reveals by the trapped flux contour map, as shown in Figure 13. Meanwhile, some

Ba-Cu-O phase is also observed in the bonding zone, as well as some pores even though Ag additive can reduce the pores in the bonding zone comparing with the Ag_2O. In order to reduce the porosity in the bonding zone, we considered to use melt YBCO/Ag sliced plate as solder.

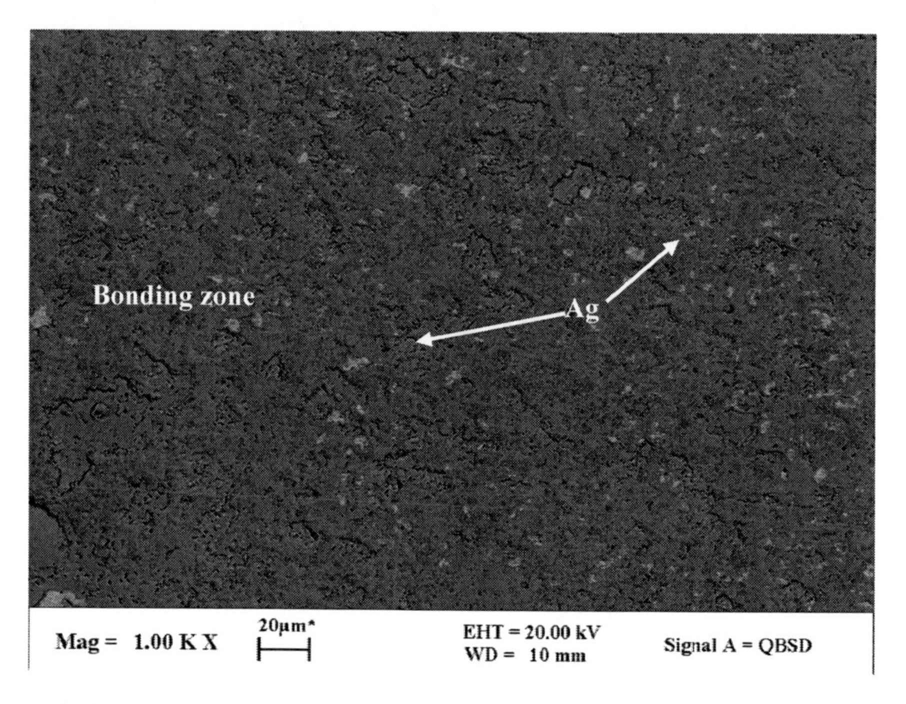

Figure 11. Distribution of Ag in the bonding zone.

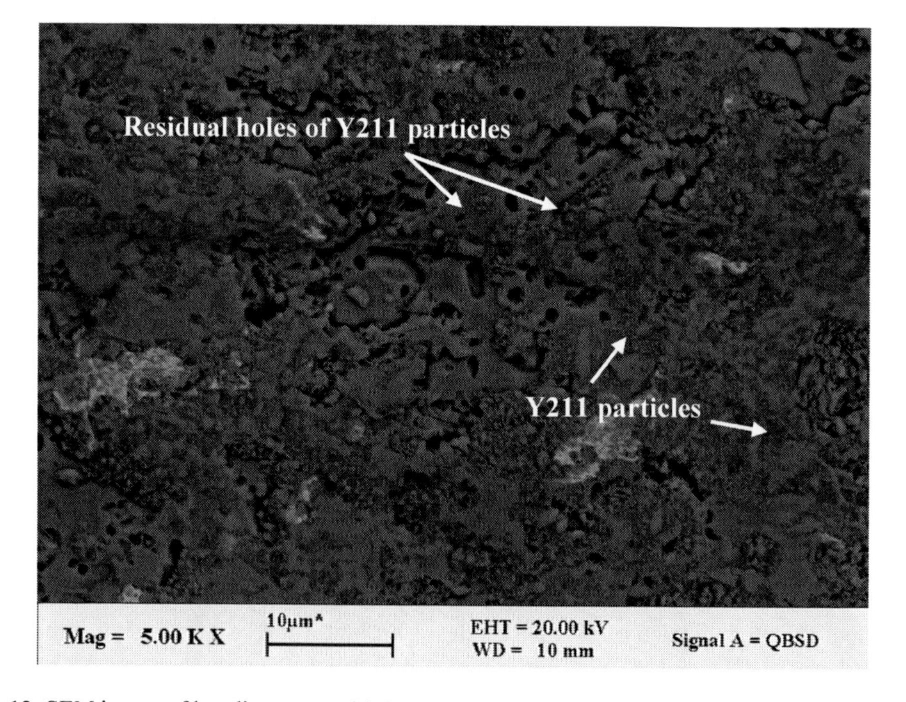

Figure 12. SEM image of bonding zone with larger magnification.

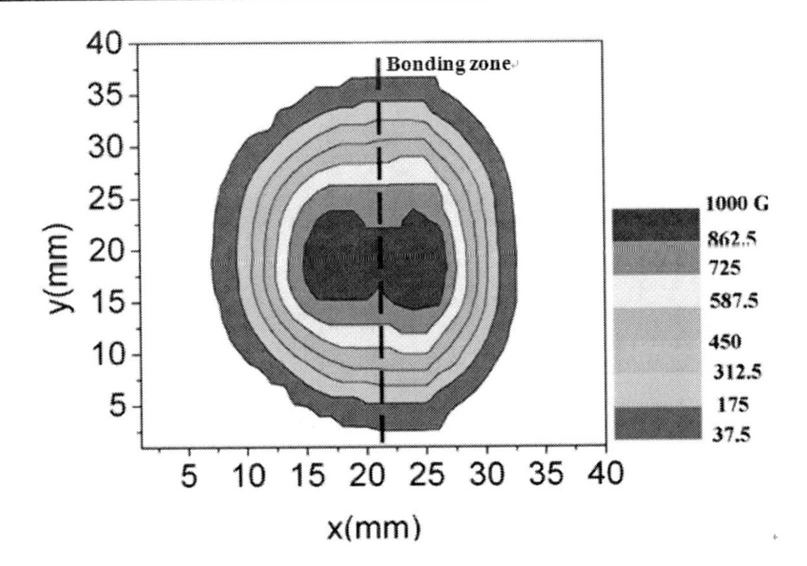

Figure 13. Trapped-flux contour map of the joint.

3.3. Joint with melt YBCO/Ag solders

The composition of YBCO/Ag solders is the same as sintered ones; the only difference between these two kinds of solders is the sintering temperature: the melt solder was prepared via the mixture powder, the powder was pressed into pellet of 28mm in diameter and 14mm in height, and the pressed pellet was sintered at 995 °C, which is above the eutectic point of YBCO and Ag system, and the solder plate was sliced by diamond saw from the pellet with the thickness of 0.3mm, about 0.1mm thinner than the sintered one. The sliced solder was evaluated by SEM, as shown in Figure 14: The white particles disturb in the figure are Ag, and one can evaluate the average diameter of the Ag particle is around 10 um, which means the Ag particles is relatively homogenous disturb in the solder and no accumulation was not happened during the heating process. These Ag particles are the decomposition product of Ag_2O, which decomposed into Ag and O_2 at about 380 °C. Meanwhile, the grey background is the mixture of Y123 phase and Y 211 phase, the distinction between Y123 phase and Y 211 phase is not very clear from this figure. However, some black holes are also observed in this figure, the diameter is ranged from about 5μm to about 20μm, these are the pores in the solder, even though the high temperature can greatly increase the density of the solder. But one have to notice that both the quantity and the pore size were much smaller than the as sintered solder which was sintered at lower temperature [27],that means the density of the melt solder is larger than the sintered one. Similar result was also reported by A.Mahmood et al. [29]

The microstructure of the bonding zone is almost the same as that in the base YBCO, so no distinct boundary exist, as shown in Figure 15. However, some Ba-Cu-O phase was found in the bonding zone, but the distribution of these black areas is discontinuous, so that the supercurrent would not be blocked. The size and distribution of Y211 particles is similar to the results of the sintered one, as indicated by Figure 16. Nonetheless, the porosity in the bonding zone is greatly reduced, so the superconductivity is better.

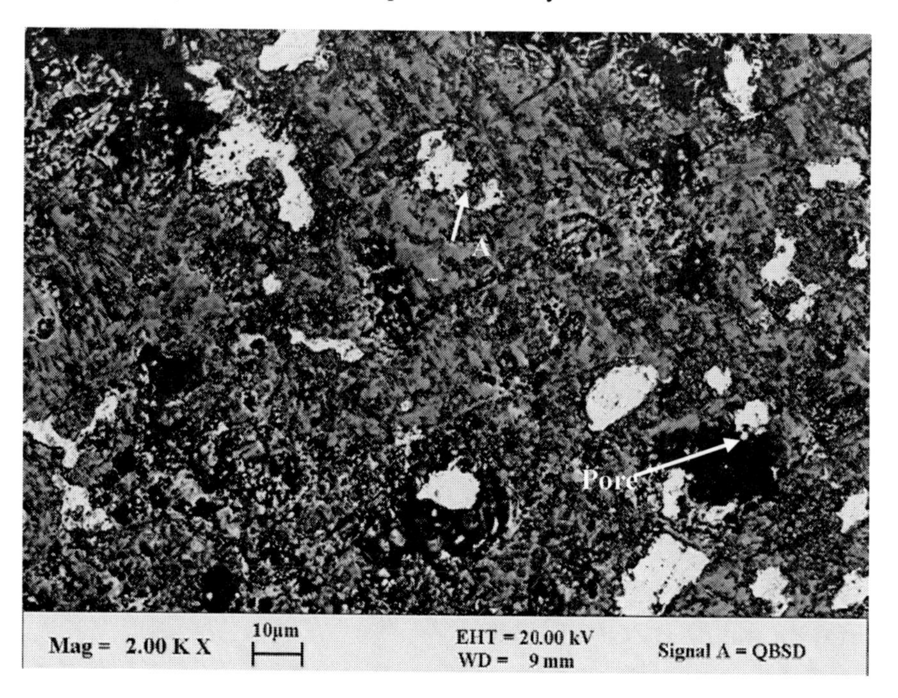

Figure 14. SEM image of the solder.

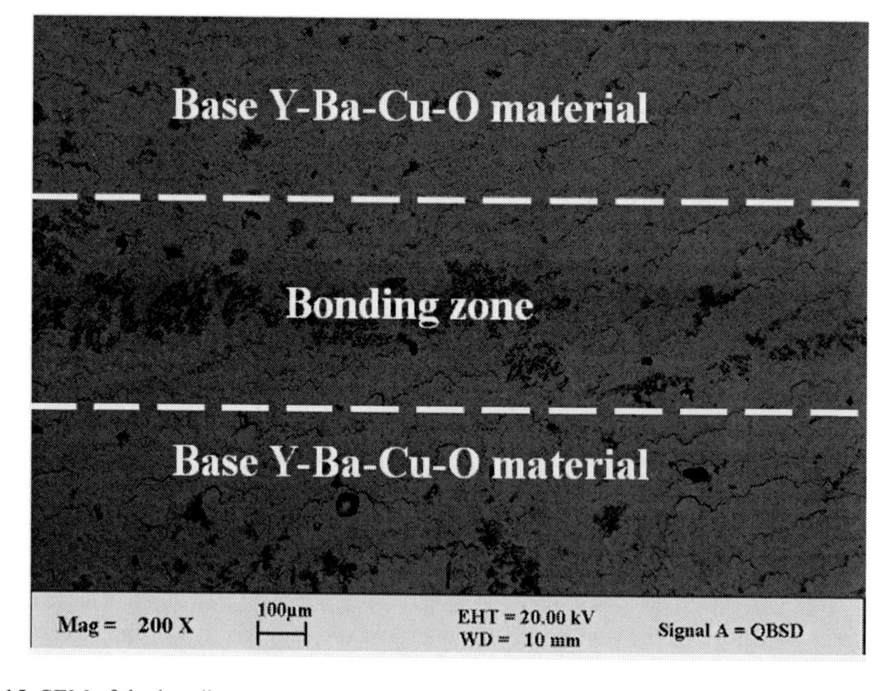

Figure 15. SEM of the bonding zone.

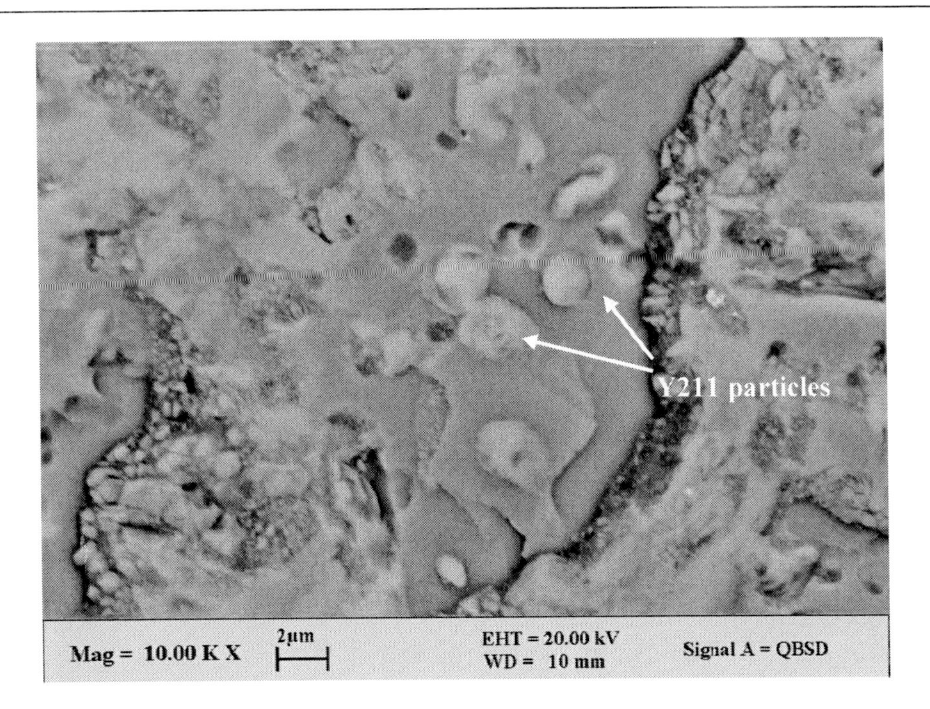

Figure 16. SEM image of distribution of Y211.

Conclusion

In a word, Ag doped YBCO solder is a promising kind of solder that can realize fast self-seeded joining of YBCO bulk, and preheat effect of the solder is very obvious, the melt YBCO/Ag solder is very promising in obtaining good superconductive joint in a relatively short time.

Acknowledgment

This research is supported by the National Natural Science Foundation of China (Grant No. 50705050), Beijing Natural Science Foundation (Grant No. 3093020)

References:

[1] B. Oswald, M. Krone, T. Straβer, K. J. Best, M. Söll, W. Gawalek, H. J. Gutt, L. Kovalev, L. Fisher, G. Fuchs, G. Krabbes, H. C. Freyhardt, *Physica C* 372-376 (2002), 1513.

[2] S. Gruss, G. Fuchs, G. Krabbes, P. Verges, G. Stöver, K. H. Müller, J. Fink, L. Schultz, *Appl. Phys. Lett.* 79 (2001), 3131.

[3] K. Matsunaga, N. Yamachi, M. Tomita, M. Murakami, N. Koshizuka, *Physica C* 392-396 (2003), 723.

[4] R. Tournier, E. Beaugnon, O. Belmont, X. Chaud, D. Bourgault, D. Isfort, L. Porcar, P. Tixador, *Supercond. Sci. Technol.* 13 (2000), 886.

[5] K. Salama, V. Selvanmanickam, *Appl. Phys. Lett.* 60(7), 17 February, 1992.

[6] D. A. Cardwell, A. D. Bradley, N. H. Babu, M. Kambara, W. Lo, *Supercond. Sci. Technol.* 15 (2002), 639.

[7] W. Lo, D. A. Cardwell, A. D. Bradley, R. A. Doyle, Y. H. Shi, S. Lloyd, *IEEE Trans. on Appl. Supercond.* vol. 9, No. 2, June 1999.

[8] S. M. Mukhopadhyay, N. Mahadev, S. Sengupta, *Physica C* 329 (2000), 95.

[9] J. G. Noudem, E. S. Reddy, E. A. Goodilin, M. Tarka, M. Noe, G. J. Schmitz, *Physica C* 372 (2002), 1187.

[10] J. G. Noudem, E. S. Reddy, M. Tarka, M. Noe, G. J. Schmitz, *Supercond. Sci. Technol.* 14 (2001), 363.

[11] T. A. Prikhna, W. Gawalek, V. E. Moshchil, L. S. Uspenskaya, R. Viznichenko, N. V. Sergienko, A. A. Kordyuk, V. B. Sverdun, A. B. Surzhenko, D. Litzkendorf, T. Habisreuther, A. V. Vlasenko, *Physica C* 392 (2003), 432.

[12] T. A. Prikhna, W. Gawalek, N. V. Novikov, V. E. Moshchil, V. B. Sverdun, N. V. Sergienko, A. B. Surzhenko, L. S. Uspenskaya, R. Viznichenko, A. A. Kordyuk, D. Litzkendorf, T. Habisreuther, S. Krachunovska, A. V. Vlasenko, *Supercond. Sci. Technol.* 18 (2005), S153.

[13] H. Zheng, M. Jiang, R. Nikolova, U. Welp, A. P. Paulikas, Y. Huang, G. W. Crabtree, B. W. Veal, H.Claus, *Physica C* 322 (1999), 1.

[14] C. Harnois, X. Chaud, I. Laffez, G. Desgardin, *Physica C* 372 (2002), 1103.

[15] C. Harnois, G. Desgardin, X. Chaud, *Supercond. Sci. Technol.* 14 (2001), 708.

[16] K. Iida, J. Yoshioka, N. Sakai, M. Murakami, *Physica C* 392 (2003), 437.

[17] K. Iida, T. Kono, T. Kaneko, K. Katagiri, N. Sakai, M. Murakami, N. Koshizuka, *Physica C* 402 (2004), 119.

[18] K. Iida, J. Yoshioka, T. Negichi, K. Noto, N. Sakai, M. Murakami, *Physica C* 378 (2002), 622.

[19] K. Iida, T. Kono, T. Kaneko, K. Katagiri, N. Sakai, M. Murakami, N. Koshizuka, *Supercond. Sci. Technol.* 17 (2004), S46-S50.

[20] S. Iliescu , S. Sena, X. Granados, E. Bartolome, T. Puig, X. Obradors, M. Carrera , J. Amoros, S. Krakunovska , T. Habisreuther, *IEEE Trans. on Appl. Supercond.* 13 (2003), 3136.

[21] S. Iliescu, X. Granados, T. Puig, X. Obradors, *J Mater Res.* 21 (2006), No. 10, 2534.

[22] S. Iliescu, X. Granados, E. Bartolome, S. Sena, A. E. Carrillo, T. Puig, X. Obradors, J. E. Evetts, *Supercond. Sci. Technol.* 17 (2004), 182.

[23] S. Iliescu, A. E. Carrillo, E. Bartolome, X. Granados, B. Bozzo, T. Puig, X. Obradors, I. Garcia, H. Walter, *Supercond. Sci. Technol.* 18 (2005), S168.

[24] T. Puig, P.Rodriguez Jr., A. E. Carrillo, X. Obradors, H. Zheng, U. Welp, L. Chen, H. Claus, B. W. Veal, G. W. Crabtree, *Physica C* 363 (2001), 75.

[25] Th. Hopfinger, R. Viznichenko, G. Krabbes, G. Fuchs, K. Nenkov, *Physica C* 398 (2003), 95.

[26] U. Wiesner, G. Krabbes, M. Ueltzen, C. Magerkurth, J. Plewa, H. Altenburg, *Physica C* 294 (1998), 17.

[27] W. Guo, G. Zou, X. Chai, A.Wu, J. He, H. Bai, J. Ren, Y. Jiao, *Physica C* 470 (2010), 482.

[28] X. Chai, G. Zou , W. Guo, A.Wu , J. He, H. Bai, L. Xiao, Y. Jiao, J. Ren, *Physica C* 470 (2010), 598.

[29] A. Mahmood, B. H. Jun, H. W. Park, C. J. Kim, *Physica C* 468 (2008), 1350.

In: Superconductivity
Editor: Vladimir Rem Romanovskii

ISBN 978-1-61324-843-0

Chapter 10

Vortices In High Temperature Superconductors: A Phenomenological Approach

Madhuparna Karmakar [1*] ***and Bishwajyoti Dey*** [2†]
[1] S. N. Bose National Center for Basic Sciences
JD-Block, Salt Lake, Kolkata-700098, India
[2] Department of Physics, University of Pune
Pune-411007, India

Abstract

It is well known that magnetic flux can penetrate a type-II superconductor in the form of Abrikosov vortices. Many properties of the vortices are well described by phenomenological Ginzburg–Landau theory. High temperature superconductor also belongs to the class of type-II superconductors. These vortices (also known as flux-line lattice) form a triangular lattice in the low-temperature type-II superconductors. But experiments have shown that in the high-temperature superconductors, these vortices form an oblique lattice. This is attributed to various characteristics of high temperature superconductors, namely, pairing state symmetries, anisotropy, planar nature of the superconducting plane, symmetry of the unit cell modulated by the presence of CuO-chain, thermal fluctuations etc. The experimentally observed softening and subsequent melting of the vortex lattice is attributed to the small shear modulus of the vortex lattice. In this review we address these problems associated with the vortex lattice of high temperature superconductors.

One of the central issues associated with the high-T_c superconductors is the pairing state symmetry of the order parameter components. Based on the experimental evidences a consensus could now be reached that these materials possesses mixed symmetry state scenario where the dominant order parameter is of d-wave symmetry along with which there is an admixture of a small s-wave order parameter component.

We have studied the properties of the high-T_c superconductors involving mixed symmetry state of the order parameters in the framework of a two-order parameter Ginzburg–Landau (GL) model, over the entire range of applied magnetic field ($H_{c1} < H < H_{c2}$) and wide range of temperature and for arbitrary GL parameter κ

*E-mail address: madhuparna@bose.res.in
†E-mail address: bdey@physics.unipune.ac.in

and vortex lattice symmetry. Using the present model the limitations of the earlier theoretical works involving the GL theory could be overcome, wherein the studies were restricted to the upper (H_{c2}) and lower (H_{c1}) critical magnetic field regions and the problem was reduced to an effective single order parameter model by using an ansatz. The present theoretical model has been further generalized to take into account the effect of in-plane anisotropy. We calculate various properties of the high temperature superconductors including vortex core radius, penetration depth, vortex lattice symmetry, upper critical magnetic field, shear modulus of the vortex lattice etc. Furthermore, the effect of the admixture of the sub-dominant s-wave order parameter on the various properties of the high-Tc cuprates have been explored. The variations of these properties with applied magnetic field and temperature are also calculated and shown to be in very good agreement with experiments on high-T_c cuprate $YBa_2Cu_3O_{7-\delta}$ superconductors [1]– [5]. We have also used the same two-order parameter Ginzburg–Landau model appropriately modified to study the properties of the two-band inter-metallic high temperature superconductor MgB_2. The results of the analytical calculations are shown to agree very well with the experimental results of MgB_2 [6].

1. Introduction

Superconductivity, as the name suggests correspond to the sudden disappearance of the electrical resistance in certain materials when their temperature is lowered. It is this very property which lead to the discovery of the phenomenon of superconductivity by Kamerlingh Onnes in 1911 when he observed that on cooling mercury (Hg) to very low temperature there is a sudden disappearance of its electrical resistivity at a temperature around 4.2K [7]. One of the highlighted features of any superconductor is the existence of long range order i. e an interaction between the electrons. It is however known that there exists coulomb repulsive interaction between the electrons which by no means is capable of giving rise to an energy gap which is a key feature of any superconducting material. The difficulty thus came up could be solved by understanding the concept of *Electron-Phonon interaction* put forward by Frölich in 1950 [8], and later extended by Bardeen, Cooper and Schrieffer (BCS) [9], [10]. According to the BCS theory the superconducting property of a material arises due to a phonon mediated attractive interaction between pair of electrons near the Fermi surface. As per this theory the critical temperature can be expressed by the relation,

$$T_c = 1.14\langle\omega\rangle exp(-1/N_0V) \tag{1}$$

where, $\langle\omega\rangle$ is the average phonon energy, N_0 is the density of states at the Fermi level and V is the pairing potential arising due to the electron-phonon interaction. Thus, depending upon the average phonon energy there is an upper limit of the critical temperature that can be attained. Based on his studies regarding the average phonon energy and electron-phonon coupling strength $\lambda = N_0V$, McMillan suggested that the maximum critical temperature that can be achieved by any superconducting material is $T_c = 30K$ [11].

The situation however changed dramatically in 1986 with the discovery of a new class of superconducting materials by Bednorz and Müller [12], which were characterized by very high critical temperature as compared to the then existing genre of superconducting materials and were hence christianized as the "High temperature superconductors" (HTS).

The HTS, with T_c ranging from $\approx 92K$ for $YBa_2Cu_3O_{7-\delta}$ to the recent Hg-based superconductors with $T_c \approx 140K$ broke the myth of the unattainability of T_c beyond $30K$ in type-II superconductors.

2. Previous Theoretical Studies

Keeping in pace with the experimental development, attempts were made to theoretically study the properties of the superconducting system. Even before the evolution of the microscopic (BCS) theory of superconductivity an entirely different approach to the problem was developed by Ginzburg and Landau in 1950 [13]. The approach was independent of the microscopic behavior of the system and was based merely on the thermodynamic property that the superconducting transition is a second order phase transition and thus, it should be possible to analyze the superconducting state on the basis of general rules of a second order phase transition. The theory failed to create much of an interest at that time and remained so till Gork'ov [14] showed that the Ginzburg–Landau (GL) theory is basically a limiting case of the BCS theory of superconductivity, applicable near all the criticals i. e critical temperature, critical magnetic field etc.

In the framework of the GL theory the theoretical investigation of the flux line lattice (FLL) of the type-II superconductors near the upper critical magnetic field (H_{c2}) was first carried out by A. A. Abrikosov [15]. In his work, Abrikosov defined the stable vortex lattice configuration by the parameter $\beta_A = \langle\psi^4\rangle/\langle\psi^2\rangle^2$, which was later termed as the Abrikosov's parameter. By calculating the free energy density for type-II superconducting systems ($\kappa \gg 1$) near the upper critical magnetic field H_{c2}, Abrikosov suggested that the stable vortex lattice configuration for such superconducting materials correspond to a triangular arrangement of the flux lines. Abrikosov's pioneering work was followed by several important theoretical studies which were carried out using the conventional GL theory in order to understand the various properties of the type-II superconducting system. In these studies the GL equations could be solved analytically under two limiting conditions viz. in the single vortex regime ($H \to H_{c1}$) where it correspond to the solutions of London theory and in the vortex lattice regime ($H \approx H_{c2}$). Both analytical and approximate numerical studies were carried out to extend the GL theory to the intermediate applied magnetic field regime [16]– [22]. The intermediate applied magnetic field regime $H_{c1} < H < H_{c2}$ however remained largely unexplored in spite of these efforts regarding the extension of the single vortex and vortex lattice solutions. This is because of the difficulty in analytically solving the coupled GL equations in this intermediate magnetic field regime.

This limitation of solving the coupled nonlinear GL equations over the entire range of applied magnetic field was sorted out by E. H. Brandt in his pioneering work [23]. In his work he developed a novel iteration technique which could numerically solve the GL equations for the ideal Abrikosov vortex lattice in type-II superconductors for arbitrary GL parameter κ ($1/\sqrt{2} \leq \kappa < \infty$) and vortex lattice symmetry. The method is applicable to wide range of magnetic field induction $10^{-3} < b < 1$, where $b = \bar{B}/B_{c2}$, with $\bar{B}$ being the spatially averaged magnetic field and $B_{c2} = \mu_0 H_{c2}$ is the upper critical magnetic field. The numerical iteration technique allows one to compute over 1000 Fourier coefficients for the order parameter $\omega(x, y)$ and magnetic field $\mathbf{B}(x, y)$ with high precision. Due to its applicability to arbitrary vortex lattice symmetry the method can compute the shear modulus (c_{66}) of the vortex lattice, an important quantity which gives information regarding the melting of the flux line lattice.

3. Unconventional Superconductivity: The High Temperature Superconductors

The term "Unconventional Superconductors" refer to the superconducting materials which possess a pairing state symmetry different from the isotropic s-wave symmetry observed in the conventional low temperature type-II superconductors. The discovery of a new class of superconducting materials by Bednorz and Müller [12] in 1986 opened up a vast new world in the field of superconductivity and condensed matter physics in general. Apart from their high critical temperature, the high-T_c superconductors (HTS) are characterized by a short coherence length and consequently a large GL parameter κ. These materials possess strong in-plane and c-axis anisotropy and their pairing symmetry is certainly different from the isotropic s-wave observed in the conventional low temperature type-II superconductors.

The various unconventional features of HTS are related to the complex structure of these cuprates. These are three-dimensional perovskite materials consisting of two-dimensional superconducting CuO_2-planes and one-dimensional CuO-chains which serve as the charge reservoirs. The superconducting property is very crucially dependent on the stoichiometry of these materials.

3.1. In-plane mass anisotropy in high temperature cuprates

As discussed in the preceding sections the high-T_c superconductors are characterized by strong in-plane anisotropy. For instance the high-T_c cuprate $YBa_2Cu_3O_{7-\delta}$ consists of an in-plane anisotropy of $\approx$ 2.4, the in-plane anisotropy is quantified by the parameter $\gamma = m_x/m_y$, where m_x and m_y are the effective electronic masses along the x and y-directions aligned respectively with the a- and b-axes of the superconducting crystal [24]. The existence of such in-plane anisotropy in these materials have been attributed to the orthorhombic distortion of the CuO_2-planes arising due to the presence of the CuO chains [25]. The presence of in-plane anisotropy in $YBCO$ has been substantiated by the microwave surface impedance measurement of penetration depth in which an in-plane anisotropy of $\gamma \approx 2.4$ has been observed [24]. Moreover, the elongated vortex core structure observed by the scanning tunneling microscopy (STM) measurements have also been attributed to the in-plane anisotropy present in these systems [26]. Furthermore, recent experiments have shown that the effect of anisotropy introduces considerable changes, such as distortion of flux line lattice, vortex lattice melting via an intermediate phase, oblique structure of vortex lattice etc [27].

In the following few sections the effect of in-plane anisotropy on the various properties of the high temperature superconducting cuprates are studied in the framework of GL theory by using the numerical iteration technique mentioned above [1], [28].

3.1.1. Theoretical formalism

For studying the effect of the in-plane mass anisotropy on the various properties of the high-T_c cuprates we begin with the two-dimensional average GL free energy density functional corresponding to the anisotropic single order parameter GL model [1], [28].

$$f = \left\langle -\omega + \omega^2/2 + \nabla\omega\mathbf{\Lambda}\nabla\omega/4\kappa_y^2\omega + \omega\mathbf{Q\Lambda Q} + (\nabla \times \mathbf{Q})^2 \right\rangle. \quad (2)$$

In Eqn. (2) ω corresponds to the dimensionless gauge invariant order parameter as, $\psi(\mathbf{r}) = \omega(\mathbf{r})^{1/2}e^{i\phi(\mathbf{r})}$ while $\mathbf{Q} = \mathbf{A} - \nabla\phi/\kappa_y$ is the supervelocity of the electrons, with the GL parameter being defined as $\kappa_y = \lambda_y/\xi_y$. The two-dimensional mass anisotropy tensor $\mathbf{\Lambda}$ is introduced in the formalism as,

$$\mathbf{\Lambda} = \begin{pmatrix} m_y/m_x & 0 \\ 0 & 1 \end{pmatrix} \quad (3)$$

where, m_x and m_y are the effective electronic masses along the x-and y-directions of the superconducting plane. Minimizing the free energy density functional (Eqn. (2)) w. r. t the order parameter ω and the supervelocity $\mathbf{Q}$ one gets the corresponding GL equations of motion. The order parameter, magnetic field and supervelocity are then expressed in the form of rapidly converging Fourier series (Eqn. (4)-(6)) with the corresponding Fourier coefficients being $a_{\mathbf{K}}$ and $b_{\mathbf{K}}$ [1], [23].

$$\omega(\mathbf{r}) = \sum_{\mathbf{K}} a_{\mathbf{K}}(1 - \cos\mathbf{K}.\mathbf{r}), \quad (4)$$

$$B(\mathbf{r}) = \bar{B} + \sum_{\mathbf{K}} b_{\mathbf{K}} \cos\mathbf{K}.\mathbf{r}, \quad (5)$$

$$\mathbf{Q}(\mathbf{r}) = \mathbf{Q}_A(\mathbf{r}) + \sum_{\mathbf{K}} b_{\mathbf{K}} \frac{\hat{z} \times \mathbf{K}}{K^2} \sin\mathbf{K}.\mathbf{r} \quad (6)$$

where, $\mathbf{r} = (x, y)$; the sums are over all the reciprocal lattice vectors $\mathbf{K}_{mn} \neq 0$. The position of the vortices in the vortex lattice are defined by $\mathbf{R} = \mathbf{R}_{mn} = (mx_1 + nx_2, ny_2)$ (m, n are integer) with x_1, x_2 and y_2 being the lattice parameters. The corresponding reciprocal lattice is defined as $\mathbf{K} = \mathbf{K}_{mn} = (2\pi/S)(my_2, nx_1 + mx_2)$, where, $S = x_1y_2$. $\mathbf{Q}_A(x, y)$ is the supervelocity of the Abrikosov B_{c2} solution. The Fourier coefficients are next determined numerically using a set of iterative equations [1], [28],

$$a_{\mathbf{K}} := \frac{-4\kappa_y^2\langle(2\omega - \omega^2 - \omega\mathbf{Q\Lambda Q} - g)\cos\mathbf{K}.\mathbf{r}\rangle}{\mathbf{K\Lambda K} + 2\kappa_y^2}, \quad (7)$$

$$a_{\mathbf{K}} := a_{\mathbf{K}}.\frac{\langle(\omega - \omega\mathbf{Q\Lambda Q} - g)\rangle}{\langle\omega^2\rangle}, \quad (8)$$

$$b_{\mathbf{K}} := \frac{-2\langle[\omega B - \bar{\omega}(B - \bar{B}) + p]\cos\mathbf{K}.\mathbf{r}\rangle}{\frac{m_x}{m_y}\mathbf{K\Lambda K} + \bar{\omega}} \quad (9)$$

where, $g = \nabla\omega\mathbf{\Lambda}\nabla\omega/4\kappa_y^2\omega$ and $p = (\nabla\omega \times \mathbf{Q}).\hat{z}$. The high precision numerical solutions of the order parameter and magnetic field thus obtained are then used to calculate the effect of the in-plane anisotropy on the various properties of the high-T_c cuprates in the framework of an anisotropic single order parameter GL model.

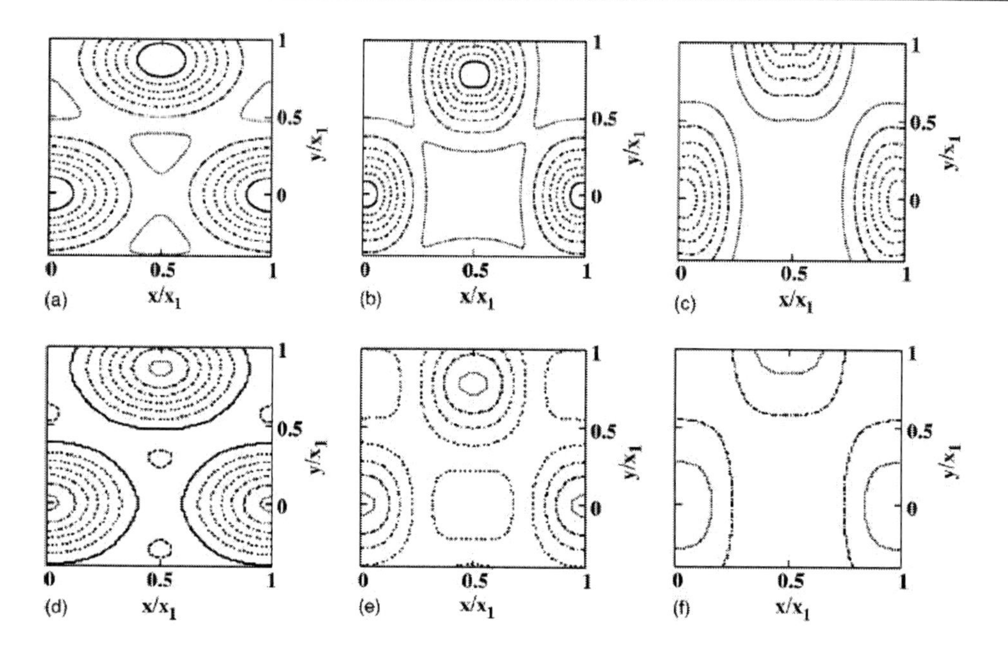

Figure 1. Contour plots showing vortex lattice of order parameter $\omega(x, y)$ for (a) $\gamma = 1$, (b) $\gamma = 5$, (c) $\gamma = 10$ and magnetic field induction $B(x, y)$ for (d) $\gamma = 1$, (e) $\gamma = 5$, (f) $\gamma = 10$, for $\bar{B}/B_{c2} = 0.4$.

3.1.2. Results and Discussions

In order to understand the effect of the in-plane mass anisotropy of the various properties of the high-T_c cuprates we begin with the determination of the stable vortex lattice configuration corresponding to the different values of the in-plane anisotropy parameter $\gamma = m_x/m_y$. This can be achieved by minimizing the free energy density w. r. t the lattice parameter y_2/x_1 (with $x_1 = 1.0$). The lattice has been found to be oblique with the exact shape being governed by the in-plane anisotropy parameter γ.

Single vortex and vortex lattice structure:- The dependence of the stable vortex lattice configuration on the in-plane anisotropy parameter γ can be understood from Figure 1 in which the order parameter and magnetic field distributions are plotted for different values of the in-plane anisotropy parameter γ [1]. The deviation of the vortex lattice structure from the triangular configuration gives a clear indication of the effect of in-plane anisotropy present in the system. Further, the single vortex shows deviation from the elliptic structure that has been observed for the isotropic case. Experimental studies based on small angle neutron scattering (SANS) [26] and scanning tunneling microscopy (STM) [29] have substantiated the presence of the elongated vortex cores and oblique vortex lattice structure in the high-T_c cuprates.

The theoretical model has further been utilized to study the local spatial behavior of the system viz. variation of widths of order parameter and magnetic field, which gives information regarding the experimentally observable quantities such as, the vortex core

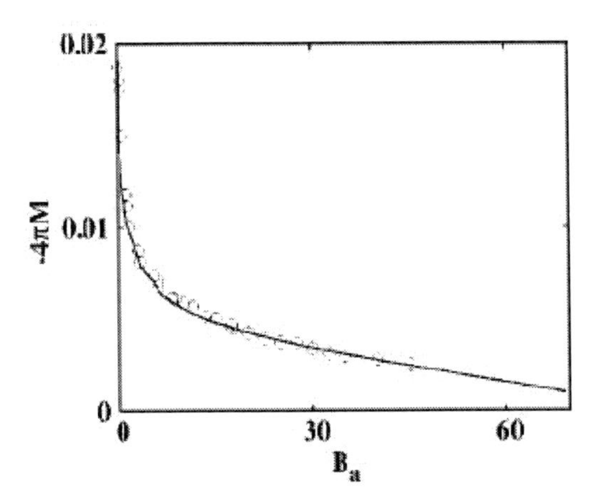

Figure 2. Reversible magnetization curve calculated for $\kappa_y = 70$ and mass anisotropy parameter $\gamma = 2$. The solid line represent numerically calculated results and the circle represents the experimental data for $YBa_2Cu_4O_8$ [31].

radius and magnetic penetration depth. It has been observed that an increase in the in-plane mass anisotropy decreases the width of both the order parameter and magnetic field profiles [1].

Reversible Magnetization:- For calculating the equilibrium applied magnetic field and thus the reversible magnetization of the system, instead of taking the numerical derivative of the free energy density functional, the virial theorem has been used [23], [30]. The corresponding equilibrium applied magnetic field B_a can be given as [1], [23], [30],

$$B_a = \frac{\langle \omega - \omega^2 + 2B^2 \rangle}{2\langle B \rangle}. \tag{10}$$

The reversible magnetization of the system can now be determined as per the relation $-4\pi M = B_a - \bar{B}$. The reversible magnetization thus calculated theoretically is compared with the experimental data corresponding to $YBa_2Cu_4O_8$ [31] and the results are plotted in Figure 2 [1].

In the present section we have discussed about the properties of the high-T_c superconductors studied in the framework of an anisotropic single order parameter GL model over the entire range of applied magnetic field. The study suggested the presence of an oblique flux line lattice, a characteristic feature of the high-T_c cuprates. The vortex core radius, magnetic penetration depth and the reversible magnetization showed profound effect of the presence of in-plane anisotropy in the system. Further, due to its applicability to arbitrary vortex lattice symmetry, the model has also been implemented to study the shear modulus of the vortex lattice (not presented here) and it is observed that a higher in-plane mass anisotropy corresponds to a harder vortex lattice and hence doesn't favor its melting.

4. Vortices in Superconducting thin Films of Finite Thickness

In the present section, we discuss about the extension of the numerical technique introduced in the previous sections for studying the properties of superconducting thin films. From the point of view of applications, the superconducting thin films make an important subject of study in the field of superconductivity. Due to the particular geometry of the thin films, with the thickness comparable to the the London penetration depth, special physical properties such as the role of surfaces on magnetic induction or defects created by various experimental techniques can be easily studied on thin films. The problem of an isolated vortex in thin films was first analytically studied by Pearl [32]. In case of thin films, the vortex-vortex interaction takes place principally through the magnetic stray fields outside the film. As a result of this, for films with thickness d smaller than the London penetration depth λ the range of vortex-vortex interaction is increased to the effective penetration depth $\Lambda = \lambda^2/d$. In the framework of the GL theory [33] and London theory [34]– [36], vortices in superconducting films of finite thickness ($d < \lambda$ and $d \geq \lambda$) and in the superconducting half space ($d >> \lambda$) have been studied. The properties of the superconducting thin films over the entire range of applied magnetic field have been studied by suitably extending Brandt's numerical iteration technique [23] to the three-dimensional case for conventional low temperature type-II superconductors [37]. In this work it has been observed that the numerical iteration technique correctly depicts the widening of the magnetic field lines as they exit the film surface. Moreover, other properties such as the variance of the order parameter and magnetic field, the reversible magnetization, the shear modulus of the vortex lattice for a superconducting thin film have also been studied.

4.1. c-axis anisotropy in high-T_c superconducting cuprates

In case of the high-T_c cuprates the large c-axis anisotropy is quantified by the parameter γ_z, where $\gamma_z^2 = m_z/m_{xy}$ and m_z and m_{xy} are the effective electronic masses along the z-direction and xy-plane respectively. For $YBa_2Cu_3O_{7-\delta}$ the c-axis anisotropy has been found to be $\gamma_z \approx 5 - 7$. Experimental studies on $YBCO$ and $BSCCO$ thin films have investigated the influence of anisotropy on transport properties such as, critical current and c-axis charge dynamics [38], [39]. From various experiments, estimate of $\gamma_z = (m_c/m_{ab})^{1/2}$ ranges from 5 to 200 [40]– [42]. Earlier theoretical work on superconducting films have considered anisotropic London model where the magnetic field is oriented parallel to the superconducting planes and normal to a crystal face of layered superconductor [43]. Also the structure of vortices within a stack of thin HTS layers is examined, where magnetic field and current distributions generated by 2D pancake vortices are calculated corresponding to Lawrence-Doniach model [44].

In the present section we will discuss about the effect of c-axis anisotropy on the various properties of the superconducting thin films corresponding to the high- T_c cuprate $YBCO$ in the framework of GL theory by suitably generalizing Brandt's numerical iteration technique developed for the isotropic superconducting thin films [37].

4.1.1. Theoretical formalism

Once again we begin with the GL free energy density functional of the system, this time in three-dimensional form. The resulting free energy density f_{tot} is the sum of the three-dimensional anisotropic free energy density and the stray magnetic field energy. In terms of the gauge invariant real quantities the free energy density can be expressed as [3],

$$f_{tot} = \langle -\omega + \frac{\omega^2}{2} + (\nabla \times \mathbf{Q})^2 + g + \omega \mathbf{Q} \Lambda \mathbf{Q} \rangle + \frac{f_{stray}}{d} \tag{11}$$

with

$$f_{stray} = 2 \int_{d/2}^{\infty} \langle B^2(\mathbf{r}) - \bar{B}^2 \rangle_{x,y} dz \tag{12}$$

where, $g = \frac{\nabla \omega \Lambda \nabla \omega}{4\kappa_{xy}^2 \omega}$ and $\mathbf{Q}(\mathbf{r}) = \mathbf{A}(\mathbf{r}) - \frac{\nabla \phi}{\kappa_{xy}}$ is the supervelocity which is related to the vector potential $\mathbf{A}$ and thus to the local magnetic field as, $\mathbf{B} = \hat{z}B = \nabla \times \mathbf{A}$, d is the film thickness. $\omega = \mid \psi \mid^2 \leq 1$ is expressed in terms of Cooper pair density or Ginzburg–Landau ψ-function as $\psi(\mathbf{r}) = \sqrt{\omega(\mathbf{r})} exp[i\phi(\mathbf{r})]$ and Λ is the three-dimensional anisotropic mass tensor given as,

$$\Lambda = \begin{pmatrix} 1 & 0 & 0 \\ 0 & 1 & 0 \\ 0 & 0 & m_{xy}/m_z \end{pmatrix} \tag{13}$$

as for uniaxial mass anisotropy $m_x = m_y = m_{xy} \neq m_z$. Here, ∇ denotes the 3D nabla operator. The reduced units are introduced as $\sqrt{\alpha/\beta}$, $\Phi_0/2\pi\xi_{xy}$ and λ_{xy} for the order parameter $\psi(\mathbf{r})$, magnetic vector potential $\mathbf{A}(\mathbf{r})$ and all lengths, respectively. The GL parameter and coherence length are defined as $\kappa_{xy} = \lambda_{xy}/\xi_{xy} = \Phi_0/(2\pi\sqrt{2}B_c\xi_{xy}^2)$ and $\xi_{xy} = \hbar/\sqrt{2m_{xy}\alpha}$ respectively, where $B_c^2/8\pi = \alpha/\beta$. The local magnetic field is expressed in units of $\sqrt{2}B_c$. The order parameter, magnetic field and the supervelocity are expressed in terms of Fourier series with the Fourier coefficients being $a_{\mathbf{K}}$ and $b_{\mathbf{K}}$ corresponding to the order parameter and magnetic field respectively. The Fourier coefficients are then determined by solving a set of iterative equations [3] using a numerical iteration technique [37].

4.1.2. Results and Discussions

The solutions of the anisotropic nonlinear GL equations are obtained for arbitrary values of the various parameters involved, such as mass anisotropy parameter γ_z, GL parameter κ_{xy}, magnetic field induction parameter $b = \bar{B}/B_{c2}$ and arbitrary film thickness d. Numerical results show that even small variation of the anisotropy parameter $\gamma_z = (m_z/m_{xy})^{1/2}$ have significant effect on several properties of the ideal vortex lattice in superconducting films of arbitrary thickness, such as the order parameter, magnetic induction profile and shear modulus.

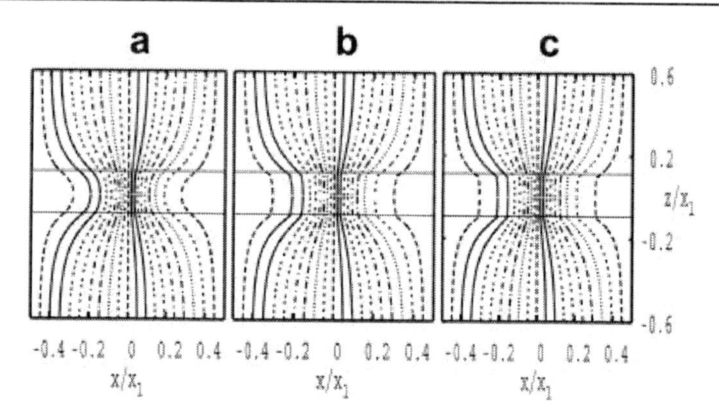

Figure 3. Magnetic field lines in one unit cell of vortex lattice plotted for a film of thickness $d/x_1 = 0.2$ and different $\gamma_z^2 = 1, 5, 10$ at $\kappa_{xy} = 1.4$.

Variance of magnetic field:- The magnetic field lines are shown in Figure 3 for film thickness $d/x_1 = 0.2$, magnetic induction parameter $b = \bar{B}/B_{c2} = 0.04$ and $\kappa_{xy} = 1.4$. Figure 3a shows the field lines for $\gamma_z^2 = 1$, Figure 3b for $\gamma_z^2 = 5$ and Figure 3c for $\gamma_z^2 = 10$, respectively [3]. The effect of γ_z^2 on magnetic field is different at the center and on the surface of the film. In the center of the film ($z = 0$), B_z decreases with increasing anisotropy. One of the known features of bulk HTS is that mass anisotropy reduces the field component in superconducting planes. The decrease in magnetic field inside the film as obtained from our calculations for uniaxial anisotropic HTS thin film supports this feature. This result is observed for lower values of magnetic induction $b < 0.1$. On the other hand, the field component on the surface increases with anisotropy. The corresponding increase or decrease of magnetic field is observed clearly in the center of vortices. These properties are confirmed by calculating the relative variance σ_z and $\sigma_\perp$ of magnetic induction defined as [37],

$$\sigma_z(z) = \left\langle [B_z(x,y,z) - \bar{B}]^2 \right\rangle_{x,y}^{1/2} / \bar{B}, \quad (14)$$

$$\sigma_\perp(z) = \left\langle B_x(x,y,z)^2 + B_y(x,y,z)^2 \right\rangle_{x,y}^{1/2} / \bar{B}. \quad (15)$$

The variance of the longitudinal and transverse components of the magnetic induction is plotted versus z. Figure 4 shows the dependence of transverse component $\sigma_\perp(z)$ on mass anisotropy parameter γ_z^2 inside and outside the film [3]. For $\gamma_z^2 = 1$ (isotropic case), it increases as one approaches the surface from inside as well as from outside the film. It attains the maximum value at the surface $z = d/2$. Whereas for anisotropic case, i. e. $\gamma_z^2 > 1$, $\sigma_\perp(z)$ shows a sharp drop at the surface inside the film. Then, it decreases gradually from surface to deep inside the film. Outside the superconducting film, it has a higher value for anisotropic case as compared to the isotropic value. The sharp fall of anisotropic $\sigma_\perp(z)$ at the surface inside the film may indicate the experimentally observed surface barrier to the flux penetration when the field is applied in a transverse direction to the superconducting planes [45]. The sharp drop in magnetic induction at the surface of the sample interpreted as Bean-Livingston barrier is experimentally observed in magneto-optical measurements of

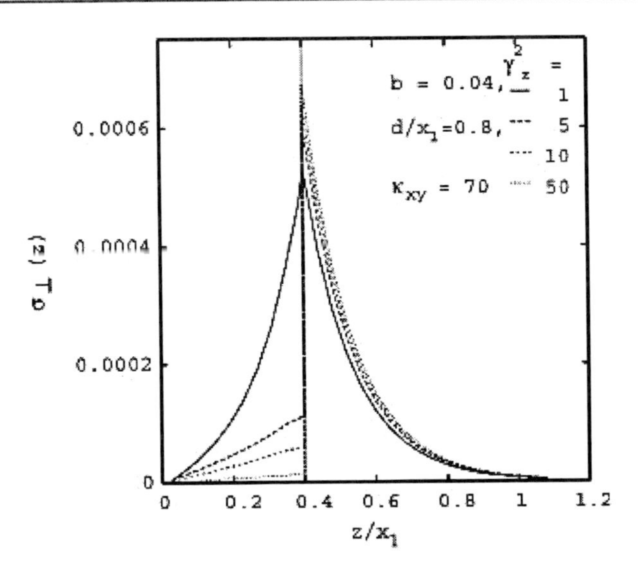

Figure 4. The transverse component of magnetic field variance $\sigma_{\perp}(z)$ plotted versus z/x_1 at reduced induction $b = 0.04$. Here $\kappa_{xy} = 70$ and $d/x_1 = 0.8$ and mass anisotropic parameter values are $\gamma_z^2 = 1, 5, 10, 50$. The vertical line at $z/x_1 = 0.4$ indicates the surface of the film.

zero field cooled $YBa_2Cu_3O_7$ single crystal [46]. Bean-Livingston barrier arises from the competition between the repulsion of vortex from the surface due to its interaction with the exponentially decreasing external field or shielding currents. The anisotropic $\sigma_{\perp}$ plots show similar behavior of magnetic induction at the surface of the film. The drop in $\sigma_{\perp}$ value at the surface inside the film is found to be $> 30\%$. This drop in the induction value at the surface depends on the anisotropy parameter γ_z^2. The barrier increases with increase in anisotropic parameter. This is purely an effect of uniaxial anisotropy, as this behavior is not observed in isotropic $\sigma_{\perp}$. Recently, similar studies on thin films by Vodolazov *et al.* have predicted the suppression of magnetic field inside the film at the surface by thickness dependent vector potential within London model [47].

The longitudinal component $\sigma_z(z)$ also decreases inside the superconductor with increase in γ_z^2. In the case of $\gamma_z^2 = 5$, the $\sigma_z(z)$ deep inside the film reduces $\approx 25\%$ than that of its isotropic value. But all the $\sigma_z(z)$ anisotropic curves inside the film cross at a point below the surface $z = d/2$. This is shown in Figure 5, where $\sigma_z(z)$ is plotted versus film thickness d along z direction and for parameter values $\kappa_{xy} = 70$, film thickness $d/x_1 = 0.8$ at reduced induction $b = 0.04$ [3]. All the anisotropic curves ($\gamma_z^2 > 1$) inside the superconductors cross at $z/x_1 = 0.3$. Above this point and outside the film, anisotropic σ_z has higher values than the isotropic σ_z. Similar to the isotropic case, $\sigma_z(z)$ and $\sigma_{\perp}(z)$ values coincide outside the film even in anisotropic case.

Current density:- To examine the dependence of current density on film thickness as well as on mass anisotropy, the current density is calculated for the entire range of magnetic induction parameter $0 < b < 1$. In most of the experiments, $100 - 1000$ nm thickness

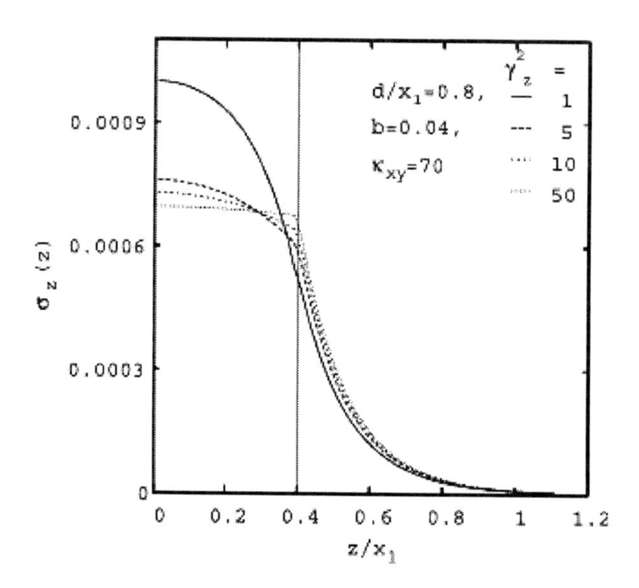

Figure 5. The variance of the longitudinal component of magnetic field $\sigma_z(z)$ plotted against z/x_1 at reduced induction $b/0.04$, $d/x_1 = 0.8$, $\kappa_{xy} = 70$ and $\gamma_z^2 = 1, 5, 10, 50$. The vertical line at $z/x_1 = 0.4$ indicates the surface of the film.

films are studied. For example, in ref. [48], the experiments on $YBa_2Cu_3O_{7-\delta}$ have used $100 - 1000$ nm thickness films. Therefore, current density is calculated within this range of film thickness. The film thickness in nanometers is introduced in numerical calculations through d/ξ_{xy} where experimental measured value for ξ_{xy} is reported to be $1.06 - 1.09$ nm [40]. Our calculations confirm the thickness dependence of **j** as depicted in Figure 6 [3]. In this figure, current density is plotted for $T = 51K, \gamma_z = 1, 10, 50$ and magnetic induction $b = 0.01$. The current density profile is consistent with the profile obtained experimentally in ref. [48] at $51K$. But decrease in j above 500 nm thick films observed experimentally for other temperatures is not obtained here. The anisotropic effect is clearly seen in small film thickness where it suppresses the j value. However, as pinning forces are neglected in this study and the discreetness of the layers of HTS films is not taken into account by anisotropic GL theory, the results may not give exact quantitative match with the experiments.

In this section discussion regarding some of the effects of c-axis anisotropy on the various properties of high temperature superconducting thin films is presented. The solutions of the coupled nonlinear GL equations corresponding to the superconducting thin films involving uniaxial anisotropy have been obtained by using a numerical iteration technique. The uniaxial anisotropy possesses significant effect on the magnetic field distribution and it has been found that that the anisotropy suppresses the magnetic field component inside the film. The variance of the magnetic field and order parameter are calculated and so is the effect of uniaxial anisotropy on these quantities. The anisotropic current density and the shear modulus of the vortex lattice is also determined numerically.

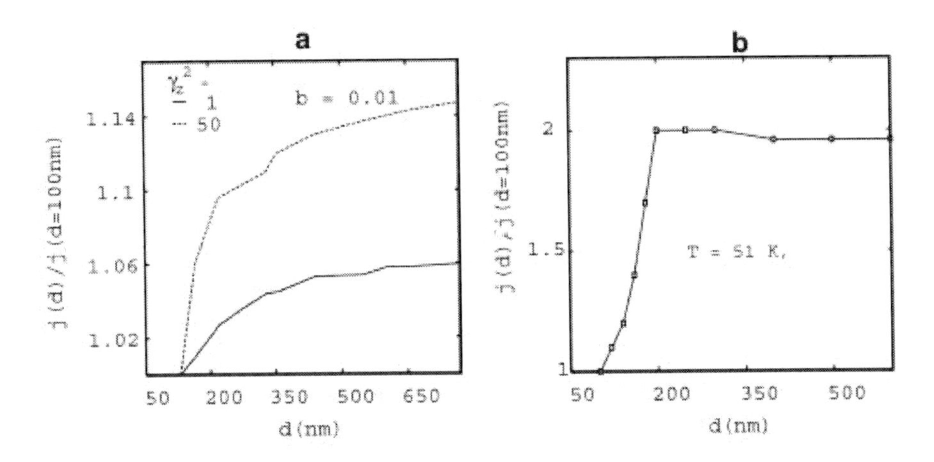

Figure 6. The current density j plotted w. r. t film thickness d. (a) Numerically obtained j versus film thickness for parameter values of $YBCO$. (b) Experimentally obtained current density plot for strongly pinned $YBa_2Cu_3O_{7-\delta}$ films of various thickness [48].

5. Mixed Pairing State Symmetry of Order Parameters: Isotropic Two-Order Parameter Ginzburg–Landau Theory

The unconventional properties of the high-T_c superconductors such as the high critical temperature, critical current etc have made these materials important for practical applications. At the same time these unconventional properties raise several questions regarding the underlying mechanism of the superconducting property and consequently the pairing state symmetry of these materials. There are several possible symmetries for the pairing state of the high-T_c superconductors [49]. However, temperature dependence of the NMR relaxation rates did provide strong evidence in support of the spin-singlet pairing state symmetry of the high-T_c superconductors, thereby restricting the pairing state to d-wave or s-wave symmetry [50]. A principal question was whether the pairing state symmetry for the high-T_c cuprates is isotropic s-wave or the d-wave i. e $d_{x^2-y^2}$ symmetry with lines of node in the energy gap. A host of experimental studies have been carried out to determine the pairing state symmetry of the high-T_c superconducting cuprates [51]. These experimental studies can broadly be classified in two categories as the (i) Non-phase-sensitive and (ii) Phase-sensitive experiments. The non-phase-sensitive experiments such as, the low temperature penetration depth measurements, thermal conductivity studies, angle resolved photo emission spectroscopy (ARPES), specific heat measurements, NMR relaxation studies, Raman scattering studies etc provided evidence in support of the presence of nodes in the energy gap of the high-T_c cuprates, which is a characteristic feature of unconventional superconductivity. These studies however, failed to identify a second characteristic feature of the unconventional superconductors viz. a sign change of the order parameter along the different k-directions. Such sign changes in the order parameter can be detected by using phase sensitive experiments such as the dc SQUID and single junction experiment, tricrystal junction experiments, c-axis tunneling experiment etc [4], [51].

The phase sensitive experiments supported the presence of a d-wave pairing state symmetry of the order parameter. However, none of these experiments could rule out the possibility of a mixed pairing state symmetry of the order parameters. The experimental and theoretical efforts devoted towards understanding the pairing state symmetry of the HTS materials led to a consensus that the high-T_c superconductor possesses an unconventional mixed symmetry state of pairing [4]. In such a mixed symmetry state scenario the dominant order parameter component is d-wave, along with which there is an admixture of a sub-dominant s-wave order parameter component. The mixed symmetry state scenario of the order parameter components is attributed to the orthorhombic distortion of the CuO_2-planes present in these materials. Such a scenario of the order parameters has been found to be a consistent answer to several of the experimentally observed unusual features of the high-T_c superconductors such as, the upward curvature in the thermodynamic $H_{c2}(T)$ vs T plot [52], the pseudogap effect in HTS [53], non-magnetic impurity effect in HTS leading to an energy gap [54], superconducting fluctuation effects in HTS [55] etc.

The anomalous magnetic field dependence of the single vortex and vortex lattice structure of the high-T_c cuprates as has been observed in the experimental studies has proved to be a point of key interest for the experimentalists and theoreticians alike. In particular the oblique structure of the flux line lattice and the four-fold structure of the single vortices indicate towards the significant deviation of the pairing state symmetry of these materials from that of the conventional low temperature superconductors. It has been discussed in the previous sections that the oblique structure of the flux line lattice can be explained by taking into account the in-plane anisotropy present in these materials. Such an anisotropic single order parameter model however could not justify the experimentally observed four-fold structure of the vortices.

Theoretical studies concerning the anomalous structure of the single vortex and vortex lattice of the high temperature superconducting cuprates have been many. By studying the symmetry properties of the vortex core of d-wave superconductors Volovik concluded the existence of mixed d-and s-wave pairing in high-T_c superconductors [56]. A relevant macroscopic approach to study the properties of the high-T_c superconducting cuprates is thus the two-order parameter Ginzburg–Landau (GL) theory consisting of two or more order parameters (such as s and d) in the GL free energy density functional, with derivative mixing terms reflecting the ionic lattice symmetry [57]. In the framework of two-order parameter GL theory the GL equations are solved numerically so as to understand the single vortex and vortex lattice structure of the high-T_c cuprates [58]– [64].

These theoretical studies though could explain the single vortex and vortex lattice structures of the high-T_c superconducting cuprates, suffer from two major limitations. Firstly, these approximate studies are limited to the regions near the upper ($H \approx H_{c2}$) and lower ($H \to H_{c1}$) critical magnetic fields corresponding to the vortex lattice and single vortex regime respectively [60]– [61], thereby leaving the experimentally relevant intermediate applied magnetic field region $H_{c1} < H < H_{c2}$ unattended [65]. Moreover, in these theoretical studies one of the order parameter components is expressed in terms of the other, thereby reducing the problem to an effective single order parameter model. In order to overcome these limitations a suitable theoretical model is required which can explain the properties of the high-T_c cuprates involving mixed symmetry state of the order parameters over the experimentally relevant applied magnetic field regime $H_{c1} < H < H_{c2}$, for arbitrary values of vortex lattice symmetry and GL parameter κ. In the present section we will discuss about the development and application of such a theoretical model.

5.1. Theoretical Formalism for the isotropic two-order parameter GL theory

To this end we begin with the free energy density of the system. In the framework of the two-order parameter GL theory the free energy density corresponding to the high-T_c superconductor involving mixed symmetry state of the order parameters can be depicted as [2], [4], [60],

$$\begin{aligned} f = \Big\langle\ & \alpha_s \mid s \mid^2 + \alpha_d \mid d \mid^2 + \beta_1 \mid s \mid^4 + \beta_2 \mid d \mid^4 + \beta_3 \mid s \mid^2 \mid d \mid^2 \\ & + \beta_4 (s^{*2} d^2 + d^{*2} s^2) + \gamma_s \mid \mathbf{\Pi} s \mid^2 + \gamma_d \mid \mathbf{\Pi} d \mid^2 \\ & + \gamma_v \Big[(\Pi_y s)^* (\Pi_y d) - (\Pi_x s)^* (\Pi_x d) + c.c \Big] + B^2/8\pi \Big\rangle \qquad (16) \end{aligned}$$

where, $s(\mathbf{r})$, $d(\mathbf{r})$ are the order parameter components and the magnetic field applied applied along z-axis is defined as $\mathbf{B} = B\hat{z} = (\nabla \times \mathbf{A})$ ($\mathbf{A}$ is the vector potential). The temperature dependent parameter $\alpha' s$ are defined as $\alpha_s = \alpha(T - T_s)$ and $\alpha_d = \alpha(T - T_d)$ with $(T_s < T_d)$ while $\langle \cdots \rangle = \frac{1}{V} \int d\mathbf{r} \cdots$ denotes the two-dimensional spatial average. The other parameters in the free energy density functional i. e β_i $(i = 1, 2, 3, 4)$, γ_j $(j = s, d, v)$ are positive quantities and the parameters γ_j are related to the effective electronic masses as $\gamma_j = \hbar^2/2m_j^*$ with $(j = s, d, v)$.

In presence of tetragonal symmetry (as considered in the present study) the major contribution of the s-wave order parameter component arises from the mixed gradient coupling term. The strength of the admixture of the s-wave order parameter component in the system is determined by the coupling parameter γ_v corresponding to the mixed gradient coupling term. Linear stability analysis have suggested that in the absence of the mixed gradient coupling term the bulk d-wave state is stable against the admixture of the s-wave order parameter component through the β_3 and β_4-terms [61]. The parameter β_4 is however the decisive factor for the relative phase between the s-and d-wave order parameter components. For $\beta_4 > 0$, $(d \pm is)$-state is stable while for $\beta_4 < 0$, $(d \pm s)$-state is favored. In the present study a relative phase of $\pi/2$ is considered between the order parameter components corresponding to the choice of the parameter $\beta_4 > 0$. The various parameters used in the formalism are interrelated by inequalities [4], [60], [61].

Next, the free energy density functional (Eqn. (16)) is expressed in terms of the gauge invariant real quantities $\omega_s(x,y)$, $\omega_d(x,y)$ and $\mathbf{Q}(x,y)$ as [2],

$$\begin{aligned} f = \Big\langle \Big[& \alpha_s\omega_s - \omega_d + \beta_1\omega_s^2 + \beta_2\omega_d^2 + \big(\beta_3 + 2\beta_4\cos(2\phi)\big)\omega_s\omega_d \\ & + (\omega_s+\omega_d)Q^2 + (\nabla\omega_s)^2/4\omega_s\kappa^2 + (\nabla\omega_d)^2/4\omega_d\kappa^2 \\ & + 2\epsilon_v\Big\{ \cos(\phi)\Big[\Big((\nabla_y\omega_s)(\nabla_y\omega_d) - (\nabla_x\omega_s)(\nabla_x\omega_d)\Big)/(4\kappa^2(\omega_s\omega_d)^{1/2}) \\ & \qquad + (Q_y^2 - Q_x^2)(\omega_s\omega_d)^{1/2}\Big] \\ & + \sin(\phi)\Big[\big(Q_y(\nabla_y\omega_s) - Q_x(\nabla_x\omega_s)\big)(\omega_d/4\kappa^2\omega_s)^{1/2} \\ & - \big(Q_y(\nabla_y\omega_d) - Q_x(\nabla_x\omega_d)\big)(\omega_s/4\kappa^2\omega_d)^{1/2}\Big]\Big\} + (\nabla\times\mathbf{Q})^2\Big]\Big\rangle \end{aligned} \quad (17)$$

where, the order parameters are expressed as, $s(x,y) = \sqrt{\omega_s(x,y)}e^{i\phi_s(x,y)}$ and $d(x,y) = \sqrt{\omega_d(x,y)}e^{i\phi_d(x,y)}$ with $\omega_s = \mid s\mid^2 \leq 1$ and $\omega_d = \mid d\mid^2 \leq 1$. The vector potential $\mathbf{A}(x,y)$ is defined in terms of the supervelocity $\mathbf{Q}(x,y)$ as, $\mathbf{Q}(x,y) = \mathbf{A}(x,y) - \nabla\phi(x,y)/\kappa$ with κ being the GL parameter. $\phi = \phi_d - \phi_s$ is the relative phase between the d-and s-wave order parameter components. It must be noted that in the above equation (Eqn. (17)) the parameters (α's and β's) are presented in terms of reduced units as, $\alpha_s = \alpha_s/\mid\alpha_d\mid$, $\beta_i = \beta_i/2\beta_2$ and $\epsilon_v = \gamma_v/\gamma_d = m_d^*/m_v^*$. The effective masses of the s-and d-wave type electrons are taken to be equal (i. e $m_s^* = m_d^*$) for the sake of simplicity, though in general they depend on the Fermi surface architecture. The order parameters are expressed in terms of $d_0 = \sqrt{\mid\alpha_d\mid/2\beta_2}$ while the free energy density is in terms of $\mid\alpha_d\mid^2/2\beta_2$. The magnetic field is measured in units of $\sqrt{2}B_c$. The parameter ϵ_v gives the strength of the admixture of s-wave order parameter in the system and for $\epsilon_v = 0$ we return to the case of pure d-wave solution. On minimizing this GL free energy density functional (Eqn. (17)) using the variational technique $\partial f/\partial\omega_i = 0$ ($i = s, d$) and $\partial f/\partial\mathbf{Q} = 0$ the corresponding GL equations are obtained [4].

The fully coupled nonlinear GL equations corresponding to the two-order parameter GL theory are solved over the entire range of applied magnetic field and wide range of temperature for arbitrary GL parameter κ and vortex lattice symmetry by using the high precision numerical iteration technique, details regarding which can be seen from ref. [2], [4]. For the numerical iteration technique the five iterative equations for the isotropic two-order parameter GL theory in the mixed symmetry states of the order parameters can be given as [2],

$$\begin{aligned} a_{\mathbf{K}}^s := \frac{2}{\left(\frac{K^2}{2\kappa^2} + c_{1s}\right)} \Big\langle \Big[& (\alpha_s - c_{1s})\omega_s + 2\beta_1\omega_s^2 + \beta_3\omega_s\omega_d + 2\beta_4\cos(2\phi)\omega_s\omega_d + g_s \\ & + \omega_s Q^2 + \epsilon_v\Big\{\cos(\phi)\Big(-\frac{\nabla_y^2\omega_d}{2\kappa^2}(\omega_s/\omega_d)^{1/2} + \frac{\nabla_x^2\omega_d}{2\kappa^2}(\omega_s/\omega_d)^{1/2} \\ & + g_{dy}(\omega_s/\omega_d)^{1/2} - g_{dx}(\omega_s/\omega_d)^{1/2} + Q_y^2(\omega_d\omega_s)^{1/2} - Q_x^2(\omega_d\omega_s)^{1/2}\Big) \\ & + 2\sin(\phi)\Big(-\frac{(\nabla_y\omega_d)Q_y}{2\kappa}(\omega_s/\omega_d)^{1/2} + \frac{(\nabla_x\omega_d)Q_x}{2\kappa}(\omega_s/\omega_d)^{1/2}\Big)\Big\}\Big] \cos\mathbf{K}.\mathbf{r}\Big\rangle \end{aligned} \quad (18)$$

where, $g_s = \frac{(\nabla\omega_s)^2}{4\kappa^2\omega_s}$, $g_{dy} = \frac{(\nabla_y\omega_d)^2}{4\kappa^2\omega_d}$, $g_{dx} = \frac{(\nabla_x\omega_d)^2}{4\kappa^2\omega_d}$.

$$a^d_{\mathbf{K}} := \frac{2}{\left(\frac{K^2}{2\kappa^2} + c_{1d}\right)}\Big\langle\Big[-(1+c_{1d})\omega_d + 2\beta_2\omega_d^2 + \beta_3\omega_s\omega_d + 2\beta_4\cos(2\phi)\omega_s\omega_d + g_d + \omega_d Q^2 + \epsilon_v\Big\{\cos(\phi)\Big(-\frac{\nabla_y^2\omega_s}{2\kappa^2}(\omega_d/\omega_s)^{1/2} + \frac{\nabla_x^2\omega_s}{2\kappa^2}(\omega_d/\omega_s)^{1/2} + g_{sy}(\omega_d/\omega_s)^{1/2} - g_{sx}(\omega_d/\omega_s)^{1/2} + Q_y^2(\omega_d\omega_s)^{1/2} - Q_x^2(\omega_d\omega_s)^{1/2}\Big) + 2\sin(\phi)\Big(\frac{(\nabla_y\omega_s)Q_y}{2\kappa}(\omega_d/\omega_s)^{1/2} - \frac{(\nabla_x\omega_s)Q_x}{2\kappa}(\omega_d/\omega_s)^{1/2}\Big)\Big\}\Big]\cos\mathbf{K}.\mathbf{r}\Big\rangle, \quad (19)$$

where, $g_d = \frac{(\nabla\omega_d)^2}{4\kappa^2\omega_d}$, $g_{sy} = \frac{(\nabla_y\omega_s)^2}{4\kappa^2\omega_s}$, $g_{sx} = \frac{(\nabla_x\omega_s)^2}{4\kappa^2\omega_s}$.

$$a^s_{\mathbf{K}} := a^s_{\mathbf{K}}\cdot\Big\langle\Big[(c_{2s}-\alpha_s)\omega_s - \beta_3\omega_s\omega_d - 2\beta_4\cos(2\phi)\omega_d\omega_s - g_s - \omega_s Q^2 - \epsilon_v\Big\{\cos(\phi)\Big(-\frac{\nabla_y^2\omega_d}{2\kappa^2}(\omega_s/\omega_d)^{1/2} + \frac{\nabla_x^2\omega_d}{2\kappa^2}(\omega_s/\omega_d)^{1/2} + Q_y^2(\omega_d\omega_s)^{1/2} - Q_x^2(\omega_d\omega_s)^{1/2} + g_{dy}(\omega_s/\omega_d)^{1/2} - g_{dx}(\omega_s/\omega_d)^{1/2}\Big) + 2\sin(\phi)\Big(-\frac{(\nabla_y\omega_d)Q_y}{2\kappa}(\omega_s/\omega_d)^{1/2} + \frac{(\nabla_x\omega_d)Q_x}{2\kappa}(\omega_s/\omega_d)^{1/2}\Big)\Big\}\Big]\Big\rangle\Big/(2\beta_1\langle\omega_s^2\rangle + c_{2s}\langle\omega_s\rangle), \quad (20)$$

$$a^d_{\mathbf{K}} := a^d_{\mathbf{K}}\cdot\Big\langle\Big[(1+c_{2d})\omega_d - \beta_3\omega_s\omega_d - 2\beta_4\cos(2\phi)\omega_d\omega_s - g_d - \omega_d Q^2 - \epsilon_v\Big\{\cos(\phi)\Big(\frac{-\nabla_y^2\omega_s}{2\kappa^2}(\omega_d/\omega_s)^{1/2} + \frac{\nabla_x^2\omega_s}{2\kappa^2}(\omega_d/\omega_s)^{1/2} + Q_y^2(\omega_d\omega_s)^{1/2} - Q_x^2(\omega_d\omega_s)^{1/2} + g_{sy}(\omega_d/\omega_s)^{1/2} - g_{sx}(\omega_d/\omega_s)^{1/2}\Big) + 2\sin(\phi)\Big(\frac{(\nabla_y\omega_s)Q_y}{2\kappa}(\omega_d/\omega_s)^{1/2} - \frac{(\nabla_x\omega_s)Q_x}{2\kappa}(\omega_d/\omega_s)^{1/2}\Big)\Big\}\Big]\Big\rangle\Big/(2\beta_2\langle\omega_d^2\rangle + c_{2d}\langle\omega_d\rangle), \quad (21)$$

$$b_{\mathbf{K}} := \frac{2}{K^2 + (\bar{\omega}_s + \bar{\omega}_d)}\Big\langle\Big[-\omega_s B - P_s - \omega_d B - P_d + (\bar{\omega}_s + \bar{\omega}_d)(B - \bar{B}) - \epsilon_v(2\cos(\phi)(\omega_s\omega_d)^{1/2}(\nabla_y Q_x + \nabla_x Q_y) + (\omega_s\omega_d)^{1/2}\{\cos(\phi)(Q_x\nabla_y + Q_y\nabla_x)\omega_d + \sin(\phi)/2\kappa[-(\nabla_x\nabla_y + \nabla_y\nabla_x)\omega_d + (\nabla_x\omega_d\nabla_y\omega_d)/\omega_d]\} + (\omega_d/\omega_s)^{1/2}\{\cos(\phi)(Q_x\nabla_y + Q_y\nabla_x)\omega_s + \sin(\phi)/2\kappa[(\nabla_x\nabla_y + \nabla_y\nabla_x)\omega_s - (\nabla_x\omega_s\nabla_y\omega_s)/\omega_s]\})\Big]\cos\mathbf{K}.\mathbf{r}\Big\rangle \quad (22)$$

where, $P_i = (\nabla\omega_i \times \mathbf{Q}).\hat{z}$ and $g_{ij} = (\nabla_j\omega_i)^2/4\kappa_y^2\omega_i$ (with $i = s, d$ and $j = x, y$). The convergence and stability of the iteration method can be enhanced by the choice of the constants c's. While the value of the constant c_{1s} is determined by an empirical relation $c_{1s} \approx \frac{8\times10^3}{(6/b)+50(1-b/4)}$, the constant c_{2s} and c_{2d} are small positive quantities, while c_{1d} is chosen to be 1. The order parameters and magnetic field thus determined numerically can now be used to study the several properties associated with the high- T_c superconducting systems.

5.2. Results and Discussions

In this section we will discuss about the various properties of the high- T_c superconducting cuprates which has been computed by using the isotropic two-order parameter GL theory. In particular, the effect of the admixture of the s-wave order parameter component in the system has been studied through the effect of the mixed gradient coupling parameter ϵ_v on the various properties.

5.2.1. Single vortex and vortex lattice structure

An important aspect of the high temperature superconductors is its unusual single vortex and vortex lattice structure. The flux line lattice configuration is significantly different from that of the conventional type-II superconductors and indicate towards the presence of unconventional pairing state symmetry in these materials. A stable vortex lattice configuration corresponds to the minimum of the free energy density functional $f(x_2 = x_1/2, y_2)$. Alternatively, one can use the minimum of the Abrikosov's parameter defined as $\beta_A = \langle\omega_s^2\rangle/\langle\omega_s\rangle^2$ to determine the stable configuration of the flux line lattice (FLL). In the present work this second approach has been used. The Abrikosov's parameter β_A is varied as a function of the lattice parameter y_2/x_1 (with $x_1 = 1.0$) and the value of the parameter y_2 which corresponds to the minimum of β_A gives the stable configuration of the vortex lattice. For various admixtures of s-wave order parameter component the structure of the FLL is found to be oblique with the exact shape being determined by the choice of the mixed gradient coupling parameter ϵ_v. The variation of the structure of the vortex lattice with the mixed gradient coupling parameter ϵ_v can be clearly observed from Figure 7 [4]. In this figure the contour plots of the s-and d-wave order parameter components have been plotted for different values of the mixed gradient coupling parameter ϵ_v mentioned in the figure caption. A closer observation of the vortices reveal that the s-wave order parameter component possesses a four-fold symmetric structure for all values of the mixed gradient coupling parameter ϵ_v, as has been interpreted experimentally [66] and observed in previous theoretical works [60], [61]. The nodal lines deviate from $k_x = \pm k_y$ expected from the pure d-wave case and the exact position of the nodes depend on the amount of admixture of the s-wave order parameter component in the system [51]. The d-wave order parameter component doesn't however manifest any special structure. The magnetic field distribution in the system shows similar behavior with the exact shape of the lattice being dependent on the amount of admixture of s-wave order parameter component in the system.

An important quantity, information regarding which can be obtained from the structure of the FLL is the angle (β) between the primitive axes of the vortex lattice. It gives an

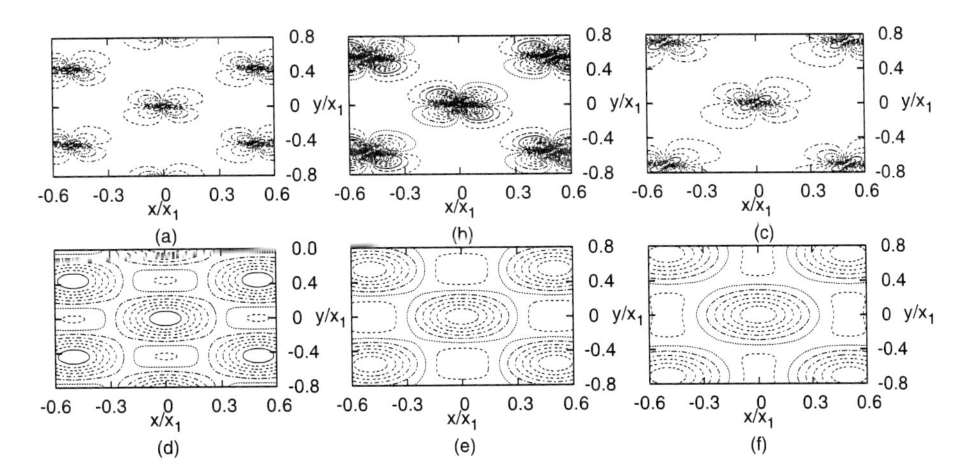

Figure 7. Contour plots of order parameter components. The parameter values are $\alpha_s / \mid \alpha_d \mid = 0.5, \beta_1/2\beta_2 = \beta_3/2\beta_2 = 1.0, \beta_4/2\beta_2 = 0.5, \phi = \phi_d - \phi_s = \pi/2, \kappa = 72$, magnetic induction parameter $b = 0.7$. Figures (a), (b), (c) shows s-wave order parameter component $\omega_s(x, y)$ for $\epsilon_v = 0.1, 0.3, 0.4$ respectively and Figures (d), (e) and (f) correspond to the plots for the d-wave order parameter component $\omega_d(x, y)$ for $\epsilon_v = 0.1, 0.3, 0.4$ respectively.

estimation of the amount of admixture of the s-wave order parameter component in the experimental samples. Small angle neutron scattering (SANS) experiments [26] and scanning tunneling microscopy (STM) measurements [29] have revealed that in case of the high-T_c superconducting cuprate $YBa_2Cu_3O_{7-\delta}$ this angle between the primitive axes of the vortex lattice amounts to $\beta \approx 77 \pm 5^o$. Thus, in order to compare with the experimental data the angle between the primitive axes of the vortex lattice is theoretically computed for different values of the mixed gradient coupling parameter ϵ_v.

The variation of the mixed gradient coupling parameter results in the variation in the structure of the flux line lattice and thus gives rise to variation in the angle β. It has been observed that the best agreement of the various theoretical results with the relevant experimental data is achieved for the mixed gradient coupling parameter of $\epsilon_v = 0.1$. For $\epsilon_v = 0.1$ the angle between the primitive axes of the vortex lattice is found to be $\beta \approx 97.3^o$ which is far from the experimentally observed value.

5.2.2. Local spatial behavior of the order parameters and magnetic field

Recent experimental studies have suggested that the presence of the admixture of more than one order parameter component in the system is manifested in the local spatial behavior of the order parameters and magnetic field. For instance, the point of inflection in the low temperature measurements of the penetration depth of the system as has been observed in μSR experiments carried out on high-T_c cuprates has been attributed to the presence of multiple order parameter components in the system, with one of the gaps being substantially smaller than the leading one [67], [68]. The local spatial behavior of the system such as the widths of the order parameter and magnetic field profiles are related to the experi-

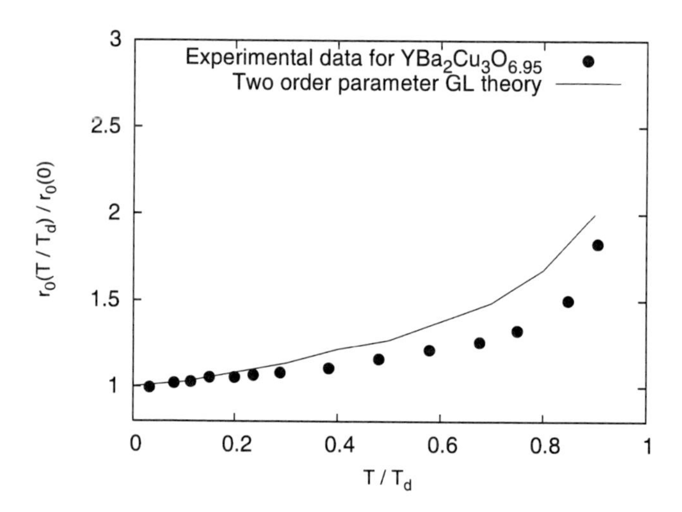

Figure 8. Temperature dependence of the vortex core radius. The solid line gives the theoretical result while the dot represents the experimental data for $YBa_2Cu_3O_{6.95}$ [71]. The parameter values used for the theoretical calculation are same as in Figure 7 with $b = 0.004$ and $\epsilon_v = 0.8$.

mentally observable vortex core radius and magnetic penetration depth of the material. The vortex core radius of the high-T_c superconducting cuprates are strongly dependent on the applied magnetic field with a shrinkage in the vortex core radius at higher magnetic field induction arising due to the vortex lattice effect [69]. The qualitative agreement between the experimental results corresponding to $YBa_2Cu_3O_{7-\delta}$ [70] and the results obtained by the present theoretical model is found to be good [4].

The temperature dependence of the vortex core radius suggests that the vortex core size shrinks with the decreasing temperature. In Figure 8 a comparison of the vortex core radius calculated by the isotropic two-order parameter GL theory with the experimental data corresponding to the high-T_c cuprate $YBa_2Cu_3O_{7-\delta}$ [71] is presented [4]. For the theoretical results the average of the vortex core radius corresponding to the s-and d-wave order parameter components has been plotted. It is worth mentioning here that since the present work is based on the Ginzburg–Landau theory, the very low temperature behavior of the system cannot be explored in this framework. An ideal approach would be to solve the BdG equations corresponding to the superconductors involving unconventional pairing state symmetry. Another relevant approach would be to extend the present phenomenological model to the low temperature regime by using Bardeen's extension of GL theory [72]. A similar approach has been used in case of conventional low temperature type-II superconductors involving an isotropic s-wave pairing state symmetry [73].

For different values of the mixed gradient coupling parameter ϵ_v the magnetic field dependence of the penetration depth has been found to remain qualitatively the same. A higher admixture of the s-wave order parameter component has been observed to correspond to a higher magnetic penetration depth [4]. Also at higher magnetic field induction a nonlinear behavior of the penetration depth is observed for higher values of the mixed gradient

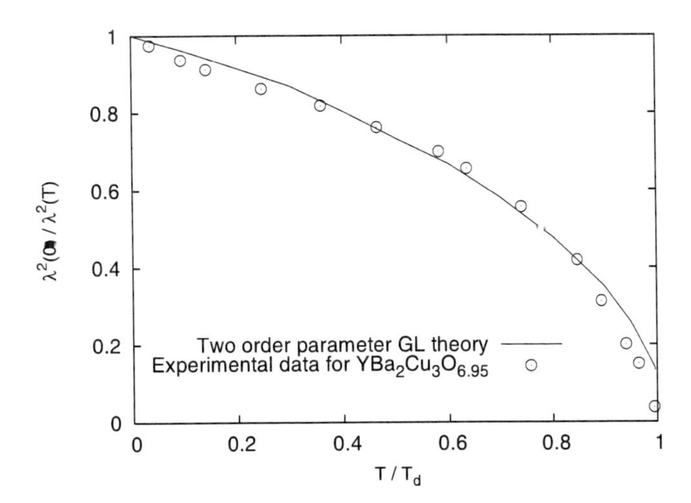

Figure 9. Variation of the ratio $\lambda^2(0)/\lambda^2(T)$ with temperature T/T_d. The solid line gives the theoretical result while the open circle corresponds to the experimental data for $YBa_2Cu_3O_{6.95}$ [76]. The parameter values used for the theoretical calculations are same as in Figure 7 with $b = 0.004$ and $\epsilon_v = 0.1$.

coupling parameter ϵ_v. In case of unconventional superconductors such nonlinear behavior can be attributed to the nonlocal effects [74]. The influence of the nonlocal effects is particularly important at higher magnetic field inductions where there is greater overlapping between the neighboring vortices. The nonlinear behavior is prominent at higher values of mixed gradient coupling parameter ϵ_v showing the influence of the admixture of the s-wave order parameter component in the system.

The temperature dependence of the magnetic penetration depth of the high-T_c superconducting cuprates is an important property. The low temperature behavior of the penetration depth gives information regarding the presence of unconventional pairing state symmetry in these materials [75]. In Figure 9 the temperature dependence of the penetration depth calculated by our isotropic two-order parameter GL theory for the mixed gradient coupling parameter $\epsilon_v = 0.1$ is compared to the corresponding experimental data for $YBa_2Cu_3O_{7-\delta}$ [76]. The excellent agreement between the theoretical and experimental result substantiates the fact that the mixed symmetry state scenario of the order parameters is a strong candidate for the underlying pairing state symmetry of the high-T_c superconducting cuprates.

5.2.3. Reversible magnetization and Upper critical magnetic field

The theoretical study of the reversible magnetization over the entire range of the applied magnetic field is required to understand the observed magnetic field dependence of the high-T_c superconducting cuprates. Earlier theoretical works carried out to determine the reversible magnetization of the conventional low temperature type-II superconductors employed a variational method [20], [21]. The results obtained from these approximate studies however involved considerable inaccuracies. Furthermore, according to the thermodynamic

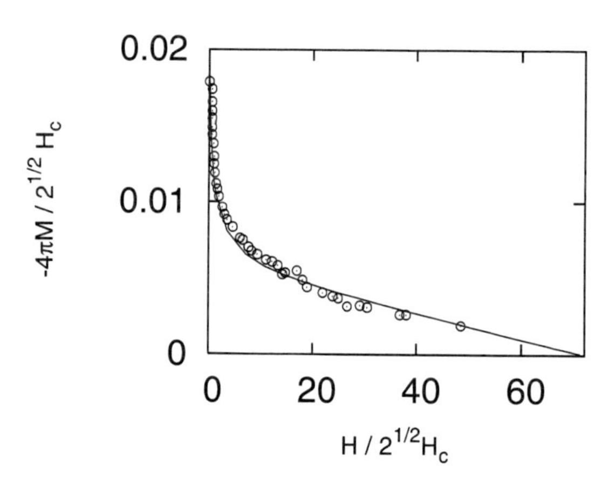

Figure 10. Reversible magnetization of the system calculated by two-order parameter GL theory in comparison with the experimental data for $YBa_2Cu_3O_{7-\delta}$ [65]. Other parameters are the same as in Figure 7 with GL parameter $\kappa = 72$ and $\epsilon_v = 0.1$.

relation $H = 4\pi(\partial f/\partial \bar{B})$, the determination of the equilibrium applied magnetic field H (and hence the reversible magnetization) requires the numerical derivative of the free energy density functional. This increases the complexity of solving the coupled nonlinear GL equations involving two order parameter components and magnetic field. A possible technique to overcome the difficulty of taking the numerical derivative of the free energy density functional is to develop a virial theorem suitable for multiple order parameter component system which establishes a relation between the average of kinetic and potential energy of the superconducting system.

In terms of the gauge invariant real quantities, ω_s, ω_d and $\mathbf{Q}$ the virial theorem for the high-T_c superconductors involving mixed symmetry state of the order parameters is expressed as [2],

$$2H\bar{B} = \left\langle -\alpha_s\omega_s + \omega_d - 2\beta_1\omega_s^2 - 2\beta_2\omega_d^2 - 2\beta_3\omega_s\omega_d - 4\beta_4\cos(2\phi)\omega_s\omega_d + 2B^2 \right\rangle. \quad (23)$$

Reversible magnetization:- Once the equilibrium applied magnetic field (H) is determined using the virial theorem (Eqn. (23)), one can now calculate the reversible magnetization of the system as per the relation $M = \bar{B} - H$. The behavior of the reversible magnetization of the system is found to be qualitatively the same for various admixtures of s-wave order parameter component in the system. The reversible magnetization of the system calculated by making use of the present theoretical model is compared with the experimental data corresponding to the high-T_c cuprate $YBa_2Cu_3O_{7-\delta}$ [65], as shown in Figure 10. For the theoretical calculation, the experimentally determined GL parameter $\kappa = 72$ relevant for the high-T_c cuprate $YBa_2Cu_3O_{7-\delta}$ is used [65]. The temperature dependence of the reversible magnetization of the system is calculated next for different magnetic field inductions and the theoretically computed results has been found to be in agreement with the experimental data corresponding to the high-T_c cuprate $YBa_2Cu_3O_{7-\delta}$, as shown in Figure 11 [4]. Our theoretical model very accurately depicts one of the fundamental char-

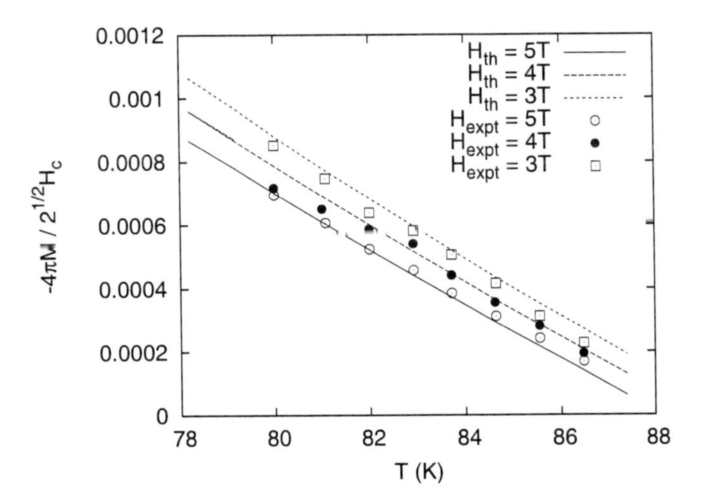

Figure 11. Temperature dependence of the reversible magnetization of the system calculated by two-order parameter GL theory for different applied magnetic field in comparison with the experimental data of $YBCO$ [21]. Other parameter values are the same as is Figure 7 with GL parameter $\kappa = 57$.

acteristics of these materials viz. the high critical temperature of $\approx 90K$.

Upper critical magnetic field:- The importance of calculating the reversible magnetization of the superconducting system lies in the fact that it gives a measure of the upper critical magnetic field (H_{c2}) of the system. The calculation of the upper critical magnetic field is specifically difficult in case of the high-T_c cuprates due to the presence of thermal fluctuation in the system. Moreover, the upper critical field derived from the resistivity ($R(T)$) curves has been found to give incorrect results. The unusual positive upward curvature of the temperature dependence of upper critical magnetic field ($H_{c2}(T)$) has been observed in the experimental studies of the high-T_c cuprates [77] and discussed theoretically in the literature [52]. The behavior is considered to be a signature of the multicomponent behavior and unconventional pairing state symmetry present in these materials. On the other hand the upward curvature observed in the results derived from the resistance measurements has been considered to be a misinterpretation of the data [78]. In the present two-order parameter GL theory the upper critical magnetic field has been computed from the analytically calculated reversible magnetization of the system as per the relation [78],

$$M(H,T) = (1/4\pi)(H_{c2}(T) - H)/(2\kappa^2 - 1)\beta_A \tag{24}$$

where, $\beta_A = \langle\omega_s^2\rangle/\langle\omega_s\rangle^2$ is the Abrikosov's parameter. The theoretically computed results are compared with the relevant experimental data [78] and has been presented in Figure 12. It is worth mentioning that the temperature dependence of the upper critical magnetic field ($H_{c2}(T)$) calculated by the present two-order parameter GL theory doesn't manifest the pronounced upward curvature that has been discussed in ref. [52]. The reason for the same can be attributed to the tetragonal symmetry of the system that has been taken into account. It

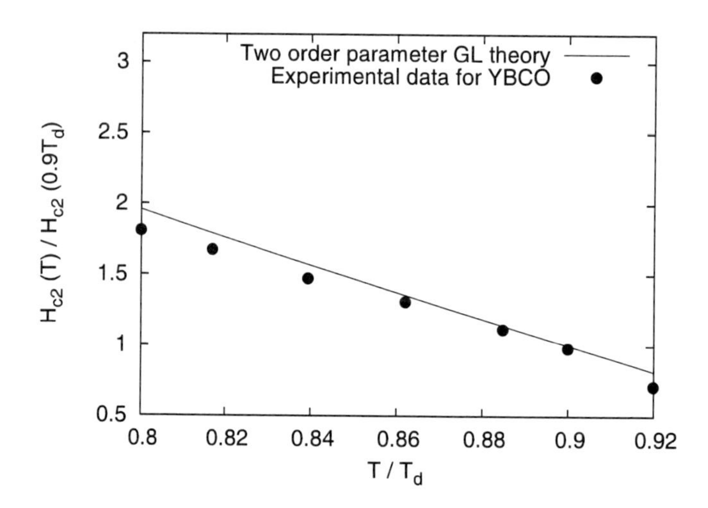

Figure 12. Comparison of the temperature dependence of the upper critical magnetic field (H_{c2}) calculated by two-order parameter GL theory with the experimental data for $YBa_2Cu_3O_{7-\delta}$ [78]. Other parameters are the same as in Figure 7 with mixed gradient coupling parameter $\epsilon_v = 0.1$.

is now clear from the experiments carried out on the high-T_c cuprates that $YBa_2Cu_3O_{7-\delta}$ is an orthorhombically distorted material. However, in the present and most of the other theoretical calculations the symmetry of the system has been approximated to be tetragonal [58], [60], [62], [64]. The presence of orthorhombic distortion in the system profoundly affects the various properties of the high-T_c cuprate $YBCO$. The vortex lattice structure shows significant magnetic field dependence and the structure of the single vortex varies from four-fold to two-fold with the increasing orthorhombic distortion. This is likely to affect the temperature dependence of the upper critical magnetic field and the upward curvature of the $H_{c2}(T)$ plot can be observed in presence of orthorhombic distortion in the system.

5.2.4. Shear modulus of the vortex lattice

An important aspect of the high-T_c superconductors is the thermal fluctuation which leads to softening of the flux line lattice (FLL) and finally its melting. The high- T_c superconducting cuprates possess high current carrying capacity and an artifact of the thermal fluctuation is the reduction of the current carrying capacity of the material. The softening of the vortex lattice gives rise to shaking of the vortices and leads to power dissipation when a large current is passed through the superconducting material. Thus, the stability of the vortex lattice and specially its dependence on the admixture of s-wave order parameter component in the system is required to be understood [79]. One such theoretically computable important quantity which gives information regarding the stability of the vortex lattice is the shear modulus (c_{66}) of the vortex lattice. The shear modulus of the vortex lattice can be theoretically computed from the difference in the free energy density between a rectangular and an oblique flux line lattice and is given by the relation [4],

$$c_{66} = 2\pi^2[y_2(\epsilon_v)/x_1]^2 \times \Big[f(x_2 = 0, y_2(\epsilon_v)) - f(x_2 = x_1/2, y_2(\epsilon_v))\Big] \tag{25}$$

where, $y_2(\epsilon_v)$ denotes the value of the unit cell height which correspond to the stable vortex lattice configuration for a particular admixture of the s-wave order parameter component in the system. The magnetic field dependence of the shear modulus of the vortex lattice for various coupling parameter ϵ_v is plotted in Figure 13 [4]. An enhanced admixture of s-wave

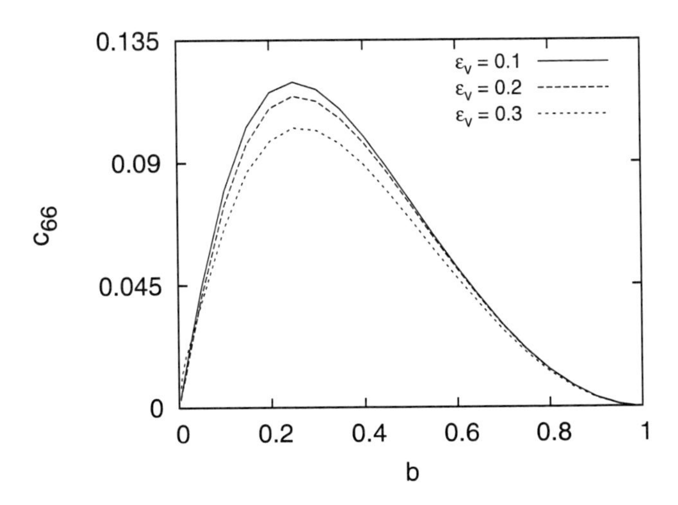

Figure 13. Variation of shear modulus of the vortex lattice with magnetic field induction b for different values of coupling parameter ϵ_v. Other parameters are same as in Figure 7.

order parameter component leads to softening of the vortex lattice. The positive value of the shear modulus for any ϵ_v corresponds to a stable vortex lattice configuration. Both at the low and high applied magnetic field inductions the vortex lattice structure is softened, at higher magnetic field the vortices are closely packed leading to an overlapping of the vortices and consequently softening of the vortex lattice. With the decreasing magnetic field the vortices drift apart and eventually goes beyond the London penetration depth λ_L. In the low magnetic field induction regime the inter-vortex interaction is exponentially small and consequently the shear modulus of the vortex lattice decreases rapidly as $c_{66} \propto exp(-a_0/\lambda_L)$ (where a_0 is of the order of lattice constant) leading to melting of the flux line lattice [79]. It has been observed that in presence of a higher admixture of the s-wave order parameter component the vortex lattice softens up while a higher magnetic field induction is required to melt the flux line lattice in this case. The magnetic field dependence of the stability of the flux line lattice has been observed in experiments carried out to study the vortex phase boundary [80]. The complete picture of the stability of the vortex lattice can be obtained by studying the temperature dependence of the shear modulus of the vortex lattice. Figure 14 shows the temperature dependence of the shear modulus of the vortex lattice calculated for various magnetic field inductions [4]. The melting of the vortex lattice is favored at high temperature as observed in experiments [80].

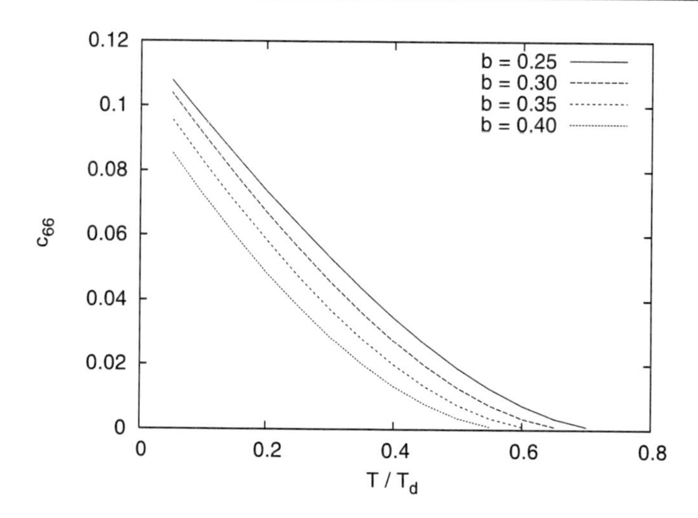

Figure 14. Temperature dependence of c_{66} for different magnetic field induction b and coupling parameter $\epsilon_v = 0.1$. Other parameters are same as in Figure 7.

The results presented in this section suggest that the isotropic two-order parameter model could successfully explain most of the experimentally observed behaviors of the high-T_c superconducting cuprates. The model could also depict the four-fold symmetric single vortex structure which could not be obtained in the framework of an anisotropic single order parameter model. However, the angle between the primitive axes of the vortex lattice derived using the isotropic two-order parameter model deviates considerably from that of the experimental observation, thus leaving scope of further improvement of the present theoretical model. One such possible improvisation of this theoretical model is to incorporate the effect of in-plane anisotropy present in the high-T_c superconducting cuprates, particularly $YBa_2Cu_3O_{7-\delta}$

6. Anisotropic Two-Order Parameter Ginzburg–Landau Theory

Of the various salient features associated with the high-T_c superconducting cuprates, a significant one is the anisotropic nature of these materials. Experimental manifestations of the in-plane anisotropy present in the high temperature superconductors are many. Raman scattering experiments carried out on $YBa_2Cu_4O_8$ [81] and $YBa_2Cu_3O_7$ [82], [83] have shown the presence of in-plane anisotropy in the energy gap (i. e $\Delta_x \neq \Delta_y$). Similar anisotropy in the energy gap has also been observed in photo emission spectra of $YBa_2Cu_3O_{7-\delta}$ [84] and in angle resolved photo emission spectroscopy experiments [66] which suggested an in-plane anisotropy of $\Delta_a/\Delta_b = 1.5$. Apart from the energy gap the effect of in-plane anisotropy in these materials have been observed in other properties as well viz., the magnetic penetration depth [24], vortex core radius [29], vortex lattice structure [26] etc. The in-plane anisotropy is attributed to the orthorhombic distortion of the CuO_2-planes and can be considered to be a signature of the mixed symmetry state of pairing present in these materials [5]. As has been discussed, the oblique structure of the flux line lattice can be depicted theoretically using an isotropic two-order parameter GL theory by taking into account only the mixed symmetry state of the order parameters even in the absence of an explicit in-plane anisotropy in the system [4]. However, such an isotropic two-order parameter GL theory cannot give the correct measure of the angle between the primitive axes of the vortex lattice. In the present section we have discussed about the effect of the in plane anisotropy on the various properties of the high-T_c superconductors involving mixed symmetry state of the order parameters.

In the present theoretical model the in-plane anisotropy in the system is the effective mass anisotropy defined by the parameter $\gamma = m_x/m_y$ and is introduced in the formulation as an anisotropic mass tensor ($\mathbf{\Lambda}$). For the sake of simplicity the effective masses of the electrons involving d-and s-wave type pairing symmetries are considered to be the same (i. e $m_s^* = m_d^* = m^*$) though in principle they are dependent on the corresponding Fermi surface architecture. For the GL parameter κ_y the experimental value of $\kappa_{avg} = \kappa_y = 72$ relevant for $YBa_2Cu_3O_{7-\delta}$ has been used [65].

6.1. Theoretical formalism, numerical results and discussions

In presence of in-plane mass anisotropy the two-dimensional average GL free energy density functional corresponding to the high-T_c superconductors involving mixed symmetry state of the order parameters can be given as [5],

$$
\begin{aligned}
f = \Bigg\langle \Bigg[& \alpha_s\omega_s - \omega_d + \beta_1\omega_s^2 + \beta_2\omega_d^2 + \beta_3\omega_s\omega_d + 2\beta_4\cos(2\phi)\omega_s\omega_d \\
& + g_s + \omega_s\mathbf{Q\Lambda Q} + g_d + \omega_d\mathbf{Q\Lambda Q} \\
& + 2\epsilon_v \Bigg[\cos(\phi)\Big\{\Big((\nabla_y\omega_s)(\nabla_y\omega_d) - (\nabla_x\omega_s)(\nabla_x\omega_d)\Big)/4\kappa_y^2(\omega_s\omega_d)^{1/2} \\
& \qquad + (Q_y^2 - Q_x^2)(\omega_s\omega_d)^{1/2}\Big\} \\
& + \sin(\phi)\Big\{\Big[Q_y(\nabla_y\omega_s) - Q_x(\nabla_x\omega_s)\Big](\omega_d/4\kappa_y^2\omega_s)^{1/2} \\
& - \Big[Q_y(\nabla_y\omega_d) - Q_x(\nabla_x\omega_d)\Big](\omega_s/4\kappa_y^2\omega_d)^{1/2}\Big\}\Bigg] + (\nabla\times\mathbf{Q})^2\Bigg]\Bigg\rangle
\end{aligned}
\tag{26}
$$

where, $g_i = (\nabla\omega_i)\mathbf{\Lambda}(\nabla\omega_i)/4\kappa_y^2\omega_i$, with $i = s, d$. As for the isotropic two-order parameter model, the coefficients α_s and β_i ($i = 1, 2, 3, 4$) are expressed in terms of reduced units and so is the free energy density functional and the order parameter components. The order parameters and vector potential (thus, the magnetic field) are expressed in terms of gauge invariant real quantities ω_s, ω_d and $\mathbf{Q}$ respectively. The in-plane anisotropy is introduced in the formulation in terms of an anisotropic mass tensor $\mathbf{\Lambda}$ defined as,

$$
\mathbf{\Lambda} = \begin{pmatrix} m_y/m_x & 0 \\ 0 & 1 \end{pmatrix} \tag{27}
$$

The magnitude of the in-plane anisotropy present in the system is defined in terms of the mass anisotropy parameter $\gamma = m_x/m_y$ and for $\gamma = 1$ the formalism reverts back to the isotropic two-order parameter GL theory. The magnetic field measured in units of $\sqrt{2}B_c$ is applied along the z-axis and is thus defined as $\mathbf{B} = B\hat{z} = (\nabla \times \mathbf{A})$. The parameter $\epsilon_v = m_y/m_v$ is the coefficient of the mixed gradient coupling term and gives the strength of the admixture of the s-wave order parameter component in the system. The mixed gradient coupling parameter is chosen to be $\epsilon_v = 0.1$ since from the isotropic two-order parameter GL model it has been found that the theoretical results computed for $\epsilon_v = 0.1$ regarding the properties of the high-T_c superconducting cuprates show very good agreement with the relevant experimental results corresponding to $YBa_2Cu_3O_{7-\delta}$ [4]. The GL equations are obtained by minimizing the anisotropic two-order parameter free energy density functional w. r. t the order parameters and supervelocity. The anisotropic nonlinear coupled GL equations are solved by using the numerical iteration technique discussed in the previous sections and the Fourier coefficients $a^s_{\mathbf{K}}$, $a^d_{\mathbf{K}}$ and $b_{\mathbf{K}}$ corresponding to the order parameters and magnetic field are determined.

6.1.1. Single vortex and vortex lattice structure

The order parameters and magnetic field computed numerically are used to study the various properties of the high-T_c cuprates in the framework of anisotropic two-order parameter GL theory. We begin with studying the effect of the in-plane anisotropy on the single vortex and vortex lattice structure of the high-T_c superconducting cuprates. In order to determine the stable vortex lattice configuration the free energy density is minimized w. r. t the lattice parameter y_2/x_1 ($x_1 = 1.0$) for different values of the in-plane anisotropy parameter γ. The magnetic field induction is chosen to be $b = 0.04$ which corresponds to the experimental applied magnetic field of $H = 5T$ [26]. The stability of the vortex lattice against the variation of the mixed gradient coupling parameter ϵ_v in presence of in-plane anisotropy has been verified. The lattice has been found to be oblique whose exact shape is determined by the magnitude of in-plane anisotropy (characterized by the parameter γ) present in the system [5]. For the remaining of the studies the mixed gradient coupling parameter has been chosen to be $\epsilon_v = 0.1$. The lattice parameter y_2/x_1 corresponding to the stable vortex lattice configuration is computed and for the experimentally determined in-plane anisotropy parameter $\gamma = 2.4$ the stable vortex lattice configuration is achieved for the lattice parameter $y_2/x_1 = 0.55$.

The angle between the primitive axes of the vortex lattice for the experimentally observed in-plane anisotropy parameter $\gamma = 2.4$ has been found to be $\beta \approx 84.5^o$ which is quite close to the experimentally determined value of $\beta \approx 77 \pm 5^o$ [26], [29]. For $\gamma = 3.0$ the angle between the primitive axes is found to be $\beta \approx 76.87^o$. The slight discrepancy between the experimental and theoretical results can be attributed to the fact that in the present theoretical model the orthorhombic distortion of the CuO_2-planes present in the high-T_c cuprates have not been taken into account explicitly. Moreover, the three-dimensional structure of the superconducting system has to be taken into account in the theoretical formalism. In case of the exclusive presence of the mixed symmetry pairing state scenario the angle between the primitive axes is $\beta \approx 97.3^o$ which is far from the experimental result. It can thus be inferred from this observation that in order to explain the experimental results corresponding to the high-T_c cuprate $YBa_2Cu_3O_{7-\delta}$ involving the mixed symmetry state of the order parameters, the presence of in-plane anisotropy in the system should be taken into ac-

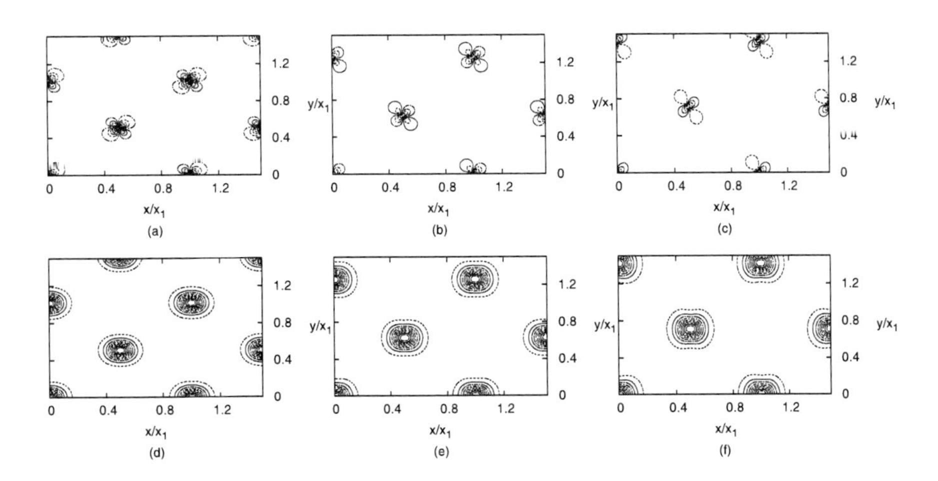

Figure 15. Contour plots of the order parameter components for different values of the mass anisotropy parameter γ. Figures (a)–(c) corresponds to the s-wave order parameter component $\omega_s(x, y)$ while Figures (d)–(f) correspond to the d-wave order parameter component $\omega_d(x, y)$ for $\gamma = 2.0, 3.0, 4.0$ respectively. Parameter values used are $\alpha_s / \mid \alpha_d \mid = 0.5$, $\beta_1/2\beta_2 = \beta_3/2\beta_2 = 1.0$, $\beta_4/2\beta_2 = 0.5$, $\kappa_y = 72$, $\phi = \phi_d - \phi_s = \pi/2$ and $\epsilon_v = 0.1$.

count. As an artifact of the presence of in-plane anisotropy in the system the single vortices have been found to be elongated with the ratio between the axes being ≈ 1.4, in agreement with the experimental observation for $YBCO$ [29]. It has been found that in presence of high in-plane anisotropy the structure of the vortices corresponding to the s-wave order parameter component shows transition from four-fold to two-fold symmetry. The transition is also observed in case of the d-wave order parameter component which shows progressively prominent tapering at both the ends with increasing in-plane anisotropy. The observation is in agreement with the one discussed by Xu *et al.* who have studied the single vortex and vortex lattice structure corresponding to the anisotropic d-wave superconductor [62]. Such a transition of the s-wave order parameter component from four-fold to two-fold symmetry has not been observed in case of an isotropic two-order parameter model [4] and hence can be attributed to the presence of in-plane anisotropy in the system. The increased prominence of the transition from four-fold to two-fold symmetry at higher γ can thus be inferred as the enhanced effect of the presence of in-plane anisotropy in the system. The single vortex structure corresponding to the s-and d-wave order parameter components for different values of the in-plane anisotropy parameter γ is presented in Figure 16 [5].

6.1.2. Local spatial behavior of the order parameters and magnetic field

In the framework of the present theoretical model the magnitude of the vortex core radius calculated along the x-and y-directions have been found to vary significantly indicating the

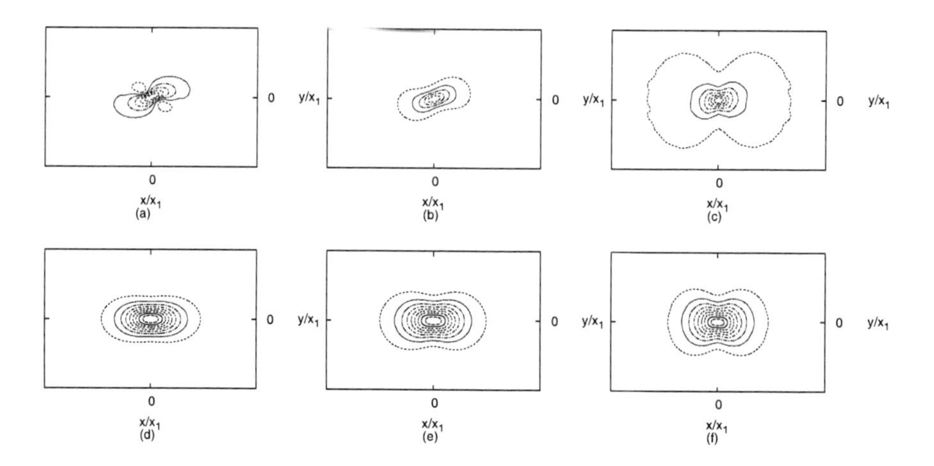

Figure 16. Single vortex solution for the order parameter components for different values of mass anisotropy parameter γ and magnetic field induction $b = 0.01$. Figures (a), (b) and (c) correspond to $\omega_s(x, y)$, while Figures (d), (e) and (f) correspond to $\omega_d(x, y)$ for $\gamma = 3.0, 5.0, 8.0$ respectively. Other parameter values are the same as Figure 15. With the increase in the mass anisotropy parameter γ the vortex structure becomes more and more two-fold symmetric.

effect of in-plane anisotropy present in the system. The magnetic field dependence of the vortex core radius calculated along the x-and y-directions for different values of the in-plane anisotropy parameter γ are plotted in Figure 17 along with the corresponding experimental data for $YBa_2Cu_3O_{7-\delta}$ [5], [71]. The qualitative behavior is the same for different values of the in-plane anisotropy parameter γ and is in agreement with the experimental results [69]. The temperature dependence of the vortex core size calculated along the x- and y-directions revealed a difference in magnitude, indicating the presence of in-plane anisotropy in the system [5]. Further, our theoretically computed results (calculated along the y-direction) for $\gamma = 2.4$ and magnetic field induction $b = 0.004$ ($H = 0.5T$) showed excellent agreement with the experimental data corresponding to $YBa_2Cu_3O_{7-\delta}$ [71] as is presented in Figure 18.

A significant effect of the presence of in-plane anisotropy in the system was also observed in case of the magnetic penetration depth which corresponds to the local spatial behavior of the magnetic field [5]. It has been observed that at higher magnetic field induction the penetration depth of the system shows a nonlinear behavior with the non linearity being enhanced with the enhancement of the in-plane anisotropy. Such a non linearity has also been observed in the experimental studies carried out on $YBa_2Cu_3O_{7-\delta}$. The nonlinear behavior of the magnetic penetration depth of the high-T_c superconducting cuprates can be attributed to the greater overlap of the vortices at higher magnetic field inductions [5]. The temperature dependence of the penetration depth of the system is studied next. In agree-

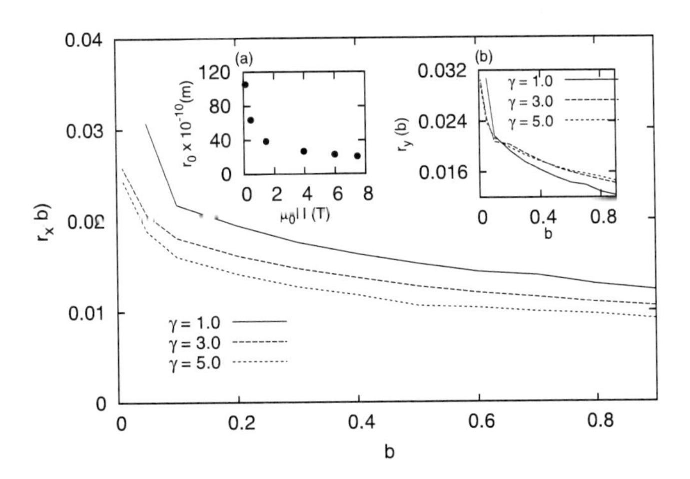

Figure 17. Variation of the vortex core radius ($r_x(b)$) calculated along the x-direction with magnetic field induction b for different values of the mass anisotropy parameter γ. Other parameters are the same as in Figure 15. The inset (a) shows the experimental data for the magnetic field dependence of the vortex core radius of high-T_c superconductor $YBa_2Cu_3O_{7-\delta}$ [71]. The inset (b) of the figure shows the magnetic field dependence of the vortex core radius ($r_y(b)$) calculated along the y-direction by our theoretical model for different values of the mass anisotropy parameter γ.

ment with the experimental data corresponding to the the high-T_c cuprate $YBa_2Cu_3O_{7-\delta}$ the magnetic penetration depth was found to increase with the increase in temperature [76]. The relevance of the results calculated by the present theoretical model can be observed from Figure 19 [5]. In this figure the theoretical results are compared with the corresponding experimental data for $YBa_2Cu_3O_{7-\delta}$ [76] and the agreement between the two has been found to be excellent.

6.1.3. Superconducting current density

The difference in the local spatial behavior of the magnetic field profile along the x-and y-directions can be attributed to the difference in the supercurrent response along these directions. Theoretically, in the framework of two-order parameter GL theory the superconducting current density can be calculated by using the third GL equation obtained by minimizing the free energy density functional w. r. t the supervelocity $\mathbf{Q}$.

The results regarding the effect of in-plane anisotropy on the superconducting current density is presented in Figure 20 wherein the current density profile calculated along the x-direction is plotted for different values of in-plane anisotropy parameter γ [5]. The magnitude of the superconducting current density gets enhanced in presence of higher in-plane anisotropy. This observation can be considered to be an evidence of the experimentally observed high current carrying capacity of the high-T_c cuprates which are known to possess significant in-plane anisotropy.

We now calculate the temperature dependence of the superconducting current density.

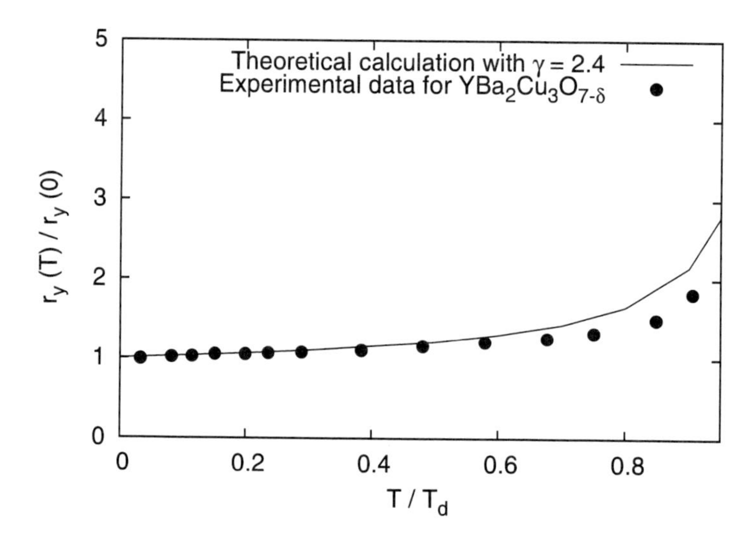

Figure 18. Temperature dependence of the vortex core radius. The solid line represents the result calculated theoretically by our anisotropic two-order parameter model for the in-plane anisotropy parameter $\gamma = 2.4$ and magnetic field induction $b = 0.004(H = 0.5T)$. Other parameter values used for the theoretical calculation are the same as in Figure 15 with $T/T_s = 0.5$. The dot gives the experimental data for high-T_c cuprate $YBa_2Cu_3O_{7-\delta}$ [71].

For the magnetic field being applied along the z-direction the superconducting current density can be resolved in two components along the x-and y-directions as $j_x(x,y)$ and $j_y(x,y)$ respectively. In Figure 21 the temperature dependence of these two components of the superconducting current density are plotted in presence of an in-plane anisotropy of $\gamma = 2.4$ and experimentally relevant field of $H = 0.5T$ ($b = 0.004$) [5]. It can be seen from the figure that there is significant difference in the magnitude of the components calculated along the x-and y-directions.

In order to ascertain the relevance of our theoretically computed results with the corresponding experimental data we next compare our results with the data corresponding to $YBa_2Cu_3O_{7-\delta}$ [48]. The observation is presented in Figure 22 where the temperature dependence of the superconducting current density calculated by the anisotropic two-order parameter GL theory is compared with the corresponding experimental results [5]. The agreement has been found to be fairly good. The slight difference can be attributed to the fact that while the present theoretical study has been carried out in presence of a small magnetic field induction $b = 0.004$, the experimental results correspond to zero magnetic field. Moreover, unlike the thin film sample used for the experimental study, the theoretical work is carried out for bulk system in which the pinning of the flux lines have not been taken into account.

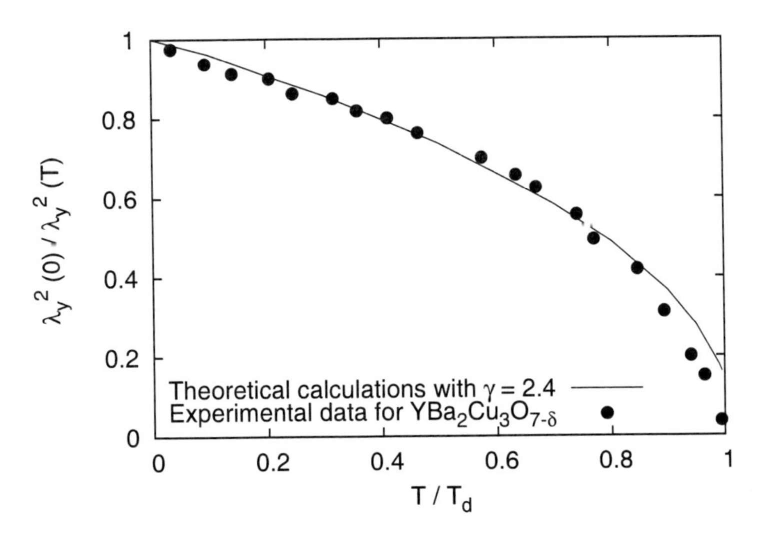

Figure 19. Temperature dependence of the penetration depth. The solid line represents the result calculated theoretically by our anisotropic two-order parameter model for the in-plane anisotropy parameter $\gamma = 2.4$ and magnetic field induction $b = 0.004(H = 0.5T)$. Other parameter values used for the theoretical calculation are the same as in Figure 15 with $T/T_s = 0.5$. The dot gives the experimental data for high-T_c cuprate $YBa_2Cu_3O_{7-\delta}$ [76].

6.1.4. Shear modulus of the vortex lattice

The quantity which can be computed by the present theoretical model and gives information regarding the softening and melting of the flux line lattice is the shear modulus (c_{66}) of the vortex lattice. We thus next compute the shear modulus of the vortex lattice (c_{66}) for different in-plane anisotropy parameter γ. In presence of the in-plane anisotropy in the system the shear modulus of the vortex lattice can be defined as [5],

$$c_{66} = 2\pi^2 [y_2(\gamma)/x_1]^2 \times \Big[f(x_2 = 0, y_2(\gamma)) - f(x_2 = x_1/2, y_2(\gamma)) \Big] \quad (28)$$

where, $y_2(\gamma)$ denotes the anisotropy parameter dependent unit cell height which corresponds to the minimum free energy density for the vortex lattice of a particular symmetry, i.e. the stable vortex lattice configuration.

It has been observed that an enhancement of the in-plane anisotropy parameter γ corresponds to a higher magnitude of the shear modulus of the flux line lattice [5]. Thus, a higher in-plane anisotropy gives rise to a harder vortex lattice and consequently the current carrying capacity of the material is enhanced [5]. A lower in-plane anisotropy leads to softening of the flux line lattice and favors its melting. A comparison with the results of the isotropic two-order parameter model showed that the enhancement in the admixture of the s-wave order parameter component in the system leads to softening of the vortex lattice. However, once the in-plane anisotropy is introduced in the system the effect of the admixture of the s-wave order parameter gets suppressed by the effect of the in-plane anisotropy in the system. In presence of a higher in-plane anisotropy a lower magnetic field induction is required to

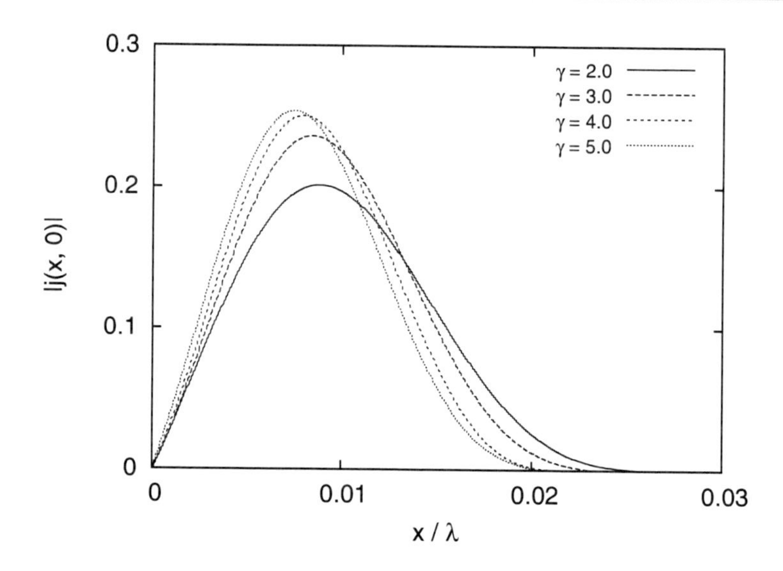

Figure 20. Super current profile for different values of the mass anisotropy parameter γ ($\gamma = 2.0, 3.0, 4.0, 5.0$) for the magnetic field induction $b = 0.5$. Other parameters are the same as in Figure 15.

melt the flux line lattice [5]. In Figure 23 we have shown the temperature dependence of the shear modulus of the vortex lattice (c_{66}) for various in-plane anisotropy parameter γ [5]. The lattice gets hardened at higher in-plane anisotropy at any given temperature. For any amount of in-plane anisotropy the lattice melts at higher temperature as has been observed in experimental studies [80].

A balance between the softening of the vortex lattice and the current carrying capacity of the high-T_c cuprates, particularly $YBa_2Cu_3O_{7-\delta}$ can be struck by controlling the amount of admixture of s-wave order parameter component in the system and the magnitude of the in-plane anisotropy. A clearer understanding can be achieved by experimentally studying the variation in the vortex phase diagram and also the current carrying capacity of the material both for different admixtures of the s-wave order parameter component and in-plane anisotropies present in the system.

The discussion presented in this section highlights the importance of the inclusion of in-plane anisotropy in the two-order parameter GL theoretical model so as to understand the various properties of the high-T_c superconducting cuprates. Apart from the various properties of these materials which could be explained in the framework of anisotropic two-order parameter GL theory, the model could also depict the experimentally observed angle between the primitive axes of the vortex lattice, which could not be correctly predicted by using the isotropic two-order parameter model.

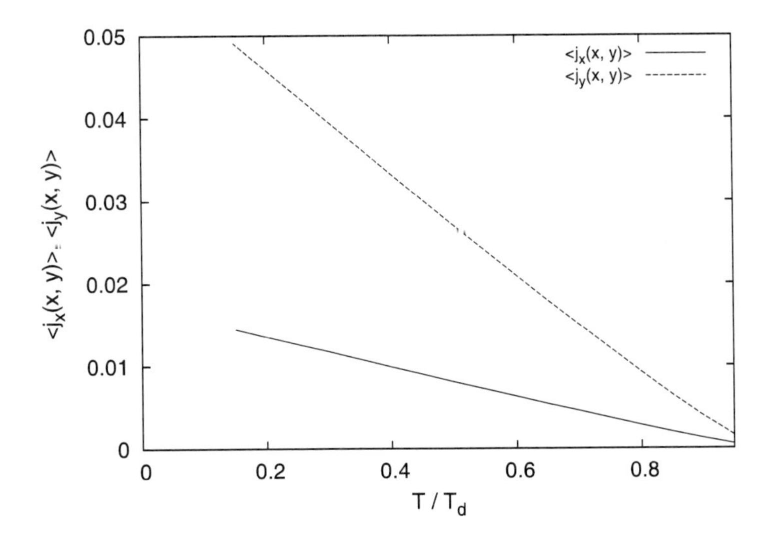

Figure 21. Temperature dependence of the different components of the superconducting current density calculated for the in-plane mass anisotropy parameter $\gamma = 2.4$. The magnetic field induction is $b = 0.004$ and the other parameters are the same as in Figure 15, with $T/T_s = 0.5$.

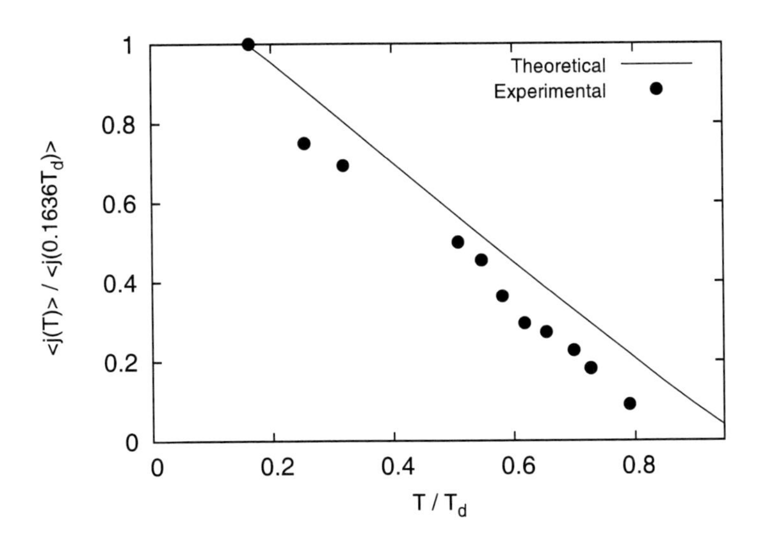

Figure 22. Comparison of the temperature dependence of current density calculated by our anisotropic two-order parameter GL theory for mass anisotropy parameter $\gamma = 2.4$ and $b = 0.004$ with the experimental data for $YBa_2Cu_3O_{7-\delta}$ thin film [48].

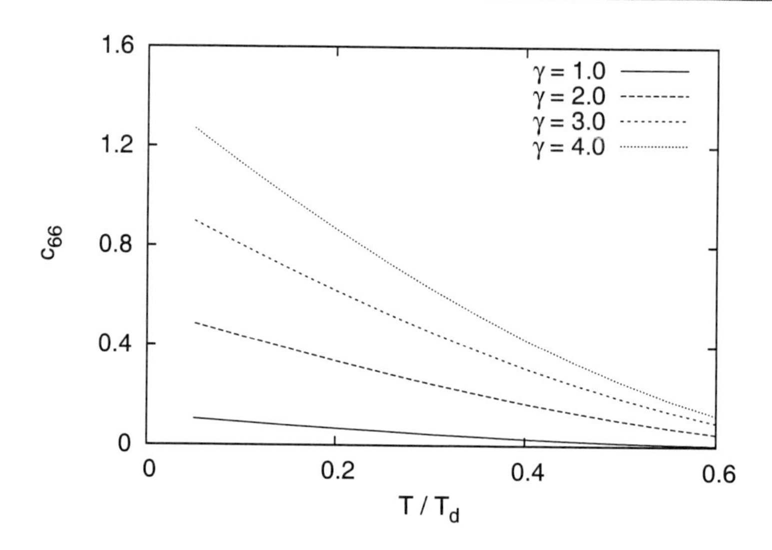

Figure 23. Temperature dependence of the shear modulus (c_{66}) of the vortex lattice for different values of the mass anisotropy parameter γ with magnetic field induction $b = 0.3$. Other parameters are the same as in Figure 15.

7. Application of the Two-Order Parameter GL Theory: The MgB_2 Problem

In the previous sections it has been shown that the two-order parameter Ginzburg–Landau (GL) model can be very successfully applied to study the properties of the high temperature superconducting cuprate $YBa_2Cu_3O_{7-\delta}$. The properties were studied over the entire range of applied magnetic field and wide range of temperature for arbitrary values of the GL parameter κ and vortex lattice symmetry. In the present section this theoretical model is used for studying the properties of another multicomponent superconducting system. For this we have chosen the recently discovered inter-metallic superconducting material MgB_2. The applicability of the two-order parameter theoretical model to probe the properties of MgB_2 shows its versatility to study the properties of multicomponent superconducting systems. Before discussing the properties of this material we present a brief introduction to this newly discovered superconducting material with significantly different properties compared to the cuprates.

7.1. Introduction

The discovery of superconductivity in the inter-metallic compound MgB_2 [85] has initiated a lot of theoretical and experimental work to study the properties of this material. The material can support a large critical current and critical temperature of $T_c \approx 39K$ despite a simple chemical composition as compared to the high temperature cuprates. Experimental studies [86], [87] have suggested that MgB_2 is an example of multiband superconductor, and the superconductivity in this material is governed by strong electron-phonon coupling as suggested by B-isotope effect on T_c [88]. As has been found by *Ab initio* calculations [89] MgB_2 is a metal possessing a layered structure. The boron atoms form honeycombed layers and the magnesium atoms are located above the center of the hexagons in between the boron planes. Superconductivity in this material is governed by the σ-and π-bonding boron orbitals. The σ-bonding orbitals are partially occupied in case of MgB_2 and are thus coupled very strongly to the in-plane vibrations of the boron planes. This gives rise to a large energy gap of magnitude $\Delta_\sigma \approx 6.8meV$. This strong electron-phonon coupling is confined to the boron plane and gives the principal contribution to the superconductivity of MgB_2. A smaller superconducting gap of magnitude $\Delta_\pi \approx 1.8meV$ originates from the weakly superconducting π-bonding states present in the remaining parts of the Fermi surface.

It has been found that owing to its two-band nature MgB_2 manifests several new and interesting features. The anisotropy of the upper critical magnetic field shows anomalous temperature dependence [90]. On the other hand the lower critical magnetic field is found to be almost isotropic indicating that in MgB_2 the coherence length and penetration depth anisotropy $\Gamma = \xi_{ab}/\xi_c = \lambda_c/\lambda_{ab}$ is field dependent and rises from an isotropic value ($\Gamma \approx 1$) at the lower magnetic field to an anisotropic value of ($\Gamma \approx 5-6$) at the higher magnetic field. The anomalous behavior is not restricted to the upper critical magnetic field only and has also been observed in case of the contribution of the σ-and π-bands towards the superconductivity of this material [91], [92]. Above a crossover field of $0.1T$ the contribution of π-band is suppressed by the magnetic field and the property of the material is then governed by the two-dimensional σ-band [93]. Below this crossover field the three dimensional π-band serves as a decisive factor for the various properties of MgB_2.

The theoretical and experimental efforts has lead to a consensus that in MgB_2, superconductivity is mainly driven by the σ-band and its interband coupling with the π-band induces a small superconductivity in the π-band [94], [95]. In the framework of Ginzburg–Landau (GL) theory attempts has been made to study the properties of MgB_2 superconductor [96]– [99]. However, in these theoretical works an effective single order parameter theory have been used in which one of the order parameter components is expressed in terms of the other. It is thus required to develop a theoretical model which can encompass the entire applied magnetic field regime and can thus explain the experimentally observed anomalous magnetic field dependence of the various properties of MgB_2.

7.2. Theoretical formalism and Numerical calculations

The two-dimensional average Ginzburg–Landau (GL) free energy density functional for MgB_2 superconductor is given as [6], [96], [97],

$$f = \Big\langle \alpha_\sigma \mid \psi_\sigma \mid^2 + \alpha_\pi \mid \psi_\pi \mid^2 + \beta_\sigma \mid \psi_\sigma \mid^4 + \beta_\pi \mid \psi_\pi \mid^4 + \gamma(\psi_\sigma^* \psi_\pi + \psi_\sigma \psi_\pi^*) + \frac{1}{2m_\sigma} \mid \mathbf{\Pi} \psi_\sigma \mid^2 + \frac{1}{2m_\pi} \mid \mathbf{\Pi} \psi_\pi \mid^2 + \gamma_1 \Big(\Pi_x \psi_\sigma \Pi_x^* \psi_\pi^* + \Pi_y \psi_\sigma \Pi_y^* \psi_\pi^* + c.c \Big) + \frac{1}{8\pi} (\nabla \times \mathbf{A})^2 \Big\rangle. \quad (29)$$

On the symmetry grounds, various types of interaction in quartic and gradient terms between the superconducting order parameters are possible and have been considered in literature [96], [97], [100], [101]. In the above free energy density functional, ψ_σ and ψ_π are the superconducting order parameter components corresponding to the σ-and π-bands respectively while $\mathbf{A}$ is the vector potential, related to the local magnetic field $\mathbf{B} = \nabla \times \mathbf{A}$, applied along the z-axis. Regarding the various parameters involved in the free energy density functional, the parameters α_σ and α_π are temperature dependent quantities and can be defined as, $\alpha_i = \alpha'(T - T_i)$, where $(i = \sigma, \pi)$. The β-parameters are positive quantities, γ is the linear interband coupling parameter and determines the interaction between the order parameter components corresponding to the σ-and π-bands, while γ_1 is the gradient interband coupling parameter. The most important parameter of the formalism is the linear interband coupling parameter γ and the sign of the parameter γ decides the stable relative phase between the order parameter components. For $\gamma < 0$ a stable phase difference of 0 is realized while for $\gamma > 0$ a relative phase of π between the order parameter components is stabilized. In case of MgB_2 the linear interband coupling term governs the properties of the material and the strength of the linear interband coupling is determined by the parameter γ.

In terms of the gauge invariant real quantities $\psi_\sigma = \sqrt{\omega_\sigma} e^{i\phi_\sigma}$, $\psi_\pi = \sqrt{\omega_\pi} e^{i\phi_\pi}$ and $\mathbf{Q} = \mathbf{A} - \nabla\phi/\kappa$ corresponding to the order parameters and supervelocity the free energy density functional (Eqn. (29)) can be given as [6],

$$
\begin{aligned}
f = \Big\langle & -\omega_\sigma + (1/2)\omega_\sigma{}^2 + \alpha\omega_\pi + \beta_\pi\omega_\pi^2 + 2\gamma\cos(\phi)(\omega_\sigma\omega_\pi)^{1/2} + g_\sigma \\
& + \omega_\sigma Q^2 + M_1 g_\pi + M_1\omega_\pi Q^2 \\
& + 2M_2\Big[\cos(\phi)\Big(Q^2(\omega_\sigma\omega_\pi)^{1/2} + (\nabla\omega_\sigma)(\nabla\omega_\pi)/4\kappa^2(\omega_\sigma\omega_\pi)^{1/2}\Big) \\
& + \sin(\phi)/2\kappa\Big(\mathbf{Q}(\nabla\omega_\pi)(\omega_\sigma/\omega_\pi)^{1/2} - \mathbf{Q}(\nabla\omega_\sigma)(\omega_\pi/\omega_\sigma)^{1/2}\Big)\Big] + (\nabla \times \mathbf{Q})^2\Big\rangle
\end{aligned}
\tag{30}
$$

with, $g_i = (\nabla\omega_i)^2/4\omega_i\kappa^2$, for $i = \sigma, \pi$. For details regarding the selection of parameters, the reader may refer [6]. The order parameters and magnetic field are determined by using the numerical iteration technique discussed in the previous sections.

7.3. Results and Discussions

The two-dimensional approach of the method makes it applicable over a wide range of applied magnetic field induction $(0.015 < b < 1)$ where the superconductivity is principally governed by the two-dimensional σ-bands. However, below a crossover applied magnetic field $H \approx 0.1T$ $(b < 0.015)$ the method ceases to be accurately applicable due to the enhanced dominance of the three-dimensional π-bands. The two-dimensional treatment to the problem is however sufficient to understand the essential physics of the system over the wide range of applied magnetic field. For its applicability to extremely low magnetic fields the method has to be generalized to three-dimension. The only free parameter in the formalism is the linear interband coupling parameter γ which has been chosen by comparing with the experimental data for the reversible magnetization of the system.

7.3.1. Vortex lattice structure

For MgB_2 a triangular vortex lattice configuration is realized in agreement to what has been suggested by several band structure calculation [102] and experimental studies [92]. The 30° orientation of the vortex lattice observed in the recent neutron scattering experiments has however not been observed in the present theoretical study [92]. The discrepancy can be attributed to the the anisotropy of the σ-and π-bands which favors different orientations of the vortex lattice [92].

7.3.2. Local spatial behaviors of order parameters and magnetic field

Next, we have studied the spatial dependence of the order parameter profiles $\omega_\sigma(x, y)$ and $\omega_\pi(x, y)$ for different values of the linear interband coupling parameter γ and in Figure 24a we have plotted the same [6]. The figure shows significant difference in the magnitude and width of the order parameter profiles corresponding to the σ-and π-bands. The order parameter profile corresponding to the π-band varies more slowly as compared to that of the σ-band. There are two different length scales in the system as ξ_σ and ξ_π, over which the σ-and π-band order parameter components vary, as suggested by the different vortex core size of the σ-and π-band order parameter components. The Figure 24a shows that with the variation in the linear interband coupling parameter γ from -0.6 to -0.05 the magnitude of the π-band order parameter component reduces by a factor of ≈ 40 while the vortex core size varies by a factor of ≈ 2. Yet another important parameter of the system is the gradient coupling parameter M_1 which gives the ratio of the effective electronic masses corresponding to the σ and π bands. In Figure 24b we have plotted the spatial variation of the order parameter components for different magnitudes of the coupling parameter M_1. From Figure 24a and (24b) one can observe an important property of the two-band superconducting systems. The vortex core size of the order parameter corresponding to the weakly superconducting band (π-band in this case) always exceeds that of the strongly superconducting band [103].

We next compare our theoretically computed magnetic field dependence of the vortex core radius with the corresponding experimental data for MgB_2 [98]. The corresponding result is depicted in Figure 25 [6]. The combined vortex core size calculated along the y-axis by the present theoretical model is defined as $r_y = 2r_{\sigma y} + r_{\pi y}$, which gives the best fit with the experimental results. The inset of the figure shows the magnetic field dependence of the relative vortex core size of the σ-and π-band order parameter components ($r_{\pi y}/r_{\sigma y}$). The behavior re-establishes the fact that for all magnetic field inductions the vortex core size corresponding to the π-band exceeds that of the σ-band with the behavior being prominent at the lower magnetic field regime.

After the vortex core radius we concentrate on the magnetic penetration depth (λ). The magnetic field dependence of the penetration depth of the system as calculated by the present theoretical model is compared with the relevant experimental data for MgB_2 as observed from SQUID measurements [104], along with the results obtained by an effective single order parameter model [98]. The corresponding observations are presented in Figure 26 [6]. The result obtained by the present theoretical model corresponds to a smaller value of the penetration depth (λ_{ab}) as compared to that obtained from SQUID measurements [104], while it is higher than the one determined by using an effective sin-

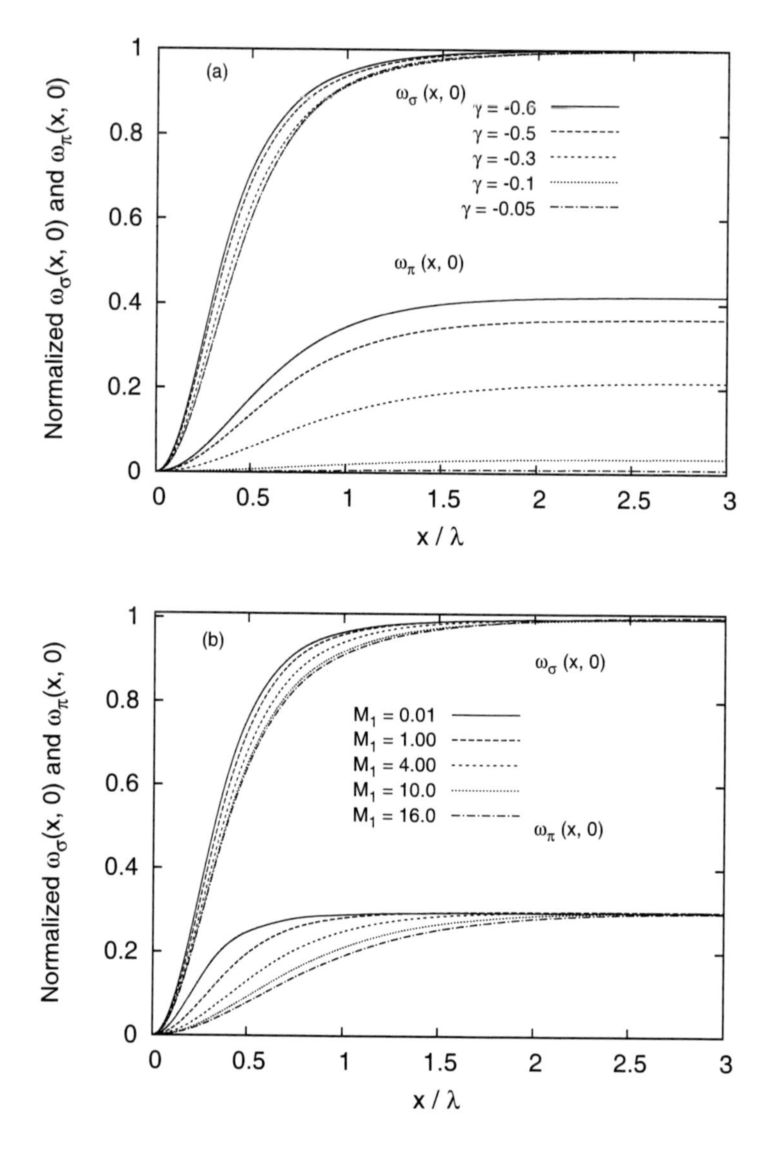

Figure 24. Variation of normalized superconducting order parameter profiles for different values of (a) the linear interband coupling parameter γ and (b) the coupling parameter M_1. The magnetic field induction is $b = 0.005$ and $(1 - T/T_\sigma) = 0.3$. For (a) the coupling parameter $M_1 = 9.0$ and for (b) the linear interband coupling parameter $\gamma = -0.4$.

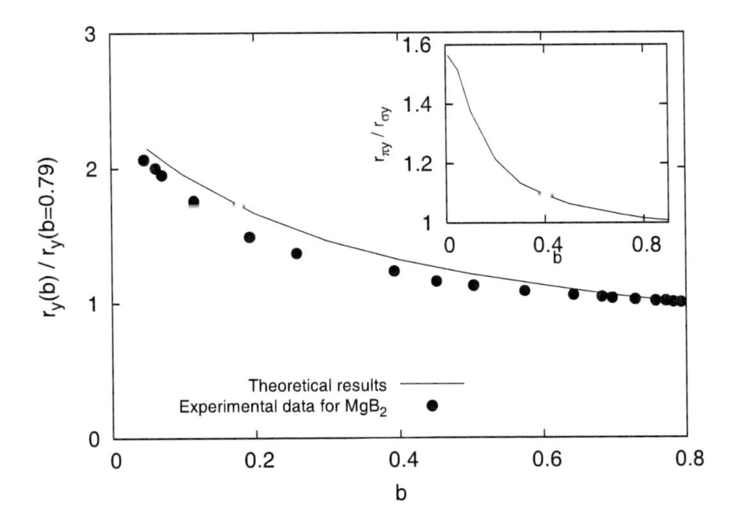

Figure 25. Comparison of the experimental data [98] for MgB_2 corresponding to the magnetic field dependence of the vortex core radius with the results calculated by the two-order parameter GL theory. The inset of the figure shows the magnetic field dependence of the relative vortex core radius of the σ-and π-band order parameter components ($r_{\pi y}/r_{\sigma y}$). Other parameters used are $\alpha_\pi / \mid \alpha_\sigma \mid = 0.66$, $\beta_\pi/2\beta_\sigma = 0.75$, $\gamma = \gamma/ \mid \alpha_\sigma \mid = -0.1$, $M_1 = m_\sigma/m_\pi = 1.06$, $\kappa = 6.4$.

gle order parameter model involving an assumed field dependent vortex core radius and penetration depth. The SQUID measurements carried out on MgB_2 [104] suggests a magnetic penetration depth of $\lambda_{ab} \approx$ 900angstrom while in the clean limit the theoretically calculated magnetic penetration depth is $\lambda_{ab} \approx$ 400angstrom, in agreement to the isotropic H_{c1} [105] and SANS measurements [106]. The effective single order parameter model discussed by Klein *et al.* shows that the penetration depth of the system varies between $\lambda_{ab} \approx 450 - 700$angstrom for the applied magnetic field varying between low to high [98]. It can be inferred from the difference in the magnetic field dependence of the penetration depth between the theoretical and experimental results that the magnetic penetration depth determined from the reversible magnetization data is grossly overestimated. The results calculated by the present two-order parameter theoretical model where the penetration depth is determined from the FWHM of the magnetic field profile suggests that the magnetic penetration depth is $\lambda_{ab} <$ 900angstrom. In the higher magnetic field regime our results are in good agreement with the one obtained by using an effective single order parameter model [98]. The discrepancy at the lower magnetic field regime can be attributed to the enhanced importance of the three-dimensionality of the π-band order parameter component which has not been taken into account in the present theoretical model. The inset of the figure shows the magnetic field dependence of the inverse square of the penetration depth of the system which decreases with increasing magnetic field induction and almost saturates at high magnetic field. This is in contrast to the result of linear dependence of the penetration depth on the magnetic field as reported earlier [98]. In Figure 27 our theoretically

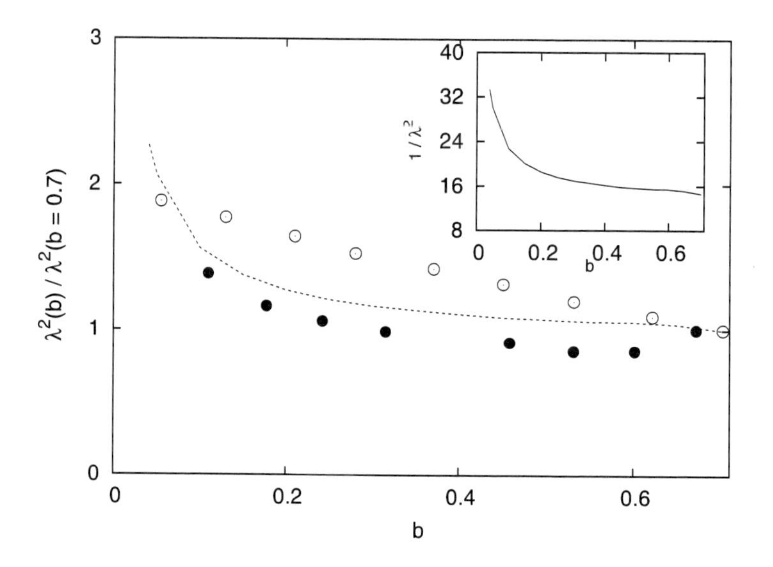

Figure 26. Magnetic field dependence of penetration depth calculated by our two-order parameter GL theory in comparison with the experimental data obtained from SQUID measurement [104] (closed circles) and the results obtained from the effective single band model [98] (open circles). The inset shows the magnetic field dependence of the inverse square of the penetration depth. The parameter values used for the theoretical calculation are the same as in Figure 25.

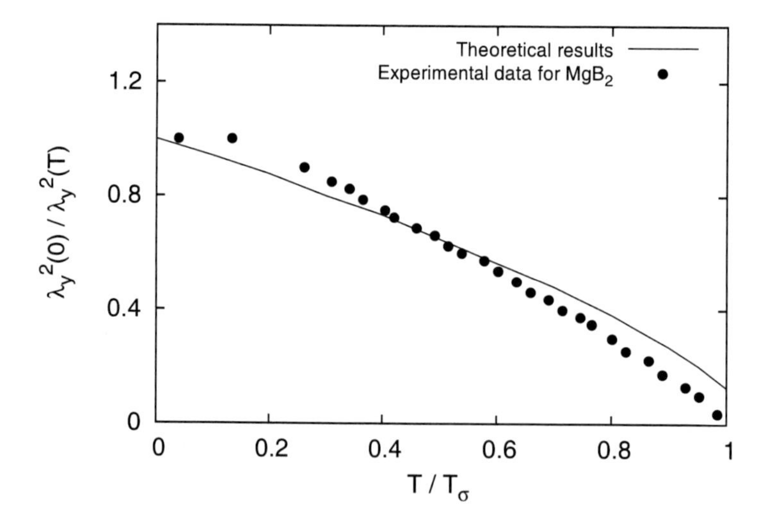

Figure 27. Temperature dependence of the penetration depth calculated by our two-order parameter model in comparison with the corresponding experimental data for MgB_2 [107]. Other parameter values are the same as in Figure 25 with magnetic field induction $b = 0.003$.

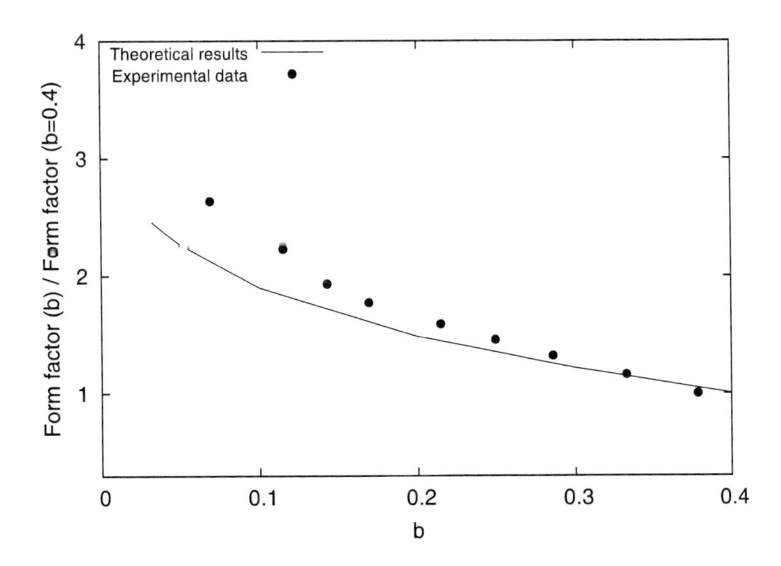

Figure 28. Comparison of magnetic form factor calculated by our two-order parameter GL theory with the corresponding experimental data for MgB_2 [98]. Other parameter values are the same as in Figure 25.

computed results [6] for the temperature dependence of magnetic penetration depth for the parameters mentioned in the figure caption are compared with the one determined experimentally [107]. The discrepancy between the theoretical and experimental result arises since for comparison with experimental data the theoretical results are calculated in the lower magnetic field regime where our model ceases to be accurate. Moreover, since the study is carried out in the framework of GL theory at very low temperatures inaccuracies tend to set in the theoretically computed results.

The magnetic form factor is an important quantity for any superconducting system as it can give information regarding the structure of the vortex lattice and is determined by SANS measurements. In the present theoretical model the magnetic form factor has been determined and its magnetic field dependence has been presented in Figure 28 [6]. The results are compared with the experimental data for MgB_2 [98]. The agreement between the theoretical and experimental results have been found to be good over a wide range of magnetic field induction.

7.3.3. Reversible Magnetization

The earlier theoretical works carried out on MgB_2 which involved an effective single order parameter model failed to explain the reversible magnetization of the system over the entire range of applied magnetic field [99]. Also when the reversible magnetization of the system is used to determine the magnetic field dependence of the penetration depth, it leads to the gross overestimation of the same. In order to compute the equilibrium applied magnetic field H and the corresponding reversible magnetization $M = \bar{B} - H$ of the system, we have used a suitable virial theorem. In terms of the gauge invariant real quantities ω_σ, ω_π and $\mathbf{Q}$

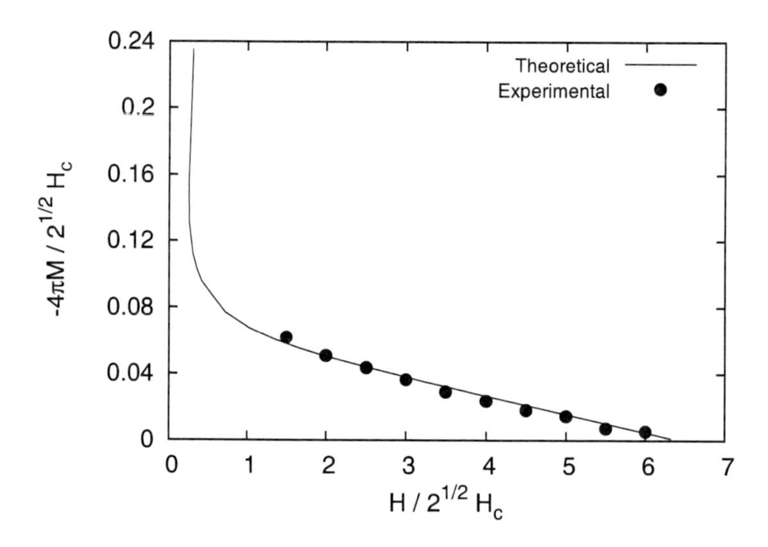

Figure 29. Comparison between the reversible magnetization result calculated by our two-order parameter model with the corresponding experimental data for MgB_2 [108]. Other parameter values are the same as in Figure 25.

corresponding to the σ, π-band order parameter components and supervelocity respectively the virial theorem corresponding to the MgB_2 problem can be expressed as [6],

$$2H\bar{B} = \Big\langle \omega_\sigma - \alpha\omega_\pi - \omega_\sigma^2 - 2\beta_\pi\omega_\pi^2 - 2\gamma\cos(\phi)(\omega_\sigma\omega_\pi)^{1/2} + 2B^2 \Big\rangle. \quad (31)$$

Using this expression for the equilibrium applied magnetic field (Eqn. (31)) the reversible magnetization of the system can be calculated as per the relation $M = \bar{B} - H$. Figure 29 shows the comparison of the theoretically calculated reversible magnetization of the MgB_2 superconducting system with the corresponding experimental data [108]. The experimental data is compared with the theoretical results calculated for different values of the linear interband coupling parameter γ and the best fit is presented in Figure 29. By comparing with the experimental data for reversible magnetization of the system the linear interband coupling parameter is fixed to be $\gamma = -0.1$. The excellent agreement between the theoretical and experimental results suggest that a two-order parameter GL theory is required to explain the properties of MgB_2 and an effective single order parameter model will not suffice.

7.3.4. Superconducting Current density

An important property of MgB_2 superconducting material is its high current carrying capacity in spite of a simple chemical composition as compared to the high-T_c cuprates. The present theoretical model has been used to carry out a detailed study of the various aspects of the superconducting current density of MgB_2. Figure 30 shows the magnetic field dependence of the current density calculated at different temperatures [6]. A lower temperature corresponds to a higher magnitude of current density. This is an expected behavior

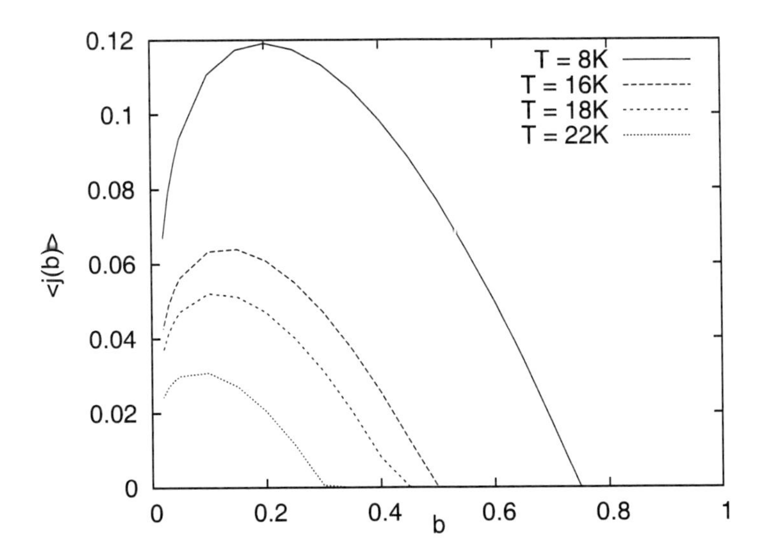

Figure 30. Variation of the supercurrent density with magnetic field induction at different temperatures. The parameters used are the same as in Figure 25.

since at higher temperatures and near the critical temperature the current carrying capacity of the material gets reduced due to the thermal fluctuations. At lower temperatures the current carrying capacity of the material is enhanced and current can be carried over a wide range of applied magnetic field. At lower applied magnetic fields the current density shows a drop in the magnitude indicating the fact that in the present theoretical model pinning of the vortices has not been taken into account. Similar behavior of the current density profile has been observed in experimental studies also [109].

The current density of a superconducting system is related to the magnetic penetration depth as $\mathbf{j}_c(T) \approx 1/\lambda^2(T)$ [110]. Thus, a higher current density should correspond to a lower magnetic penetration depth of the system and vice verse. The behavior has been observed from our theoretical calculations also. The theoretically calculated superconducting current density of the system is found to be in good agreement with the relevant experimental data for MgB_2 [111]. The corresponding results are presented in Figure 31.

7.3.5. Shear modulus of the vortex lattice

On calculating the shear modulus (c_{66}) of the vortex lattice it has been observed that a higher linear interband coupling between the order parameters correspond to a lower magnitude of the shear modulus of the vortex lattice indicating the softening of the flux line lattice [6]. The observation is in agreement with our previous results on the dependence of the magnetic penetration depth on the linear interband coupling parameter γ. The shear modulus of the vortex lattice is inversely proportional to the magnetic penetration depth of the system ($c_{66} \approx 1/\lambda^2$). For any particular magnetic field induction the shear modulus of the vortex lattice decreases with increasing temperature. This indicates that the melting of the flux line lattice is favored near T_c, in agreement with the experimental observations of melting of the

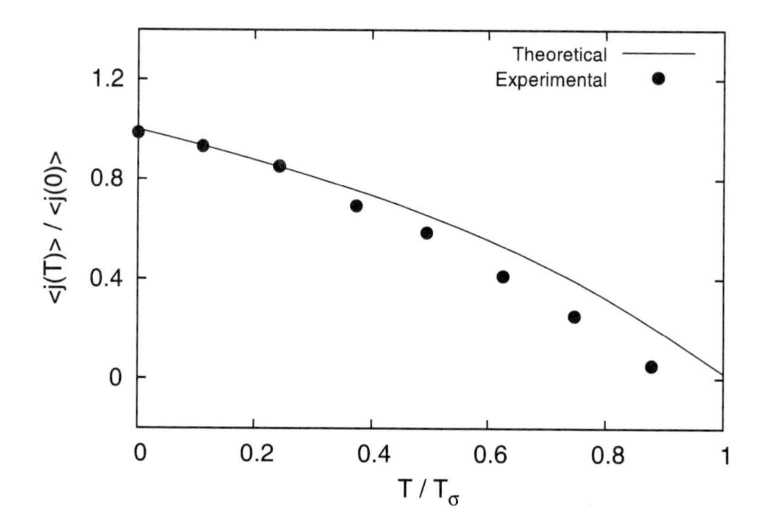

Figure 31. Temperature dependence of the supercurrent density calculated theoretically by our two-order parameter GL theory in comparison with the corresponding experimental data for MgB_2 [111]. The parameter values are the same as in Figure 25, with the magnetic field induction $b = 0.003$.

flux line lattice of MgB_2 [112].

In this section the results obtained by theoretically studying the various properties of the two-band superconductor MgB_2 over the entire range of applied magnetic field $H_{c1} < H < H_{c2}$ and wide range of temperature is presented. The study has been carried out in the framework of a two-order parameter GL theory. The spatial distribution of the order parameters reveals two different length scales corresponding to the σ and π-bands in agreement with the recent STM experiments on MgB_2. Both the vortex core radius and magnetic penetration depth is observed to be field dependent and this is due to the change in supercarrier density as superconductivity in the π-band is suppressed by the applied field. Apart from the local spatial behaviors viz. the vortex core radius and magnetic penetration depth, other properties such as, the reversible magnetization, shear modulus of the vortex lattice, superconducting current density of the system has also been calculated and the effect of the variation of the linear interband coupling parameter γ on these properties is discussed.

8. Conclusions

In this chapter, various aspects of the high temperature superconductors have been studied and discussed in the framework of Ginzburg–Landau theory. The key issue of the article is to theoretically explore the experimentally relevant applied magnetic field region ($H_{c1} < H < H_{c2}$). The studies are carried out for arbitrary values of GL parameter κ and vortex lattice symmetry. This has been achieved by using a high precision numerical iteration technique. The work could overcome the limitations of the earlier theoretical works carried out in the framework of GL formalism. The work can be subdivided in four parts, the first part comprises of studying the effect of in-plane mass anisotropy on two-dimensional bulk high-T_c cuprate $YBCO$ in the framework of a single order parameter GL theory. In the second part we have discussed the effect of c-axis anisotropy on the properties of superconducting thin films. The third part of the article is devoted in developing and application of a two-order parameter GL theory for studying the properties of the high- T_c cuprates involving complex order parameters. Finally the last part discusses the application of the two order parameter GL theory to study the properties of two-band intermetallic superconductor MgB_2.

Both the two-dimensional anisotropic single order parameter model and the isotropic two-order parameter model have confirmed the oblique structure of the vortex lattice as has been observed experimentally. The four-fold symmetric structure of the vortices as has been found experimentally however could not be observed through the anisotropic single order parameter model. This experimentally observed behavior could be interpreted using a two-order parameter GL theory. This indicates the fact that the mixed symmetry state scenario is a suitable candidate for studying the properties of the high- T_c superconducting cuprates, particularly $YBCO$. Apart from the oblique structure of the vortex lattice and four-fold structure of the single vortices the isotropic two-order parameter GL model could further depict the magnetic field and temperature dependence of the vortex core radius and magnetic penetration depth. In order to calculate the reversible magnetization of the system, instead of taking the numerical derivative of the free energy density functional we have derived a virial theorem suitable for the two-order parameter systems. The applicability of the model to arbitrary vortex lattice symmetry has been utilized to calculate the shear modulus (c_{66}) of the flux line lattice. We have compared our theoretical result corresponding to various properties of the high-T_c cuprates to the relevant experimental data and the agreement has been found to be very good.

The isotropic two-order parameter model though could depict the oblique structure of the vortex lattice failed to predict the experimentally observed angle between the primitive axes of the lattice. In order to obtain this, along with the mixed symmetry state of the order parameters we have introduced the effect of in-plane anisotropy in the system. The presence of in-plane anisotropy in the system can be attributed to the orthorhombic distortion of the CuO_2-planes arising due to the presence of CuO chains in the high temperature superconductors. Along with the other properties of the high- T_c cuprate $YBCO$ the anisotropic two-order parameter model could theoretically justify the experimentally observed angle between the primitive axes of the vortex lattice for the experimentally determined value of in-plane anisotropy parameter $\gamma = 2.4$. The model further suggested that the single vortex structure undergoes a transition from the four-fold to two-fold symmetry with increasing in-plane anisotropy. An interesting behavior has been observed regarding the shear modulus of the flux line lattice. While a higher admixture of s-wave order parameter component tends to softening of the lattice and thereby favor its melting, a higher in-plane anisotropy makes the lattice hard and difficult to melt. In presence of both the mixed symmetry state of the order parameters and in-plane anisotropy, the subdominant effect of the admixture of the s-wave order parameter in the system is suppressed by the dominant effect of the in-plane anisotropy. A harder lattice supports a larger critical current, thus a balance between the softening of the vortex lattice and the current carrying capacity of the high-T_c cuprates can be achieved by controlling the amount of admixture of s-wave order parameter component in the system and the amount of in-plane anisotropy.

As an application of the two-order parameter model we have used this model to calculate the properties of the two-band superconductor MgB_2. Superconductivity in this material is principally governed by the two-dimensional σ-band and thus for a wide range of magnetic field our two-dimensional approach to the problem gives very good agreement with the experimentally observed behaviors. The study showed that the material is characterized by two length scales as ξ_σ and ξ_π corresponding to the σ and π-band order parameter components. As a characteristic of two-band superconductors the order parameter component corresponding to the passive π-band varies more slowly as compared to that of the active band. The variation of various properties of the system with the linear interband coupling parameter γ have been studied in depth along with the magnetic field and temperature dependence of the various properties. A larger interband coupling favors the melting of the flux line lattice and reduces the current carrying capacity of the material.

As a generalization of the single-order parameter model, the two-dimensional method has been extended to study the effect of c-axis anisotropy in three-dimensional superconducting thin films. The profiles of order parameter, magnetic field distribution inside and outside the film are calculated for arbitrary thickness of the HTS films. As an artifact of the uniaxial anisotropy the magnetic field inside the film gets suppressed. The longitudinal and transverse variance outside the film shows the enhancement in the field component near the film surface, indicating that the uniaxial anisotropy leads to screening currents on the surface of the film which causes excess magnetic field on the surface. The sharp drop in the value of transverse component of variance of magnetic field exactly at the surface inside the film gives an indication of presence of surface barrier to the flux penetration. This barrier is observed to be increasing with increase in mass anisotropy ratio. The higher magnetic field on the surface of the film reduces the order parameter near the surface of the film. Also this increase in magnetic field on the surface results in enhancement of the stray field energy. This is important to know since the interaction of vortices in thin films basically depends on the stray field outside the film. Since surface energy originates mainly from magnetic stray field, the surface energy shows higher values for anisotropic case than that of the isotropic values. The anisotropic current density obtained numerically for $YBa_2Cu_3O_{7-\delta}$ films of a few nanometer thickness is qualitatively compared with the experimental current density curves. Experimentally observed increase in current density within the range $50 - 300$nm thick films is more clearly observed in the anisotropic current density curves as compared to the isotropic curves. The shear modulus calculations indicate that the rigidity of vortex lattice caused by film surface for thick films depend on anisotropy parameter γ_z as well as on film thickness d.

9. Acknowledgments

Financial assistance from BCUD, University of Pune, Pune, India through a research grant is gratefully acknowledged.

References

[1] Achalere, A. ; Dey, B. *Phys. Rev. B 2005* 2005, **71**, 224505.

[2] Karmakar, M.; Dey, B. *Phys. Rev. B* 2006, **74**, 172508.

[3] Achalere, A.; Dey, B. *Physica C* 2008, **468**, 2241.

[4] Karmakar, M.; Dey, B. *J. Phys. Cond. Mat.* 2008, **20**, 255218, and the references therein.

[5] Karmakar, M.; Dey, B. *J. Phys. Cond. Mat.* 2009, **21**, 405702, and the references therein.

[6] Karmakar, M.; Dey, B. *J. Phys. Cond. Mat.* 2010, **22**, 205701, and the references therein.

[7] Onnes, H. Kamerlingh *Leiden Comm.* 1911, **120b**, **122b**, **124c**.

[8] Frölich, H. *Phys. Rev.* 1950, **79**, 845.

[9] Cooper, L. N. *Phys. Rev.* 1956, **104**, 1189.

[10] Bardeen, J.; Cooper, L. N.; Schrieffer, J. R. *Phys. Rev.* 1957, **108**, 1175.

[11] McMillan, W. L. *Phys. Rev.* 1968, **167**, 331.

[12] Bednorz, J. G.; Müller, K. *Z. Physik B* 1986, **64**, 189.

[13] Ginzburg, V. L.; Landau, L. D. *Zh. Eksperim. i. Teor. Fiz.* 1950, **20**, 1064.

[14] Gor'kov, L. P. *Zh. Eksperim. i. Teor. Fiz.* 1959, **36**, 1918 [*Soviet Phys. - JETP* 1959, **9**, 1364].

[15] Abrikosov, A. A. *Sov. Phys. JETP* 1957, **5**, 1174.

[16] Eilenberger, G. *Z. Phys.* 1964, **180**, 32.

[17] Ihle, D. *Phys. Stat. Solidi (b)* 1971, **47**, 423.

[18] Saint-James, D.; Sarma, G.; Thomas, E. *Type II superconductors* (Oxford Univ. Press, Newyork, 1969); Oliviera, I. G.; Thompson, A. M. *Phys. Rev. B* 1998, **57**, 7477; Fetter, A. L. *Phys. Rev.* 1966, **147**, 153.

[19] Clem, J. R. *J. Low. Temp. Phys.* 1975, **18**, 427

[20] Hao, Z.; Clem, J. R. *Phys. Rev. Lett.* 1991, **67**, 2371.

[21] Hao, Z.; Clem, J. R.; Mc Elfresh, M. W.; Civale, L.; Malozemof, A. P.; Holtzberg, F. *Phys. Rev. B* 1991, **43**, 2844.

[22] Brandt, E. H. *Phys. Status Solidi (b)* 1972, **51**, 354.

[23] Brandt, E. H. *Phys. Rev. Lett.* 1997, **78**, 2208.

[24] Zhang, K.; Bonn, D. A.; Kamal, S.; Liang, R.; Baar, D. J.; Hardy, W. N.; Basov, D.; Timusk, T. *Phys. Rev. Lett.* 1994, **73**, 2484.

[25] Lake, B.; Rnnow, H. M.; Christensen, N. B.; Aeppli ,G.; Lefmann, K.; Mcmorrow, Vorderwisch, P.; Smeibidl, P.; Mangkorntong, N.; Sasagawa, T.; Nohara, M.; Takagi, H.; Mason, T. E. *Nature* 2002, **415**, 299.

[26] Keimer, B.; Shih, W. Y.; Erwin, R. W.; Lynn, J. W.; Dogan, F.; Aksay, I. A. *Phys. Rev. Lett.* 1994, **73**, 3459; *J. Appl. Phys.* 1994, **76**, 6778.

[27] Kealey, P. G.; Charamlambous, D.; Forgan, E. M.; Lee, S. L.; Johnson, S. T.; Schleger, P.; Cubitt, R.; McK. Paul, D.; Aegerter, C. M.; Tajima, S.; Rykov, A. *Phys. Rev. B* 2001, **64**, 174501; Carlson, E. W.; Castro Neto, A. H.; Campbell, D. K. *Phys. Rev. Lett.* 2003, **90**, 087001.

[28] de Oliveira, I. G.; Doria, M. M.; Brandt, E. H. *Physica C* 2000, **341-348**, 1069.

[29] Maggio-Aprile, I.; Renner, Ch.; Erb, A.; Walker, E.; Fischer, ø *Phys. Rev. Lett.* 1995, **75**, 2754.

[30] Doria, M. M.; Gubernatis, J. E.; Rainer, D. *Phys. Rev. B* 1989, **39**, 9573.

[31] Sok, J.; Xu, M.; Chen, W.; Suh, B. J.; Gihng, J.; Finnemore, D. K.; Kramer, M. J.; Schwartzkopf, L. A.; Dabrowski, B. *Phys. Rev. B* 1995, **51**, 6035.

[32] Pearl, J. *Appl. Phys. Lett.* 1964, **5**, 65.

[33] Fritz, O.; Wulfert, M.; Hug, H. J.; Thomasm, H.; Guntherodt, H. J. *Phys. Rev. B* 1993, **47**, 384.

[34] Irx, D. Yu.; Ryzhov, V. N.; Tareyeva, E. E. *Phys. Lett. A* 1995, **374**, 207.

[35] Wei, J. -C.; Yang, T. -J. *Jpn. J. Appl. Phys.* Part 1 1996, **35**, 5696 .

[36] Carneiro, G.; Brandt, E. H. *Phys. Rev. B* 2000, **61**, 6370.

[37] Brandt, E. H. *Phys. Rev. B* 2005, **71**, 014521.

[38] Durell, J. H.; Events, J. E. *Appl. Phys. Lett.* 2003, **83**, 4999.

[39] Tajima, S.; Schutzman, J.; Miyamoto, S.; Terasaki, I.; Sato, Y.; Hauff, R. *Phys. Rev. B* 1997, **55**, 6051.

[40] Triscone, G.; Khoder, A. F.; Opagiste, C.; Genoud, J. Y.; Graf, T.; Janod, E.; Tsukamoto, T.; Couach, M.; Junod, A.; Muller, J. *Physica C* 1994, **224**, 263.

[41] Farell, D. E.; Bonhman, S.; Foster, J.; Chang, Y. C.; Chang, P. Z.; Jiang, P. Z.; Vandervoot, K. G.; Lam, D. J.; Kogan, V. G. *Phys. Rev. Lett.* 1989, **63**, 782.

[42] Martinez, J. C.; Brongersma, S. H.; Koshelev, A.; Ivlev, B.; Kes, P. H.; Griessen, R. P.; de Groot, D. G.; Tarnavski, Z.; Menovsky, A. A. *Phys. Rev. Lett.* 1992, **69**, 2276.

[43] Kirtley, J. R.; Kogan, V. G.; Clem, J. R.; Moler, K. A. *Phys. Rev. B* 1999, **59**, 4343.

[44] Clem, J. R. *Phys. Rev. B* 1991, **43**, 7837.

[45] Brandt, E. H. *Rep. Prog. Phys.* 1995, **58**, 1465.

[46] Dorosinskii, L. A.; Nikitenko, V. I.; Polyanskii, A. A. *Phys. Rev. B* 1994, **50**, 501.

[47] Vodolazov, D. Yu.; Maksimov, I. L. *Physica C* 2001, **349**, 125.

[48] Van der Beek, C. J; Konczykowski, M.; Abal'oshev, A.; Abal'osheva, I.; Gierlowski, P.; Lewandowski, S. J.; Indenbom, M. V.; Barbanera, S. *Phys. Rev. B* 2002, **66**, 024532.

[49] Van Harlingen, D. J. *Rev. Mod. Phys.* 1995, **67**, 515.

[50] Barrett, S. E.; Martindale, J. A.; Durand, D. J.; Pennington, C. P.; Slicher, C. P.; Friedmann, T. A.; Rice, J. P.; Ginsberg, D. M. *Phys. Rev. Lett.* 1991, **66**, 108.

[51] Tsuei, C. C.; Kirtley, J. R. *Rev. Mod. Phys.* 2000, **72**, 969.

[52] Joynt, R. *Phys. Rev B* 1990, **41**, 4271.

[53] Chakravarty, S.; Laughlin, R. B.; Morr, D. K.; Nayak, C. *Phys. Rev. B* 2001, **63**, 094503.

[54] Beal-Monod, M. T.; Maki, K. *Europhys. Lett.* 1996, **33**, 309.

[55] Curras, S. R.; Ferro, G.; Gonzalez, M. T.; Ramallo, M. V.; Ruibal, M.; Veira, J. A.; Wagner, P.; Vidal, F. *Phys. Rev. B* 2003, **68**, 094501; Ramallo, M. V.; Pomar, A.; Vidal, F. *Phys. Rev. B* 1996, **54**, 4341.

[56] Volovik, G. E. *Pis'ma Zh. Eksp. Teor. Fiz.* 1993, **58**, 457.

[57] Affleck, I.; Franz, M.; Amin, M. H. S *Phys. Rev. B* 1997, **55**, R704.

[58] Ren, Y.; Xu, J.; Ting, C. S. *Phys. Rev. Lett.* 1995, **74**, 3680; Xu, J.; Ren, Y.; Ting, C. S. *Phys. Rev B* 1995, **52**.

[59] Soininen, P. I.; Kallin, C.; Berlinsky, A. J. *Phys. Rev B* 1994, **50**, 13883.

[60] Franz, M.; Kallin, C.; Soininen, P. I.; Berlinsky, A. J.; Fetter, A. L. *Phys. Rev. B* 1996, **53**, 5795.

[61] Heeb, R.; van Otterlo, A.; Sigrist, M.; Blatter, G. *Phys. Rev. B* 1996, **54**, 9385.

[62] Xu, J.; Ren, Y.; Ting, C. S. *Phys.Rev. B* 1996, **53**, R2991.

[63] Ichioka, M.; Enomoto, N.; Hayashi, N.; Machida, K. *Phys.Rev. B* 1996, **53**, 2233.

[64] Mel'nikov, A. S.; Nefedov, I. M.; Ryzhov, D. A.; Shereshevskii, I. A.; Vysheslavtsev, P. P. *Phys. Rev. B* 2000, **62**, 11820.

[65] Gohng, J.; Finnemore, D. K. *Phys. Rev. B* 1992, **46**, 398.

[66] Smilde, H. J. H.; Golubov, A. A.; Ariando, Rjinders, G.; Dekkers, J. M.; Harkema, S.; Blank, D. H. A.; Rogalla, H.; Hilgenkamo, H. *Phys. Rev. Lett.* 2005, **95**, 257001.

[67] Khasanov, R.; Strassle, S.; Di Castro, D.; Masui, T.; Miyasaka, S.; Tajima, S.; Bussmann-Holder, A.; Keller, H. *Phys. Rev. Lett.* 2007, **99**, 237601.

[68] Khasanov, R.; Shengelaya, A.; Maisuradze, A.; La Mattina, F.; Bussmann-Holder, A.; Keller, H.; Müller, K. A. *Phys. Rev. Lett.* 2007, **98**, 057007.

[69] Shan, L.; Huang, Y.; Ren, C.; Wen, H. H. *Phys. Rev. B* 2006, **73**, 134508.

[70] Sonier, J. E.; Brewer, J. H.; Kiefl, R. F.; Morris, G. D.; Miller, R. I.; Bonn, D. A.; Chakhalian, J.; Heffner, R. H.; Hardy, W. N.; Liang, R. *Phys. Rev. Lett.* 1999, **83**, 4156.

[71] Sonier, J. E.; Brewer, J. H.; Kiefl, R. F. *Rev. Mod. Phys.* 2000, **72**, 769.

[72] Bardeen, J. *Phys. Rev.* 1954, **94**, 554.

[73] Lipavský, P.; Koláček, J.; Morawetz, K.; Brandt, E. H. *Phys. Rev. B* 2002, **65**, 144511.

[74] Amin, M. H. S.; Affleck, Ian; Franz, M. *Phys. Rev. Lett.* 2000, **84** 5864.

[75] Prozorov, Ruslan; Giannetta, Russell W. *Sup. Sci. and Technol.* 2006, **19**, R41-R67.

[76] Hardy, W. N.; Bonn, D. A.; Morgan, D. C.; Liang, Ruixing; Kuan, Zhang *Phys. Rev. Lett.* 1993, **70**, 3999).

[77] Worthington, T. K.; Gallagher, W. J.; Dinger, T. R. *Phys. Rev. Lett.* 1987, **59**, 1160.

[78] Landau, I. L.; Ott, H. R. *Phys. Rev. B* 2002, **66**, 144506.

[79] Blatter, G.; Feigel'man, M. V.; Geshkenbein, V. B.; Larkin, A. I.; Vinokur, V. M. *Rev. Mod. Phys.* 1994, **66**, 1125.

[80] Ferrell, D. E.; Rice, J. P.; Ginsberg, D. M. *Phys. Rev. Lett.* 1991, **67**, 1165.

[81] Heyen, E. T.; Cardona, M.; Karpinski, J.; Kaldis, E.; Rusiecki, S. *Phys. Rev. B* 1991, **43**, 12958.

[82] Limonov, M. F.; Rykov, A. I.; Tajima, S.; Yamanaka, A. *Phys. Rev. Lett.* 1998, **80**, 825.

[83] Limonov, M. F.; Rykov, A. I.; Tajima, S.; Yamanaka, A. *Phys. Rev. B* 2000, **61**, 12412.

[84] Lu, D. H.; Feng, D. L.; Armitage, N. P.; Shen, K. M.; Damascelli, A.; Kim, C.; Ronning, F.; Shen, Z. X.; Bonn, D. A.; Liang, R.; Hardy, W. N.; Rykov, A. I.; Tajima, S. *Phys. Rev Lett.* 2001, **86**, 4370.

[85] Nagamatsu, J.; Nakagawa, N.; Muranaka, T.; Zenitani, Y.; Akimatsu, J. *Nature* 2001, **410**, 63.

[86] Giubileo, F.; Roditchev, D.; Sacks, W.; Lamy, R.; Thanh, D. X.; Klein, J.; Miraglia, S.; Fruchart, D.; Marcus, J.; Monod, Ph. *Phys. Rev. Lett.* 2001, **87**, 177008.

[87] Iavarone, M.; Karapetrov, G.; Koshelev, A. E.; Kwok, W. K.; Crabtree, G. W.; Hinks, D. G.; Kang, W. N.; Choi, Eun-mi; Kim, Hyun Jung; Kim, Hyeong-Jin; Lee, S. I. *Phys. Rev. Lett.* 2002, **89**, 187002.

[88] Bud'ko, S. L. *et al., Phys. Rev. Lett.* 2001, **86**, 1877.

[89] Choi, H. J.; Roudny, D.; Sun, H.; Cohen, M. L.; Louie, S. G. *Nature* 2002, **418** 758.

[90] Lyard, L. *et al. Phys. Rev. B* 2002, **66**, 180502(R); Bud'ko, S. L.; Kogan, V. G.; Canfield, P. C. *ibid.* 2001, **64**, 180506; Angst, M.; Puzniak, R.; Wisniewski, A.; Jun, J.; Kazakov, S. M.; Karpinski, J.; Roos, J.; Keller, H. *Phys. Rev. Lett.* 2002, **88**, 167004; Welp, U. *et al. Phys. Rev. B* 2003, **67**, 012505.

[91] Szabo, P. *et al. Phys. Rev. Lett.* 2001, **87**, 137005; Samuely, P. *et al. Physica C* 2003, **385**, 244.

[92] Cubitt, R.; Eskildsen, M. R.; Dewhurst, C. D.; Jun, J.; Kazakov, S. M.; Karpinski, J. *Phys. Rev. Lett.* 2003, **91**, 047002.

[93] Kunchur, M.; Saracila, G.; Arcos, D. A.; Cui, Y.; Pogrebnyakov, A.; Orgiani, P.; Xi, X. X.; Adams, P. W.; Young, D. P. *Physica C* 2006, **437-438**, 171-175.

[94] Pissas, M.; Papavassiliou, G.; Karayanni, M.; Fardis, M.; Maurin, I.; Margiolaki, I.; Prassides, K.; Christides, C. *Phys. Rev. B* 2002, **65**, 184514; Papavassiliou, G.; Pissas, M.; Fardis, M.; Karayanni, M.; Christides, C. *Phys. Rev. B* 2002, **65**, 012510.

[95] Jin, R.; Paranthaman, M.; Zhai, H. Y.; Christen, H. M.; Christen, D. K.; Mandrus, D. *Phys. Rev. B* 2001, **64**, 220506(R).

[96] Zhitomirsky, M. E.; Dao, V. H. *Phys. Rev. B* 2004, **69**, 054508.

[97] Betouras, J. J.; Ivanov, V. A.; Peeters, F. M. *Eur. Phys. J. B* 2003, **31**, 349.

[98] Klein, T.; Lyard, L.; Marcus, J.; Holanova, Z.; Marcenat, C. *Phys. Rev. B* 2006, **73**, 184513.

[99] Zehetmeyer, M.; Eisterer, M.; Jun, J.; Kazakov, S. M.; Karpinski, J.; Weber, H. W. *Phys. Rev. B* 2004, **70**, 214516.

[100] Gurevich, A.; Vinokour, V. M. *Phys. Rev. Lett.* 2003, **90**, 047004.

[101] Doh, H.; Sigrist, M.; Cho, B. K.; Lee, Sunk-Ik *Phys. Rev. Lett.* 1999, **83**, 5350.

[102] Dahm, T.; Schopohl, N. *Phys. Rev. Lett.* 2003, **91**, 017001; Eskildsen, M. R.; Kugler, M.; Tanaka, S.; Jun, J.; Kazakov, S. M.; Karpinski, J.; Fischer, ø. *Phys. Rev. Lett.* 2002, **89**, 187003.

[103] Golubov, A. A.; Brinkman, A.; Dolgov, O. V.; Kortus, J.; Jepsen, O. *Phys. Rev. B* 2002, **66**, 054524.

[104] Angst, M.; DiCastro, D.; Eschenko, D. G.; Khasanov, R.; Kohout, S.; Savic, I. M.; Shengelaya, A.; Budko, S. L.; Canfield, P. C.; Jun, J.; Karpinski, J.; Kazarov, S. M.; Ribeiro, R. A.; Keller, H. *Phys. Rev. B* 2004, **70**, 224513.

[105] Lyard, L.; Szabo, P.; Klein, T.; Marcus, J.; Marcenat, C.; Kim, K. H.; Kang, B. W.; Lee, H. S.; Lee, S. I. *Phys. Rev. Lett.* 2004, **92**, 057001.

[106] Pal, D.; De Beer-Schmitt, L.; Bera, T.; Cubitt, R.; Dewhurst, C. D.; Jun, J.; Zhigadlo, N. D.; Karpinski, J.; Kogan, V. G.; Eskildsen, M. R. *Phys. Rev. B* 2006, **73**, 012513.

[107] Golubov, A. A.; Brinkman, A.; Dolgov, O. V.; Kortus, J.; Jepsen, O. *Phys. Rev. B* 2002, **66**, 054524.

[108] Kang, B.; Kim, H. J.; Park, M.; Kim, K. H.; Lee, S. *Phys. Rev. B* 2004, **69**, 14451.

[109] Nishida, A.; Taka, C.; Chromik, S.; Durny, R. *Physica C* 2005, **426-431**, 340-344.

[110] Askerzade, I. N. *Physica C* 2003, **390**, 281-285.

[111] Sahin, H.; Askerzade, I. N. *Eur. Phys. J. Appl. Phys.* 2007, **36**, 267-270.

[112] Nie, Qing-Miao; Lv, Jian-Ping; Chen, Qing-Hu *Phys. Lett. A* 2009, **374**, 655-658.

In: Superconductivity
Editor: Vladimir Rem Romanovskii
ISBN 978-1-61324-843-0

Chapter 11

CHARGE DISTRIBUTION AND HYPERFINE INTERACTIONS IN $GdBa_2Cu_3O_7$ FROM FIRST PRINCIPLES

Maciej Łuszczek *
Faculty of Applied Physics and Mathematics
Gdańsk University of Technology
G. Narutowicza 11/12, 80-233 Gdańsk, Poland

Abstract

The electronic band structure, charge distribution and hyperfine interactions, namely, electric field gradients (EFG) and contact terms of hyperfine fields (HFF) in stoichiometric $GdBa_2Cu_3O_7$ (Gd123) high-temperature superconductor have been calculated on the first-principles basis using the *full-potential linearized augmented plane wave* (FP-LAPW) method. The *generalized gradient approximation* (GGA) was employed to treat the exchange and correlation effects. The Hubbard correction U was applied for $4f$ electrons to account for the strong on-site Coulomb repulsion. The calculated electronic structure and charge distribution is typical for $(rare - earth)123$ class of compounds. It was shown that the evaluated strongly localized Gd magnetic moment, which is comparable with the one obtained from the neutron diffraction data, could not effectively influence the neighbors resulting in small itinerant moments in the CuO_2 planes. The calculated EFG and HFF parameters are consistent with the Mössbauer measurements. Our results indicate that in the highly correlated $Gd123$ system the applied computational method yields reliable charge distribution to which the EFG and HFF are very sensitive.

PACS 74.72.-h, 71.20.-b, 71.15.Mb, 31.30.Gs.
Keywords: Cuprate superconductors, electronic structure, density functional theory, hyperfine interactions.

1. Introduction

It was demonstrated that in the $YBa_2Cu_3O_{7-\delta}$ ($Y123$) high-temperature superconductor (HTSC) the Y^{3+} ion can be replaced by most of the rare-earth R^{3+} ions [1] without any

*E-mail address: maclu@mif.pg.gda.pl

significant changes in its superconducting properties. The correct understanding of the electronic structure upon the rare-earth substitutions is still a challenge because the coexistence of the magnetism and the superconductivity in this class of compounds is completely different from the behavior of the conventional superconductors.

The charge distribution appears to be a crucial quantity in the analysis of the superconducting properties of the $Y123$–like systems. The hole concentration in the CuO_2 layers, which are commonly believed to be involved in the superconductivity, depends on the filling of certain Cu $3d$ and O $2p$ states. Unfortunately, in such complex materials a detailed charge distribution cannot be obtained directly from the experiment (e.g. from the x-ray diffraction data). However, hyperfine interactions, such as electric field gradient (EFG) and hyperfine field (HFF), can be measured by various techniques and provide information on the interaction of a nucleus with the surrounding charge. An interpretation of these measurements allows for precise insight into the electronic and magnetic structure. Therefore, the EFG and HFF measurements seem to be ideally suited for the study of charge distribution in this class of high–T_c superconductors. On the other hand, the accurate electronic structure calculations within *density functional theory* (DFT) provide the basis for computing either EFG or HFF parameters, what enables direct comparison between theory and experiment.

The local ground state electronic structure can be nicely probed through the EFG at a given nuclear site. In the solid-state physics, this quantity measures the rate of change of the electric field at an atomic nucleus generated by the electronic charge distribution and the other nuclei. The EFG couples with the nuclear electric quadrupole moment of quadrupolar nuclei (with spin quantum number greater than one-half) to generate an effect which can be measured using several spectroscopic methods, such as nuclear magnetic resonance (NMR), nuclear quadrupole resonance (NQR), Mössbauer spectroscopy or perturbed angular correlation (PAC) method. The EFG is defined as the second derivative of the electrostatic potential written as a traceless tensor. The diagonal terms of the EFG tensor with respect to the crystallographic a, b and c axes are V_{aa}, V_{bb} and V_{cc}, respectively. By ordering these components according to their magnitudes we can define $|V_{zz}| \geq |V_{yy}| \geq |V_{xx}|$. The EFG is commonly characterized by the largest component V_{zz} and the anisotropy parameter $\eta = (V_{xx} - V_{yy})/V_{zz}$.

The hyperfine field is a precise and essential probe of the magnetic state of a solid and of the quality of theoretical core wave functions. The HFF on a nuclei of an atom may be written as $B_{hf} = B_c + B_{dip} + B_{orb} + B_{lat}$, where B_c is the Fermi contact term, B_{dip} is the dipolar field from the on-site spin density, B_{orb} is the field associated with the on-site orbital moment. B_{lat} is classical dipolar field from all other atoms in the system that carry the magnetic moment. The most important contribution is the Fermi contact term B_c, which is the sum of the contributions of valence and core electrons. Usually, nuclear hyperfine fields are associated with local magnetic moments.

In this Chapter we concentrate on the electronic band structure, charge distribution and hyperfine interactions in $GdBa_2Cu_3O_7$ high-temperature superconductor because the reliable experimental results are available [2], [3], [4], [5] and so the direct comparison with our computational approach is possible.

2. Computational Details

All calculations were performed within the *generalized gradient approximation* (GGA) [6], which provides an improved description of exchange and correlation effects in solids in comparison with the standard *local-spin density approximation* (LSDA). The local orbital extension (LO) of the *full-potential linearized augmented-plane wave* (FP - LAPW) method [7], [8] was applied. The well-known WIEN2k code [9], [10] was used for the electronic structure calculations. Approximately 1400 LAPW functions plus LOs were used. The crystallographic data for $GdBa_2Cu_3O_{7-\delta}$ with $Pmmm$ symmetry, collected in Ref. [11], were used as the input. Atomic sphere radii of 2.30, 2.30, 1.90, and 1.55 $a.u.$ were taken for Gd, Ba, Cu, and O, respectively. The plane-wave cutoff $R_{min}K_{max} = 6.0$ was chosen, where R_{min} denotes the smallest atomic sphere radius and K_{max} gives the magnitude of the largest ***K*** vector in the plane-wave expansion. Self-consistency was obtained using a set of 16 special ***k*** points in the irreducible Brillouin zone (IBZ) with the charge density taken as a criteria of convergence (the "charge distance" of the order of $10^{-5}e$ between the successive iterations was reached).

The core and valence states were treated self-consistently, the former relativistically in an atomic-like approximation [12] and the later — scalar relativistically with spin-orbit interactions included in a second-variational procedure [13]. The starting (free-atom) electron configurations of the constituent atoms were as follows: Gd $[Xe]$ $4f^7$ $5d^1$ $6s^2$, Ba $[Xe]$ $6s^2$, Cu $[Ar]$ $3d^9$ $4s^2$ and O $[He]$ $2s^2$ $2p^4$. The GGA plus Hubbard parameter U(GGA+U) approach [14] was employed to account for the strong on-site Coulomb repulsion of the $4f$ electrons in Gd. The $U_{Gd} = 0.4$ Ry and $J_{Gd} = 0.07$ Ry parameters were applied (the same values were successfully used for $4f$ states in the previous study of the $DyBa_2Cu_3O_7$ and $TbBa_2Cu_3O_7$ systems [15], [16]). It was reported in the literature that conventional band structure calculations can quite properly reproduce many physical properties of the metallic (doped) HTSC and U_{Cu} parameter could play an important role mainly in the case of the underdoped materials [17]. In this study we assumed the optimally doped (metallic) character of the $Gd123$ system and thus this correction for Cu $3d$ states was not applied. The modified tetrahedron method of Blöchl *et al.*[18] was used to determine the density of states (DOS). The electric field gradients (EFG) can be calculated directly from the self-consistent charge density by solving Poisson's equation without further approximations, as it was firstly demonstrated by Blaha *et al.*[19]. The hyperfine field (HFF) calculations were performed using relativistic formula given by Breit [20]. The employed implementation of FP-LAPW method enable to determine the relativistic contact contribution B_c, obtained by the averaging of spin over a small region (with the size described by the Thomson radius $r_T = Ze^2/mc^2$) around the nucleus.

3. Results and Discussion

The calculated electronic structure of $GdBa_2Cu_3O_7$, with the dominant $Cu(1)$ $3d$ – $O(4)$ $2p$ (CuO chain-derived) and $Cu(2)$ $3d$ – $O(2,3)$ $2p$ (CuO_2 plane-derived) bands crossing the Fermi level, exhibits the metallic character, as expected, and the bands are similar to those established in the $Y123$ system [21], [22]. The total density of states (DOS) of $Gd123$ is presented in Figure 1. One can notice some narrow high-density peaks, corresponding to $Gd\ 4f$-derived bands, which are shown separately in Figure 2. Noteworthy, there is no fractional occupation of $Gd\ 4f$ states at the Fermi level (E_F) in the GGA calculations, what was recently observed for $Dy\ 4f$ states in $Dy123$ system [15], resulting in an atomic-like Gd^{3+} ion behavior. The U_{Gd} parameter produces the additional Hubbard splitting between the occupied and empty $4f$ states. In the both methods the $Gd\ 4f$ minority spin states are entirely above the Fermi energy.

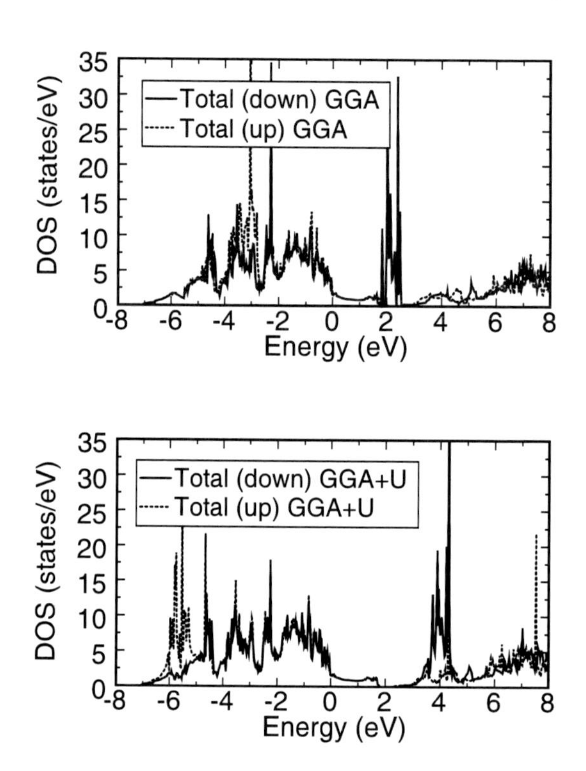

Figure 1. The total density of states (DOS) for majority (up) and minority (down) spin states in $GdBa_2Cu_3O_7$ system from GGA and GGA+U calculations. The Fermi level was put at $0\ eV$.

The details of the total DOS in the nearest vicinity of E_F obtained from GGA and GGA+U calculations are shown in Figure 3. At a first glance there are some differences between the both approaches. As can be seen, the noticeable spin splitting of bands at the Fermi level occurs only in the GGA calculations. The GGA+U calculations also give the splitting of DOS but it extends roughly below $-50\ meV$ and is not supposed to affect the superconductivity (it is well known that the energy gap 2Δ in the high-T_c copper oxide superconductors is roughly equal $50\ meV$ [23]). The same effect was previously reported for $Dy123$ system in the analogous GGA+U study [15]. What is more, the $Gd123$ system exhibits slightly higher value of density of states at the Fermi level than $Dy123$, that is,

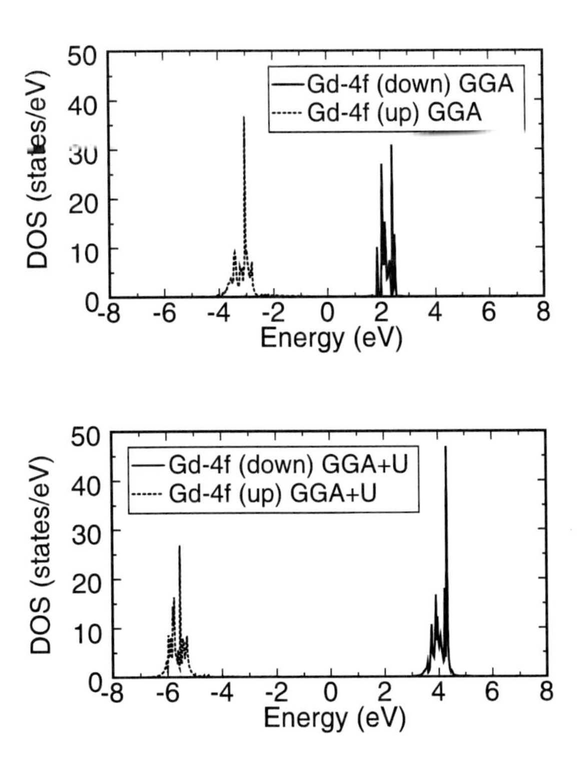

Figure 2. The positions of majority (up) and minority (down) Gd $4f$ bands from GGA and GGA+U calculations. The Fermi level was put at $0\ eV$.

$N_{Gd123}(E_F) = 3.50\ states/eV$ and $N_{Dy123}(E_F) = 3.16\ states/eV$ what could imply slightly better superconducting properties of this compound. Indeed, it is known from the experiment that the superconducting transition temperature T_c in $R123$ shows a week increase with the size of the R^{3+} ion [24]. Taking the above facts into considerations, the use of GGA+U seems to be a better choice in the case of $Gd123$ system.

The atomic-decomposed (partial) DOS characteristics from the GGA+U approach are presented in Figure 4. As stated before, the partial DOS of Gd in $Gd123$ is dominated by $4f$ peaks. Ba atoms generate relatively small DOS in the valence bands and unoccupied states are shifted upwards by approximately $3.5\ eV$ above E_F. The plane $Cu(2)$ density of states at the Fermi level is significantly higher than that of chain $Cu(1)$ with the considerable weight centered around $-1\ eV$. All the Cu bands extend to very similar positions at the top of the bands ($2\ eV$). However, the chain-derived $Cu(1)$ bands extend down only to (approximately) $-5.5\ eV$, while the plane-derived $Cu(2)$ bands extend to $-7\ eV$. As could be seen from Figure 4, the plane $O(2)$ and $O(3)$ partial densities are practically identical, indicating that orthorhombicity arising from the chains has very little effect on the electronic structure of the CuO_2 planes. The plane $O(2,3)$ valence bands extend from roughly

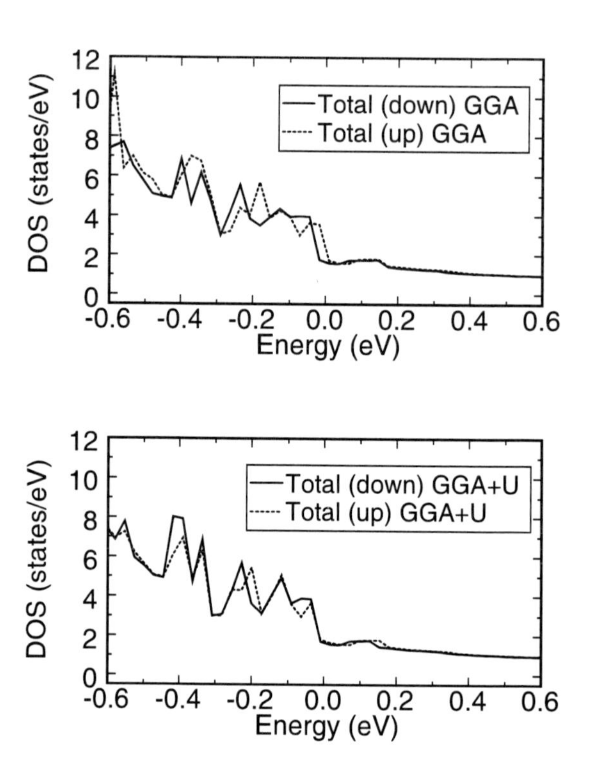

Figure 3. Details of total DOS of $GdBa_2Cu_3O_7$ systems in the nearest vicinity of the Fermi level for the majority (up) and minority (down) spin states obtained from the GGA and GGA+U calculations. The Fermi level was put at $0\ eV$.

$-7\ eV$ to $2\ eV$, that is, in the same range as $Cu(2)$ bands. As the chain oxygens $O(4)$ are considered, the partial DOS is lower than in the case of the $O(2)$ and $O(3)$ oxygens from the planes. These results are generally consistent with the previously reported calculations of $Y123$-like systems [15], [16], [25], [26].

In the applied method of calculations the electronic charge inside the unit cell is distributed either in the interstitial region or in the atomic spheres giving so-called partial charges. Thus, the symmetry decomposition according to lm numbers can be made [9], [10]. Partial charges (in electrons) corresponding to $Gd(5s, 5p, 5d, 4f)$, $Ba(6s, 5p, 5d)$, $Cu(4s, 4p, 3d)$ and $O(2s, 2p)$ majority and minority spin valence states in $GdBa_2Cu_3O_7$ are collected in Table 1 and the symmetry decomposition of p and d states is presented in Table 2.

As could be seen from Table 1, only small leakage of $Gd\ 4f$ charge from the atomic sphere is visible, what means that $4f$ states are quite well spatially confined and therefore they cannot effectively hybridize with the other valence electrons. The partial charges of Gd correspond to almost nominal $4f^7\ 5s^2\ 5p^6$ configuration of isolated Gd^{3+} ion with about

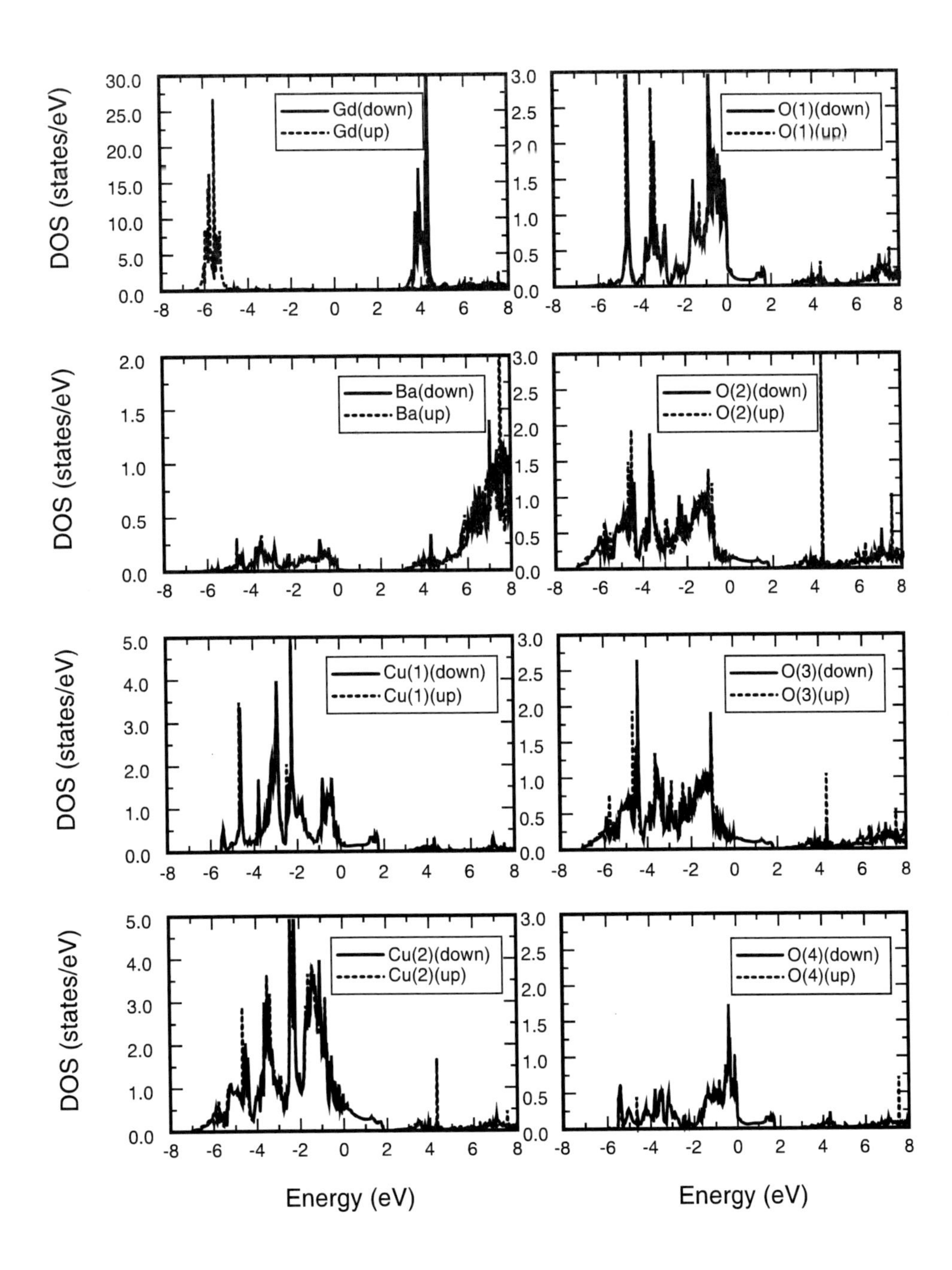

Figure 4. Partial DOS of $GdBa_2Cu_3O_7$ system for the majority (up) and minority (down) spin states in the GGA+U approach. The Fermi level was put at $0\ eV$.

Table 1. Calculated partial charges in $GdBa_2Cu_3O_7$ system. All data is given in electrons (e).

	s	p	d	f
Gd	1.99	5.51	0.39	6.97
Ba	1.79	4.55	0.12	
$Cu(1)$	0.22	6.16	8.31	
$Cu(2)$	0.20	6.16	8.66	
$O(1)$	1.55	3.38		
$O(2)$	1.54	3.42		
$O(3)$	1.54	3.42		
$O(4)$	1.55	3.35		

0.4 e shifted to d orbitals. One can notice that p_x, p_y and p_z states of Gd are filled without significant anisotropy and that the charges of d_{z^2} and d_{yz} symmetries are significantly smaller than that of the $d_{x^2-y^2}$ and d_{xy}, d_{xz} (see Table 2). In the Ba atomic sphere some part of the valence charge is moved to $5d$ states, as well, when compared with the $5s^25p^6$ configuration of Ba^{2+} ion. For both copper sites, the highest partial charges correspond to d_{xy}, d_{xz} and d_{yz} orbitals, which are nearly filled, what means that their interactions with the neighbors are weak. The chain $Cu(1)$ atoms gather the smallest charge in the d_{z^2} and $d_{x^2-y^2}$ symmetries as a result of strong interactions with p orbitals of $O(1)$ and $O(4)$, whereas the $Cu(2)$ atoms (from the planes) have less charge in the $d_{x^2-y^2}$ state than in d_{z^2} due to strong interactions with the $O(2)$ and $O(3)$ atoms. The $O(2)$ sphere is involved in strong interaction mainly in x direction (the p_x partial charge is the smallest one). On the other hand, for the $O(3)$ position the interactions have mostly y character. Consequently, the strongest interactions can be determined along the z axis for the apical $O(1)$ atoms what is indicated by the smallest value of charge with the p_z symmetry. In the case of the chain $O(4)$ atoms the smallest p component is visible for y direction. These results are consistent with the previously reported charge distribution for the $Y123$ [27] and $Dy123$ [15] HTSC. One should be aware, however, that the quoted charges depend on the atomic sphere radii chosen to calculate them and they may not reflect correctly the ionic charges associated with particular ions. This is mainly because of the s and p states for which a large fraction of charge lies outside the atomic sphere.

Table 3 summarizes the EFG tensor, the principal component V_{zz} and the asymmetry parameter η. The value of $V_{zz} = -8.41 \cdot 10^{21}\ V/m^2$ calculated for Gd position is larger than the experimental values of $V_{zz} = -(5.1 \pm 0.2) \cdot 10^{21}\ V/m^2$ [4] and $V_{zz} = -4.9 \cdot 10^{21}\ V/m^2$ [5] derived from Mössbauer measurements. On the other hand, the asymmetry parameter $\eta = 0.41$ obtained from our calculations is equal (within the experimental error) to the experimental value of $\eta = 0.42 \pm 0.03$ from Ref. [4] and is slightly smaller than $\eta = 0.6 \pm 0.1$ and $\eta = 0.57 \pm 0.02$ reported for the orthorhombic phase in Refs. [2] and [5], respectively. As the copper and oxygen positions are concerned, the calculated EFG parameters are generally consistent with those either calculated or found experimentally in the parent high-temperature superconductor $YBa_2Cu_3O_7$ (see Ref. [27] and references therein). It was demonstrated that the anisotropy in the charge density around the oxygen atoms is directly proportional to the corresponding EFG. The principal component is positive for all oxygens

Table 2. Symmetry decomposition of p and d valence states in $GdBa_2Cu_3O_7$. All data is given in electrons (e).

	p_x	p_y	p_z	d_{z^2}	$d_{x^2-y^2}$	d_{xy}	d_{xz}	d_{yz}
Gd	1.835	1.833	1.840	0.040	0.099	0.037	0.108	0.102
Ba	1.526	1.517	1.509	0.025	0.016	0.037	0.024	0.017
$Cu(1)$	2.019	2.062	2.082	1.402	1.645	1.846	1.845	1.863
$Cu(2)$	2.066	2.062	2.029	1.732	1.417	1.854	1.827	1.826
$O(1)$	1.185	1.211	0.981					
$O(2)$	1.007	1.225	1.191					
$O(3)$	1.223	1.004	1.191					
$O(4)$	1.187	0.923	1.242					

and points in the direction of the largest anisotropy. As can be noticed from Table 3, the largest anisotropy parameter is found for the chain $O(4)$ position yielding the highest EFG among oxygens and pointing in the b direction. This is because the $O(4)$ site has copper neighbors only in the y direction (i.e. along the b axis) leading to large V_{bb} and relatively small $\eta = 0.28$. The V_{zz} of plane $O(2)$ and $O(3)$ positions point into a and b directions, respectively. The smallest anisotropy parameter corresponds to the $O(1)$ site for which the main EFG component points into the c direction. For the $Cu(1)$ position the c component V_{cc} is very small, while V_{aa} and V_{bb} have similar values with opposite signs leading to large asymmetry parameter $\eta = 0.80$. On the other hand, the EFG on $Cu(2)$ site points into the c direction while V_{aa} and V_{bb} are similar giving small value of $\eta = 0.14$.

The interaction between nuclei and electrons can be also described by an effective field called hyperfine field (HFF). The values of the Fermi contact term B_c on different sites and the corresponding magnetic moments are collected in Table 4. In our calculations Gd magnetic moment of $\mu_{Gd} \simeq 6.93\mu_B$ is found, which is in a very good agreement with the experimental value of $\mu_{Gd} = (7.4 \pm 0.6)\mu_B$ from the neutron diffraction data [28]. The $4f$ spin moment polarizes the valence bands, what gives rise to small itinerant moments in the system, as well. As can be noticed, the strong localized magnetic moment of Gd is responsible for the generation of relatively small magnetic interactions, mainly

Table 3. Calculated electric field gradient (EFG) components (in 10^{21} V/m^2) for $GdBa_2Cu_3O_7$ superconductor.

Position	V_{zz}	η	V_{aa}	V_{bb}	V_{cc}
Gd	-8.41	0.41	2.48	5.93	-8.41
Ba	-8.80	0.98	-8.80	0.070	8.73
$Cu(1)$	-6.42	0.80	-6.42	5.78	0.64
$Cu(2)$	-5.23	0.14	2.99	2.24	-5.23
$O(1)$	13.09	0.19	-5.32	-7.77	13.09
$O(2)$	12.45	0.21	12.45	-7.53	-4.92
$O(3)$	12.44	0.20	-7.48	12.44	-4.96
$O(4)$	17.83	0.28	-6.45	17.83	-11.38

Table 4. The calculated Fermi contact term B_c of HFF (in units of kG) and the corresponding magnetic moments (in μ_B) in the atomic spheres in $GdBa_2Cu_3O_7$.

Position	B_c	μ
Gd	268.117	6.9291
Ba	-0.494	0.0000
$Cu(1)$	-1.167	-0.0003
$Cu(2)$	-0.424	-0.0039
$O(1)$	0.279	-0.0018
$O(2)$	-0.384	-0.0093
$O(3)$	-0.515	-0.0096
$O(4)$	-0.582	-0.0014

in the CuO_2 planes. The calculated total contact contribution at the Gd position is $B_c \simeq 268\ kG$ and it is slightly underestimated when compared with the experimental values of $B_{hf} = (290 \pm 30)kG$ [3], [4] or $B_{hf} \simeq 302\ kG$ [2]. This difference might originate from the smaller theoretical Gg magnetic moment (as compared with the experiment) and the neglected dipolar and orbital contributions to the HFF. Noteworthy, it is obvious that the assumption of the simple proportionality of the hyperfine field to the local magnetic moments seems not to be justified in the cases of the investigated $GdBa_2Cu_3O_7$ system.

4. Conclusion

In this work the electronic band structure, charge distribution and hyperfine interactions in stoichiometric $GdBa_2Cu_3O_7$ high-temperature superconductor have been calculated on the first-principles basis. The electronic structure of $Gd123$ is typical for this class of compounds. Also charge distribution is similar to analogous $R123$ systems [15], [27]. The presented results show that the evaluated strong on-site magnetic moment of Gd, which is in a good agreement with the neutron diffraction data [28], could not effectively influence the neighbors resulting in small itinerant moments in the CuO_2 planes. The calculated values of electric field gradient (EFG) and hyperfine field (HFF) parameters, which may be treated as a measure of the correctness of the applied theoretical approach, are quite well reproduced and consistent with the Mössbauer measurements [2], [3], [4], [5].

In conclusion, a good agreement between the *ab initio* calculations and the experiment has been found, what indicates that in the highly correlated $GdBa_2Cu_3O_7$ system the applied computational method yields reliable charge distribution to which the EFG and HFF are very sensitive.

Acknowledgements

The Academic Computer Center in Gdańsk (TASK), where all the calculations were performed, is acknowledged.

References

[1] Tarascon, J. M.; McKinnon, W. R.; Greene, L. H.; Hull, G. W.; Vogel, E. M. Phys. Rev. B 1987, 36, 226–234.

[2] Cashion, J. D.; Fraser, J. R.; McGrath, A. C.; Mair, R. H.; Driver, R. Hyp. Int. 1988, 42, 1253.

[3] Taylor, R. D.; Willis, J. O.; Fisk, Z. Hyp. Int. 1988, 42, 1257.

[4] Wortmann, G.; Kolodziejczyk, A.; Bergold, M.; Stadermann, G.; Simmons, C. T.; Kaindl, G. Hyp. Int. 1989, 50, 555–568.

[5] Wang, X. Z.; Hellebrand, B.; Bäuerle, D.; Strecker, M.; Wortmann, G.; Lang, W. Physica C 1995, 242, 55–62.

[6] Perdew, J. P.; Burke, K.; Ernzerhof, M. Phys. Rev. Lett. 1996, 77, 3865–3868.

[7] Sjöstedt, E.; Nordström, L.; Singh, D. J. Solid State Commun. 2000, 114, 15–20.

[8] Madsen, G. K. H.; Blaha, P.; Schwarz, K.; Sjöstedt, E.; Nordström, L. Phys. Rev. B 2001, 64, 195134-195142.

[9] Blaha, P.; Schwarz, K.; Madsen, G. K. H.; Kvasnicka, D.; Luitz, J. WIEN2K, an augmented plane wave+local orbitals program for calculating crystal properties, Techn. Universität Wien, Getreidemarkt 9/156 A, 1060 Wien, Austria, 2001.

[10] Schwarz, K. J. Solid State Chem. 2003, 176, 319–328.

[11] Gladyshevskii, R.; Galez, P. In Handbook of Superconductivity; Poole Jr., C.P.; Ed.; Academic Press, A Harcourt Science and Technology Company, San Diego, 2000, pp. 267–413.

[12] Desclaux, J. P. Comp. Phys. Commun. 1969, 1, 216–222.

[13] MacDonald, A. H.; Picket, W. E.; Koelling, D. D. J. Phys. C 1980, 13, 2675–2683.

[14] Anisimov, V. I.; Zaanen, J.; Andersen, O. K. Phys. Rev. B 1991, 44, 943–954.

[15] Łuszczek, M. Physica C, 2009, 469, 1892–1897.

[16] Łuszczek, M. Phys. Stat. Sol. B 2010, 247, 104–108.

[17] Blaha, P.; Schwarz, K.; Novak, P. Int. J. Quantum Chem. 2004, 101, 550–556.

[18] Blöchl, P. E.; Jepsen, O.; Andersen, O. K. Phys. Rev. B 1994, 49, 16223–16233.

[19] Blaha, P.; Schwarz, K. Phys. Rev. Lett. 1985, 54, 1192-1195.

[20] Breit, G. Phys. Rev. 1930, 35, 1447–1451.

[21] Pickett, W. E. Rev. Mod. Phys. 1989, 61, 433–512.

[22] Pickett, W. E.; Cohen, R. E. Phys. Rev. B 1990, 42, 8764–8767.

[23] Li, Y.; Lieber, C. M. Mod. Phys. Lett. B 1993, 7, 143–153.

[24] Williams, G. V. M.; Tallon, J. L. Physica C 1996, 258, 41–46.

[25] Biagini, M.; Calandra, C.; Ossicini, S. Phys. Rev. B 1995, 52, 10468–10473.

[26] Łuszczek, M.; Laskowski, R. Phys. Stat. Sol. B 2003, 239, 361–366.

[27] Schwarz, K.; Ambrosch-Draxl, C.; Blaha, P. Phys. Rev. B 1990, 42, 2051–2061.

[28] Paul, D. McK.; Mook, H. A.; Hewat, A. W.; Sales, B. C.; Boatner, L. A.; Thompson, J. R.; Mostoller, M. Phys. Rev. B 1988, 37, 2341-2344.

In: Superconductivity
Editor: Vladimir Rem Romanovskiĭ

ISBN 978-1-61324-843-0

Chapter 12

JOINING OF HIGH TEMPERATURE SUPERCONDUCTING BSCCO/AG TAPES

***Guo Wei*[*,1,2], *Zou Guisheng*[2], *Wu Aiping*[2] *and Ren Jialie*[2]**
[1]The School of Mechanical Engineering and Automation, Beijing University of Aeronautics & Astronautics, Beijing, China
[2]Department of Mechanical Engineering, Tsinghua University and Key Laboratory for Advanced Manufacturing by Materials Processing Technology, Ministry of Education of P. R., Beijing, China

ABSTRACT

High temperature superconducting BSCCO/Ag tapes have been fabricated commercially by PIT methods and many applications of the tapes need to join the tapes. There are several joining methods developed to implement the aim. Soldering technology is conventional joining methods and the joint owns larger resistance. Some superconducting joining methods also have been put out. First of all, the cold-press and post heat treatment technique according to the fabrication process of the tape is used to made superconducting joint with hundreds of hours and the joint is of lower superconducting property. Second, Diffusion bonding without an interlayer (directly) or with superconducting powders methods are developed to join the BSCCO/Ag tapes in shorter time, and these joints are of excellent superconducting properties comparing with the cold-press and post heat treatment method.

Keywords: Bi-2223/Ag, soldering, diffusion bonding, microstructures

[*]Corresponding author: Dr. Guo Wei, Postal address: The School of Mechanical Engineering and Automation, Beijing University of Aeronautics & Astronautics, China, Beijing 100191, China, E-mail address: gwei@buaa.edu.cn (Guo Wei)

INTRODUCTION

Up to now, thousands of superconducting materials have been discovered and synthesized. According to the critical temperature (Tc) value, the superconducting materials are divided into low temperature superconducting (LTS) materials and High temperature superconducting (HTS) materials, which are generally used at boiling point of liquid helium temperature of 4.2 K and boiling point of liquid nitrogen temperature of 77 K, respectively.

Up to now, HTS materials are regarded to copper oxide superconductors or cuprates superconductors, mainly including La-based, Y-based, Tl-based, Bi-based, Hg-based cuprates. In these HTS materials, BSCCO/Ag tape (Ag-sheathed Bi-2223($Bi_2Sr_2Ca_2Cu_3O_{7-\delta}$)) is one of important HTS material, which has been fabricated commercially using power-in-tube (PIT) technology by Japan, USA, China and some other countries. Moreover, many application projects of the HTS BSCCO/Ag tapes, such as power cables, transformers, motors, etc., are ongoing in many countries [1-3]. However, in these applications, bonding technology is needed to obtain interconnections between various parts of superconducting apparatus. And these joints are often required to possess mechanical and electric properties.

In general, the bonding technology of HTS BSCCO/Ag includes mainly soldering, cold-press and post heat treatment, diffusion bonding directly and diffusion bonding with a superconducting interlayer. Moreover, these joints are divided two kinds of types, that is, the resistance joint and the superconducting joint. Soldering is main resistance bonding technique and cold-pressure and post heat treatment, diffusion bonding directly and diffusion bonding with a superconducting interlayer are superconducting bonding technique.

In this paper, the Bi-2223/Ag superconducting tapes used in the experiments are provided by Beijing Innova Superconductor Technology Co. Ltd. The critical current (at 77 K in zero magnetic field) is about 90A. The tape contains 61 superconducting filaments and is 4.3mm wide and 0.22mm thick.

In all kinds of joints, the critical current(I_C) is used to evaluated the superconducting properties. The voltage-current (*V-I*) curves of the joints were measured by standard four probe method with a 1μV/cm criterion in liquid nitrogen under zero magnetic field. The system was connected to a computer for data acquisition. Here we defined critical current ratio (CCR) as the ratio of the critical current of the joint to that of original tape.

1. SOLDERING

As an economical, reliable and facilitate method, soldering is still a widely used one to connect the superconducting tapes [4-6]. In the process, the soldered joint is fabricated at low temperature, and thus the superconducting property of Bi-2223/Ag-sheathed tape does not degrade greatly. For soldering, the low melting filler materials such as Bi-Pb-Sn-Cd, Sn-Pb, Sn-Cu-Ag, Sn-Pb-Ag and Sn-Pb-Sb alloys, etc., are utilized.

A series of soldering experiments with 63wt%Sn-34wt%Pb-1wt%Bi-2wt%Ag alloy paste solder were performed in a special soldering stove in air, and the soldering temperature was 200□ which is a little higher than the melting point of the solder (about 183°C). Meanwhile, an approximate 1MPa uniaxial press was applied on the superconducting tape.

1.1. Electrical Properties of the Soldered Joints

The *V-I* curves of the soldered joints with 20mm overlap length are shown in Fig. 1, and three curves are shown in the figure. The relations of the current and the voltage of these joints, i.e., joint 1, joint 2 and joint 3 are approximately linear, which indicates the joints accord with ohms law. Thus, the electrical properties of the joints are just like traditional conductors. The average electrical resistance of the joints is about $5.8\times10^{-8}\Omega$ derived from ohms law. Fig. 2 shows the effects of the overlap length on the CCR and the resistance of the soldered joints. It can be seen that the CCR of joint is direct proportional to the overlap length of joint. In contrary, the resistance of the joints decreases with the increase of the overlap length. Note that the magnitude of the resistance is as small as the order of $10^{-8}\ \Omega$.

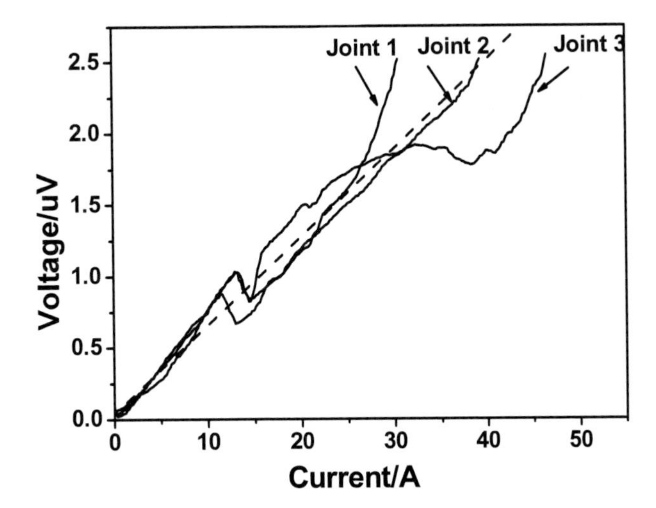

Figure 1. The I_C of the soldering joints

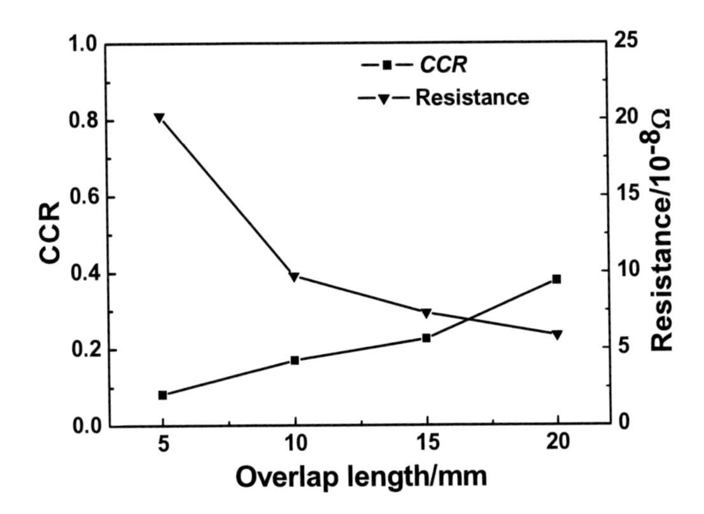

Figure 2. The CCR and resistance of the soldering joints.

1.2. Microstructures of the Joint

Figs. 3a and b are microstructures of the transverse and longitudinal sections of soldered joints, respectively. From Fig. 3b, the upper and lower superconducting tapes have been connected due to intermetallic compound created between silver sheathed layer and Sn-Pb-Bi-Ag solder. The thickness of the diffusion reaction layer is about 40μm.

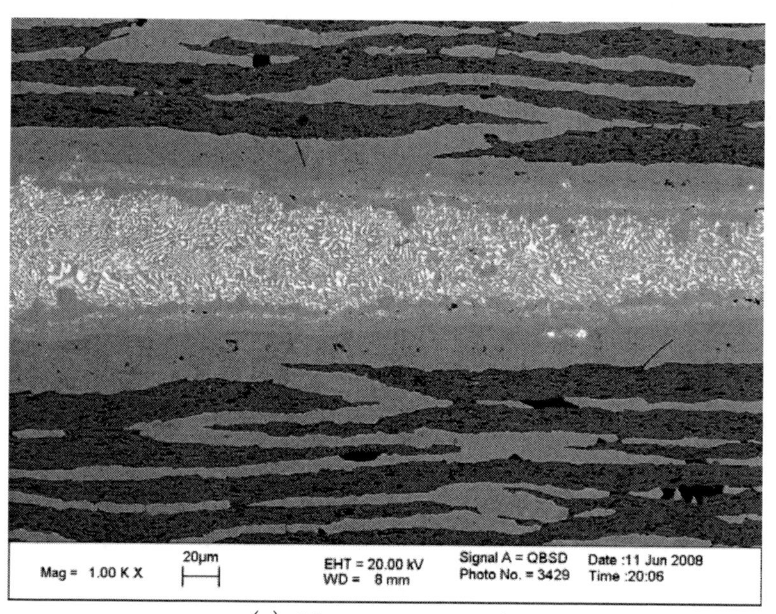

(a) The transverse section

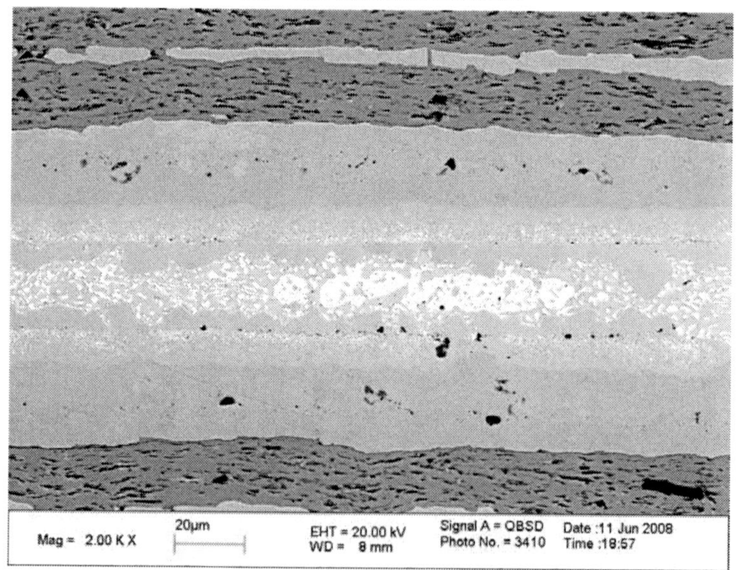

(b) The longitudinal section

Figure 3. The microstructures of the transverse and longitudinal sections of soldered joints.

2. COLD-PRESS AND POST HEAT TREATMENT

Soldering technology is resistance bonding technique, that is, the soldering joint has a larger resistance comparing with the original tape. So some researcher put out a new bonding technology, that is, cold-press and post heat treatment, based on the mechanism of PIT-fabricating method for BSCCO tape itself, which consists of first cold-weld under a high pressure at room temperature to form a joint and then high temperature reaction-annealing under no pressure without or with 1~3 times intermediate press at room temperature to improve the superconductivity of the above joint. And this method is used in the bonding mono-filamentary tape and multi-filamentary one. First of all, Tkaczyk et al. [7] described a method to join partially reacted monofilamentary Bi-2223/Ag tape consisting of mechanical peeling off the Ag sheath and overlapping the superconducting cores, and the joint shows approximately 50% current capability of the unjoined region of the tape, whose superconducting property is lower than about 10-20% of the original HTS tape due to temperature cycle. Subsequently, many other groups used similar methods to join monofilamentary or small quantity filaments HTS tapes. Jaimoo Yoo et al. [8] reported that monofilamentary Bi-2223/Ag tape lap-joints were fabricated by a single or multiple cold presses firstly and post-annealing at about 840°C for 50-150h in air, and the transition critical currents (joint) reached about 60% of the original tapes. And Hee-Gyoun Lee[9] fabricated butt-joints with 9 filaments Bi-2223/Ag tapes using a load of 2 ton and post-annealing at 830°C for 100h, and the joint showed about 75% current carrying capacity of the tapes without joining part. Jung Ho Kim et al. [10] reported that lap-joints of 37 filaments Bi-2223/Ag superconducting tapes with or without steps were carried out with 2000MPa and post-annealing at 840°C for 50h, and the joint attained 25%-58% of current capability of the unjoined regions of the tape. In these works, the common characteristics were cold press with about 140-4000MPa pressure and post-annealing at about 830-850°C for as long as 50-400h in order to healing the damage caused by the cold press process. Thus, the process of fabricating a joint was as long as hundreds of hours. So it was necessary to find a more efficient way to fabricate the joints. The microstructures of the joint are shown in Fig. 4.

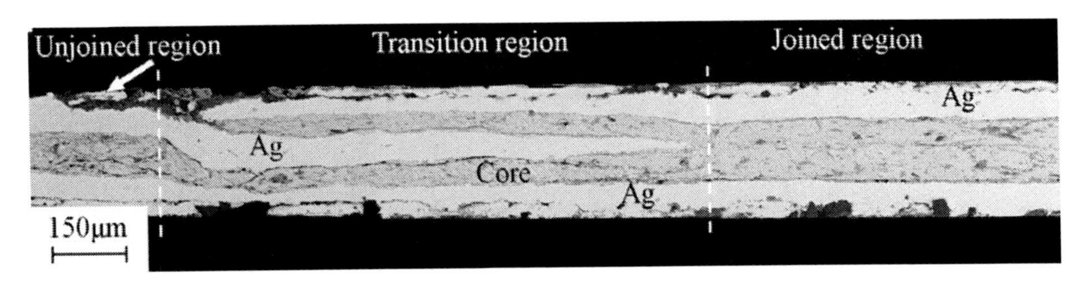

Figure 4. The microstructures of the cold-press and post heat treatment joint.

3. DIFFUSION BONDING DIRECTLY

Due to the disadvantage of the traditional cold-press and post heat treatment method (too longer fabricating time and lower electric property of the joint), Diffusion bonding directly is used to join the BSCCO/Ag types.

Before the bonding experiment, a certain area of superconducting cores as a joining window was exposed by etching method with a mixture of $NH_3{\cdot}H_2O$ (30ml) and H_2O_2 (10ml), and the lap-joint with overlap length 20mm was used. The sketch map of lap-joint and its dimension were shown in Fig. 5.

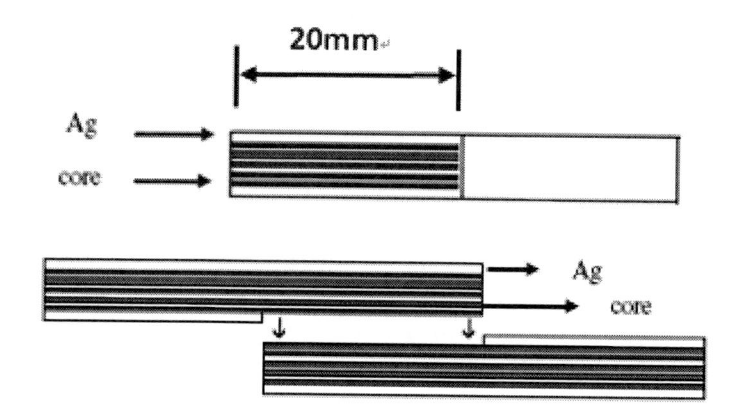

Figure 5. The sketch map of lap-joint and its dimension during the diffusion bonding process.

The temperature parameter of bonding process was shown in Fig. 6. And a certain uniaxial pressure (3MPa) was applied by ZrO_2 powder throughout the whole bonding process, and the bonding time is 120min.

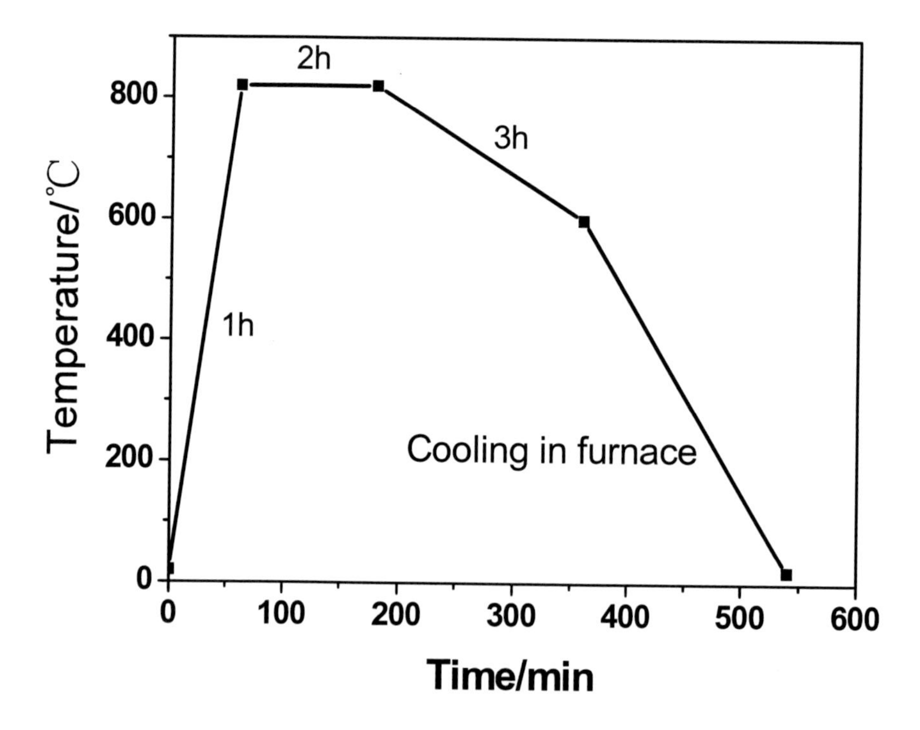

Figure 6. The temperature parameter during the bonding process.

3.1. Electric Property of Superconducting Joints

Ten joints have been fabricated. And the *V-I* curves of these joints and the original tape were tested measured, and the results are displayed in Fig. 7. It can be seen that these diffusion bonding joints sustain superconducting properties shown by the profiles of *V-I* curves. Comparing with the original tape, the critical currents of these joints are smaller.

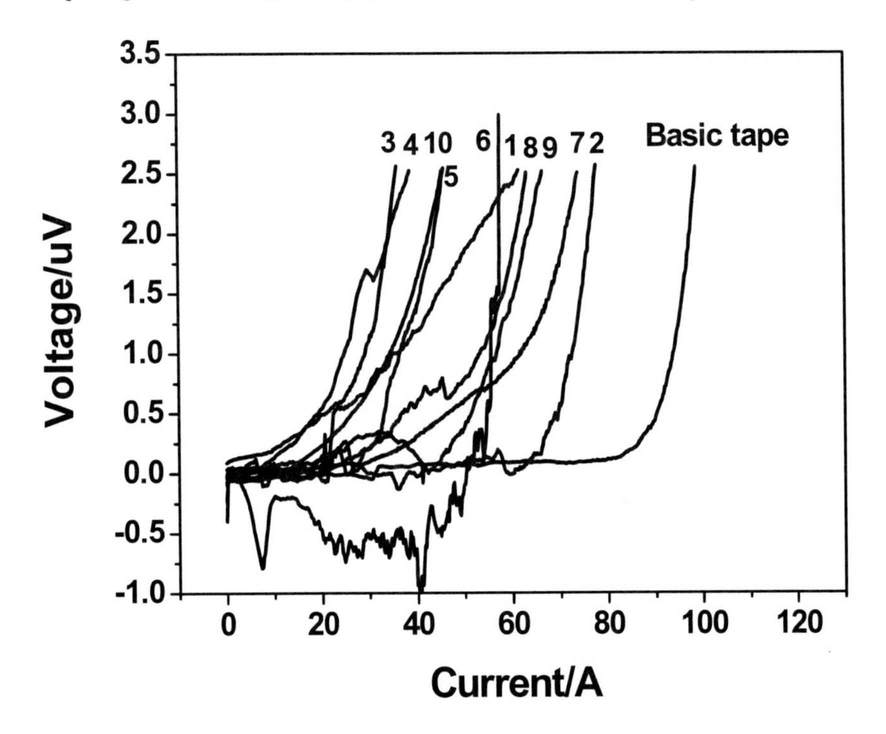

Figure 7. The *V-I* curves of these joints and the original tape.

The critical current values of these joints are listed in Table 1. According to the experimental data, the highest CCR_O of the joints arrives at nearly 80%, while all CCR_Os are in the range of 35%-80%. In comparison with the traditional joining methods, diffusion bonding can produce a joint with better superconducting property. However, a fact needs to be noticed is that the CCR_O values of all joints distribute in a wide range, which is similar to the traditional method. There are two main reasons for this dispersive distribution of the CCR_O values. First, the bonding interfaces are composed of several superconducting filaments (7-10) which are made of brittle oxygen ceramic powders by PIT method. These oxygen ceramic have low atom diffusion ability, so it is difficult to directly join by diffusion bonding. Second, there are some grooves between the superconducting filaments after the Ag sheath are etched off, so the bonding interface is uneven before joining. During the process of diffusion bonding, the brittle filaments are pushed to fill up the concave parts on the interface by bonding pressure at high temperature, so that some micro-cracks often generate on the interface. If the cracks can't be healed by the atom interdiffusion during diffusion bonding process, some of them will retain on the interface after the completion of joint.

3.2. Microstructures of the Joint

The typical micrographs of longitudinal section of the joint are displayed in Fig. 8. Comparing with the superconducting cores in upper and lower tapes, a new and thicker core has been formed on the interface. No discontinuities exist on the interface of the joint. The superconducting cores near the interface present the fine texture orientation which is similar to that in original tape. Sometimes, there are cracks in the bonding interface due to brittle ceramic superconducting materials, shown in Fig. 9.

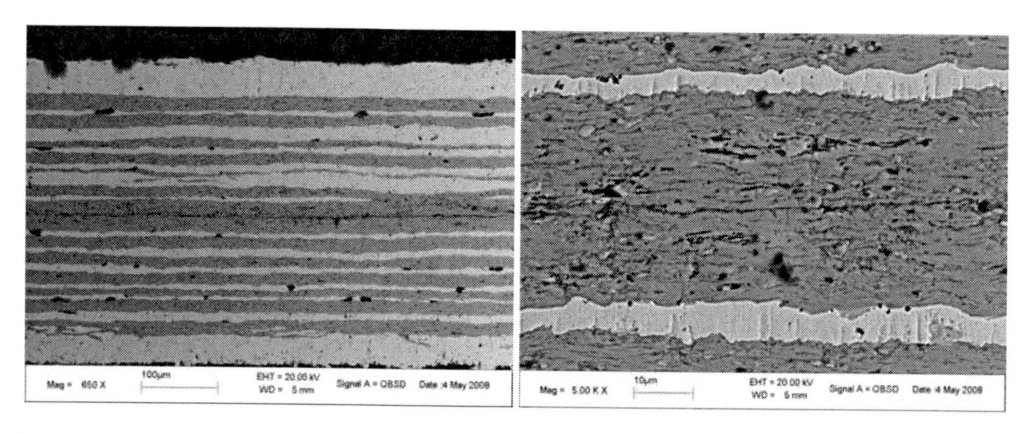

Figure 8. The microstructures of longitudinal section of the joint.

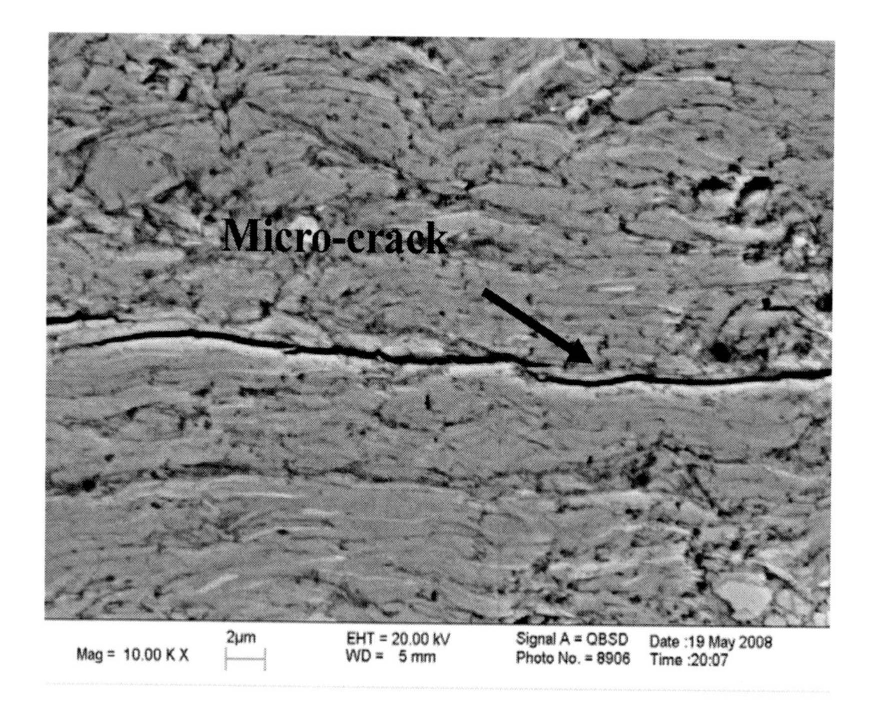

Figure 9. The cracks exist in the bonding interface.

4. DIFFUSION BONDING WITH SUPERCONDUCTING POWDERS

There is other superconducting bonding technique, that is, diffusion bonding with superconducting powders. There are main two kinds of bonding types, which are the joint with Bi-2223 superconducting powder interlayer(Bi-2223 51%) or the joint with Bi-2212 superconducting powder interlayer(prepowder). In the two types of joints, the outside silver sheath was peeled off completely by etching with a mixture of $NH_3{\cdot}H_2O$ (30 ml) and H_2O_2 (10 ml) before diffusion bonding process. The schematic diagrams of three types of joints are shown in Fig. 10.

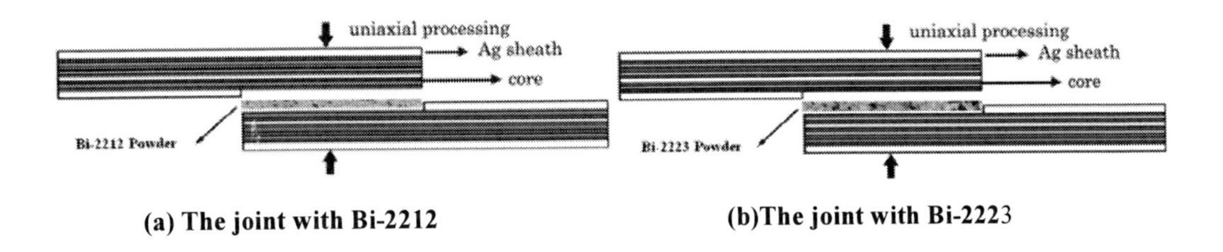

Figure 10. The schematic diagrams of the joint with superconducting powders.

Fig. 11a and b are microstructures of the joints with Bi-2212 superconducting powders and Bi-2223 superconducting powders respectively. Fig .11a shows the microstructure of joint with Bi-2212 superconducting powders. Due to existence of the Bi-2212 superconducting phase in the joint zone, this kind of joint also possesses but smaller superconducting property and the critical current only reaches about 38% of original tape. Fig .11b shows that the microstructure of the joint with Bi-2223 superconducting phase interlayer. And the two superconducting tapes have been bonded together completely. Additionally, no point defect and cracks exist in the joint zone. So the superconducting property of this joint is better than that of joint with Bi-2212 phase and lower than that of direct bonding joint. The critical current of this type of joint reaches about 48% of the basic tape.

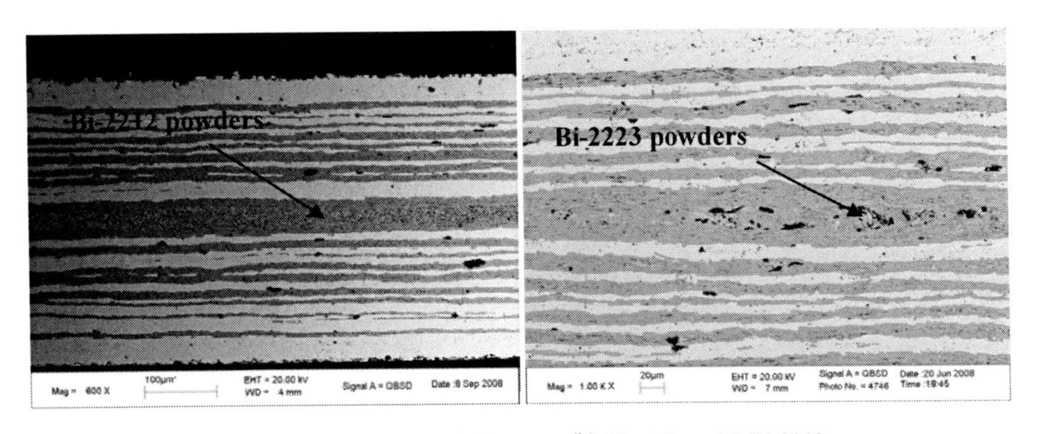

(a) The joint with Bi-2212 (b) The joint with Bi-2212

Figure 11. The microstructures of the joints with superconducting powders.

Conclusion

In a word, HTS BSCCO/Ag tape is one of promising superconducting materials. And there are several bonding technology to join the tape, Soldering provides a conventional resistance bonding technique. Cold-press and post heat treatment method could realize the superconducting joining, however this technique spends hundreds of hours and the joint owns worse electric properties. The diffusion bonding directly could finish the superconducting joint with much short time comparing with the cold-press and post heat treatment method and the joint owns better electric property. At last, the diffusion bonding with superconducting powders provides another choice to fabricate a superconducting joint.

Acknowledgments

This research is supported by the Fundamental Research Funds for the Central Universities and the National Natural Science Foundation of China (Grant No. 50705050, 50575114 and 50635050).

References

[1] H. Takigawa , H. Yumura, T. Masuda, M. Watanabe, Y. Ashibe, H. Itoh, *Physica C,* 463-465(2007) 1127.

[2] T. Masuda, T. Kato , H. Yumura ,M. Watanabe, Y. Ashibe , K. Ohkura , *Physica C* 378–381 (2002) 1174.

[3] S. Kato , T. Hashimoto, H. Hasegawa, S. Hirano, S. Nagasawa, *Physica C* 378–381 (2002) 1486.

[4] C. Gu, C. Zhang, T.M. Qu, Z. Han, *Pgysica C* 426-431 (2005) 1385.

[5] J.H. Kim, J. Joo, *Supercond. Sci. Technol.* 13 (2000) 237.

[6] J.H. Kim, K.T. Kim, J.H. Joo, W. Nah, *Physica C* 372 (2002) 909.

[7] J. E. Tkaczyk, R. H.Arendt, P. J. Bednarczyk, M. F. Garbauskas, B.A. Jones, R. J. Kilmer, K. W. Lay, *IEEE Trans. Appl. Supercond.* 3 (1993) 946.

[8] J. Yoo, H. Chung, J. Ko, H. Kim, *J. Sha, Physica* C 267 (1996) 53.

[9] H. G. Lee, Il. H. Kuk, G. W. Hong, E. A. Kim, K. S. No, W. Goldacker, *Physica C* 259 (1996) 69.

[10] J. H. Kim, K. T. Kim, J. H. Joo, W. Nah, *Physica C* 372-376 (2002) 909.

In: Superconductivity
Editor: Vladimir Rem Romanovskiĭ

ISBN 978-1-61324-843-0

Chapter 13

MASS PRODUCTION OF LOW-COST LRE-123 BULK SUPERCONDUCTORS: PROCESSING, FLUX PINNING AND ITS APPLICATION POTENTIAL UP TO LIQUID ARGON TEMPERATURE

Muralidhar Miryala and Masaru Tomita
Railway Technical Research Institute (RTRI), Applied Superconductivity,
Materials Technology Division, Tokyo, Japan

1. INTRODUCTION

Recent progress in technology of melt-textured $REBa_2Cu_3O_y$ "RE-123" high-T_c superconductors has brought applications of superconducting permanent magnets in variety of industrial, medical, public, and research applications. When the superconducting pellet is magnetized to a high magnetic field, part of this field is trapped in the pellet and we get a superconducting permanent magnet or, shortly, super-magnet [1-13]. Such a name is fully justified as high-T_c superconductors can trap magnetic field by order of magnitude higher than the best hard ferromagnets nowadays known [14]. The trapped field depends on the critical current density, J_c of the melt textured material and on the size of the single grain. Therefore, to achieve a large B_T, one needs to enhance both J_c and the bulk material size. A further improvement of the critical current density and fabrication of large homogeneous good-quality single-grain superconducting disks capable of trapping high magnetic fields are fundamental issues for many industrial applications. In recent years, the major emphasis has been devoted to fabricating single-grain, high-performance LRE-$Ba_2Cu_3O_y$ "LRE-123" (LRE: Nd, Sm, Eu, Gd) pellets by means of the oxygen-controlled melt-growth process [15-40]. Melt-processed bulk disks as large as 140 mm in diameter were produced [41]. As a result of a continuous improvement in the pinning defect efficiency, this material showed very high critical current density even at 90.2 K, the boiling temperature of liquid oxygen [42-60]. Nowadays the performance reached the level necessary for industrial applications and we believe that bulk superconducting magnets will in a close future enter the market as basic parts of various industrial applications [61,62]. In all these cases a high number of pieces with equally high quality are required. Batch processing of LRE-123 materials with uniform properties is the necessary step in this process.

Batch processing is a principal step towards economical production of these materials. For a successful batch processing the following requirements are necessary: (i) the ability to process a large number of grains with the lowest possible loss fraction, (ii) For all processed blocks the final product quality must be satisfactory within each batch process. The seed crystal represents a key issue in this direction. In the batch melt growth a cold seeding process is used. The seed crystal is placed on the precursor pellet at room temperature. Normally, for the batch-processed melt-textured Y-123 pellets Nd-123 or Sm-123 crystals have been used as seeds [63-65]. However, in the case of LRE-123 the peritectic temperature is higher and these kinds of seeds get partially melted, degrading thus superconducting performance of the samples. In order to fabricate LRE-123 bulks using cold seeding method, the Cambridge group led by D. Cardwell introduced the Mg-doped LRE-123 crystals as a seed material [66]. These crystals possess higher melting temperature than pure LRE-123 ones. With generic Mg-doped Nd-123 seeds they succeeded in growing single-grain samples of 20 mm in diameter in air [67]. For elimination of a multiple nucleation in the case of LRE-123 bulk growth, one has to increase the maximum temperature that the seed can withstand to about 80 ^{o}C above the peritectic temperature of the particular LRE-123 [68]. For this, the recently discovered superheating effect of LRE-123 thin films has offered a genuine solution. Oda et al. [69] reported use of Sm-123 thin film seed grown on MgO crystal for growing Sm-123 block in the cold seeding process with a reduced oxygen atmosphere, taking advantage of the superheating phenomenon [70]. With this effect, the YBCO/MgO thin film grown by liquid phase epitaxy could be used as seeds for melt processing of high-temperature-melting NdBCO, [70] (in the bulk form YBCO has much lower peritectic temperature than Nd-123). Similarly, Sm-123/MgO thin films were shown to exhibit superheating effect, being stable up to 1100 ^{o}C, so that they can be used as seeds for growing any LRE-123 bulk [71]. Recently, we succeeded in batch processing of Gd-123 pellets 24 mm in size, 12 samples in one batch, using the Nd-123 thin film seeds [72-73].

In this contribution, we report on successful growth of several batches of LRE-123 pellets in air and partial oxygen pressure using the new class of Nd-123/MgO seeds. The emphasis was focused on producing high-performance large size single grains at a reduced price. The superconducting performance, microstructure, as well as the trapped magnetic field measurements at liquid nitrogen temperature and liquid argon temperature are reported. Finally, taking advantage of the batch processed material, we constructed home made child levitation disk, which is capable of levitating a mass greater than 35 kg. Our experimental results clearly indicate that using the new class of seeds, LRE-123 can scale up from laboratory to industrial production.

2. Low-Cost LRE-123 Material

In order to reduce price and to improve the performance of the LRE-123 system, the Gd-123 and Gd-211 powders were carefully prepared by us ourselves. This reduced the price more than 10 times as compared to the commercial powders of a comparable quality. Further, we processed the Gd-123 pellets in air instead of the oxygen-reduced atmosphere and reduced Pt content from the common 0.5 wt% to 0.1wt% [74]. At present, the LRE-123 melt-textured blocks are offered by only a few suppliers and are very expensive. In Japan, the commercial 45 mm in size Gd-123 is available for around Yen 300,000 (US$ 3250). To reduce the LRE-123 price, batch process is essential. For more details of the low-cost Gd-123 see Table 1.

Table 1. Some ways how to reduce costs of preparing high-performance $LREBa_2Cu_3O_y$ bulks

	Normal Process	Low-cost Process
1	melt-growth with hot seeding (one sample)	melt-growth with cold seeding in the batch process (12 to 20 samples)
2	processing in Ar-1% O_2 (Yen 38,000/puck)	processing in Air (Ba excess)
3	commercial Gd-123 (Yen 180,000/kg)	self-made Gd-123, starting from oxide powders (Yen 11,000/kg)
4	commercial Gd-211 (Yen 120,000/kg)	self-made Gd-211, starting from oxide powders (Yen 12,900/kg)
5	Pt addition (Yen 3500/gram)	CeO_2 addition (Yen 16/gram)
6	Nd-123 single crystal or MT seed (low melting temperature)	commercial Nd-123 thin film seed on MgO single crystal (high melting temperature) (Yen 250 for 1 seed)

3. EXPERIMENTAL

The commercial Nd-123 thin film seeds grown on MgO crystal, single-grain $GdBa_2Cu_3O_y$, $(Nd,Eu,Gd)Ba_2Cu_3O_y$, and $SmBa_2Cu_3O_y$ superconductors were prepared by batch processing. For the batch process self made Gd-123, Sm-123, NEG-123, Gd-211, Sm-211 powders were used.

3. (a) Processing of Nd-123 Thin Film Seeds on MgO Crystal

The films were prepared by simultaneous thermal evaporation of Nd, Ba, and Cu from three different boats. The substrate was heated to 680 °C and the oxygen was supplied just next to the substrate. The partial pressure during oxidation was around $5x10^{-3}$ mbar. All films showed T_c (onset) of 94.4 K and the critical current density about 3.7 MA/cm^2 at liquid nitrogen temperature. We tested the preferential orientation of the NdBaCuO films by x-ray $\theta/2\theta$ diffraction with CuKα radiation. Fig. 1 presents the typical X-ray $\theta/2\theta$ diffraction pattern of the $NdBa_2Cu_3O_y$ seed crystal. Notably, only (0 0 l) reflections were observed, indicating that the Nd-123 films were *c*-axis oriented. On the other hand, the in-plane orientation of the Nd-123 films was evaluated by x-ray Φ-scan by using (1 0 2) plane of Nd-123. Four peaks in the Φ-scan were clearly observed, indicating four fold symmetry of the Nd-123 film (see in Fig. 2). We could thus conclude that the Nd-123 films grew epitaxially on MgO substrate.

The scanning electron micrograph (SEM) of the smooth, high quality surface of the Nd-123 film was reported elsewhere. [88].

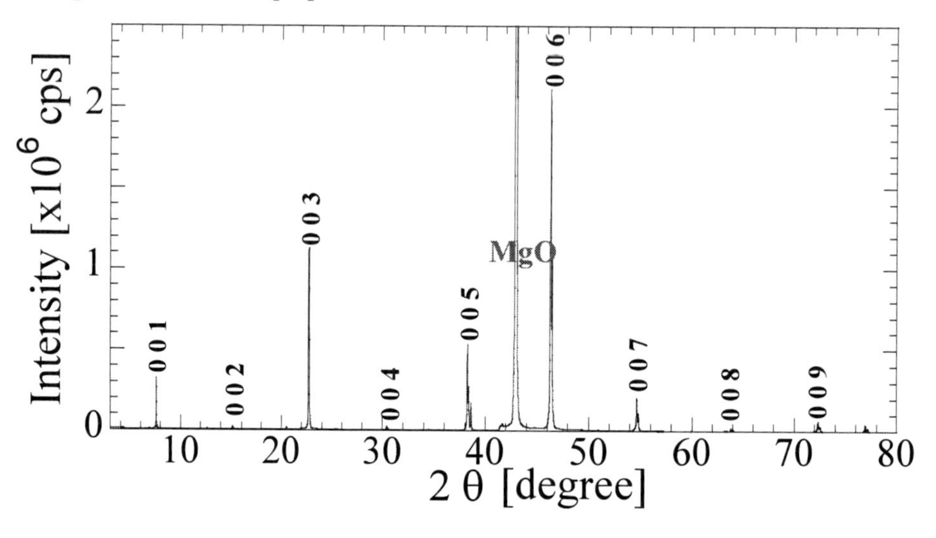

Figure. 1. The x-ray diffraction pattern of the Nd-123 thin film grown on the MgO substrate. Note that only (0 0 l) reflections from the Nd-123 films indicating the *c*-axis orientation.

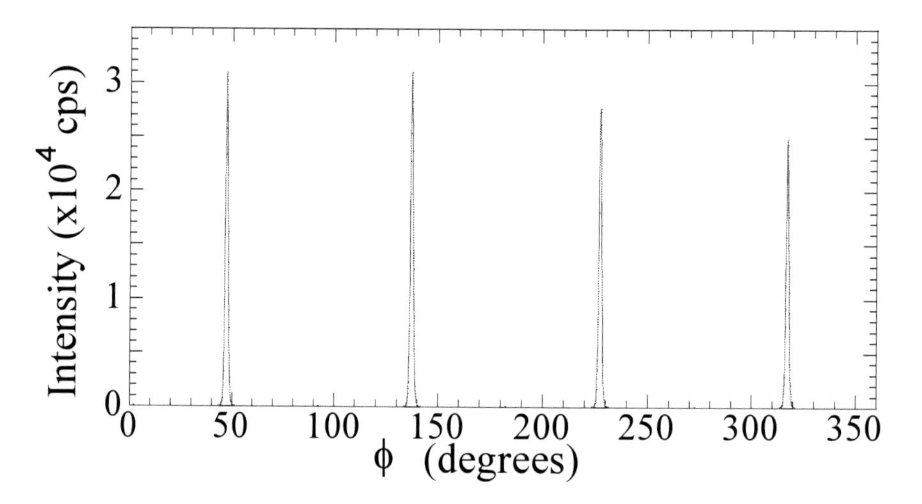

Figure. 2. An in plane crystal alignment characterized by using XRD Φ–scan using A (1 0 2) plane of Nd-123.

3. (b) Batch Processing of Gd-123 Material in Air

High-purity commercial powders of Gd_2O_3, $BaCO_3$ and CuO were mixed in a nominal composition of $GdBa_2Cu_3O_y$. The starting powders were thoroughly ground and calcined at 880°C and 900°C for 20 hours and then pressed into pellets. Sintering was performed at 925° C for 15 hours. This process was repeated twice, in air. In parallel, the powders of Gd_2O_3, BaO_2, and CuO were mixed in the nominal composition of Gd_2BaCuO_5 and calcined three times at 840° C, 870° C and 900° C for 10 h. The average particle size of the Gd-211 powders was less than 1 μm, as determined by the BET specific area measurements [75]. For the batch process Gd-123 and Gd-211 powders were mixed in the molar ratios of Gd123:Gd211 = 10:5. In order to suppress coarsening of the Gd-211 particles during the melt process, 0.1 wt% Pt and 1 mol% CeO_2 were added. In this experiment we used less of Pt and more CeO_2, to reduce the cost. In principle, one can use only CeO_2. The CeO_2 dopant allows decreasing size of the RE-211 inclusions up to micron and sub-micron size. CeO_2 modifies RE-211 morphology during the melt process. It probably acts on interfacial energy and on viscosity of the melt and thus decreases RE-211 coarsening.

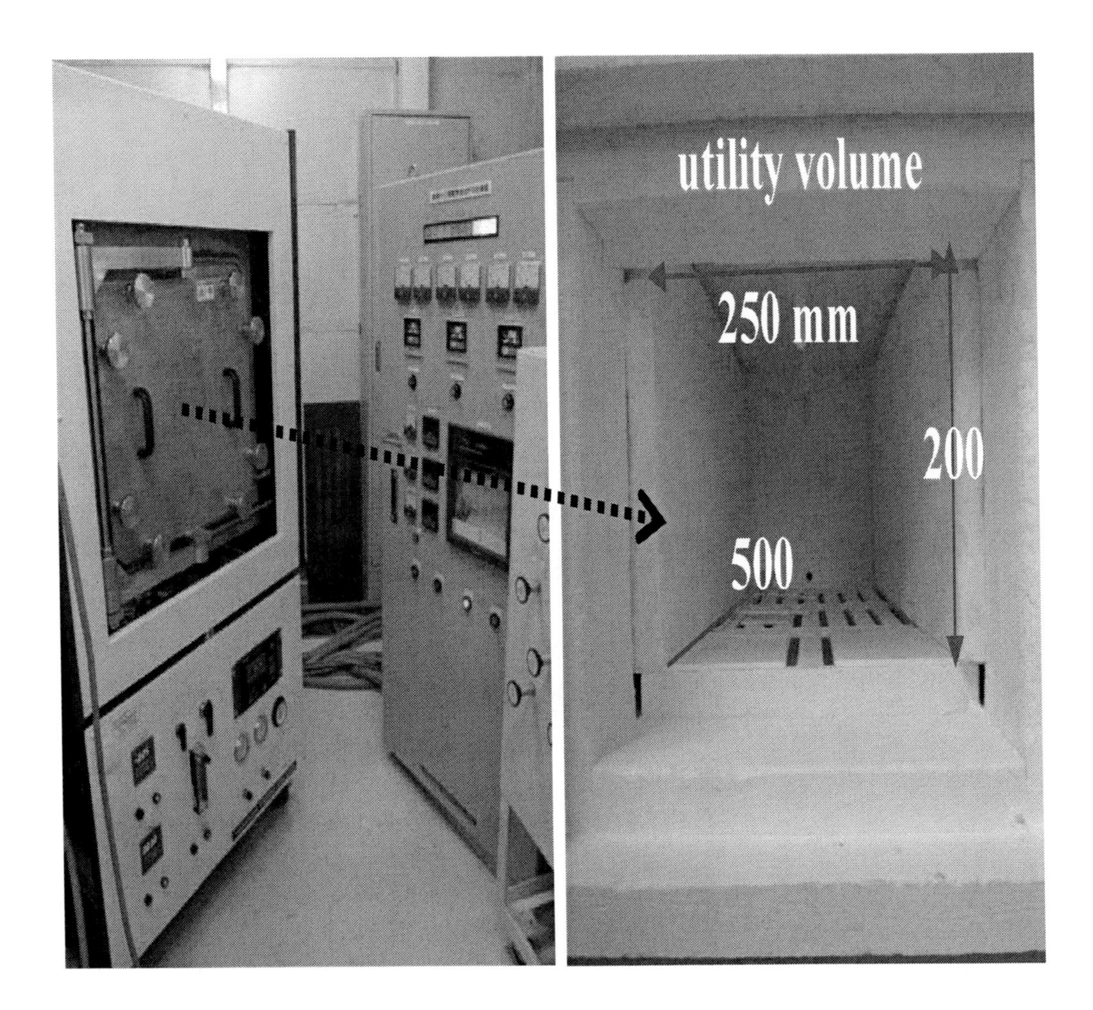

Figure 3. Special box type furnace for batch production of LRE-123 material in air and partial oxygen pressure.

The effect of CeO_2 and PtO_2 dopants on microstructure was recognized long time ago [76, 77] with the conclusion that both oxides act on microstructure in the same way, namely that the size of RE-211 particles in the RE123/RE211 composite is effectively refined. In the present case, besides CeO_2 also 1wt% of BaO_2 was added to suppress Gd/Ba substitution in the superconducting phase composition. To improve the mechanical performance of the bulk, 20 wt% of Ag_2O was also added. All constituents were thoroughly mixed for 5 hours in a milling machine incorporating an electrical mortar and pestle. The powder mixture was pressed into pellets of 30 mm to 55 mm in diameter and subject to a cold isostatic pressing under 200 MPa. Finally, commercial 700 nm thin Nd-123 films evaporated on 0.2 mm MgO <100> substrates were used as seed crystals.

We used a special box furnace with the utility volume of 20x50x25 cm^3 in which a vertical temperature gradient could be created (see in Fig. 3). Twelve Gd-123 pellets were placed on yttrium-stabilized ZrO_2 rods inside the furnace. The MgO crystals covered by Nd-123 thin films were placed at the top center of each pellet, which were then melt-grown in air. The heat treatment profile used in the present experiment was as follows. The sample was first heated in 5 h to T_p+90 °C (T_p is the peritectic temperature) and held there for 50 min. Then the temperature was reduced during 30 minutes to T_p+2 °C and then decreased by 25 °C with the cooling rate of 0.25 °C/h. Finally, the temperature was reduced with the cooling rate of 20 °C/h to 100 °C and then the furnace was left to cool down to room temperature. The melt-textured samples with the final diameters of 24 mm and 32 mm were annealed at 400-450 °C in the flowing pure O_2 for 250 hours and 310 hours, respectively.

3. (c) Batch Processing of Sm-123 Material in air

For the development of Sm-123 batch process, Sm-123 and Sm-211 powders were mixed in the molar ratios of Sm-123:Sm-211 = 10:5. In order to suppress coarsening of the Sm-211 particles during the melt process, 0.1 wt% Pt and 1 mol% CeO_2 were added, similar to the Gd-123 material. Moreover, 0wt% to 3wt% of BaO_2 was added to optimize the best composition and suppress Gd/Ba solid state substitution in the superconducting phase composition. 20wt% of Ag_2O was also added to improve the mechanical performance of the material. Finally, well-mixed powders were pressed into pellets of 20 mm diameter and 15 mm thickness, which were consolidated by cold isostatic pressing with a pressure of 200 MPa. The differential thermal analysis (DTA) measurements were performed in air to determine the peritectic decomposition temperature, T_p. This temperature was then used to schedule the heat treatment profile of the melt growth process. Eventually, commercial 700 nm thin Nd-123 films evaporated on 0.2 mm MgO <100> substrates were used as seed crystals. Six Sm-123 pellets were placed on yttrium stabilized ZrO_2 rods inside the small box furnace and melt processed in Air. The heat treatment profile used in the present experiment was as follows. The sample was heated in 5 h to the temperature 80°C above the peritectic temperature (T_p) and held there for 50 min. Then the temperature was reduced during 30 min to 2°C above T_p, and slowly decreased by 25°C with a cooling rate of 0.20°C/h. Finally, the temperature was reduced a cooling rate of 20°C/h to 100 °C and then the furnace was left to cool down to room temperature. The melt textured samples were annealed at 350-400°C for 250 hours in flowing pure O_2 gas.

3. (d) Batch Processing of NEG-123 Material in Ar-1% O_2

For batch process of NEG-123, high-purity commercial powders of Nd_2O_3, Eu_2O_3, Gd_2O_3, $BaCO_3$ and CuO were mixed in the ratio corresponding to the nominal composition of $(Nd_{0.33},Eu_{0.28},Gd_{0.38})Ba_2Cu_3O_y$ "NEG-123". The starting powders were thoroughly ground and calcined at 880°C for 24 h with intermediate grinding, then pressed into pellets. Sintering was carried out at 900°C for 15 h. This process was repeated three times under oxygen partial pressure (pO_2) of 1% O_2. For the batch process NEG-123 and Gd-211 powders were mixed in the molar ratios of NEG-123:Gd-211 = 10:4. More details for the oxygen controlled melt growth (OCMG) process and oxygenation found elsewhere [78].

3. (e) Characterization of Batch Processed LRE-123 Material

Measurements of critical temperature (T_c) and magnetization hysteresis loops (M-H_a loops) in fields from –2 to +6 T were measured at 77 K using a commercial SQUID magnetometer (Quantum Design, model MPMS7). J_c values were estimated using the extended Bean's critical state model for a rectangular sample [79]. X-ray diffraction measurements were carried out using CuKα radiation to confirm the orientation (Rigaku, Rint-2000). The measurements of trapped magnetic field were carried out by magnetizing the bulk samples in a 10 T superconducting magnet. The bulk samples were cooled to the liquid nitrogen temperature in magnetic field of 1.5 T applied parallel to the c-axis and kept 15 min. in this field. After switching the external field off, the profile of the trapped magnetic flux density was measured by scanning Hall probe sensor. The total gap between the top surface of the sample and the active area of the Hall sensor was 1.2 mm, including the sensor mold thickness, 0.7 mm.

4. Batec Production of LRE-123

4. (a) Gd-123 Batch Production in Air

An important feature of the new Nd-123 thin film seed is the high melting temperature > 1100 °C. As a result, one can use these seeds in a cold seeding process to grow any class of LRE-123 system. This is why the high performance batch processing of LRE-123 became possible. Figure 4 presents the result - the photograph of as grown samples after the melt-growth in air. 12 pieces of samples with the final diameter 24 mm per one batch have been repeatedly produced (see figure 4, left). The four-fold growth facet lines clearly visible on the top surface of all pellets indicate that the crystal growth was perfectly controlled in all positions in the furnace. In other words, we succeeded in governing the batch production of the Gd-123 samples. In another batch process we fabricated 7 pieces of Gd-123 material with the final diameter varying from 24 mm to 45 mm have been repeatedly produced, as seen in Fig.4, right. These results clearly indicate that one can produce samples with a large diameter using the cold seeding with Nd-123/MgO seeds. Based on our long-term experience, we do not find any principal difference in fabricating Gd-123 and other LRE-123 blocks. Therefore, we can extend the present experience with Gd-123 to all the LRE-123 family.

Figure 4. A photograph of batch processed as-grown Gd-123 samples prepared by the cold-seeding method using a Nd/MgO film as a seed crystal. Note that same size or varies sizes can produced in the single run.

4. (b) NEG-123 Batch Production in Ar-1% O_2 Atmosphere

In order to check performance of the new seed crystals in the Ar-1% O_2 atmosphere, we prepared a batch of six NEG-123 samples. In this process we used normal Gd-211 particles with size around 1 micrometer. The details are described in the experimental section. Figure 5 shows the as grown NEG-123/Gd-211 samples, melt processed in Ar-1% O_2 atmosphere. The traces of the growth facets on the surface reveal that single-grain compact NEG-123 bulks were successfully fabricated (see figure 5, left). The final sample size was 24 mm in diameter and 16 mm in thickness. These results proved again that use of the new Nd-123 seeds enables batch fabrication of any LRE-123 material with the same uniform growth orientation with (a,b) plane parallel to the pellet surface.

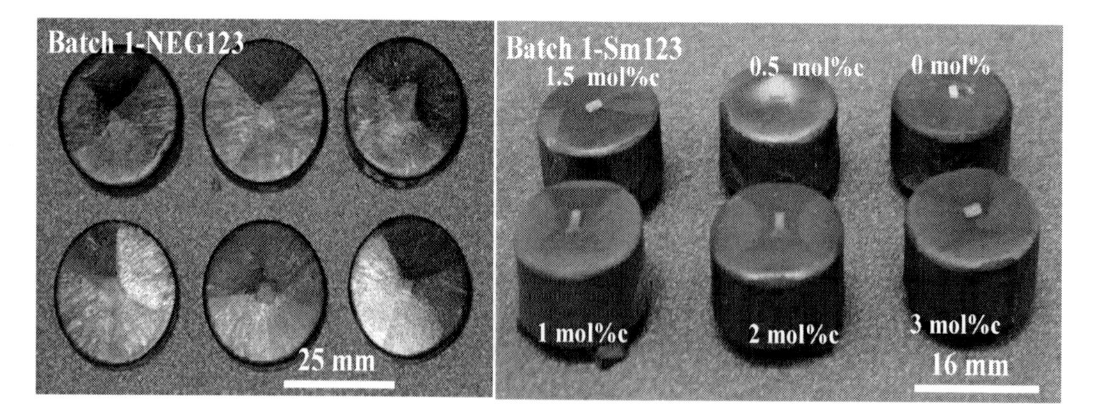

Figure 5. A photograph of batch processed as-grown LRE-123 samples prepared by the cold-seeding method using a Nd-123/MgO film as a seed crystal. As-grown NEG-123 samples processed in Ar-1% O_2 atmosphere (left), As-grown Sm-123 with varying content of BaO_2 (0wt%, 1wt%, 1.5wt%, 2wt%, 2.5wt% and 3wt%) processed in Air (right).

4. (c) Sm-123 Batch Production in Air

To develop the batch process of Sm-123 material in air for the best performance, the optimization of BaO_2 content is necessary to suppress Sm/Ba solid state substitution in the superconducting phase composition. For this we made small pellets with initial size of 20 mm in diameter and varied the BaO_2 content 0 wt%, 1wt%, 1.5wt%, 2wt%, 2.5wt%, and 3wt% and made single run in a box furnace using the Nd-123/MgO seed crystal. The result of as grown samples is shown in Fig.5, right. It is clear that the top surface of the samples has four-fold growth facet lines clearly visible similar to the Gd-123 or NEG-123 systems. Note that in this experiment the maximum temperature was 1105 °C (T_p+80°C) during the melt growth. It is evident that one can use the new class of cold seeding process above 1100 °C, make it possible to produce the batch process of the any kind of LRE-123 material.

5. MAGENTIZATION MEASUREMENTS

5. (a) Superconducting Properties of Batch Processed LRE-123 Material

In general, superconducting properties of large bulks vary from place to place. In particular, properties of the pellet top usually differ from its bottom, partially due to the bottom surface reaction with the supporting material, but also due to different dominant growth orientations (on the top *a*-axis sectors prevail, while the bottom lies mostly in the *c*-axis growth sector. Therefore, we mapped the local properties within the large bulk. The positions of the specimens selected from the bulk are shown in the inset of Fig. 6. Figure 6 displays superconducting transition of the Gd-123/Gd-211 (10:5) sample measured after zero-field cooling (ZFC) in magnetic field of 1 mT. All the specimens, MT, MM, ME, TE, and BE, exhibited a sharp superconducting transition (around 1 K wide) with the onset T_c around 93.5 K. The onset T_c slightly decreased from 93.5 to 93 K in the bottom edge sample (BE).

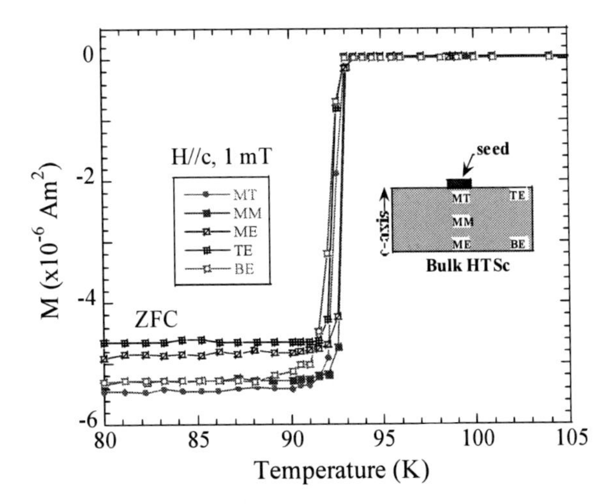

Figure 6. Superconducting transitions for specimens cut from various locations of the single-domain batch-processed GdBaCuO superconductor prepared in air by cold seeding method using an Nd-123/MgO film as a seed crystal.

The reason might be a slight contamination of the bottom surface from ZrO_2 supporting rod. The transition width ΔT_c and the transition temperature are similar to the oxygen-controlled melt-grown Gd-123 material [75]. Evidently, a proper quantity of BaO_2 (Ba excess) enables to control the Gd/Ba solid state solution in the superconducting phase.

5. (b) Critical Current Density Characteristics of Batch Processed LRE-123 Material

The critical current density, J_c, at 77 K in the field applied parallel to the *c*-axis, calculated from the *M-H* curves using the extended Bean model formula, is shown in Figure 7. In all measured samples the remnant critical current density reached a value of 70 kA/cm^2. In MT (middle top), MM, and TE (top edge) samples the critical current density reached 24 kA/cm^2 at 1.8 T. This peak value decreased in bottom samples, especially in the BE sample.

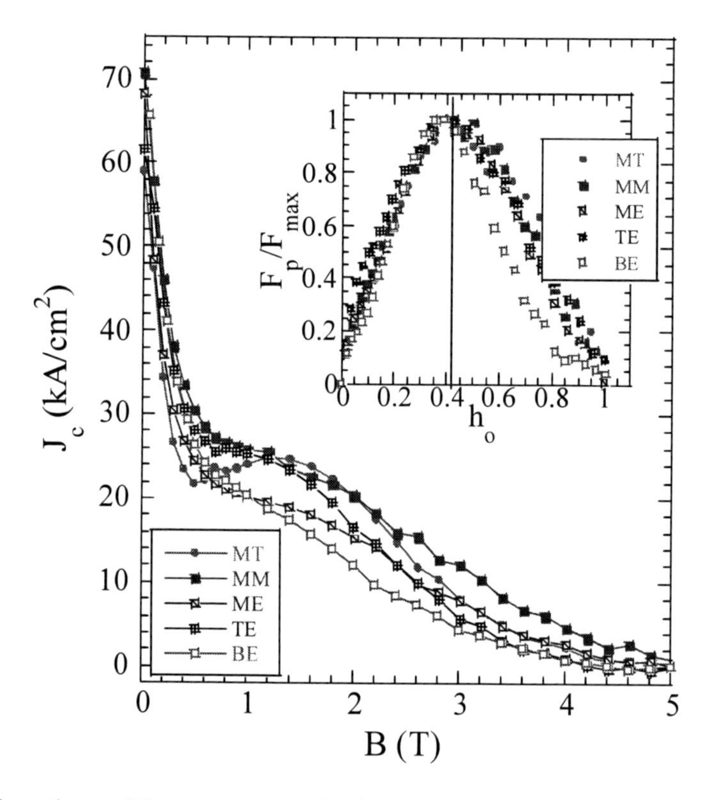

Figure 7. Field dependence of the super-current density at 77 K for specimens cut from various locations in the batch-processed single-domain GdBaCuO superconductor prepared in air by cold seeding method using an Nd-123/MgO film as a seed crystal. The inset shows the normalized volume pinning force density, $f_p = F_p/F_{pmax}$, as a function of the reduced field, $h_0 = H_a/H_{irr}$.

The T_c (onset) also slightly decreased in this specimen, which we attribute to contamination from the supporting ZrO_2 rod. The self-field current density in the top samples are slightly lower than in the bottom samples, which can be explained by variation in fine Gd-211 particles distribution in the 123 matrix [80]. The high J_c value at the top centre of the pellet is favorable for reaching high trapped fields. All the J_c values presented here are similar to those of the OCMG processed Gd-123 bulks [75]. The normalized volume pinning force density, $f_p = F_p/F_{pmax}$, as a function of the reduced field, $h = H_a/H_{irr}$, is frequently used as a measure of the pinning structure effectiveness. We checked performance of the present air-processed bulk samples also in terms of the pinning force density. For this, the irreversibility field, H_{irr}, was determined from magnetization loops with the criterion of 100 A/cm^2. The $f_p(h)$ curves of the MT, MM, ME, TE, and BE samples are presented in the inset of Fig. 7. For all the samples from various locations, the $f_p(h)$ dependence peaked close to 0.41. This is similar to the OCMG processed LRE-123 materials [81]. This again proves that an optimum BaO_2 concentration can effectively suppress Gd/Ba substitution.

6. Microstructure Analysis

In melt-textured LRE-123 materials some portion of a superconducting phase always exists. However, for a reasonable pinning at low fields the concentration of such natural defects is not sufficient and additional LRE-211 or Nd-422 (in the case of Nd-123) have to be introduced intentionally. As these particles are normal and relatively large, they are strong pinning sites because a vortex trapping on them is associated with a high condensation energy gain.

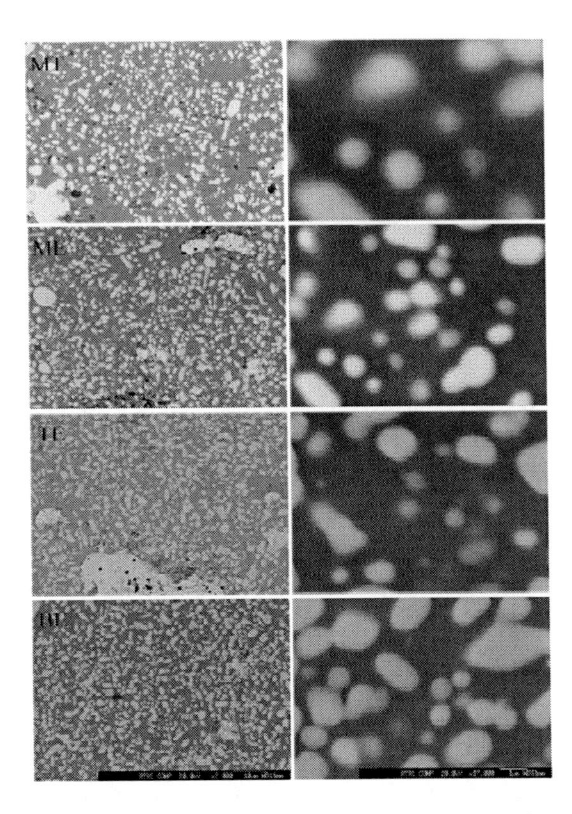

Figure 8. SEM images at different positions of batch processed single domain GdBaCuO superconductor prepared in air.

This is particularly true in the regime of individual vortex pinning and therefore these particles are most effective at low magnetic fields. Their efficiency is inversely proportional to a small power of their dimension, the actual power value depending on the particle geometry [82, 83]. In any case, size reduction of the secondary phase is an important task. In principle, each externally added 211 particle acts as a nucleation site. A slight amount of Pt is usually added to any LRE-123 composite in order to keep the LRE-211 phase dispersion in the LRE-123 superconductive parent phase fine and more stable [84,85].

Morphology of the final secondary phase network in the Gd-123 matrix was studied by scanning electron microscopy on batch processed Gd-123 sample. The results are shown in Fig. 8 that presents both the low and high magnification images of the MT, ME, TE, and BE test samples. In this pellet no large cracks were observed, in accord with the trapped field measurements. The microstructure was similar on the top and bottom of the pellet and found the number of submicron size particles are more in ME and BE respectively. The optimum calcinations process and the mesh used to make the particle size uniform for both 211 as well as 123, influence the microstructure. Similar results are also observed in Gd-123 system processing in partial oxygen pressure [75]. The Gd-211 particles were of sub-micrometer size, which made them the main source of the improved trapped field at 77K.

7. Trapped Field Measurements

Trapped field measurements were performed on all the samples to confirm the single-grain nature of the material. For this, the samples were cooled in the field of 1.5 T to 77 K. After switching off the applied field, the trapped field distribution was detected by scanning the surface with Hall sensors at the distance of 1.2 mm. The more experimental details are presented in the experimental section and the results are shown in Fig. 9. The field profiles had a single peak at the center in all samples of the batch, indicating absence of weak links. The maximum trapped field of 0.8 to 0.9 T was routinely obtained for all samples in the batch, for the molar ratio Gd-123/Gd-211 = 10:5. The maximum observed value of the trapped field reached 0.9 T at 1.2 mm above the sample surface and around 1 T when measured "directly" on the bulk surface. This value is similar to that obtained in the 30 mm Gd-123 single-grain bulk processed by the hot seeding method in a low partial oxygen pressure [86]. The trapped field also exceeded the recently reported result in 26 mm GdBCO/Ag single grain containing Gd-2411 (Nb) fabricated in air [87].

These results indicate that samples are perfectly grown in the form of single grains. Obviously, the trapped magnetic flux density increases with increasing diameter of the material. For this we increased the Gd-123 material size to show the performance of the new seeds. The 45 mm sample was cooled in the field of 3 to 77 K. After switching off the applied field, and waiting for 30 s the trapped field distribution was detected by scanning the surface with Hall sensors at the distance of 1.2 mm. 3D trapped field profile of the best 24 mm sample is presented along with the 45 mm sample in Fig. 10. The maximum trapped field of 1.35 T was recorded at the center of the sample. The result demonstrates the quality of the large-size Gd-123 pellet produced by the batch process.

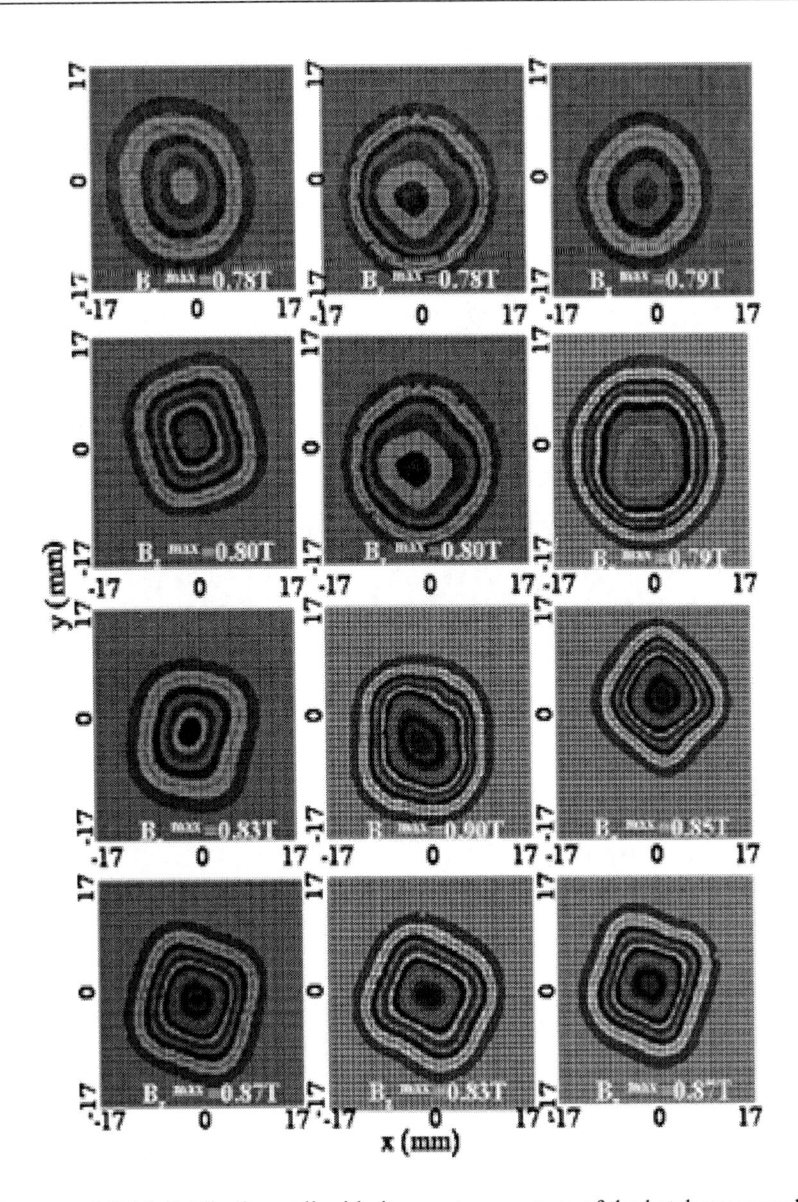

Figure 9. The trapped field distribution at liquid nitrogen temperature of the batch processed Gd-123 bulks processed in air by cold seeding method using a Nd/MgO film as a seed crystal.

To further explore the quality of the material at higher temperatures, we measured the trapped field performance at liquid argon (87.3 K) temperature for both 24 mm and 45 mm single grain Gd-123 material and results are presented in Fig. 11. All experimental conditions were similar to the 77 K measurement described above. The maximum trapped fields at the center for the samples of 24 mm and 45 mm were 0.19 and 0.35 T, respectively, enough for levitation applications.

Figure 12 displays the trapped field distribution at 77 K of the six NEG-123 samples produced in a single run. The trapped field at its maximum reached 0.9 T to 1.05 T at 1.2 mm above the sample surface. The best sample, measured "directly" on the bulk surface, showed

1.2 T. It is worth of noting that the whole batch of NEG-123 samples revealed single-grain nature and equal high quality. The trapped field values exceeded the values recently reported on a 24 mm single grain of Gd-123 processed in air [88], and also that of the 26 mm single grain of Gd-123 material with fine $Gd_2Ba_4CuNbO_y$ phasc [87].

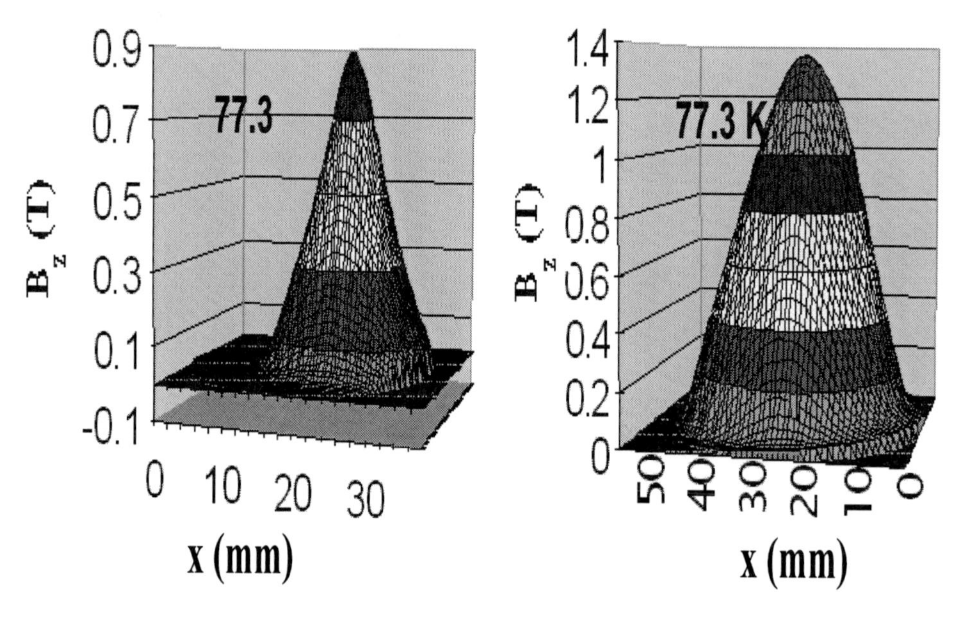

Figure 10. The trapped field profile at 77 K for 24 mm in diameter (left) and 45 mm in diameter (right) Gd123 single grain samples processed in air by cold seeding method using an Nd-123/MgO film as a seed crystal.

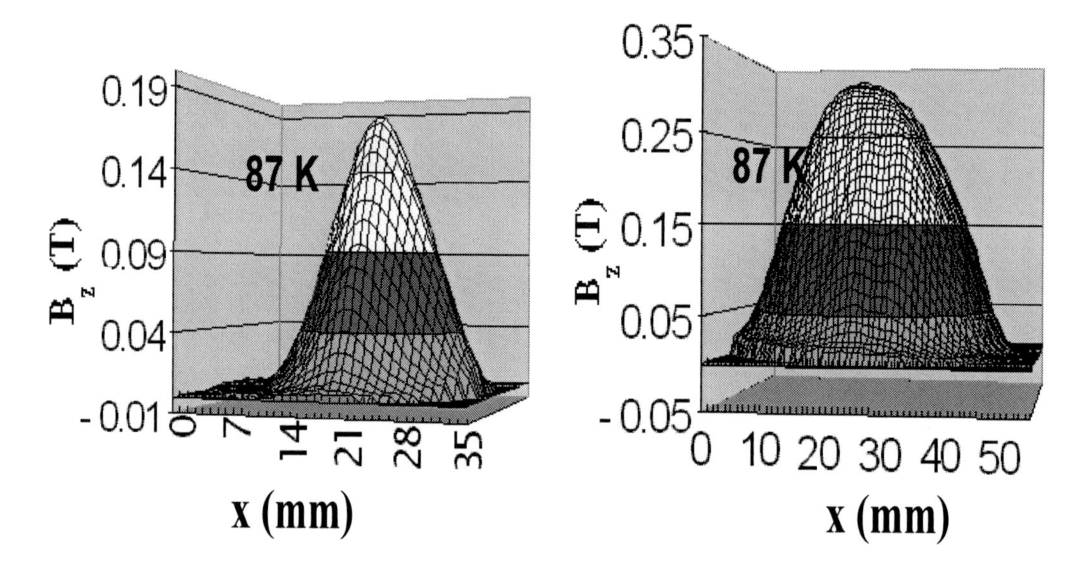

Figure 11. The trapped field profile at 87 K for 24 mm in diameter (left) and 45 mm in diameter (right) Gd123 single grain samples processed in air by cold seeding method using an Nd-123/MgO film as a seed crystal. Not that one can use these material even higher temperatures.

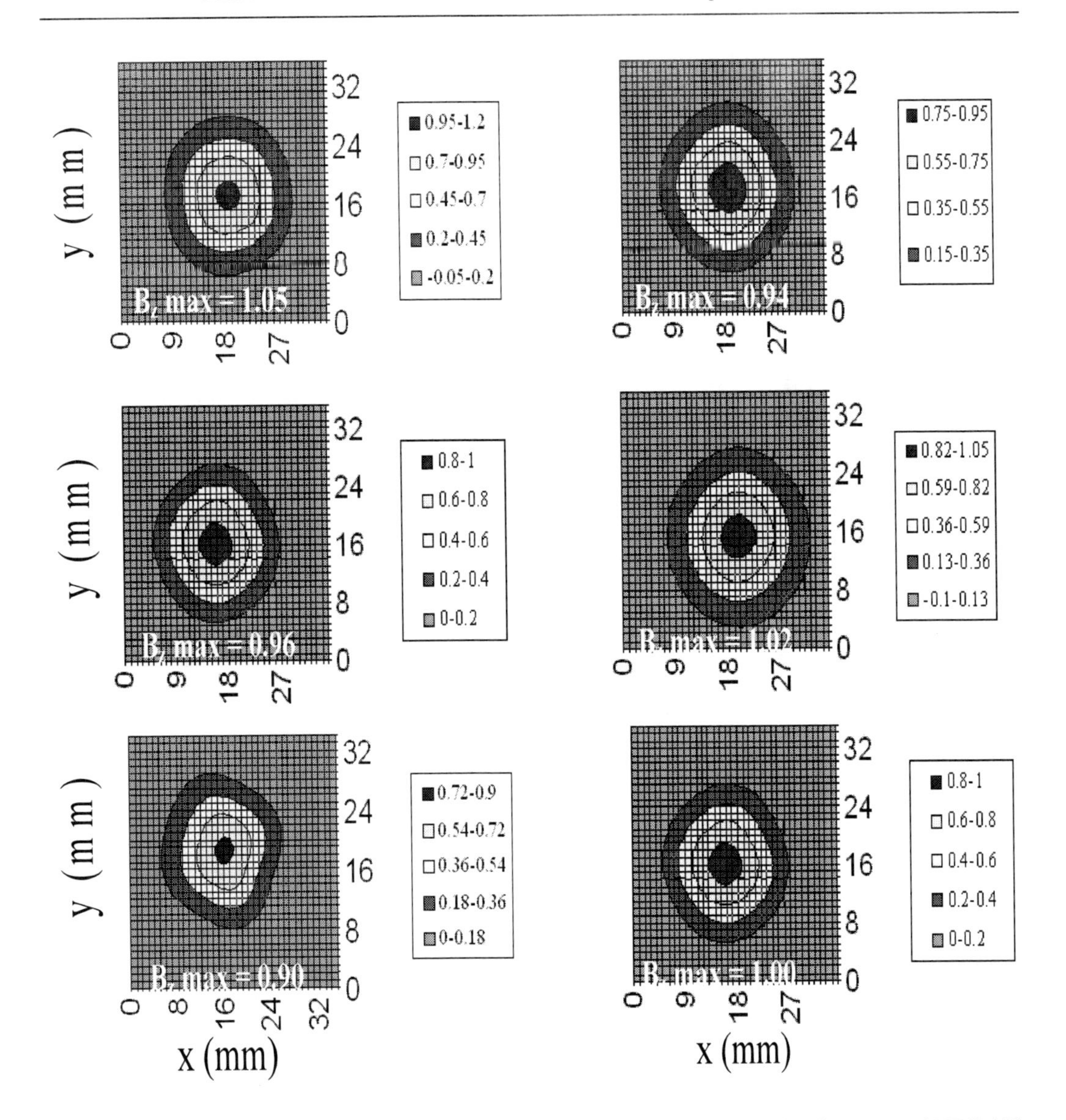

Figure 12. The trapped field distribution at liquid nitrogen temperature of the batch processed NEG-123 bulks processed in Ar-1% O_2 atmosphere by cold seeding method using a Nd/MgO film as a seed crystal.

The trapped field measurements were performed on all Sm-123 samples with varying contents of BaO_2. After the melt growth the final size of the Sm-123 bulks is 16 mm in diameter and 10 mm in thickness. For trapped field measurement, the samples were cooled in the field of 1 T to 77.3 K. After switching off the applied field, the trapped field distribution was detected by scanning the surface with Hall sensors at the distance of 1.2 mm. The more experimental details are presented in the experimental section and the Sm-123 sample with 0, 1, and 3 wt% of BaO_2 added sample results are shown in Fig. 13. It is clear that field profiles have a single peak at the center in all samples, indicate that samples are grown in single grain. Further, the trapped field values increases 0.08 T to 0.23 T with increasing the BaO_2 content in Sm-123 system from 0 to 3 wt%, thus indicating that a proper quantity of BaO_2 (Ba excess) enables to control the Sm/Ba solid state solution in the superconducting phase resulting improved performance. Further experiments are needed since in present experimental results, we do not observe any saturation or reduction and the values are continuously increasing with increasing the BaO_2 in the Sm-123 matrix. This proves again that Nd-123 thin film seeds on MgO single crystals possess a big potential for use in batch process of LRE-123 bulks, reducing the production time and costs.

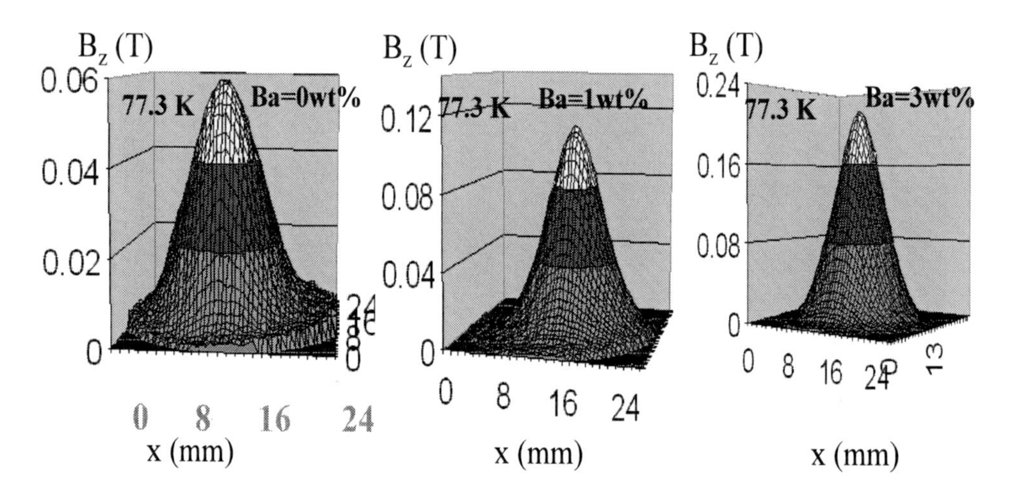

Figure 13. The trapped field profile at 77.3 K for 16 mm diameter Sm-123 single grain processed in air by a cold seeding method using an Nd-123/MgO film as a seed crystal. Note that trapped field of 0.09 T, 0.11T, and 0.23 T was recorded Sm-123 sample with 0wt%, 1wt% and 3 wt% for BaO_2 respectively.

8. Construction of a Child Levitation Disk

The reproducible quality and quantity of the batch processed LRE-123 bulks allowed us to construct a child levitation disk. It is one of the necessary steps on the way to potentially replace superconducting coils of Maglev vehicle (working at liquid helium temperature) by bulk super-magnets cooled by far more economic liquid nitrogen or cryogen-free refrigeration. The main parts of this levitation set-up were: (i) a metallic disk with attached high-performance permanent magnets; (ii) high-performance superconducting super-magnets; and (iii) a liquid nitrogen vessel.

8. (a) The Metallic Disk with Permanent Magnets

Figure 14 shows a permanent magnet disk constructed at railway technical research institute (RTRI). The levitating disk is an iron plate 350 mm in diameter with NdFeB permanent magnets arranged into two concentric rings. The inner permanent magnet ring has outer diameter of 119 mm, inner diameter of 51 mm, and 5 mm thickness. The outer permanent magnet ring has outer diameter of 219 mm, the inner diameter is 151 mm, and thickness is 5 mm. The total weight of the disk with magnets is 5.55 kg.

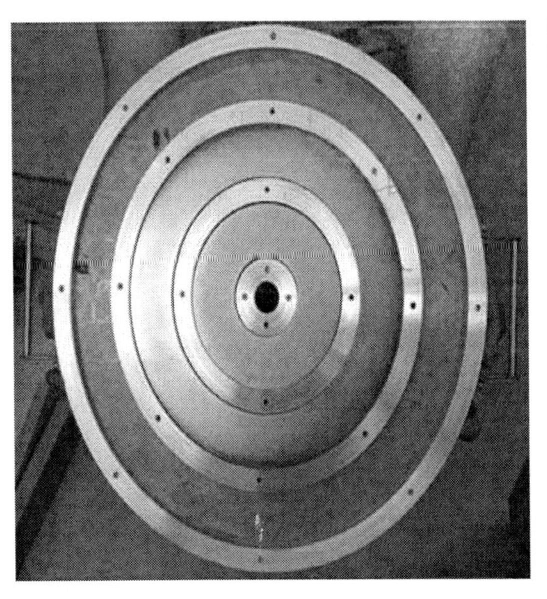

Figure. 14. Construction of home made child levitation disk; home made permanent magnet disk (left), home made FRP cool vessel and home made batch processed LRE-123 material (right).

Figure 15. Levitation of home made magnetic disk using the home made batch processed LRE-123 material. One can see large gap around 15 mm between the disk and sample surface (left). Levitation of child using a repulsive force between home made batch processed $LREBa_2Cu_3O_y$ pellets and home made permanent magnet disk (right).

8. (b) The Liquid Nitrogen Vessel And Superconducting Material

The liquid nitrogen vessel was constructed again at RTRI using the fiber-reinforced plastic (FRP) an insulated material to minimize the liquid nitrogen losses [89]. More then 125 batch-processed Gd-123 pellets were used for this equipment (see Fig. 14(right)). Most of the samples were 24 mm or 32 mm in diameter. Few samples were also between 32 mm and 45 mm diameter. In the first step all samples were arranged in the FRP cool vessel and cooled to liquid nitrogen temperature. Then the disk with permanent magnets was put above the cooled superconductors and successfully levitated above them (see Fig. 15 (left)). The disk was used for the public, during the open day of rail way technical research institute (RTRI) on October 9, 2010 [90]. More then 150 children stood on the disk and enjoyed the experience of levitation (see Fig. 15 (right)). We also loaded a maximum weight of around 35 kg with which the gap maintained around 5 mm gap. Further improvements on the levitation disk (e.g. by adding an additional ring of magnets) are under way, which will be able to levitate adult human beings. This again proved that rather large, and high-performance melt-processed LRE-123 superconductors can be fabricated using the batch process with Nd-123/MgO thin film seeds set on the pellets by cold seeding.

CONCLUSION

In this contribution, a new class of thin film Nd-123 seeds grown on MgO crystals are used to develop a batch production for fabrication of $LREBa_2Cu_3O_y$ (LRE: Sm, Gd, NEG) "LRE-123" pellets in air and Ar-1% O_2. The scanning electron microscopy (SEM) and x-ray diffraction (XRD) results conformed that the quality and orientation of the seed crystals are excellent. On the other hand, new seeds can withstand temperatures >1100°C, as a result, the cold seeding process was applied even to grow Sm-123 material in Air.

XRD analysis confirmed that all the bulks were *c*-axis oriented. The superconducting and magnetic performance of the pellets was checked on several small test samples cut out at various standard positions within the bulk. The values were reasonably uniform and the performance was similar to the oxygen-controlled melt-grown Gd-123 samples. The trapped field measurements showed that the samples were single-domain and of good pinning performance. A large size Gd-123 bulk sample, 45 mm in diameter could also be fabricated by cold seed method in air. This leads to the maximum trapped field value of 1.35 T and 0.35 T at 77.3 K and 87.3 K, respectively. Further, 24 mm in diameter NEG-123 sample showed the maximum trapped field of 1.2 T at the sample surface. Moreover, the maximum trapped field of 0.23 T at 77.3 K was recorded 16 mm in diameter Sm-123 with 3 mol% BaO_2 addition. Initial tests with a home made child levitation disk using the batch processed LRE-123 bulks were performed and tested during open day of RTRI. The batch processed LRE-123 composites possess a big potential for industrial super-magnet applications and present technology opening way from a laboratory to industrial production.

ACKNOWLEDGMENTS

The authors would like to record thanks to Prof. S. Tanaka, the former Director of ISTEC-SRL for his encouragement. We also acknowledge the stimulating discussions with Dr. U. Balachandran (Argonne), Prof. David A Cardwell (University of Cambridge), Dr. Shunichi KUBO (RTRI), Prof. M. Murakami (SIT), Prof. V. Hari Babu (Osmania University), Dr. A. Das (Canada), Dr. M. R. Koblischka (Germany), and Dr. P. Diko (SAS,Slovakia).

REFERENCES

[1] G. Fuchs, P. Schatzle, G. Krabbes, S. Gruβ, P. Verges, K. H. Muller, J. Fink & S. Schultz, *Appl. Phys. Lett.,* 76 (2000) 2107.

[2] M. Muralidhar, N. Sakai, M. Jirsa, M. Murakami & N. Koshizuka, *IEEE Trans. Appl. Supercond.,* 14 (2004) 1206.

[3] F. C. Moon, P. Z. Chang, *Appl. Phys. Lett.* 56 (1990) 22.

[4] K. Salama, V. Selvamanickam, L. Gao, and K. Sun, *Appl. Phys. Lett.* 54 (1989) 2352.

[5] A. M. Campbell, and D. A. Cardwell, Cryogenics 37 (1997) 567.

[6] M. Muralidhar, M. Jirsa, N. Sakai and Murakami, M, *Supercond. Sci. Technol.* 15 (2002) R1.

[7] K. Yokoyama, T. Oka, H. Okada & K. Noto, *Physica C* 392-396 (2003) 739.

[8] S. Wang, *et al.*, IEEE Trans. *Applied Superconductivity* 13 (2003) 2134.

[9] M. Murakami, *Supercond. Sci. Technol.* 5 (1992) 185.

[10] R. Weinstein, I. G. Chen, J. Liu, J. Xu, Y. Obot, C. Foster, *J. Appl. Phys.* 73 (1993) 6533.

[11] J. R. Hull, *Supercond. Sci. Technol.* 13 (2001) R1.

[12] T. Ohara, H. Kumakura, H. Wada, *Physica C* 13 (2001) 1272.

[13] C. Day *et al.*, *Supercond. Sci. Technol.* 15 (2002) 838.

[14] M. Tomita & M. Murakami, *Nature* 421 (2003) 517.

[15] Murakami, M. in: *Melt processed high temperature superconductors, M. Murakami (Ed.)*, World Scientific Publisher Co. Singapore (1992).

[16] N. Chikumoto, S. Ozawa, S. I. Yoo, N. Hayashi, M. Murakami, *Physica* C 278 (1997)187.

[17] T. Egi, J. G. Wen, K. Kurada, N. Koshizuka, S. Tanaka, *Appl. Phys. Lett.* 67 (1995) 2406.

[18] S. I. Yoo, N. Sakai, H. Takaichi, M. Murakami, Appl. Phys. Lett. 65, 633 (1994).

[19] M. Murakami, N. Sakai, T. Higuchi, S. I. Yoo, *Supercond. Sci. Technol.* 9, 1015 (1996).

[20] W. Ting, T. Egi, K. Kurada, K. Koshizuka, S. Tanaka, *Appl. Phys. Lett.* 70, 770 (1997).

[21] G. Osabe, S. I. Yoo, N. Sakai, T. Higuchi, T. Takizawa, K. Yasohama, M. Murakami, *Supercond. Sci. Technol.* 13, 637 (2000).

[22] M. Muralidhar, M. Murakami, K. Segawa, K. Kamada, and T. Saitho, *United States Patent,* (2000) Patent Number 6,063,753.

[23] Muralidhar M, Koblischka M R, and Murakami M, in *Studies of high temperature superconductors* (edited by A. V. Narlikar) 31 (2000) 89.

[24] Muralidhar M, Koblischka M R, Diko P, and Murakami M 2000 *Appl., Phys. Lett.* 76 91.

[25] A. K. Pradhan, M. Muralidhar, M. R. Koblischka, M. Murakami, K. Nakao and N. Koshizuka *Appl. Phys. Lett.* 75 (1999) 253.

[26] M. R. Koblischka, M. Muralidhar, M. Murakami, *Appl. Phys. Lett.* 73 (1998) 2351.

[27] M. M. Muralidhar, T. Saitoh, K. Sagawa, M. Murakami, *J. of Applied Superconductivity* 6 (1998) 139.

[28] M. Muralidhar, M. Murakami, *Applied Supercond.* C 5 (1997) 127.

[29] M. Muralidhar, H.C. Chauhan, T. Saitoh, K. Segawa, M. Murakami, *Supercond. Sci. Technol.* 10 (1997) 663.

[30] M. Muralidhar & M. Murakami, *Physica C* 309 (1998) 39.

[31] M. Muralidhar & M. Murakami, *Physica C* 309 (1998) 43.

[32] M. Muralidhar, H.C. Chauhan, T. Saitoh, K. Segawa, K. Kamada, & M. Murakami, *Physica C* 282-287 (1997) 503.

[33] Das, S. Koshikawa, T. Fukuzaki, M. Muralidhar and M. Murakami *Applied Superconductivity*, 6 (1998) 193.

[34] M. Zamboni, M. Muralidhar, & M. Murakami, *Supercond. Sci. Technol.*, 13 (2000) 811.

[35] M. Muralidhar, H.C. Chauhan, T. Saitoh, K. Segawa, M. Murakami, *Physica C* 280 (1997) 200

[36] M. Muralidhar, M. R. Koblischka and M. Murakami, *Supercond. Sci. Technol.* 12 (1999) 555.

[37] M. Muralidhar, M. R. Koblischka, M. Murakami, *Supercond. Sci. Techno.* 12 (1999) 105.

[38] M. Muralidhar, M. R. Koblischka, M. Murakami, *Physica Status Solid (a)* 171 (1999) R7.

[39] M. Muralidhar, N. Sakai, M. Murakami, K. Segawa, K. Kamada, and T. Saitho, *Japanese Patent*, date of filling: July 5 (2002).

[40] M. Muralidhar, M. Jirsa, N. Sakai, M.Murakami, *Appl. Phys. Lett.* 79 (2001) 3107.

[41] N. Sakai, S. Nariki, K. Nagashima, T. Miyazaki, M. Muralidhar & I. Hirabayashi, *Physica C* 305-309 (2007) 305.

[42] M. Muralidhar, M. Jirsa, N. Sakai, I. Hirabayashi, M. Murakami, 2006 in A. V. Narlikar (Edt.,) Studies of High Temperature Superconductors, (Nova Science Publishers, New York) 50 229.

[43] M. Muralidhar, N. Sakai, N. Chikumoto, M. Jirsa, T. Machi, M. Nishiyama, Y. Wu, & M. Murakami, Phys. Rev. Lett. 89 (2002) 237001.

[44] M. Muralidhar, N. Sakai, M. Jirsa, M. Murakami, & N. Koshizuka, *Supercond. Sci. Technol.* 16 (2003) L46.

[45] M. Muralidhar, N. Sakai, M. Jirsa, N. Koshizuka & M. Murakami, *Appl. Phys. Lett.* 85 (2004) 3504.

[46] M. Muralidhar, N. Sakai, M. Murakami, K. Segawa, T. Saitho & T. Ona, *Japanese Patent*, KN02B317-2, date of filling: June 15 (2003).

[47] M. Muralidhar, N. Sakai, M. Jirsa, M. Murakami, & I. Hirabayashi, *Int. J. of Condensed Matter, Advanced Materials and Superconductivity Research*, 6 (2008) 269.

[48] M. Muralidhar, N. Sakai, M. Jirsa, M. Murakami & N. Koshizuka, *Supercond. Sci. Technolo.*, 17 (2004) S66.

[49] M. Muralidhar, N. Sakai, M. Jirsa, M. Murakami, & N. Koshizuka, *Trans. MRS-J*, 29 (2004) 1305.

[50] M. Jirsa, M. Muralidhar, M. Murakami, K. Noto, T. Nishizaki, and N. Kobayashi, Supercond. Sci. Technol. 14 (2001) 50.

[51] M. Muralidhar, N. Sakai, M. Jirsa, M. Murakami & N. Koshizuka, *Supercond. Sci. Technolo.,* 18 (2005) S47.

[52] M. Jirsa & M Muralidhar, *Czechoslovak Journal of Physics*, 54 (2004).

[53] M. Muralidhar, M. Jirsa, N. Sakai, M. Murakami, & I. Hirabayashi, *J. Mater. Sci. Eng. B* 151 (2008) 90.

[54] M. Muralidhar, M. Jirsa, Y. Wu, N. Sakai, & M. Murakami, *J. Mater. Res*. 18 (2003) 1073.

[55] M. Muralidhar, N. Sakai, M. Jirsa, M. Murakami, & N. Koshizuka, *Supercond. Sci. Technolo.,* 18 (2005) L9.

[56] M. Muralidhar, N. Sakai, M. Jirsa, M. Murakami, & N. Koshizuka, *Physica C* 412-414 (2004) 575.

[57] M. Muralidhar, N. Sakai, M. Jirsa, M. Murakami, & N. Koshizuka, *Physica C* 412-414 (2004) 739.

[58] M. Muralidhar, M. Jirsa, N. Sakai, & M. Murakami, *Recent Res. Devel. Applied Phys.* 6 (2003) 813.

[59] M. Muralidhar, N. Sakai, M. Jirsa, M. Murakami, & N. Koshizuka, *Supercond. Sci. Technolo.,* 17 (2004) 1129.

[60] M. Muralidhar, N. Sakai, M. Jirsa, N. Koshizuka & M. Murakami, *Appl. Phys. Lett.* 83 (2003) 5005.

[61] H. Hayashi, K. Tsutsumi, N. Saho, N. Nishizima & K. Asano, *Physica C* 392-396 (2003) 745.

[62] C. A. Luongo, J. Masson, T. Nam, D. Mavris, H. D. Kim, G. V. Brown & M. W. David Hall, *IEEE Trans. Appl. Supercond.* 19 (2009) 1055.

[63] H. T. Ren, L. Xiao, Y. L. Jiao & M. H. Zheng, *Physica C* 412-414 (2004) 597.

[64] W. Gawalek, T. Habisreuther, M. Zeisberger, D. Litzkendorf, O. Surzhenko, S. Kracunovska, T. A. Prikhana, B. Oswald, L. K. Kovalev, W. Canders, *Supercond. Sci. Technol.* 17 (2004) 1185.

[65] D. Litzkendorf, T. Habisreuther, J. Bierlich, O. Surzhenko, M. Zeisberger, S. Kracunovska, & W. Gawalek, *Supercond. Sci. Technol.* 18 (2005) S206.

[66] Y. Shi, N. H. Babu, & D. A. Cardwell, *Supercond. Sci. Technol.* 18 (2005) L13.

[67] N. H. Babu, Y. Shi, K. Iida, & D. A. Cardwell, *Nature Mater*. 4 (2005) 476.

[68] K. Iida, N. H. Babu, Y. Shi, D. A. Cardwell & M. Murakami, *Supercond. Sci. Technol.* 19 (2006) 641.

[69] M. Oda, X. Yao, Y. Yoshida & H. Ikuta, *Supercond. Sci. Technol.* 22 (2009) 075012.

[70] X. Yao, J. Hu, T. Izumi, & Y. Shiohara, *J. Condens. Matter*. 16 (2004) 3816.

[71] L. Cheng, C.Y.Tang, X. Q. Xu, L. J. Sun, W. Li, X. Yao, Y. Yoshida & H. Ikuta, *J. Phys. D: Appl. Phys*. 42 (2009) 175303.

[72] M. Muralidhar & M. Tomita, *Japanese Patent No 2010-244160*, date of filling: October 29 (2010)

[73] M. Muralidhar, K. Suzuki, M. Jirsa, Y. Fukumoto, & A. Ishihara, M. Tomita *Supercond. Sci. Technol.* 23 (2010) 045033.

[74] M. Muralidhar, M. Tomita, K. Suzuki, & Y. Fukumoto, *Physica C*. 470 (2010) 1158.

[75] S. Nariki, N. Sakai & M. Murakami, *Supercond. Sci. Technol.* 18 (2005) S126.
[76] M. Muralidhar, M. Jirsa, S. Nariki, & M. Murakami, *Supercond. Sci. Technol.*, 14 (2001) 832.
[77] M. Muralidhar & M. Murakami, *in Studies of High Temperature Superconductors, edited by A. V. Narlikar* (NovaScience Publishers, New York, vol. 41) (2002)106.
[78] M. Muralidhar, N. Sakai, M. Jirsa, M. Murakami, & I. Hirabayashi, *Appl. Phys. Lett.* 92 (2008) 162512.
[79] D.X. Chen, R. B. Goldfarb, *J. Appl. Phys.* 66 (1989) 2489.
[80] R. Cloots, T. Koutzarova, J-P. Mathieu & M. Ausloos, *Supercond. Sci. Technol.* 18 (2005) R9.
[81] M. R. Koblischka, A. L. van Dalen, T. Higuchi, S. I Yoo & M. Murakami, *Phys. Rev. B* 58 (1998) 2863.
[82] P. Diko, *in Studies of High Temperature Superconductors* (*edit. by A. Narlikar*), Nova Science Publishers, New York, 28 (1991) 1.
[83] M. Muralidhar & M. Murakami, *Journal of Advances in Cryogenic Engineering.*, 48 (2002) 669.
[84] M. Muralidhar, M. R. Koblischka & M. Murakami, *Supercond. Sci. Technolo.*, 13 (2000) 693.
[85] Y. Zhang, V. Selvamanickam, D.F. Lee, & K. Salama, *Jpn. J. Appl. Phys.* 33 (1994) 3419.
[86] M. Oda & H. Ikuta, *Physica C* 460-462 (2007) 301.
[87] Y. Shi, N. H. Babu, K. Iida, W. K. Yeoh, A. R. Dennis & D. A. Cardwell, *Supercond. Sci. Technol.* 22 (2009) 075025.
[88] M. Muralidhar, K. Suzuki, Y. Fukumoto, A. Ishihara, & M. Tomita, *J. of IEEE Tran. On Appl. Supercond.* (2011) at press.
[89] M. Muralidhar, K. Suzuki, A. Ishihara, M. Jirsa, Y. Fukumoto, & M. Tomita *Supercond. Sci. Technol.* 23 (2010) 124003.
[90] M. Muralidhar, *Railway Technology Avalanche* 33 (2010) 196.

In: Superconductivity: Theory, Materials and Applications ISBN 978-1-61324-843-0
Editor: Vladimir Rem Romanovskii

Chapter 14

ALTERNATING-CURRENT SUSCEPTIBILITY, CRITICAL-CURRENT DENSITY, LONDON PENETRATION DEPTH, AND EDGE BARRIER OF TYPE-II SUPERCONDUCTING FILM

***D.-X. Chen*[1,2], *C. Navau*[2], *N. Del-Valle*[2] *and A. Sanchez*[2]**
[1]Institució Catalana de Recerca i Estudis Avançats (ICREA)
Barcelona, Spain
[2]Departament de Física, Universitat Autònoma de Barcelona
08193 Bellaterra, Barcelona
Spain

Abstract

From accurate measurements of ac susceptibility, χ, of a square $YBa_2Cu_3O_{7-\delta}$ superconducting film of sides $2a = 4$ mm and thickness $t = 0.25$ μm as a function of temperature T at different values of ac field amplitude H_m and frequency f, important superconducting properties are extracted and analyzed. The London penetration depth $\lambda(T)$ is determined from low-H_m $\chi(T)$ measurements after the Meissner susceptibility of the square film is calculated numerically by minimizing relevant Gibbs potential. The critical-current density $J_c(T)$ is extracted from the measured $\chi(T, H_m, f)$ based on the critical-state and flux-creep models. Having $\lambda(T)$ and $J_c(T)$ determined, properly normalized χ vs H_m curves are further converted from the measured $\chi(T, H_m, f)$ to be compared directly with the calculated critical-state curve. It is concluded that vortex dynamics is dominated by collective creep, J_c is practically independent of local flux density B, and the contributions from edge barriers to $\chi(T, H_m, f)$ are important especially when T approaches T_c. For a more quantitative explanation of the observed $\chi(T, H_m, f)$, type-II superconductivity theory for thin films in a perpendicular magnetic field is developed. Important anomalous phenomena are found and remain open for further study.

1. Introduction

The technique of ac susceptibility, $\chi = \chi' - j\chi''$, is a very useful and widespread tool for studying electromagnetic properties of superconductors. When this technique is used for measuring high-temperature superconducting (HTS) films, the field is normally applied in the direction perpendicular to the film surface, so that $|\chi|$ is very high owing to a strong demagnetizing effect superimposed on the diamagnetism. The first application of χ measurements was to determine the superconducting transition temperature T_c. After χ of round films (thin disks) was calculated from the critical-state (CS) model assuming a constant critical-current density J_c (i.e., the Bean model) or a power-law $E(J)$ [1–3], J_c and its mechanism could be quantitatively studied by χ measurements in $YBa_2Cu_3O_{7-\delta}$ (YBCO) films [4–10]. The theoretical CS χ for round films was recently converted accurately to rectangular films by numerical calculations [11, 12], so that further quantitative experimental study on χ becomes possible and necessary. On the other hand, previous χ calculations were performed with a neglected London penetration depth λ, but our recent calculation for a thin strip shows that an effective penetration depth λ_{eff} is large enough to influence significantly its low-field χ [13]. Thus, it would be interesting to calculate λ_{eff} of a square film so that the λ of superconducting films can be determined by χ measurements. In this way, not only intrinsic properties such as Cooper-pair density, critical fields, and depairing current density can be studied for each state-of-the-art HTS films, but also some features of χ itself may be better understood.

We study in the present work both the London penetration depth λ and the ac susceptibility χ of a typical HTS film. The sample is a square film of sides $2a = 4$ mm and thickness $t = 250$ nm, and its χ as a function of temperature T at different values of ac field amplitude H_m and frequency f is measured with a high-quality ac susceptometer, as described in Sec. II. The London penetration depth $\lambda(T)$ is determined from low-H_m $\chi(T)$, after the Meissner susceptibility of a square film is calculated numerically, as described in Sec. III. In Sec. IV, $\chi(T)$ curves measured in a large range of H_m and f are presented, from which $J_c(T)$ is obtained and $\chi(T, H_m, f)$ is converted to a normalized version to be compared directly with theoretical CS susceptibility. The vortex dynamics is discussed in Sec. V and associated effects of thermal equilibrium magnetization and edge barrier are discussed in Sec. VI. Conclusions are presented in Sec. VII.

2. Sample and Measurements

The sample being studied was an epitaxial YBCO film grown by chemical solution deposition on a 5×5 mm^2 $LaAlO_3$ single crystal [14]. The crystallographic c axis of YBCO was perpendicular to the film surface and the a and b axes were parallel to the edges with both in-plane and out-of-plane misalignment angles being approximately 0.5 °. The film was patterned into a square shape by optical lithography, in order to have a well defined geometry and to avoid any defects on the edges. Final film dimensions were $2a = 4.00 \pm 0.01$ mm in sizes and $t = 0.250 \pm 0.025$ μm in thickness (the latter error signifies the surface roughness), which were measured by using an optical microscope and a profilometer, respectively.

The ac susceptibility of the film after zero-field cooling was measured as a function of T at different values of H_m and f with an ac susceptometer of Quantum Design PPMS, which had been calibrated by using a copper cylinder [15, 16]. A magnetic moment magnitude correction was made by multiplying a constant factor, so that the measured $\chi(T, H_m)$ became very accurate at $f = 1111$ Hz, and the errors for the magnitude and phase of χ were within $\pm 0.7\%$ and $\pm 0.3°$, respectively, in a f range between 111 and 1111 Hz. The magnitude error can reach 1% and the phase error can reach $-0.6°$ at $T < 20$ K for f beyond this range. The maximum H_m value was 800 A/m in the T-dependence measurements with this commercial ac susceptometer. For H_m up to 80 kA/m at 77 K, a home-made high field ac susceptometer was used [17].

3. London Penetration Depth

3.1. Calculation of effective London penetration depth

For a superconducting film of thickness $t < \lambda$, the relevant two-dimensional screening length is the Pearl length [18]

$$\Lambda = \lambda^2/t. \tag{1}$$

For a YBCO film of $t \approx 0.3$ μm with $\lambda \approx 0.3$ μm at $T = 77$ K, we have $\Lambda \approx 0.3$ μm, which is 10^{-4} times the usual sample sizes of several millimeters. Since the susceptibility of a completely shielded square film of sides $2a \times 2a \times t$ is [11]

$$\chi_0(0) = -0.9094a/t, \tag{2}$$

where (0) means the susceptibility for the completely shielded state of $\Lambda/a = 0$, the detection of a reduction in the half width by an effective penetration depth $\lambda_{\text{eff}} = \Lambda$ would require the detection of a variation of 10^{-4} times χ. This is beyond the accuracy of any ac susceptibility technique. Fortunately, our calculation for a thin strip showed that λ_{eff} is not Λ but $\sim 10\Lambda$, so that it becomes detectable by high-quality χ measurements [13]. In the present work, we will extend our calculation to square films and show how λ can be determined by χ measurements.

The studied square film occupies a region of $-a \leq x, y \leq a$ and $-t/2 \leq z \leq t/2$ $(t < \lambda)$ with an applied field H_a in the z direction and a circulating surface current of density $\mathbf{K}(x, y)$. As done in [19–21], a scalar function $g(x, y)$, which is zero at the film edges, is defined on the xy plane by $\mathbf{K}(x, y) = -\hat{\mathbf{z}} \times \nabla g(x, y)$, where $\hat{\mathbf{z}}$ is the unit vector in the z direction. The total magnetic moment of the film is an integration over the entire film surface, $\mathbf{m} = \hat{\mathbf{z}} \int_S g(x, y) ds$. The function $g(x, y)$ in the Meissner state corresponds to a minimum of the Gibbs potential, $G = E_{\text{mag}} + E_{\text{kin}} + E_{\text{ext}}$, where the internal magnetic and kinetic energies and the interaction energy with external field are, respectively,

$$E_{\text{mag}} = \frac{\mu_0}{8\pi} \int_S \int_{S'} \frac{\nabla g(x, y) \cdot \nabla g(x', y')}{\sqrt{(x - x')^2 + (y - y')^2}} \, ds \, ds', \tag{3}$$

$$E_{\text{kin}} = \frac{\mu_0 \Lambda}{2} \int_S [\nabla g(x, y)]^2 \, ds, \tag{4}$$

$$E_{\text{ext}} = \mu_0 H_a \int_S g(x, y) \, ds. \tag{5}$$

For numerical calculations, the film is divided into $N \times N$ square cells, each having an area of $A_c = 4a^2/N^2$. In this case, a continuous $g(x, y)$ is approximated by an $(N + 1) \times$

$(N+1)$ array of $g(i,j)$, where indices $i,j = 0,1,2,...,N$ correspond to the node positions and $g(i,j) = 0$ holds if $i,j = 0$ or N. Starting with $g(i,j)A_c = 0$ everywhere, we apply a positive value of H_a and look for the position (i,j), with $i,j \neq 0$ or N, at which decreasing $g(i,j)A_c$ by a given negative moment increment $\Delta g A_c$ would yield the maximum decrease in G. When this place is found we change the $g(i,j)A_c$ there by adding $\Delta g A_c$. This process is repeated until G cannot be further minimized, so that a set of $g(i,j)A_c$ is found for this value of H_a, from which the Meissner susceptibility $\chi_0(\Lambda/a)$ is calculated by

$$\chi_0(\Lambda/a) = \frac{1}{H_a N^2 t} \sum_{i,j=0}^{N} g(i,j). \tag{6}$$

As described in [13], the calculated $\chi_0(\Lambda/a)$ changes monotonically with increasing H_a and N, which result from a reduction of systematic discretization errors, and two linear extrapolations to $1/H_a = 0$ and $1/N = 0$ are needed to get χ_0 with accuracy better than 0.1%. This was done for obtaining Eq. (2) for the case of $\Lambda/a = 0$, as explained in [11]. In the present work, we fix $N = 201$ and only make one extrapolation with respect to $1/H_a$. In this way, the calculated $-\chi_0(0)$ is 0.3% smaller than that given in Eq. (2), but $\chi_0(\Lambda/a)/\chi_0(0)$ can be calculated accurate enough. Similarly to the case of a thin strip [13], λ_{eff} for the present case is defined by $\chi_0(\Lambda/a) = -0.9094(a - \lambda_{\text{eff}})/t$ according to Eq. (2), so that

$$\lambda_{\text{eff}}/a = 1 - \chi_0(\Lambda/a)/\chi_0(0). \tag{7}$$

Thus, λ_{eff}/a is calculated as a function of Λ/a. For the purpose of λ determination, an inverse function is plotted in Figure 1 by open circles, which are well fitted by a curve expressed analytically by

$$\Lambda/a = 0.0735\lambda_{\text{eff}}/a + 0.22(\lambda_{\text{eff}}/a)^{1.7} + 1.1(\lambda_{\text{eff}}/a)^{5.2}. \tag{8}$$

The fitting error is less than 0.3% and 1% for $\lambda_{\text{eff}}/a < 0.1$ and < 0.4, respectively. For comparison, the results for a thin strip calculated in [13] are also plotted in Figure 1.

3.2. Determination of London penetration depth by ac susceptibility measurements

To determine λ of a square film of sides $2a$ and thickness t, we should measure $\chi_0(\Lambda/a)$ in Eq. (7), where $\chi_0(0)$ is expressed by Eq. (2), to get λ_{eff}/a, from which Λ/a and then λ are obtained from Eq. (8) and Eq. (1).

The measured χ at $H_m = 8$ A/m and $f = 1111$ Hz as a function of T is shown in Figure 2(a). The finite $H_m = 8$ A/m is already large enough to cause a critical-current penetration, as seen from the χ'' peak when $T > 85$ K. According to [1], there is a relation for thin disks between $\chi_0(0)$ and CS χ at H_m much lower than the field for full penetration of critical current,

$$\chi_0(0) = \chi' - 15\pi\chi''/32. \tag{9}$$

This relation is assumed to be approximately valid for the square film with $\Lambda/a > 0$, so that $\chi_0(T)$ is obtained from the measured $\chi'(T)$ and $\chi''(T)$. Using this $\chi_0(T)$ to replace $\chi_0(\Lambda/a)$ in Eq. (7), we obtain λ as a function of T, as shown in Figure 2(b) by open circles.

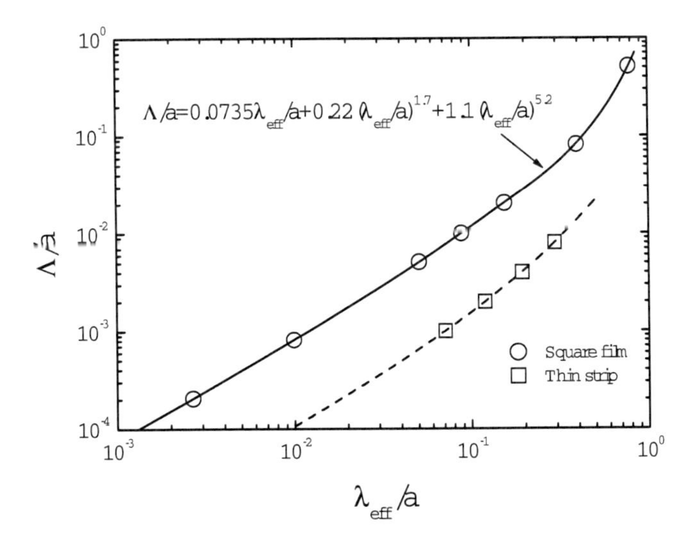

Figure 1. Λ/a as a function of λ_{eff}/a for a square superconducting film of sides $2a$. Circular symbols are numerically calculated results and solid curve is their fitting expressed by Eq. (8). Square symbols and dashed curve are calculated for a thin strip of width $2a$, included for comparison.

We should emphasize that in ac measurements, the Meissner susceptibility $\chi_0(\Lambda/a)$ is defined as $\chi'(H_m < H_{c1})$, where the lower-critical field H_{c1} is practically zero owing to the strong demagnetizing effects for any actual superconducting film (see Figure 14), but for calculating $\chi_0(\Lambda/a)$ in Sec. III A, H_a can be so large that extrapolation to $1/H_a = 0$ can be made, because the Meissner state is already guaranteed by minimizing the G expressed above.

3.3. The $\lambda(T)$ dependence

$\lambda(T)$ of YBCO films or single crystals was previously studied by a number of groups using high-frequency techniques [22–29]. It was shown for YBCO superconductors that λ is highly anisotropic. λ_c and λ_{ab} are defined for Meissner currents flowing along the c axis and within the ab plane, respectively. One has $\lambda_c \gg \lambda_{ab}$ at any T and $d\lambda_c/dT = 0$ and $d\lambda_{ab}/dT > 0$ at $T = 0$. It is commonly accepted that $d\lambda_{ab}/dT > 0$ at $T = 0$ reflects a d-wave pairing mechanism. In a wide T interval below T_c, the following relation is approximately valid:

$$\lambda(T) = \lambda(0)[1 - (T/T_c)^p]^{-0.5}, \tag{10}$$

where p is a constant. For our sample, $T_c = 90.3$ K is determined by the onset of diamagnetism and $\lambda(0) = 0.18$ μm is used as determined for a similar film [29]. We see in Figure 2(b) that the determined data points can be fitted by a solid curve with $p = 1.7$, which is far above the dashed curve for $p = 4$ describing an empirical law (or the two-fluid dependence) for some low-temperature superconductors [30, 31]. The measured λ is actually λ_{ab} since the Meissner currents flow within the ab plane. The value of $p = 1.7$ is located between two typical vales of 1.84 and 1.36. $p = 1.84$ was obtained from the average of

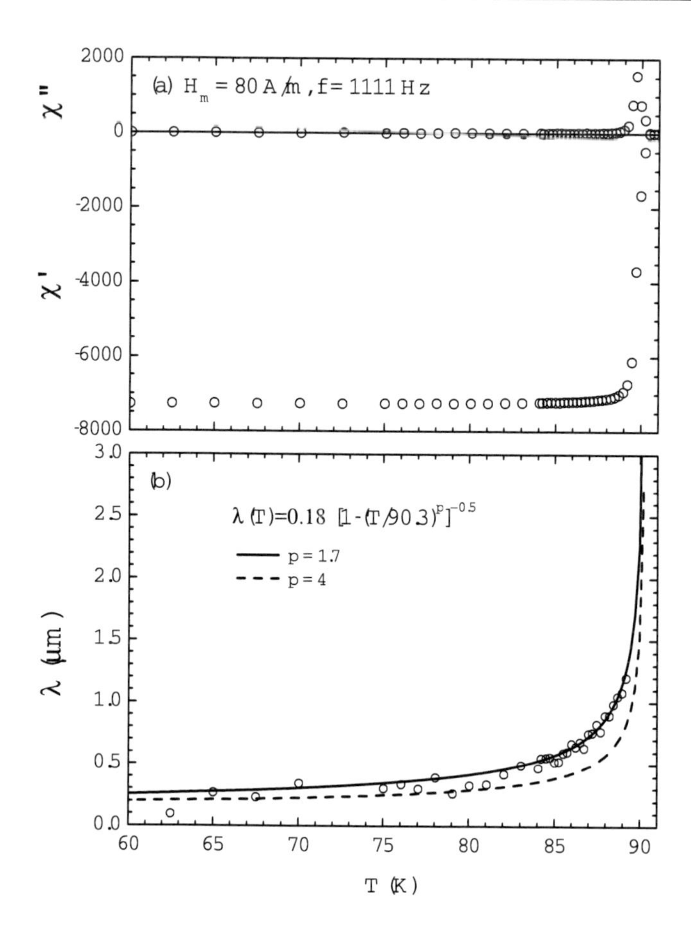

Figure 2. (a) The χ' and χ'' of the YBCO film measured at $H_m = 8$ A/m and $f = 1111$ Hz as functions of temperature T. (b) λ as a function of T (symbols) deduced from the measured $\chi'(T)$ and $\chi''(T)$ based on Eqs. (1), (7), (8), and (9). The curves are calculated from Eq. (10) with $\lambda(0) = 0.18$ μm, $T_c = 90.3$ K, and $p = 1.7$ and 4.

previously measured data of $\lambda_a(T)$ and $\lambda_b(T)$ (Figure 2 of [25]), corresponding to a d-wave pairing with $d_{x^2-y^2}$ symmetry. $p = 1.36$ was obtained from the calculated d-wave pairing $\lambda_{ab}^2(0)/\lambda_{ab}^2(T)$ versus T/T_c curves in Figure 1 of [32].

4. Ac Susceptibility

4.1. Temperature and ac field dependence of χ

The portions around $\chi'' = \chi''_m$ of the measured χ' and χ'' as functions of T at $f = 11, 33, 111, 333, 1111, 3333$, and 10000 Hz are plotted in Figures 3(a), 3(b), and 3(c) for $H_m = 800, 80$, and 8 A/m, respectively. Only the lines connecting the data points are plotted in these figures, except for the cases used for determining J_c (see below), for which data points are also plotted. It can be seen that at fixed H_m and f with increasing T, negative χ' increases to zero at $T = T_c = 90.3$ K, accompanied by a positive χ'' peak. At a fixed

H_m with increasing f, both χ' and χ'' curves shift to higher T, and the maximum χ'', χ''_m, increases slightly. The T at $\chi'' = \chi''_m$, $T(\chi''_m)$, increases with decreasing H_m. All these features are rather common for superconductors.

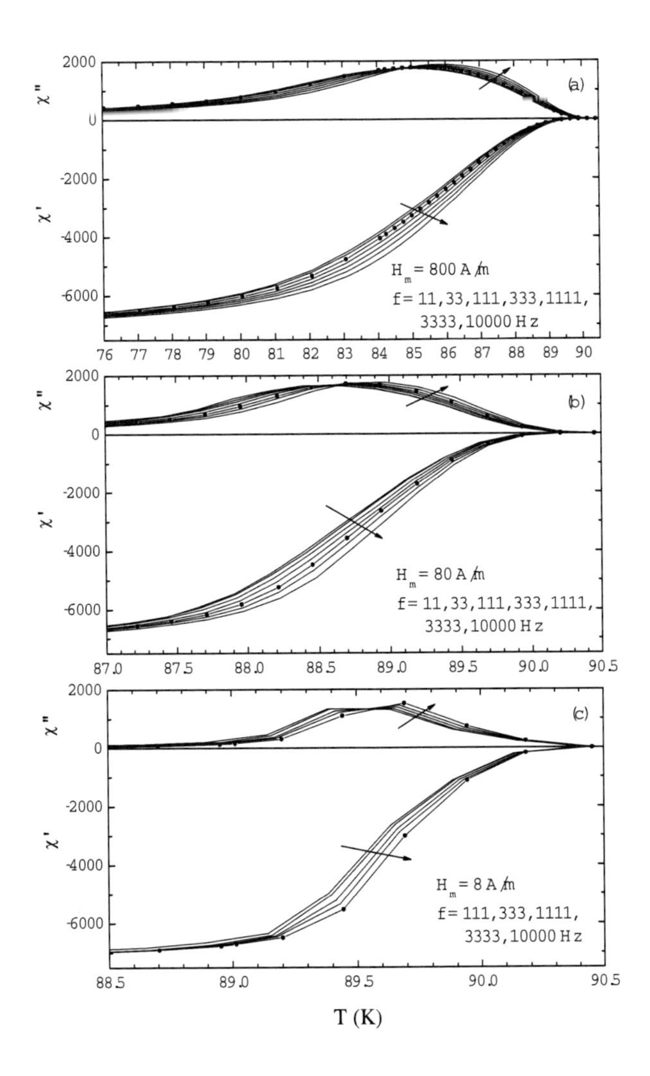

Figure 3. The χ' and χ'' of the YBCO film studied, measured at $H_m = 800$ (a), 80 (b), and 8 A/m (c) and f between 11 and 10000 Hz as functions of temperature T. The $H_m(\chi''_m, T, f)$ functions with data points for $f = 111, 1111$, and 10000 Hz at $H_m = 800, 80$, and 8 A/m, respectively, are used for determining $J_c(T)$.

The $\chi(T)$ curves shown in Figures 3(a), 3(b), and 3(c) cannot be compared directly with the calculated CS $\chi(H_m)$ curves, but the $\chi''(\chi')$ curves (Cole-Cole plots) can be compared with their CS counterpart for $\Lambda/a = 0$ and constant J_c, as shown in Figures 4(a), 4(b), and 4(c) for $H_m = 800, 80$, and 8 A/m, respectively. Such a comparison was often made in the literature [5, 10, 33, 34]. We see that the best coincidence between the experimental and calculated curves occurs for the case of $H_m = 800$ A/m, although the experimental χ' at $\chi'' = \chi''_m$, $\chi'(\chi''_m)$, is somewhat lower than its CS value and χ''_m increases from its

CS value with increasing f. For the case of $H_m = 8$ A/m, both the experimental $\chi'(\chi''_m)$ and χ''_m are significantly lower than their CS values. The case of $H_m = 80$ A/m lies in between. Since $-\chi_0(\Lambda/a)$ decreases with increasing T, such a comparison cannot be made accurately at least for the case of $H_m = 8$ A/m, for which the main portion of the $\chi''(\chi')$ curve is associated with few temperatures very close to T_c.

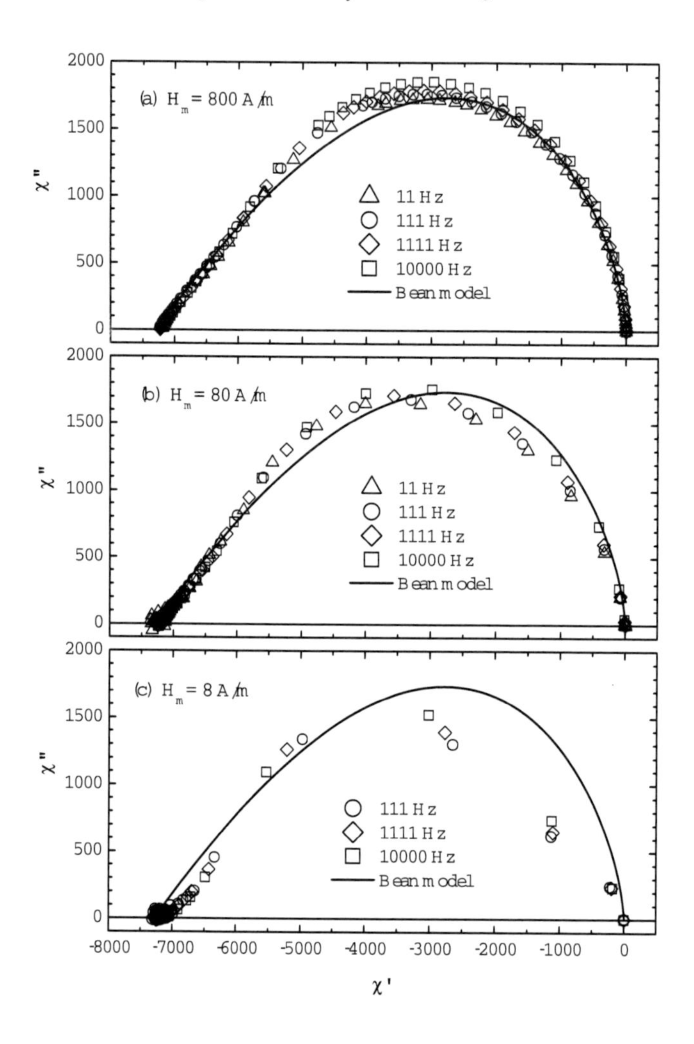

Figure 4. χ'' vs χ' corresponding to Figure 3 for $H_m = 800$ (a), 80 (b), and 8 A/m (c). The lines are calculated from the Bean model with $\chi_0(0) = -7275$.

The results for $\chi(H_m)$ measured at $T = 77$ K and $f = 10, 30, 90$, and 270 Hz will be described below.

4.2. Determination of $J_c(T)$

In the CS calculations, J_c is defined as the value of current density $|J|$ below which the electrical field $E = 0$ and above which $E \to \infty$. As a result, the obtained ac susceptibility is frequency independent. In any real material, however, $|E|$ is finite and increasing with

$|J|$, so that J_c has to be defined as the J where $E = E_c$, E_c being a criterion E on the order of 1 μV/cm. Thus, the experimental determination of J_c requires the measurements of both J and E. Using the ac susceptibility technique at a fixed T, one measures $\chi(H_m)$ at several values of f to get $H_m(\chi''_m)$, from which $J(T, f)$ is calculated from a CS formula and $E(T, f)$ is calculated from a formula for emf. Two examples for HTS films measured at $T = 77$ K were given in [4, 8]. In the present case, instead of $\chi(H_m)$, $\chi(T)$ curves have been measured at fixed values of H_m and f, so that an alternative technique should be developed.

The CS formula is Eq. (20) in [12] and should be modified for each set of $\chi(T, f)$ at a fixed H_m as follows. J_c is written $J(T, f)$, being the time and position averaged J measured at frequency f and temperature T where $\chi'' = \chi''_m$, $-\chi_0$ is written $\chi_0(\Lambda/a)$, being the Meissner susceptibility at the same T and the corresponding Λ, and $H_m(\chi''_m)$ is written $H_m(\chi''_m, T, f)$, being the fixed H_m itself. Thus, the modified formula becomes

$$J(T, f) = -1.21\chi_0(\Lambda/a)H_m(\chi''_m, T, f)/a. \tag{11}$$

The corresponding time and position averaged $E(T, f)$ may be approximated by [3]

$$E(T, f) = fa\mu_0 H_m(\chi''_m, T, f)/\sqrt{2}. \tag{12}$$

We find that $E(T, f) \approx 1$ μV/cm may be calculated from Eq. (12) for $H_m = 800$, 80, and 8 A/m with $f = f_c = 111, 1111$, and 10000 Hz, respectively. Therefore, $E_c \approx 1$ μV/cm is used for the present case, and $J_c(T) \equiv J(T, f_c)$ is calculated from Eq. (11) to be 0.349, 0.0344, and 0.00334 MA/cm^2 at $T = 85.17, 88.76$, and 89.67 K, respectively. In order to get a $J_c(T)$ function more accurately, $J_c = 1.3$ MA/cm^2 has been obtained by $\chi(H_m)$ measurements at $T = 77$ K and $f = 30$ Hz. The four data points are plotted in Figure 5.

To get a $J_c(T)$ function with high accuracy in a T range near T_c, we plot $1/J_c$ vs $q \equiv 1/(1 - T/T_c)$ in Figure 5(b), and get a fitting formula as

$$J_c(T)^{-1} = 0.0744q^{1.2} + 0.000104q^{2.98}, \tag{13}$$

in which the unit of J_c is MA/cm^2. The curve in Figure 5(a) is plotted also according to Eq. (13).

We note that J_c determined from Eqs. (11) and (12) is a few percent under-estimated, since a correction factor $k_J(n) = 1 + 0.086[1 - \exp(-40/n)]$, which was obtained in [3], is not included in Eq. (11). The value of n in this factor has to be determined iteratively after obtaining the first approximation of $E(J)$ relation as done in Sec. V. We will not make the iterations, since such an approximation will not influence the main conclusions of the present work.

4.3. Normalized susceptibility and field amplitude

In the CS susceptibility calculation, London penetration depth is assumed to be zero and J_c is assumed to be constant (the Bean model), and the results are expressed by χ normalized to $-\chi_0(0)$ as a function of H_m/J_ct [1, 12]. In order to compare our T dependence curves with the field dependence curves for the CS model, H_m should be normalized to $J_c(T)t$,

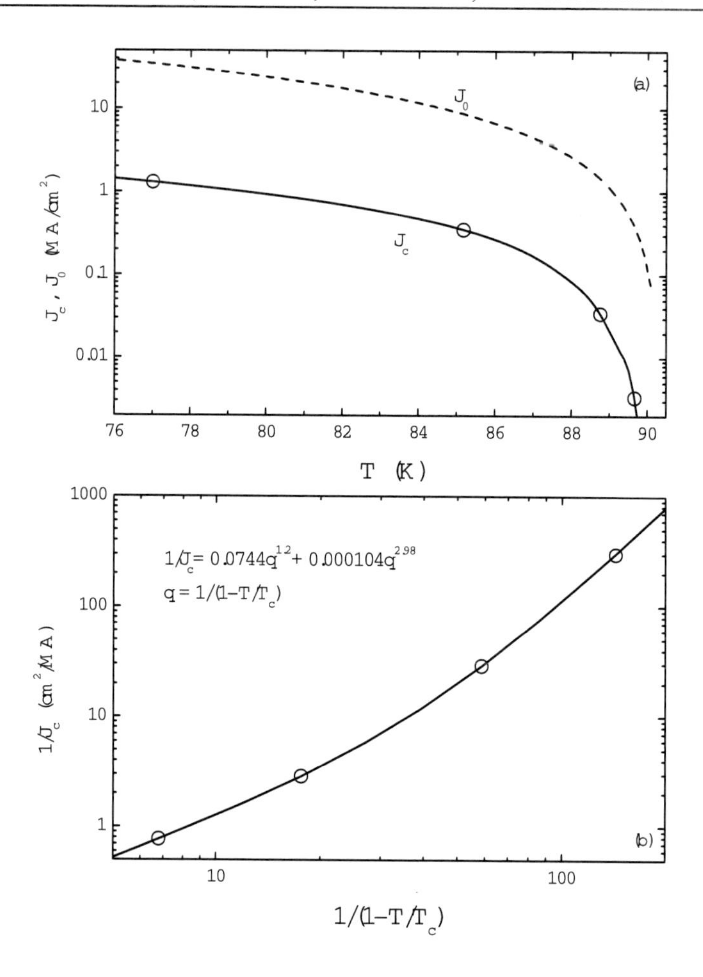

Figure 5. J_c determined at $E_c \approx 1\ \mu$V/cm at four values of T (circles) and the continuous $J_c(T)$ curve (a) and $1/J_c$ as a function of $1/(1 - T/T_c)$ (b). The depairing current density $J_0(T)$ is shown in (a) for comparison.

where $J_c(T)$ is calculated from Eq. (13) with the unit converted to A/m ^{2}. The normalized $-\chi/\chi_0(0)$ should be replaced by $-\chi/\chi_0(\Lambda/a)$ by multiplying $(1 - \lambda_{\rm eff}/a)^{-1}$ according to Eq. (7). $\lambda_{\rm eff}/a$ is calculated from Λ/a by

$$\lambda_{\rm eff}/a = 16.9\Lambda/a - 20.2(\Lambda/a)^{1.2}, \tag{14}$$

which is an inverted function of Eq. (8) and valid in an enough-large range $\lambda_{\rm eff}/a < 0.2$. Λ/a in Eq. (14) is calculated by

$$\Lambda/a = \lambda(T)^2/at = [\lambda(0)^2/at][1 - (T/T_c)^{1.7}]^{-1} \tag{15}$$

according to Eqs. (1) and (10).

By such a normalization, the results of T-dependent actual χ shown in Figures 3 and 4 are converted to Figures 6 and 7 for each set of experimental results at $H_m = 800, 80$, and 8 A/m. The same normalization has been made for the H_m-dependent actual χ measured at $T = 77$ K, and the results are given in Figure 8.

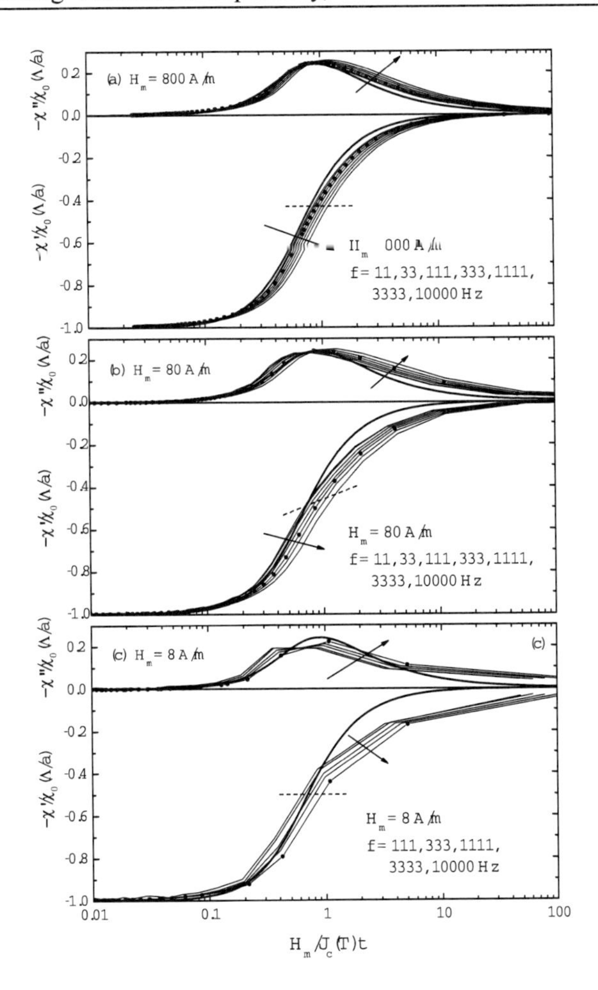

Figure 6. Normalized susceptibility $-\chi/\chi_0(\Lambda/a)$ as a function of normalized field amplitude $H_m/J_c(T)t$ for $H_m = 800$ (a), 80 (b), and 8 A/m (c). The cases with data points for determining $J_c(T)$ corresponding to those in Figures 3 may be directly compared with the thick curves of $-\chi/\chi_0(0)$ vs H_m/J_ct for the Bean model.

4.4. Comparison with critical-state susceptibility

The thick curves for the relations among $-\chi'/\chi_0(0)$, $-\chi''/\chi_0(0)$, and H_m/J_ct in Figures 6, 7, and 8 are the results for a thin film calculated from the CS model with a constant J_c [1,12], which can be compared with the normalized measured results of the film studied here, especially with the data points for the cases from which $J_c(T)$ is determined.

Some general features are shown in Figures 6(a), 6(b), and 6(c) and Figure 8(a). The normalized χ' increases from -1 to 0 with increasing the normalized H_m, accompanied by a positive peak for the normalized $\chi''(H_m)$. The curve of $-\chi/\chi_0(\Lambda/a)$ vs $H_m/J_c(T)t$ shifts to higher $H_m/J_c(T)t$ with increasing f, just like the curve of χ vs T shifts to higher T in Figures 3(a), 3(b), and 3(c). For the T dependence measurements, the increments of the shift increases with decreasing H_m, as seen in Figures 6(a), 6(b), and 6(c). The $-\chi''/\chi_0(\Lambda/a)$ peak and the $-\chi'/\chi_0(\Lambda/a)$ transition from -1 to 0 are wider than their

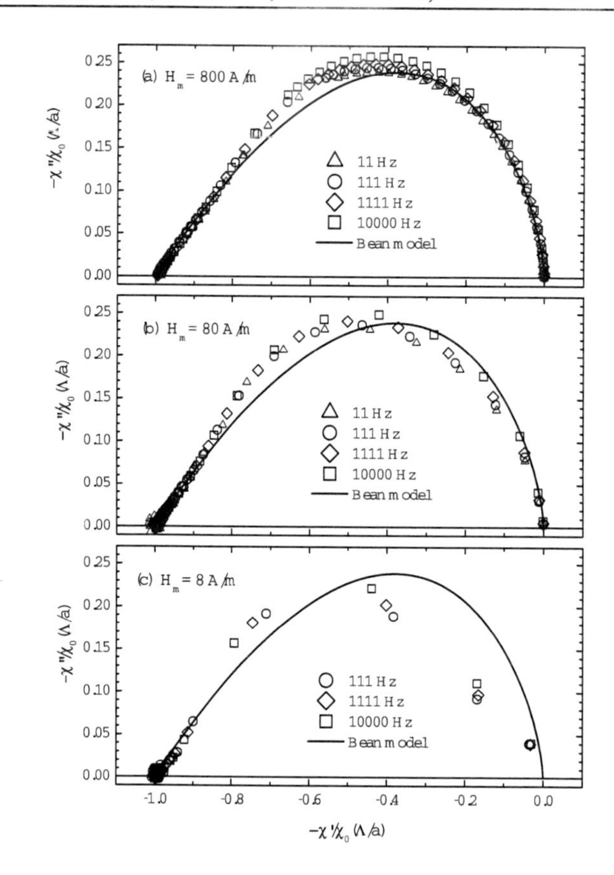

Figure 7. $-\chi''/\chi_0(\Lambda/a)$ as a function of $-\chi'/\chi_0(\Lambda/a)$ for $H_m = 800$ (a), 80 (b), and 8 A/m (c). The curves are calculated from the Bean model.

CS counterpart, and the difference between both increases with decreasing H_m for the T dependence measurements.

In order to describe the last feature quantitatively, we compare the experimental susceptibility values for the cases with data points, whose χ'' peak is centered at the same field as that for the CS. At a typical value of $H_m/J_c(T)t = 10$, $|\chi'/\chi_0(\Lambda/a)|$ and $|\chi''/\chi_0(\Lambda/a)|$ are larger than their CS counterparts by factors of 1.35 and 1.35 for $H_m = 800$ A/m, of 1.9 and 3 for $H_m = 80$ A/m, and of 1.6 and 8 for $H_m = 8$ A/m, as seen from Figures 6(a), 6(b), and 6(c), respectively. For the results obtained at $T = 77$ K, however, both are the same within experimental errors, although systematic differences can still be observed at $H_m/J_c(T)t = 2$.

It is shown in Figures 7(a), 7(b), and 7(c) that $-\chi''_m/\chi_0(\Lambda/a)$ increases with increasing f for all the values of H_m. The values of $-\chi''_m/\chi_0(\Lambda/a)$ depart from the CS value 0.2408 by a maximum $\pm 7\%$ for all the values of H_m and f. The right side of $-\chi''_m/\chi_0(\Lambda/a)$ vs $-\chi'_m/\chi_0(\Lambda/a)$ peaks are sheared to lower $-\chi'_m/\chi_0(\Lambda/a)$ for $H_m = 80$ and 8 A/m. A weaker f dependence is seen in Figure 8(b) for the H_m dependence measurements at $T = 77$ K.

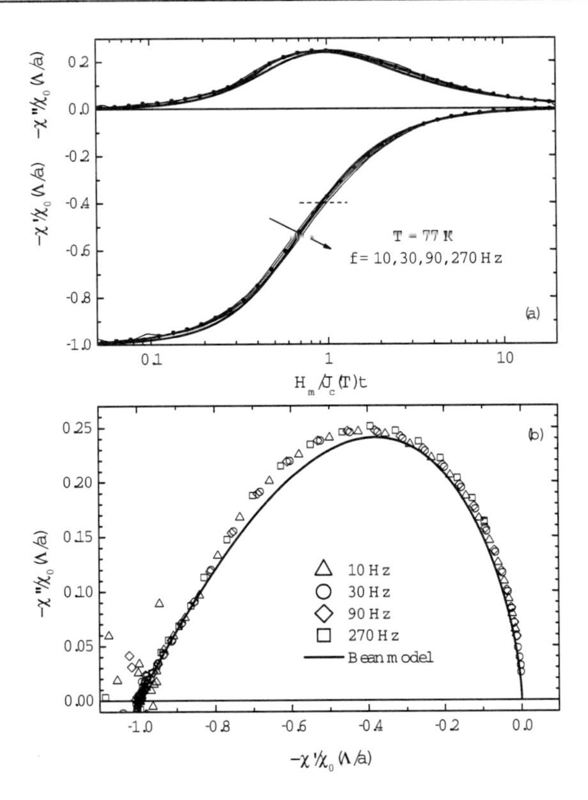

Figure 8. (a) Normalized susceptibility $-\chi/\chi_0(\Lambda/a)$ as a function of normalized field amplitude $H_m/J_c(T)t$ for $T = 77$ K. The cases with data points for determining $J_c(T)$ may be directly compared with the thick curve of $-\chi/\chi_0(0)$ vs H_m/J_ct for the Bean model. (b) $-\chi''/\chi_0(\Lambda/a)$ as a function of $-\chi'/\chi_0(\Lambda/a)$ for $T = 77$ K. The curve is calculated from the Bean model.

5. Vortex Dynamics

5.1. $E(T)/E_c$ vs $J(T)/J_c(T)$ relation

The relation of $E(T)/E_c$ vs $J(T)/J_c(T)$ is obtained using Eq. (12) and Eq. (11), which is modified to

$$J(T)/J_c(T) = [1.21\chi_0(T)t/a][H_m(\chi''_m)/J_c(T)t] \tag{16}$$

for the present purpose, from the $H_m(\chi''_m)/J_c(T)t$ vs f relation for $H_m = 800, 80$, and 8 A/m shown in Figure 6. As done in [3,4], the values of $H_m(\chi''_m)/J_c(T)t$ are determined from $-\chi'/\chi_0(\Lambda/a) = -0.43$, -0.5 to -0.44, and -0.5 for $H_m = 800, 80$, and 8 A/m, respectively, following the dashed lines plotted in Figures 6(a), 6(b), and 6(c). The resulting $E(T)/E_c$ vs $J(T)/J_c(T)$ curves are plotted in Figure 9(a). It can be seen that they are not linear in logarithmic scales for all values of H_m. The average slope $n_{1,2}$ between each pair of neighboring f values, f_1 and f_2, for any given value of H_m is calculated by

$$n_{1,2} = \frac{\log f_2 - \log f_1}{\log \frac{H_m}{J_c(T)t}(\chi''_m, f_2) - \log \frac{H_m}{J_c(T)t}(\chi''_m, f_1)}. \tag{17}$$

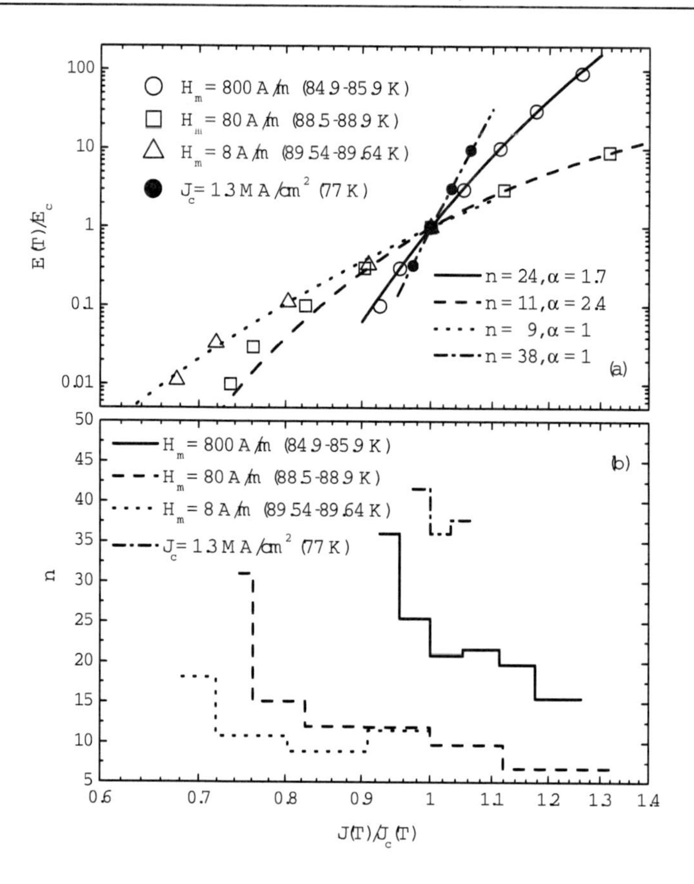

Figure 9. $E(T)/E_c$ (a) and n (b) as functions of $J(T)/J_c(T)$. Curves in (a) are calculated from Eqs. (19) and (20).

The $n_{1,2}$ as a function of $J(T)/J_c(T)$ is shown in Figure 9(b) for $H_m = 800, 80$, and 8 A/m, where the subscripts "1,2" are omitted.

Similar determinations are made from Figure 8(a) for the H_m-dependence measurements. The $E(T)/E_c$ vs $J(T)/J_c(T)$ curve determined using Eqs. (11) and (12) is plotted in Figure 9(a), showing the maximum slope compared with the others. Equation (17) should be replaced by

$$n_{1,2} = 1 + \frac{\log f_2 - \log f_1}{\log \frac{H_m}{J_c(T)t}(\chi''_m, f_2) - \log \frac{H_m}{J_c(T)t}(\chi''_m, f_1)} \tag{18}$$

for calculating $n_{1,2}$, since in this case, H_m itself increases with f in the E calculation by Eq. (12). The results of n are plotted in Figure 9(b).

5.2. Flux-creep and flux-flow $E(J)$

In Bean's CS model, a constant J_c is assumed as a property of hard superconductors regardless of its origin [35]. In this case, the local electric field is $E = 0$ when the magnitude of current density $|J| < J_c$ occurs, whereas certain finite values of E can be induced when

$|J| = J_c$. As a result, the magnetic properties deduced from the CS model are hysteretic without time or frequency dependence. Such a dependence will result from specific flux motions associated with certain $E(J)$ functions.

A general function of $E(J)$ for the thermally activated flux creep may be written as [2,36,37]

$$E(J) = \text{sgn}(J)E_c \exp[-U(J)/kT], \tag{19}$$

where $U(J)$ is a current-density-dependent activation energy for vortex depinning,

$$U(J) = U_0(|J_c/J|^\alpha - 1)/\alpha, \tag{20}$$

α being a case dependent parameter.

Thermally activated creep of vortices depinned from defects of sizes comparable to the vortex core diameter was a mechanism for the CS in hard type-II superconductors first proposed by Kim and Anderson [38,39]. As derived in [36,37,39], the Kim-Anderson type of vortex creep corresponds to $\alpha = -1$ in Eq. (20) with

$$U(J) = U_0(1 - |J/J_c|), \tag{21}$$

$$E(J) = \text{sgn}(J)E_c \exp[m(|J/J_c| - 1)], \tag{22}$$

where U_0 is the pinning barrier for a vortex (or a vortex-bundle) [39], J_c is the critical-current density for vortex depinning at $T = 0$, and $m = U_0/kT$.

Since high-temperature superconductors are doped insulators rather than conventional metals, their pinning centers are mainly provided by point defects, e.g., oxygen vacancies. As a result, the AV pinning is weak, the critical current density J_c is much smaller than the depairing current density J_0 [$J_c/J_0 < 0.02$ for the present case as shown in Figure 5(a)], and a collective creep can take place. The simplest $E(J)$ for collective creep is obtained by assuming $\alpha \to 0$ in Eq. (20) as [2,36,37]

$$U(J) = U_0 \ln |J_c/J|, \tag{23}$$

$$E(J) = \text{sgn}(J)E_c|J/J_c|^n, \tag{24}$$

where

$$n = U_0/kT, \tag{25}$$

U_0 being pinning barrier constant, and J_c may be conveniently defined as the J when the electrical field meets the criterion $E = E_c$. Parameter α may take small positive values for other cases of collective creep.

On the other hand, flux flow may occur when the pinning barrier is overcome, with a linear $E(J)$ in the simplest case as

$$E = 0 \quad (|J| \leq J_c), \tag{26}$$

$$E = [J - \text{sgn}(J)J_c]/\sigma_f \quad (|J| > J_c), \tag{27}$$

where σ_f is flux-flow conductivity.

E/E_c as a function of J/J_c calculated from Eqs. (22), (24), and (27), as well as from Eqs. (19) and (20) with $\alpha = 1$, is shown in Figure 10, where only the portion for $0.01 < E/E_c < 100$ is plotted, $m = n = 10$ are assumed for the flux creep models, and $J_c = 10\sigma_f E_c$ is assumed for the flux flow model.

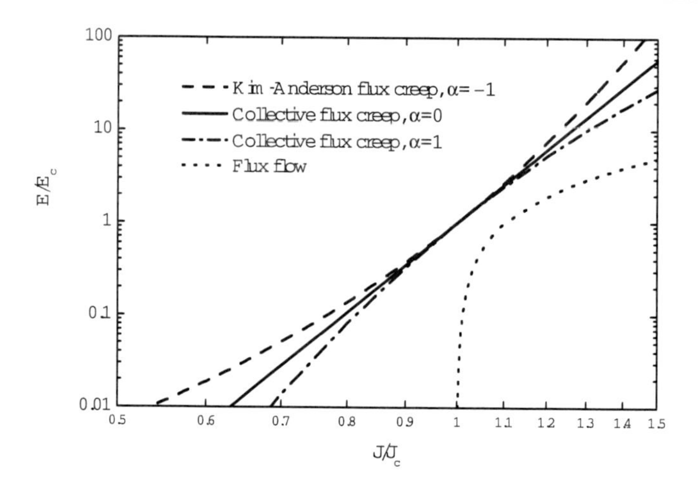

Figure 10. E/E_c as a function of J/J_c for the Kim-Anderson flux creep ($\alpha = -1$), collective flux creep ($\alpha = 0$ and 1), and flux flow models with $m = n = 10$ (for flux creep) and $J_c = 10\sigma_f E_c$ (for flux flow).

5.3. Flux-creep and flux-flow $\chi(H_m, f)$

The $\chi(H_m/J_c a, f)$ of a long cylinder of radius a and critical-current density J_c has been calculated from the Kim-Anderson flux creep, the collective flux creep with $\alpha \to 0$, and flux flow models in [3,8,40]. A common feature is that the calculated $\chi(H_m/J_c a)$ shifts to higher H_m with increasing f. However, there are significant differences in $\chi(H_m/J_c a, f)$ among the three models, which are clearly shown in the χ'' vs χ' curve. For the Kim-Anderson flux creep model, χ''_m decreases with increasing f, as shown in Figure 11(a) for $m = 10$. For the collective flux creep model with $\alpha \to 0$, however, there is an important scaling law for $\chi(H_m, f)$ calculated from Eq. (24): This $\chi(H_m, f)$ with constant E_c, J_c, and n is invariant when multiplying simultaneously f and H_m by any positive constant C and $C^{1/(n-1)}$, respectively. Thus, the χ'' vs χ' curve is independent of f, as shown in Figure 11(b) for $n = 5, 10, 20$, and 40. For flux flow model, χ''_m increases with increasing $\log f$ at an increasing rate, as shown in Figure 11(c).

The CS model is the high-m, high-n, or low-σ_f limit of the flux-creep or flux-flow models. The χ'' vs χ' curves for the flux creep and flux flow models are above the CS curve, as shown in Figure 11 for the case of Bean model.

Up to now, the J_c in the CS model has been assumed to be constant, but in general J_c decreases with increasing the local magnetic induction B. For example, $J_c(B)$ can be of a Kim type or an exponential type [41,42]. For the flux flow model, $-\chi''_m/\chi_0$ may either increase or decrease with increasing f to reach its eddy-current value. For a constant J_c, the low-f CS value of $-\chi''_m/\chi_0$ between 0.21 and 0.24 is lower than the eddy-current $-\chi''_m/\chi_0$ value between 0.35 and 0.44 for different sample shapes [16,43,44], so that $-\chi''_m/\chi_0$ increases with increasing f, as shown in Figure 11(c). For J_c that decreases with increasing flux density B, the low-f CS value of $-\chi''_m/\chi_0$ may be larger than the eddy-current value, so that $-\chi''_m/\chi_0$ decreases with increasing f. Examples for both cases can be found in [45,46] for a polycrystalline $LaFeAsO_{0.94}F_{0.06}$ sample and a sintered YBCO sample,

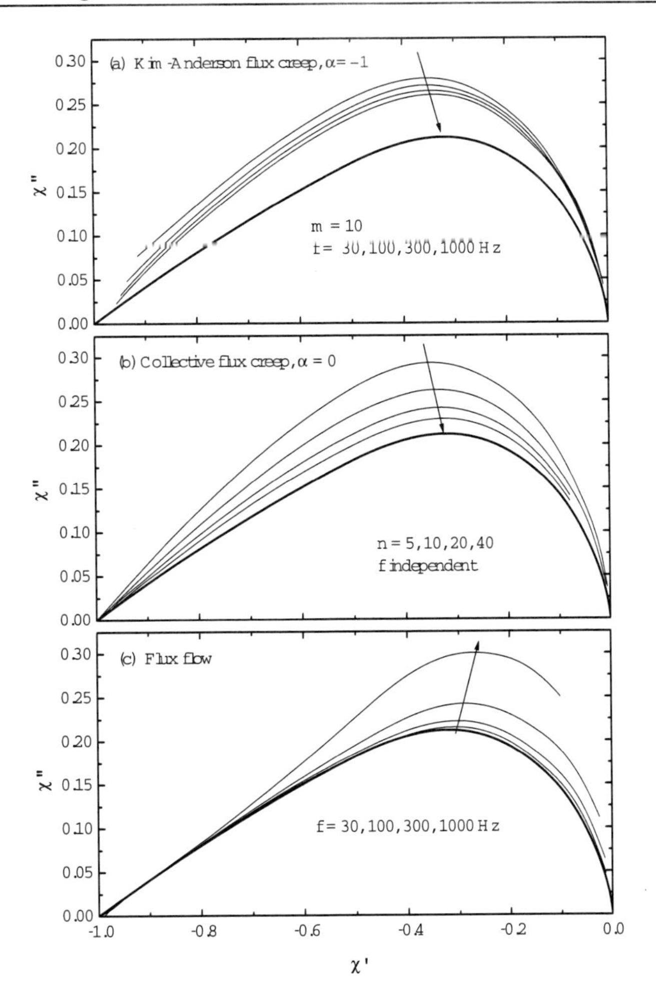

Figure 11. χ'' as a function of χ' of a cylinder calculated for the Kim-Anderson flux creep (a), collective flux creep with $\alpha = 0$ (b), and flux flow models (c), compared with the Bean model (thick curves).

whose intergranular J_c is not determined by Josephson vortex pinning but by the net dc Josephson currents (across neighboring grains) themselves. The net dc Josephson currents may decrease slowly or quickly with increasing B when Josephson junctions are short or long, resulting in nearly constant or highly B-dependent J_c and low or high intergranular χ''_m at low frequencies.

The behavior for flux creep with $\alpha = 1$ is not shown in Figure 11. In this case, χ''_m should increase steadily with increasing $\log f$ for any given value of n, just opposite to the case of Kim-Anderson flux creep.

In [47], $\chi(T, f)$ functions have been calculated for some cases of the Kim-Anderson creep with $\alpha = -1$ and the collective creep (referred to as vortex glass in [47]) with $\alpha = 1/7$. The features for the Kim-Anderson creep are similar to those for $\chi(H_m/J_c a, f)$; and for the the collective creep, both χ''_m and $T(\chi''_m)$ increase steadily with increasing $\log f$.

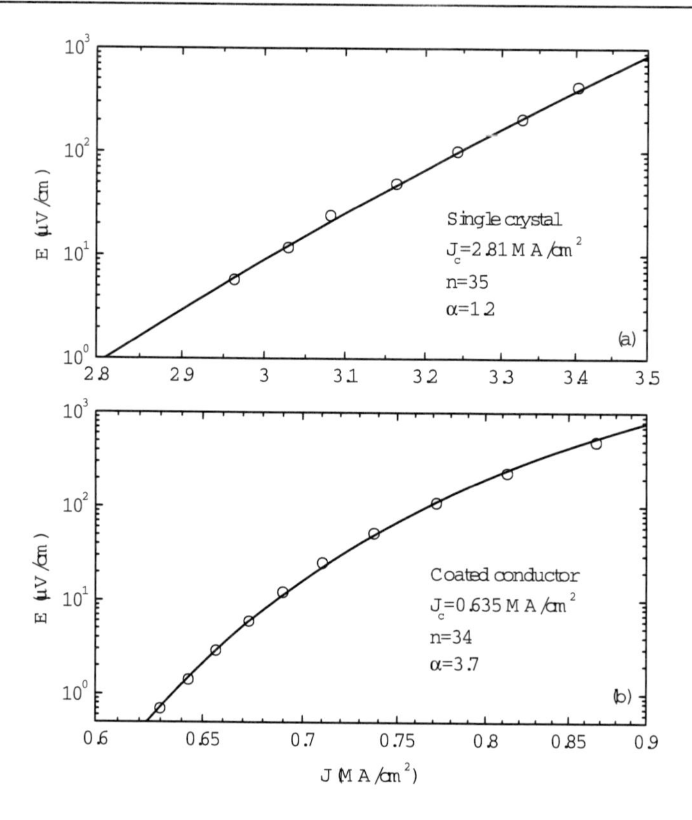

Figure 12. Experimental E vs J data for (a) single crystal and (b) coated conductor YBCO films reported in [4] fitted by Eq. (19).

5.4. Vortex dynamics for the studied film

Figures 6, 7, and 8 show that $-\chi''_m/\chi_0(\Lambda/a)$ increases with increasing f, so that the dominant vortex dynamics is not the Kim-Anderson flux creep but the collective flux creep and/or flux flow in all cases.

Comparing Figure 9 with Figure 10, one can see that the experimental $E(T)/E_c$ vs $J(T)/J_c(T)$ is similar to that of the collective flux creep with a positive α. In fact, the experimental $E(T)/E_c$ vs $J(T)/J_c(T)$ curves may be well fitted by the curves in Figure 9(a) calculated from Eqs. (19) and (20) with $\alpha = 1, 1.7, 2.4,$ and 1 and $n = 38, 24, 11,$ and 9 for the cases of $T = 77$ K and $H_m = 800$, 80, and 8 A/m, respectively.

The susceptibility curves in Figure 8 measured at 77 K are similar to those for a single crystal YBCO film prepared by pulsed laser deposition (PLD) on a $SrTiO_3$ (STO) substrate reported in [4], where its E vs J was well fitted by a power-law $E(J)$ so that a mechanism of collective creep of Pearl vortices (PVs) was proposed. PVs in films are different from the Abrikosov vortices (AVs) in bulk superconductors by having a large characteristic size, Pearl length $\Lambda \equiv \lambda^2/t$, where λ is the London penetration length relevant to AV and t is the film thickness [18]. For a coated conductor, however, the E vs J could not be well fitted by a power-law $E(J)$, which led to a conclusion that the flux flow may also occur for Abrikosov-Josephson vortices (AJVs) located along the grain boundaries (GBs) [4].

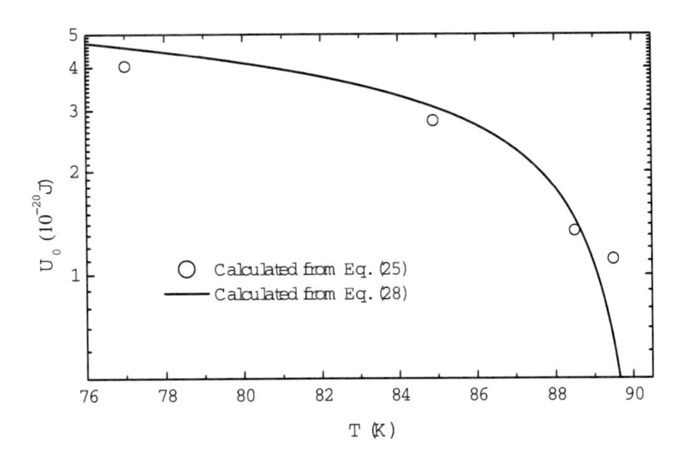

Figure 13. Comparison between the U_0 vs T functions calculated from Eqs. (25) and (28).

In fact, the E vs J data in [4] may be better fitted by Eq. (19) with $\alpha = 1.2$ and 3.7 and $n = 35$ and 34 for the single crystal and coated conductor, respectively, as shown in Figure 12. This implies that the film being studied here is basically very similar to the single crystal studied in [4], and the vortex dynamics of the coated conductor studied in [4] can still be regarded as collective flux creep with a larger α.

According to [36], $\alpha \approx 0.1$ occurs when AVs undergo creep individually and $\alpha \approx 1$ occurs if AVs are grouped in bundles. Thus, for the film studied here and the single crystal film studied in [4], PV-bundles are the units undergoing creep at 77 K. In the coated conductor mentioned above with $\alpha = 3.7$, PVs and AJVs may be coupled together to undergo creep or there may be a coupling between flux creep of PVs and flux flow of AJVs. Similar effects may occur for the cases of $H_m = 800$ and 80 A/m with $\alpha = 1.7$ and 2.4. The case of $H_m = 8$ A/m at T near T_c with $\alpha = 1$ is special, which will be discussed below.

5.5. Vortex pinning barrier

Using the values of $n = 38, 24, 11,$ and 9 occurring at $T = 77, 84.9, 88.5,$ and 89.5 K, respectively, to characterize the collective flux creep, we can easily calculate using Eq. (25) the pinning barrier constant $U_0 = 4.04 \times 10^{-20}$, 2.81×10^{-20}, 1.34×10^{-20}, and 1.11×10^{-20} J, as shown in Figure 13.

On the other hand, U_0 may be calculated by [36]

$$U_0 \approx 4\pi\mu_0 H_c^2 \xi^3 (J_c/J_0)^{1/2}, \tag{28}$$

where ξ is the coherence length and H_c and J_0 are the thermodynamic critical field and the depairing current density expressed by

$$H_c = \frac{\sqrt{2}\Phi_0}{4\pi\mu_0\lambda\xi}, \tag{29}$$

$$J_0 = H_c/\lambda. \tag{30}$$

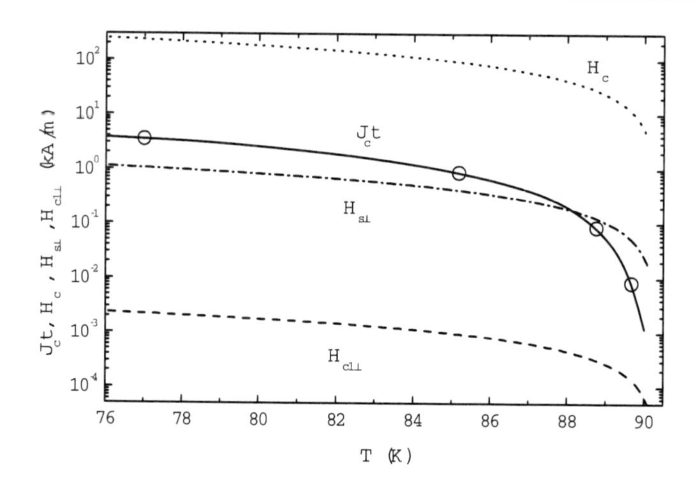

Figure 14. J_ct, $H_{s\perp}$, and $H_{c1\perp}$ as functions of T for the studied film, compared with $H_c(T)$ for bulk superconductor.

If $\kappa = \lambda/\xi$ is independent of T at $T \geq 77$ K [48], then we have from Eqs. (28)-(30) that $U_0 \propto (J_c\lambda)^{1/2}$.

Using Eqs. (10) and (13) for $\lambda(T)$ and $J_c(T)$ and assuming $\kappa = 57$ as measured in [48], H_c, J_0, and U_0 as functions of T are calculated and plotted in Figures 14, 5, and 13, respectively.

We observe in Figure 13 that the values of U_0 calculated in both ways are consistent except near T_c, where U_0 calculated using Eq. (25) is significantly higher.

We should mention that the values of U_0 and α are sensitive to the choice of the value of E_c, by which J_c is defined. However, this does not change the feature for U_0 calculated from Eq. (28) to be significantly lower than that calculated from Eq. (25) at high T, which will be discussed below.

6. Extension of Critical-State Model

6.1. B dependence of J_c and other effects

For collective flux creep with $\alpha \to 0$ and J_c independent of B, $-\chi''_m/\chi_0$ is larger than its Bean CS counterpart $-\chi''_{m,\text{Bean}}/\chi_0$ and the ratio $\chi''_m/\chi''_{m,\text{Bean}}$ increases from 1 roughly linearly with increasing $1/n$ from 0, as shown in Figure 2 of [3]. Using this rule we can estimate from the low-f data in Figures 8(b), 7(a), and 7(b) that if there were a CS, then the CS χ''_m would be smaller than that for the Bean model by 5, 7, and 12% for $T = 77, 84.9,$ and 88.5 K, respectively.

As mentioned above, the CS $J_c(B)$ is generally a decreasing function of B, which results in a position-dependent J_c in the magnetized superconductor. As calculated in [43] for long superconductors, $-\chi''_m/\chi_0$ always increases with increasing this spatial nonuniformity of J_c in the sample owing to the B dependence of J_c. Thus, the CS $-\chi''_m/\chi_0$ being lower than its Bean value would require J_c to increase with increasing B, which is contradicted to the vortex pinning and any other mechanisms of the CS.

In order to solve this problem, we first explain that the Bean CS is the low B limit for most actual B dependences of J_c. For example, the Kim type of $J_c(B)$ is actually a simplification of constant J_c at low B, where the vortex density is smaller than the pinning center density, followed by $J_c \propto 1/B$ at high B [49]. For thin films with small $J_c t$, the positional nonuniformity of J_c caused by $J_c(B)$ is small, so that Bean model should work well even if J_c is weakly B dependent.

In fact, the electromagnetic properties of type-II superconductors are influenced by their thermal equilibrium magnetization and surface barriers. An extended model considering CS and these effects was proposed in [50] and successfully used for explaining dc magnetization curves of sintered HTSs [51–53]. This model was also used for analyzing $\chi(H_m)$ of HTS grains [54, 55], leading to conclusions that χ''_m is lowered and it occurs at lower χ' by the effects of equilibrium magnetization and surface barrier. Since these features are found for the present film, both effects should also occur in it.

6.2. Equilibrium magnetization and surface barrier for bulk samples

The equilibrium magnetization curve of a type-II superconducting long cylinder of radius a is calculated by minimizing the Gibbs potential with respect to AV density [30]. As a result, with increasing the applied field H from zero to the lower-critical field H_{c1} [56],

$$H_{c1} = \frac{H_c}{\sqrt{2}\kappa}(\ln \kappa + 0.497)\ (\kappa \gg 1), \tag{31}$$

the magnetization M decreases linearly with slope $M/H = -1$, and then increases first quickly and later slowly to zero at the higher-critical field H_{c2},

$$H_{c2} = \sqrt{2}\kappa H_c. \tag{32}$$

On the other hand, there is a Bean-Livingston surface barrier to the AV entry, so that the first AV entry occurs at a surface pinning field H_s larger than H_{c1} [30, 57]

$$H_s = H_c/\sqrt{2}. \tag{33}$$

In fact, there is another field H_n for first AV nucleation on the surface, derived by Fink [58],

$$H_n = \frac{\sqrt{5}H_c}{3}[1 + (2\kappa)^{-1/2}], \tag{34}$$

so that the first AV entry should occur at $H = H_n > H_s > H_{c1}$ if the sample surface is perfectly smooth. The existence of H_n for $\kappa \sim 2$ has been displayed by numerical calculations based on time-dependent Ginzburg-Landau equations (huge numerical errors occur for larger κ) [59–62]. A similar situation has been found for Josephson vortices in a resistively shunted square-bar Josephson junction array up to a large κ [49, 63, 64], for which

$$H_n \approx \sqrt{\pi/2}H_c. \tag{35}$$

It may be obtained for YBCO superconductors at 77 K that $H_{c1} = 5.8$ kA/m, $H_s = 165$ kA/m, and $H_n = 191$ kA/m, using Eqs. (10) and (31)-(34) with $\kappa = 57$. All these values are comparable with the full penetration field H_p ($= J_c a$ for the cylinder), so that the effects of equilibrium magnetization and surface barrier on the magnetization curve of bulk superconductor, or on the $\chi(H_m)$ of superconducting grains, may be significant.

6.3. Equilibrium magnetization and edge barrier for films

For a type-II superconducting film in a perpendicular field H, H_{c2} should be approximated by Eq. (32) [65], which is reasonable since $M = 0$ at H_{c2} so that demagnetizing field is zero. We estimate the H_{c1}, H_s, and H_n of a perpendicularly magnetized thin superconducting strip, $H_{c1\perp}$, $H_{s\perp}$, and $H_{n\perp}$, based on the London equation.

For a type-II superconducting film in a perpendicular field H, the lower-critical field $H_{c1\perp}$ may be derived as follows. As calculated by De Gennes for bulk superconductor [30], $H_{c1\perp}$ for a film of thickness t and other sizes much larger than Λ may be derived from the increment of the Gibbs potential with respect to the Meissner state,

$$\Delta G = n_{\mathrm{PV}} E_{\mathrm{PV}} - \mu_0 M t H, \tag{36}$$

where M is the magnetization provided by Pearl vortices (PVs) [18], whose centers are far away from each other so that the interaction energy is negligible, with surface density n_{PV} and E_{PV} is the energy of one isolated PV [18],

$$E_{\mathrm{PV}} = \frac{\Phi_0^2}{8\mu_0\Lambda}\left[H_0\left(\frac{\xi}{2\Lambda}\right) - Y_0\left(\frac{\xi}{2\Lambda}\right) + \frac{1}{2\pi}\right], \tag{37}$$

where H_i and Y_i are the Struve function and the second-kind Bessel function of the ith order and the term $1/2\pi$ is added by considering the condensation energy $\mu_0 H_c^2 \pi\xi^2 t/2$ in the core. Since $\Delta G = 0$ in the Meissner state, $H_{c1\perp}$ is the smallest H that causes $\Delta G < 0$ and is obtained from Eq. (36) as

$$H_{c1\perp} = n_{\mathrm{PV}} E_{\mathrm{PV}} / \mu_0 M t. \tag{38}$$

We have $n_{\mathrm{PV}} = B/\Phi_0$, since each PV carries Φ_0, and $B = -\mu_0 M/\chi_0(0)$ when $-M \gg H$. Considering these, Eq. (38) becomes

$$\begin{aligned} H_{c1\perp} &= -\frac{\pi\xi H_c}{2\sqrt{2}\chi_0(0)\lambda}\left[H_0\left(\frac{\xi}{2\Lambda}\right) - Y_0\left(\frac{\xi}{2\Lambda}\right) + \frac{1}{2\pi}\right] \\ &= -\frac{\xi H_c}{\sqrt{2}\chi_0(0)\lambda}[\ln(2\Lambda/\xi) + 0.37], \end{aligned} \tag{39}$$

where the error of the second equality is less than 1% when $\Lambda/\xi > 13$.

Surface barrier in bulk samples is reduced to edge barrier in films. The characterizing field for edge barrier is the edge pinning field $H_{s\perp}$. Modifying the approaches of Bean and Livingston and De Gennes for bulk superconductor to the case of films [30, 57], we assume that the film is in the xy plane occupying $x \geq 0$ and a PV is located near the film edge with its center at $x = \xi$ and $y = 0$. We then calculate the current density J_s at the vortex center produced by the image PV centered at $x = -\xi$ and $y = 0$, and balance it by the current density J_{edg} at the edge of a thin strip of width $2a$ in the Meissner state under a perpendicular applied field $H_{s\perp}$.

J_s is obtained from the J of PV at radius $r = 2\xi$ [18],

$$J_s = \frac{\pi \xi H_c}{2\sqrt{2}\lambda\Lambda}\left[H_1\left(\frac{\xi}{\Lambda}\right) - Y_1\left(\frac{\xi}{\Lambda}\right) - \frac{2}{\pi}\right]. \tag{40}$$

$J_{\rm edg}/H$ for a thin strip of width $2a$ is obtained as [66]

$$\frac{J_{\rm edg}t}{H} = \left(\frac{a/\Lambda}{1/\pi + \Lambda/a}\right)^{1/2}, \tag{41}$$

which is consistent with results obtained numerically from minimizing Gibbs potential [13]. Replacing $J_{\rm edg}$ and H by J_s and $H_{s\perp}$, we have

$$\begin{aligned} H_{s\perp} &= \frac{\pi\sqrt{t}\xi H_c}{2\sqrt{2a}\Lambda}\left[H_1\left(\frac{\xi}{\Lambda}\right) - Y_1\left(\frac{\xi}{\Lambda}\right) - \frac{2}{\pi}\right]\left(\frac{1}{\pi} + \frac{\lambda^2}{at}\right)^{1/2} \\ &= \frac{\sqrt{t}H_c}{\sqrt{2a}}\left(\frac{1}{\pi} + \frac{\lambda^2}{at}\right)^{1/2}, \end{aligned} \tag{42}$$

where the error of the second equality is less than 1% when $\Lambda/\xi > 100$.

Assuming that at $H = H_n$, $J_{\rm edg}$ of the strip equals the depairing current density J_0 expressed by Eq. (30), we have from Eq. (41) that

$$H_{n\perp} = \frac{\sqrt{t}H_c}{\sqrt{a}}\left(\frac{1}{\pi} + \frac{\lambda^2}{at}\right)^{1/2} = \sqrt{2}H_{s\perp}. \tag{43}$$

Approximating the studied film by a strip of $a = 2$ mm and at $T = 77$ K, we have $H_{c1} \approx 2$ A/m and $H_s \approx 1000$ A/m, calculated from Eqs. (39) and (42). H_c, $H_{s\perp}$, $H_{c1\perp}$, and J_ct as functions of T are shown in Figure 14. We see that $H_{c1\perp}$ is always negligible compared with J_ct but $H_{s\perp}$ is comparable with J_ct although both have different T dependences.

6.4. Anomalous phenomena and previous explanations

The fact that χ''_m in the $\chi''(T)$ curve of superconductors measured at low H_m is below its Bean value is an anomalous phenomenon frequently reported in the literature [5, 10, 33, 34, 67]. For the YBCO film studied here, this anomaly has been extended to several phenomena with respect to the CS and flux creep as summarized below. (i) The low-f limit $-\chi''_m/\chi_0(\Lambda/a)$ decreases with increasing T and it is lower than its Bean value when T approaches T_c, as shown in Figures 7 and 8. (ii) The $-\chi'/\chi_0(\Lambda/a)$ vs $-\chi''/\chi_0(\Lambda/a)$ curve is sheared to the left with increasing T as shown in Figure 7. (iii) The pinning barrier U_0 for vortex creep calculated from Eq. (25) is higher than that from Eq. (28) at $T \sim T_c$. (iv) A sudden drop occurs for the value of α defined in Eq. (20) at $T \sim T_c$.

The high-T over-low χ''_m in $\chi(T)$ measurements of YBCO films at low H_m was attributed by the authors of [5] to the growing dominance of screening, i.e., the decrease in $-\chi_0(\Lambda/a)$ in terms of the present work, rather than to a change towards a different loss mechanism. This explanation may be discarded after normalizing χ to $-\chi_0(\Lambda/a)$ as in Figure 7. On the other hand, the author of [10] attributed the same phenomenon simply to the breakdown of CS at low H_m, without discussing possible mechanism for it. We should mention that theoretical CS $\chi''(\chi')$ curves given in both papers were questionable, so that no correct comparison between the measured and CS results could be carried out.

Some authors have owed the reduced χ''_m of superconducting YBCO or lead films at $T \sim T_c$ to the elastic motion of the vortices in the so-called Campbell regime in which the low driving force is too weak to depin vortices [33, 34, 67].

For an ideal type-II superconductor without defects in a vortex state in a dc field above H_{c1}, the ac magnetization at small H_m follows the equilibrium magnetization curve with a paramagnetic (positive) χ. According to Campbell and Evetts [68], $\chi' < 0$ and $\chi'' = 0$ occur at a fixed dc field and a small ac field amplitude H_m if vortices are pinned by defects with elastic motions, and χ' takes a value somewhat larger (smaller in magnitude) than the Meissner χ_0 treated in the present work. However, the χ at larger H_m and the resultant χ''_m are not involved in the Campbell regime, and if the magnetization process after depinning may be well described by the CS model in this case and the normalized χ and H_m are used, as done in the present work, then the over-low χ''_m is still not explained. Thus, the anomaly of over-low χ''_m is a phenomenon with unknown mechanism.

6.5. Effects of equilibrium magnetization and edge barrier

Since J_c and H_c decrease with increasing T roughly in proportion to $1 - T/T_c$ and $1 - (T/T_c)^{1.7}$, respectively, the ratios of $H_{c1\perp}$ and $H_{s\perp}$ of the film to $J_c t$ will increase with increasing T and take maximum values at $T \sim T_c$. Thus, possible effects of equilibrium magnetization and surface barrier on the ac susceptibility of the film should be enhanced at $T \sim T_c$, which is indeed the case of our results. Several possible effects are described below.

The $-\chi'/\chi_0(\Lambda/a)$ vs $-\chi''/\chi_0(\Lambda/a)$ curve should be sheared to the left with increasing T. This is explained by a narrow-waist $M(H)$ loop resulting from the enhanced $H_{s,n}(T)/J_c(T)t$ and $H_{c1}(T)/J_c(T)t$, as shown in [54, 55] for ac susceptibility of YBCO grains. However, such an effect requires the first field for vortex entry to be larger than the last field for vortex exit. This may be realized by considering the image effect during both vortex entry and exit in combination with inner vortex pinning.

U_0 calculated from Eq. (25) should be higher than that from Eq. (28) at $T \sim T_c$. This is because of the fact that Eq. (28) is valid for the flux creep of vortices depinned from the volume defects, but a part of ac losses comes from edge pinning, corresponding to a partial χ with a weak f dependence. As a result, the effective n should be larger than that when there is volume-defects pinning only, leading to over-estimated U_0 calculated from Eq. (25).

If $\alpha < 1$ is, as explained above, attributed to the existence of AJVs, then the reduction of α to 1 when T approaches T_c is consistent with the enhanced edge barrier effects, which makes the AJV effect relatively smaller.

The reduced low-f limit $-\chi''_m/\chi_0(\Lambda/a)$ at high T is still a difficult phenomenon to explain. It requires a sufficiently large $H_{c1}(T)/J_c(T)t$, since in the limit of $J_c(T) = 0$, χ'' should remain zero with increasing χ' from $\chi_0(T)$ to zero. However, this effect should be negligible since $H_{c1}(T)/J_c(T)t < 0.02$ even at T as high as 89.7 K.

One possibility to explain this is to assume a B dependent χ_0 as for the Campbell regime. In this case, the normalization to the Meissner susceptibility $\chi_0(\Lambda/a)$ in Figure

7 should be replaced by that to the Campbell susceptibility $\chi_0(\Lambda/a, B)$, which increases with increasing B (so increasing χ'), so that the normalized χ''_m and χ'' at higher χ' can be increased to their Bean values.

7. Conclusion

The complex ac susceptibility $\chi = \chi' - j\chi''$ of a square YBCO film of sides $2a = 4$ mm and thickness $t = 0.25$ μm has been measured in ac field of different amplitude H_m and frequency f applied perpendicularly to the film plane as a function of temperature T. After its Meissner susceptibility $\chi_0(\Lambda/a)$, where $\Lambda \equiv \lambda^2/t$ is the Pearl length, is calculated numerically by minimizing the Gibbs potential, the London penetration depth $\lambda(T)$ is determined from the measured $\chi(T)$ at $H_m = 8$ A/m and $f = 1111$ Hz. Assuming a constant $\kappa = 57$, this $\lambda(T)$ leads to the determination of thermodynamic fields, $H_{c1}(T)$, $H_c(T)$, and $H_{c2}(T)$, surface pinning and vortex nucleation fields, $H_s(T)$ and $H_n(T)$, and the depairing current density $J_0(T)$ for the corresponding bulk YBCO superconductor.

Based on $\chi(H_m/J_c t)$ accurately calculated from the critical-state model with a constant critical-current density J_c (the Bean model) and from collective flux creep model with a power-law $E(J)$, $J_c(T)$ is extracted from the measured $\chi(T, H_m, f)$, so that a normalized $-\chi/\chi_0(\Lambda/a)$ vs $H_m/J_c(T)t$ function is obtained and may be quantitatively compared with theoretical results for the Bean model and the flux creep model.

It is found for the studied film that J_c is practically independent of local flux density B and the f dependence may be explained quantitatively by the collective flux creep model with a vortex pinning barrier $U_0(T)$. There are significant effects on $\chi(T, H_m, f)$ and apparent $U_0(T)$ from thermal equilibrium magnetization and edge-pinning and/or vortex nucleation, especially at T approaching T_c. However, unlike the case of bulk superconductors, such effects have not been well studied theoretically for films in a perpendicular field. The fields $H_{c1\perp}$, $H_{s\perp}$, and $H_{n\perp}$ for a perpendicularly magnetized superconducting thin strip are derived from the London equation, and $H_{c1\perp}(T)$ and $H_{s\perp}(T)$ for the studied film are compared with $J_c(T)t$. It is concluded that the anomalous behavior of $\chi(T, H_m, f)$ found at $T \sim T_c$ should be related to edge barrier effects that are comparable with the volume pinning. However, further work is needed for understanding the anomalies in detail.

Acknowledgements

We thank V. Skumryev for high-quality measurements and E. H. Brandt, J. R. Clem, E. Pardo, A. Palau, E. Bartolome, T. Puig, X. Obradors, J. Gutierrez, and A. V. Silhanek for discussions and help. Financial support from Consolider Project CSD2007-00041 is acknowledged.

References

[1] J. R. Clem and A. Sanchez, *Phys. Rev. B* **50**, 9355 (1994).

[2] E. H. Brandt, *Phys. Rev. B* **55**, 14513 (1997).

[3] D.-X. Chen and E. Pardo, *Appl. Phys. Lett.* **88**, 222505 (2006).

[4] D.-X. Chen, E. Pardo, A. Sanchez, A. Palau, T. Puig, and X. Obradors, *Appl. Phys. Lett.* **85**, 5646 (2004).

[5] Th. Herzog, H. A. Radovan, P. Ziemann, and E. H. Brandt, *Phys. Rev. B* **56**, 2871 (1997).

[6] T. Puig, A. Palau, X. Obradors, E. Pardo, C. Navau, A. Sanchez, Ch. Jooss, K. Guth, and H. C. Freyhardt, *Supercond. Sci. Technol.* **17**, 1283 (2004).

[7] D.-X. Chen, E. Pardo, A. Sanchez, S.-S. Wang, Z.-H. Han, E. Bartalome, T. Puig, and X. Obradors, *Phys. Rev. B* **72**, 052504 (2005).

[8] D.-X. Chen, E. Pardo, A. Sanchez, M. N. Iliev, S.-S. Wang, and Z.-H. Han, *J. Appl. Phys.* **101**, 073905 (2007).

[9] A. Palau, T. Puig, and X. Obradors, *J. Appl. Phys.* **102**, 073911 (2007).

[10] A. A. Elabbar, Physica C **469**, 147 (2009).

[11] D.-X. Chen, C. Navau, N. Del-Valle, and A. Sanchez, *Appl. Phys. Lett.* **92**, 202503 (2008).

[12] D.-X. Chen, C. Navau, N. Del-Valle, and A. Sanchez, *Physica C* **470**, 89 (2010).

[13] D.-X. Chen, C. Navau, N. Del-Valle, and A. Sanchez, *Supercond. Sci. Technol.* **21**, 105010 (2008).

[14] X. Obradors, T. Puig, A. Pomar, F. Sandiumenge, S. Piñol, N. Mestres, O. Castaño, M. Coll, A. Cavallaro, A. Palau, J. Gázquez, J. C. González, J. Gutiérrez, N. Romà, S. Ricart, J. M. Moretó, M. D. Rossell, and G. van Tendeloo, *Supercond. Sci. Technol.* **17**, 1055 (2004).

[15] D.-X. Chen and V. Skumryev, *Rev. Sci. Instrum.* **81**, 025104 (2010).

[16] D.-X. Chen and C. Gu, *IEEE Trans. Magn.* **41**, 2436 (2005).

[17] D.-X. Chen, *Meas. Sci. Technol.* **15**, 1195 (2004)

[18] J. Pearl, *Appl. Phys. Lett.* **5**, 65 (1964).

[19] E. H. Brandt, *Phys. Rev. Lett.* **74**, 3025 (1995).

[20] C. Navau, A. Sanchez, N. Del-Valle, and D.-X. Chen, *J. Appl. Phys.* **103**, 113907 (2008).

[21] C. Navau, D.-X. Chen, A. Sanchez, and N. Del-Valle, *Appl. Phys. Lett.* **94**, 242501 (2009).

[22] W. N. Hardy, D. A. Bonn, D. C. Morgan, R. Liang, and K. Zhang, *Phys. Rev. Lett.* **70**, 3999 (1993).

[23] J. Mao, D. H. Wu, J. L. Peng, R. L. Greene, and S. M. Anlage, *Phys. Rev. B* **51**, 3316 (1995).

[24] C. C. Homes, T. Timusk, D. A. Bonn, R. Liang, and W. N. Hardy, *Physica C* **254**, 265 (1995).

[25] A. Hosseini, S. Kamal, D. A. Bonn, R. Liang, and W. N. Hardy, *Phys. Rev. Lett.* **81**, 1298 (1998).

[26] S. M. Anlage, B. W. Langley, G. Deutscher, J. Halbritter, and M. R. Beasley, *Phys. Rev. B* **44**, 9764 (1991).

[27] Z. Ma, R. C. Taber, L. W. Lombardo, A. Kapitulnik, M. R. Beasley, P. Merchant, C. B. Eom, S. Y. Hou, and J. M. Phillips, *Phys. Rev. Lett.* **71**, 781 (1993).

[28] E. Farber, G. Deutscher, J. P. Contour, and E. Jerby, *Eur. Phys. J. B* **5**, 159 (1998).

[29] S. Djordjevic, E. Farber, G. Deutscher, N. Bontemps, O. Durand, and J. P. Contour, *Eur. Phys. J. B* **25**, 407 (2002).

[30] P. G. de Gennes, *Superconductivity of Metals and Alloys*, Addison-Wesley Publ., 1989, p. 26.

[31] H. W. Lewis, *Phys. Rev.* **102**, 1508 (1956).

[32] R. J. Radtke, V. N. Kostur, and K. Levin, *Phys. Rev.* **53** R522 (1996).

[33] S. Raedts, A. V. Silhanek, and V. V. Moshchalkov, *Phys. Rev. B* **73**, 174514 (2006).

[34] M. M. Ozer, J. R. Thompson, and H. H. Weitering, *Phys. Rev. B* **74**, 235427 (2006).

[35] C. P. Bean, *Phys. Rev. Lett.* **8**, 250 (1962).

[36] G. Blatter, M. V. Feigel'man, V. B. Geshkenbein, A. I. Larkin, and V. M. Vinokur, *Rev. Mod. Phys.* **66**, 1125 (1994).

[37] E. H. Brandt, *Phys. Rev. B* **58**, 6506 (1998).

[38] Y. B. Kim, C. F. Hempstead, and A. R. Strnad, *Phys. Rev. Lett.* **9**, 306 (1962).

[39] P. W. Anderson, *Phys. Rev. Lett.* **9**, 309 (1962).

[40] D.-X. Chen, E. Pardo, and A. Sanchez, *Appl. Phys. Lett.* **86**, 242503 (2005).

[41] D.-X. Chen and R. B. Goldfarb, *J. Appl. Phys.* **66**, 2489 (1989).

[42] D.-X. Chen, A. Sanchez, and J. S. Munoz, *J. Appl. Phys.* **67**, 3430 (1990).

[43] D.-X. Chen and A. Sanchez, *J. Appl. Phys.* **70**, 5463 (1991).

[44] E. H. Brandt, *Phys. Rev. B* **50**, 4034 (1994).

[45] G. Bonsignore, A. A. Gallitto, M. L. Vigni, J. L. Luo, G. F. Chen, Z. Li, N. L. Wang, and D. V. Shovkun, *Low Temp. Phys.* **162**, 40 (2011).

[46] D.-X. Chen, E. Pardo, A. Sanchez, and E. Bartolome, *Appl. Phys. Lett.* **89**, 072501 (2006).

[47] M. Polichetti, M. G. Adesso, and S. Pace, *Physica A* **339**, 119 (2004).

[48] Z. Hao, J. R. Clem, M. W. McElfresh, L. Civale, A. P. Malozemoff, and E. Holtzberg, *Phys. Rev. B* **43**, 2844 (1991).

[49] D.-X. Chen, J. J. Moreno, A. Hernando, and A. Sanchez, in: *Studies of High Temperature Superconductors* **40**, ed. A. Narlikar (Nova Science, New York, 2002) p. 1.

[50] D.-X. Chen, R. W. Cross, and A. Sanchez, *Cryogenics* **33**, 695 (1993).

[51] D.-X. Chen, R. B. Goldfarb, R. W. Cross, and A. Sanchez, *Phys. Rev. B* **48**, 6426 (1993).

[52] D.-X. Chen, A. Hernando, C. Conde, J. Ramirez, and J. M. Gonzales-Calbet, *J. Appl. Phys.* **75**, 2578 (1994).

[53] D.-X. Chen, A. Varela, A. Hernando, J. M. Gonzales-Calbet, and M. Vallet, *Phys. Rev. B* **53**, 5160 (1996).

[54] D.-X. Chen and A. Sanchez, in: *Magnetic Susceptibility of Superconductors and Other Spin Systems*, edited by R. A. Hein et al. (Plenum, New York, 1991) p. 259.

[55] D.-X. Chen and A. Sanchez, *Phys. Rev. B* **45**, 10793 (1992).

[56] C. R. Hu, *Phys. Rev. B* **6**, 1756 (1972).

[57] C. P. Bean and J. D. Livingston, *Phys. Rev. Lett.* **12**, 14 (1964).

[58] H. J. Fink and A. G. Presson, *Phys. Rev.* **182**, 498 (1969).

[59] R. Kato, Y. Enomoto, and S. Maekawa, *Phys. Rev. B* **47**, 8016 (1993).

[60] A. Hernandez and D. Dominguez, *Phys. Rev. B* **65**, 144529 (2002).

[61] D. Y. Vodolazov, I. L. Maksimov, and E. H. Brandt, *Physica C* **384**, 211 (2003).

[62] Y. Enomoto and K. Okada, *J. Phys.: Condens. Matter* **9**, 10203 (1997).

[63] D.-X. Chen, J. J. Moreno, and A. Hernando, *Phys. Rev. B* **53**, 6579 (1996).

[64] D.-X. Chen, J. J. Moreno, and A. Hernando, *Phys. Rev. B* **56**, 2364 (1997).

[65] M. Tinkham, *Phys. Rev.* **129**, 2413 (1963).

[66] B. L. T. Plourde, D. J. van Harlingen, D. Yu. Vodolazov, R. Besseling, M. B. S. Hesselberth, and P. H. Kes, *Phys. Rev. B* **64**, 014503 (2001).

[67] G. Pasquini, P. Levy, L. Civale, G. Nieva, and H. Lanza, *Physica C* **274**, 165 (1997).

[68] A. M. Campbell and J. E. Evetts, *Adv. Phys.* **50**, 1249 (2001) [reprinted from **21**, 199 (1972)].

INDEX

A

B

C

D

I

J

L

M

N

O

P

Q

R

S

T

U

V

W

Y

Z